*Edited by*
*Daniel A. Erlanson and*
*Wolfgang Jahnke*

**Fragment-based Drug Discovery**

*Methods and Principles in Medicinal Chemistry*
Edited by R. Mannhold, H. Kubinyi, G. Folkers


## *Previous Volumes of this Series:*

Urbán, László/Patel, Vinod F./Vaz, Roy J. (Eds.)

**Antitargets and Drug Safety**

2015
ISBN: 978-3-527-33511-4
Vol. 66

Keserü, György M./Swinney, David C. (Eds.)

**Kinetics and Thermodynamics of Drug Binding**

2015
ISBN: 978-3-527-33582-4
Vol. 65

Pfannkuch, Friedlieb/Suter-Dick, Laura (Eds.)

**Predictive Toxicology**

From Vision to Reality

2014
ISBN: 978-3-527-33608-1
Vol. 64

Kirchmair, Johannes (Ed.)

**Drug Metabolism Prediction**

2014
ISBN: 978-3-527-33566-4
Vol. 63

Vela, José Miguel/Maldonado, Rafael/Hamon, Michel (Eds.)

**In vivo Models for Drug Discovery**

2014
ISBN: 978-3-527-33328-8
Vol. 62

Liras, Spiros/Bell, Andrew S. (Eds.)

**Phosphodiesterases and Their Inhibitors**

2014
ISBN: 978-3-527-33219-9
Vol. 61

Hanessian, Stephen (Ed.)

**Natural Products in Medicinal Chemistry**

2014
ISBN: 978-3-527-33218-2
Vol. 60

Lackey, Karen/Roth, Bruce (Eds.)

**Medicinal Chemistry Approaches to Personalized Medicine**

2013
ISBN: 978-3-527-33394-3
Vol. 59

Brown, Nathan (Ed.)

**Scaffold Hopping in Medicinal Chemistry**

2013
ISBN: 978-3-527-33364-6
Vol. 58

Hoffmann, Rémy/Gohier, Arnaud/Pospisil, Pavel (Eds.)

**Data Mining in Drug Discovery**

2013
ISBN: 978-3-527-32984-7
Vol. 57

Dömling, Alexander (Ed.)

**Protein-Protein Interactions in Drug Discovery**

2013
ISBN: 978-3-527-33107-9
Vol. 56

*Edited by Daniel A. Erlanson and Wolfgang Jahnke*

# Fragment-based Drug Discovery

## Lessons and Outlook

WILEY-VCH
Verlag GmbH & Co. KGaA

**Series Editors:**

***Prof. Dr. Raimund Mannhold***
Rosenweg 7
40489 Düsseldorf
Germany
mannhold@uni-duesseldorf.de

***Prof. Dr. Hugo Kubinyi***
Donnersbergstr. 9
67256 Weisenheim am Sand
Germany
kubinyi@t-online.de

***Prof. Dr. Gerd Folkers***
Collegium Helveticum
ETH-Zentrum/STW
Schmelzbergstr. 25
8092 Zürich
Switzerland
gerd.folkers@collegium.ethz.ch

**Volume Editors:**

***Dr. Daniel A. Erlanson***
Carmot Therapeutics, Inc.
409 Illinois Street
San Francisco, CA 94158
USA

***Dr. Wolfgang Jahnke***
Novartis Pharma AG
NIBR/DT/PSU
WSJ- 88 904
4002 Basel
Switzerland



**Library of Congress Card No.:** applied for

**British Library Cataloguing-in-Publication Data**
A catalogue record for this book is available from the British Library.

**Bibliographic information published by the Deutsche Nationalbibliothek**
The Deutsche Nationalbibliothek lists this publication in the Deutsche Nationalbibliografie; detailed bibliographic data are available on the Internet at <http://dnb.d-nb.de>.



**Print ISBN:** 978-3-527-33775-0
**ePDF ISBN:** 978-3-527-68361-1
**ePub ISBN:** 978-3-527-68362-8
**Mobi ISBN:** 978-3-527-68363-5
**oBook ISBN:** 978-3-527-68360-4

**Cover Design** Grafik-Design Schulz, Fußgönheim, Germany
**Typesetting** Thomson Digital, Noida, India
**Printing and Binding** Markono Print Media Pte Ltd

Printed on acid-free paper

# Contents

## Contributors

*Chris Abell*
University of Cambridge
Department of Chemistry
Lensfield Road
Cambridge
CB2 1EW
UK

*Michelle R. Arkin*
University of California, San Francisco
School of Pharmacy
Small Molecule Discovery Center

and

Department of Pharmaceutical Chemistry
UCSF Mission Bay Campus
1700 4th Street
San Francisco
CA 94158
USA

*Christoph Arkona*
Freie Universität Berlin
Institut für Pharmazie
Königin-Luise-Str. 2 +4
14195 Berlin
Germany

*Mahendra Awale*
University of Berne
Department of Chemistry and Biochemistry
Freiestrasse 3
3012 Berne
Switzerland

*Justin Bower*
Cancer Research UK Beatson Institute
Garscube Estate
Switchback Road
Bearsden
Glasgow
G61 1BD
UK

*Alexander L. Breeze*
AstraZeneca R&D
Discovery Sciences
Alderley Park
Macclesfield
SK10 4TG
UK

and

University of Leeds
Astbury Centre for Structural Molecular Biology
Faculty of Biological Sciences
Leeds
LS2 9JT
UK

***Peter J. Brown***
2 Structural Genomics Consortium
7th Floor, MaRS South Tower
101 College Street
Toronto
ON M5G 1L7
Canada

***Stacie L. Bulfer***
University of California, San
Francisco
School of Pharmacy
Small Molecule Discovery Center

and

Department of Pharmaceutical
Chemistry
UCSF Mission Bay Campus
1700 4th Street
San Francisco
CA 94158
USA

***John A. Christopher***
Heptares Therapeutics Ltd
BioPark, Broadwater Road
Welwyn Garden City
AL73AX
UK

***Miles Congreve***
Heptares Therapeutics Ltd
BioPark, Broadwater Road
Welwyn Garden City
AL73AX
UK

***Jared N. Cumming***
Merck Research Laboratories
Structural Chemistry
2015 Galloping Hill Road
Kenilworth
NJ 07033
USA

***Thomas G. Davies***
Astex Pharmaceuticals
436 Cambridge Science Park
Milton Road
Cambridge
CB4 0QA
UK

***Ben J. Davis***
Vernalis Research
Granta Park
Cambridge
CB21 6GB
UK

***Martin Drysdale***
Cancer Research UK Beatson
Institute
Garscube Estate
Switchback Road
Bearsden
Glasgow
G61 1BD
UK

***Minh-Dao Duong-Thi***
Nanyang Technological University
School of Biological Sciences
60 Nanyang Drive
637551
Singapore

***György G. Ferenczy***
Hungarian Academy of Sciences
Research Center for Natural
Sciences
Magyar tudósok krt. 2
H-1117 Budapest
Hungary

**_Stephen W. Fesik_**
Professor of Biochemistry
Pharmacology and Chemistry
Vanderbilt University School of Medicine
2215 Garland Ave.
607 Light Hall
Nashville
TN 37232-0146
USA

**_Anthony M. Giannetti_**
Google[x]
1600 Amphitheatre Parkway
Mountain View
CA 94043
USA

**_Laurie E. Grove_**
Wentworth Institute of Technology
Department of Sciences
Boston
MA 02115
USA

**_Roderick E. Hubbard_**
Vernalis Research
Granta Park, Cambridge
CB21 6GB
UK

and

YSBL
University of York
Heslington, York
YO10 5DD
UK

**_Sean A. Hudson_**
University of California, San Francisco
Department of Pharmaceutical Chemistry
UCSF Mission Bay Campus
1700 4th Street
San Francisco
CA 94158
USA

**_Aman Iqbal_**
Proteorex Therapeutics Inc.
40 King Street West
Toronto
ON M5H 3Y4
Canada

**_Mike Jaegle_**
Freie Universität Berlin
Institut für Pharmazie
Königin-Luise-Str. 2 + 4
14195 Berlin
Germany

**_Frantz Jean-Francois_**
University of California, San Francisco
School of Pharmacy
Small Molecule Discovery Center

and

Department of Pharmaceutical Chemistry
UCSF Mission Bay Campus
1700 4th Street
San Francisco
CA 94158
USA

***Harren Jhoti***
Astex Pharmaceuticals
436 Cambridge Science Park
Milton Road
Cambridge
CB4 0QA
UK

***György M. Keserű***
Hungarian Academy of Sciences
Research Center for Natural Sciences
Magyar tudósok krt. 2
H-1117 Budapest
Hungary

***Dima Kozakov***
Boston University
Department of Biomedical Engineering
Boston
MA 02215
USA

***Jürgen Maibaum***
Novartis Pharma AG
Novartis Institutes for Biomedical Research
Novartis Campus
4002 Basel
Switzerland

***Eric Nawrotzky***
Freie Universität Berlin
Institut für Pharmazie
Königin-Luise-Str. 2 + 4
14195 Berlin
Germany

***Sten Ohlson***
Nanyang Technological University
School of Biological Sciences
60 Nanyang Drive
Singapore 637551
Singapore

***Puja Pathuri***
Astex Pharmaceuticals
436 Cambridge Science Park
Milton Road
Cambridge
CB4 0QA
UK

***Angelo Pugliese***
Cancer Research UK Beatson Institute
Garscube Estate
Switchback Road
Bearsden
Glasgow
G61 1BD
UK

***Jörg Rademann***
Freie Universität Berlin
Institut für Pharmazie
Königin-Luise-Str. 2 + 4
14195 Berlin
Germany

***T. Justin Rettenmaier***
University of California, San Francisco
Department of Pharmaceutical Chemistry
UCSF Mission Bay Campus
1700 4th Street
San Francisco
CA 94158
USA

***Jean-Louis Reymond***
University of Berne
Department of Chemistry and Biochemistry
Freiestrasse 3
3012 Berne
Switzerland

***Simon Rüdisser***
Novartis Pharma AG
Novartis Institutes for Biomedical Research
Novartis Campus
4002 Basel
Switzerland

***Gordon Saxty***
Fidelta Ltd.
Prilaz baruna Filipovića 29
10000 Zagreb
Croatia

***Duncan E. Scott***
University of Cambridge
Department of Chemistry
Lensfield Road
Cambridge
CB2 1EW
UK

***Christina Spry***
University of Cambridge
Department of Chemistry
Lensfield Road
Cambridge
CB2 1EW
UK

and

The Australian National University
Research School of Biology
Linnaeus Way
Canberra
ACT 2601
Australia

***Andrew W. Stamford***
Merck Research Laboratories
126 East Lincoln Avenue
Rahway
NJ 07065
USA

***Corey O. Strickland***
Merck Research Laboratories
Structural Chemistry
2015 Galloping Hill Road
Kenilworth
NJ 07033
USA

***Sandor Vajda***
Boston University
Department of Biomedical Engineering and
Department of Chemistry
Boston
MA 02215
USA

***Eric Vangrevelinghe***
Novartis Pharma AG
Novartis Institutes for Biomedical Research
Novartis Campus
4002 Basel
Switzerland

***Ricardo Visini***
University of Berne
Department of Chemistry and Biochemistry
Freiestrasse 3
3012 Berne
Switzerland

***Feng Wang***
Vanderbilt University School of Medicine
Department of Biochemistry
2200 Pierce Ave.
802/804 RRB
Nashville
TN 37232-0146
USA

***Richard A. Ward***
AstraZeneca R&D
Oncology iMED
Alderley Park
Macclesfield
SK10 4TG
UK

***James A. Wells***
University of California, San Francisco
Department of Pharmaceutical Chemistry
UCSF Mission Bay Campus
1700 4th Street
San Francisco
CA 94158
USA

***Glyn Williams***
Astex Pharmaceuticals
436 Cambridge Science Park
Milton Road
Cambridge
CB4 0QA
UK

***Jon Winter***
AstraZeneca R&D
Oncology iMED
Alderley Park
Macclesfield
SK10 4TG
UK

***Ee Lin Wong***
Freie Universität Berlin
Institut für Pharmazie
Königin-Luise-Str. 2 + 4
14195 Berlin
Germany

***Daniel F. Wyss***
Merck Research Laboratories
Structural Chemistry
2015 Galloping Hill Road
Kenilworth
NJ 07033
USA

# Preface

Just two decades ago, Stephen Fesik initiated fragment-based ligand design by developing an NMR-based method to search for small, low-affinity ligands in adjacent binding pockets of a protein and to link them to a high-affinity ligand [1]. A broader use of this approach was hindered both by its limitation to relatively small proteins and by a patent application. However, within short time alternative methods emerged, originally based on different NMR techniques, later using protein crystallography. Thus, structure-based design was not any longer restricted to "large" molecules: libraries of much smaller fragment-type compounds were tested experimentally or screened in silico, with the advantage that a small ligand has a much better chance to fit a certain binding site. In further steps, the ligand can grow into the environment of its pocket or can be linked to an adjacent fragment. The only critical step in fragment combination is the search for a linker that combines the fragments in a relaxed, bioactive conformation, optimally stabilizing this favorable conformation.

Ten years later, in 2006, time was already ripe to review the techniques and the accumulated experience in fragment-based ligand design: Wolfgang Jahnke and Daniel Erlanson edited the very first book on this topic [2]. Now, another 10 years later, the discipline has significantly developed and a major number of drug candidates resulted from its use. Thus, we are very grateful that both experts agreed to edit not only a new edition but also a completely new book on fragment-based design. In its introductory section, leading scientists of this area review the role of fragment-based approaches in lead finding and the selection of appropriate targets. Next, an overview on chemical space is provided. The second section discusses library design and various screening techniques, together with a major number of issues that are relevant in fragment-based ligand discovery. The last section presents a significant number of success stories, providing evidence for the broad applicability of fragment-based design in drug research.

As last time, we are very grateful to the editors Daniel Erlanson and Wolfgang Jahnke for assembling such a unique collection of important topics, as well as to all chapter authors for their excellent work. Last but not least we thank the publisher Wiley-VCH, in particular Waltraud Wüst and Frank Weinreich, for their valuable contributions to this project and the entire series.

Düsseldorf — *Raimund Mannhold*
Weisenheim am Sand — *Hugo Kubinyi*
Zürich — *Gerd Folkers*

October 2015

## Reference

1 Shuker, S.B., Hajduk, P.J., Meadows, R.P. and Fesik, S.W. (1996) Discovering high-affinity ligands for proteins: SAR by NMR. *Science*, **274**, 1531–1534.

2 Jahnke, W. and Erlanson, D. eds., (2006) *Fragment-based Approaches in Drug Discovery (Volume 34 of Methods and Principles in Medicinal Chemistry*, eds.), R. Mannhold, H. Kubinyi, and G. Folkers Wiley-VCH Verlag GmbH, Weinheim.

# A Personal Foreword

For the great things are not done by impulse, but by a series of small things brought together.

*Vincent Van Gogh, 1888*

When Wiley-VCH asked us whether we would be willing to edit a new book on fragment-based drug discovery, our first reaction was panic. Editing a book is a daunting task, and having done it once already we knew well what was in store.

Our second reaction was to ask whether a new book was really needed. Since the very first book on fragment-based drug discovery was published by Wiley-VCH in 2006, six more books have appeared, along with dedicated journal issues and dozens of reviews. Was there anything new to say?

Happily, as you will soon discover, the answer is an emphatic yes! This is clearly illustrated by a search for publications containing the phrase "fragment-based drug discovery" in SciFinder®, as seen in the figure.

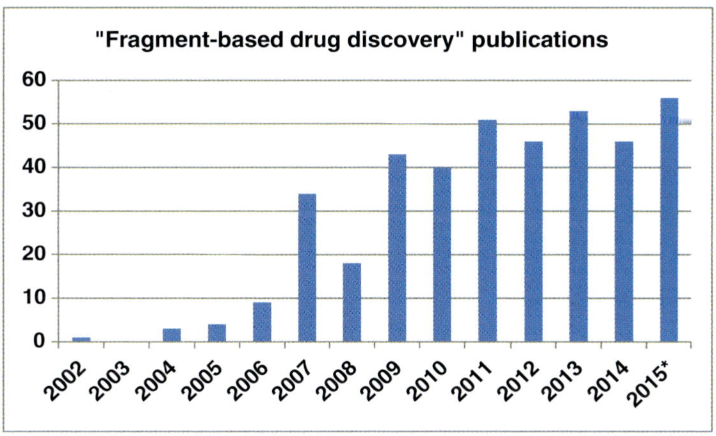


(*2015 numbers through August 25 only)

The past few years have seen a bumper crop of papers on the topic, and given that this search was run in August of 2015 this trend looks set to continue if not accelerate. From its origins as a niche technique, fragment-based approaches have spread throughout the world to organizations large and small and are embraced by biologists, biophysicists, chemists, modelers, and more. More than 30 drugs derived from fragments have entered the clinic (http://practicalfragments.blogspot.com/2015/01/fragments-in-clinic-2015-edition.html), and one (vemurafenib) has already been approved. This book is a comprehensive view of where the field stands – and where it is going.

We would like to thank Wiley-VCH, especially Frank Weinreich and Waltraud Wüst, for encouraging us to undertake this project and patiently working with us through the inevitable but nonetheless frustrating difficulties and delays. We would also like to thank our contributors, all of whom are extraordinarily busy and accomplished scientists. We are thrilled with the response we received to our invitations and with the depth and quality of the chapters. Finally, we would like to thank you for reading. We hope that you will find something useful to apply to your own research: each of our fragmentary efforts advances the great human enterprise of drug discovery.

San Francisco
Basel
August 2015

*Daniel A. Erlanson*
*Wolfgang Jahnke*

# Part I
# The Concept of Fragment-based Drug Discovery

# 1
# The Role of Fragment-based Discovery in Lead Finding

*Roderick E. Hubbard*

## 1.1
## Introduction

Fragment-based lead discovery (FBLD) is now firmly established as a mature collection of methods and approaches for the discovery of small molecules that bind to protein or nucleic acid targets. The approach is being successfully applied in the search for new drugs, with many compounds now in clinical trials (see summary in [1]) and with the first fragment-derived compound now treating patients [2]. The approach has also had a number of other impacts such as providing starting points for lead discovery for challenging, unconventional targets such as protein–protein interactions [3–5], increasing the use of biophysics to characterize compound binding and properties, and providing small groups, particularly in academia, with access to the tools to identify chemical probes of biological systems [6,7].

The other chapters in this book will discuss the details and new advances in the methods and provide examples of how fragments have been used in specific projects. In this chapter, I will draw on my own experiences and view of the literature to discuss three main areas. First, I will review current practice in FBLD, highlighting how and when fragments have an impact on the drug discovery process. Second, I will then review how the ideas have developed, with particular emphasis on the past 10 years. I will discuss how fragment methods and thinking have been extended and refined and how these developments have affected the lead discovery process in drug discovery. Finally, I will discuss some of the areas where we can see that improvements in fragment methods could have further impact on discovery.

The discussion will focus on fragment-based discovery against protein targets. Although there are a few examples of fragments being used against RNA [8–10] and DNA [11] targets, the majority of reported campaigns are against proteins. Two types of protein target will be considered. The first shall be called conventional targets. These are proteins such as kinases where although it is never straightforward to achieve the required selectivity and balance of physicochemical properties in the compound, the proteins usually behave in most of the

*Fragment-based Drug Discovery: Lessons and Outlook*, First Edition. Edited by Daniel A. Erlanson and Wolfgang Jahnke.

experiments and assays. Crystal structures are usually readily obtained, large amounts of pure, homogeneous, and functional proteins can be generated for biophysical studies, and the activity assays are robust and well understood. The second class of target shall be called unconventional targets. There are two types here – the first are protein–protein interaction targets such as the proapoptotic Bcl-2 family or Ras, where experience over the years has eventually established reasonably robust assays and although crystal structures take some time to determine and the protein does not always behave in biophysical assays, it is possible to establish structure-based discovery. The main challenge here is the nature of the binding sites, with often large, hydrophobic, and sometimes flexible sites. The second type of unconventional targets are the results of recent advances in our understanding of mammalian disease biology and consist of new classes of enzymes (such as the ubiquitin processing machinery [12]), disrupting multiprotein complexes, and proteins that are intrinsically disordered in some way (such as the one described in [13]). Here, the primary challenges are often in producing sufficient, homogeneous, functional protein for study, knowing what the post-translational modification state or even which complex is the true target and establishing robust assays to report on activity or binding. This last issue is often not appreciated – it can take a long time to establish the assays on new classes of target, not only because there is intrinsic variability in the behavior of the system but also because there is often not a tool compound available with which to validate the assay.

## 1.2 What is FBLD?

There are two distinctive features of fragment-based discovery compared to other approaches to lead finding. The first is that the discovery process begins with screening a small (usually 1–2000 member) library of low molecular weight (typically less than 20 heavy atom) compounds for binding to a particular site on the target. Key is the molecular weight of the fragments – they are big enough to probe interactions in the protein but small enough to minimize chances of unfavorable interactions. The second distinctive feature lies in the approach to optimizing these hits to lead compounds, either through careful, usually structure-guided, growth of the fragment or through merging information from fragments and elsewhere to generate optimized hits.

In many ways, fragments can be viewed as a state of mind – an approach to use the fragments as chemical tools to dissect what the requirements are for the chemical matter that affects a particular target in the desired way (affinity, selectivity) and using a combination of rational, usually structure-guided, and often biophysics-based methods for generation of the optimized compounds. We can define a fragment approach as one of intent – and that intent affects the strategy, methods, and thinking that is applied during the early parts of a discovery project. Detection and characterization of such weakly binding compounds can be problematic for some classes of target, with concerns over false positive and false negative hits, changes in binding mode, and so on. So, fragment methods engender a questioning, problem-solving approach to research. This is carried through into the usually structure-guided evolution of the initial fragment hits, which allows careful assembly of compounds that bind with high efficiency combined with suitable compound properties.

## 1.3 FBLD: Current Practice

Figure 1.1 and its legend summarize the contemporary approach to fragment-based discovery followed by most practitioners. There are five main components to a fragment platform: a fragment library, a method for finding which fragments bind, characterizing how the fragments bind by determining structure and biophysical measurements, exploring fragment SAR to identify the best fragment(s) to progress, and using the fragment(s) to generate lead compounds. Figure 1.1 also emphasizes how information about binding motifs is combined with information from HTS hits, literature compounds, or virtual screening hits. Other chapters in this book will provide detail on each of these different areas. In this chapter, I am focusing on the impact fragments have had on the lead discovery process. This is best done with some examples.

### 1.3.1 Using Fragments: Conventional Targets

Conventional targets are ones with well-defined active sites (such as most enzymes) where structural information is readily available. It is usual to get a large number of fragment hits for such targets – at Vernalis our experience has been 50–150 validated hits from screening a library of about 1500 fragments [14,15]. A lower hit rate can indicate there may be issues with progressing compounds against the target as discussed later. Modeling of the binding of these fragments can be helpful, but the most effective fragment to hit to lead optimization campaigns uses the detailed information available from experimental structures determined by X-ray crystallography (preferred) or if necessary by NMR. The main issue with NMR is the time it takes to generate structures. A suitable crystal form can generate many hundreds of crystal structures during the early months of a project, whereas it takes at best a few days for NMR methods to generate models for binding. In addition, NMR models rarely have the resolution to give confidence in some of the subtleties of binding mode necessary for design of selective compounds (such as for kinases).

The three main ways of using fragments are growing, merging, and linking. Figure 1.2a–c shows some representative examples that we can use to describe the essential features of each approach.

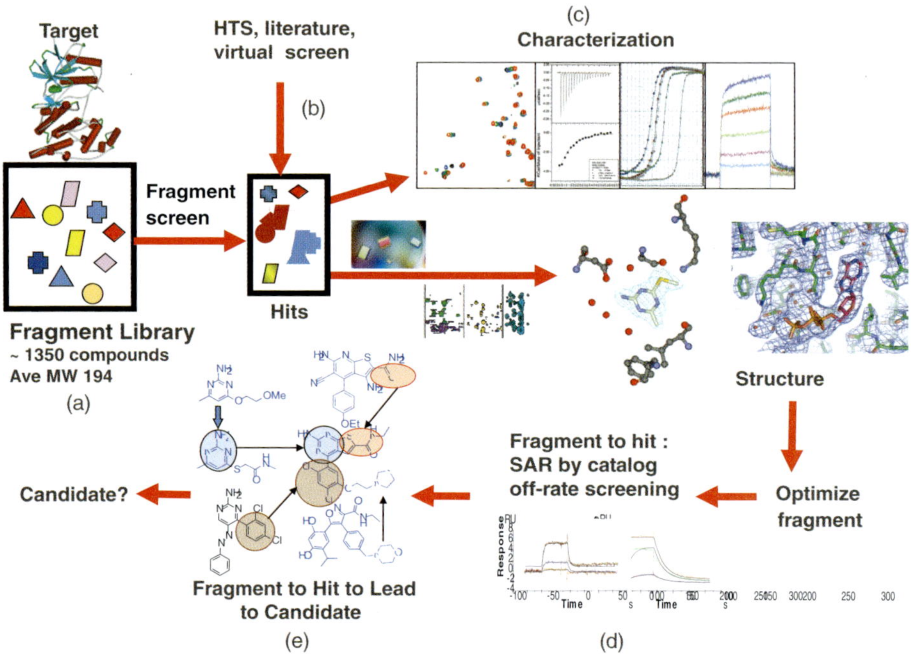


**Figure 1.1** the FBLD process. There are five main components to a fragment platform. (a) Fragment library: there is an extensive literature on the design of fragment libraries [26,31,32,41]. The choice of compounds is constrained both by the demands of the screening methods (solubility, detection) and by the need to evolve the compounds (elaboration vectors, synthetic tractability) as well as avoiding reactive or toxic substructures. Key is the number of heavy atoms in the compounds. Analyses by Reymond [38,39] suggest that the number of possible lead-like compounds (chemical space) increases by around eightfold for each heavy atom. There are many approximations but this means that a fragment library of 1000 compounds of average MW 190 is equivalent to $10^8$ compounds of MW 280 and $10^{18}$ compounds of MW 450. (b) Fragment screening: Table 1.1 summarizes the experiences at Vernalis over the years that are variously described elsewhere [15]. For all techniques, the main limitations are whether the protein target can be prepared in a suitable format for screening and whether the fragments are sufficiently soluble. The most robust method of screening is ligand-observed NMR, which has the dynamic range (typically from 5 mM to 100 nM) seen for fragment binding and particularly important for unconventional targets, as the integrity of the ligand and protein is checked at each experiment. (c) Characterizing fragment binding: for conventional targets, it is often possible to rapidly determine crystal structures of the fragment binding to the protein and, if the biochemical or binding assay is not suitable, use a biophysical method to validate and if possible quantify potency. For unconventional targets, this step is particularly important as the targets can have challenging binding sites, where conformational flexibility or large hydrophobic surfaces can challenge reliable detection of fragment binding. NMR methods can be used for unconventional targets, ranging from binding site localization (HSQC) to NMR-guided models (measuring NOE distances from ligand atoms to protein residues) and full structure determination. These are constrained by the size of the protein and requirement for isotope labeling. (d) Fragment SAR and optimization: there are two well-established methods – (1) SAR by catalog where features of the fragment are used to identify commercially available compounds for purchase and assay and

**Table 1.1** A summary of the characteristics of the most widely used fragment screening methods.

| Method | Sensitivity | Issues |
|---|---|---|
| Ligand-observed NMR – a number of NMR experiments (STD [82], Water-Logsy [83], and CPMG [84]) detect binding of a ligand to the protein | 10 mM–100 nM | Requires large amounts of protein (many 10 s mgs) but the most robust method for detecting weak binding. Each experiment confirms that the ligand and protein maintain their integrity in solution; the use of a competitor ligand to displace the fragment can identify nonspecific binding. These features make the technique particularly suitable for weak binding to challenging targets. Requires careful design to identify allosteric or cryptic binding sites |
| Protein-observed NMR – HSQC experiment detects changes in the local environment of $^{15}N$ or $^{13}C$ nuclei as ligand added | 5 mM–100 nM | Requires isotopic labeling of the protein; limited to proteins<35 kDa; can titrate ligand onto protein and determine $K_D$; pattern of changes in spectra can confirm the same binding site for different ligands and identify allosteric sites; assignment of spectrum allows localization of site |
| X-ray crystallography: either cocrystallization (crystals formed from the preformed protein–ligand complex) or soaking (high concentrations of ligand added to apo crystals) | All affinities | Cocrystallization can require different crystal conditions for each ligand. Soaking of apo crystals requires crystal form with accessible protein binding site; depending on crystal form can identify cryptic sites Crystal structure provides information-rich description of protein–ligand interactions ready for design |
| Surface plasmon resonance [47]; monitor molecular weight change as one component flows past the other attached to a surface | 500 μM lower limit | Two modes – direct binding (protein attached, ligand flows) allows kinetics ($k_{on}$ and $k_{off}$) to be measured; indirect, or affinity in solution, where tool compound attached and protein (in the presence of possible fragment) is flowed past. Main issue is immobilization and integrity of protein on surface |

(*continued*)

(2) detailed design of bespoke compounds to optimize the fragment itself and explore potential vectors for elaboration. More recently, there have been new methods such as off-rate screening [16] that allow rapid profiling of compounds where substituents have been added to particular positions on the fragment, prospecting for suitable vectors for fragment evolution. This can be particularly important when limited structural information is available. (e) Fragment to candidate: medicinal chemistry optimization, supported where possible by rapid crystal structure determination, to bring together information from the portfolio of fragments, hits, HTS, literature, and so on to design and optimize lead compounds.

**Table 1.1** (*Continued*)

| Method | Sensitivity | Issues |
|---|---|---|
| Enzyme/binding assays | 100 μM lower limit usually | The high concentrations of ligand interfere with most formats of assay preventing detection of mM binding fragments; effective for some assay formats and for well-defined active sites – for example, kinases |
| Isothermal titration calorimetry (ITC) [85] | 1 mM–10 nM | Requires too much protein and ligand to be useful for screening, but the most robust method for measuring $K_D$ as long as the interaction involves a change in $\Delta H$ |
| Mass spectrometry | 100 μM | Requires protein/buffer system that "flies" in the mass spectrometer and an interaction that can survive in the gas phase. Effective for covalent interactions –too variable for weakly binding fragments |
| Weak-affinity chromatography [51] – immobilize the target on a silica column, then use LC–MS to identify retained ligands | 1 μM upper limit | A cheap way of measuring weak interactions (using simple LC–MS equipment). As for SPR, the main limitation is attachment of protein to surface and behavior of the fragments on the surface |
| Thermal shift analysis (TSA) [86] – measure the melting temperature of the target by monitoring the increase in fluorescence as the target is heated up in the presence of a dye plus and minus the ligand | 500 μM lower limit | A relatively reliable technique for detecting binding of ligands that bind better than 10 μM, but many false positives and negatives in detecting fragment binding – the change in melting temperature is too small to measure. Uses small amounts of material and inexpensive instrumentation |

Fragment linking is a conceptually very attractive idea – find two fragments that bind in adjacent sites and achieve a high-affinity compound by linking them together.This was the basis of the initial SAR by NMR approach, but with a few exceptions, only the Abbott group (such as summarized in Figure 1.2a, see also Table 1.3), and the follow-on work by Fesik at Vanderbilt (see later), has

**Figure 1.2** (a) Evolving fragments – linking. The SAR by NMR approach was developed by the Abbott group in the 1990s [22] (see also the reviews [63,64]). Protein-observed NMR screening of a library identifies the first site binder (screen 1) that can then be optimized (optimize 1). The second screen (screen 2) is then performed in the presence of an excess of the optimized first site binder to identify the second site binder that can also be optimized (optimize 2). NMR structure determination identifies appropriate vectors for linking the two fragments (link) to give a compound that can then be optimized. The first disclosed example was for FKBP [22]; the first drug discovery project was on stromelysin [65] and arguably the most successful was for the Bcl-2 family of proteins [66–69]. For stromelysin, compound **1** was not from screening but is a known metalloprotease binding motif. Screening in the presence of **1** identified compounds such as **2** that after

(a)

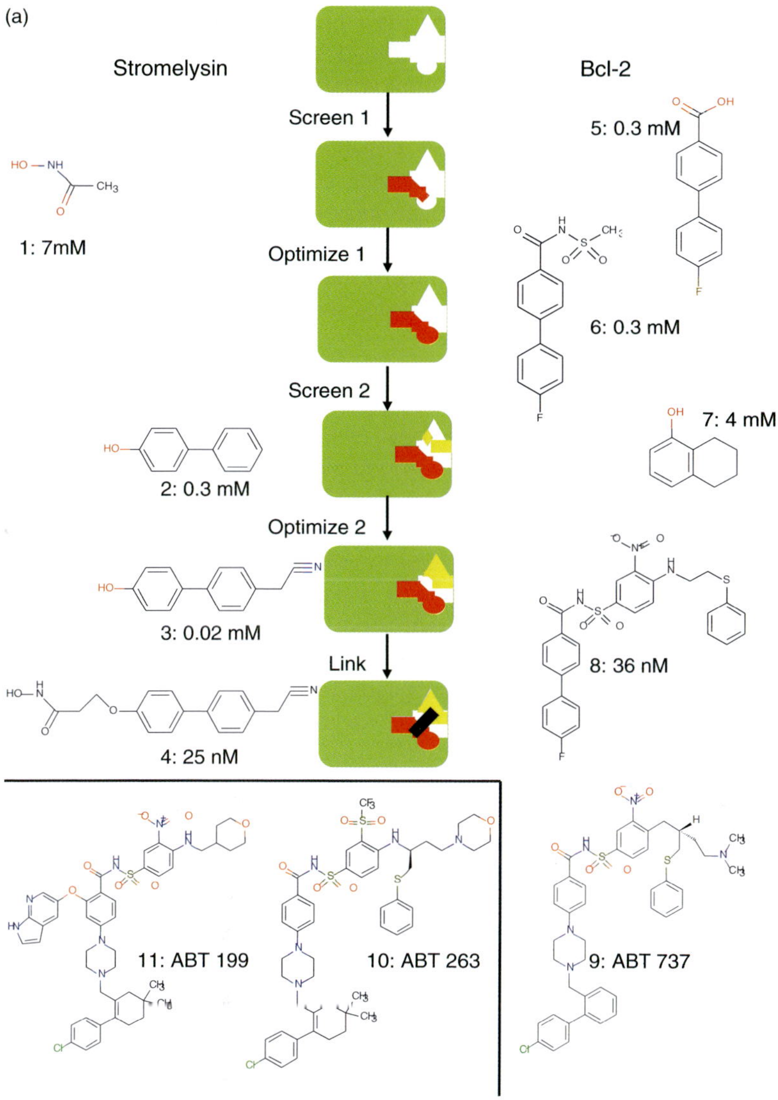


**Figure 1.2** (a) (***Continued***) optimization gave **3**. Combining these in **4** very neatly demonstrates the power of the method – a large increase in potency, clearly retaining the two weakly binding fragments. For the Bcl-2 family, the evolution from the two site binding fragments **5** and **6** is less obvious in compound **7**, although the method did provide starting points for chemistry where conventional HTS failed. A considerable amount of medicinal chemistry optimization was needed to generate ABT-737 [66] that briefly entered clinical trials, followed by ABT-263 [70] with better drug-like properties though still with a dual Bcl-2/Bcl-$x_L$ profile that can give undesired pharmacology. This has recently been succeeded in the clinic by the more Bcl-2 selective ABT-199 [71]. With few exceptions [72], the continued champion of the linking approach is Fesik, now at Vanderbilt (see Figure 1.2d). Most other practitioners find it difficult to identify such multiple sites and commit such dedicated chemistry resources to a linking strategy.

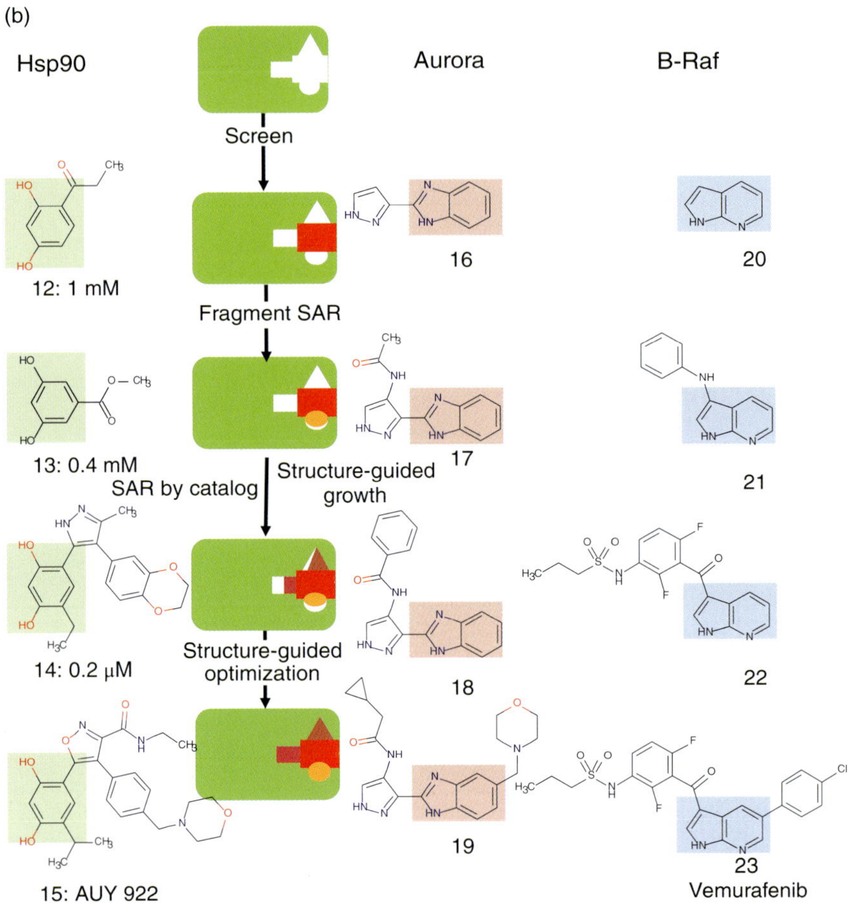


**Figure 1.2** (b) Evolving fragments – growing. There are two widely used approaches for growing fragments. The first is SAR by catalog, where features of the bound fragment are used to search a database of accessible compounds that are then assayed. An example is the HSP90 program at Vernalis that led to **15**, AUY922, currently in multiple phase II clinical trials for various cancers. A ligand-observed screen identified the resorcinol **12** and near-neighbor fragments such as **13**. Search of available compounds for resorcinols that were subsequently triaged with pharmacophore-biased docking, identified compounds such as **14**, which show good affinity. Structure-guided optimization led to **15**. A summary of the HSP90 discovery project is available [73] as well as more details on AUY922 [74,75]. The second approach is growing by careful structure-guided ligand design. The Aurora example from Astex [76] is a particularly good example where fragment **16** was identified from a crystallographic screen of fragments against CDK2; exploration (**17**) identified good vectors for optimization, leading to the hit **18** that was subsequently optimized to the clinical candidate **19**. A similar chemogenomic approach (i.e., transferring knowledge about chemotypes that bind to a particular family) can be seen in the B-Raf kinase example from Plexxikon [2], where **20** was initially characterized binding to Pim-1 kinase, with the related **21** studied in FGFR1 kinase leading to **22** as a potent hit against B-Raf that was optimized to Vemurafenib, **23**, the first fragment-derived compound on the market.

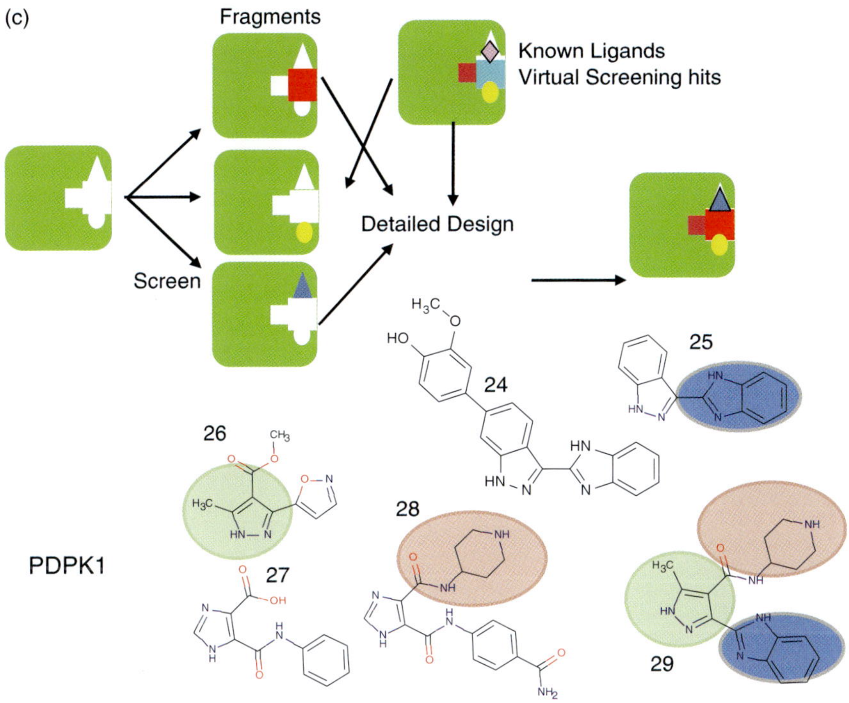


**Figure 1.2** (c) Evolving fragments – merging. This is an approach where information from fragments, derived hits, literature, and HTS hits is all combined together to generate the new lead compound. The approach relies heavily on multiple crystal structures to identify the subtle opportunities provided by binding modes for novel scaffolds and achieving selectivity. The example shown is for PDPK1 [56]. Compound **24** is a known promiscuous kinase inhibitor from which the fragment **25** was derived and the crystal structure determined bound to PDPK1. A ligand-observed screen with staurosporine as competitor identified more than 80 fragment hits, for 50 of which crystal structures were determined. The crystal structures of fragments **26** and **27** showed distinctive binding modes. The crystal structure of an inhibitor from a published study on CDK2, bound to PDPK1, identified a hydrophobic region adjacent to the carboxylic acid of **27**. A search of the available chemicals identified a compound that with small optimization gave **28** with 1 μM affinity for PDPK1. Superposition of the structures of the hits suggested a number of combinations of scaffolds and features. One of these is taking the highlighted features from **25**, **26**, and **28** to generate **29** that showed low nM affinity with good selectivity against other kinases, but importantly affected *in vivo* PD markers in mouse xenograft models.

succeeded in making this work. Most practitioners have found that either the binding site of the target does not contain appropriate features or that they have found it difficult to either design or resource the chemistry required to link the two fragments and preserve the binding mode of the initial hits.

The growing approach is summarized in Figure 1.2b. This has been the most widely used with a number of variants. The first is to use the idea of SAR by

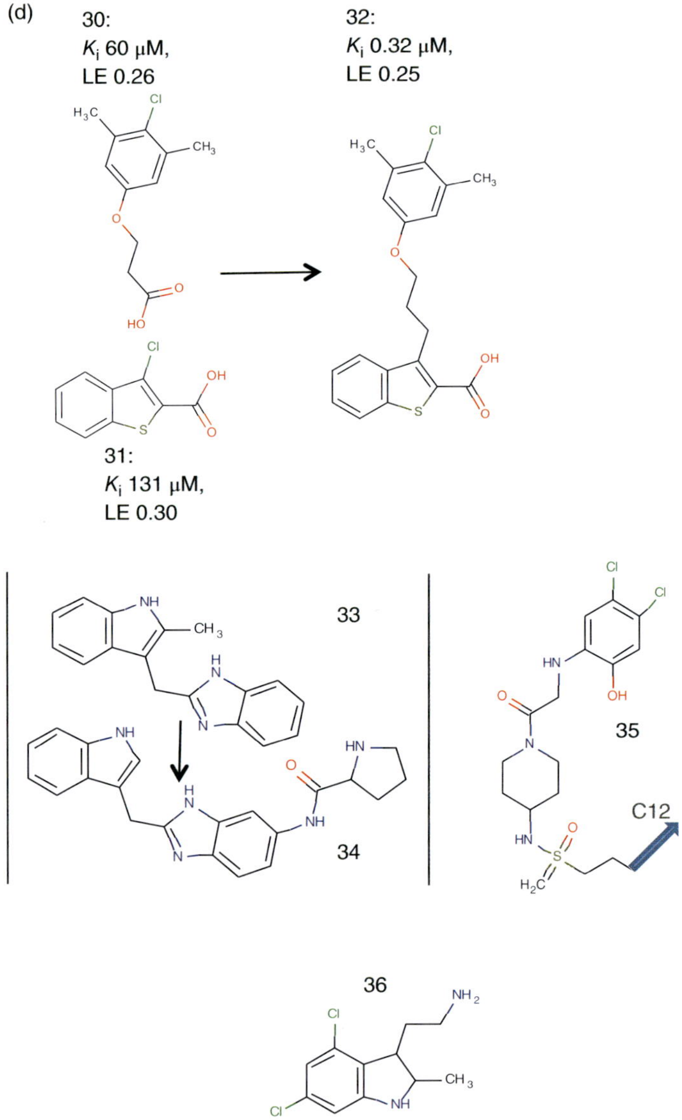


**Figure 1.2** (d) Recent protein–protein interaction FBLD projects. A number of studies have been published recently on the use of fragments for therapeutically important protein–protein interaction targets that have been studied extensively but with little published success through conventional drug discovery efforts. The Bcl-2 family member Mcl-1 is such an example. The Fesik group at Vanderbilt used SAR by NMR to identify **30** and **31** binding in adjacent sites that when linked (**32**) gave a useful starting lead compound for further evolution [77]. The same group used a similar approach to identify **33** that when grown gave **34** as an inhibitor of Ras [4], a key oncogene. Other recently published Ras projects are tethering through the G12C mutant as in **35** from a group at UCSF [5] and identification of another binding pocket for fragment **36** at Genentech [3].

catalog – that is, where the fragment hit provides a central scaffold with which to search for near-neighbor compounds from available compounds (either the corporate collection or that can be purchased). The second is to generate limited libraries of compounds based on the fragment to explore vectors for affinity or selectivity. Here, a recent innovation is to look at changes in the off-rate of binding as a marker for improvements in binding (the so-called off-rate screening [16]). For both of these approaches, the key value is in characterizing possible vectors for elaboration and the types of functional groups that could be used – developing SAR around the hit fragment. It is important to characterize the binding as it is probable that the evolved compound will include additional atoms that are not optimal – and these should be removed before further optimization is attempted. The final and most widely used method for growing fragments is to use detailed structure-based design to iteratively grow the fragment a few atoms at a time to pick up specific interactions with the binding site.

In the merging approach, insights from the binding mode of the fragments are combined with information from literature, virtual screening, or HTS hits to design new scaffolds. The PDPK1 in Figure 1.2c is one example; others include some of the series developed at Pfizer against biotin carboxylase [17] and the oral compound BEP800 [18] against HSP90 designed at Vernalis. For this type of approach, the main requirements are multiple crystal structures and confidence from the medicinal chemists to embark on such radical compound redesign. This is one of the approaches that is being used more frequently as fragment-based approaches are being embraced by large pharmaceutical companies and integrated with HTS. It is fascinating to see how a medicinal chemist with experience of fragment-based discovery approaches HTS hits – their first instinct is to dissect the compound down to the core binding motifs, and once that has been identified at the fragment level and then optimized, the functionalization from the HTS hit can be reassembled.

### 1.3.2
### Using Fragments: Unconventional Targets

One of the striking advantages of FBLD is that fragment hits can be found for nearly all targets (see Figure 1.5 later). This is particularly valuable for more challenging unconventional targets such as protein–protein interactions, multi-protein complexes, intrinsically disordered proteins, and new classes of targets (such as ubiquitin-specific proteases [12]) where HTS screening of conventional libraries fails to identify tractable hits. The early, high-profile example of this was the discovery of ABT-737 (and subsequently, ABT-263 and ABT-199) as seen in Figure 1.2a; additional examples are shown in Figure 1.2d such as the work on Mcl-1 by the Vanderbilt group and that at Genentech and UCSF in identifying hit compounds that affect the activity of K-Ras (referenced in the figure legends). These proteins are often difficult to work with in terms of solubility, homogeneity, and folding. In addition, the binding sites are often quite shallow or diffuse, which means that binding affinity (and thus ligand efficiency, see Table 1.3) is low. The main published successes have, therefore, not surprisingly, used NMR methods to detect and characterize binding, a technique that not only has the necessary sensitivity but also importantly has the in-built quality control to assess the protein state for each binding experiment.

## 1.4
## What do Fragments Bring to Lead Discovery?

Later in this chapter, I will discuss some of the details of how our ideas and practice of FBLD has evolved over the past 10 years. Here, the major features will be summarized of how FBLD is used and has had an impact on drug discovery, with a somewhat arbitrary separation of comments against both conventional and unconventional classes of targets.

For conventional targets:

- Fragments can sample the chemical space of what will bind to a binding site. There are still not many examples where this has been analyzed in detail (see [19] for an analysis on HSP90 compounds), but fragments usually recapitulate the key binding features seen in compounds derived by other methods (such as HTS or natural product derivatives).
- Fragments can show selectivity even for closely related proteins and even when a fragment binds to many similar targets, it can adopt different binding modes (– see the example for kinases in Table 1.4 and Figure 1.7 discussed later).
- Where crystal structures are available, the important first step in assimilating the set of fragment hits is to categorize the fragments on binding mode.
- The selection of which fragments to take forward for evolution is as much about opportunity (such as IP, selectivity, and chemical tractability) as the current affinity of a specific fragment. Often, there will be regions of a fragment that are not optimal or required for binding. For this reason, it is important to explore the SAR of the initial fragment(s) before optimization, identifying which binding modes and potential vectors offer the opportunity to gain selectivity and affinity.
- It is usually the case that the central core of the fragment does not change the binding mode as the fragment is evolved. If it does, then it can be a sign that the initial fragment was not optimal or that the additional atoms added to the fragment have challenged the binding efficiency – as seen in the evolution of the same fragment in three different kinases [20].
- A concern voiced by some is how it is possible to achieve novelty when most are screening very similar fragment libraries against the same targets. As discussed later (and shown in Figure 1.8), even where the same fragments are found, the medicinal chemists will optimize differently.

For unconventional targets:

- Fragments provide the opportunity to assess challenging targets for chemical matter that binds; the hit rate can be an indication of how difficult it is going to be to progress compounds against a target (see Figure 1.5 later).
- It is usually not an issue in identifying fragments that bind to such targets.
- The major challenge is establishing robust, validated assays – both for establishing binding and for activity. An issue is that, often, there are not validated tool compounds available and so it can take some time (and iteration with evolving fragments) to establish an assay that can be relied on.
- The use of multiple (sometimes called orthogonal) binding experiments can be crucial to success – helping to validate the binding and binding mode.
- A major issue is the time it takes to generate lead compounds – it takes commitment to the long haul. A project can spend many years in the exploratory phase, establishing robust assays and validated starting points before a drug discovery program can begin. This time in fragment and early hit space does not necessarily require large resources, but it can be long. The key to continuation of the project is maintaining confidence in the target and hits and that the next steps for the project are clear.

For all targets:

- A fragment screen is a rapid way of assessing how difficult it will be to find new chemical matter that will bind to a particular target – a low fragment hit rate does not necessarily mean the target is undruggable, but it can indicate that finding high affinity, selective compounds could be a challenge.
- Start the fragment campaign early enough in the project cycle
  - Many large companies have found they needed to establish dedicated teams that promoted the fragment approaches – this required top-down implementation by management to ensure the resources were applied (and staff objectives suitably adjusted).
  - The main issues are cultural – most organizations have the different technologies/capabilities in place to perform fragment screening. Key is integration alongside more conventional HTS type of methods and building a culture of seeing the techniques as complementary and not a competition.
- The focus (and required assays) is on binding rather than activity in the early part of lead identification.
- An important feature of fragments is that the optimization process starts with a core that is small. Careful optimization can maximize the ligand efficiency of evolving compounds and ensure incorporation of the optimal properties. A drop in ligand efficiency on optimization should always be questioned. In addition, maintaining a high ligand efficiency during early to mid lead optimization allows that efficiency to be spent in fine-tuning

the physicochemical properties and efficacy of the compound in the later stages. Overall, fragments provide the opportunities and scope for the medicinal chemist to develop compounds with optimal properties.

- For most targets, fragments provide choice – this gives potential for many different chemical series as starting points, providing backups, ideas on key interactions and binding modes that can be exploited in optimization or the potential for scaffold hopping if issues appear during optimization (such as physicochemical properties, CYP inhibition, or hERG).

In addition to explicit FBLD campaigns, the fragment ideas have permeated conventional lead optimization. The most recent example I have come across is the work at Kaken Pharmaceuticals on PDE7 [21], where a novel HTS hit was dissected back to a fragment and then evolved – there are many other examples emerging in the literature that show how fragment thinking has infected many areas of medicinal chemistry.

## 1.5
## How did We Get Here?

### 1.5.1
### Evolution of the Early Ideas and History

It is now nearly 20 years since the first publication described a fragment-based approach to discover potent lead compounds [22] and almost 10 years since the publication of the first edition of this book [23]. Table 1.2 provides a summary of the ideas and methods that made an important contribution to the development of the first use of fragments in ligand discovery. Table 1.3 summarizes the early developments in the field that led to the publication of the first edition of this book in 2006.

### 1.5.2
### What has Changed Since the First Book was Published in 2006?

In 2006, the publication of the first edition of this book announced the arrival of fragment-based discovery as a thriving area of method development and application in the pharmaceutical industry.

The book contained 16 chapters which can be summarized as follows:

1) Introduction with a brief sketch of the field
2) The idea of multivalency
3) Ideas in entropy of binding and combination in fragment binding
4) Multiple solvent crystal soaking (MSCS) as a method for experimental solvent mapping
5) Cheminformatics analysis of fragments in literature ligands

**Table 1.2** The ideas and concepts that underpinned the emergence of the first demonstration of fragment-based lead discovery in 1996.

| |
|---|
| Double the $\Delta G$; square the $K_D$: papers by Jencks [87] from the early 1970s remind the community that $\Delta G = -RT\ln K$ – so combining two weak interactions gives a strong association. Also that the first ligand binding overcomes rotational and translational entropy, so additional interactions are stronger. |
| High-throughput crystallography: Perutz and coworkers [88] demonstrated in the early 1980s the benefit of multiple crystal structures in analyzing protein–drug interactions |
| Functional group efficiency: Andrews, Craik, and Martin [89] developed the idea in the early 1980s that particular functional groups make a distinct average contribution to binding |
| Functional group binding – computational: Goodford [90] developed GRID in the early 1980s to map the predicted binding of single-point probes to an active site with impact on drug discovery such as Relenza [91]. The MCSS [92] approach extended this to larger functional groups and LUDI [93] derived interactions from crystal structures |
| Functional group binding – experimental: the first experimental mapping of a binding site was by crystallography, with the MCSC approach pioneered by Ringe and coworkers [94] in the early 1990s and developed by others [95,96] |
| Fragment linking – computational: Approaches such as HOOK [97], Caveat [98], and LUDI [99] were developed in the early 1990s to link functional groups; the main challenges were predicting binding affinity and design of synthetically tractable molecules |

6) Analysis of fragments in marketed drugs
7) Multiple copy simultaneous search (MCSS) for computational solvent mapping
8) Using NMR both to identify fragments that bind and to guide fragment assembly
9) SAR by NMR from the Abbott group
10) The Astex platform with a focus on high-throughput crystallography and library design
11) The platform developed at SGX using high-throughput crystallography with some examples and discussion of fragment evolution strategies
12) The use of NMR and X-ray crystallography for some particular examples of using fragments at Abbott
13) The use of mass spectrometry
14) Tethering
15) Click chemistry
16) Dynamic combinatorial libraries

It is fascinating to look critically at these chapters from this distance of 10 years and how it represented the state of FBLD at that time. Many of the methods described in the chapters on click chemistry, dynamic combinatorial libraries, MCSS, and multiple solvent crystal structure (MCSC) have found little application in practical FBLD. However, quite significant parts of Chapters 8–10 on the central methods and Chapter 11 on a crystallographic platform and fragment

**Table 1.3** The developments in FBLD between 1996 and 2005.

| |
|---|
| SAR by NMR: the phrase "SAR by NMR" was coined to describe the approach developed at Abbott that saw the first publication on fragment-based drug discovery [22]. Protein-observed NMR ($^{15}$N-$^{1}$H HSQC) spectroscopy is used to screen for binding of a fragment that is then optimized. A second screen is then performed in the presence of the optimized fragment to identify second site binders, which are then linked together. The optimization and linking of fragments is aided by structure determination by NMR. The first paper demonstrated proof of concept in identifying nanomolar inhibitors of the FK506 binding protein [22]; subsequent early disclosures were on stromelysin [65], E2 from papilloma virus [100], urokinase [101], and phosphatases [29] |
| Combinatorics: an early idea to emphasize the power of the method is as follows [65]: screening of a 1000 fragment library for binding to each of two sites and then trying 10 different linkers in a total of 2010 experiments, which samples the $1000 \times 1000 \times 10$ (10 million) possible compounds |
| Detecting weak binding: techniques established and widely used by 2005 were protein-observed NMR [22] ($^{15}$N-$^{1}$H HSQC), ligand-observed NMR [26] (STD, Water-Logsy, CPMG), X-ray structure of fragment mixtures soaked into apo crystals. Enzyme/binding assays dismissed by most (Plexikkon scaffolds were the exception) [102] |
| High-throughput crystallography: Nienaber at Abbott was the first to publish screening by crystallography [103]; there was considerable development and promotion by companies such as Syrrx [104], SGX [105], and Astex [106] |
| Fragment evolution: two main approaches had been established by the mid-2000s – linking and growing [107]. Using the target protein as a template for self-assembly is an attractive idea that was discussed frequently, such as a chapter in the first edition of this book and mentioned in this review [107]. Also of great promise was the idea of tethering [107–110] from Sunesis and colleagues |
| Hann complexity [111]; this is a concept whose significance has grown over the years. This was a theoretical analysis arguing that the chance of finding a hit is a balance between enough complexity in the molecule to bind with sufficient affinity to register in an assay and being too complex that the compound no longer will fit into the binding site |
| Ligand efficiency [112]: this concept (that the efficiency of binding is the amount of free energy of binding for each heavy atom) built on earlier ideas about the maximum affinity obtainable by ligands [113]. The metric was rapidly adopted by the fragment community, with slower uptake by more conventional medicinal chemists. It is a persuasive way of demonstrating that a small (10 heavy atom) compound with a $K_D$ of 1 mM is equivalent in "quality of binding" as a drug-like compound (30 heavy atom) with a $K_D$ of 1 nM. |
| Fragment library design: the essential details were established of the approach to identifying fragments with defined criteria for exclusion (toxicity/reactivity) and with desired properties and opportunities for evolution, together with the rigorous QC for library curation [26]. The so-called "rule of 3" was identified by Astex [114] and perhaps overused in the community (people find security in rules – it saves thinking) |

This is a somewhat selective precis of the origin of the underlying principles and early applications of fragment-based methods (see also [81]).

evolution would not require much updating to be included in this book. The chapter on tethering is relatively up to date; the approach remains intellectually appealing, and although there have been some high-profile applications, the approach [5,24] has not been widely used. Apart from these chapters, the book lacked in real examples of what we would today call FBLD in practice. There is some screening data presented and linking portions of molecules; however, most of these were small compounds identified by HTS. At first consideration, this is a striking omission. However, looking back over the publication timelines, the first examples of success at companies such as Astex and Vernalis were being presented at conferences but had not been published at the time the book was assembled.

The book did not contain a comprehensive summary of the developments across the FBLD field – it carried individual reports from particular practitioners where they tend to stress their own platforms. Combining the book with two representative papers from Astex [25] and Vernalis [26] allows a summary of the contemporary state of FBLD in the mid-2000s from which to begin our discussion of how the methods and approach have changed.

What is striking is that most of the essential ideas in fragment-based discovery had made the transition from the SAR-by-NMR approach to be essentially the same approach as followed today. The major difference has been in the extent of application and demonstration of success. Other articles in this book will describe those successes. What I want to focus on in the next section are the small differences in emphasis and demonstration of ideas and principles that emphasize the strength and opportunity given by the approach.

## 1.6 Evolution of the Methods and Their Application Since 2005

Most new technologies follow the same cycle illustrated in Figure 1.3. When a new technology arrives (the technology trigger), there is great expectation of the impact this technology will have. These expectations are usually overhyped (perhaps they have to be to realize the investment for developing and implementing the methods). The methods are then often implemented and applied inexpertly or to the wrong problems. This leads to disillusionment – until the practitioners become more experienced with the methods and how and when to apply them. The methods then contribute to the general productivity of the area. There have been many such hype cycles in drug discovery – molecular modeling, combinatorial chemistry, genomics, proteomics, and so on. Fragment technologies have to some extent gone through such a cycle – the early introduction of SAR by NMR led to a number of large companies attempting to replicate the success of the Abbott group and failing. The advantage fragments have had is that the methods do not require very large investments and so small, structure-based companies such as Astex, Vernalis, Plexikkon, SGX, and others were able to develop, implement, and refine the methods with a focus that comes from within small companies.

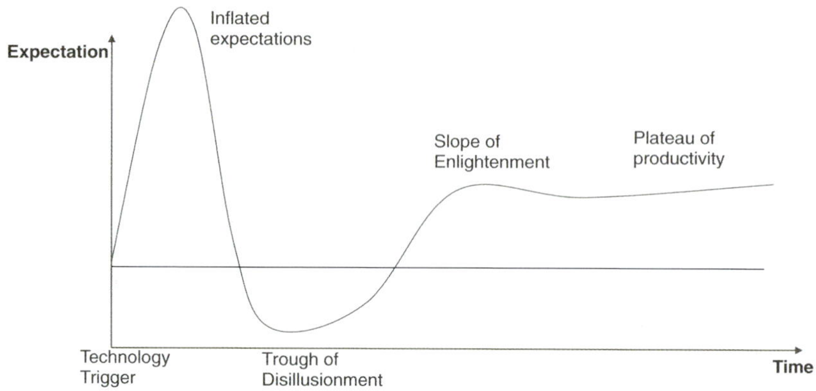


**Figure 1.3** The technology hype cycle. The first reference I can find to the use of this was in an editorial introducing fuzzy systems by Bezdek [78]; it has subsequently been branded by Gartner management consultants and can be applied to most new technologies; the first use I saw for drug discovery was by Mark Murcko.

The successes mean that all accept FBLD has reached that plateau of productivity, at least for conventional targets. This is evident in the rise in the number of publications describing projects where FBLD was the main hit-finding approach. Figure 1.4 shows an analysis by Derek Lowe (personal communication) of the number of papers in the *Journal of Medicinal Chemistry* each year. Between 1996 and 2004, there was 1 review [27] and 3 projects (IL2 [28],

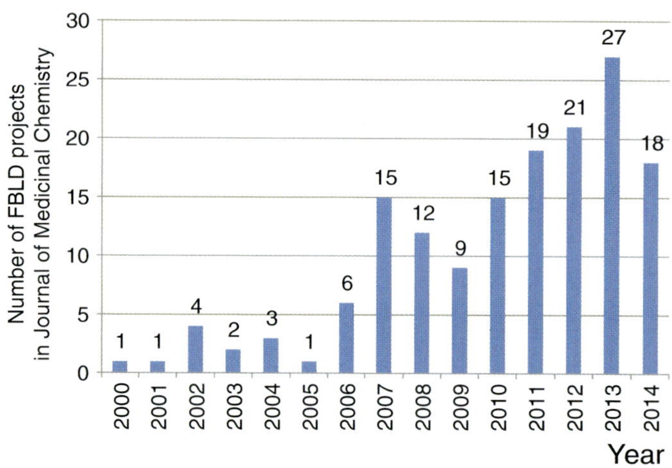


**Figure 1.4** Publications on FBLD. This plot uses data provided by Derek Lowe from Vertex who counted the number of publications on fragment-based discovery that appeared in different years in the *Journal of Medicinal Chemistry* (up to August 2014). The method was to search for fragments as a keyword and then check from the abstract that they were relevant.

PTP1B [29], and Src SH2 [27,30]), whereas in the first 6 months of 2014, there were 13 relevant papers.

So, what are the main differences between 2005 and now? The following is a rather personal and anecdotal summary of the emergence and demonstration of some of the key ideas and methods that underpin and inform current practice in FBLD. The approach has been to go through the various talks I have given at fragment-based discovery meetings over the past 10 years and summarize the main results that I was presenting under the general title of fragmentology (a rather obvious pun on the tendency of many in the United States to corrupt my first name). The timing of these disclosures was sometimes limited by confidentiality issues, but it is a useful reflection of the emerging ideas and practices.

### 1.6.1
### Developments in Fragment Libraries

The essential principles of how to design a fragment library were established in the mid-2000s and there have been a succession of papers describing only small variations in this basic approach from various organizations [31–33], including one personal example of the design of a library that maximally represents available compounds to aid in SAR by catalog [34]. Perhaps, a distinctive lesson to be learnt for a large organization is to ensure the whole team who need to use the fragments are involved in the design of the library and thus have ownership of its use [31]. One recent development is to have a library of fragments containing fluorine to exploit the sensitivity of $^{19}$F NMR [35]. There is an issue of the limitations of which compounds can be synthesized or are available containing such an atom. Another way to exploit $^{19}$F is to screen a generic fragment library for displacement of a $^{19}$F-labeled probe molecule [36].

The notion that a fragment potentially samples a huge chemical space has been around from the early days of fragment library design. From the early days of FBLD, the chemical universe was cited as containing up to $10^{60}$ compounds with less than 30 heavy atoms [37]. An exhaustive analysis by Reymond showed that using known chemistries, the number of compounds of up to 11 heavy atoms increases by about eightfold per heavy atom [38,39]. This suggests that a library of 1000 compounds of MW 190 (13 heavy atoms) is equivalent to $10^9$ compounds of MW 280 (20 HA) and $10^{20}$ compounds of MW 450 (32 HA). More recent analyses [39] from this group were less exhaustive and gave a smaller average rate of growth (about fivefold per HA), but this is still a powerful demonstration for fragments being as small as possible and the size of compounds in the libraries of most practitioners have reflected this. The Vernalis library is currently average MW 190 Da; that for Astex is anecdotally closer to MW 170 Da. However, as shown recently, it is still likely that screening a library of a few thousand fragments will leave gaps in chemical space coverage, and perhaps virtual screening can help to identify these [40].

There has been some analysis of the properties of the fragments that are hits versus not hits against any targets [14,41,42], which have concluded there are no distinctive features or reasons to exclude a particular compound from the library – in some senses, the nonhits are fragments waiting for the right target. In general, the hit rates from larger compounds (i.e., above 20 or so heavy atoms) is seen to be lower, for the obvious reasons of complexity. One of the few trends is that fragments that are hits against protein–protein interaction targets are slightly larger than those for conventional targets and have a slightly higher lipophilicity. This is not surprising, given the nature of those binding sites and that the library compounds need to be highly soluble for screening and thus have many polar groups. During the late 2000s, a theme emerging from analysis of developability of drugs was that an increase in 3D character is useful [43] and there has been some discussion of generating 3D fragment libraries. To date, this has remained a hypothesis – there are not yet the compounds available to test if having more 3D fragments (usually requiring more $sp^3$ character in the core scaffold) has any advantage.

### 1.6.2
### Fragment Hit Rate and Druggability

There have been a number of papers suggesting that the hit rate from a fragment screen reflects how difficult it will be to generate high-affinity ligands against a target. The first publications [44,45] were followed by others [14,41]; a more recent analysis of the Vernalis experiences (see Figure 1.5) [15] suggests that the

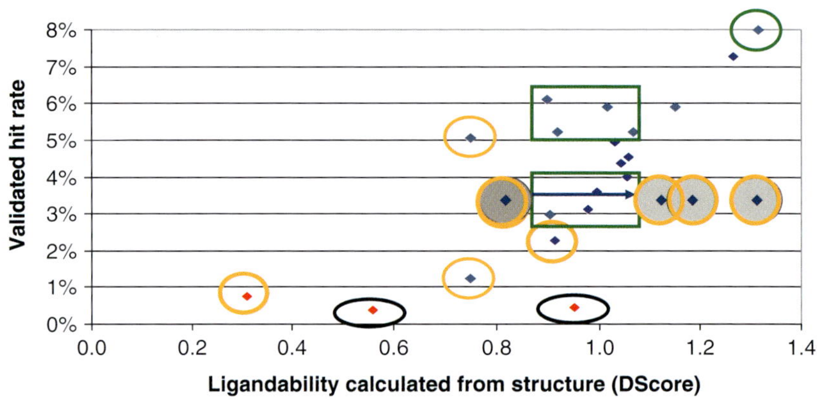


**Figure 1.5** Druggability and conformational change. This plot is an update of an analysis published in 2009 [14] on the number of validated hits identified binding to different targets by ligand-observed NMR screening of the Vernalis fragment library. Each point represents a different target with kinases highlighted with a green box and protein–protein interaction targets with a yellow circle; the *x*-axis is the calculated druggability for a representative structure calculated with the program Sitemap. The red points are targets for which it proved difficult to progress in lead optimization (from left to right, β-catenin, Pin1, and Hsp70). The blue arrow and shaded yellow circles reflect how the shape of the binding site changed as leads were optimized for Bcl-2.

initial hit rate from fragment screening gives an indication of the "ligandability," with two main additional observations. The first is that for some targets (such as protein–protein interaction targets), the binding surface is quite malleable and that the right ligands can induce or select a conformation of the protein that is able to bind a ligand more strongly; this is a demonstration of a principle discussed by others [46]. The second is that a fragment screen hit rate that is lower than that expected on the basis of the known structure of the target can indicate possible issues in evolving ligands against that target.

### 1.6.3 Developments in Fragment Screening

This issue will be discussed at length in other chapters in this book. The most striking change since 2005 is the widespread use of surface plasmon resonance (SPR). This is primarily because there is a better understanding and wider expertise in how to use the method [47,48] as well as improvements in the sensitivity of the hardware. There continue to be proposals for new methods. Some are variations on immobilization of the protein to allow detection of ligand binding by optical [49], NMR [50], or mass spectrometry methods [51], others are monitoring changes on a fluorescent labeled protein [52] or exploiting the sensitivity of $^{19}$F in NMR spectroscopy [53].

### 1.6.4 Ways of Evolving Fragments

Again, there will be many examples in this book of fragment growing, merging, and perhaps linking. One new idea for exploring the potential for optimization of a fragment is to exploit SPR to screen rapidly (and cheaply) small libraries of compounds that exploit a particular vector on a fragment. The essential idea is that the increase in binding affinity typically seen on evolving a fragment from 100 s to single-digit μM is a change in the off-rate of binding [16].

### 1.6.5 Integrating Fragments Alongside Other Lead-Finding Strategies

All of these successes have led to increased investment in the methods within most pharmaceutical companies. There has been a tendency for FBLD to be the method of last resort for unconventional targets that fail in HTS. For these targets, the main issue is to allow the flexible, multidisciplinary and problem-solving approach the time necessary to get FBLD methods to work. Where there is still a real challenge in some companies is how to organize project teams and processes to effectively use fragment methods for more conventional targets such as kinases. There are real opportunities in large companies to combine the insights that can be given by fragments with information provided by HTS hits, either from a fragment screen or by designing fragments from the HTS hit. However, a fragment optimization campaign can spend much longer working with relatively low-affinity compounds (typically in the hundreds of μM range) compared to HTS hits (which often begin with tens of nM affinity). If affinity is the metric used to drive the project, then the HTS hit will displace the fragment hit. However, the fragments can often give hits with better properties – better not necessarily faster.

### 1.6.6 Fragments Can be Selective

An early concern was that the same fragments would emerge as hits for multiple examples of the same protein family. The protein kinases represent a major area of activity for the past 10 years; a surprising degree of selectivity is seen in the fragment hits found against such targets. Two main points emerge from an unpublished Vernalis analysis of screening the same library against eight different protein kinases. The first is that the number of fragments that are just a hit for a single kinase (the leading diagonal of Table 1.4) is quite high; the pattern of selectivity (the off-diagonal numbers) is also quite marked. Second, there are remarkably few fragments that are hits against more than two kinases (the numbers are 41, 26, 14, 14, 8, and 2 fragments for 3, 4, 5, 6, 7, and 8 kinases, respectively). One of the fragments always seen binding is compound **37** but as shown in Figure 1.6, it has the capability of binding in many different ways. It is not like a conventional "frequent hitter" – the promiscuity is real and not an assay artifact.

**Table 1.4** Fragments are selective.

| Total number of hits | 28<br>CDK2 | 102<br>DYRK1A | 55<br>Kin1 | 37<br>Jnk3 | 51<br>Pak1 | 42<br>Pak4 | 32<br>PDPK1 | 42<br>Stk33 |
|---|---|---|---|---|---|---|---|---|
| CDK2 | 6 | 14 | 11 | 9 | 8 | 5 | 8 | 13 |
| DYRK1A | | 30 | 35 | 19 | 18 | 21 | 20 | 29 |
| Kin1 | | | 9 | 18 | 20 | 12 | 14 | 12 |
| Jnk3 | | | | 7 | 13 | 9 | 12 | 10 |
| Pak1 | | | | | 20 | 12 | 7 | 7 |
| Pak4 | | | | | | 13 | 8 | 7 |
| PDPK1 | | | | | | | 4 | 12 |
| Stk33 | | | | | | | | 8 |

The table summarizes the results at Vernalis for screening a series of kinase targets by ligand-observed NMR. The experiments were performed over a 10 year period when the fragment library consisted of between 1100 and 1500 fragments. As the library has been updated and has evolved over the years [14], only 565 fragments were screened against all 8 kinase targets. Of these, 297 were never a hit against any kinase. Ninety-six of the fragments hit a single kinase; this is not a function of the threshold for binding. In general, kinases bind fragments with relatively high affinity (up to 100 nM) so the fragments are showing considerable selectivity. The table summarizes how many of the 565 fragments were hits against each of the kinases and how many were hits against two kinases. For example, the highlighted boxes show that although 12 fragments were hits for both Pak1 and Pak4, there were 20 and 13 fragments out of the 565 that were hits for Pak1 and Pak4, respectively, and not for any of the other kinases.

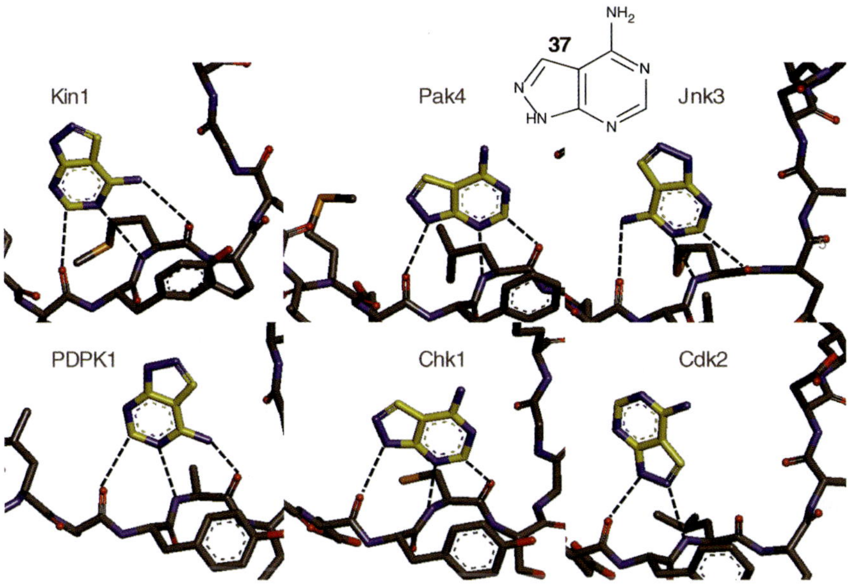


**Figure 1.6** Binding modes of fragments. Details of the crystal structures of **37** bound to six different kinases. The hinge binding region of the ATP binding site of each kinase is shown in stick with carbon atoms in gray, oxygen in red, nitrogen in blue, and sulfur in yellow; the carbon atoms of **37** are shown in yellow with hydrogen bonds to the hinge backbone shown as dashed lines.

## 1.6.7
## Fragment Binding Modes

As fragments became established, there was a flurry of activity disassembling known compounds. Not surprisingly, the fragments sometimes adopted the same binding mode as the parent ligand, sometimes not; it depends on the

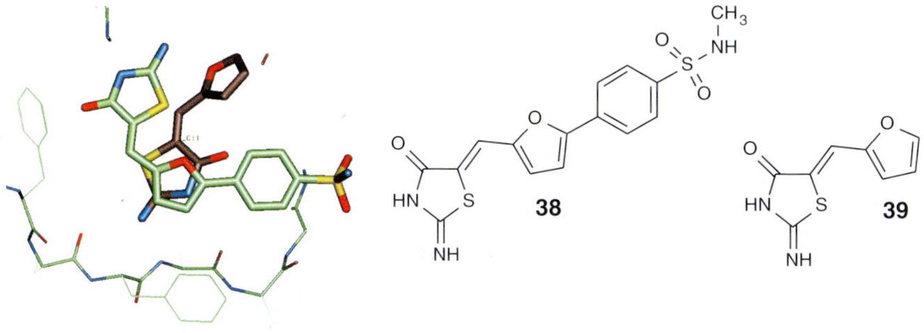


**Figure 1.7** Deconstruction of hits to fragments. Overlay of the crystal structure of compounds in stick **38** (light green carbons) and **39** (dark brown carbons) bound to the hinge region of CDK2 (lines).

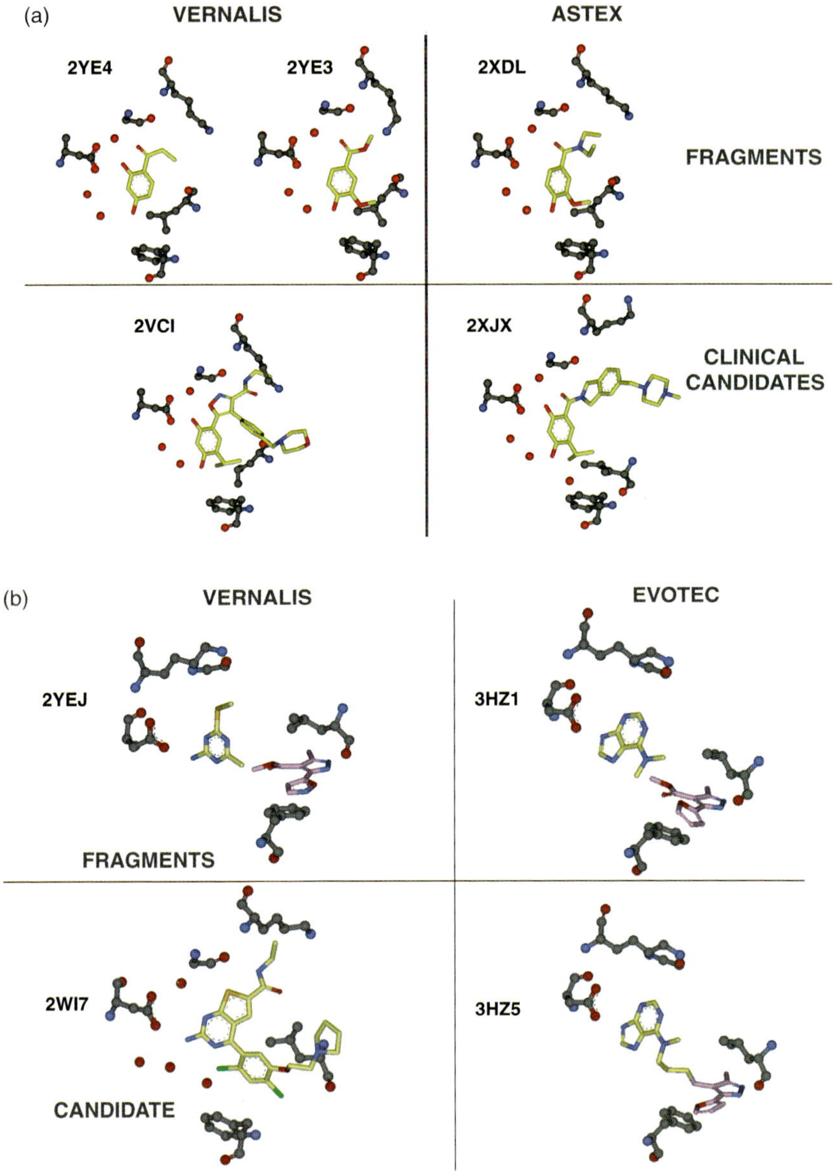


**Figure 1.8** Chemical space and novelty. Crystal structures (PDB codes shown) of various compounds (stick with yellow or purple carbons) bound to HSP90 (ball and stick, key residues shown). The top panels of (a) show fragments discovered at Vernalis (left) [73,74] and Astex (right) [79]; clinical candidates derived from the fragments at each company are shown in the bottom panels. The top panels in (b) show a pair of fragments discovered at Vernalis (left) [18,73] and Evotec [72,80] (right) with the resulting optimized compounds from each company in the bottom panels.

efficiency of binding and the compromises required to bind the full ligand [54,55]. An example that reinforces this is from a Vernalis screen against CDK2 where there is very different binding mode (Figure 1.7) for the initial virtual screening hit **38** with that seen for the derived fragment **39**.

### 1.6.8
### Fragments, Chemical Space, and Novelty

Another issue that is often raised is that there are similarities in the fragment libraries in different organizations and so similar, if not the same, fragments are likely to emerge from screening of the same target. As Figure 1.8 demonstrates for HSP90, that can be the case, but the fragment evolution depends on the medicinal and structural chemistry involved in optimization. Figure 1.8a summarizes the initial phenol/resorcinol fragments found at Vernalis and subsequently at Astex; there are some similarities in the final clinical candidates but they reside in sufficiently different IP space. By contrast, the dual fragment binding seen at Vernalis and then at Evotec (Figure 1.8b) resulted in very different candidates being developed. At Vernalis, features were merged to give the final compound that entered preclinical trials, whereas Evotec attempted a linking strategy to give a lead compound that was not pursued.

## 1.7
## Current Application and Impact

In addition to the impact on drug discovery, I would like to emphasize three other significant impacts of FBLD methods.

The first is the ability to rapidly develop tool compounds that are potent and selective enough to ask questions about targets. This is very important for assessing targets in pharmaceutical discovery, such as the work at Vernalis on PDPK1 [56]. It is also proving extremely valuable in academia and there are an increasing number of examples emerging from such laboratories on protein–protein interactions [5,57], exploring potential new antibiotic targets [6] and the recent discovery of enzyme activators with potential for improving industrial enzymes [7].

Second, FBLD has helped to promote (and required) the development of biophysical methods in drug discovery. The rational approach to FBLD has increased the focus on compound properties and provides routes to resolving any issues. This has allowed some of the lessons learnt from analysis of the past decade or so of drug discovery to be considered during optimization.

Finally, the methods have emphasized the advantage of a nonindustrialized approach to drug discovery. FBLD campaigns against conventional targets such as kinases require some modification of the metrics usually applied – the project can spend considerably longer working on lower affinity compounds while suitable fragments are identified and suitable SAR developed to allow incorporation of selectivity and affinity. But for many such projects, the eventual time taken to achieve an advanced lead series is not dissimilar from starting with an HTS and then trying to fix the problems in the HTS hit. Where the strain really appears in large companies is setting the expectation that to be successful on an unconventional target can take many years – some of the recent projects such as the BH3 mimetics have taken 5–7 years to identify clinical candidates.

## 1.8
## Future Opportunities

So, what are the areas of FBLD where we can expect some new developments in the next 10 years? What additional impacts will fragment-based methods have on lead discovery? There will continue to be improvements in the methods: for identifying fragment hits (reduced cost in terms of sample requirements and instrument as well as methods for robust characterization of very weak, mM small fragments [51,58]); for characterizing the kinetics and thermodynamics of binding (wider range of faster, cheaper approaches); and in additional ideas such as off-rate screening by SPR [16] to allow rapid elaboration of fragments. In terms of application, we will see an increased deployment and expertise base in the academic community, able to use the methods to not only identify chemical tools to probe biology but also bring about further integration within the pharmaceutical industry, realizing the full potential synergy between FBLD, HTS, and other hit- and lead-finding technologies (DNA-encoded libraries, tethering, etc.). Alongside this, there will be improvements in the methods, perhaps most striking will be the engagement with academic synthetic organic chemists, devising novel chemistries, such as seen recently [5,59].

An area that has not developed as rapidly as (I) expected is the use of computational methods in support of the process. There are some attempts at docking and simulation methods to augment experimental screening [60–62], but as yet there is no substantial body of research that exploits the vast amounts of detailed structural information available on protein–fragment complexes – be that analyzing the origins of binding thermodynamics and kinetics or devising ways of merging fragment information together to design new compounds.

There are still issues with effective integration of the methods in some organizations: fragments are sometimes deployed only when HTS or other hit-finding methods have failed, the fragment screening is not initiated early enough for the fragment hits to have an impact on the choice of lead for progression, or the metrics applied in the early hit-finding phase do not give time for the fragments to be properly considered. For example, some organizations have defined schedules for progression from hit to lead or have a definition of a hit that requires biological activity. Such rigid criteria are perhaps needed to prioritize resources across many competing projects – such an industrialized approach allows managers to manage. But research, even for a conventional target, needs to respond to the data to ensure the best possible compounds are taken forward.

In conclusion, although the essential features of FBLD methods were established in the early 2000s, the surge in application of the methods and demonstration of success has happened only since the publication of the first edition of this book. The methods and their impact on lead discovery have continued to develop; it will be fascinating to see if another edition of this book is justified in a space of a decade or if, as we are beginning to see, the methods become sufficiently integrated into the general lexicon of drug discovery that it is no longer seen as a separate field.

## References


1 Erlanson, D.A. and Zartler, E. (2013) Practical Fragments Blog. http://practicalfragments.blogspot.com/2015/01/fragments-in-the-clinic-2015.html (August 2015).

2 Bollag, G., Tsai, J., Zhang, J., Zhang, C., Ibrahim, P., Nolop, K., and Hirth, P. (2012) Vemurafenib: the first drug approved for BRAF-mutant cancer. *Nature Reviews Drug Discovery*, **11**, 873–886.

3 Maurer, T., Garrenton, L.S., Oh, A., Pitts, K., Anderson, D.J., Skelton, N.J., Fauber, B.P., Pan, B., Malek, S., Stokoe, D., Ludlam, M.J., Bowman, K.K., Wu, J., Giannetti, A.M., Starovasnik, M.A., Mellman, I., Jackson, P.K., Rudolph, J., Wang, W., and Fang, G. (2012) Small-molecule ligands bind to a distinct pocket in Ras and inhibit SOS-mediated nucleotide exchange activity. *Proceedings of the National Academy of Sciences of the United States of America*, **109**, 5299–5304.

4 Sun, Q., Burke, J.P., Phan, J., Burns, M.C., Olejniczak, E.T., Waterson, A.G., Lee, T., Rossanese, O.W., and Fesik, S.W. (2012) Discovery of small molecules that bind to K-Ras and inhibit sos-mediated activation. *Angewandte Chemie International Edition*, **51**, 6140–6143.

5 Ostrem, J.M., Peters, U., Sos, M.L., Wells, J.A., and Shokat, K.M. (2013) K-Ras (G12C) inhibitors allosterically control GTP affinity and effector interactions. *Nature*, **503**, 548–551.

6 Yin, Z., Whittell, L.R., Wang, Y., Jergic, S., Liu, M., Harry, E.J., Dixon, N.E., Beck, J.L., Kelso, M.J., and Oakley, A.J. (2014) Discovery of lead compounds targeting the bacterial sliding clamp using a fragment-based approach. *Journal of Medicinal Chemistry*, **57**, 2799–2806.

7 Darby, J.F., Landstrom, J., Roth, C., He, Y., Davies, G.J., and Hubbard, R.E. (2014) Discovery of selective small-molecule activators of a bacterial glycoside hydrolase. *Angewandte Chemie*, **53**, 13419–13423.

8 Seth, P.P., Miyaji, A., Jefferson, E.A., Sannes-Lowery, K.A., Osgood, S.A., Propp, S.S., Ranken, R., Massire, C., Sampath, R., Ecker, D.J., Swayze, E.E., and Griffey, R.H. (2005) SAR by MS: discovery of a new class of RNA-binding small molecules for the hepatitis C virus: internal ribosome entry site IIA subdomain. *Journal of Medicinal Chemistry*, **48**, 7099–7102.

9 Warner, K.D., Homan, P., Weeks, K.M., Smith, A.G., Abell, C., and Ferre-D'Amare, A.R. (2014) Validating fragment-based drug discovery for biological RNAs: lead fragments bind and remodel the TPP riboswitch specifically. *Chemistry and Biology*, **21**, 591–595.

10 Garavis, M., Lopez-Mendez, B., Somoza, A., Oyarzabal, J., Dalvit, C., Villasante, A., Campos-Olivas, R., and Gonzalez, C. (2014) Discovery of selective ligands for telomeric RNA G-quadruplexes (TERRA) through F-19-NMR based fragment screening. *ACS Chemical Biology*, **9**, 1559–1566.

11 Nasiri, H.R., Bell, N.M., McLuckie, K.I.E., Husby, J., Abell, C., Neidle, S., and Balasubramanian, S. (2014) Targeting a c-MYC G-quadruplex DNA with a fragment library. *Chemical Communications*, **50**, 1704–1707.

12 Komander, D. and Rape, M. (2012) The ubiquitin code. *Annual Review of Biochemistry*, **81**, 203–229.

13 Wang, N., Majmudar, C.Y., Pomerantz, W.C., Gagnon, J.K., Sadowsky, J.D., Meagher, J.L., Johnson, T.K., Stuckey, J.A., Brooks, C.L. 3rd; Wells, J.A., and Mapp, A.K. (2013) Ordering a dynamic protein via a small-molecule stabilizer. *Journal of the American Chemical Society*, **135**, 3363–3366.

14 Chen, I.J. and Hubbard, R.E. (2009) Lessons for fragment library design: analysis of output from multiple screening campaigns. *Journal of Computer-Aided Molecular Design*, **23**, 603–620.

15 Hubbard, R.E. and Murray, J.B. (2011) Experiences in fragment-based lead discovery. *Methods in Enzymology*, **493**, 509–531.

16 Murray, J.B., Roughley, S.D., Matassova, N., and Brough, P.A. (2014) Off-rate screening (ORS) by surface plasmon resonance. An efficient method to kinetically sample hit to lead chemical space from unpurified reaction products. *Journal of Medicinal Chemistry*, **57** (7), 2845–2850.

17 Mochalkin, I., Miller, J.R., Narasimhan, L., Thanabal, V., Erdman, P., Cox, P.B., Prasad, J.V., Lightle, S., Huband, M.D., and Stover, C.K. (2009) Discovery of antibacterial biotin carboxylase inhibitors by virtual screening and fragment-based approaches. *ACS Chemical Biology*, **4**, 473–483.

18 Brough, P.A., Barril, X., Borgognoni, J., Chene, P., Davies, N.G., Davis, B., Drysdale, M.J., Dymock, B., Eccles, S.A., Garcia-Echeverria, C., Fromont, C., Hayes, A., Hubbard, R.E., Jordan, A.M., Jensen, M.R., Massey, A., Merrett, A., Padfield, A., Parsons, R., Radimerski, T., Raynaud, F.I., Robertson, A., Roughley, S.D., Schoepfer, J., Simmonite, H., Sharp, S.Y., Surgenor, A., Valenti, M., Walls, S., Webb, P., Wood, M., Workman, P., and Wright, L. (2009) Combining hit identification strategies: fragment-based and *in silico* approaches to orally active 2-aminothieno[2,3-d]pyrimidine inhibitors of the Hsp90 molecular chaperone. *Journal of Medicinal Chemistry*, **52**, 4794–4809.

19 Roughley, S.D. and Hubbard, R.E. (2011) How well can fragments explore accessed chemical space? A case study from Heat Shock Protein 90. *Journal of Medicinal Chemistry*, **54** (12), 3989–4005.

20 Hubbard, R.E., Davis, B., Chen, I., and Drysdale, M.J. (2007) The SeeDs approach: integrating fragments into drug discovery. *Current Topics in Medicinal Chemistry*, **7**, 1568–1581.

21 Kawai, K., Endo, Y., Asano, T., Amano, S., Sawada, K., Ueo, N., Takahashi, N., Sonoda, Y., Nagai, M., Kamei, N., and Nagata, N. (2014) Discovery of 2-(cyclopentylamino)thieno[3,2-d]pyrimidin-4(3H)-one derivatives as a new series of potent phosphodiesterase 7 inhibitors. *Journal of Medicinal Chemistry*, **57**, 9844–9854.

22 Shuker, S.B., Hajduk, P.J., Meadows, R.P., and Fesik, S.W. (1996) Discovering high-affinity ligands for proteins: SAR by NMR. *Science*, **274**, 1531–1534.

23 Erlanson, D.A. and Jahnke, W. (2006) *The Concept of Fragment-based Drug Discovery. Fragment-based Approaches in Drug Discovery*, Wiley-VCH Verlag GmbH, Weinheim, pp. 1–10.

24 Wilson, C.G. and Arkin, M.R. (2013) Probing structural adaptivity at PPI interfaces with small molecules. *Drug Discovery Today: Technologies*, **10**, e501–e508.

25 Carr, R.A.E., Congreve, M., Murray, C.W., and Rees, D.C. (2005) Fragment-based lead discovery: leads by design. *Drug Discovery Today*, **10**, 987–992.

26 Baurin, N., Aboul-Ela, F., Barril, X., Davis, B., Drysdale, M., Dymock, B., Finch, H., Fromont, C., Richardson, C., Simmonite, H., and Hubbard, R.E. (2004) Design and characterization of libraries of molecular fragments for use in NMR screening against protein targets. *Journal of Chemical Information and Computer Science*, **44**, 2157–2166.

27 Erlanson, D.A., McDowell, R.S., and O'Brien, T. (2004) Fragment-based drug discovery. *Journal of Medicinal Chemistry*, **47**, 3463–3482.

28 Raimundo, B.C., Oslob, J.D., Braisted, A.C., Hyde, J., McDowell, R.S., Randal, M., Waal, N.D., Wilkinson, J., Yu, C.H., and Arkin, M.R. (2004) Integrating fragment assembly and biophysical methods in the chemical advancement of small-molecule antagonists of IL-2: an approach for inhibiting protein–protein interactions. *Journal of Medicinal Chemistry*, **47**, 3111–3130.

29 Liu, G., Xin, Z., Pei, Z., Hajduk, P.J., Abad-Zapatero, C., Hutchins, C.W., Zhao, H., Lubben, T.H., Ballaron, S.J., Haasch, D.L., Kaszubska, W., Rondinone, C.M., Trevillyan, J.M., and Jirousek, M.R. (2003) Fragment screening and assembly: a highly efficient approach to a selective and cell active protein tyrosine phosphatase 1B inhibitor. *Journal of Medicinal Chemistry*, **46**, 4232–4235.

30 Lesuisse, D., Lange, G., Deprez, P., Benard, D., Schoot, B., Delettre, G., Marquette, J.P., Broto, P., Jean-Baptiste, V., Bichet, P., Sarubbi, E., and Mandine, E. (2002) SAR and X-ray. A new approach combining fragment-based screening and rational drug design: application to the discovery of nanomolar inhibitors of Src SH2. *Journal of Medicinal Chemistry*, **45**, 2379–2387.

31 Lau, W.F., Withka, J.M., Hepworth, D., Magee, T.V., Du, Y.J., Bakken, G.A., Miller, M.D., Hendsch, Z.S., Thanabal, V., Kolodziej, S.A., Xing, L., Hu, Q., Narasimhan, L.S., Love, R., Charlton, M.E., Hughes, S., van Hoorn, W.P., and Mills, J.E. (2011) Design of a multi-purpose fragment screening library using molecular complexity and orthogonal diversity metrics. *Journal of Computer-Aided Molecular Design*, **25**, 621–636.

32 Doak, B.C., Morton, C.J., Simpson, J.S., and Scanlon, M.J. (2013) Design and evaluation of the performance of an NMR screening fragment library. *Australian Journal of Chemistry*, **66** (12), 1465–1472. (http://dx.doi.org/10.1071/CH13280.)

33 Albert, J.S., Blomberg, N., Breeze, A.L., Brown, A.J., Burrows, J.N., Edwards, P.D., Folmer, R.H., Geschwindner, S., Griffen, E.J., Kenny, P.W., Nowak, T., Olsson, L.L., Sanganee, H., and Shapiro, A.B. (2007) An integrated approach to fragment-based lead generation: philosophy, strategy and case studies from AstraZeneca's drug discovery programmes. *Current Topics in Medicinal Chemistry*, **7**, 1600–1629.

34 Schulz, M.N., Landstrom, J., Bright, K., and Hubbard, R.E. (2011) Design of a fragment library that maximally represents available chemical space. *Journal of Computer-Aided Molecular Design*, **25**, 611–620.

35 Dalvit, C., Mongelli, N., Papeo, G., Giordano, P., Veronesi, M., Moskau, D., and Kummerle, R. (2005) Sensitivity improvement in 19F NMR-based screening experiments: theoretical considerations and experimental applications. *Journal of the American Chemical Society*, **127**, 13380–13385.

36 Dalvit, C., Fagerness, P.E., Hadden, D.T.A., Sarver, R.W., and Stockman, B.J. (2003) Fluorine-NMR experiments for high-throughput screening: theoretical aspects, practical considerations, and range of applicability. *Journal of the American Chemical Society*, **125**, 7696–7703.

37 Bohacek, R.S., McMartin, C., and Guida, W.C. (1996) The art and practice of structure-based drug design: a molecular modeling perspective. *Medicinal Research Reviews*, **16**, 3–50.

38 Fink, T. and Reymond, J.L. (2007) Virtual exploration of the chemical universe up to 11 atoms of C, N, O, F: assembly of 26.4 million structures (110.9 million stereoisomers) and analysis for new ring systems, stereochemistry, physicochemical properties, compound classes, and drug discovery. *Journal of Chemical Information and Modeling*, **47**, 342–353.

39 Ruddigkeit, L., van Deursen, R., Blum, L.C., and Reymond, J.L. (2012) Enumeration of 166 billion organic small molecules in the chemical universe database GDB-17. *Journal of Chemical Information and Modeling*, **52**, 2864–2875.

40 Barelier, S., Eidam, O., Fish, I., Hollander, J., Figaroa, F., Nachane, R., Irwin, J.J., Shoichet, B.K., and Siegal, G. (2014) Increasing chemical space coverage by combining empirical and computational fragment screens. *ACS Chemical Biology*, **9**, 1528–1535.

41 Hubbard, R.E., Chen, I., and Davis, B. (2007) Informatics and modeling challenges in fragment-based drug discovery. *Current Opinion in Drug Discovery and Development*, **10**, 289–297.

42 Hajduk, P.J. (2006) Fragment-based drug design: how big is too big? *Journal of Medicinal Chemistry*, **49**, 6972–6976.

43 Lovering, F., Bikker, J., and Humblet, C. (2009) Escape from flatland: increasing saturation as an approach to improving clinical success. *Journal of Medicinal Chemistry*, **52**, 6752–6756.

44 Hajduk, P.J., Huth, J.R., and Fesik, S.W. (2005) Druggability indices for protein targets derived from NMR-based screening data. *Journal of Medicinal Chemistry*, **48**, 2518–2525.

45 Hajduk, P.J., Huth, J.R., and Tse, C. (2005) Predicting protein druggability. *Drug Discovery Today*, **10**, 1675–1682.

46 Wells, J.A. and McClendon, C.L. (2007) Reaching for high-hanging fruit in drug discovery at protein–protein interfaces. *Nature*, **450**, 1001–1009.

47 Giannetti, A.M. (2011) From experimental design to validated hits a comprehensive walk-through of fragment lead identification using surface plasmon resonance. *Methods in Enzymology*, **493**, 169–218.

48 Giannetti, A.M., Koch, B.D., and Browner, M.F. (2008) Surface plasmon resonance based assay for the detection and characterization of promiscuous inhibitors. *Journal of Medicinal Chemistry*, **51**, 574–580.

49 Rich, R.L., Quinn, J.G., Morton, T., Stepp, J.D., and Myszka, D.G. (2010) Biosensor-based fragment screening using FastStep injections. *Analytical Biochemistry*, **407**, 270–277.

50 Siegal, G. and Hollander, J.G. (2009) Target immobilization and NMR screening of fragments in early drug discovery. *Current Topics in Medicinal Chemistry*, **9**, 1736–1745.

51 Meiby, E., Simmonite, H., le Strat, L., Davis, B., Matassova, N., Moore, J.D., Mrosek, M., Murray, J., Hubbard, R.E., and Ohlson, S. (2013) Fragment screening by weak affinity chromatography: comparison with established techniques for screening against HSP90. *Analytical Chemistry*, **85**, 6756–6766.

52 Pollack, S.J., Beyer, K.S., Lock, C., Muller, I., Sheppard, D., Lipkin, M., Hardick, D., Blurton, P., Leonard, P.M., Hubbard, P.A., Todd, D., Richardson, C.M., Ahrens, T., Baader, M., Hafenbradl, D.O., Hilyard, K., and Burli, R.W. (2011) A comparative study of fragment screening methods on the p38alpha kinase: new methods, new insights. *Journal of Computer-Aided Molecular Design*, **25**, 677–687.

53 Vulpetti, A. and Dalvit, C. (2012) Fluorine local environment: from screening to drug design. *Drug Discovery Today*, **17**, 890–897.

54 Babaoglu, K. and Shoichet, B.K. (2006) Deconstructing fragment-based inhibitor discovery. *Nature Chemical Biology*, **2**, 720–723.

55 Fry, D.C., Wartchow, C., Graves, B., Janson, C., Lukacs, C., Kammlott, U., Belunis, C., Palme, S., Klein, C., and Vu, B. (2013) Deconstruction of a nutlin: dissecting the binding determinants of a potent protein–protein interaction inhibitor. *ACS Medicinal Chemistry Letters*, **4**, 95–100.

56 Hubbard, R.E. (2008) Fragment approaches in structure-based drug discovery. *Journal of Synchrotron Radiation*, **15**, 227–230.

57 Scott, D.E., Ehebauer, M.T., Pukala, T., Marsh, M., Blundell, T.L., Venkitaraman, A.R., Abell, C., and Hyvonen, M. (2013) Using a fragment-based approach to target protein–protein interactions. *Chembiochem*, **14**, 332–342.

58 Buratto, R., Mammoli, D., Chiarparin, E., Williams, G., and Bodenhausen, G. (2014) Exploring weak ligand–protein interactions by long-lived NMR states: improved contrast in fragment-based drug screening. *Angewandte Chemie*, **53**, 11376–11380.

59 Hung, A.W., Ramek, A., Wang, Y., Kaya, T., Wilson, J.A., Clemons, P.A., and Young, D.W. (2011) Route to three-dimensional fragments using diversity-oriented synthesis. *Proceedings of the National Academy of Sciences of the United States of America*, **108**, 6799–6804.

60 Alvarez-Garcia, D. and Barril, X. (2014) Molecular simulations with solvent competition quantify water displaceability and provide accurate interaction maps of protein binding sites. *Journal of Medicinal Chemistry*, **57**, 8530–8539.

61 Yu, W., Lakkaraju, S.K., Raman, E.P., and MacKerell, A.D. Jr. (2014) Site-identification by ligand competitive saturation (SILCS) assisted pharmacophore modeling. *Journal of Computer-Aided Molecular Design*, **28**, 491–507.

62 Hall, D.R., Ngan, C.H., Zerbe, B.S., Kozakov, D., and Vajda, S. (2012) Hot spot analysis for driving the development of hits into leads in fragment-based drug discovery. *Journal of Chemical Information and Modeling*, **52**, 199–209.

63 Hajduk, P.J. (2006) SAR by NMR: putting the pieces together. *Molecular Interventions*, **6**, 266–272.

64 Hajduk, P.J. and Greer, J. (2007) A decade of fragment-based drug design: strategic advances and lessons learned. *Nature Reviews Drug Discovery*, **6**, 211–219.

65 Olejniczak, E.T., Hajduk, P.J., Marcotte, P.A., Nettesheim, D.G., Meadows, R.P., Edalji, R., Holzman, T.F., and Fesik, S.W. (1997) Stromelysin inhibitors designed from weakly bound fragments: effects of linking and cooperativity. *Journal of the American Chemical Society*, **119**, 5828–5832.

66 Oltersdorf, T., Elmore, S.W., Shoemaker, A.R., Armstrong, R.C., Augeri, D.J., Belli, B.A., Bruncko, M., Deckwerth, T.L., Dinges, J., Hajduk, P.J., Joseph, M.K., Kitada, S., Korsmeyer, S.J., Kunzer, A.R., Letai, A., Li, C., Mitten, M.J., Nettesheim, D.G., Ng, S., Nimmer, P.M., O'Connor, J.M., Oleksijew, A., Petros, A.M., Reed, J.C., Shen, W., Tahir, S.K., Thompson, C.B., Tomaselli, K.J., Wang, B.L., Wendt, M.D., Zhang, H.C., Fesik, S.W., and Rosenberg, S.H. (2005) An inhibitor of Bcl-2 family proteins induces regression of solid tumours. *Nature*, **435**, 677–681.

67 Petros, A.M., Dinges, J., Augeri, D.J., Baumeister, S.A., Betebenner, D.A., Bures, M.G., Elmore, S.W., Hajduk, P.J., Joseph, M.K., Landis, S.K., Nettesheim, D.G., Rosenberg, S.H., Shen, W., Thomas, S., Wang, X., Zanze, I., Zhang, H., and Fesik, S.W. (2006) Discovery of a potent inhibitor of the antiapoptotic protein Bcl-xL from NMR and parallel synthesis. *Journal of Medicinal Chemistry*, **49**, 656–663.

68 Tse, C., Shoemaker, A.R., Adickes, J., Anderson, M.G., Chen, J., Jin, S., Johnson, E.F., Marsh, K.C., Mitten, M.J., Nimmer, P., Roberts, L., Tahir, S.K., Xiao, Y., Yang, X., Zhang, H., Fesik, S., Rosenberg, S.H., and Elmore, S.W. (2008) ABT-263: a potent and orally bioavailable Bcl-2 family inhibitor. *Cancer Research*, **68**, 3421–3428.

69 Petros, A.M., Huth, J.R., Oost, T., Park, C.M., Ding, H., Wang, X., Zhang, H., Nimmer, P., Mendoza, R., Sun, C., Mack, J., Walter, K., Dorwin, S., Gramling, E., Ladror, U., Rosenberg, S.H., Elmore, S.W., Fesik, S.W., and Hajduk, P.J. (2010) Discovery of a potent and selective Bcl-2 inhibitor using SAR by NMR. *Bioorganic and Medicinal Chemistry Letters*, **20**, 6587–6591.

70 Park, C.-M., Bruncko, M., Adickes, J., Bauch, J., Ding, H., Kunzer, A., Marsh, K.C., Nimmer, P., Shoemaker, A.R., Song, X., Tahir, S.K., Tse, C., Wang, X., Wendt, M.D., Yang, X., Zhang, H., Fesik, S.W., Rosenberg, S.H., and Elmore, S.W. (2008) Discovery of an orally bioavailable small molecule inhibitor of prosurvival B-cell lymphoma 2 proteins. *Journal of Medicinal Chemistry*, **51**, 6902–6915.

71 Souers, A.J., Leverson, J.D., Boghaert, E.R., Ackler, S.L., Catron, N.D., Chen, J., Dayton, B.D., Ding, H., Enschede, S.H., Fairbrother, W.J., Huang, D.C., Hymowitz, S.G., Jin, S., Khaw, S.L., Kovar, P.J., Lam, L.T., Lee, J., Maecker, H.L., Marsh, K.C., Mason, K.D., Mitten, M.J., Nimmer, P.M., Oleksijew, A., Park, C.H., Park, C.M., Phillips, D.C., Roberts, A.W.,

Sampath, D., Seymour, J.F., Smith, M.L., Sullivan, G.M., Tahir, S.K., Tse, C., Wendt, M.D., Xiao, Y., Xue, J.C., Zhang, H., Humerickhouse, R.A., Rosenberg, S.H., and Elmore, S.W. (2013) ABT-199, a potent and selective BCL-2 inhibitor, achieves antitumor activity while sparing platelets. *Nature Medicine*, **19**, 202–208.

72 Barker, J.J., Barker, O., Courtney, S.M., Gardiner, M., Hesterkamp, T., Ichihara, O., Mather, O., Montalbetti, C.A., Muller, A., Varasi, M., Whittaker, M., and Yarnold, C.J. (2010) Discovery of a novel Hsp90 inhibitor by fragment linking. *ChemMedChem*, **5**, 1697–1700.

73 Roughley, S., Wright, L., Brough, P., Massey, A., and Hubbard, R.E. (2012) Hsp90 inhibitors and drugs from fragment and virtual screening. *Topics in Current Chemistry*, **317**, 61–82.

74 Brough, P.A., Aherne, W., Barril, X., Borgognoni, J., Boxall, K., Cansfield, J.E., Cheung, K.M., Collins, I., Davies, N.G., Drysdale, M.J., Dymock, B., Eccles, S.A., Finch, H., Fink, A., Hayes, A., Howes, R., Hubbard, R.E., James, K., Jordan, A.M., Lockie, A., Martins, V., Massey, A., Matthews, T.P., McDonald, E., Northfield, C.J., Pearl, L.H., Prodromou, C., Ray, S., Raynaud, F.I., Roughley, S.D., Sharp, S.Y., Surgenor, A., Walmsley, D.L., Webb, P., Wood, M., Workman, P., and Wright, L. (2008) 4,5-Diarylisoxazole Hsp90 chaperone inhibitors: potential therapeutic agents for the treatment of cancer. *Journal of Medicinal Chemistry*, **51**, 196–218.

75 Eccles, S.A., Massey, A., Raynaud, F.I., Sharp, S.Y., Box, G., Valenti, M., Patterson, L., de Haven Brandon, A., Gowan, S., Boxall, F., Aherne, W., Rowlands, M., Hayes, A., Martins, V., Urban, F., Boxall, K., Prodromou, C., Pearl, L., James, K., Matthews, T.P., Cheung, K.M., Kalusa, A., Jones, K., McDonald, E., Barril, X., Brough, P.A., Cansfield, J.E., Dymock, B., Drysdale, M.J., Finch, H., Howes, R., Hubbard, R.E., Surgenor, A., Webb, P., Wood, M., Wright, L., and Workman, P. (2008) NVP-AUY922: a novel heat shock protein 90 inhibitor active against xenograft tumor growth, angiogenesis, and metastasis. *Cancer Research*, **68**, 2850–2860.

76 Howard, S., Berdini, V., Boulstridge, J.A., Carr, M.G., Cross, D.M., Curry, J., Devine, L.A., Early, T.R., Fazal, L., Gill, A.L., Heathcote, M., Maman, S., Matthews, J.E., McMenamin, R.L., Navarro, E.F., O'Brien, M.A., O'Reilly, M., Rees, D.C., Reule, M., Tisi, D., Williams, G., Vinkovic, M., and Wyatt, P.G. (2009) Fragment-based discovery of the pyrazol-4-yl urea (AT9283), a multitargeted kinase inhibitor with potent aurora kinase activity. *Journal of Medicinal Chemistry*, **52**, 379–388.

77 Friberg, A., Vigil, D., Zhao, B., Daniels, R.N., Burke, J.P., Garcia-Barrantes, P.M., Camper, D., Chauder, B.A., Lee, T., Olejniczak, E.T., and Fesik, S.W. (2013) Discovery of potent myeloid cell leukemia 1 (Mcl-1) inhibitors using fragment-based methods and structure-based design. *Journal of Medicinal Chemistry*, **56**, 15–30.

78 Bezdek, J.C. (1993) Fuzzy models&#x2014: what are they, and why? [Editorial]. *IEEE Transactions on Fuzzy Systems*, **1**, 1–6.

79 Murray, C.W., Carr, M.G., Callaghan, O., Chessari, G., Congreve, M., Cowan, S., Coyle, J.E., Downham, R., Figueroa, E., Frederickson, M., Graham, B., McMenamin, R., O'Brien, M.A., Patel, S., Phillips, T.R., Williams, G., Woodhead, A.J., and Woolford, A.J. (2010) Fragment-based drug discovery applied to Hsp90. Discovery of two lead series with high ligand efficiency. *Journal of Medicinal Chemistry*, **53**, 5942–5955.

80 Barker, J.J., Barker, O., Boggio, R., Chauhan, V., Cheng, R.K., Corden, V., Courtney, S.M., Edwards, N., Falque, V.M., Fusar, F., Gardiner, M., Hamelin, E.M., Hesterkamp, T., Ichihara, O., Jones, R.S., Mather, O., Mercurio, C., Minucci, S., Montalbetti, C.A., Muller, A., Patel, D., Phillips, B.G., Varasi, M., Whittaker, M., Winkler, D., and Yarnold, C.J. (2009) Fragment-based identification of Hsp90 inhibitors. *ChemMedChem*, **4**, 963–966.

81 Chessari, G. and Woodhead, A.J. (2009) From fragment to clinical candidate: a

historical perspective. *Drug Discovery Today*, **14**, 668–675.

82 Mayer, M. and Meyer, B. (1999) Characterization of ligand binding by saturation transfer difference NMR spectroscopy. *Angewandte Chemie International Edition*, **38**, 1784–1788.

83 Dalvit, C., Fogliatto, G., Stewart, A., Veronesi, M., and Stockman, B. (2001) WaterLOGSY as a method for primary NMR screening: practical aspects and range of applicability. *Journal of Biomolecular NMR*, **21**, 349–359.

84 Meiboom, S.G.D. (1958) Modified spin-echo method for measuring nuclear relaxation times. *Review of Scientific Instruments*, **29**, 688–691.

85 Ladbury, J.E., Klebe, G., and Freire, E. (2010) Adding calorimetric data to decision making in lead discovery: a hot tip. *Nature Reviews Drug Discovery*, **9**, 23–27.

86 Kranz, J.K. and Schalk-Hihi, C. (2011) Protein thermal shifts to identify low molecular weight fragments. *Methods in Enzymology*, **493**, 277–298.

87 Jencks, W.P. (1981) On the attribution and additivity of binding energies. *Proceedings of the National Academy of Sciences of the United States of America*, **78**, 4046–4050.

88 Abraham, D.J., Perutz, M.F., and Phillips, S.E. (1983) Physiological and X-ray studies of potential antisickling agents. *Proceedings of the National Academy of Sciences of the United States of America*, **80**, 324–328.

89 Andrews, P.R., Craik, D.J., and Martin, J.L. (1984) Functional-group contributions to drug receptor interactions. *Journal of Medicinal Chemistry*, **27**, 1648–1657.

90 Goodford, P.J. (1985) A computational procedure for determining energetically favorable binding sites on biologically important macromolecules. *Journal of Medicinal Chemistry*, **28**, 849–857.

91 Vonitzstein, M., Wu, W.Y., Kok, G.B., Pegg, M.S., Dyason, J.C., Jin, B., Phan, T.V., Smythe, M.L., White, H.F., Oliver, S.W., Colman, P.M., Varghese, J.N., Ryan, D.M., Woods, J.M., Bethell, R.C., Hotham, V.J., Cameron, J.M., and Penn, C.R. (1993) Rational design of potent sialidase-based inhibitors of influenza-virus replication. *Nature*, **363**, 418–423.

92 Miranker, A. and Karplus, M. (1991) Functionality maps of binding sites: a multiple copy simultaneous search method. *Proteins*, **11**, 29–34.

93 Bohm, H.J. (1992) LUDI: rule-based automatic design of new substituents for enzyme inhibitor leads. *Journal of Computer-Aided Molecular Design*, **6**, 593–606.

94 Allen, K.N., Bellamacina, C.R., Ding, X.C., Jeffery, C.J., Mattos, C., Petsko, G.A., and Ringe, D. (1996) An experimental approach to mapping the binding surfaces of crystalline proteins. *Journal of Physical Chemistry*, **100**, 2605–2611.

95 English, A.C., Done, S.H., Caves, L.S., Groom, C.R., and Hubbard, R.E. (1999) Locating interaction sites on proteins: the crystal structure of thermolysin soaked in 2% to 100% isopropanol. *Proteins*, **37**, 628–640.

96 English, A.C., Groom, C.R., and Hubbard, R.E. (2001) Experimental and computational mapping of the binding surface of a crystalline protein. *Protein Engineering*, **14**, 47–59.

97 Eisen, M.B., Wiley, D.C.;., Karplus, M., and Hubbard, R.E. (1994) HOOK: a program for finding novel molecular architectures that satisfy the chemical and steric requirements of a macromolecule binding site. *Proteins*, **19**, 199–221.

98 Bartlett, P.A., Shea, G.T., Waterman, S., and Telfer, S.J. (1991) Caveat – a vector-based approach to computer-aided molecule design. *Abstracts of Papers of the American Chemical Society*, **202**, 44–COMP.

99 Bohm, H.J. (1992) The computer program LUDI: a new method for the *de novo* design of enzyme inhibitors. *Journal of Computer-Aided Molecular Design*, **6**, 61–78.

100 Hajduk, P.J., Dinges, J., Miknis, G.F., Merlock, M., Middleton, T., Kempf, D.J., Egan, D.A., Walter, K.A., Robins, T.S., Shuker, S.B., Holzman, T.F., and Fesik, S.W. (1997) NMR-based discovery of lead inhibitors that block DNA binding of the

human papillomavirus E2 protein. *Journal of Medicinal Chemistry*, **40**, 3144–3150.

101 Hajduk, P.J., Boyd, S., Nettesheim, D., Nienaber, V., Severin, J., Smith, R., Davidson, D., Rockway, T., and Fesik, S.W. (2000) Identification of novel inhibitors of urokinase via NMR-based screening. *Journal of Medicinal Chemistry*, **43**, 3862–3866.

102 Card, G.L., Blasdel, L., England, B.P., Zhang, C., Suzuki, Y., Gillette, S., Fong, D., Ibrahim, P.N., Artis, D.R., Bollag, G., Milburn, M.V., Kim, S.H., Schlessinger, J., and Zhang, K.Y. (2005) A family of phosphodiesterase inhibitors discovered by cocrystallography and scaffold-based drug design. *Nature Biotechnology*, **23**, 201–207.

103 Nienaber, V.L., Richardson, P.L., Klighofer, V., Bouska, J.J., Giranda, V.L., and Greer, J. (2000) Discovering novel ligands for macromolecules using X-ray crystallographic screening. *Nature Biotechnology*, **18**, 1105–1108.

104 Kuhn, P., Wilson, K., Patch, M.G., and Stevens, R.C. (2002) The genesis of high-throughput structure-based drug discovery using protein crystallography. *Current Opinion in Chemical Biology*, **6**, 704–710.

105 Blaney, J., Nienaber, V., and Burley, S.K. (2006) Fragment-based lead discovery and optimization using X-ray crystallography, computational chemistry, and high-throughput organic synthesis, in *Fragment-based Approaches in Drug Discovery*, Wiley-VCH Verlag GmbH, Weinheim, pp. 215–248.

106 Sharff, A. and Jhoti, H. (2003) High-throughput crystallography to enhance drug discovery. *Current Opinion in Chemical Biology*, **7**, 340–345.

107 Rees, D.C., Congreve, M., Murray, C.W., and Carr, R. (2004) Fragment-based lead discovery. *Nature Reviews Drug Discovery*, **3**, 660–672.

108 Arkin, M.R., Randal, M., DeLano, W.L., Hyde, J., Luong, T.N., Oslob, J.D., Raphael, D.R., Taylor, L., Wang, J., McDowell, R.S., Wells, J.A., and Braisted, A.C. (2003) Binding of small molecules to an adaptive protein–protein interface. *Proceedings of the National Academy of Sciences of the United States of America*, **100**, 1603–1608.

109 Arkin, M.R. and Wells, J.A. (2004) Small-molecule inhibitors of protein–protein interactions: progressing towards the dream. *Nature Reviews Drug Discovery*, **3**, 301–317.

110 Erlanson, D.A., Wells, J.A., and Braisted, A.C. (2004) Tethering: fragment-based drug discovery. *Annual Review of Biophysics and Biomolecular Structure*, **33**, 199–223.

111 Hann, M.M., Leach, A.R., and Harper, G. (2001) Molecular complexity and its impact on the probability of finding leads for drug discovery. *Journal of Chemical Information and Computer Science*, **41**, 856–864.

112 Hopkins, A.L., Groom, C.R., and Alex, A. (2004) Ligand efficiency: a useful metric for lead selection. *Drug Discovery Today*, **9**, 430–431.

113 Kuntz, I.D., Chen, K., Sharp, K.A., and Kollman, P.A. (1999) The maximal affinity of ligands. *Proceedings of the National Academy of Sciences of the United States of America*, **96**, 9997–10002.

114 Congreve, M., Carr, R., Murray, C., and Jhoti, H. (2003) A 'rule of three' for fragment-based lead discovery? *Drug Discovery Today*, **8**, 876–877.

# 2
# Selecting the Right Targets for Fragment-Based Drug Discovery

*Thomas G. Davies, Harren Jhoti, Puja Pathuri, and Glyn Williams*

## 2.1
## Introduction

Fragment-based drug discovery (FBDD) has been actively pursued by groups both in biotechnology and pharmaceutical companies and in academia, for more than 15 years. During this time, one fragment-derived drug, Vemurafenib [1], has reached the market and a number of others are in late-stage clinical trials [2]. The advantages of fragment-based approaches have been well documented by its practitioners, but the emergence of FBDD has also had a broader impact within the lead generation community as more attention has become focused on generating lead compounds with improved physicochemical properties [3,4]. For example, a recent comparison of compounds derived from high-throughput screening (HTS) and fragment methods suggests that there are advantages of using FBDD, particularly when combined with structure-guided optimization during hit elaboration [5]. This wider appreciation of the impact and benefits of FBDD has led to its adoption as a standard part of many screening cascades, and partly as a consequence, against an increasing range of therapeutic target classes. As the field has matured, literature has appeared that not only details the successes but also describes the challenges and limitations of a fragment approach [6–9]. Choosing targets to enter a company's discovery portfolio is a challenging step, regardless of the screening modality, but some of these difficulties may arise from selecting targets that are less appropriate for FBDD. In this chapter, we will outline some of the principles that we have found useful to apply in this area and provide examples of targets where the advantages of using a fragment-based approach have been clearly demonstrated.

Irrespective of whether FBDD or HTS methods are used to pursue a particular target, assessment of the potential of either inhibition or activation to drive the desired biological effect within the context of a disease is of primary importance. However, in many cases this question can be confidently answered only when a suitable tool compound has been generated. Here we take a "tractability-centric" view of target selection, but in practice this is usually applied after careful

*Fragment-based Drug Discovery: Lessons and Outlook*, First Edition. Edited by Daniel A. Erlanson and Wolfgang Jahnke.

consideration of the strength of its biological validation and potential patient selection strategies.

The experimental approaches to FBDD, at least in the early stages of the process, rely heavily on biophysical methods for detecting ligand binding to the protein target, measuring the binding affinities of hits, and generating structural models to guide the elaboration of fragment-sized hits. Thus, an initial filter when choosing suitable targets for FBDD is often to select proteins to which the greatest number of biophysical methods can be applied. The availability of X-ray or NMR data on the target or a close homolog is also an encouraging sign, even if the amount of structural information is limited. In our experience, the structure-driven progression of a target from initial screening to lead optimization will typically require more than 100 mg of purified protein, so proteins that express well in bacteria or in insect cell systems also have a significant practical advantage. However, technological advances continue to be made that can increase the availability of stable, correctly folded protein [10] or improve the sensitivity and widen the applicability of biophysical methods [11–13]. As methods evolve, views on which targets are tractable by FBDD and which provide a clear advantage for FBDD will change. The challenge for research groups involved in drug discovery is to recognize and incorporate these developments within an existing framework.

An early assessment of likely druggability is also essential for the prioritization of potential disease targets [14]. We define druggability as the ability of a protein target to bind a selective, low molecular weight compound (200–500 Da) with high affinity (typically $K_d < 10$ nM), and in a way that can modulate target activity *in vivo*. A therapeutic target that is druggable is likely to have one or more binding sites characterized by well-defined pockets or clefts containing a balance of polar and nonpolar functionality, in order to bind a small molecule with appropriate physicochemical properties. These pockets must be large enough to contain drug-sized molecules; however, depending on the target, they may change shape or even be formed only upon ligand binding. We discuss the assessment of target druggability in more detail in Section 2.3 below.

One of the key strengths of FBDD is its ability to sample chemical space in a highly efficient manner, which implies that it will have a higher chance of finding molecules that can occupy protein pockets or clefts than conventional approaches. FBDD may also allow the identification of binding sites that modulate the function of a protein in a more unexpected manner, given that the detection of fragment hits typically relies on some type of direct binding measure and not a preselected functional readout. In addition, the potential to identify more diverse hit matter from fragment screening may broaden the range of targets for which intervention with a drug-like molecule becomes possible. Selecting the right target for FBDD should include an assessment of the probable impact of these advantages, and we present a more detailed discussion of these points in the following sections.

## 2.2
## Properties of Targets and Binding Sites

While FBDD can be applied to many targets, is it necessarily the best method? For targets where a robust bioassay exists or can be developed, HTS remains an attractive option and is routinely applied prior to or in parallel with fragment screening. Improvements in assay design and the availability of more selective readouts based on fluorescence proximity assays (e.g., FRET or alpha screen) have reduced reliance on coupled enzymatic assays, which can be prone to interference and the generation of false positive hits. However, the extent to which the available HTS screening library explores useful chemical space remains a critical issue, with a downstream impact on the choice of targets for which HTS is most appropriate.

HTS is usually successful when applied to druggable targets and where the chemical library contains examples of chemotypes that are known to bind to the active sites of related targets. The improved coverage of chemical space by fragment-sized compounds (typically containing less than 16 nonhydrogen atoms or with molecular weights less than 200 Da) gives fragment screening a clear advantage to identify new chemotypes or hits for novel or poorly druggable sites [15,16]. It has been estimated that the number of possible lead-like molecules of a given size increases by approximately eightfold for each nonhydrogen atom (C, N, O, P, S, and Halogen) that is added [17]. If, on average, compounds in an HTS library are larger by seven atoms (approximately 100 Da) than compounds in a fragment library, then the size of chemical space that the HTS library must explore is around $2 \times 10^6$-fold larger than that of the fragment library. Even though typical HTS libraries contain $10^6$ compounds while fragment libraries generally contain only $10^3$ fragments, this still gives the fragment library a considerable ($10^3$-fold) sampling advantage.

In the case of fragment libraries, considerable effort is typically spent in selecting members in a way that will generate valuable hit matter [18], including attention not only to compound diversity but also to molecular properties that favor the biophysical detection method(s) that will be used for screening and chemical properties that facilitate elaboration. This will certainly include good aqueous solubility, but other factors such as the presence of aromatic protons or $^{19}$F for NMR detection and the presence of a range of synthetically accessible vectors may also be considered. By contrast, HTS libraries are often populated by compounds derived from previous drug discovery campaigns and may be biased toward finding hits for druggable targets. FBDD libraries may, therefore, allow access to a greater range of targets, including those which are novel, or that have not proved tractable using HTS methods.

There are several examples of challenging targets for which progressable compounds have been identified by FBDD. These include beta-secretase (BACE) [19–21], lactate dehydrogenase A (LDA) [22,23], protein kinase B (PKB) [24], the BCR fusion with Abelson kinase (BcrAbl) [25], cyclin-dependent kinase 4 (CDK4) [26], and PIM kinase (PIM) [27,28]. In some cases, wider coverage of chemical space by the fragment library was clearly an advantage, leading to the identification of what were, at the time, novel or less well-precedented chemotypes, for example, dihydroisocytosines for BACE, pyrazoles for PKB, and cinnamic acids for PIM. Identification of such chemotypes by FBDD has subsequently been used to improve the coverage of lead-like chemical space in some HTS libraries.

In other cases, the success of FBDD has relied upon the identification of hits at remote sites that can exert an allosteric effect on protein function. These include farnesyl pyrophosphate synthase (FPPS) [29], BcrAbl [25], hepatitis C virus non-structural protein 3 (HCV-NS3) [30], pyruvate kinase M2 (PKM2) [31], and soluble adenylate cyclase (SolAC) [32]. These sites were either previously unknown (FPPS, HCV-NS3, PKM2) or not precedented for lead-like molecules (BcrAbl, SolAC), and some are discussed in more detail below. The identification of allosteric hits remains a challenge for many HTS methods since the hit must not only bind to the target protein but also significantly alter the rate-determining step of the biochemical assay, under conditions (substrate or cofactor concentration, etc.) that may be far removed from the physiological situation. In contrast, binding alone is normally sufficient to register a molecule as a hit in an FBDD screen. Any allosteric effects on activity are not usually investigated until the hit is selected for further validation or follow up. The general utility of allosteric hits has been a matter of debate [33]. However, there are a number of marketed drugs that act allosterically (e.g., maraviroc, diazepam) and allosteric sites offer the possibility that, even when fully occupied, they may not fully inhibit the target protein thus potentially reducing or modifying the effects of on-target toxicity. In our (limited) experience, remote fragment binding pockets generally do exert some allosteric functional effect. This may be due to either the stabilization of a conformational state that modulates substrate or cofactor binding or by competition with a protein–protein interaction that is either transiently present in the cell or may be required for folding or transport [34].

The accumulation of a large body of X-ray structural data on fragment binding to a wide range of protein targets at Astex has given some insight into the occurrence of remote binding sites. These data have been used to determine the number of nonoverlapping sites capable of supporting fragment binding for each protein studied. The data set included all X-ray structures of in-house protein targets for which at least 100 different fragments had been screened using crystallographic methods and comprised 5590 structures from 25 different protein targets, containing 4950 distinct compounds. All X-ray structures used to detect fragment binding had resolutions better than 2.8 Å and fragments that bound in interstitial sites formed by crystal packing were excluded from the analysis, as were inorganic compounds, metal clusters, solvent molecules, and covalently bound molecules. Complex structures were obtained from soaking experiments at high ligand concentrations, typically 20–50 mM, so the observation of a bound ligand with reasonable occupancy implies a binding affinity of better than approximately 10 mM in the crystal environment.

Analysis of the set revealed that about 60% of liganded protein structures have at least one additional ligand bound at a secondary site, with some having several additional binding pockets [manuscript in preparation]. In one example, CDK2, two copies of the same ligand were seen: one binding to the ATP-site and the other binding within the C-terminal lobe, far away from the active site. In another case, that of the tetrameric enzyme PKM2, five different binding sites were observed, including the catalytic site and a previously unidentified binding site of an allosteric activator that is described in greater detail below. Overall, 55 binding sites were observed across 25 targets, with an average of $2.2 \pm 1.3$ sites per target. This value represents a lower limit for the number of binding sites. It is likely that additional sites exist in some cases, but cannot be observed in the X-ray experiment when these sites participate in protein–protein contacts or are not accessible via solvent channels in the particular crystal system being used.

Observation of fragment binding does not guarantee that any site will ultimately prove druggable, since it may not need to support tight binding ligands if it is the target of a small molecule, peptide or protein, which is at high local concentration within the cell. While improvements in biophysical screening methods and the chemical properties of fragment libraries have allowed the identification of more fragment hits in a wider range of sites, this has created an urgent need to be able to assess the druggability of each site and, ideally, each chemical series before further resources are spent. In part, this is being addressed using structure-guided computational methods that are discussed in the next section. In addition, it will be highly desirable to develop robust biophysical and biochemical methods that can detect target engagement and the functional effects of fragment-derived leads in cells at relatively low affinities (10–100 μM or weaker).

## 2.3 Assessing Druggability

With the increased activity in pursuing more challenging targets, such as those involving protein–protein interactions, it is critical for companies to elucidate the druggability of a target quickly and efficiently [35]. We begin by noting that experimental hit rates from fragment screening have emerged as a good indicator of druggability, and typically there is a good correlation between the number of fragment hits from a biophysical screen and the probability of producing high-affinity drug leads [36,37]. The use of screening itself to assess druggability relies on the superior sampling of chemical space by fragments when coupled with highly sensitive detection techniques [38] and may be particularly useful in assessing challenging protein targets. In this context, it is clearly important to use a fragment screening library that is large and diverse enough to draw valid conclusions about a target's druggability and to employ a well-characterized screening cascade to filter out false positives while minimizing the number of false negatives. In our hands, X-ray crystallography is often the most sensitive screening technique, but it is used in conjunction with other biophysical approaches including nuclear magnetic resonance (NMR), surface plasmon resonance (SPR), and thermal unfolding (Tm) or differential scanning fluorimetry (DSF) assays in order to increase the number of fragments screened.

Although direct evidence from experimental fragment screening is an attractive approach, and may give information on the viability of progression into the hit-to-lead phase, it is not usually practical to screen all potential targets of interest as a means to assess druggability up front. Predictions and judgments are, therefore, normally required in order to prioritize and triage targets that enter a company's pipeline. Where the three-dimensional structure of the target is known, simple inspection of the protein surface and architecture by an experienced molecular modeler or structural biologist can be a powerful means to assess druggability. In addition, and particularly where multiple targets must be considered and compared, a large number of computational approaches have been developed to score and rank targets based on the predicted ability of surface cavities to support the binding of fragments or other small molecules (for a full review, see Ref. [39]). In particular, where the three-dimensional structure is known, geometric algorithms (e.g., LigSite [40], PocketDepth [41], and PocketPicker [42]) and/or probe- or energy-based algorithms (e.g., GRID [43], FTMap [44], SiteMap [45], QSiteFinder [46], AutoLigand [47], and ICM-PocketFinder [48]) can be used to analyze known or novel binding sites for their potential to bind fragments or small molecules.

The computational methods above are less likely to be successful if the three-dimensional structure of the protein target itself is not available or if known structures do not reveal the full extent of conformational changes that might occur on small molecule binding. In the former situation, inferring the nature of binding sites based on closely related homolog structures within a superfamily may be feasible (e.g., kinases), perhaps in conjunction with homology modeling. However, such bioinformatics approaches may be less appropriate for identifying novel binding sites on a protein target if these are poorly conserved among family members. The scenario where existing protein structures do not reveal the full conformational changes that are possible upon small molecule binding is particularly common for protein–protein interaction targets [49]. In these cases, ligand binding frequently leads to deeper pockets than anticipated from apo structures or from complex structures with the partner protein. Finally, we note that assessing and establishing the druggability of a protein target is no longer restricted to pharmaceutical companies. The relatively small size and cost of fragment libraries (compared to HTS screening decks) means that academic groups can now assess the druggability of a protein target by screening commercially available fragment libraries in biophysical experiments (see Chapter 10). In addition, novel computational programs like DruGUI and DrugFeature are making it easier and faster to calculate and predict a target's druggability [50,51].

## 2.4
## Properties of Ligands and Drugs

More than 90% of all compounds that enter clinical trials fail for reasons of safety or efficacy [52]. Failure of the compound due to safety concerns can often be traced to off-target or nonspecific effects, caused by lack of selectivity for its intended target or inappropriate physicochemical properties that lead to promiscuous protein, nucleotide, or membrane binding.

The ability of FBDD to generate multiple, chemically diverse starting points for drug design enables the design of lead-like molecules with a broader range of physicochemical properties, from which drug-like chemical series can be selected for lead optimization. This may be of particular importance for those targets where existing drug candidates are known to have issues relating to high molecular weight and lipophilicity. Often, such targets have active sites that have evolved to process hydrophobic substrates (lipids, steroids, collagen, acetyl-lysine, etc.) and FBDD has shown some promise here in the development of more polar inhibitors ([53], see also Figure 2.2 of reference [3]).

At the other extreme, the discovery of potent inhibitors of targets whose binding partners are highly polar acids, for example, phosphotyrosine and phosphoserine (SH2 domains and phosphatases) or sugars (lectins, glycosidases, and cell-surface receptors), has proved very challenging. Recent papers have demonstrated the ability of fragments to find alternative interactions in two cases: CD44 that binds the carbohydrate hyaluronan [54] and the mixed phosphatase, CDC25B, whose substrates contain the charged, polar residues phosphoserine, phosphothreonine, or phosphotyrosine [55]. In the case of CD44, the fragment only partially overlaps the carbohydrate binding site and also makes hydrophobic interactions in a new, induced pocket. In general, the ability of fragment screening to identify hit matter with novel representations of binding pharmacophores (and hence alternative physicochemical properties) may allow FBDD to access target classes that appear intractable based on considerations of the properties of known inhibitors or natural substrates and binding partners.

While the number and variety of fragment hits provide more options for limiting lipophilicity during optimization, the ability to increase the potency of those hits, while minimizing the molecular weight of the eventual lead and also improving its selectivity, requires both structural information and affinity data. Finding an efficient route to the lead, and reducing the number of compounds that must be synthesized, also requires structural knowledge (or at least a good model) of the proteins over which selectivity must be achieved.

## 2.5 Case Studies

Some examples are described below that illustrate the ability of FBDD to find novel hit matter with improved chemical properties or novel binding sites with a potential for allosteric modulation of target activity. These case studies are drawn from our own projects that have combined X-ray screening and structure determination with NMR screening, thermal shift screening, and computational methods, followed by a structure-guided medicinal chemistry hits-to-leads and lead optimization campaign. Some of the key principles that we follow when selecting optimal targets for FBDD are highlighted in these examples.

### 2.5.1 Case Study 1: Inhibitors of Apoptosis Proteins (IAPs)

One of the mechanisms by which cancer cells avoid apoptosis and continue to proliferate is by overexpressing members of the IAP (inhibitors of apoptosis) protein family. IAPs contain small protein modules, known as baculovirus inhibitors of apoptosis repeats (BIR domains), whose function is to bind to the N-termini of caspase enzymes, inactivating the enzyme and preventing caspase-mediated apoptosis. The N-terminal sequence that is recognized by BIR domains contains the motif $A^1x^2P^3x^4$. This interaction can be antagonized in cells by a peptide (the second mitochondrial activator of caspase or SMAC) containing the sequence $A^1V^2P^3I^4$, which causes the release of caspase and subsequent cell death. There are a number of peptidomimetic antagonists in the clinic and all contain an essential alanine warhead (Figure 2.1). These compounds show a strong (greater than 30-fold) selectivity for some members of the IAP family

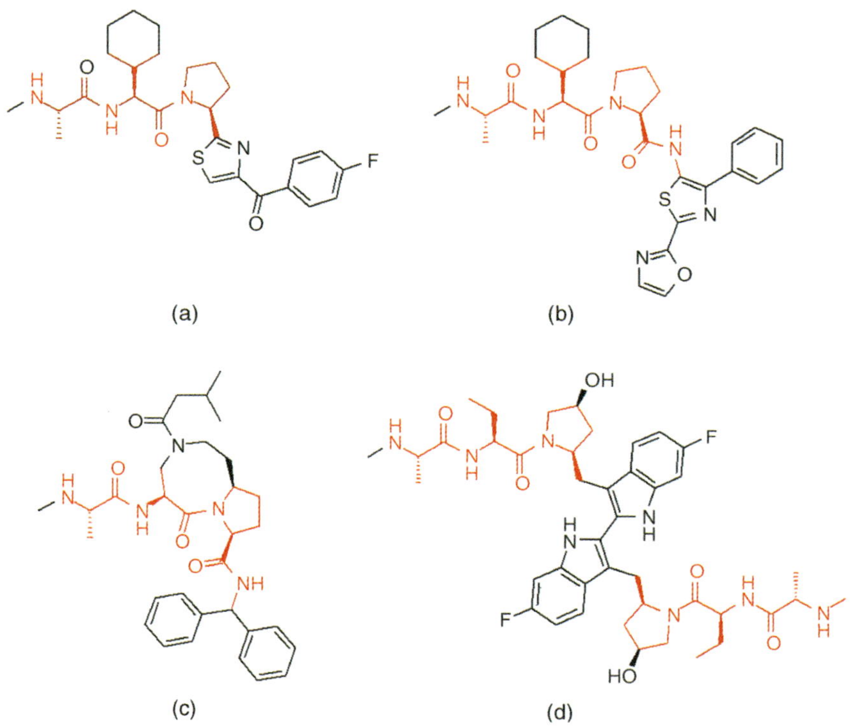


**Figure 2.1** Inhibitors of the IAP BIR3 domain in clinical trials (January 2015). All contain a peptide backbone (highlighted in red) with a terminal alanine residue. This pharmacophore gives potent inhibitors of cIAP but with >30-fold selectivity over XIAP (Astex, unpublished data). (a) LCL-161 from Novartis, (b) CUDC-427 from Curis, (c) Debio-1143 from Debiopharma, and (d) Birinapant from Tetralogic.

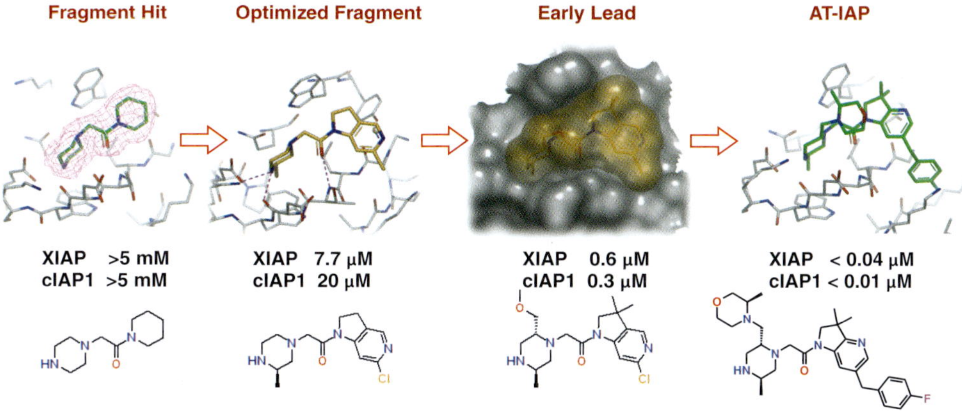


**Figure 2.2** X-ray and NMR fragment screening of XIAP identified piperazine hits that were elaborated into potent cell active inhibitors of both XIAP and cIAP. Numbers represent IC50 values as determined by XIAP and cIAP1 fluorescence polarization assays.

(cellular IAP 1; cIAP1) over others (X-linked IAP: XIAP). However, recent data suggests that a true dual antagonist of both cIAP1 and XIAP will promote more effective apoptotic response through the generation of death-inducing ripoptosome complexes, with resultant caspase activation.

Astex set out to find a nonpeptidic dual inhibitor of XIAP and cIAP by carrying out crystallographic and NMR fragment screens on XIAP. Both methods yielded relatively few hits, suggesting that this would be a challenging target, and many of these hits were primary amines that recapitulated the binding mode of alanine. However, a number of secondary amine hits were also detected, including piperazines, and these showed a more balanced affinity profile between XIAP and cIAP. These hits were subsequently elaborated to generate potent, nonpeptidic dual inhibitors of XIAP and cIAP1 (Figure 2.2) with low nanomolar potencies in cells ([56,57]).

Some nonamine hits were also detected, and NMR data indicated that these bound in the region of XIAP that is occupied by the AVP peptide of SMAC. However, these hits could not be crystallographically characterized and were not pursued. This illustrates one of the challenges of defining the best targets for FBDD. While there are many potential methods for identifying fragment hits, the probability of deriving a drug from this information is usually limited by the ability to direct chemical synthesis to improve affinity or generate detectable functional effects. Thus, ligandability is a necessary but not sufficient criterion for selecting a druggable target. Equally important though are the biochemical insights that can be generated from these fragment ligands, which may include the ability to model their interactions based on homologies with known ligands, their affinity for related proteins, isoforms, or single-site mutants and the availability of sensitive and robust bioassays to generate functional SAR data.

### 2.5.2
### Case Study 2: HCV-NS3

HCV-NS3 consists of two domains with protease and helicase activities. A crystallographic fragment screen has revealed a previously unidentified pocket at the interface between the protease and the helicase domains [30], in addition to a fragment binding site within the DNA binding groove of the helicase domain. The binding of ligands in the protease–helicase interface is expected to stabilize the crystallographic conformation, which represents an autoinhibited state of HCV-NS3 in which the C-terminal peptide of the helicase domain occupies the protease active site (Figure 2.3). The observation of several fragments binding at the interdomain site indicated that it was, at the least, a ligandable site and, coupled with the observation that this site was conserved across all HCV genotypes, led to a medicinal chemistry campaign to design molecules that could bind tightly here. In addition, molecules that target this site are expected to generate a different resistance profile from that of existing protease inhibitors and may, therefore, lead to new combination therapies for HCV infection.

One of the initial fragment leads ($K_d \sim 500\,\mu M$) is shown in Figure 2.4a, which also illustrates its progression to a lead compound with an $EC_{50}$ of 0.4 µM in a cell-based replicon assay (Figure 2.4c–e). Further optimization has yielded an optimized lead with an $EC_{50}$ of 8 nM in the replicon assay. The relevance of this autoinhibited conformation in cells was supported by the observation that

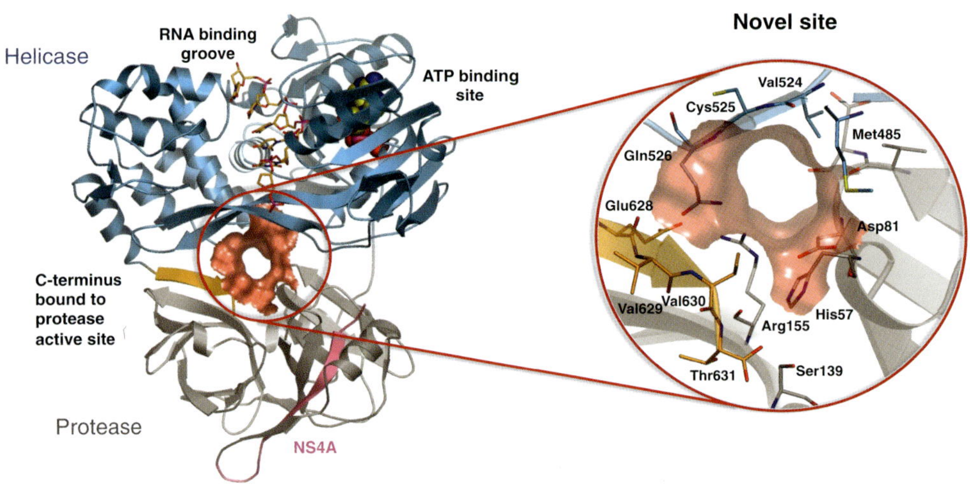


**Figure 2.3** Ribbon diagram of the crystal structure of full-length NS3-4a protein, with the new allosteric site (red surface) [30]. Shown are the protease domain (gray), the NS4a cofactor peptide (pink), and the helicase domain (blue). The C-terminus of the helicase domain (orange) occupies the protease active site in this autoinhibited conformation. Inset is a magnified view of the allosteric pocket showing residues in the protease–helicase interface as well as the protease catalytic triad (Ser139, His57, and Asp81).

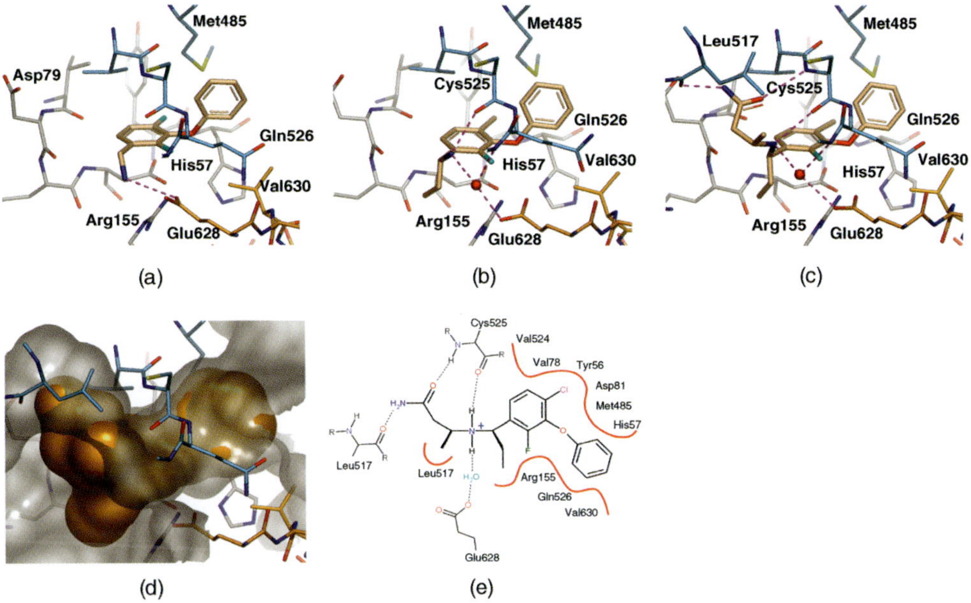


**Figure 2.4** Progression from the fragment hit to a cell active lead within the new allosteric site of HCV-NS3 [30]. Hydrogen bonds are indicated by dashed lines and conserved water molecules are indicated by red spheres. (a) Flanking fluorine substituents have been introduced into the initial fragment hit to improve hydrophobic interactions and stabilize a twisted conformation that is observed when bound to the protein. (b) The 4-fluoro substituent was replaced with a larger chlorine atom to maximize van der Waals interactions. An ethyl group was also introduced at the benzylic position to restrict the rotation of the aminomethyl side chain, which now interacts with Glu628 via a water molecule and also forms a hydrogen bond to the backbone carbonyl of Cys525. (c) The X-ray crystal structure of the complex shown in (b) was used to design a small set of compounds to probe a pocket formed by Tyr516, Leu517, Val524, and Cys525. This resulted in the addition of a small amide side chain, which makes key interactions with residues from the protease domain and the helicase domain, including the C-terminus. (d) Surface representation of the compound shown in (c) and the allosteric pocket. (e) Two-dimensional representation of the compound shown in (c) and its interactions with protein when bound in the allosteric site. Hydrophobic contacts are indicated by solid red lines.

prolonged low-level dosing of cell lines containing the replicon led to the generation of resistant mutations at Met485 and Val630 within this interdomain pocket. Colonies expressing these mutations showed decreased susceptibility to compound 5 but not to Telaprevir that binds at the protease active site.

### 2.5.3
### Case Study 3: PKM2

The example of HCV-NS3 described above illustrates the potential of FBDD to stabilize inactive states of the target protein. In the case of NS3, the inactive

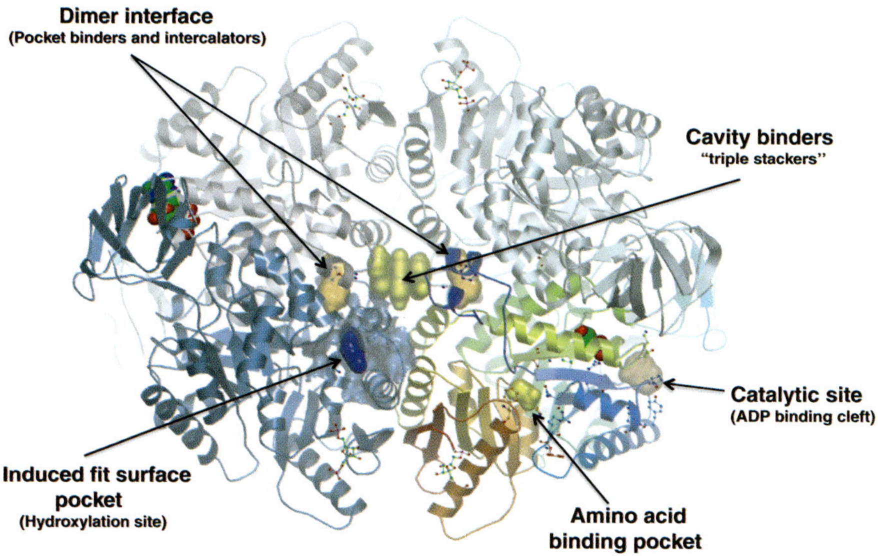


**Figure 2.5** Ribbon diagram of the crystal structure of homotetrameric PKM2. Fragment binding was observed at five different sites within the tetramer, including the catalytic site, and a dimer interface site that is the target of previously described allosteric activators [58]. Fragments were also observed to bind in the central cavity of the tetramer and an inducible pocket on the protein surface, as well as a previously unknown allosteric site that preferentially binds serine, a physiological activator of PKM2 [31].

conformation is probably also present to a significant extent in the absence of the ligand. The second potential advantage of drug binding at sites that are remote from catalytic or cofactor sites is that binding may modulate (increase or reduce) target activity without eliminating it.

Pyruvate kinase M2 (PKM2), a tetrameric enzyme involved in controlling the fate of molecules passing through the glycolytic pathway, illustrates the potential for complex allosteric control by ligand binding. Figure 2.5 shows the architecture of PKM2; in this view, the dimer–dimer interface runs vertically and is formed by a number of helix–helix contacts. Formation of the active tetramer is assisted by binding four molecules of fructose bis-phosphate (FBP) at the dimer–dimer interface, leading to stabilization of the helix–helix contacts. In a fragment screen conducted using X-ray crystallography against FBP-bound tetrameric PKM2, five additional fragment binding sites were identified and these are described in Figure 2.5. One site, here labeled the dimer interface site, is the binding site for "Cantley activators," a previously described series of quinolone sulphonamides that increase PKM2 activity but display a different pharmacology from that of FBP [58]. Fragments were also observed to bind both in the ADP binding cleft of the catalytic site and in the central cavity of the tetramer where they formed multimeric stacks and in a small surface pocket that was formed only in the presence of suitable ligands. Finally, small amino acids (glycine, alanine, serine, cysteine, and threonine) were found to bind to PKM2 at a site, labeled amino-acid binding pocket, which is distinct not only from the catalytic site but also the FBP and Cantley activator sites [31]. Further studies using the 20 naturally occurring amino acids showed that only the binding of serine caused a change in PKM2 activity, leading to a sixfold increase at saturation, which is comparable to the level of activation caused by FBP. Regulation of PKM2 activity by serine is probably linked to its biosynthesis, which requires 3-phosphoglycerate (3-PG); increasing levels of serine increase the activity of PKM2 that deplete the pool of 3-PG by diverting it into the TCA cycle [59].

### 2.5.4
### Case Study 4: Soluble Adenylate Cyclase

The second messenger, 3′,5′-cyclic adenosine monophosphate(cAMP), is produced by adenylate or adenylyl cyclases and its cellular concentration is tightly regulated. Mammalian cells contain a cytosolic enzyme, soluble adenylate cyclase (solAC), which is responsible for generating cAMP in cellular microdomains and which is regulated by both $Ca^{2+}$ and bicarbonate ions [60]. Due to its predominance in testes and spermatozoa, human solAC (hsolAC) was considered as a potential drug target for male contraception. The discovery of its presence in a wider range of tissues and its importance in intracellular signaling has recently led to interest in this target for a broader spectrum of potential therapeutic indications.

A fragment screen, conducted using a thermal unfolding assay, generated a number of hits [32]. Fragment hits were validated using X-ray crystallography, leading to the identification of their binding sites and interactions. The results showed that fragments occupied three overlapping regions, in a cleft containing the ATP-substrate binding site of hsolAC (Figure 2.6a).

A majority of the fragment hits were carboxylic acids or weakly acidic heterocycles and occupied the lower portion of the cleft, approximately 12 Å from the ATP binding site itself. Comparison of the structures of several complexes showed that the fragment binding cleft demonstrated a structural plasticity that depended on the binding partner. In particular, the middle of the cleft contained an induced pocket that was formed only in the presence of suitable fragments. The allosteric activator, bicarbonate ion, was also shown to bind in the lower pocket, consistent with the hypothesis that this comprises a functionally relevant site. It should be noted that this was the first time that hsolAC had been crystallized [61] and that the bicarbonate site had not been identified in earlier structures of related cytosolic adenylate cyclases. Bicarbonate binding induces movement of the side chain of Arg 176 that prevents its interaction with a peptide loop adjacent to the ATP binding region, and this change is hypothesized to be responsible for the pharmacology of bicarbonate as an activator. The fragment hits obtained from screening mimic some of the interactions between bicarbonate and hsolAC but were found to be inhibitory and do not lead to the

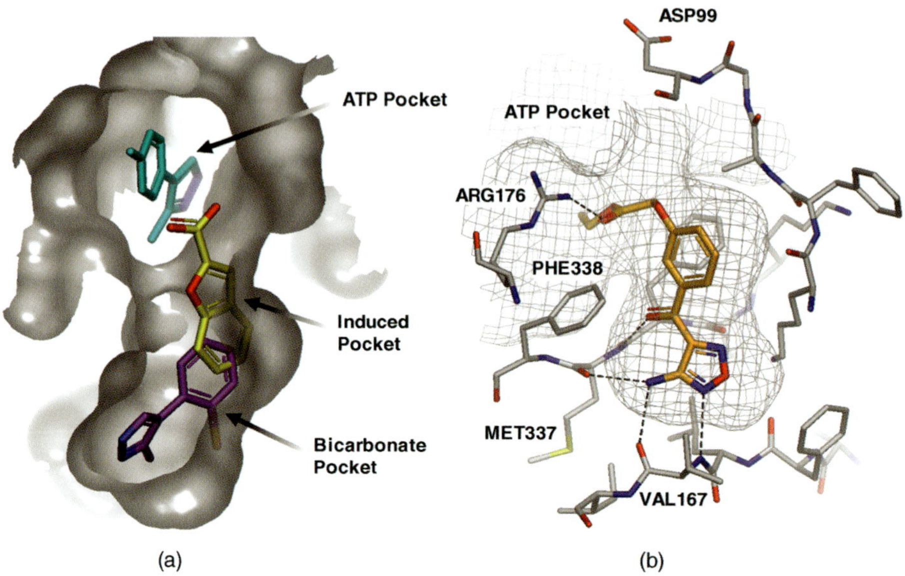


**Figure 2.6** Fragment binding in hSolAC. (a) Overlay of three different crystal structures showing fragments binding in multiple regions of the hsolAC catalytic and allosteric pockets. A phenyl pyrazole (cyan) binds in the ATP pocket, an aromatic carboxylic acid (yellow) induces a new pocket, and another pyrazole (magenta) binds in the allosteric bicarbonate pocket. (b) X-ray structure of the complex of hSolAC with an advanced lead compound that spans the bicarbonate and ATP binding sites. Hydrogen bonds are shown as dashed lines.

movement of Arg176 that is observed with bicarbonate binding. After medicinal chemistry elaboration of the initial screening hits, the most advanced lead compound exploited the geometry of Arg176 as well as additional conformational changes that were generated by ligand binding. This compound occupies both the bicarbonate and the ATP sites and is an inhibitor of hsolAC with an affinity of 0.36 μM (Figure 2.6b).

## 2.6 Conclusions

As highlighted in this chapter and elsewhere in this book, FBDD has emerged as a powerful approach for finding new lead compounds. Of course, it is only one approach in the tool kit of lead discovery and should therefore be utilized in a considered manner. One pleasing observation is that it is becoming less often used only as a "salvage" option when HTS has failed. This emphasizes the progress that has been made by the FBDD community. However, there are still many examples where the method has not been applied with sufficient care, either by

**Table 2.1** Areas where FBDD offers potential advantages.

| Potential areas of benefit | Opportunities for FBDD |
|---|---|
| Poorly druggable targets | Improved exploration of chemical space increases the probability of finding hits |
| Screening of targets with low/no biochemical activity | Hits can be detected that bind to and stabilize a less active/inactive conformation of the target |
| Existing drugs have suboptimal PK/PD properties | More hits are generated, with better physicochemical properties and more options during elaboration |
| Existing drugs show on-target toxicity or resistance | Direct binding methods allow detection of novel, potentially allosteric sites |

selecting an inappropriate target or using libraries in which the compounds resemble large leads and not fragments.

General areas of current need where FBDD may have particular advantages are summarized in Table 2.1.

Target selection for FBDD remains one of the biggest challenges and in this chapter we have attempted to highlight some of the key principles that we have found useful. Ensuring that the target can be screened using multiple biophysical techniques and having a strategy to employ structure-driven optimization of the fragment into a lead compound are two key principles. However, it is also clear that FBDD techniques must continue to develop as we tackle more complex disease targets and mechanisms.

## References


1 Bollag, G., Tsai, J., Zhang, J., Zhang, C., Ibrahim, P., Nolop, K., and Hirth, P. (2012) Vemurafenib: the first drug approved for *BRAF*-mutant cancer. *Nature Reviews Drug Discovery*, **11**, 873–886.

2 Baker, M. (2013) Fragment-based lead discovery grows up. *Nature Reviews Drug Discovery*, **12**, 5–7.

3 Hopkins, A.L., Keserü, G.M., Leeson, P.D., Rees, D.C., and Reynolds, C.H. (2014) The role of ligand efficiency metrics in drug discovery. *Nature Reviews Drug Discovery*, **13**, 105–121.

4 Jhoti, H., Williams, G., Rees, D.C., and Murray, C.W. (2014) The 'rule of three' for fragment-based drug discovery: where are we now? *Nature Reviews Drug Discovery*, **12**, 644.

5 Leeson, P.D. and St-Gallay, S.A. (2011) The influence of the 'organizational factor' on compound quality in drug discovery. *Nature Reviews Drug Discovery*, **10**, 749–765.

6 Barelier, S., Pons, J., Marcillat, O., Lancelin, J.-M., and Krimm, I. (2012) Fragment-based deconstruction of Bcl-xL inhibitors. *Journal of Medicinal Chemistry*, **53**, 2577–2588.

7 Fry, D.C., Wartchow, C., Graves, B., Janson, C., Lukacs, C., Kammlott, U., Belunis, C., Palme, S., Klein, C., and Vu, B. (2013) Deconstruction of a nutlin: dissecting the binding determinants of a potent protein–protein interaction inhibitor. *ACS Medicinal Chemistry Letters*, **4**, 660–665.

8 Dias, D.M., Van Molle, I., Baud, M.G.J., Galdeano, C., Geraldes, C.F.G.C., and Ciulli, A. (2013) Is NMR fragment screening fine-tuned to assess druggability

of protein–protein interactions? *ACS Medicinal Chemistry Letters*, **4**, 660–665.

9 Dragovich, P.S., Zhao, G., Baumeister, T., Bravo, B., Giannetti, A.M., Ho, Y.-C., Hua, R., Li, G., Liang, X., Ma, X., O'Brien, T., Oh, A., Skelton, N.J., Wang, C., Wang, W., Wang, Y., Xiao, Y., Yuen, P.-W., Zak, M., Zhao, Q., and Zheng, X. (2014) Fragment-based design of 3-aminopyridine-derived amides as potent inhibitors of human nicotinamide phosphoribosyltransferase (NAMPT). *Bioorganic and Medicinal Chemistry Letters*, **24**, 954–962.

10 Andrews, S.P., Brown, G.A., and Christopher, J.A. (2014) Structure-based and fragment-based GPCR drug discovery. *ChemMedChem*, **9**, 256–275.

11 Jerabek-Willemsen, M., André, T., Wanner, R., Roth, H.M., Duhr, S., Baaske, P., and Breitsprecher, D. (2014) MicroScale thermophoresis: interaction analysis and beyond. *Journal of Molecular Structure*, **1077**, 101–113.

12 Buratto, R., Mammoli, D., Chiarparin, E., Williams, G., and Bodenhausen, G. (2014) Exploring weak ligand–protein interactions by long-lived NMR states: improved contrast in fragment-based drug screening. *Angewandte Chemie International Edition*, **53**, 11376–11380.

13 Marx, V. (2014) Structural biology: 'seeing' crystals the XFEL way. *Nature Methods*, **11**, 903–908.

14 Cheng, A.C., Coleman, R.G., Smyth, K.T., Cao, Q., Soulard, P., Caffrey, D.R., Salzberg, A.C., and Huang, E.S. (2007) Structure-based maximal affinity model predicts small-molecule druggability. *Nature Biotechnology*, **25**, 71–75.

15 Boyd, S.M. and de Kloe, G.E. (2010) Fragment library design: efficiently hunting drugs in chemical space. *Drug Discovery Today*, **7**, 173–180.

16 Hall, R.J., Mortenson, P.N., and Murray, C.W. (2014) Efficient exploration of chemical space by fragment-based screening. *Progress in Biophysics and Molecular Biology*, **116**, 82–91.

17 Fink, T. and Reymond, J.-L. (2007) Virtual exploration of the chemical universe up to 11 atoms of C, N, O, F: assembly of 26.4 million structures (110.9 million stereoisomers) and analysis for new ring systems, stereochemistry, physicochemical properties, compound classes, and drug discovery. *Journal of Chemical Information and Modeling*, **47**, 342–353.

18 Kutchukian, P.S., Vasilyeva, N.Y., Xu, J., Lindvall, M.K., Dillon, M.P., Glick, M., Coley, J.D., and Brooijmans, N. (2012) Inside the mind of a medicinal chemist: the role of human bias in compound prioritization during drug discovery. *PLOS One*, **7** (11), e48476.

19 Stamford, A. and Strickland, C. (2013) Inhibitors of BACE for treating Alzheimer's disease: a fragment-based drug discovery story. *Current Opinion in Chemical Biology*, **17**, 320–328.

20 Efremov, I.V., Vajdos, F.F., Borzilleri, K.A., Capetta, S., Chen, H., Dorff, P.H., Dutra, J.K., Goldstein, S.W., Mansour, M., McColl, A., Noell, S., Oborski, C.E., O'Connell, T.N., O'Sullivan, T.J., Pandit, J., Wang, H., Wei, B., and Withka, J.M. (2012) Discovery and optimization of a novel spiropyrrolidine inhibitor of β-secretase (BACE1) through fragment-based drug design. *Journal of Medicinal Chemistry*, **55**, 9069–9088.

21 Albert, J.S. and Edwards, P.D. (2008) Identification of high-affinity-secretase inhibitors using fragment-based lead generation, in *Fragment-Based Drug Discovery* (eds E.R. Zartler and M. Shapiro), John Wiley & Sons, Ltd, Chichester, UK, pp. 261–280.

22 Ward, R.A., Brassington, C., Breeze, A.L., Caputo, A., Critchlow, S., Davies, G., Goodwin, L., Hassall, G., Greenwood, R., Holdgate, G.A., Mrosek, M., Norman, R.A., Pearson, S., Tart, J., Tucker, J.A., Vogtherr, M., Whittaker, D., Wingfield, J., Winter, J., and Hudson, K. (2012) Design and synthesis of novel lactate dehydrogenase A inhibitors by fragment-based lead generation. *Journal of Medicinal Chemistry*, **55**, 3285–3306.

23 Dragovich, P.S., Fauber, B.P., Corson, L.B., Ding, C.Z., Eigenbrot, C., Ge, H.-X., Giannetti, A.M., Hunsaker, T., Labadie, S., Liu, Y., Malek, S., Pan, B., Peterson, D., Pitts, K., Purkey, H.E., Sideris, S., Ultsch, M., Vander Porten, E., Wei, B.-Q., Xu, Q., Yen, I., Yue, Q., Zhang, H., and Zhang, X. (2013) Identification of substituted

2-thio-6-*oxo*-1,6-dihydropyrimidines as inhibitors of human lactate dehydrogenase. *Bioorganic and Medicinal Chemistry Letters*, **23**, 3186–3194.

24 Saxty, G., Woodhead, S.J., Berdini, V., Davies, T.G., Verdonk, M.L., Wyatt, P.G., Boyle, R.G., Barford, D., Downham, R., Garrett, M.D., and Carr, R.A. (2007) Identification of inhibitors of protein kinase B using fragment-based lead discovery. *Journal of Medicinal Chemistry*, **50**, 2293–2296.

25 Jahnke, W., Grotzfeld, R.M., Pelle, X., Strauss, A., Fendrich, G., Cowan-Jacob, S. W., Cotesta, S., Fabbro, D., Furet, P., Mestan, J., and Marzinzik, A.L. (2010) Binding or bending: distinction of allosteric Abl kinase agonists from antagonists by an NMR-based conformational assay. *JACS*, **132**, 7043–7048.

26 Cho, Y.S., Angove, H., Brain, C., Chen, C.H.-T., Cheng, H., Cheng, R., Chopra, R., Chung, K., Congreve, M., Dagostin, C., Davis, D.J., Feltell, R., Giraldes, J., Hiscock, S.D., Kim, S., Kovats, S., Lagu, B., Lewry, K., Loo, A., Lu, Y., Luzzio, M., Maniara, W., McMenamin, R., Mortenson, P.N., Benning, R., O'Reilly, M., Rees, D.C., Shen, J., Smith, T., Wang, Y., Williams, G., Alison Woolford, A.J.-A., Wrona, W., Xu, M., Yang, F., and Howard, S. (2012) Fragment-based discovery of 7-azabenzimidazoles as potent, highly selective, and orally active CDK4/6 inhibitors. *ACS Medicinal Chemistry Letters*, **3**, 445–449.

27 Schulz, M.N., Fanghanel, J., Schafer, M., Badock, V., Briem, H., Boemer, U., Nguyen, D., Husemann, M., and Hillig, R.C. (2011) A crystallographic fragment screen identifies cinnamic acid derivatives as starting points for potent Pim-1 inhibitors. *Acta Crystallographica*, **D67**, 156–166.

28 Xiang, Y., Hirth, B., Asmussen, G., Biemann, H.-P., Bishop, K.A., Good, A., Fitzgerald, M., Gladysheva, T., Jain, A., Jancsics, K., Liu, J., Metz, M., Papoulis, A., Skerlj, R., Stepp, J.D., and Wei, R.R. (2011) The discovery of novel benzofuran-2-carboxylic acids as potent Pim-1 inhibitors. *Bioorganic and Medicinal Chemistry Letters*, **21**, 3050–3056.

29 Jahnke, W., Rondeau, J.-M., Cotesta, S., Marzinzik, A., Pelle, X., Geiser, M., Strauss, A., Gotte, M., Bitsh, F., Hemmig, R., Henry, C., Lehmann, S., Glickman, J.F., Roddy, T.P., Stout, S.J., and Green, J.R. (2010) Allosteric non-bisphosphonate FPPS inhibitors identified by fragment-based discovery. *Nature Chemical Biology*, **6**, 660–666.

30 Saalau-Bethell, S.M., Woodhead, A.J., Chessari, G., Carr, M.G., Coyle, J., Graham, B., Hiscock, S.D., Murray, C.W., Pathuri, P., Rich, S.J., Richardson, C.J., Williams, P.A., and Jhoti, H. (2012) Discovery of an allosteric mechanism for the regulation of HCV NS3 protein function. *Nature Chemical Biology*, **8**, 920–925.

31 Chaneton, B., Hillmann, P., Zheng, L., Martin, A.C.L., Maddocks, O.D.K., Chokkathukalam, A., Coyle, J.E., Jankevics, A., Holding, F.P., Vousden, K.H., Frezza, C., O'Reilly, M., and Gottlieb, E. (2012) Serine is a natural ligand and allosteric activator of pyruvate kinase M2. *Nature*, **491**, 458–462.

32 Saalau-Bethell, S.-M., Berdini, V., Cleasby, A., Congreve, M., Coyle, J.E., Lock, V., Murray, C.W., O'Brien, M.A., Rich, S.J., Sambrook, T., Vinkovic, M., Yon, J.R., and Jhoti, H. (2014) Crystal structure of human soluble adenylate cyclase reveals a distinct, highly flexible allosteric bicarbonate binding pocket. *ChemMedChem*, **9**, 823–832.

33 Kar, G., Keskin, O., Gursoy, A., and Nussinov, R. (2010) Allostery and population shift in drug discovery. *Current Opinion in Pharmacology*, **10**, 715–722.

34 Zimmermann, G., Schultz-Fademrecht, C., Küchler, P., Murarka, S., Ismail, S., Triola, G., Nussbaumer, P., Wittinghofer, A., and Waldmann, H. (2014) Structure guided design and kinetic analysis of highly potent benzimidazole inhibitors targeting the PDEδ prenyl binding site. *Journal of Medicinal Chemistry*, **57**, 5435–5448.

35 Egner, U. and Hillig, R.C. (2008) A structural biology view of target druggability. *Expert Opinion on Drug Discovery*, **3**, 391–401.

36 Hajduk, P.J., Huth, J.R., and Fesik, S.W. (2005) Druggability indices for protein

targets derived from NMR-based screening data. *Journal of Medicinal Chemistry*, **48**, 2518–2525.

37 Hajduk, P.J., Huth, J.R., and Tse, C. (2005) Predicting protein druggability. *Drug Discovery Today*, **10**, 1675–1682.

38 Edfeldt, F.N.B., Folmer, R.H.A., and Breeze, A.L. (2011) Fragment screening to predict druggability (ligandability) and lead discovery success. *Drug Discovery Today*, **16**, 284–287.

39 Pérot, S., Sperandio, O., Miteva, M.A., Camproux, A., and Villoutreix, B.O. (2010) Druggable pockets and binding site centric chemical space: a paradigm shift in drug discovery. *Drug Discovery Today*, **15**, 656–667.

40 Hendlich, M., Rippmann, F., and Barnickel, G. (1997) LIGSITE: automatic and efficient detection of potential small molecule-binding sites in proteins. *Journal of Molecular Graphics and Modelling*, **15**, 359–363.

41 Kalidas, Y. and Chandra, N. (2008) PocketDepth: a new depth based algorithm for identification of ligand binding sites in proteins. *Journal of Structural Biology*, **161**, 31–42.

42 Weisel, M., Proschak, E., and Schneider, G. (2007) PocketPicker: analysis of ligand binding-sites with shape descriptors. *Chemistry Central Journal*, **13**, 1–7.

43 Goodford, P.J. (1985) A computational procedure for determining energetically favorable binding sites on biologically important macromolecules. *Journal of Medicinal Chemistry*, **28**, 849–857.

44 Brenke R, Kozakov D, Chuang GY, Beglov D, Hall D, et al. (2009) Fragment-based identification of druggable 'hot spots' of proteins using Fourier domain correlation techniques. Bioinformatics 25: 621–627.

45 Halgren, T. (2007) New method for fast and accurate binding-site identification and analysis. *Chemical Biology and Drug Design*, **69**, 146–148.

46 Laurie, A.L.T. and Jackson, R.M. (2005) Q-SiteFinder: an energy-based method for the prediction of protein–ligand binding sites. *Bioinformatics*, **21**, 1908–1916.

47 Harris, R., Olson, A.J., and Goodsell, D.S. (2008) Automated prediction of ligand-binding sites in proteins. *Proteins*, **70**, 1506–1517.

48 An, J., Totrov, M., and Abagyan, R. (2005) Pocketome via comprehensive identification and classification of ligand binding envelopes. *Molecular and Cellular Proteomics*, **4**, 752–761.

49 Arkin, M.R., Tang, Y., and Wells, J.A. (2014) Small-molecule inhibitors of protein–protein interactions: progressing toward the reality. *Chemistry and Biology*, **21**, 1102–1114.

50 Bakan, A., Meireles, L.M., and Bahar, I. (2011) ProDy: protein dynamics inferred from theory and experiments. *Bioinformatics*, **27**, 1575–1577.

51 Liu, T. and Altman, R.B. (2014) Identifying druggable targets by protein microenvironments matching: application to transcription factors. *CPT: Pharmacometrics and Systems Pharmacology*, **3**, 1–9.

52 Plenge, R.M., Scolnick, E.M., and Altshuler, D. (2013) Validating therapeutic targets through human genetics. *Nature Reviews Drug Discovery*, **12**, 581–594.

53 Murray, C.W., Verdonk, M.L., and Rees, D.W. (2012) Experiences in fragment-based drug discovery. *Trends in Pharmacological Sciences*, **33**, 224–232.

54 Liu, L.-K. and Finzel, B.C. (2014) Fragment-based identification of an inducible binding site on cell surface receptor CD44 for the design of protein–carbohydrate interaction inhibitors. *Journal of Medicinal Chemistry*, **57**, 2714–2725.

55 Lund, G., Dudkin, S., Borkin, D., Ni, W., Grembecka, J., and Cierpicki, T. (2014) Inhibition of CDC25B phosphatase through disruption of protein–protein interaction. *ACS Chemical Biology*. doi: 10.1021/cb500883h.

56 Crawford, N., Stasik, I., Holohan, C., Majkut, J., McGrath, M., Johnston, P.G., Chessari, G., Ward, G.A., Waugh, D.J., Fennell, D.A., and Longley, D.B. (2013) SAHA overcomes FLIP-mediated inhibition of SMAC mimetic-induced apoptosis in mesothelioma. *Cell Death*

*and Disease*, **4**, e733. doi: 10.1038/cddis.2013.258.

57 Chessari G, Buck IM, Day JE, Day PJ, Iqbal A, Johnson CN, Lewis EJ, Martins V, Miller D, Reader M, Rees DC, Rich SJ, Tamanini E, Vitorino M, Ward GA, Williams PA, Williams G, Wilsher NE, Woolford AJ.(2015) Fragment-Based Drug Discovery Targeting Inhibitor of Apoptosis Proteins: Discovery of a Non-Alanine Lead Series with Dual Activity Against cIAP1 and XIAP. Journal of Medicinal Chemistry 58(16): 6574–6588

58 Kung, C., Hixon, J., Choe, S., Marks, K., Gross, S., Murphy, E., DeLaBarre, B., Cianchetta, G., Sethumadhavan, S., Wang, X., Yan, S., Gao, Y., Fang, F., Wei, W., Jiang, F., Wang, S., Qian, K., Saunders, J., Driggers, E., Koon Woo, H., Kunii, K., Murray, S., Yang, H., Yen, K., Liu, W., Cantley, L.C., Vander Heiden, M.G., Su, S.M., Jin, S., Salituro, F.G., and Dang, L. (2012) Small molecule activation of PKM2 in cancer cells induces serine auxotrophy. *Chemistry and Biology*, **19**, 1187–1198.

59 Ye, J., Mancuso, A., Tong, X., Ward, P.S., Fan, J., Rabinowitz, J.D., and Thompson, C.B. (2012) Pyruvate kinase M2 promotes *de novo* serine synthesis to sustain mTORC1 activity and cell proliferation. *Proceedings of the National Academy of Sciences*, **109**, 6904–6909.

60 Litvin, T.N., Kamenetsky, M., Zarifyan, A., Buck, J., and Levin, L.R. (2003) Kinetic properties of "soluble" adenylyl cyclase. *Journal of Biological Chemistry*, **278**, 15922–15926.

61 Saalau-Bethell, S.M., Cleasby, A., Sambrook, T.A., Coyle, J.E., and Vinkovic, M. (2007) PCT Int. Appl. WO 2007010285 A2 20070125.

# 3
# Enumeration of Chemical Fragment Space

*Jean-Louis Reymond, Ricardo Visini, and Mahendra Awale*

## 3.1
## Introduction

In fragment-based drug discovery (FBDD), screening small collections of only a few thousand fragments, defined as very small organic molecules that are structurally rigid and carry only few functional groups, for weak binding to a protein target is used to identify hit compounds that can later be optimized by adding substituents or other fragments [1,2]. The successes of FBDD show that relatively small sets of fragments, which are often selected as substructures of known drugs, are sufficient as starting points to develop new drugs. One may, therefore, conclude that there is no need to look for new fragments as alternative starting points. In fact, new fragments might be difficult to find because existing compound collections in catalogs or public databases are not very abundant for fragment-sized molecules. For example, there are less than a million fragment-like molecules in the public database PubChem that contains over 50 million entries [3].

We believe that the above conclusion is too restrictive and misses an enormous opportunity for chemical innovation. To take a closer look at small molecule chemical diversity, our group has performed an exhaustive computational enumeration of the small molecule chemical space restricted only by simple rules of chemical stability and synthetic feasibility, and produced the chemical universe databases GDB-11 (*G*enerated *D*ata*B*ase up to 11 atoms of C, N, O, and F; 26.4 million structures) [4,5], GDB-13 (up to 13 atoms of C, N, O, S, and Cl; 977 million structures) [6], and GDB-17 (up to 17 atoms of C, N, O, S, and halogen; 166.4 billion structures) [7,8]. These databases contain extremely large numbers of possible fragment-sized molecules, several orders of magnitude more than those available in known collections. GDB-derived molecules cover a much broader property range than known molecules, in particular in terms of molecular shape and functional group combinations. Following up on previous accounts on the GDB [9–12], including a recent perspective on fragrances that share properties of fragments [13], we herein review the generation and analysis of GDB and its use for drug discovery that led to the development of several visualization and virtual screening tools available at www.gdb.unibe.ch. We then

*Fragment-based Drug Discovery: Lessons and Outlook*, First Edition. Edited by Daniel A. Erlanson and Wolfgang Jahnke.

discuss how to use these virtual screening tools to identify new fragments in GDB.

## 3.2 The Enumeration of Chemical Space

### 3.2.1 Counting and Sampling Approaches

The enumeration of chemical space goes back to Cayley, the nineteenth-century inventor of graph theory, who used the graph theory to calculate the number of possible acyclic hydrocarbons as a function of size [14–16], an approach later followed for other topologies [17,18]. Estimates of the size of chemical space have also been reported, for example, $10^{60}$ for all molecules obeying Lipinski's rule of five for oral bioavailability [19,20], and $10^{20}$–$10^{24}$ for all molecules up to 30 atoms [21,22].

Beyond these counting exercises, many programs have been written capable of computationally generating molecular structures in a specific context. For example, MOLGEN [23,24] enumerates molecules fitting predefined criteria of elemental composition, mass, and the presence or absence of functional groups to enable computer-aided structure elucidation [25–28]. Furthermore, structure generators such as SPROUT [29] evolve organic molecules for maximum docking to a target protein or highest similarity to a reference molecule using a genetic algorithm [30]. In these algorithms, molecules are often assembled from known building blocks using known coupling reactions, an approach also used to enumerate virtual combinatorial libraries [31–34], for example, the Pfizer Global Virtual Library that corresponds to 10 trillion molecules, although molecules are enumerated only in response to specific searches [35].

Genetic algorithms can also be used to sample chemical space stochastically, for example, by performing single atom and bond changes within molecules to generate new structures [36–38]. In our "chemical space travel" algorithm [39], such structural mutations on a starting molecule are coupled to selection by similarity to a target molecule, which by iterative cycles allows one to generate libraries of hundreds of thousands of intermediates representing the chemical space between start and target. When performed in undirected manner, such mutations can be used to explore chemical space through the generation of random molecules [40].

### 3.2.2 Enumeration of the Chemical Universe Database GDB

Most of the above stochastic approaches start from known molecules and building blocks to sample chemical space. By contrast, the philosophy of the GDB project was to exhaustively enumerate all possible molecules taking only minimal constraints in consideration (ring strain, stable functional groups) and no specific known molecule as starting point, such as to gain an unbiased insight into chemical space and possibly discover new structural types. To realize this enumeration, we started from the analogy between organic molecules and mathematical graphs noted by Cayley, with atoms corresponding to graph nodes and chemical bonds to graph edges. An exhaustive list of molecules was derived from an exhaustive list of graphs provided by the program GENG [41] through the following operations: (1) select graphs corresponding to a geometrically possible hydrocarbon considering topology and ring strain criteria; (2) expand each graph to various skeletons by selectively introducing double and triple bonds following valency and ring strain rules; and (3) enumerate molecules corresponding to each skeleton by substituting heteroatoms for carbons systematically, taking valency and functional group stability rules as well as tautomer equilibria into account. This approach was used to generate the databases GDB-11 [4,5], GDB-13 [6], and most recently GDB-17 that contains 166.4 billion molecules up to 17 atoms of C, N, O, S, and halogens and represents by far the largest database of explicitly enumerated small molecules reported to date (Figure 3.1a) [7,8]. GDB molecules are stored as SMILES representing 2D structures [42], which can be expanded to stereoisomers using the 3D generator CORINA [43], which increases database size by at least five- to sixfold.

### 3.2.3 GDB Contents

For each additional heavy atom in molecule size, the number of molecules enumerated in the GDB increases approximately by a factor corresponding to the heavy atom count (HAC), which is more than exponential. To partly compensate for this effect in the case of GDB-17 and keep the total number of molecules enumerated to a manageable size (<100 billion), increasingly drastic selection criteria were applied for graphs, skeletons, and functional groups as functions of molecule size above 11 atoms (Figure 3.1b). Even under these restrictions, however, the increase in database size as a function of molecule size is much steeper than that observed in databases of experimental molecules such as PubChem, ChEMBL, and DrugBank, indicating the increasingly scarce sampling of chemical space in experimental databases for larger molecules.

GDB molecules have comparable topologies to molecules in experimental databases, with only a slight shift toward polycyclic compounds, bicyclic molecules being the most abundant in the database (Figure 3.1c). On the other hand, GDB differs very strongly from experimental databases in terms of molecular categories, and is largely dominated by nonaromatic heterocycles, while aromatic molecules are comparatively rare although they dominate experimental databases (Figure 3.1d). This difference also occurs in GDB-17 even though small rings (3- and 4-membered rings), which are not used to form aromatic rings, were reduced to maximum one per molecule at 16 atoms and entirely eliminated at 17 atoms to contain the combinatorial explosion of possibilities (Figure 3.1e).

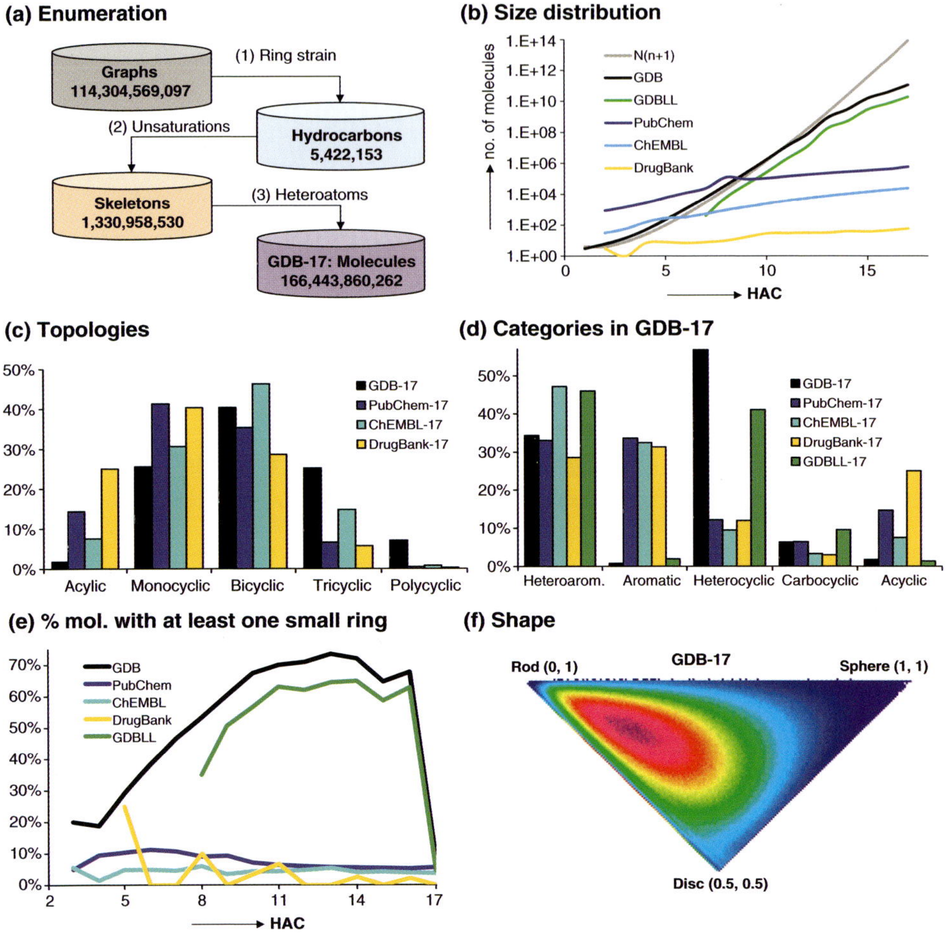


**Figure 3.1** Enumeration and properties of GDB-17 and its lead-like (LL) subset compared to molecules up to 17 atoms in PubChem, ChEMBL, and DrugBank. (a) Enumeration strategy. (b) Number of molecules per database as function of the heavy atom count (HAC = nonhydrogen atoms). The $N(n+1)$ curve is obtained as number of molecules at $\text{HAC}(n+1)$ = number of molecules at $\text{HAC}(n) \times (n+1)$ starting with $N = \text{HAC}(1) = 4$. (c and d) Percentages of molecular topologies and categories in the different databases. (e) Percentages of molecules containing small rings (3- or 4-membered rings) as a function of HAC in the different databases. (f) Distribution of GDB-17 molecules in the triangular shape plot.

One striking consequence of the abundance of nonaromatic heterocycles and small rings in GDB is that GDB molecules show very substantial 3D shape as estimated by their distribution in the shape triangle computed from the normalized principal moments of inertia, which separates rod-like, disk-like, and sphere-like molecules (Figure 3.1f) [44]. This stands in stark contrast to experimental databases that are strongly dominated by planar, rod-like, and disk-like molecules. The abundance of 3D-shaped molecules in GDB compared to experimental databases should be seen as favorable considering the fact that 3D shape may correlate with clinical success in drug candidates [45]. The current relative scarcity of 3D-shaped molecules in experimental compound databases probably reflects the very strong dominance of aromatic chemistry in synthetic methodology.

Due to their small size, essentially all GDB molecules follow Lipinski's rule of five for bioavailability [19]. Almost all of the 977 million molecules in GDB-13 (98.9%) are "lead-like" as defined by a polarity range $1 < \text{clog}P < 3$ and size range $100 < \text{MW} < 350$ Da [46]. Further restriction to Congreve's "rule of three" for fragment likeness (HBA and HBD $\leq 3$, $\text{clog}P < 3$, $\text{RBC} < 3$) [47] still leaves 45.1% of GDB-13. In the case of the larger molecules in GDB-17, only approximately 32 billion (19.2%) are fragment-like. The breakdown stems from an abundance of densely functionalized molecules exceeding the number of allowed hydrogen bond acceptor (HBA) and donor (HBD) atoms, rather than from apolar molecules that are comparatively rare in GDB since low functionalization does not give rise to many combinatorial possibilities in the systematic enumeration (see Section 3.4.2 for details).

## 3.3 Using and Understanding GDB

### 3.3.1 Drug Discovery

The GDBs contain so many more molecules than experimental databases that almost all GDB molecules have never been synthesized to date. Note that the many millions of molecules already synthesized in the context of medicinal chemistry mostly cover the size range of 20–30 atoms, which is yet untouched by the GDB project [18,19]. As a consequence, any list of selected GDB molecules such as close analogs of known drugs identified by virtual screening (see below) points to many new molecules. We have reported several examples of drug discovery with GDB by combining fingerprint-based similarity searches and docking with chemical synthesis and experimental testing (Figure 3.2). Dipeptide **1** and diketopiperazine **2** are glycine analogs from GDB-11 acting as inhibitors of the glycine site of the NMDA receptor [50,51]. Norbornane aspartic acid *rac*-**3** is an aspartate analog derived from GDB-11 acting as a selective inhibitor of the glutamate transporter GLT-1 [52]. Amine **4** is a competitive inhibitor of the α7 nicotinic acetylcholine receptor (nAChR) and *rac*-**5**, *rac*-**6, 7,** and (*R*)-**8** are positive allosteric modulators (PAM) of the α3β2 nAChR discovered as analogs of the known drug PNU-282,987 by replacing its 3-amino-quinuclidine core by related diamines from GDB-11 and GDB-13 [53–56]. Searching for nicotine analogs in GDB-13 similarly led to the known neonicotine *rac*-**9** and

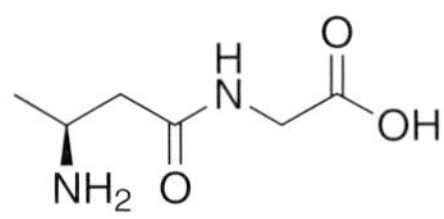

**1** (NMDA glycine site, $IC_{50}$ = 28 μM)

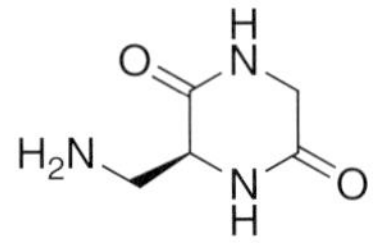

**2** (NMDA glycine site, $IC_{50}$ = 370 μM)

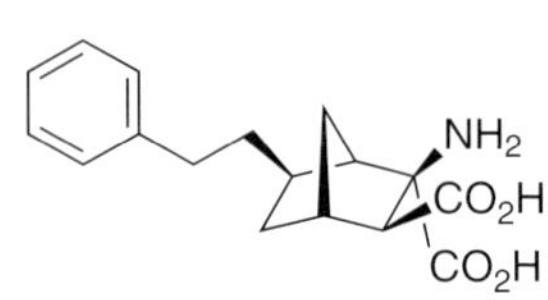

*rac*-**3** (GLT-1, $IC_{50}$ = 1.4 μM)

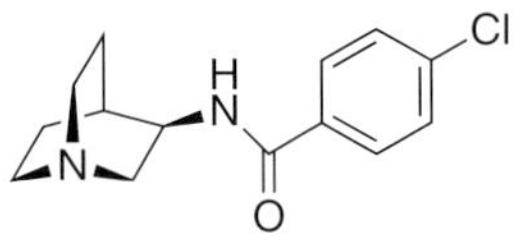

PNU-282,987

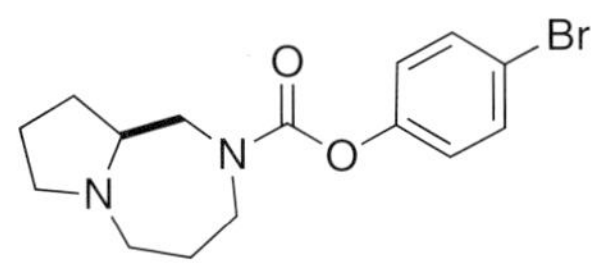

**4** (a7 nAChR, $IC_{50}$ = 5.6 μM)

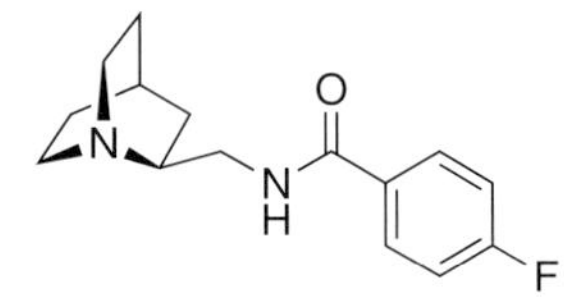

*rac*-**5** (α3β2 nAChR PAM)

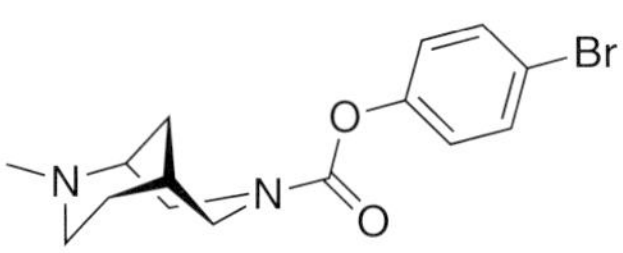

*rac*-**6** (α3β2 nAChR PAM)

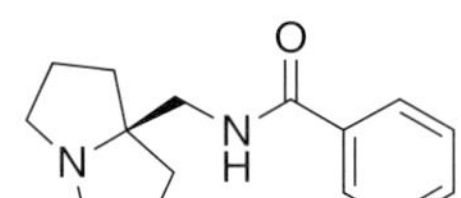

**7**(α3β2 nAChR PAM)

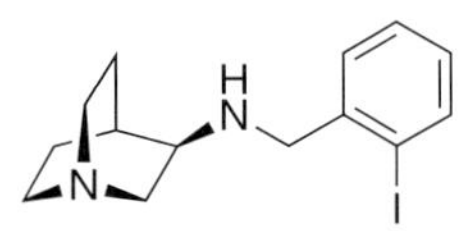

(*R*)-**8** (α3β2 nAChR PAM)

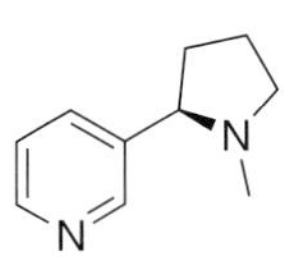

nicotine

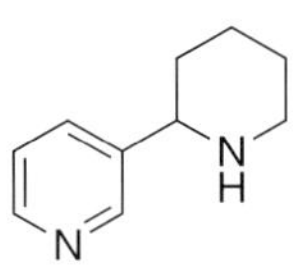

*rac*-**9** ((±)-neonicotine)

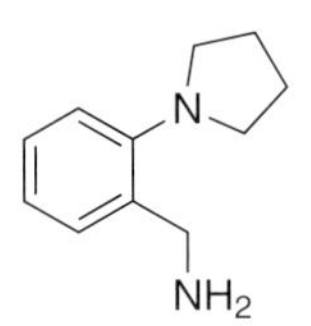

**10**(α7 nAChR, $IC_{50}$ = 7.0 μM)

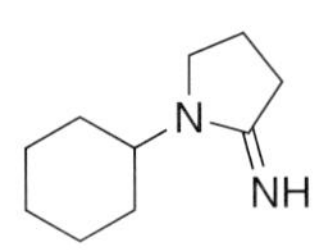

**11**(α7 nAChR, $IC_{50}$ = 9.2 μM)

**Figure 3.2** Fragment-sized bioactive compounds selected from GDB. The α3β2 nAChR PAMs enhance the signal induced by 50 μM acetylcholine by 1.5–3-fold at 10 μM (**5–7**) and 10 fold at 1 μM (**8**).

previously unknown inhibitors such as **10** and **11** [57]. All of these examples are fragment-sized compounds.

### 3.3.2 The MQN System

Identifying bioactive molecules in GDB required virtual screening methods for handling very large databases, which we have developed starting from known cheminformatic principles [58–61]. In particular, we have created a "periodic system for molecules" using the concept of property space as a general method to understand molecular diversity. In a property space, molecules are assigned numerical descriptor values, collected in the so-called fingerprint [62], and placed at the corresponding coordinates of a multidimensional space where each dimension represents one of the descriptors [63]. Our system uses 42 molecular quantum numbers (MQN) counting features important for chemical structure and biological activity (Figure 3.3a) [6,64]. In the corresponding 42-dimensional MQN space, spatial proximity between molecules measures similarity. Similarity is calculated using the city–block distance $CBD_{MQN}$, which is simply the sum of the 42 absolute differences between MQN values of two molecules. Using CBD as similarity measure makes it possible to preorganize the GDBs for extremely fast similarity searching. While CBD is different from the more commonly used Tanimoto coefficient, both similarity measures give almost the same molecules as closest analogs and perform similarly well in typical analog recovery studies with various fingerprints [64–66].

Most importantly, the MQN classification system groups molecules with similar bioactivities together. For example, nicotinic receptor modulators up to 13 atoms documented in ChEMBL [67] are closer to the reference compound nicotine in MQN space than average GDB-13 molecules (Figure 3.3b) [68]. Nearest neighbor searches to identify close analogs of any molecule in the various GDBs can be performed within seconds using the MQN browsers, which are Web-based tools freely accessible at www.gdb.unibe.ch [12,69]. For example, 31 504 analogs of nicotine in GDB-13 are found within a city–block distance of $CBD_{MQN} \leq 11$ and following a list of restricted criteria (43.7 M subset: no small rings, aldehydes, carbonates, esters, epoxides, aziridines, nonaromatic CC double bonds, and NN and NO bonds), and are retrieved in less than one second using the online browser (Figure 3.3c and d) [57].

Another interesting aspect of property spaces is that principal component analysis (PCA) allows one to represent the property space by projection into the principal component plane (PC1, PC2). Projecting GDB from MQN space shows that molecules are organized in MQN space according to structural rigidity (cyclic atoms, Figure 3.3e) and molecular categories (Figure 3.3f). MQN maps provide a logical and elegant visualization of MQN space, which can be accessed interactively using the MQN mapplet, a freely downloadable Java application [70].

### 3.3.3
### Other Fingerprints

We have extended the principle of MQN search and visualization tools to the SMILES fingerprint (SMIfp), which counts 34 characters appearing in the SMILES of a molecule [66]. This classification system is somewhat similar to the MQN system with the important difference that aromatic and nonaromatic molecules, which are not singled out in MQN space, are strongly separated in SMIfp space because upper case letter element symbols (nonaromatic) and lower case element symbols (aromatic) are counted separately.

Other useful fingerprints so far cannot be implemented for fast-searchable versions of GDB due to fingerprint complexity, although they have been implemented for smaller databases such as ZINC [71]. These include the 1024-bit

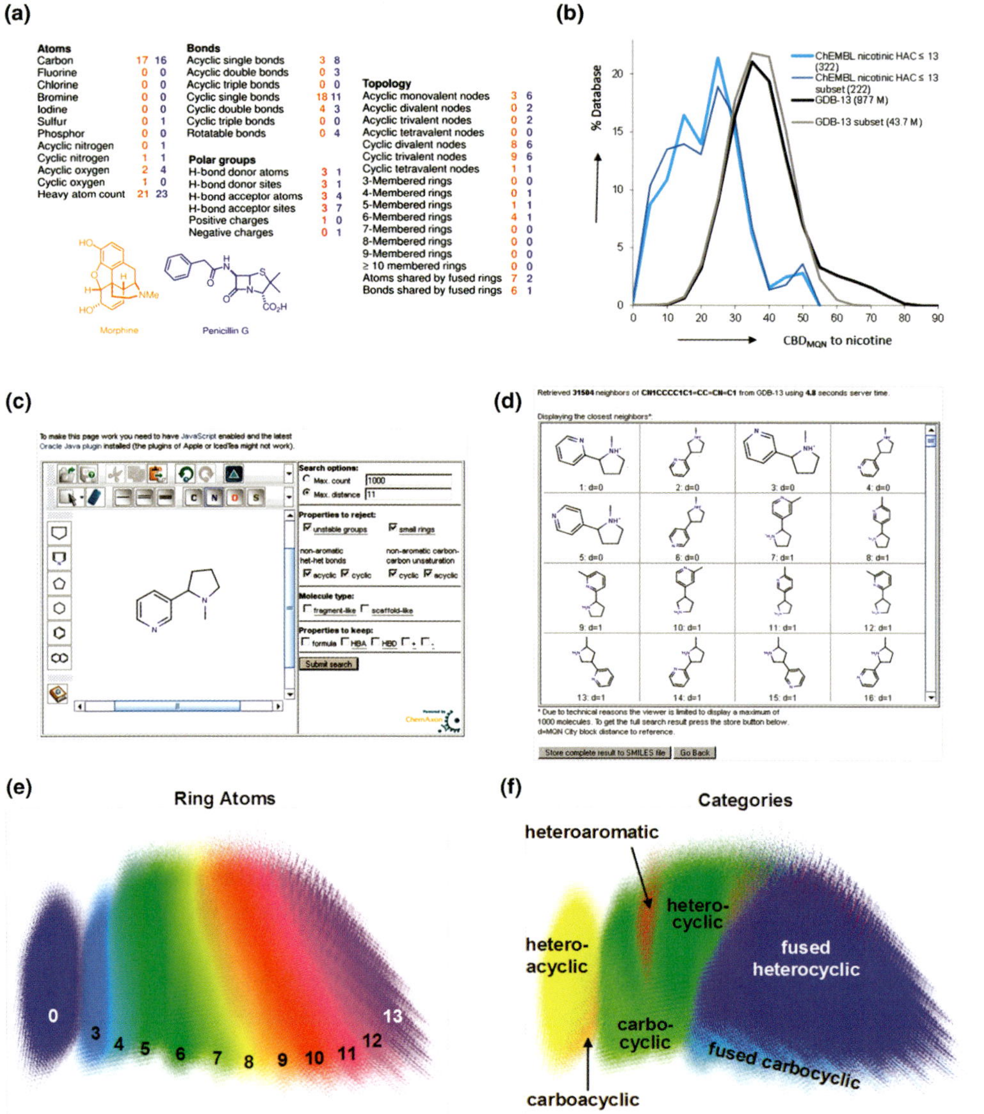


**Figure 3.3** The MQN system for searching and visualizing the GDB, illustrated for GDB-13 and nicotine. (a) The 42 MQNs with the values for morphine and penicillin G as examples. (b) Histogram as function of the distance from nicotine (in CBD$_{MQN}$) for documented nicotinic receptor modulators up to 13 atoms in ChEMBL (blue lines) and GDB-13 molecules (black lines) for the entire GDB-13 (977 M molecules, thick lines) or a restricted subset (thin lines, subset without 3- or 4-membered rings, esters, aldehydes, epoxides, aziridines, nonaromatic C=C double bonds and nonaromatic N–N or N–O bonds, 43.7 M molecules). (c and d) MQN-browser exemplified with a nicotine analog search. (e and f) Color-coded MQN maps of the 43.7 M subset of GDB-13. Ring atoms in (e) are the number of heavy atoms that are part of rings in the molecules. In (f) GDB molecules are assigned to one category following the priority rule heteroaromatic > aromatic>fused heterocyclic > heterocyclic > fused carbocyclic > carbocyclic > heteroacyclic (interrupted carbon chain) >carboacyclic (continuous carbon chain).

binary Daylight type substructure fingerprint (Sfp) [72], the extended connectivity fingerprint (ECfp4) [73], and the atom pair fingerprint (APfp) encoding molecular shape and its related category extended atom pair fingerprint (Xfp) encoding pharmacophores, which are particularly efficient for shape-based virtual screening [65]. The strategy of choice to use these fingerprints consists in screening GDB based on MQN or SMIfp similarity to a reference molecule to obtain a subset of a few million molecules. These smaller subsets can later be analyzed closer using these more complex fingerprints or other scoring functions such as docking, an approach we followed in most drug discovery examples discussed above.

## 3.4 Fragments from GDB

### 3.4.1 Fragment Replacement

As discussed above, the GDBs contain only fragment-sized molecules; however, only a fraction of those follow fragment-likeness criteria (GDB-13: 45.1%; GDB-17: 5%) due to the high frequency of highly functionalized molecules in these databases. Despite these unfavorable percentages, the GDB provide a very large reservoir of fragments in absolute numbers.

One of the simplest approaches to access new fragments from GDB consists in taking an existing fragment collection of a few thousand compounds, and systematically replacing every fragment with an analog identified by MQN-proximity searching in GDB. In this manner, the properties of the original fragment collection are preserved but the GDB-derived compounds are all new compounds. A remarkable feature of the MQN-similarity search is that over 97% of all shape-similar analogs of any reference compound occur within the distance $CBD_{MQN} \leq 12$ from the reference, a distance constraint that also preserves the overall polarity, structural rigidity, and pharmacophore range of the compound. Applying the $CBD_{MQN} \leq 12$ distance constraint to typical fragments shows that thousands to millions of such analogs are available in GDB depending on the molecular size and complexity of the starting fragment, which leaves the possibility to apply additional search criteria to further restrict the search.

For example, searching GDB-13 for analogs of a commercially available oxetane fragment can be carried out with our online browser using the $CBD_{MQN}$ $\leq 12$ distance constraint, the fragment-likeness condition [47], as well as locking the neutral charge, and removing unstable groups and nonaromatic double bonds and NN or NO bonds, which gives 646 possible analogs from which new fragments can be selected (Figure 3.4). While many of the examples shown may seem trivial and are indeed already known, a systematic analysis of the complete set readily shows that many yet unknown fragments are present in the list that can constitute valuable synthetic targets to enhance a fragment collection.

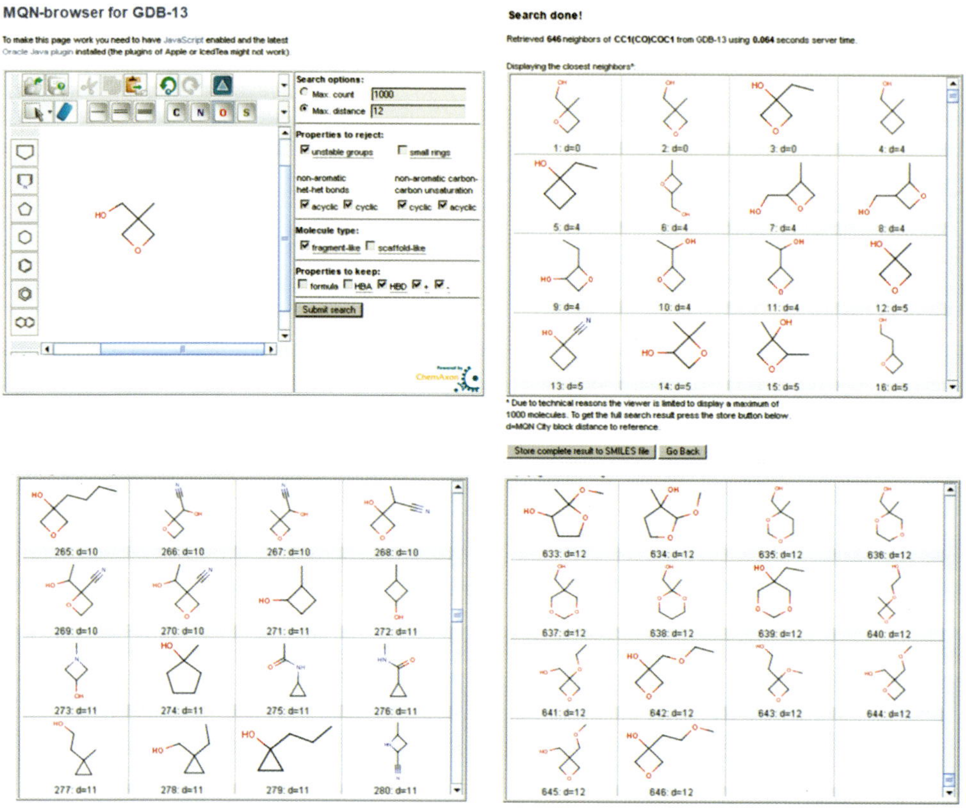


**Figure 3.4** Using the MQN browser for GDB-13 to identify new analogs of an oxetane fragment.

### 3.4.2 Shape Diversity of GDB Fragments

The oxetane example above features a typical GDB-type molecule: a saturated heterocycle containing a small ring, a quaternary center, and therefore having a globular shape. By contrast, existing fragment collections are mostly composed of substituted benzene rings with planar shape. The differences between GDB molecules and commercial fragments are illustrated here by comparing 41 000 fragments collected from various commercially available fragment sets with the public 50 M random subset of GDB-17. Additionally, the 50 M GDB-17 cpds were filtered according to the "rule of three" fragment likeness criteria [47]. Overall 98% passed MW $\leq$ 300, 25% passed HBA $\leq$ 3, 86% passed HBD $\leq$ 3, 74% passed RBC $\leq$ 3, and 98% passed clog$P \leq 3$, which left a 9.6 M (19.2%) GDB-17 fragment-like subset. This was further reduced to a 1 M subset by choosing randomly from this fragment-like subset similar numbers of molecules in the size ranges HAC = 1–13, 14, 15, 16, and 17.

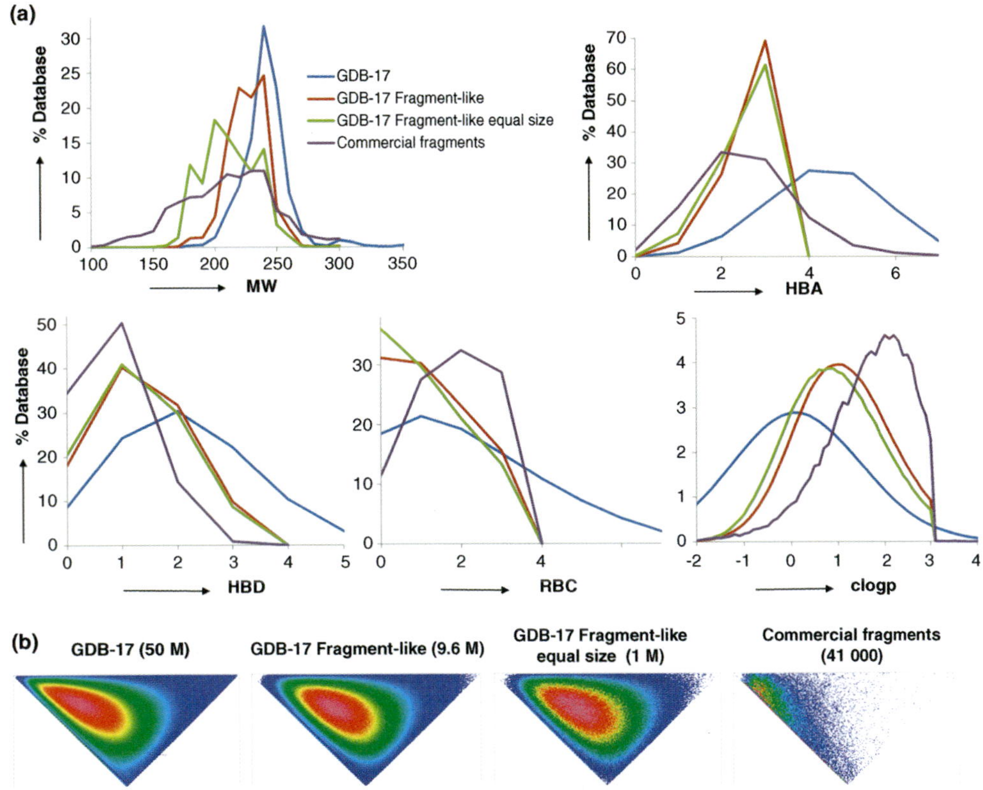


**Figure 3.5** Comparing GDB-17 (50 M cpds), the fragment-like subset (9.6 M cpds), and a size-equalized fragment-like subset (1 M cpds) with commercial fragments (41000 cpds). (a) Property histograms. (b) Molecular shape distribution according to normalized principal moments of inertia, see Figure 3.1.

In terms of the various fragment property histograms (Figure 3.5a), GDB-17 and its subsets have more molecules with high HBA and HBD values, but offer relatively more rigid molecules (RBC < 2) compared to commercial fragments. Furthermore, the polarity range of GDB-17 molecules extends to negative clog*P* values due to the presence of many multiple charged and highly polar molecules. The higher frequency of polar molecules in GDB-17 and its subsets compared to commercial fragments is mostly a consequence of the exhaustive enumeration of chemical space that produces many more combinatorial possibilities at higher ratios of heteroatom to carbon. On the other hand, the higher abundance of rigid fragments reflects a diversity of fused ring systems stemming from the graph enumeration but which are rarely present in commercial fragments due to their challenging syntheses. Most strikingly, the shape triangle analysis shows that GDB-17 and its derived fragments subsets cover a very broad molecular shape range, in contrast to commercial fragments that are in majority rod-like and disk-like due to their aromatic nature (Figure 3.5b).

### 3.4.3
### Aromatic Fragments from GDB

Although the high frequency of planar aromatic molecules in bioactive hits from fragment screens might simply reflect the shape distribution of fragment sets, strong binding to protein pockets often occurs by means of specific interactions possible only with aromatic rings such as π-stacks, XH-π, and CH-π interactions [74]. The preference for aromatic fragments might, therefore, reflect a functional selection process. In this perspective, it must be pointed out that although planar molecules such as aromatic compounds are relatively rare in GDB, their number is very large compared to experimental databases. For instance, the 0.8% aromatic molecules in GDB-17 correspond to 1.3 billion structures, which is larger than the entire GDB-13 or any collection of known molecules including enumerated aromatic heterocycle collections designed to explore the specific chemical space of aromatic molecules [75,76]. Therefore, the fragment replacement approach discussed above (Figure 3.4) would certainly be successful even for a collection consisting mostly of planar aromatic compounds. In summary, the size of GDB is sufficient to satisfy needs for chemical diversity even within narrowly defined subsets.

## 3.5
## Conclusions and Outlook

The exhaustive enumeration of chemical space led to the chemical universe databases GDB-11, GDB-13, and GDB-17, which all contain molecules within the size range of fragments. These enumerations revealed that chemical space is orders of magnitude larger than the known chemical space even when applying restrictive enumeration rules allowing only unstrained rings and stable functional groups. The GDBs offer an extremely large reservoir of new molecular structures with emphasis on molecular shape diversity and bicyclic saturated heterocycles.

Beyond enumeration, proof-of-principle examples of using GDB for drug discovery have been achieved in the area of neurotransmitter receptors, in particular the nicotinic acetylcholine receptor. These application projects led to the development of virtual screening and visualization methods suitable for very large databases, in particular the MQN and SMIfp system freely accessible at www.gdb.unibe.ch that allow very fast similarity searches and interactive visualization of the GDBs. These search tools can be utilized readily to search for new fragments within the GDB, for example, by substituting individual fragments of existing collections with yet unknown analogs from GDB identifiable in an MQN-proximity search.

The key question in applying GDB for fragment design lies in the choice of molecule type. One possibility might be to evenly sample the molecular shape space offered by GDB to obtain entirely novel fragment collections very rich in nonplanar molecules. Alternatively, systematic fragment replacement of existing collections would produce GDB-derived fragment sets rich in planar and aromatic structures but still entirely novel. Such aromatic sets might possibly be more favorable to identify good protein binders exploiting specific aromatic interactions. In any approach chosen to identify fragments, the GDBs are large enough to exhaust synthetic curiosity. Any selected fragment must then be reduced to practice by chemical synthesis, which requires a substantial investment in resources. The public availability of virtual screening and visualization tools to explore the GDB should facilitate this project, which is currently ongoing in our laboratory.

## Acknowledgment

This work was supported financially by the University of Berne, the Swiss National Science Foundation, and the NCCR TransCure.

## References

1 Foloppe, N. (2011) The benefits of constructing leads from fragment hits. *Future Medicinal Chemistry*, **3**, 1111–1115.
2 Murray, C.W. and Rees, D.C. (2009) The rise of fragment-based drug discovery. *Nature Chemistry*, **1**, 187–192.
3 van Deursen, R., Blum, L.C., and Reymond, J.L. (2011) Visualisation of the chemical space of fragments, lead-like and drug-like molecules in PubChem. *Journal of Computer-Aided Molecular Design*, **25**, 649–662.
4 Fink, T., Bruggesser, H., and Reymond, J.L. (2005) Virtual exploration of the small-molecule chemical universe below 160 daltons. *Angewandte Chemie International Edition*, **44**, 1504–1508.
5 Fink, T. and Reymond, J.L. (2007) Virtual exploration of the chemical universe up to 11 atoms of C, N, O, F: assembly of 26.4 million structures (110.9 million stereoisomers) and analysis for new ring systems, stereochemistry, physicochemical properties, compound classes, and drug discovery. *Journal of Chemical Information and Modeling*, **47**, 342–353.
6 Nguyen, K.T., Blum, L.C., van Deursen, R., and Reymond, J.-L. (2009) Classification of organic molecules by molecular quantum numbers. *ChemMedChem*, **4**, 1803–1805.
7 Ruddigkeit, L., van Deursen, R., Blum, L.C., and Reymond, J.L. (2012) Enumeration of 166 billion organic small molecules in the chemical universe database GDB-17. *Journal of Chemical Information and Modeling*, **52**, 2864–2875.
8 Ruddigkeit, L., Blum, L.C., and Reymond, J.L. (2013) Visualization and virtual screening of the chemical universe database GDB-17. *Journal of Chemical Information and Modeling*, **53**, 56–65.
9 Reymond, J.L., van Deursen, R., and Blum, L.C., and Ruddigkeit, L. (2010) Chemical space as a source for new drugs. *MedChemComm*, **1**, 30–38.
10 Reymond, J.L., Ruddigkeit, L., Blum, L.C., and van Deursen, R. (2012) The enumeration of chemical space. *WIREs Computational Molecular Science*. doi: 10.1002/wcms.1104
11 Amarouch, M.Y., Syam, N., and Abriel, H. (2013) Biochemical, single-channel, whole-cell patch clamp, and pharmacological analyses of endogenous TRPM4 channels in HEK293 cells. *Neuroscience Letters*, **541**, 105–110.

12 Reymond, J.L., Blum, L.C., and van Deursen, R. (2011) Exploring the chemical space of known and unknown organic small molecules at www.gdb.unibe.ch. *Chimia*, **65**, 863–867.

13 Ruddigkeit, L., Awale, M., and Reymond, J. L. (2014) Expanding the fragrance chemical space for virtual screening. *Journal of Cheminformatics*, **6**, 27–39.

14 Cayley, E. (1875) Ueber die analytischen Figuren, welche in der Mathematik Bäume genannt werden und ihre Anwendung auf die Theorie chemischer Verbindungen. *Chemische Berichte*, **8**, 1056–1059.

15 Schiff, H. (1875) Zur statistik chemischer verbindungen. *Chemische Berichte*, **8**, 1542–1547.

16 Henze, H.R. and Blair, C.M. (1931) The number of isomeric hydrocarbons of the methane series. *Journal of the American Chemical Society*, **53**, 3077–3085.

17 Brinkmann, G., Caporossi, G., and Hansen, P. (2003) A survey and new results on computer enumeration of polyhex and fusene hydrocarbons. *Journal of Chemical Information and Computer Science*, **43**, 842–851.

18 Dias, J.R. (2010) The polyhex/polypent topological paradigm: regularities in the isomer numbers and topological properties of select subclasses of benzenoid hydrocarbons and related systems. *Chemical Society Reviews*, **39**, 1913–1924.

19 Lipinski, C.A., Lombardo, F., Dominy, B. W., and Feeney, P.J. (1997) Experimental and computational approaches to estimate solubility and permeability in drug discovery and development settings. *Advanced Drug Delivery Reviews*, **23**, 3–25.

20 Bohacek, R.S., McMartin, C., and Guida, W.C. (1996) The art and practice of structure-based drug design: a molecular modeling perspective. *Medicinal Research Reviews*, **16**, 3–50.

21 Ertl, P. (2003) Cheminformatics analysis of organic substituents: identification of the most common substituents, calculation of substituent properties, and automatic identification of drug-like bioisosteric groups. *Journal of Chemical Information and Computer Science*, **43**, 374–380.

22 Kirkpatrick, P. and Ellis, C. (2004) Chemical space. *Nature*, **432**, 823–823.

23 Wieland, T., Kerber, A., and Laue, R. (1996) Principles of the generation of constitutional and configurational isomers. *Journal of Chemical Information and Computer Science*, **36**, 413–419.

24 Buchanan, B.G., Smith, D.H., White, W.C., Gritter, R.J., Feigenbaum, E.A., Lederberg, J., and Djerassi, C. (1976) Applications of artificial intelligence for chemical inference. 22. Automatic rule formation in mass spectrometry by means of the meta-DENDRAL program. *Journal of the American Chemical Society*, **98**, 6168–6178.

25 Lederberg, J., Sutherland, G.L., Buchanan, B.G., Feigenbaum, E.A., Robertson, A.V., Duffield, A.M., and Djerassi, C. (1969) Applications of artificial intelligence for chemical inference. I. Number of possible organic compounds. Acyclic structures containing carbon, hydrogen, oxygen, and nitrogen. *Journal of the American Chemical Society*, **91**, 2973–2976.

26 Warr, W.A. (1993) Computer-assisted structure elucidation. Part II. Indirect database approaches and established systems. *Analytical Chemistry*, **65**, 1087A–1095A.

27 Steinbeck, C. (2004) Recent developments in automated structure elucidation of natural products. *Natural Products Reports*, **21**, 512–518.

28 Elyashberg, M., Blinov, K., Molodtsov, S., Smurnyy, Y., Williams, A.J., and Churanova, T. (2009) Computer-assisted methods for molecular structure elucidation: realizing a spectroscopist's dream. *Journal of Cheminformatics*, **1**, 3.

29 Gillet, V.J., Newell, W., Mata, P., Myatt, G., Sike, S., Zsoldos, Z., and Johnson, A.P. (1994) SPROUT: recent developments in the *de novo* design of molecules. *Journal of Chemical Information and Computer Science*, **34**, 207–217.

30 Schneider, G. and Fechner, U. (2005) Computer-based *de novo* design of drug-like molecules. *Nature Reviews Drug Discovery*, **4**, 649–663.

31 Danziger, D.J. and Dean, P.M. (1989) Automated site-directed drug design: a general algorithm for knowledge

acquisition about hydrogen-bonding regions at protein surfaces. *Proceedings of the Royal Society of London B: Biological Sciences*, **236**, 101–113.

32 Lewell, X.Q., Judd, D.B., Watson, S.P., and Hann, M.M. (1998) RECAP-retrosynthetic combinatorial analysis procedure: a powerful new technique for identifying privileged molecular fragments with useful applications in combinatorial chemistry. *Journal of Chemical Information and Computer Science*, **38**, 511–522.

33 Leach, A.R. and Hann, M.M. (2000) The *in silico* world of virtual libraries. *Drug Discovery Today*, **5**, 326–336.

34 Patel, H., Bodkin, M.J., Chen, B., and Gillet, V.J. (2009) Knowledge-based approach to *de novo* design using reaction vectors. *Journal of Chemical Information and Modeling*, **49**, 1163–1184.

35 Hu, Q., Peng, Z., Kostrowicki, J., and Kuki, A. (2011) LEAP into the Pfizer Global Virtual Library (PGVL) space: creation of readily synthesizable design ideas automatically. *Methods in Molecular Biology*, **685**, 253–276.

36 Brown, N., McKay, B., Gilardoni, F., and Gasteiger, J. (2004) A graph-based genetic algorithm and its application to the multiobjective evolution of median molecules. *Journal of Chemical Information and Computer Science*, **44**, 1079–1087.

37 Brown, N., McKay, B., and Gasteiger, J. (2004) The *de novo* design of median molecules within a property range of interest. *Journal of Computer-Aided Molecular Design*, **18**, 761–771.

38 Lameijer, E.W., Kok, J.N., Back, T., and IJzerman, A.P. (2006) The molecule evoluator. An interactive evolutionary algorithm for the design of drug-like molecules. *Journal of Chemical Information and Modeling*, **46**, 545–552.

39 van Deursen, R. and Reymond, J.L. (2007) Chemical space travel. *ChemMedChem*, **2**, 636–640.

40 Virshup, A.M., Contreras-Garcia, J., Wipf, P., Yang, W., and Beratan, D.N. (2013) Stochastic voyages into uncharted chemical space produce a representative library of all possible drug-like compounds. *Journal of the American Chemical Society*, **135**, 7296–7303.

41 McKay, B.D. (1981) Practical graph isomorphism. *Congressus Numerantium*, **30**, 45–87.

42 Weininger, D. (1988) Smiles, a chemical language and information-system.1. Introduction to methodology and encoding rules. *Journal of Chemical Information and Computer Science*, **28**, 31–36.

43 Sadowski, J. and Gasteiger, J. (1993) From atoms and bonds to 3-dimensional atomic coordinates – automatic model builders. *Chemical Reviews*, **93**, 2567–2581.

44 Sauer, W.H. and Schwarz, M.K. (2003) Molecular shape diversity of combinatorial libraries: a prerequisite for broad bioactivity. *Journal of Chemical Information and Computer Science*, **43**, 987–1003.

45 Lovering, F., Bikker, J., and Humblet, C. (2009) Escape from flatland: increasing saturation as an approach to improving clinical success. *Journal of Medicinal Chemistry*, **52**, 6752–6756.

46 Teague, S.J., Davis, A.M., Leeson, P.D., and Oprea, T. (1999) The design of leadlike combinatorial libraries. *Angewandte Chemie International Edition in English*, **38**, 3743–3748.

47 Congreve, M., Carr, R., Murray, C., and Jhoti, H. (2003) A rule of three for fragment-based lead discovery? *Drug Discovery Today*, **8**, 876–877.

48 Mayr, L.M. and Bojanic, D. (2009) Novel trends in high-throughput screening. *Current Opinion in Pharmacology*, **9**, 580–588.

49 Renner, S., Popov, M., Schuffenhauer, A., Roth, H.J., Breitenstein, W., Marzinzik, A., Lewis, I., Krastel, P., Nigsch, F., Jenkins, J. *et al.* (2011) Recent trends and observations in the design of high-quality screening collections. *Future Medicinal Chemistry*, **3**, 751–766.

50 Nguyen, K.T., Syed, S., Urwyler, S., Bertrand, S., Bertrand, D., and Reymond, J. L. (2008) Discovery of NMDA glycine site inhibitors from the chemical universe database GDB. *ChemMedChem*, **3**, 1520–1524.

51 Nguyen, K.T., Luethi, E., Syed, S., Urwyler, S., Bertrand, S., Bertrand, D., and Reymond, J.L. (2009) 3-(Aminomethyl)

piperazine-2,5-dione as a novel NMDA glycine site inhibitor from the chemical universe database GDB. *Bioorganic and Medicinal Chemistry Letters*, **19**, 3832–3835.

52 Luethi, E., Nguyen, K.T., Burzle, M., Blum, L.C., Suzuki, Y., Hediger, M., and Reymond, J.L. (2010) Identification of selective norbornane-type aspartate analogue inhibitors of the glutamate transporter 1 (GLT-1) from the chemical universe generated database (GDB). *Journal of Medicinal Chemistry*, **53**, 7236–7250.

53 Bodnar, A.L., Cortes-Burgos, L.A., Cook, K.K., Dinh, D.M., Groppi, V.E., Hajos, M., Higdon, N.R., Hoffmann, W.E., Hurst, R. S., Myers, J.K. *et al.* (2005) Discovery and structure–activity relationship of quinuclidine benzamides as agonists of alpha 7 nicotinic acetylcholine receptors. *Journal of Medicinal Chemistry*, **48**, 905–908.

54 Garcia-Delgado, N., Bertrand, S., Nguyen, K.T., van Deursen, R., and Bertrand, D., and Reymond, J.-L. (2010) Exploring α 7-nicotinic receptor ligand diversity by scaffold enumeration from the chemical universe database GDB. *ACS Medicinal Chemistry Letters*, **1**, 422–426.

55 Brethous, L., Garcia-Delgado, N., Schwartz, J., Bertrand, S., Bertrand, D., and Reymond, J.L. (2012) Synthesis and nicotinic receptor activity of chemical space analogues of *N*-(3*R*)-1-Azabicyclo[2.2.2]oct-3-yl-4-chlorobenzamide (PNU-282,987) and 1,4-Diazabicyclo[3.2.2]nonane-4-carboxylic acid 4-bromophenyl ester (SSR180711). *Journal of Medicinal Chemistry*, **55**, 4605–4618.

56 Burgi, J.J., Awale, M., Boss, S.D., Schaer, T., Marger, F., Viveros-Paredes, J.M., Bertrand, S., Gertsch, J., Bertrand, D., and Reymond, J.L. (2014) Discovery of potent positive allosteric modulators of the alpha3beta2 nicotinic acetylcholine receptor by a chemical space walk in ChEMBL. *ACS Chemical Neuroscience*, **5**, 346–359.

57 Blum, L.C., van Deursen, R., Bertrand, S., Mayer, M., Burgi, J.J., and Bertrand, D., and Reymond, J.L. (2011) Discovery of alpha7-nicotinic receptor ligands by virtual screening of the chemical universe database GDB-13. *Journal of Chemical Information and Modeling*, **51**, 3105–3112.

58 Chen, W.L. (2006) Chemoinformatics: past, present, and future. *Journal of Chemical Information and Modeling*, **46**, 2230–2255.

59 Akella, L.B. and DeCaprio, D. (2010) Cheminformatics approaches to analyze diversity in compound screening libraries. *Current Opinion in Chemical Biology*, **14**, 325–330.

60 Rosen, J., Gottfries, J., Muresan, S., Backlund, A., and Oprea, T.I. (2009) Novel chemical space exploration via natural products. *Journal of Medicinal Chemistry*, **52**, 1953–1962.

61 Singh, N., Guha, R., Giulianotti, M.A., Pinilla, C., Houghten, R.A., and Medina-Franco, J.L. (2009) Chemoinformatic analysis of combinatorial libraries, drugs, natural products, and molecular libraries small molecule repository. *Journal of Chemical Information and Modeling*, **49**, 1010–1024.

62 Willett, P. (2006) Similarity-based virtual screening using 2D fingerprints. *Drug Discovery Today*, **11**, 1046–1053.

63 Pearlman, R.S. and Smith, K.M. (1998) Novel software tools for chemical diversity. *Perspectives in Drug Discovery and Design*, **9–11**, 339–353.

64 van Deursen, R., Blum, L.C., and Reymond, J.L. (2010) A searchable map of PubChem. *Journal of Chemical Information and Modeling*, **50**, 1924–1934.

65 Awale, M. and Reymond, J.L. (2014) Atom pair 2D-fingerprints perceive 3D-molecular shape and pharmacophores for very fast virtual screening of ZINC and GDB-17. *Journal of Chemical Information and Modeling*, **54**, 1892–1897.

66 Schwartz, J., Awale, M., and Reymond, J.L. (2013) SMIfp (SMILES fingerprint) chemical space for virtual screening and visualization of large databases of organic molecules. *Journal of Chemical Information and Modeling*, **53**, 1979–1989.

67 Gaulton, A., Bellis, L.J., Bento, A.P., Chambers, J., Davies, M., Hersey, A., Light, Y., McGlinchey, S., Michalovich, D.,

Al-Lazikani, B. *et al.* (2012) ChEMBL: a large-scale bioactivity database for drug discovery. *Nucleic Acids Research*, **40**, D1100–D1107

68 Khalifa, A.A., Haranczyk, M., and Holliday, J. (2009) Comparison of nonbinary similarity coefficients for similarity searching, clustering and compound selection. *Journal of Chemical Information and Modeling*, **49**, 1193–1201.

69 Blum, L.C., van Deursen, R., and Reymond, J.L. (2011) Visualisation and subsets of the chemical universe database GDB-13 for virtual screening. *Journal of Computer-Aided Molecular Design*, **25**, 637–647.

70 Awale, M., van Deursen, R., and Reymond, J.L. (2013) MQN-mapplet: visualization of chemical space with interactive maps of DrugBank, ChEMBL, PubChem, GDB-11, and GDB-13. *Journal of Chemical Information and Modeling*, **53**, 509–518.

71 Awale, M. and Reymond, J.L. (2014) A multi-fingerprint browser for the ZINC database. *Nucleic Acids Research*. doi: 10.1093/nar/gku1379

72 Hagadone, T.R. (1992) Molecular substructure similarity searching: efficient retrieval in two-dimensional structure databases. *Journal of Chemical Information and Computer Science*, **32**, 515–521.

73 Rogers, D. and Hahn, M. (2010) Extended-connectivity fingerprints. *Journal of Chemical Information and Modeling*, **50**, 742–754.

74 Kadam, R.U., Garg, D., Schwartz, J., Visini, R., Sattler, M., Stocker, A., Darbre, T., and Reymond, J.L. (2013) CH-pi "T-Shape" interaction with histidine explains binding of aromatic galactosides to *Pseudomonas aeruginosa* lectin LecA. *ACS Chemical Biology*, **8**, 1925–1930.

75 Ertl, P., Jelfs, S., Mühlbacher, J., Schuffenhauer, A., and Selzer, P. (2006) Quest for the rings. *In silico* exploration of ring universe to identify novel bioactive heteroaromatic scaffolds. *Journal of Medicinal Chemistry*, **49**, 4568–4573.

76 Pitt, W.R., Parry, D.M., Perry, B.G., and Groom, C.R. (2009) Heteroaromatic rings of the future. *Journal of Medicinal Chemistry*, **52**, 2952–2963.

# 4
# Ligand Efficiency Metrics and their Use in Fragment Optimizations

*György G. Ferenczy and György M. Keserű*

## 4.1
## Introduction

The optimization of fragment hits to development candidates typically includes a significant increase in affinity that is accompanied by important changes in other molecular properties relevant in drug discovery. Among them, the pharmacokinetic profile is of primary importance. Although the pharmacokinetic profile signifies a set of complex properties, it can be effectively characterized by simple descriptors such as molecular weight and lipophilicity. The primary objective of early-phase fragment optimizations is to increase affinity toward the target, and the optimization almost inevitably results in an increase in both molecular weight and lipophilicity. The excessive increase of these latter properties, however, leads to suboptimal pharmacokinetic profiles. Therefore, their control in the course of optimization is highly desirable in order to ensure balanced optimization of affinity, molecular weight, and lipophilicity. Various efficiency indices have been proposed for the simultaneous monitoring of these properties. Efficiency indices are typically composite measures that are able to characterize the balance of the above properties. They are especially useful at decision points like hit and lead selection and in the early phases of optimizations when compound properties from various series have to be compared and the result of significant structural modifications has to be assessed.

## 4.2
## Ligand Efficiency

Probably, ligand efficiency is the most widely used index among efficiency indices. It represents the ratio of the affinity and ligand size; the latter is most often expressed in terms of heavy atom count (number of non-hydrogen atoms, HA). The most frequent definitions are shown in Table 4.1.

The question of how much affinity can be achieved as a function of ligand size was first addressed in Ref. [4]. It was shown that the maximal observed binding

*Fragment-based Drug Discovery: Lessons and Outlook*, First Edition. Edited by Daniel A. Erlanson and Wolfgang Jahnke.

**Table 4.1** Definitions of ligand efficiency: ligand efficiency (LE), binding efficiency index (BEI), and size-independent ligand efficiency (SILE).

| Definition | Reference |
|---|---|
| $LE = \Delta G/HA \sim 1.37pIC_{50}/HA \sim pIC_{50}/HA$ | [1] |
| $BEI = pIC_{50}/MW$ | [2] |
| $SILE = pIC_{50}/HA^{0.3}$ or $pK_i/HA^{0.3}$ | [3] |

affinity for noncovalent complexes increases with ligand size; apart from some singletons, highest affinity compounds are those around HA = 40 (see also Refs. [5,6]). On the other hand, the maximum binding affinity per heavy atom decreases with increasing heavy atom number. This quantity was proposed as an index called ligand efficiency (LE) that can be advantageously used in lead selection since it includes both affinity and ligand size and it can be considered as potency normalized to molecular size [1]. It is a better selection criterion than potency itself and it is used to compare compounds from different series.

Although LE is meant to compare compounds with different sizes, one has to observe that the maximal available ligand efficiency does depend on the molecular size as was pointed out in Ref. [4] and was analyzed in detail in Refs. [3,5,7]. Affinities and ligand efficiencies determined in thermodynamic measurements, namely, by isothermal titration calorimetry (ITC), were shown to exhibit similar behavior [6]. Moreover, ITC data provide us with the enthalpy and entropy components of binding that will be used in our subsequent analysis of other indices. Binding thermodynamics data for ligand–protein complexes were collected from databases and ligands that are not of the interest of drug discovery programs were removed; ligands containing other than C, N, O, P, S, H, and halogen atoms were discarded just as highly polar compounds (see Ref [6] for details). Figure 4.1 shows ligand affinities and Figure 4.2 shows maximal available

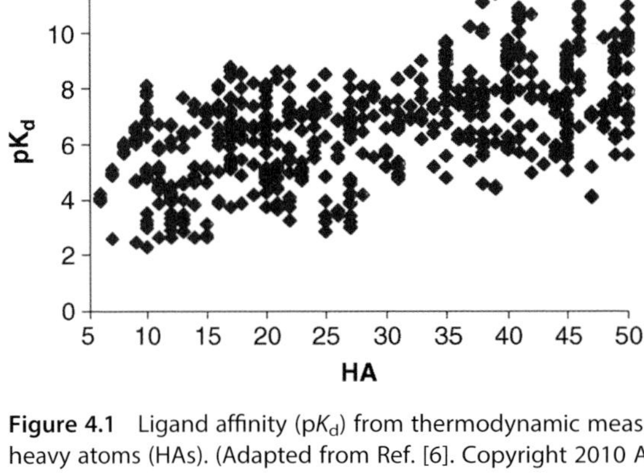


**Figure 4.1** Ligand affinity ($pK_d$) from thermodynamic measurements versus the number of heavy atoms (HAs). (Adapted from Ref. [6]. Copyright 2010 American Chemical Society.)

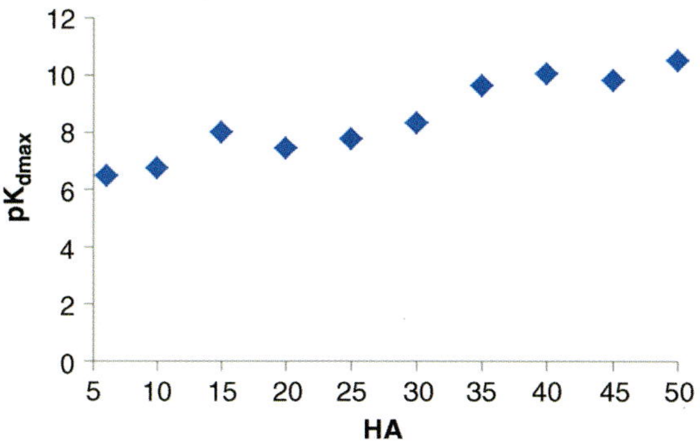


**Figure 4.2** $pK_d$ of the most potent ligands versus the number of heavy atoms.

affinities both as functions of molecular size expressed as heavy atom count. The latter plot was generated by dividing the data of Figure 4.1 into bins of HA = 5 and selecting the value of the 95th percentile. This procedure smoothens the function in Figure 4.2, but it does not affect the trend as it was checked by varying binning and percentiles.

Ligand efficiencies are shown in Figure 4.3. The significant dependence of the maximal available ligand efficiency on ligand size biases the comparison of different sized ligands and this bias is pronounced in fragment optimizations where the size of the hits is increased by 10–25 heavy atoms. A size-independent version of ligand efficiency (SILE) was defined by Nissink [3] as $\text{SILE} = \text{pIC}_{50}/\text{HA}^{0.3}$ or $pK_i/\text{HA}^{0.3}$. The maximal value of SILE is approximately independent of ligand size and this makes the comparison of different size ligands more meaningful. Indeed, we found that LE values most often do not exhibit a clear trend in optimizations, while SILE shows a significant increase in successful optimizations and thus it can be effectively used both in hit and lead selection and in monitoring optimizations (see examples in Section 4.6).

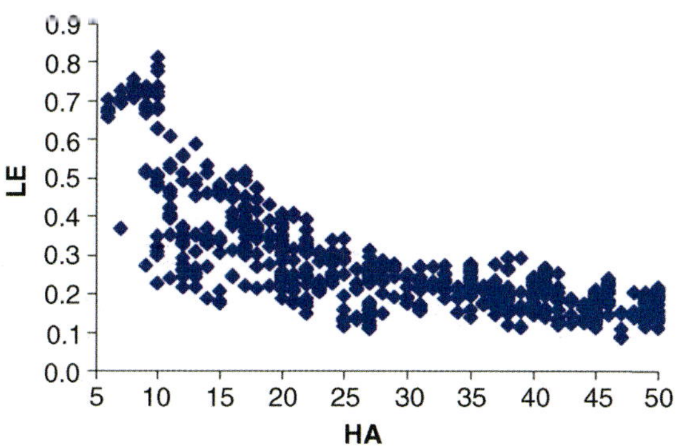


**Figure 4.3** Ligand efficiency ($\text{LE} = pK_d/\text{HA}$) from thermodynamic measurements versus the number of HAs. (Adapted from Ref. [6]. Copyright 2010 American Chemical Society.)

LE shows an overall decreasing trend except for compounds having not more than 10 heavy atoms where the increase of LE with increasing ligand size is observed (Figure 4.3). This behavior of the function is specific for very small ligands not typical even in fragment-based approaches. The increase of maximal LE with ligand size is the consequence of the sharp increase of the maximal affinity with ligand size in this region. The steepness of the affinity–ligand size function up to 10–15 heavy atoms is apparent in Figure 4.1. The structural background of this phenomenon is discussed in Section 4.3.

## 4.3 Binding Thermodynamics and Efficiency Indices

Drug discovery programs typically achieve higher ligand affinity by adding more atoms to the molecule and thus increasing the molecular size. This method is justified by the observation that the available maximum affinity increases with ligand size as shown in Figures 4.1 and 4.2. However, excessive size increase may deteriorate other properties such as pharmacokinetic profile. This is more so if a size increase is accompanied by an increase in lipophilicity. Lipophilic indices are designed to characterize molecular lipophilicity in relation to other properties such as affinity and molecular size and they are subject of Section 4.5. Here, we investigate the connection of these molecular properties with binding thermodynamics by rationalizing the size dependence of structural and thermodynamical features of ligand–protein binding.

The size dependence of ligand affinity and efficiency is discussed in Section 4.2. Ligand affinity is equivalent to the free energy change when the ligand binds to the protein and this can be decomposed into enthalpic and entropic components. The size dependence of these components does not follow that of the free energy. The enthalpic component of the same set of complexes appearing in Figure 4.1 is shown in Figure 4.4. The enthalpic component is expressed as $\mathrm{p}K_\mathrm{H} = -\Delta H/2.303RT$. (Later $\mathrm{p}K_\mathrm{S} = \Delta S/2.303R$ will also be used. Note that $\mathrm{p}K_\mathrm{d} = \mathrm{p}K_\mathrm{H} + \mathrm{p}K_\mathrm{S}$ [6]). The maximal enthalpic components are shown in Figure 4.5.

The maximal available enthalpy appears to decrease with increasing ligand size except for the smallest compounds. This trend is the opposite of that observed for the affinity (cf. Figure 4.2). Considering now compounds with maximal available affinity and decomposing their affinity into enthalpic and entropic components, similar trends are observed (Figure 4.6).

The same trends, that is, increasing potency and entropic benefit and decreasing enthalpic benefit are found for the averages of these quantities as functions of heavy atom count (not shown). These relations show that fragment binding is typically enthalpy dominated while favorable entropy contribution increases with increasing molecular size.

The changes of the binding thermodynamic profile with ligand size can be rationalized by investigating the structural requirements of ligand binding and this, in turn, facilitates the interpretation of efficiency indices.

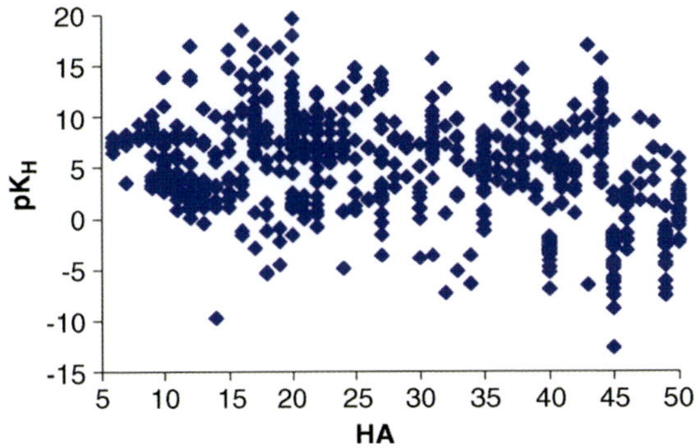


**Figure 4.4** The enthalpic component of ligand binding (p$K_H$; see text) versus the number of heavy atoms. (Adapted from Ref. [6]. Copyright 2010 American Chemical Society.)

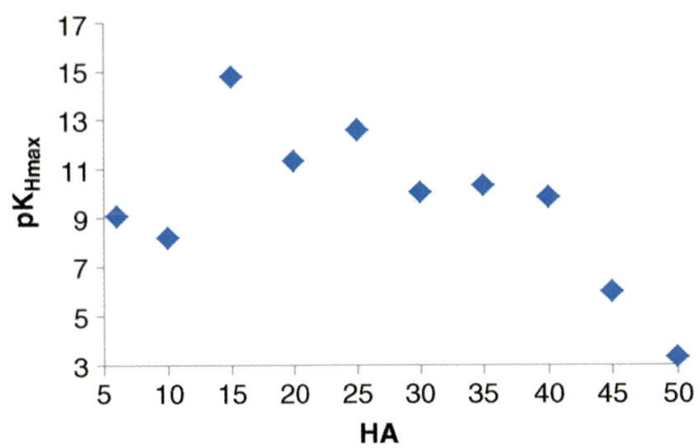


**Figure 4.5** p$K_H$ of the most enthalpic binders versus the number of heavy atoms.

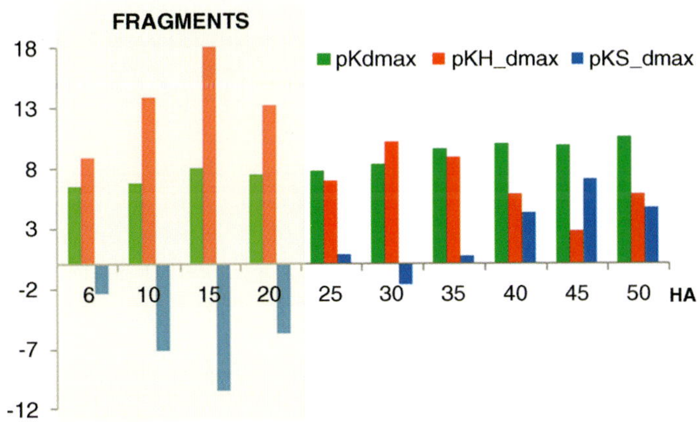


**Figure 4.6** Affinity and its enthalpic and entropic components of the most potent ligands versus the number of heavy atoms.

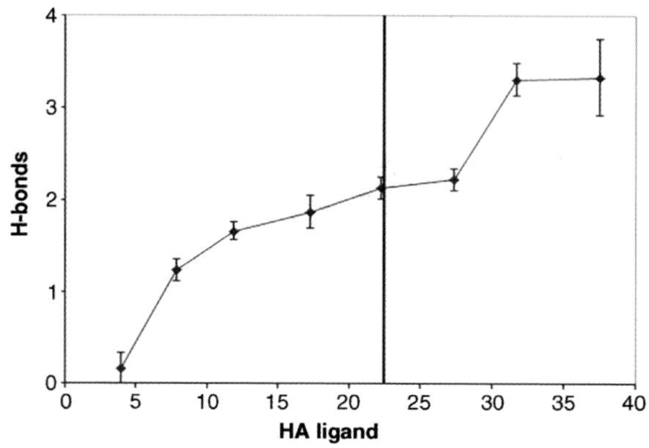


**Figure 4.7** Number of optimal H-bonds in protein–ligand complexes as a function of the number of ligand heavy atoms. Vertical bars represent standard errors. The vertical line indicates the fragment limit [12]. Copyright 2012 American Chemical Society.

It has been established [8–10] that the favorable enthalpic contribution to ligand–protein binding primarily stems from advantageous polar interactions while the entropy gain is dominated by the desolvation of apolar ligand (and protein) moieties that are buried in the complex upon binding. The enthalpy-dominated binding of fragment-sized compounds is in line with our finding that the number of optimal H-bonds in ligand–protein complexes deposited in the Protein Data Bank [11] is close to two on average and this number hardly changes with increasing ligand size (Figure 4.7) [12]. It should be emphasized that only optimal geometry H-bonds whose donor–acceptor separation is not larger than 3 Å and whose D–H . . . A angle is not smaller than 160° are counted. These optimal geometry H-bonds are expected to favorably contribute to binding free energy primarily by enthalpic gain.

Figure 4.7 suggests that the enthalpy gain originating from optimal geometry H-bonds contributes to the binding of small ligands, but it does not increase with increasing ligand size except for the smallest compounds. In contrast, the dominantly entropic apolar desolvation component potentially increases linearly with increasing molecular size as shown in Figure 4.8 [12].

These findings provide structural interpretation for the observation that the enthalpy-dominated binding of fragment-sized compounds becomes entropy-dominated binding for larger compounds. Hydrogen bonds are primarily responsible for the binding of small compounds and the average number of H-bonds sharply increases with increasing ligand size up to about 10 heavy atoms. The corresponding sharp increase in enthalpy-dominated affinity results in increasing LE as shown in Figure 4.3. The sensitivity of the binding free energy to the geometry of interacting polar groups explains why it is difficult to achieve affinity increase by introducing further polar interactions, although they would render good physicochemical properties and beneficial pharmacokinetic profile to molecules. On the contrary, it is less difficult to increase affinity by introducing

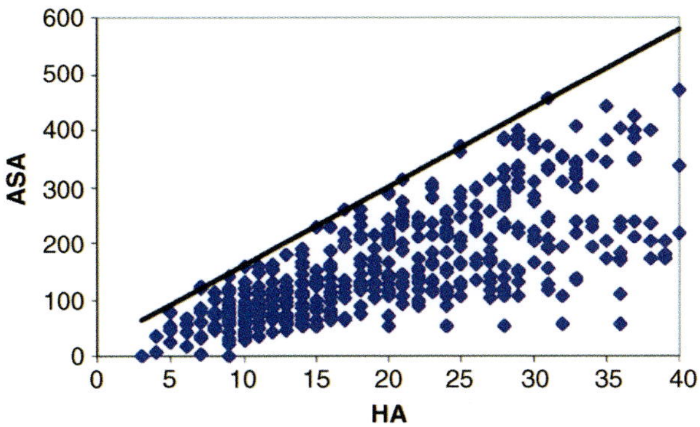


**Figure 4.8** Apolar surface area (ASA) versus number of heavy atoms for 571 distinct ligands of 1297 complexes from PDB [12]. Straight line indicates an upper limit for available ASA. Copyright 2012 American Chemical Society.

appropriate apolar moieties that render suboptimal physicochemical properties to molecules. Various indices like ligand efficiencies, enthalpic efficiencies, and lipophilic indices are well suited to simultaneously monitor affinity and physicochemical properties.

## 4.4 Enthalpic Efficiency Indices

Binding thermodynamics supports the interpretation of the relation between ligand affinities, their enthalpic and entropic components, and physicochemical properties. It suggests that an affinity increase achieved with a higher favorable enthalpy component advantageously affects the physicochemical properties of ligands. This led to the proposal of monitoring the enthalpic component of binding by the enthalpic efficiency index defined as $EE = \Delta H/Q$, where $Q$ is either the number of heavy atoms or the molecular mass [13]. This index measures the change in binding enthalpy relative to the size of the ligand. An alternative index is specific $EE = \Delta H/N_{pol}$ that measures the enthalpy change relative to the number of polar atoms [10]. This latter index directly measures the quality of polar interactions and their contribution to binding.

It was later recognized that the maximal available enthalpy depends on the size of the ligand [6,14]. The available favorable binding enthalpy shows a decreasing trend with increasing ligand size except for the smallest ligands as shown in Figure 4.5. Both maximal available EE (Figure 4.9) and specific EE (not shown) decrease monotonically with increasing ligand size.

Size-independent enthalpic efficiency was defined as $SIHE = (pK_H/40)HA^{0.3}$ [6]. The formula for SIHE was derived from the log–log plot of maximal $pK_H$ versus heavy atom count in analogy to the size-independent ligand efficiency derivation

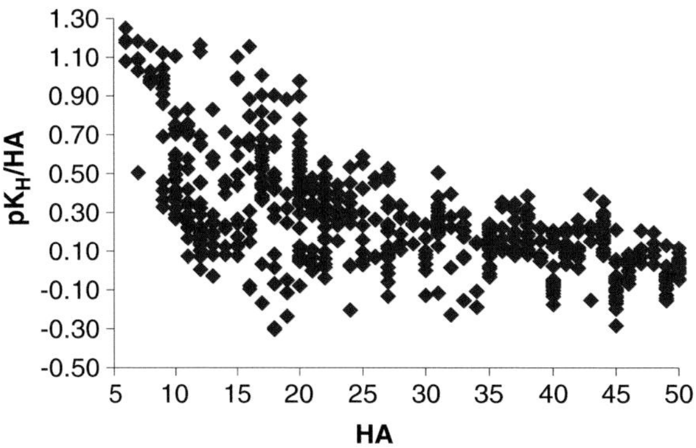


**Figure 4.9** p$K_H$/HA (proportional to EE defined in the text) versus the number of heavy atoms.

by Nissink [3]. The largest available value of SIHE is close to 1 and it is designed to support the comparison of the enthalpic components of different sized ligands. The formula for SIHE is obtained by using the maximal p$K_H$ values in the heavy atom range of 10–50. However, the p$K_H$–HA plot in Figure 4.5 suggests that it is worth considering the HA 15–45 range separately. Ligands with less heavy atom count exhibit an increasing p$K_H$ trend in line with the increasing number of average H-bonds and the corresponding increase of the favorable enthalpic component of binding (Figure 4.7) [12]. Fitting a line on the p$K_{Hmax}$–HA plot in the HA 15–45 range gives the equation p$K_{Hmax} = -0.147\text{HA} + 17.0$. (Note that in contrast to the derivation of SILE and SIHE, no log–log plot is used here.) This allows us to define a linear size-independent enthalpic efficiency index SIHEI = (p$K_H$ − 17)/HA + 1.147. The first term ensures a constant maximal value, while the second term normalizes this value to 1. The SIHEI–HA plot is shown in Figure 4.10.

It is important to realize that the application of SIHEI is legitimate in the HA 15–45 range and smaller fragments cannot achieve SIHEI values close to 1. In the range of HA 15–45, SIHEI is well suited to compare the enthalpic efficiency of different sized ligands.

Enthalpic efficiency indices are appropriate to measure the enthalpic contribution to binding relative to the ligand size and they support the assessment of the balance between affinity and physicochemical properties. They require the knowledge of the binding enthalpy that can be reliably obtained either from isothermal titration calorimetry or from the determination of binding constants at various temperatures followed by the application of the van't Hoff equation (see, for example, Ref. [9]). An alternative to the use of enthalpic indices is that of the lipophilic indices that are related to the entropic component of binding. An advantageous feature of lipophilic indices discussed in Section 4.5 is that they are generated from calculated properties.

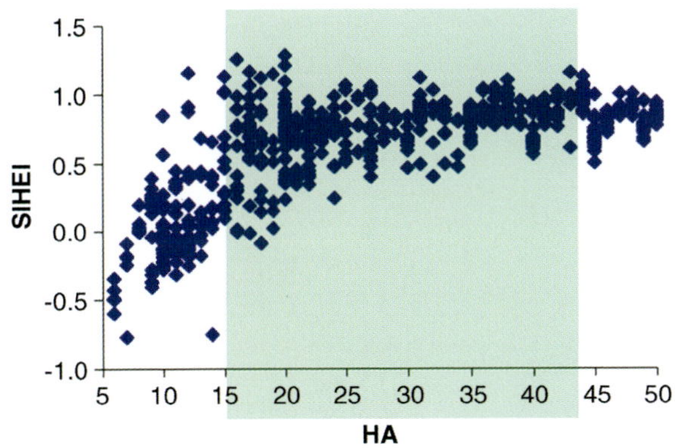


**Figure 4.10** Linear size-independent enthalpic efficiency (SIHEI) from thermodynamic measurements versus the number of heavy atoms. Shaded HA range indicates where the application of SIHEI is recommended.

## 4.5 Lipophilic Efficiency Indices

Lipophilicity is a key physicochemical parameter that influences the affinity, selectivity, and absorption, distribution, metabolism, excretion, and toxicological (ADMET) properties of compounds. In general, binding affinity increases [15] while selectivity and specificity typically decrease with increasing lipophilicity [16]. The relationship of lipophilicity and ADMET properties is more complex. Low lipophilicity is advantageous since compounds are more soluble [17], less impacted by liver metabolism and hepatic clearance [18], and less toxic [19]. On the other hand, they show low permeability across biological membranes [20] and are highly affected by renal clearance [21]. All of these and other observations, therefore, suggest that lipophilicity has an optimal range to achieve during drug discovery optimizations. In fact, analyzing large-scale discovery data sets, we identified the medicinal chemistry sweet spot in the molecular weight – lipophilicity space [22]. Although Waring's analysis suggested a somewhat wider range defined by 0.8 and 3.2 log*P* units as low- and high-lipophilicity thresholds [23], our results suggested that the target range of lipophilicity is even smaller, between 2 and 3.2 log*P* units. It should be noted that both analyses have certain limitations since they were focused on oral drugs and did not consider the dependence of optimal lipophilicity on target classes. The narrow range of optimal lipophilicity has serious consequences for both the identification and the optimization of viable chemical starting points. Considering the well-documented increase of lipophilicity during optimization programs [16,24], low-lipophilicity starting points (such as fragments and lead-like compounds) are clearly preferred. The other conclusion on the optimization strategy is the strict control on lipophilicity that was identified as one of the key success factors in

**Table 4.2** Changes in potency and physicochemical properties during drug discovery optimizations.

| Process | *N* | *p*Pot[a)] change | MW change | log*P* change | Reference |
|---|---|---|---|---|---|
| Fragment opt. | 145 | 2.71 | 185.1 | 1.3 | 25 |
| Fragment opt. –successful[b)] | 13 | 3.05 | 168.1 | 0.7 | 25 |
| HTS-based opt. | 319 | 1.39 | 51.5 | 0.27 | 24 |
| Lead opt. – successful[c)] | 60 | 2.08 | 89.9 | 0.05 | 26 |
| Lipo opt.[d)] | 59 | 1.23 | 45.0 | −0.77 | 28 |

a) *p*Pot is −log(potency), in which potency is either a measure of binding ($K_d$) or biological activity ($K_i$, $IC_{50}$).
b) Fragment opt. – successful are published fragment optimizations, providing clinical candidates with known structures.
c) Lead opt. – successful are published lead optimizations, resulting in marketed drugs.
d) Lipo opt. are optimizations where lipophilic efficiency indices has been explicitly used in the optimization process.

both fragment to candidate [25] and lead to drug optimizations [26] (see Table 4.2).

Medicinal chemistry optimizations should, therefore, be balanced and truly multidimensional, which requires efficient metrics rather than defining preferred ranges to control lipophilicity. Lipophilic ligand efficiency (LLE) suggested by Leeson and Springthorpe [16] was the first index that combined potency and lipophilicity (Table 4.3).

Based on the properties of an average oral drug with cLog*P* ~2.5 and potency in the range ~1–10 nM the authors set an LLE target of ~5–7 or greater. From these parameters, however, it seems that reduced complexity compounds such as fragments or lead-like compounds would never reach the target value of LLE. Viable leads have typically log*P* smaller than 3 and potency in the micromolar range, while fragments have even smaller log*P* and potency values that prevent achieving the target range of LLE. LLE is, therefore, best suited for lead optimization programs and earlier stages would need other metrics that deal with the size of the ligands. Lipophilicity corrected ligand efficiency (LELP) is the first realization of this concept [24]. LELP describes the price of ligand efficiency

**Table 4.3** Definitions of lipophilic efficiency: lipophilic ligand efficiency (LLE), lipophilicity-corrected ligand efficiency (LELP), and LLE adjusted for heavy atom count ($LLE_{AT}$).

| Definition | Reference |
|---|---|
| $LLE = pIC_{50} - \log P$ | 26 |
| $LELP = \log P / LE$ | 24 |
| $LLE_{AT} = 0.111 + 1.37(LLE/HA)$ | 27 |

paid in log*P*. Considering the widely accepted lower limit of ligand efficiency (0.3) and the lipophilicity range for lead-like compounds ($-3 < \log P < 3$), the optimal LELP value in lead discovery should be in the range $-10 < \text{LELP} < 10$. More recently, Astex published an alternative metric for fragment-based lead discovery program $\text{LLE}_{\text{AT}}$ that scales LLE with the heavy atom count of the ligand [27].

The concept of lipophilic efficiency is exemplified by the Astex optimization of an initial fragment known as a respiratory stimulant drug to a HSP90 inhibitor (AT13387) that reached the clinic in oncology indications (Table 4.4). The

**Table 4.4** Fragment optimization to AT13387.

| Stage | Structure | Potency (μM) | *p*Pot | MW | $N_{hev}$ [a)] | log*P* | LE | LLE | LELP | $\text{LLE}_{\text{AT}}$ |
|---|---|---|---|---|---|---|---|---|---|---|
| Start | 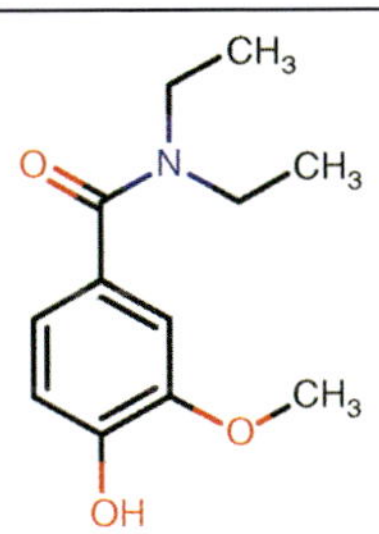 | 790 | 3.10 | 223.12 | 16.00 | 1.52 | 0.27 | 1.58 | 5.74 | 0.25 |
| 1 | 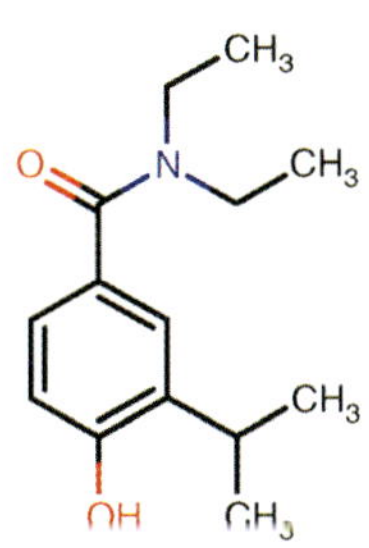 | 7 | 5.15 | 235.16 | 17.00 | 2.93 | 0.42 | 2.23 | 7.04 | 0.29 |
| 2 | 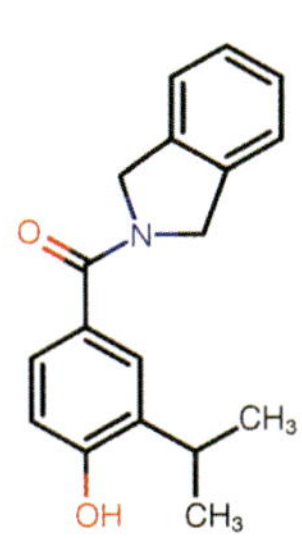 | 0.068 | 7.17 | 281.14 | 21.00 | 3.69 | 0.47 | 3.48 | 7.89 | 0.34 |

(*continued*)

**Table 4.4** (*Continued*)

| Stage | Structure | Potency (μM) | *p*Pot | MW | $N_{hev}$[a)] | log*P* | LE | LLE | LELP | $LLE_{AT}$ |
|---|---|---|---|---|---|---|---|---|---|---|
| 3 |  | 0.00054 | 9.27 | 297.14 | 22.00 | 4.03 | 0.58 | 5.23 | 6.99 | 0.44 |
| End | 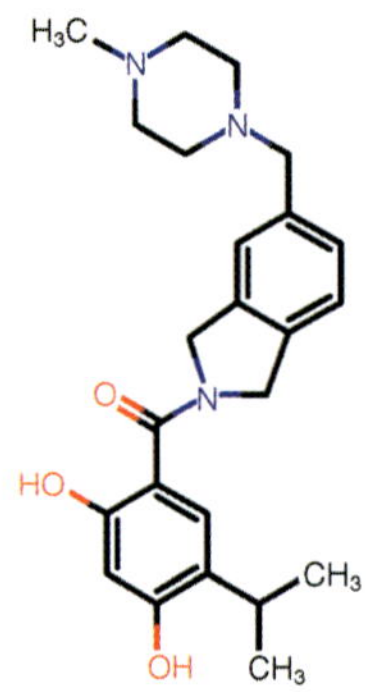 | 0.00071 | 9.15 | 409.24 | 30.00 | 3.19 | 0.42 | 5.96 | 7.64 | 0.38 |

a) $N_{hev}$ is the number of non-hydrogen atoms.

methoxy group of the starting point was first changed to the somewhat larger and obviously more lipophilic isopropyl group. Since this resulted in an almost two orders of magnitude improvement in potency, LE, LLE, and $LLE_{AT}$ improved, but LELP was getting worse in line with the almost two-fold increase in log*P*. In the next step, the team increased the lipophilicity further, which improved the potency again by two orders of magnitude. Consequently, LE, LLE, and $LLE_{AT}$ improved and LELP did not change too much. The introduction of an additional hydroxyl substituent on the central aromatic ring resulted in a sub-nanomolar HSP90 inhibitor with exceptional LE and improved lipophilic efficiency metrics. LLE and LELP reached their suggested values and $LLE_{AT}$ improved a lot. In the final step, a solubilizing piperazine group was added that improved the log*P* but decreased the potency a bit. As a result, LE and $LLE_{AT}$ dropped, LELP did not change significantly; however, LLE improved further and the candidate is located in the most favorable region of the LLE–LELP space.

A number of real-life drug discovery programs demonstrated that lipophilic efficiency measures such as LLE and LELP would support balanced optimization of potency and ADMET properties [28]. In Table 4.5, we summarized those

**Table 4.5** Fragment optimizations using lipophilic efficiency metrics.

| Company | Target | Δ*p*Pot | Δ Mol Wt | Δ c*Log**P* | Δ LLE | Δ LELP | Δ $LLE_{AT}$ | Ref. |
|---|---|---|---|---|---|---|---|---|
| AstraZeneca | 11 beta HSD | −0.11 | 1.95 | 1.53 | −1.64 | 5.27 | −0.08 | 30 |
| Heptares | β1AR | 1.97 | −12.01 | −0.45 | 2.42 | −2.13 | 0.24 | 31 |
| Pfizer | AKT kinase | 2.56 | 183.16 | 1.71 | 0.85 | 4.24 | −0.19 | 32 |
| AstraZeneca | PI3 K p110b | 1.12 | 112.13 | 1.58 | −0.46 | 4.91 | −0.17 | 33 |
| Pfizer | Progesterone | 0.68 | 95.10 | −1.85 | 2.53 | −2.86 | 0.07 | 34 |
| Uni. S. Denmark | FFA1/GPR40 | 1.51 | 57.03 | −0.44 | 1.95 | −1.01 | 0.09 | 35 |
| Roche | mGlu2/3 | 2.60 | 147.10 | 1.33 | 1.28 | 4.18 | 0.01 | 36 |
| Genentech | JAK1 | −0.16 | 55.04 | −2.67 | 2.50 | −3.89 | 0.08 | 37 |
| Dainippon | Epoxide hydrolase | 1.82 | 92.14 | −0.25 | 2.08 | −0.55 | 0.06 | 38 |
| AstraZeneca | 11 beta HSD | 1.70 | 141.17 | 0.89 | 0.81 | 3.94 | −0.03 | 39 |
| SK Biopharma | mGluR5 | 2.53 | −12.39 | −0.38 | 2.91 | −1.97 | 0.19 | 40 |
| Wyeth | BACE1 | 2.98 | 104.11 | −0.31 | 3.28 | −3.17 | 0.13 | 41 |
| Vertex | PKCθ | 3.31 | 112.10 | 0.96 | 2.36 | 1.77 | 0.02 | 42 |
| Pfizer | HIV integrase | 1.22 | 109.17 | 1.16 | 0.06 | 4.39 | −0.06 | 43 |
| ICL | *N*-Myristoyl-transferase | −0.44 | 1.94 | −1.95 | 1.51 | −4.73 | 0.09 | 44 |
| Pfizer | HIV RT | 0.28 | −41.91 | −1.63 | 1.91 | −4.33 | 0.12 | 45 |
| Average | | 1.47 | 71.61 | −0.05 | 1.52 | 0.25 | 0.04 | |

examples that were started from fragments as defined by the rule of three [29]. Although the dependency on the target is obvious, we can see that in general lipophilic efficiency metrics resulted in potency improvements with significantly less increase in lipophilicity. LLE improved significantly while the already beneficial LELP and $LLE_{AT}$ values of the fragments were kept virtually constant on average.

The most important conclusion from these data is that the application of lipophilic efficiency metrics contributed significantly to keep the lipophilicity controlled.

More recently, it has been shown that these metrics have direct connection to a number of ADMET and safety end points including membrane permeability, metabolic clearance, and Pgp substrates and CYP inhibition, hERG inhibition, and cell viability, respectively [46]. LLE was found to be sensible for the development stage but does not prefer early optimization and LELP has the advantage of this latter stage (Figure 4.11). LELP was able to discriminate HTS and fragment hits, successful leads and all lead molecules involved in the study, and more importantly compounds passed phase I trials as those having acceptable ADMET profile in humans. Considering the suggested thresholds of LLE and LELP, we demonstrated that only compounds with promising human pharmacokinetics (i.e., marketed drugs and phase II compounds) fulfill both criteria on

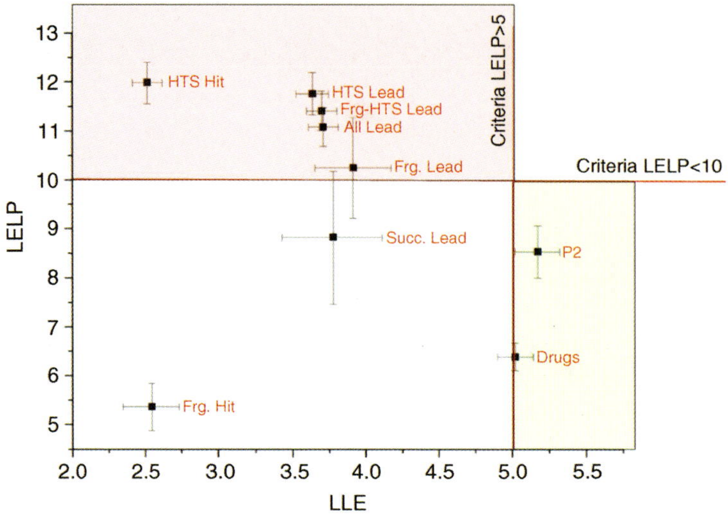


**Figure 4.11** Discrimination of compounds with different development stage in the LLE–LELP space [46]. LELP versus LLE for the different subsets including fragment hits, corresponding leads, HTS hits and corresponding leads, successful leads, compounds that entered phase II trials, development candidates, and marketed drugs. Fragment hits ($N = 100$) and leads ($N = 95$) are from Ref. [25] and HTS hits ($N = 319$) and HTS leads ($N = 319$) are from Ref. [24]. Successful leads (Succ. Lead) were reported by Perola ($N = 60$) [26]. Drugs and phase II (P2) compounds, which entered P2 trials, were collected from Thomson Reuters Integrity database. All data are mean values with error bars representing the standard error values. Copyright 2012 American Chemical Society.

lipophilic efficiency that suggests the usefulness of depicting compounds in the LLE–LELP space. In addition, an independent Pfizer study showed that unlike LE and LLE, development candidates and marketed drugs were successfully discriminated by LELP [47].

## 4.6 Application of Efficiency Indices in Fragment-Based Drug Discovery Programs

The typical starting points of fragment-based drug discovery programs are low-affinity, low-complexity polar compounds that should be optimized to viable chemical leads having improved potency, promising selectivity, and well-balanced physicochemical and ADMET properties. The early optimization of fragments is usually, therefore, potency driven that increases the size and lipophilicity of initial fragments significantly. Analyzing a total of 145 fragment optimizations reported in the biomedical literature up to 2011, we investigated changes in affinity, physicochemical properties, and ligand efficiency metrics [25].

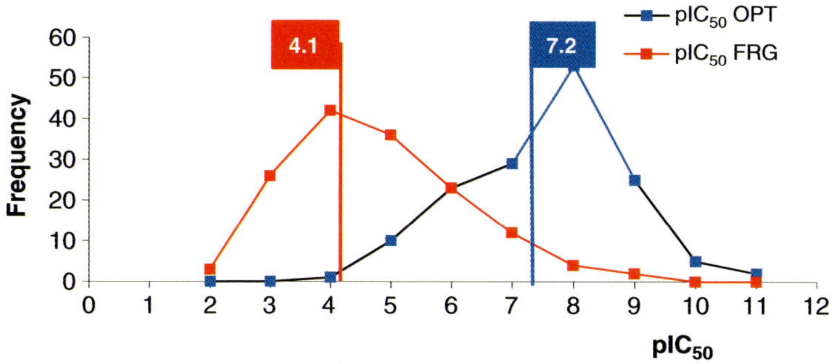


**Figure 4.12** Affinity distribution of fragment hits and optimized compounds. Median values are indicated by vertical lines. (Adapted from Ref. [25]. Copyright 2013 American Chemical Society.)

First, we investigated the improvement in potency achieved during the optimization programs. We found that the typical starting fragment has low milimolar, high micromolar potency having the median value of $pIC_{50} = 4.10$. Due to the highly efficient optimization protocols reported, this affinity improves about three orders of magnitude reaching the median value of $pIC_{50} = 7.22$. Figure 4.12 shows the affinity distribution of fragment starting points (FRG) and corresponding optimized compounds (OPT).

As expected, this remarkable improvement in potency is associated with significant changes in physiocochemical parameters, such as molecular weight, size, and lipophilicity (Figure 4.13a, b and c, respectively).

Although the median values calculated for the optimized compounds are well within the drug-like limits, we observed that the distributions are rather wide indicating very different optimization strategies. In fact, considering the molecular weight and lipophilicity about one-third (29% and 31%, respectively) of the optimizations resulted in compounds breaking at least one Lipinski criterion. Analyzing the distributions of heavy atoms, we first observed that fragment starting points have typically 10–20 non-hydrogen atoms that fits perfectly in the results of recent polls at Practical Fragments indicating the minimum and maximum size of fragments as $N_{\text{heavy}}$ of 7–8 and 18–20, respectively [48,49]. More importantly, however, we found that the number of heavy atoms was almost doubled. That in light of the demonstrated size dependency of ligand efficiency has serious consequences when comparing the ligand efficiency indices of molecules with different sizes. Investigating this effect, we analyzed both the size-dependent and the size-independent versions of ligand efficiency and lipophilic ligand efficiency.

Changes in ligand efficiency were monitored using LE and its size-independent version SILE (Figure 4.14).

As was reported earlier, ligand efficiency did not change too much during the optimization; that is in line with the conventional FBDD strategy keeping LE constant when improving the potency [50]. On the other hand, improvements in ligand efficiency could be more straightforwardly followed using SILE that

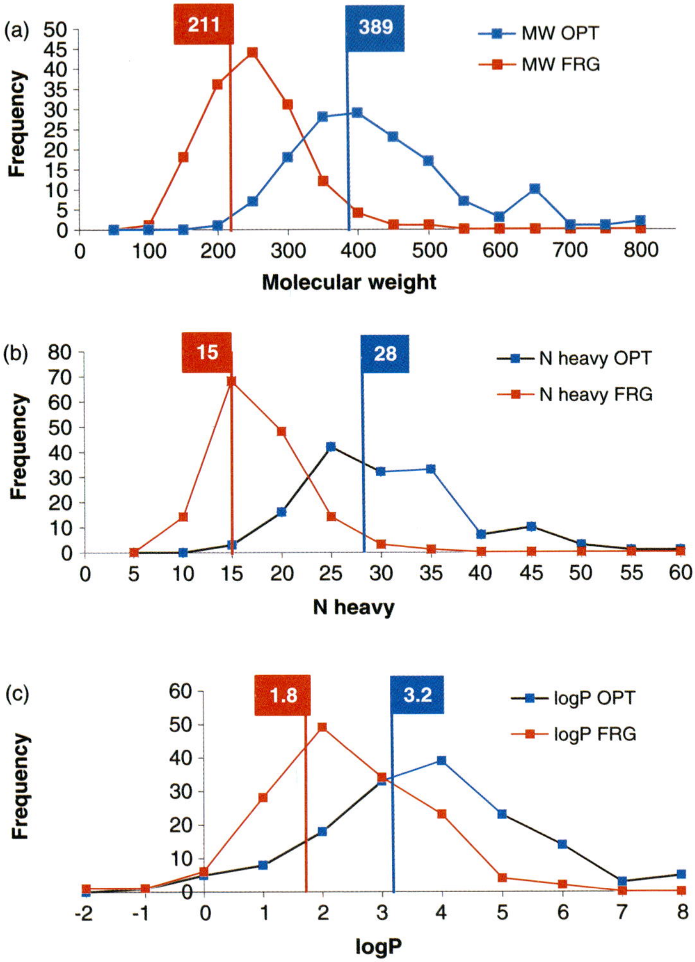


**Figure 4.13** Distribution of (a) molecular weight, (b) number of heavy atoms, and (c) log*P* for fragment hits and optimized compounds. Median values are indicated by vertical lines. ((a) and (c) adapted from Ref. [25]. Copyright 2013 American Chemical Society.)

improves significantly during successful fragment optimizations. In terms of lipophilic efficiency metrics, we investigated LLE, LELP, and $LLE_{AT}$; the latter two consider size information (Figure 4.15a, b and c, respectively).

The distribution of LLE demonstrates that lipophilic efficiency improved during this set of fragment optimizations from the median LLE = 2.1 to 4.1. Although the median LLE value is lower than expected for drug-like compounds, it is interesting to note that the distribution reaches the maximum around LLE = 5–6 that is close to its optimal value [16]. The favorable median of LELP = 4.6 became less favorable reaching the median LELP = 9.0; however,

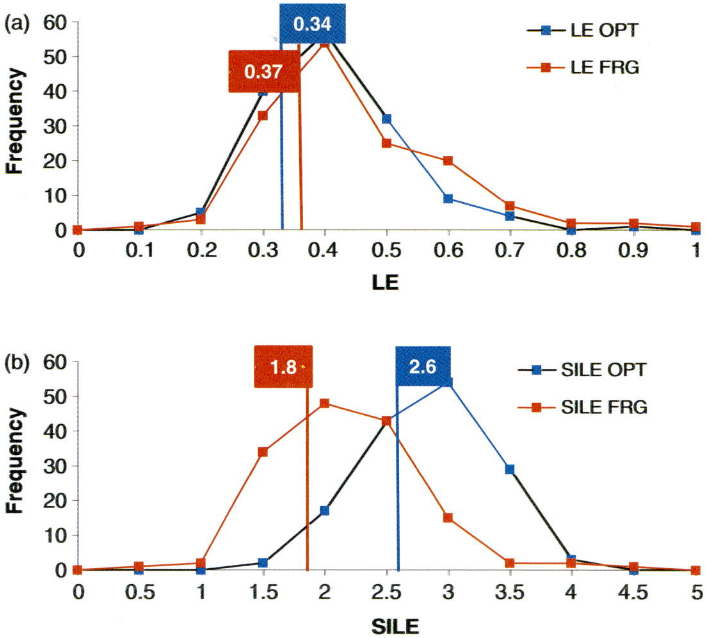


**Figure 4.14** Ligand efficiency (a) and size-independent ligand efficiency. (b) Distributions for starting fragments and optimized compounds. Median values are indicated by vertical lines. (Adapted from Ref. [25]. Copyright 2013 American Chemical Society.)

both values are well below the suggested threshold of LELP = 10 [24]. The disfavored increase of LELP reflects the undesired shift in lipophilicity. The wide range of LELP values for the optimized compounds, however, suggests significant differences in the control of lipophilicity. Finally, $LLE_{AT}$ did not change very much during the optimization programs. Due to its normalization, in this respect $LLE_{AT}$ behaves similar to LE and consequently it should be kept virtually constant when improving the potency.

Our analysis on size-dependent and size-independent metrics suggests that it is important to consider size information especially for compounds with very different numbers of heavy atoms. We think that metrics such as SILE, LELP, and $LLE_{AT}$ help medicinal chemistry teams control molecular obesity [22] in fragment-based drug discovery programs. This trend can be followed in the recent poll at Practical Fragments comparing the use of ligand efficiency metrics in 2011 and 2014 (Figure 4.16). As can be seen, LE and LLE are the most widely used metrics since they could effectively support lead optimizations. Early optimizations, however, are usually associated with large change in molecular size that might require the use of size-independent metrics. In fact, SILE and LELP usage increased significantly reaching 10% in 2014, while LELP and $LLE_{AT}$ represent about 20% together as lipophilic efficiency metrics considering size information.

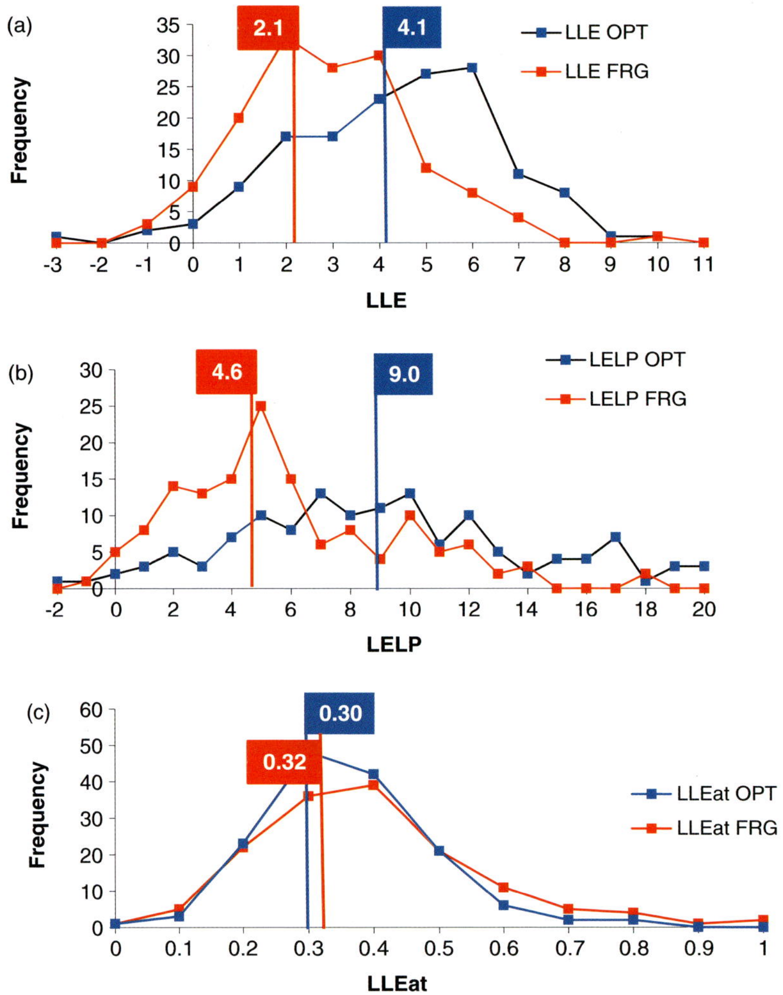


**Figure 4.15** Distribution of lipophilic efficiency metrics (a) LLE, (b) LELP, and (c) $LLE_{AT}$ for starting fragments and optimized compounds. Median values are indicated by vertical lines. ((a) and (b) adapted from Ref. [25]. Copyright 2013 American Chemical Society.)

The last objective of our analysis was comparing changes in affinity, physicochemical properties, and ligand efficiency metrics found in fragment optimizations with those observed in successful cases, that is, optimizations that provided clinical candidates. At the time of our analysis, there were 18 clinical candidates identified in FBDD programs out of which 13 had starting points and candidates with known structures. Table 4.6 summarizes the characteristics of these optimizations in comparison to corresponding median values obtained for the 145 fragment-based optimization programs analyzed.

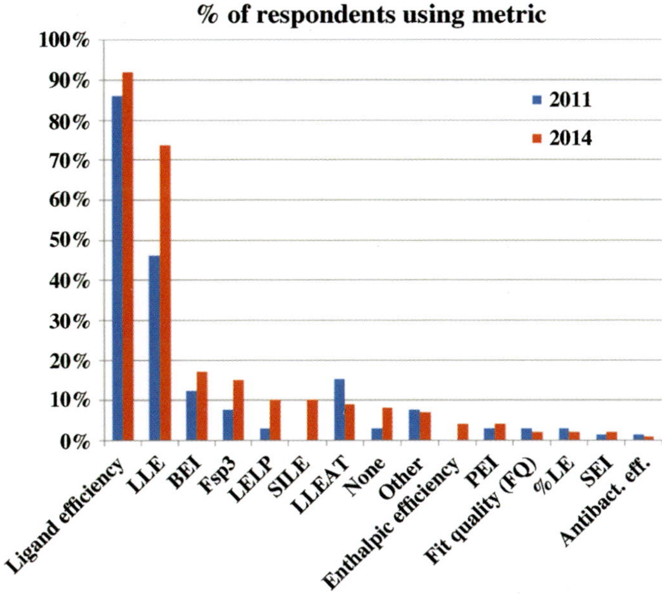


**Figure 4.16** The use of ligand efficiency metrics in 2011 and 2014 evaluated at Practical Fragments Web site in June, 2014 [51].

The properties of all compounds optimized from fragment hits were compared with the properties of those compounds that reached clinic to see if certain properties are appropriate to indicate the success of optimizations. Our data show that clinical candidates originated from hits with similar size and lipophilicity as hits from other programs. Due to their higher affinity accompanied with a size and lipohilicity similar to average optimizations, the size-independent ligand efficiency, SILE of candidates, and lipophilic efficiencies, LLE, LELP, and

**Table 4.6** Median properties for starting fragments and for optimized compounds in general and successful fragment optimizations.

| | Starting fragment | | Optimized compound | |
|---|---|---|---|---|
| | General opt. | Successful opt. | General opt. | Successful opt. |
| $pIC_{50}$ | 4.10 | 4.00 | 7.22 | 7.85 |
| HA | 15 | 16 | 28 | 28 |
| Mol. weight | 216 | 211 | 380 | 389 |
| log*P* | 1.79 | 1.99 | 3.19 | 2.97 |
| LE | 0.37 | 0.39 | 0.34 | 0.40 |
| SILE | 1.75 | 1.91 | 2.64 | 2.89 |
| LLE | 2.15 | 2.15 | 4.13 | 4.71 |
| LELP | 4.62 | 5.71 | 8.47 | 7.57 |
| $LLE_{AT}$ | 0.32 | 0.31 | 0.30 | 0.35 |

$LLE_{AT}$ of compounds from successful fragment optimizations are significantly better. In the course of optimizing their ligand efficiency, LE hardly changes, SILE improves, and log$P$ increases modestly, and thus LLE, LELP, and $LLE_{AT}$ exhibit significantly more beneficial changes than do average optimizations. These observations clearly show that lipophilic efficiency monitoring, similar to drug discovery programs in general [46], has fundamental importance in fragment-based optimizations, as well.

## 4.7 Conclusions

Efficiency indices are designed to support the generation of development candidates with balanced affinity, physicochemical properties, and ADMET profiles. Although there was some debate on the mathematics and the use of ligand efficiency metrics [52], it has been demonstrated that they are mathematically valid and their role is more than obvious in the effective control of molecular size and lipophilicity during medicinal chemistry optimizations [53]. There is a vast amount of publicly available data published by major pharma companies indicating that the inflation of physicochemical properties – often cited as molecular obesity – results in serious complications along the development path of discovery compounds [22]. Since ligand efficiency indices are composite metrics that typically measure the improvement in affinity relative to some other quantities most often related to molecular size and lipophilicity, they are simple and effective tools reducing compound-related attrition. Efficiency indices are more appropriate than ligand affinity and other individual properties to compare the quality of different ligands. They are useful at decision points like hit and lead selection and also to monitor multiparametric optimizations. The maximal value of indices may depend on ligand size and this introduces ambiguity in the comparison of different sized ligands. Size-independent indices (SILE, SIHE) eliminate this ambiguity and thus they are particularly useful in fragment optimizations accompanied with significant increases in ligand size. Similarly, lipophilic efficiency indices containing size information, such as LELP and $LLE_{AT}$, are better suited for supporting early-phase fragment optimization programs. Analyses of reported fragment optimizations show that efficiency indices are able to distinguish successful optimizations resulting in clinical candidates. Considering that the characteristics of the binding sites and the target class have an impact on ligand efficiency metrics (e.g., protein–protein interactions versus enzyme inhibitors) and similarly, different mechanisms of action should be treated separately for ligand efficiency (e.g., reversible versus covalent inhibitors), we do not suggest hard standardized target values but rather to evaluate the changes along the optimization path. Focusing to hit identification and hit-to-lead settings, we advocate to improve or keep ligand efficiency constant and obviously avoid large drops in efficiency metrics. Importantly, however, we underline that larger, more optimized compounds may have smaller ligand efficiencies, which is due in part to the fact that such compounds have been optimized for properties other than binding to the target (e.g., pharmacokinetics, solubility, selectivity, or cell permeability).

## References


1 Hopkins, A.L., Groom, C.R., and Alex, A. (2004) Ligand efficiency: a useful metric for lead selection. *Drug Discovery Today*, **9**, 430–341.

2 Abad-Zapatero, C. (2007) Ligand efficiency indices for effective drug discovery. *Expert Opinion on Drug Discovery*, **2**, 469–488.

3 Nissink, J.W.M. (2009) Simple size-independent measure of ligand efficiency. *Journal of Chemical Information and Modeling*, **49**, 1617–1622.

4 Kuntz, I.D., Chen, K., Sharp, K.A., and Kollman, P.A. (1999) The maximal affinity of ligands. *Proceedings of the National Academy of Sciences of the United States of America*, **96**, 9997–10002.

5 Reynolds, C.H., Bembenek, S.D., and Tounge, B.A. (2007) The role of molecular size in ligand efficiency. *Bioorganic and Medicinal Chemistry Letters*, **17**, 4258–4261.

6 Ferenczy, G.G. and Keserü, G.M. (2010) Enthalpic efficiency of ligand binding. *Journal of Chemical Information and Modeling*, **50**, 1536–1541.

7 Reynolds, C.H., Tounge, B.A., and Bembenek, S.D. (2008) Ligand binding efficiency: trends, physical basis, and implications. *Journal of Medicinal Chemistry*, **51**, 2432–2438.

8 Freire, E. (2008) Do enthalpy and entropy distinguish first in class from best in class? *Drug Discovery Today*, **13**, 869–874.

9 Ferenczy, G.G. and Keserü, G.M. (2010) Thermodynamics guided lead discovery and optimization. *Drug Discovery Today*, **15**, 919–932.

10 Ladbury, J.E., Klebe, G., and Freire, E. (2010) Adding calorimetric data to decision making in lead discovery: a hot tip. *Nature Reviews Drug Discovery*, **9**, 23–27.

11 Berman, H.M., Westbrook, J., Feng, Z., Gilliland, G., Bhat, T.N., Weissig, H., Shindyalov, I.N., and Bourne, P.E. (2000) The Protein Data Bank. *Nucleic Acids Research*, **28**, 235–242.

12 Ferenczy, G.G. and Keserü, G.M. (2012) Thermodynamics of fragment binding. *Journal of Chemical Information and Modeling*, **52**, 1039–1045.

13 Ladbury, J.E. (2007) Enthalpic efficiency and the role of thermodynamic data in drug development: possibility or a pipeline dream. *European Pharmaceutical Review*, **12**, 59–62.

14 Reynolds, C.H. and Holloway, M.K. (2011) Thermodynamics of ligand binding and efficiency. *ACS Medicinal Chemistry Letters*, **2**, 433–437.

15 Gleeson, M.P., Hersey, A., Montanari, D., and Overington, J. (2011) Probing the links between *in vitro* potency, ADMET and physicochemical parameters. *Nature Reviews Drug Discovery*, **10**, 197–208.

16 Leeson, P.D. and Springthorpe, B. (2007) The influence of drug-like concepts on decision-making in medicinal chemistry. *Nature Reviews Drug Discovery*, **6**, 881–890.

17 Kramer, C., Heinisch, T., Fligge, T. *et al.* (2009) A consistent dataset of kinetic solubilities for early-phase drug discovery. *ChemMedChem*, **4**, 1529–1536.

18 Johnson, T.W., Dress, K.R., and Edwards, M. (2009) Using the Golden Triangle to optimize clearance and oral absorption. *Bioorganic and Medicinal Chemistry Letters*, **19**, 5560–5564.

19 Hughes, J.D., Blagg, J., Price, D.A. *et al.* (2008) Physiochemical drug properties associated with *in vivo* toxicological outcomes. *Bioorganic and Medicinal Chemistry Letters*, **18**, 4872–4875.

20 Waring, M.J. (2009) Defining optimum lipophilicity and molecular weight ranges for drug candidates: molecular weight dependent lower log*D* limits based on permeability. *Bioorganic and Medicinal Chemistry Letters*, **19**, 2844–2851.

21 Varma, M.V.S., Feng, B., Obach, R.S. *et al.* (2009) Physicochemical determinants of human renal clearance. *Journal of Medicinal Chemistry*, **52**, 4844–4852.

22 Hann, M.M. and Keserü, G.M. (2012) Finding the sweet spot: the role of nature and nurture in medicinal chemistry. *Nature Reviews Drug Discovery*, **11**, 355–365.

23 Waring, M.J. (2010) Lipophilicity in drug discovery. *Expert Opinion on Drug Discovery*, **5**, 235–248.

24 Keserű, G.M. and Makara, G.M. (2009) The influence of lead discovery strategies on the properties of drug candidates. *Nature Reviews Drug Discovery*, **8**, 203–212.

25 Ferenczy, G.G. and Keserű, G.M. (2013) How are fragments optimized? A retrospective analysis of 145 fragment optimizations. *Journal of Medicinal Chemistry*, **56**, 2478–2486.

26 Perola, E. (2010) An analysis of the binding efficiencies of drugs and their leads in successful drug discovery programs. *Journal of Medicinal Chemistry*, **53**, 2986–2997.

27 Paul, N., Mortenson, P.M., and Murray, C. W. (2011) Assessing the lipophilicity of fragments and early hits. *Journal of Computer-Aided Molecular Design*, **25**, 663–667.

28 Hopkins, A.L., Keserü, G.M., Leeson, P.D., Rees, D.C., and Reynolds, C.H. (2014) The role of ligand efficiency metrics in drug discovery. *Nature Reviews Drug Discovery*, **13**, 105–121.

29 Congreve, M., Carr, R., Murray, C., and Jhoti, H. (2003) A 'rule of three' for fragment-based lead discovery? *Drug Discovery Today*, **8**, 876–877.

30 Scott, J.S., deSchoolmeester, J., Kilgour, E., Mayers, R.M., Packer, M.J., Hargreaves, D., Gerhardt, S., Ogg, D.J., Rees, A., Selmi, N., Stocker, A., Swales, J.G., and Whittamore, P.R.O. (2012) Novel acidic 11β-hydroxysteroid dehydrogenase type 1 (11β-HSD1) inhibitor with reduced acyl glucuronide liability: the discovery of 4-[4-(2-adamantylcarbamoyl)-5-*tert*-butyl-pyrazol-1-yl]benzoic acid (AZD8329). *Journal of Medicinal Chemistry*, **55**, 10136–10147.

31 Christopher, J.A. *et al.* (2013) Biophysical fragment screening of the β1⁻ adrenergic receptor: identification of high affinity arylpiperazine leads using structure-based drug design. *Journal of Medicinal Chemistry*, **56**, 3446–3455.

32 Freeman-Cook, K.D., Autry, C., Borzillo, G., Gordon, D., Barbacci-Tobin, E., Bernardo, V., Briere, D., Clark, T., Corbett, M., Jakubczak, J., Kakar, S., Knauth, E., Lippa, B., Luzzio, M.J., Mansour, M., Martinelli, G., Marx, M., Nelson, K., Pandit, J., Rajamohan, F., Robinson, S., Subramanyam, C., Wei, L., Wythes, M., and Morris, J. (2010) Design of selective, ATP-competitive inhibitors of Akt. *Journal of Medicinal Chemistry*, **53** 4615–4622.

33 Giordanetto, F. *et al.* (2012) Discovery of 4-morpholino-pyrimidin-6-one and 4-morpholino-pyrimidin-2-one-containing phosphoinositide 3-kinase (PI3K) p110β isoform inhibitors through structure-based fragment optimisation. *Bioorganic and Medicinal Chemistry Letters*, **22**, 6665–6670.

34 Dack, K.N., Skerratt, S., Johnson, P.S., Bradley, P.A., and Marsh, I.R. (2010) Optimisation of a pyrazole series of progesterone antagonists. Part 1. *Bioorganic and Medicinal Chemistry Letters*, **20**, 3384–3386.

35 Christiansen, E. *et al.* (2013) Discovery of TUG-770: a highly potent free fatty acid receptor 1 (FFA1/GPR40) agonist for treatment of type 2 diabetes. *ACS Medicinal Chemistry Letters*, **4**, 441–445.

36 Woltering, T.J., Wichmann, J., Goetschi, E., Knoflach, F., Ballard, T.M., Huwyler, J., and Gatti, S. (2010) Synthesis and characterization of 1,3-dihydro-benzo[b][1,4] diazepin-2-one derivatives. Part 4. *In vivo* active potent and selective non-competitive metabotropic glutamate receptor 2/3 antagonists. *Bioorganic and Medicinal Chemistry Letters*, **20**, 6969–6974.

37 Zak, M. *et al.* (2013) Identification of C-2 hydroxyethyl imidazopyrrolopyridines as potent JAK1 inhibitors with favorable physicochemical properties and high selectivity over JAK2. *Journal of Medicinal Chemistry*, **56**, 4764–4785.

38 Tanaka, D., Tsuda, Y., Shiyama, T., Nishimura, T., Chiyo, N., Tominaga, Y., Sawada, N., Mimoto, T., and Kusunose, N. (2011) A practical use of ligand efficiency

indices out of the fragment-based approach: ligand efficiency-guided lead identification of soluble epoxide hydrolase inhibitors. *Journal of Medicinal Chemistry*, **54**, 851–857.

39 Scott, J.S., Bowker, S.S., Deschoolmeester, J., Gerhardt, S., Hargreaves, D., Kilgour, E., Lloyd, A., Mayers, R.M., McCoull, W., Newcombe, N.J., Ogg, D., Packer, M.J., Rees, A., Revill, J., Schofield, P., Selmi, N., Swales, J.G., and Whittamore, P.R. (2012) Discovery of a potent, selective, and orally bioavailable acidic 11β-hydroxysteroid dehydrogenase type 1 (11β-HSD1) inhibitor: discovery of 2-[(3*S*)-1-[5-(cyclohexylcarbamoyl)-6-propylsulfanylpyridin-2-yl]-3-piperidyl] acetic acid (AZD4017). *Journal of Medicinal Chemistry*, **55**, 5951–5964.

40 Chae, E. *et al.* (2013) Discovery of biological evaluation of pyrazole/imidazole amides as mGlu5 receptor negative allosteric modulators. *Bioorganic and Medicinal Chemistry Letters*, **23**, 2134–2139.

41 Nowak, P., Cole, D.C., Aulabaugh, A., Bard, J., Chopra, R., Cowling, R., Fan, K.Y., Huc, B., Jacobsen, S., Jani, M., Jin, G., Lo, M., Malamas, M.S., Manas, E.S., Narasimhan, R., Reinhart, P., Robichaud, A.J., Stock, J.T., Subrath, J., Svenson, K., Turner, J., Wagner, E., Zhou, P., and Ellingboe, J.W. (2010) Discovery and initial optimization of 5,50-disubstituted aminohydantoins as potent b-secretase (BACE1) inhibitors. *Bioorganic and Medicinal Chemistry Letters*, **20**, 632–635.

42 Jimenez, J.-M. *et al.* (2013) Design and optimization of selective protein kinase Cθ (PKCθ) inhibitors for the treatment of autoimmune diseases. *Journal of Medicinal Chemistry*, **56**, 1799–1810.

43 Johnson, T.W., Tanis, S.P., Butler, S.L., Dalvie, D., DeLisle, D.M., Dress, K.R., Flahive, E.J., Hu, Q., Kuehler, J.E., Kuki, A., Liu, W., McClellan, G.A., Peng, Q., Plewe, M.B., Richardson, P.F., Smith, G.L., Solowiej, J., Tran, K.T., Wang, H., Yu, X., Zhang, J., and Zhu, H. (2011) Design and synthesis of novel *N*-hydroxy-dihydronaphthyridinones as potent and orally bioavailable HIV-1 integrase inhibitors. *Journal of Medicinal Chemistry*, **54**, 3393–3417.

44 Goncalves, V., Brannigan, J.A., Whalley, D., Ansell, K.H., Saxty, B., Holder, A.A., Wilkinson, A.J., Tate, E.W., and Leatherbarrow, R.J. (2012) Discovery of plasmodium vivax *N*-myristoyltransferase inhibitors: screening, synthesis, and structural characterization of their binding mode. *Journal of Medicinal Chemistry*, **55**, 3578–3582.

45 Mowbray, C.E., Corbau, R., Hawes, M., Jones, L.H., Mills, J.E., Perros, M., Selby, M.D., Stupple, P.A., Webster, R., and Wood, A. (2009) Pyrazole NNRTIs 3: optimisation of physicochemical properties. *Bioorganic and Medicinal Chemistry Letters*, **19**, 5603–5606.

46 Tarcsay, A., Nyíri, K., and Keserű, G.M. (2012) Impact of lipophilic efficiency on compound quality. *Journal of Medicinal Chemistry*, **55**, 1252–1260.

47 Wager, T.T., Chandrasekaran, R.Y., Hou, X., Troutman, M.D., Verhoest, P.R., Villalobos, A., and Will, Y. (2010) Defining desirable central nervous system drug space through the alignment of molecular properties, *in vitro* ADME, and safety attributes. *ACS Chemical Neuroscience*, **1**, 420–434.

48 http://practicalfragments.blogspot.hu/2013/06/poll-results-how-small-are-your.html, access date: 18 January 2015.

49 http://practicalfragments.blogspot.hu/2012/06/poll-results-how-big-are-your-fragments.html, access date: 18 January 2015.

50 Hajduk, P.J. and Greer, J. (2007) A decade of fragment-based drug design: strategic advances and lessons learned. *Nature Reviews Drug Discovery*, **6**, 211–9.

51 http://practicalfragments.blogspot.hu/2014/06/metric-poll-results-plus-ca-change.html, access date: 18 January 2015.

52 Shultz, M.D. (2014) Improving the plausibility of success with inefficient metrics. *ACS Medicinal Chemistry Letters*, **5**, 2–5.

53 Murray, C.W., Erlanson, D.A., Hopkins, A.L., Keserű, G.M., Leeson, P.D., Rees, D.C., Reynolds, C.H., and Richmond, N.J. (2014) Validity of ligand efficiency metrics. *ACS Medicinal Chemistry Letters*, **5**, 616–618.

# Part II
# Methods and Approaches for Fragment-based Drug Discovery

# 5
# Strategies for Fragment Library Design

*Justin Bower, Angelo Pugliese, and Martin Drysdale*

## 5.1
## Introduction

Fragment-based drug discovery (FBDD) methods are a proven strategy for hit identification, providing diverse, high-quality start points for numerous drug discovery programs across a broad spectrum of disease areas. Many fragment hits have been optimized and transitioned to the clinical setting with one, namely, vemurafenib, gaining FDA approval to treat late-stage V600E mutant B-Raf-driven malignant melanoma [1–3].

Critical for successful fragment screening is a high-quality compound library and the ability to detect and validate binding, preferentially through multiple orthogonal techniques, to the target being screened. While fragment screening techniques have evolved over the years [4], increasing in sensitivity and throughput, fragment library molecules have received less attention. Target agnostic fragment libraries are typically composed of low molecular weight (<300 Da), low complexity molecules with good physicochemical properties [5], rendering them highly water soluble, a critical characteristic of this hit identification strategy. In addition to the excellent physicochemical characteristics, a well selected, diverse fragment library is able to sample the available chemical space more effectively than a much larger HTS library, which in general contains compounds with higher molecular weights.

Many fragment libraries are chemically diverse and have been selected based on a good balance of properties. However, they all tend to have limited three-dimensional diversity, typically being composed of flat $sp^2$-rich aromatic and heteroaromatic compounds. This is due in part to the types of biological targets that fragments have historically been screened against and the rise in the widespread use of catalytic $sp^2$–$sp^2$ coupling reactions [6]. This move toward "flatter" molecules not only reduces the sampling of chemical space available for fragments but also fails to capitalize on the additional attributes of $sp^3$ character. Recently, two groups have independently published data showing the improvements in profile and project progression by, for example, increasing the proportion of $sp^3$ centers contained in molecules

*Fragment-based Drug Discovery: Lessons and Outlook*, First Edition. Edited by Daniel A. Erlanson and Wolfgang Jahnke.

or reducing lipohilicity (log*P*) [7–9]. While questions have been raised over the way that some of the statistics used in these papers has been presented [10], there are multiple examples in the literature reporting the positive impact of increasing $sp^3$-ness or three-dimensionality through inducing conformational twist [11].

While significant effort has been expended in improving the profiles of HTS libraries, relatively little has been published concerning the introduction of three-dimensionality into fragment screening collections [12,13]. Introducing a descriptor of shape for fragment molecules will add another degree of diversity to the fragment library, not only providing more choice when selecting molecules to optimize but also offering different opportunities when screening against "old" and "new" targets such as protein–protein interactions (PPI) where complementarity to a three-dimensional interface of one protein may be important in disrupting its interaction with a partner protein. This chapter will discuss our efforts toward this goal.

## 5.2 Aims

- Analyze commercial fragment space to assemble a virtual library of "chemically attractive" and readily available fragments.
- Redesign our fragment library to enhance chemical diversity and pharmacophoric content using commercial and in-house chemical space.
- Enrich the fragment screening library with three-dimensionally enhanced molecules in order to increase diversity.
- Maintain excellent physicochemical properties and chemical tractability without increasing molecule complexity.
- Synthesize molecules to fill holes in unexploited chemical and three-dimensional space.

## 5.3 Progress

In order to achieve our aim of enhancing the chemical and shape diversity profile of our fragment library through the introduction of new fragments, we have developed a three-stage strategy:

- *Stage 1*: Undertake an analysis of commercial fragment space and purchase those compounds that meet our primary criteria (fragment space compliant, diverse (chemical, pharmacophore, and shape), and attractive chemical start points).
- *Stage 2*: Identify noncommercial fragments that meet our primary criteria (enhanced chemical and shape diversity over stage 1) – these molecules will

form the basis of future synthetic chemistry efforts and delivery will be achieved through collaborating with other groups.

- *Stage 3*: Identify an optimal set of natural product-inspired fragments using a DOS (diversity-oriented synthesis) approach, prioritize for synthesis, and deliver.

### 5.3.1 BDDP Fragment Library Design: Maximizing Diversity

The Beatson Drug Discovery Programme (BDDP) fragment screening collection currently consists of 1014 compounds that conform to the well-established guidelines for selecting fragment molecules [5]. Our main aim was to improve the overall diversity of our fragment screening set, using descriptors that are pharmacophore-, chemical-, and shape-based. Maximizing the overall diversity of a library enables a more efficient representation of the available chemical space (greater chemical diversity correlates with more singletons in a screening library). This provides increased chemical diversity at the hit ID and hits-to-leads stages, translating to more choice in lead optimization and greater chances of success due to reduced chemistry-related attrition. We have addressed the chemical diversity of our fragment library using well-known dissimilarity and clustering methods based on 2D fingerprints.

Fragment binding is a molecular recognition event driven by specific interactions with the target. Consequently, maximizing the number of fragment recognition patterns is required in order to cover as much interaction space as possible. Fragment recognition patterns can be simulated by (encoded in) 3D pharmacophore features and the diversity achieved by exploiting the number of these diverse features. Therefore, recognition diversity can be addressed using dissimilarity methods based on 2D or 3D (three- or four-point) pharmacophore fingerprints.

### 5.3.2 Assessing Three-Dimensionality

The first step to incorporating three-dimensionality in compound selection is selecting a set of descriptors to define the shape of a molecule. The principal moments of inertia (PMI) introduced by Sauer and Schwartz [14] provides a way of describing the "low-resolution" shape of a molecule in the form of a 2D triangular plot. The PMI parameters of a molecule, normalized to the molecule's longest dimension, display rod-shaped molecules in the top left, disk shaped at the bottom and spherical at the top right of the triangle. One- and two-dimensional molecules occupy the extreme left-hand diagonal axis with increasingly three-dimensional compounds extending to the right (Figure 5.1). A plot of the 1014-membered Beatson fragment library is shown in Figure 5.1a indicating a preponderance of flat molecules. Interestingly, if we compare the Beatson library with a set of fragmented molecules generated from the InMan subset (all molecules

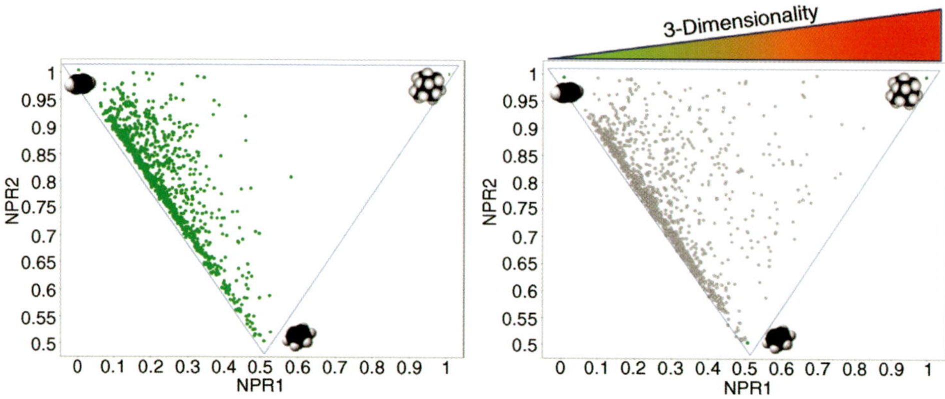


**Figure 5.1** Shape diversity of the Beatson fragment library (a) and fragmented biologically active molecules (b) showing the increase in three-dimesionality going from the left to the right of the PMI plot.

that reached clinical trials including marketed drugs) of the ZINC database [15] using the RECAP method of fragmentation [16], we note a greater spread of three-dimensionality (Figure 5.1b).

One advantage of the PMI descriptors is that they are fast to calculate and, in combination with modern 3D structure generators, can effectively profile the three-dimensionality of large commercial libraries. From this plot, a histogram or pie chart can easily be generated to profile the fraction of 2D compounds by plotting the proportion of compounds in a slice close to the left-hand axis and this can be used as a filter to remove 2D compounds if desired.

An additional attractive feature of PMI plots is that they also provide a grid-based shape space allowing unrepresented "holes" in a compound collection to be identified. This feature enabled us to rapidly identify areas to focus our initial commercial searches and those areas where synthetic chemistry would be needed to generate the required compounds. There are more detailed shape descriptors available that provide a "higher resolution" description of a molecule's shape; for example, Gaussian overlay methods such as ROCS [17], triplet alignment techniques in phase shape [18], or orientation-independent approaches such as USR [19]. While these methods are sufficiently detailed to be effective in virtual screening and can also be used for dissimilarity comparisons, they are not as amenable to cell-based representation as the PMI method because of their higher dimensionality and the conventional metrics for the (dis)similarity in these approaches is either a shape Tanimoto or a Manhattan distance.

### 5.3.3
### 3DFrag Consortium

Prior to undertaking a thorough shape analysis of our own in-house fragment library, we have identified a small set of fragments (190 molecules) that covers a

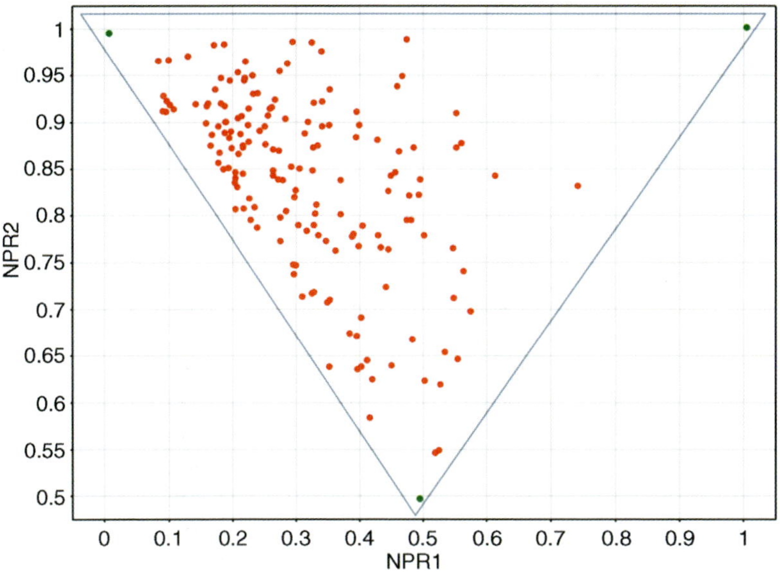


**Figure 5.2** 3DFrag Consortium foundation fragment set.

wide range of three-dimensional space. This work was undertaken in collaboration with other not-for-profit drug discovery groups, collectively known as the 3D Fragment Consortium [20]. This has enabled us to access a set of three-dimensionally enriched compounds rapidly, and these molecules have been added to our current fragment library. The overall goal of the consortium is to increase the hit rate and value for drug discovery projects through building a high-quality fragment library that includes core components found in many existing libraries and to add a complementary set of analogs possessing greater three-dimensional shapes. The consortium is committed to sharing fragment hit data for this foundation set of molecules at the target class level and it is envisaged that this data will help shape the design of future fragment libraries. The foundation set PMI plot is shown in Figure 5.2, presenting a good spread of shape diversity with excellent physicochemical properties.

### 5.3.4 Commercial Fragment Space Analysis

In order to expand the diversity of our library, we have selected approximately 236 000 commercially available (in stock) molecules from three major vendor files that comply with fragment property criteria using the process outlined in Figure 5.3. As can be seen from the process in Figure 5.3a, application of fragment property space filters removes the vast majority of molecules from the 20 million compound set, demonstrating the more "drug-like" and lead-like nature of commercial vendor collections. This set of fragment molecules acted not only as a source of compounds for our analysis but also as a readily searchable database from which we can undertake fast near-neighbor and substructure searches of fragment hits. The shape-diversity PMI plot of this filtered set (~236 000) binned by the range of sum of normalized principal moments of inertia (NPR1 + NPR2 or "flat distance") is shown in Figure 5.3b and c, indicating a good spread of shape diversity from which to select diverse fragments. A similar process to the one outlined above has been used to identify approximately 2 million lead-like molecules (virtual lead-like library) that will be used for future virtual screening campaigns against suitable drug discovery targets. As can be seen from this analysis, approximately 75% of commercial fragment property space is populated with "flat" molecules, defined as having a PMI-sum or flat distance of <1.1. While this may still provide chemically diverse molecules, it limits the potential to access chemically attractive, three-dimensionally biased molecules.

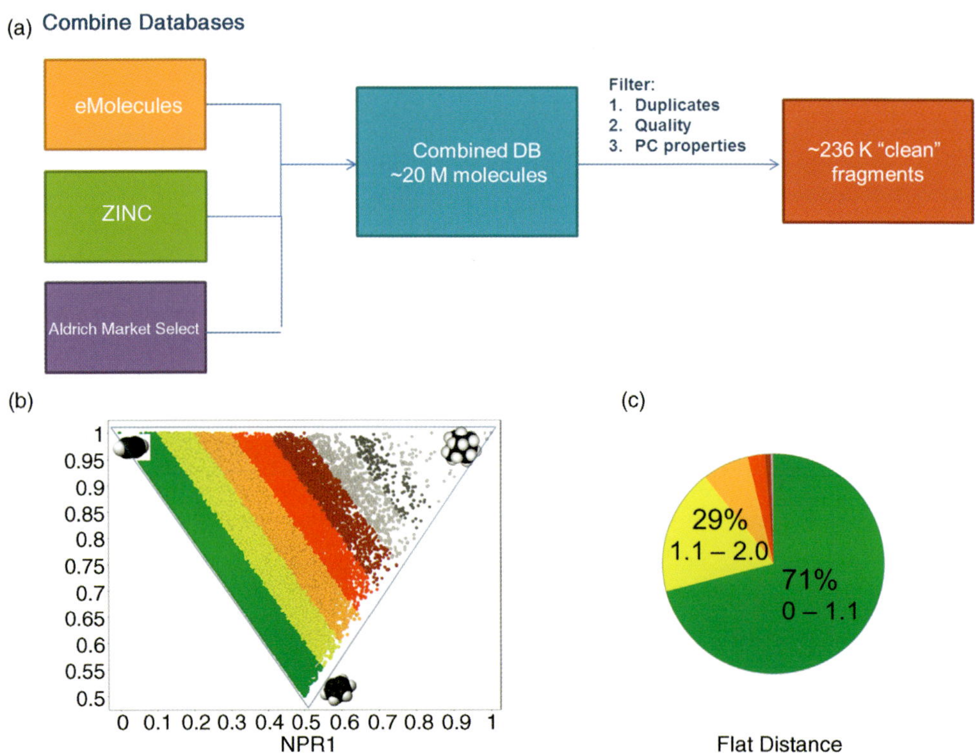


**Figure 5.3** (a) Process defined to select a "clean" fragments database. PMI plot (b) binned by PMIsum or flat distance (NPR1 + NPR2), and pie chart of flat distance distribution (c) of the ~236 000 commercially available fragments. (d) PipelinePilot protocol used for the filtering process.

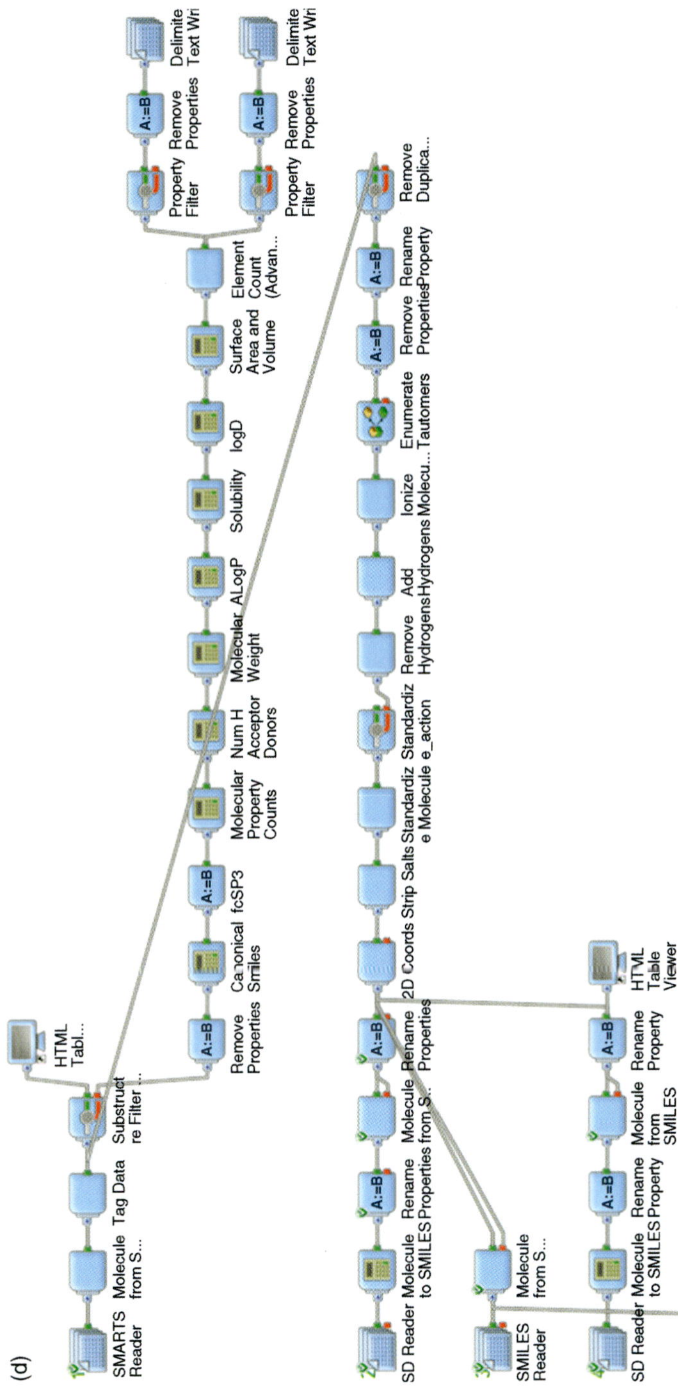


**Figure 5.3** (*Continued*)

### 5.3.5
### BDDP Fragment Library Design

The first step we undertook to enhance our fragment screening library was to improve the intraset chemical diversity. In order to maintain a "usable" fragment screening library, we undertook a phased approach to compound selection and initially focused our analysis on ~50% of the library, named FS2 (fragment set 2). Following comparison of FS2 with the 3D fragment foundation library, clustering, and a subsequent diversity selection, 181 compounds were selected to be part of the new, enhanced fragment library. Additional compounds to be added to these sets were selected from the 236 000 commercially available fragment set as outlined in Figure 5.4. In order to maintain a balanced shape distribution of our fragment library, we selected chemically diverse, pharmacophore-rich, and chemically attractive molecules from across the shape-diversity space. In part, we achieved this through restricting shape space to a PMIsum < /=1.5 (in practice, molecules from commercial sources with PMIsum>1.5 trended toward poor physicochemical profiles and failed our lipophilicity cutoff criteria as well as generally lacking attractive features for medicinal chemistry optimization), which also enabled us to maintain an acceptable level of complexity for the selected molecules. In addition to selecting molecules fitting the above criteria, we also developed Pipeline Pilot protocols to identify fragments that have a higher number of commercially available elaborated analogs in our virtual lead-like library. After removing all compounds too similar to the FS2 and 3D fragment foundation library followed by medicinal chemistry eyeballing, 198 compounds were selected. Following further diversity refinement and the addition of in-house synthesized fragments, this was boosted to 290 molecules. In total, 661 fragments (Figure 5.4a) have been selected to enhance the BDDP fragment screening set.

As can be seen in the PMI plots in Figure 5.4, the three-dimensional space exemplified by the enhanced fragment library (661 fragments, PMI plot in (c)) is more comprehensive than the initial BDDP fragment library (PMI plot in (b)) with approximately one-third (33%) of the fragments having a flat distance >1.1, compared to only 13% in the original set.

In addition to improving the shape-diversity profile of our library, we have also enhanced the overall chemical diversity (diversity coefficient calculated considering the average of the nearest neighbor fingerprint-based similarities – diversity coefficient (Maccs Keys) of 0.72, compared to 0.82 for the entire original fragment library and 0.80 for the original FS2 only), providing more singletons and thus greater coverage of chemical space. Application of radial fingerprints (ECFP4, more feature independent compared to Maccs Keys) also indicates an improved diversity coefficient (0.24) over the original set (0.37) with a similar difference in similarity values distribution. A maximum similarity (MaxSim) plot [21] enables the diversity of a library to be visualized. Nearest neighbor fingerprint-based similarities are calculated and the

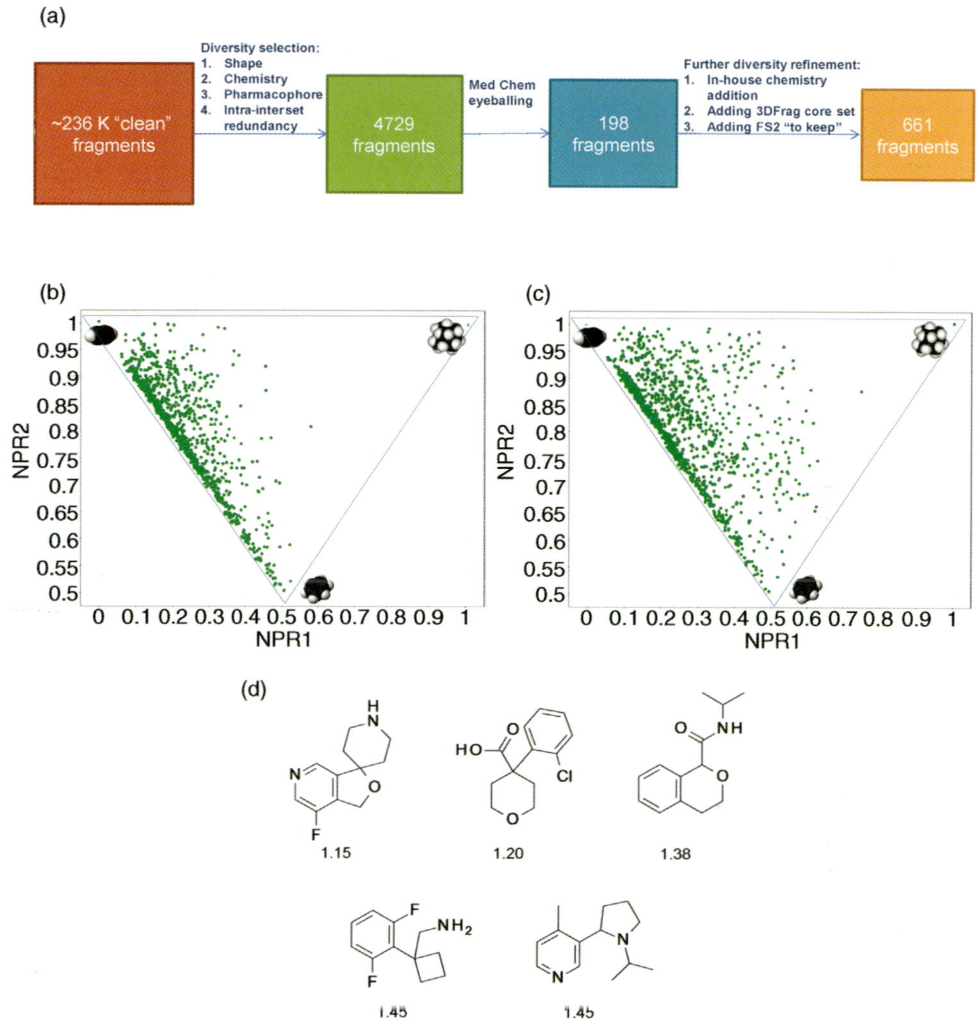


**Figure 5.4** (a) Process defined to select fragments to enhance diversity. PMI plots highlighting shape space covered by initial BDDP library (b) and diversity enhanced library (c). (d) Examples of compounds selected to enhance the library (taken from PMI plot in (c)); flat distances are reported under the structures.

frequency (count) plotted; the histogram for a library with a high degree of internal similarity will have most of the library at the greatest similarity, whereas a library with a low similarity will have a large part of the library at smaller similarities. Figure 5.5 shows the MaxSim plots for our original FS2 library (a) and our enhanced version (b); the original FS2 visibly presents the greatest similarity.

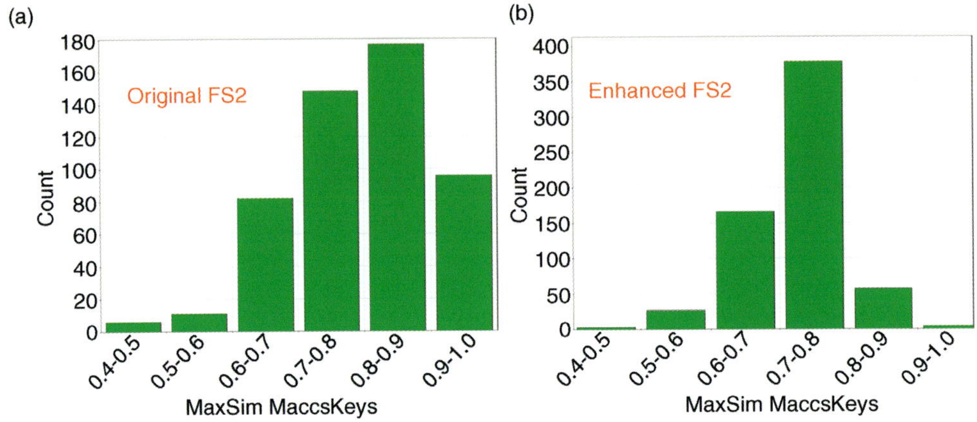


**Figure 5.5** Chemical diversity of the original FS2 library (a) and the diversity enhanced library (b).

Pharmacophore diversity of the enhanced FS2 (based on 3D pharmacophore fingerprints) was also improved with a diversity coefficient of 0.4 compared to a value of 0.51 for the original entire set. Chemical and pharmacophore diversity were also favorable compared to a commercially available "diversity" library composed of 1000 fragments. Table 5.1 shows the comparison between the two libraries, where three different diversity metrics (coefficients) are considered and the enhanced FS2 profile is favorable in all measures of diversity.

The shape distribution of the commercially available fragment diversity library was also considered and is shown in Figure 5.6a. The PMI space is not as well covered as compared to the BDDP enhanced fragment library, the vast majority having flat motifs (Figure 5.6b).

We also analyzed a commercially available 3D-biased fragment library and in this case, the majority (134, 65%) of the 205 fragments composing the set had a flat distance >1.1 but this is achieved at the expense of lower chemical diversity and greater molecule complexity (see fragment complexity section). In fact, the fingerprint-based (using Maccs Keys) diversity coefficient is 0.82 compared to 0.72 of our enhanced FS2.

**Table 5.1** Comparison of diversity coefficients between the BDDP enhanced library and a commercially available fragment diversity library.

| Diversity | Enhanced FS2 | Commercial library |
|---|---|---|
| Chemical – Maccs Keys | 0.72 | 0.78 |
| Chemical – ECFP4 | 0.24 | 0.33 |
| 3D Pharmacophore | 0.40 | 0.44 |

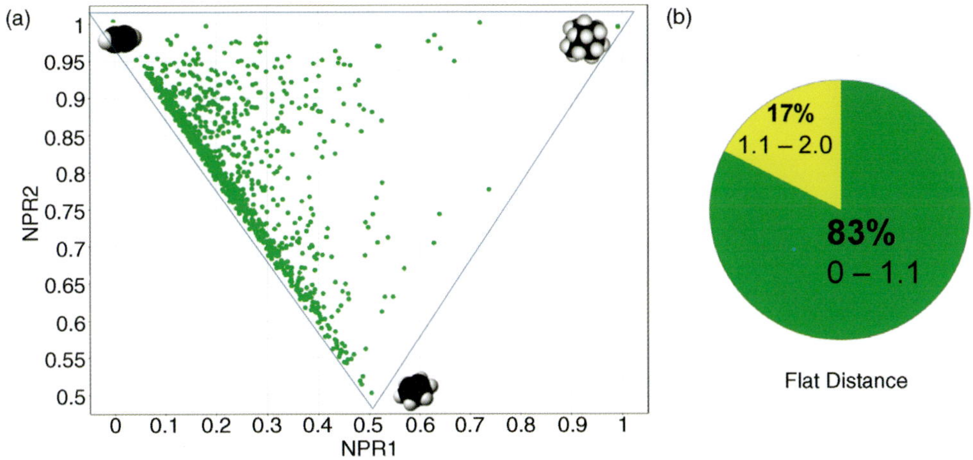


**Figure 5.6** Shape distribution of a commercially available "diversity" fragment library (a) and pie chart of flat distance distribution (b).

## 5.3.6 Fragment Complexity

Increasing the diversity of a fragment library has clear advantages, and we were keen to improve this facet but not at the expense of dramatically increasing molecule complexity. It is well known that reduced "interaction" complexity correlates with the increased probability of achieving binding to a chosen target [22]. We measured the complexity of our libraries using the method of Nilakantan [23] that is independent of atom, bond types, and connectivity and follows the instinctive view of molecular complexity. Figure 5.7 indicates this complexity measure plotted against the flat distance of four libraries: the original BDDP fragment library (a), the enhanced fragment library (b), a commercially available diversity fragment library (c), and a commercially available shape-biased fragment library (d). As can be seen, although we have increased shape diversity, this has not significantly increased molecule complexity (BDDP enhanced mean complexity score 42.4) compared to our original fragment set (BDDP mean complexity score 38.0) and this new library is considerably less complex than a commercially available shape-biased library (mean complexity score 59.4).

A recent publication highlights data for 145 fragment hits against a range of targets and the subsequent elaboration of these hits into leads [24]. We have calculated the complexity of these published fragment hits (removing molecules lying outside of the fragment property space in order to maintain consistency with our fragment library). Figure 5.8 highlights the complexity distribution of both sets of compounds indicating that our enhanced library (Figure 5.8a) has a

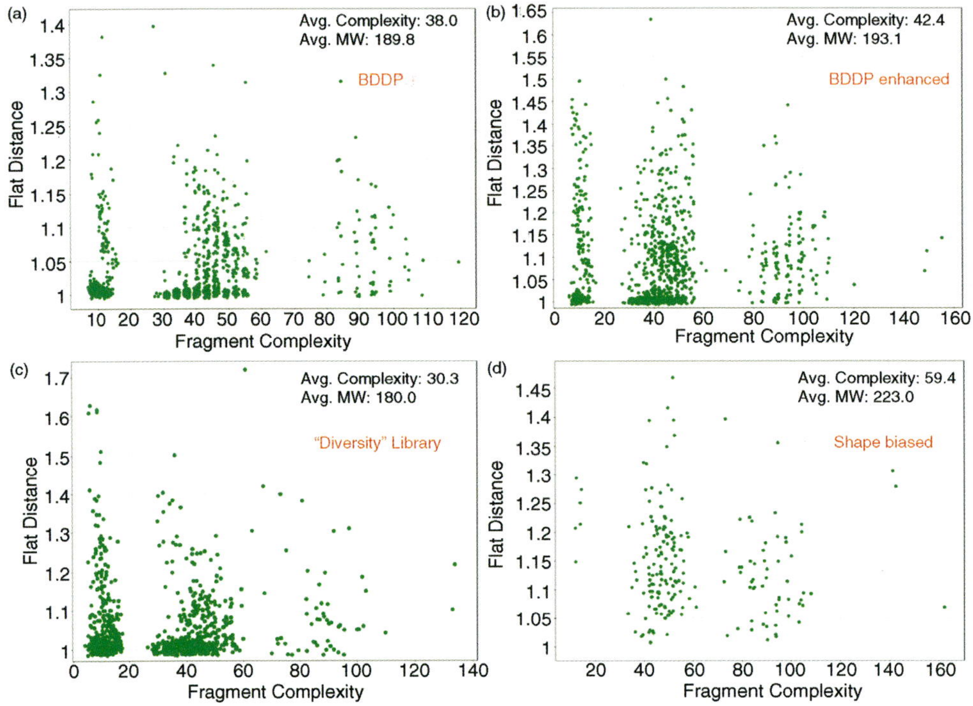


**Figure 5.7** The complexity measure of the BDDP fragment library (a) compared with the BDDP enhanced library (b), a commercially available diversity library (c), and a commercially available shape-biased library (d).

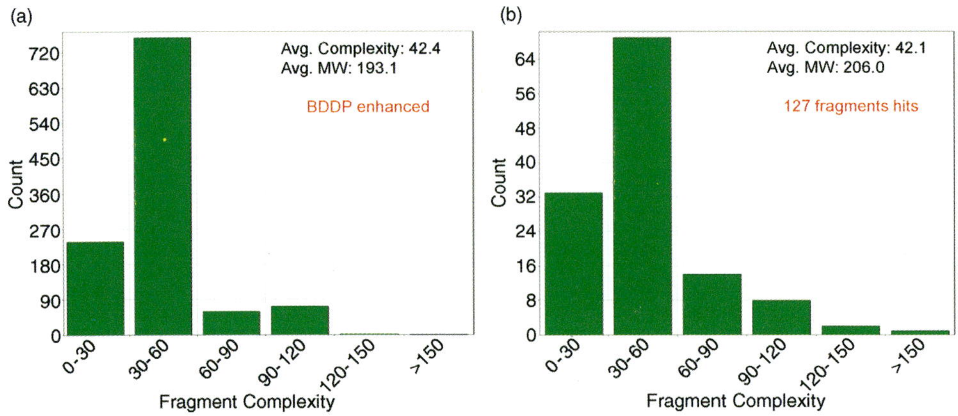


**Figure 5.8** Complexity of the diversity enhanced Beatson fragment library (a) compared with published fragment hits (b).

favorable profile versus quantified fragment hits against a broad range of drug discovery targets (Figure 5.8b).

#### 5.3.6.1 Diversity-Oriented Synthesis-Derived Fragment-Like Molecules

Perhaps, the greatest source of three-dimensional molecules is nature itself, providing a broad range of molecular scaffolds, active against numerous disease-relevant targets [25]. A key attribute of biologically validated natural products is that they are often rich in stereogenic centers, enabling them to populate areas of chemical space not accessed by many synthetic drugs. This attribute presents synthetic chemistry challenges that cannot easily be addressed in the timescales of a typical drug discovery effort. Diversity-oriented synthesis (DOS) may help to address this challenge through the synthesis of diverse molecules with natural product-like features in a very efficient manner [26]. DOS library compounds typically have more $sp^3$ character enabling them to explore more chemical space than standard lead-like HTS collections while maintaining good physicochemical properties. Identifying fragments from DOS-derived libraries presents a unique opportunity to access three-dimensional space not readily available through other means. Damian Young at the Centre for Drug Discovery at Baylor College of Medicine has published [27] on the synthesis of three-dimensional fragments using diversity-oriented synthesis (Figure 5.9). He and his team have used the build/couple/pair (B/C/P) algorithm to create structurally and stereochemically diverse bicyclic and spirocyclic fragments managing to keep the size of the compounds low (<300 Da). Despite the small molecular weight, the chemical complexity of the compounds was increased. Cheminformatics helped them to assess the shape and properties of the library showing that the property space covered was within the desired range of known

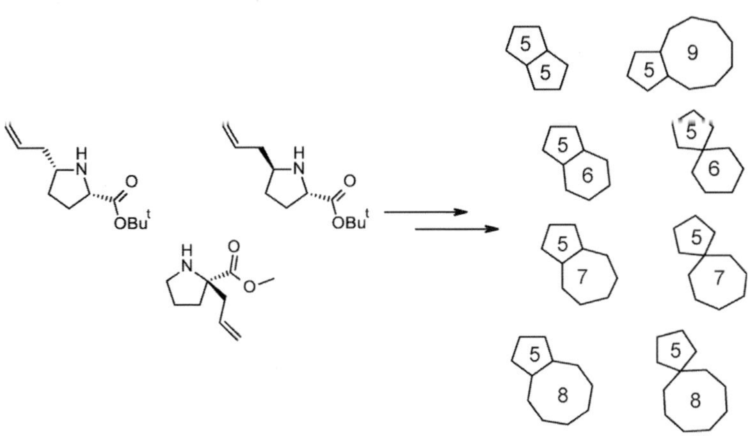


**Figure 5.9** Three-dimensional fragments prepared from mutual antecedents. Each of the fused rings and spirocycles can contain heteroatoms.

fragments; therefore, the set was amenable to a fragment-based screening approach.

## 5.4
## Future Plans

We have completed a thorough analysis of commercial fragment space and purchased a set of molecules that not only enhances the chemical diversity of our fragment library but also the three-dimensional diversity. While we have been able to purchase a number of diverse compounds, it is clear that commercially available libraries are limited in scope and in order to address this, we have initiated an analysis of compounds that report biological activity. The output from this work will form the basis of future synthetic chemistry work that will be undertaken in collaboration with other members of the 3D Fragment Consortium in order to facilitate compound delivery.

With this in mind, we have compiled a set of filtered compounds from ChEMBL (an open access database of biologically active drug-like small molecules) [28], ZINC "InMan" subset (an open access database of approved drugs and experimental compounds in man) [15] and SURECHEM (compounds from patent literature) [29], totaling around 2.5 million compounds. This set was fragmented using the RECAP algorithm, which is based on predefined synthesis-aware cleavage rules. The resulting fragments were filtered, removing unwanted functionalities and restricting the physicochemical property space and biasing to compounds with some "3D character," leaving approximately 7000 scaffolds that were subsequently enumerated (capping basic amine, acid, and alcohol functionalities) resulting in 660 000 structures that after another round of filtering were further reduced to ∼57 000 fragments. From this, a diverse set of 1000 molecules was selected using the process as outlined in Figure 5.10 that was reduced to 730 diverse, chemically attractive fragments following eye-balling. The PMI plot in Figure 5.11 shows the shape coverage of the set, visibly highlighting how fragments derived from biologically active compounds cover a vast range of shape space, considerably more than that covered by a commercial "diversity" and a shape-biased library. These molecules (730) will form the basis of future synthetic chemistry work to be undertaken both in-house and through collaboration across the 3D Consortium network. A paper detailing this work is currently being prepared.

Compounds produced through synthesis and commercial sources will be added to the Beatson fragment library and screened against our portfolio of drug discovery targets. The above strategies are designed to complement our current fragment screening set and on-going enhancement activities, increasing the library size to ∼2000 diverse molecules. In addition, as our knowledge and understanding of the features required to bind to and inhibit, for example, PPIs increases we will be able to use this information to shape the design of our future fragment screening sets.

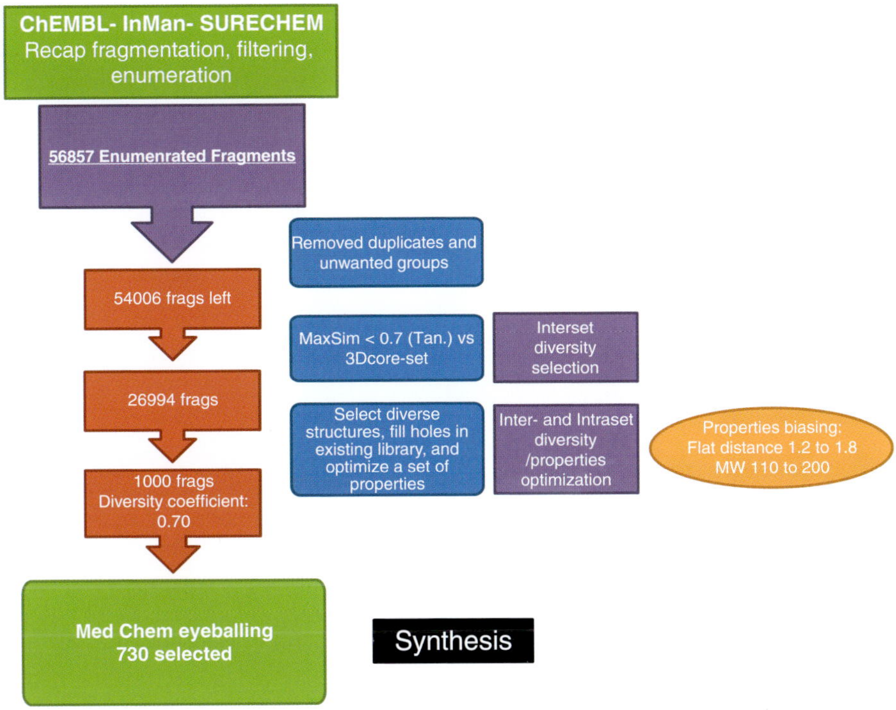


Figure 5.10 Process defined to identify fragments from "biologically active" space.

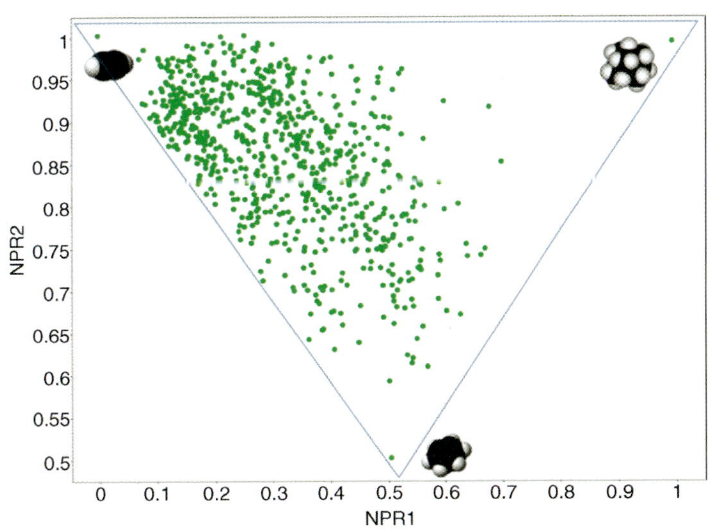


Figure 5.11 Shape diversity coverage of 730 fragments identified from "biologically active" scaffolds.

## 5.5
## Summary

In summary, we have been able to enhance the diversity of our fragment library through employing a number of descriptors for diversity, including one for shape, a feature not typically described when selecting fragments. This has enabled us to broaden our representation of "fragment chemical space" and will provide greater choice for future hit validation efforts and beyond, not only for the Beatson Drug Discovery Programme but also as an available fragment screening set for CRUK.

## 5.6
## Key Achievements

- Constructed a database of "filtered" commercially available fragment molecules covering a spectrum of chemical and shape-diversity space.
- Built a fragment set with enhanced chemical, pharmacophore, and shape diversity compared to our current library and a typical "diverse" commercial library.
- Shape coverage comparable with a shape-biased commercial library but achieving greater chemical diversity.
- Pharmacophore rich but size and molecular complexity not compromised. Designed and implemented *in silico* protocols to achieve all the above.

## References

1 Murray, C.W., Verdonk, M.L., and Rees, D.C. (2012) Experiences in fragment-based drug discovery. *Trends in Pharmacological Sciences*, **33**, 224–232.

2 Scott, D.E., Coyne, A.G., Hudson, S.A., and Abell, C. (2012) Fragment-based approaches in drug discovery and chemical biology. *Biochemistry*, **51**, 4990–5003.

3 Hubbard, R.E., Davis, B., Chen, I., and Drysdale, M.J. (2007) The SeeDs approach: integrating fragments into drug discovery. *Current Topics in Medicinal Chemistry*, **7**, 1568–1581.

4 Winter, A., Higueruelo, A.P., Marsh, M., Sigurdardottir, A., Pitt, W.R., and Blundell, T.L. (2012) Biophysical and computational fragment-based approaches to targeting protein–protein interactions: applications in structure-guided drug discovery. *Quarterly Reviews of Biophysics*, **45**, 383–426.

5 Congreve, M., Carr, R., Murray, C., and Jhoti, H. (2003) A 'rule of three' for fragment-based lead discovery? *Drug Discovery Today*, **8**, 876–877.

6 Walters, W.P., Green, J., Weiss, J.R., and Murcko, M.A. (2011) What do medicinal chemists actually make? A 50-year retrospective. *Journal of Medicinal Chemistry*, **54**, 6405–6416.

7 Ritchie, T.J. and Macdonald, S.J. (2009) The impact of aromatic ring count on compound developability: are too many aromatic rings a liability in drug design? *Drug Discovery Today*, **14**, 1011–1020.

8 Lovering, F., Bikker, J., and Humblet, C. (2009) Escape from flatland: increasing saturation as an approach to improving clinical success. *Journal of Medicinal Chemistry*, **52**, 6752–6756.

9 Lovering, F. (2013) Escape from Flatland 2: complexity and promiscuity. *MedChemComm*, **4**, 515.

10 Kenny, P.W. and Montanari, C.A. (2013) Inflation of correlation in the pursuit of drug-likeness. *Journal of Computer-Aided Molecular Design*, **27**, 1–13.

11 Stepan, A.F., Subramanyam, C., Efremov, I.V., Dutra, J.K., O'Sullivan, T.J., DiRico, K. J., McDonald, W.S., Won, A., Dorff, P.H., Nolan, C.E. *et al.* (2012) Application of the bicyclo[1.1.1]pentane motif as a nonclassical phenyl ring bioisostere in the design of a potent and orally active gamma-secretase inhibitor. *Journal of Medicinal Chemistry*, **55**, 3414–3424.

12 Hung, A.W., Ramek, A., Wang, Y., Kaya, T., Wilson, J.A., Clemons, P.A., and Young, D.W. (2011) Route to three-dimensional fragments using diversity-oriented synthesis. *Proceedings of the National Academy of Sciences of the United States of America*, **108**, 6799–6804.

13 Biofocus shape enhanced fragment libraries, www.biofocus.com/home.htm (accessed 25 October 2012).

14 Sauer, W.H. and Schwarz, M.K. (2003) Molecular shape diversity of combinatorial libraries: a prerequisite for broad bioactivity. *Journal of Chemical Information and Computer Sciences*, **43**, 987–1003.

15 ZINC, "InMan" subset, http://zinc.docking.org/subsets/zim (accessed 5 May 2012).

16 Lewell, X.Q., Judd, D.B., Watson, S.P., and Hann, M.M. (1998) RECAP–retrosynthetic combinatorial analysis procedure: a powerful new technique for identifying privileged molecular fragments with useful applications in combinatorial chemistry. *Journal of Chemical Information and Computer Sciences*, **38**, 511–522.

17 Rush, T.S. 3rd, Grant, J.A., Mosyak, L., and Nicholls, A. (2005) A shape-based 3-D scaffold hopping method and its application to a bacterial protein–protein interaction. *Journal of Medicinal Chemistry*, **48**, 1489–1495.

18 Sastry, G.M., Dixon, S.L., and Sherman, W. (2011) Rapid shape-based ligand alignment and virtual screening method based on atom/feature-pair similarities and volume overlap scoring. *Journal of Chemical Information and Modeling*, **51**, 2455–2466.

19 Ballester, P.J. and Richards, W.G. (2007) Ultrafast shape recognition to search compound databases for similar molecular shapes. *Journal of Computational Chemistry*, **28**, 1711–1723.

20 Morley, A., Pugliese, A., Birchall, K., Bower, J., Brennan, P., Brown, N., Chapman, T., Drysdale, M., Gilbert, I.H., Hoelder, S., Jordan, A., Ley, S.V., Merritt, A., Miller, D., Swarbrick, M.E., and Wyatt, P.G. (2013) Fragment-based hit identification: thinking in 3D. *Drug Discovery Today*,. **18**, 1221–1227.

21 Canvas, version 1.5, Schrödinger, LLC, New York, NY (2012).

22 Hann, M.M., Leach, A.R., and Harper, G. (2001) Molecular complexity and its impact on the probability of finding leads for drug discovery. *Journal of Chemical Information and Computer Sciences*, **41**, 856–864.

23 Nilakantan, R., Nunn, D.S., Greenblatt, L., Walker, G., Haraki, K., and Mobilio, D. (2006) A family of ring system-based structural fragments for use in structure–activity studies: database mining and recursive partitioning. *Journal of Chemical Information and Modeling*, **46**, 1069–1077.

24 Ferenczy, G.G. and Keseru, G.M. (2013) How are fragments optimized? A retrospective analysis of 145 fragment optimizations. *Journal of Medicinal Chemistry*, **56**, 2478–2486.

25 Koehn, F.E. and Carter, G.T. (2005) The evolving role of natural products in drug discovery. *Nature Reviews Drug Discovery*, **4**, 206–220.

26 Schreiber, S.L. (2000) Target-oriented and diversity-oriented organic synthesis in drug discovery. *Science*, **287**, 1964–1969.

27 Hung, A.W., Ramek, A., Wang, Y., Kaya, T., Wilson, J.A., Clemons, P.A., and Young, D.W. (2011) Route to three-dimensional fragments using diversity-oriented synthesis. *Proceedings of the National Academy of Sciences of the United States of America*, **108**, 6799–6804.

28 ChEMBL, http://www.ebi.ac.uk/chembl/ (accessed 20 October 2012).

29 SureChem, sureche.com (accessed 7 December 2012).

# 6
# The Synthesis of Biophysical Methods In Support of Robust Fragment-Based Lead Discovery

*Ben J. Davis and Anthony M. Giannetti*

## 6.1
## Introduction

The successful application of fragment-based lead discovery (FBLD) requires the ability to robustly and rapidly detect and quantitate the association of low affinity, low molecular weight interactions. Fragment molecules generally contain fewer than 15–18 nonhydrogen atoms and often have dissociation constants ($K_D$) in the range of high micromolar to millimolar. Detection of protein/small molecule interactions in this affinity range can be challenging and is typically the domain of biophysical techniques. While high-concentration enzyme screening can work, it is frequently plagued by high false positive rates [1,2]. The collection of biophysical data, especially near the limit of detection, has historically been a slow and painstaking process, usually carried out by experts in that technique. However, to have impact in a pharmaceutical discovery environment, any technique needs to be fast, reliable, and trustworthy enough to impact medicinal chemistry design in real time. Since the mid-1990s, extensive effort has gone into the development of biophysical techniques to support FBLD within drug discovery environments. In particular, significant advances in NMR, SPR, and crystallography methods, software, and hardware, as well as garnering increased familiarity and acceptance by chemists, have allowed these techniques to emerge as the primary supporters of FBLD efforts. Additional technologies such as thermal melt, weak affinity chromatography (WAC), isothermal titration calorimetry (ITC), microscale thermophoresis, and others are being explored, improving or finding routine implementation as well. Extensive reviews on the specifics of each technique, especially as applied to FBLD, are detailed elsewhere [3–15] and are beyond the scope of this chapter, which will focus on the integration of multiple techniques in advancing FBLD.

Historically, most FBLD campaigns have been driven by one biophysical technique (e.g., NMR or SPR) in conjunction with X-ray crystallography for structure determination. As FBLD was relatively new, most organizations, both academic and industrial, did not choose to invest in large-scale acquisition and deployment of multiple expensive specialized hardware platforms and the teams to run

*Fragment-based Drug Discovery: Lessons and Outlook*, First Edition. Edited by Daniel A. Erlanson and Wolfgang Jahnke.

them, instead often favoring fewer techniques and smaller teams. This has allowed for specialists within each area to refine and push their technologies in the directions demanded by FBLD. In the first few years at Genentech, for example, we were able to prosecute over 40 screening campaigns by SPR allowing for large-scale analysis of hit selection behaviors, refinement of screening cascades, methods, software, and hardware. With increasing experience, the various capabilities and limitations of each method as they relate to FBLD have become more clear, in particular the challenges associated with working at the edge of detection for most methods.

While these siloed screens were useful for establishing technology platforms, we believe that the combination and examination of results from multiple platforms is a substantially more useful activity. Each platform has strengths and weaknesses near the limit of detection, and combining the results allows a more comprehensive picture of the system's behavior to emerge. This is particularly important early in a discovery program during assay development to address concerns regarding nonspecific binding, assignment of mechanism of action for hits, choice of screening buffers/conditions, cross-target specificity, and especially comparing the behavior of mutant, surrogate, or other nonwild-type engineered systems. The most frequent challenge in FBLD is that most biophysical techniques do not give binary "yes or no" answers regarding binding, but instead generate rich and complex data that requires expert interpretation. SPR gives time-resolved data, including apparent stoichiometry, even when that binding does not look "clean." Similarly, NMR gives complex spectra where small deviations between spectra can be meaningful. Working near a technique's detection limits combined with intrinsically feature-rich data can mean that features appear in the data that can be hard to interpret but may indicate that something might not be "right" with the experimental setup. In these cases, more techniques, expertise, and care in asking the right question can yield insights into the protein system, and consequently the ligand screen, which can be extremely enlightening. However, care and time must be spent understanding these apparent artifacts in order to achieve the goal of robust, high throughput, assays for supporting screening and hit-to-lead generation.

The complexity of the data and difficulty in assigning hit/nonhit categories makes the comparison of hits from screens using different techniques (or where one technique has characterized the screening hits from a different technique) challenging. If the screens are uncoordinated, the variance in conditions, constructs, and so on may introduce real and meaningful differences in hit sets. When hits agree between techniques [2], the call on what to do next is straightforward; however, the choices of what to do when there are discrepancies is more challenging. In some cases, a fragment may be undetectable by one technique, a random false negative, or real information may be offered by the new technique. For example, SPR may identify a binder where NMR would identify that the sample has degraded and the ligand does not have the expected chemical structure. In such a case, there may be a real hit in play, but additional understanding and refinement is necessary. This was described in a case where an old, orange-colored, high-concentration stock of a compound exhibited binding to HCV NS5B by SPR and crystallography [16]. However, the initial electron density was not consistent with the expected structure, and fresh colorless samples from powder had no observable binding by SPR at any tested concentration. By aging a fresh sample under fluorescent light until it turned orange, binding was restored. The analytics group determined the structure of the parent and degraded compounds for final interpretation of the X-ray structure. The degraded compound was chemically unattractive but defined a novel binding motif for the protein. Computational chemists designed a series of new, more medicinally relevant, scaffolds capable of preserving the novel interaction and developed new series from there. By doing the assay development for all platforms, both screening and analytical, in an open and coordinated manner, the screening groups could focus on the questions related to getting a reliable screen that identifies chemical matter of a desired mechanism of action, under conditions where binding was meaningful to all techniques involved.

Through the rest of this chapter, we will describe examples where the interplay of multiple techniques and data early in the project helped identify and further characterize difficulties with the protein system or lead molecules. In each case, one technique may have been sufficient to identify an issue with the assay, but in all cases the detailed coordination of activities from all experiments and biophysical/screening experts was needed to push the project forward. Given the emerging utility of FBLD for developing not only drugs but also tool compounds, it is our hope that more organizations will drive toward integrated broad biophysical approaches not only for FBLD but also for understanding small molecule binding and inhibition for all systems.

## 6.2 Fragment-Based Lead Discovery on a Difficult Kinase

Inhibition of kinases is generally considered straightforward to achieve, given the large number of known kinase inhibitors, kinase binding motifs, and known determinants of specificity. In practice though, some kinases exhibit resistance to lead finding and produce low hit rates and yield hits with low affinities and ligand efficiencies (LE) in high-throughput screening, FBLD, and other lead-finding approaches. We sought to develop inhibitors of an undisclosed kinase, referred to here as Kin1, but were met with low hit rates and low hit quality in HTS and FBLD despite identifying a number of potent, though unselective, inhibitors through kinase selectivity panels. An SPR-based fragment screen of 2500 Genentech fragments yielded approximately 100 binders. However, none of them had an affinity better than 150 μM or LE significantly greater than 0.3. This stands in contrast to all our other kinase screens, which typically resulted in observed LEs closer to 0.4–0.5, with single-digit micromolar affinities. Crystallographic success in obtaining fragment-bound structures was also poor with >90% of soaks observed to be empty. At this stage, we sought to reevaluate the

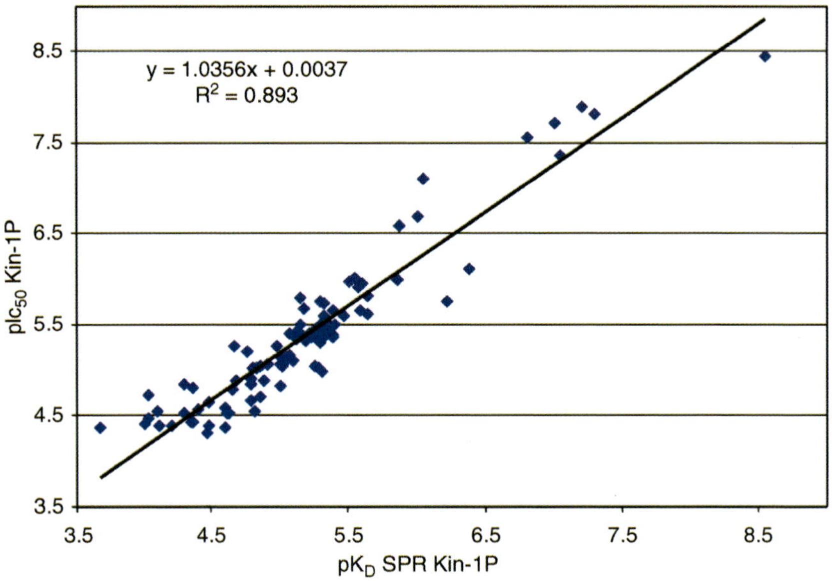


**Figure 6.1** High degree of correlation between SPR binding affinity and biochemical inhibition for Kin-1P compounds.

protein platform and assays by multiple biophysical and biochemical approaches to understand the source of the challenges with the system.

Our high-throughput cloning and expression approaches produced a variety of Kin1 constructs. Biochemical assay development was initiated before other activities and used highly purified Kin1-1P (wild-type kinase domain with a single phosphorylation site). SPR was also carried out using this protein to maintain assay alignment between the biophysical (binding) experiment and the biochemical (inhibition) experiment. Figure 6.1 shows the correlation of ~60 compounds based on their SPR binding (X-axis) versus their observed $IC_{50}$ in an enzyme inhibition reaction. This establishes a tight correlation over several orders of magnitude between the binding ATP-competitive molecules and the degrees of inhibition. Similarly, the SPR assay robustness was assessed by measuring 40 molecules in duplicate (different machine, operator, and protein prep) (Figure 6.2) and finding an extremely tight correlation indicating assay reproducibility.

Due to challenges with toxicity when expressing Kin1at large scale, X-ray and NMR activities were performed using the Kin1-DN constructs, which harbors an aspartate to asparagine substitution. This less toxic protein expresses and purifies easily in large quantities. However, when SPR testing was performed a systematic shift toward weaker binding for the Kin1-DN mutant was observed over Kin1-1P for ~40 compounds. If the variance in the SPR data was due to random noise, we would expect some compounds to be more potent on the mutant; however, all compounds were weaker or only close to equipotent

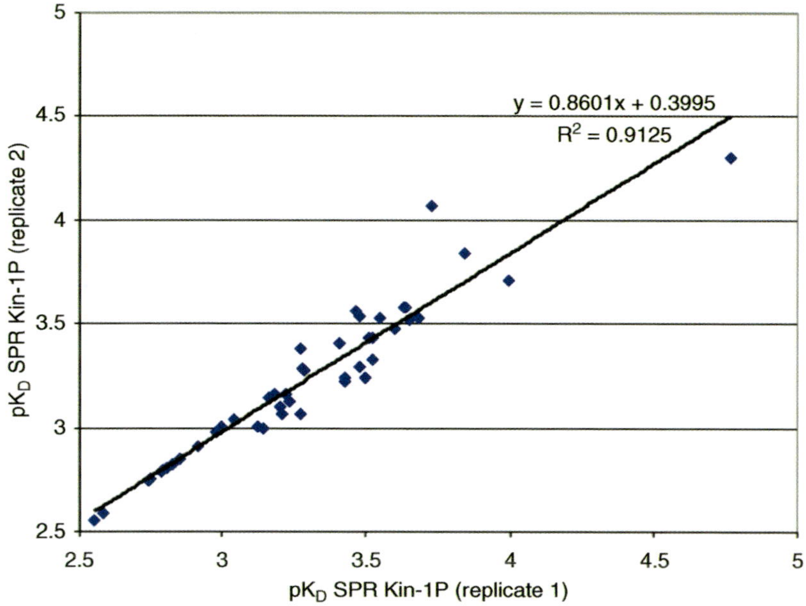


**Figure 6.2** Correlation between replicate measurements of compounds by SPR on Kin-1P.

suggesting overall weaker binding to this construct; this was observed across multiple different protein preparations (Figure 6.3). To further refine this result, we carried out $IC_{50}$ experiments using the Kin1-DN mutant in place of the Kin1-1P in the enzymatic assay. In this case as well, inhibition was systematically weaker relative to Kin-1P $IC_{50}$ or SPR $K_D$ measurements (data not shown). The use of the Kin1-DN protein in crystallization may explain the high empty structure rate for that protein. The already weak fragment hits are further weakened by several fold on the Kin1-DN protein making production of fully occupied complexes for crystallization that much more difficult to achieve.

Owing to the difficulty in producing large quantities of Kin1-1P for NMR, a pilot NMR fragment screen was attempted using the Kin1-DN protein. Three independent NMR measurements were made (STD, water-LOGSY, and $T_2$-filtered) to minimize false positives and negatives and observe the system from several perspectives [8]. Several issues were quickly noted. The hit rate appeared to be extremely high with multiple compounds in the mixtures of 12 appearing to bind simultaneously. This is at odds with the SPR observation of an unusually low hit rate. Additionally, control compounds, such as adenine, showed weak binding in the STD experiment relative to more wild-type constructs (Figure 6.4). Most perplexing was the "inversion" of the LOGSY NMR spectra (Figure 6.5) such that resonances in the LOGSY spectrum that are expected to get more negative after competition become more positive, while STD and CPMG NMR experiments indicate reduced binding. While the meaning of a shift in LOGSY

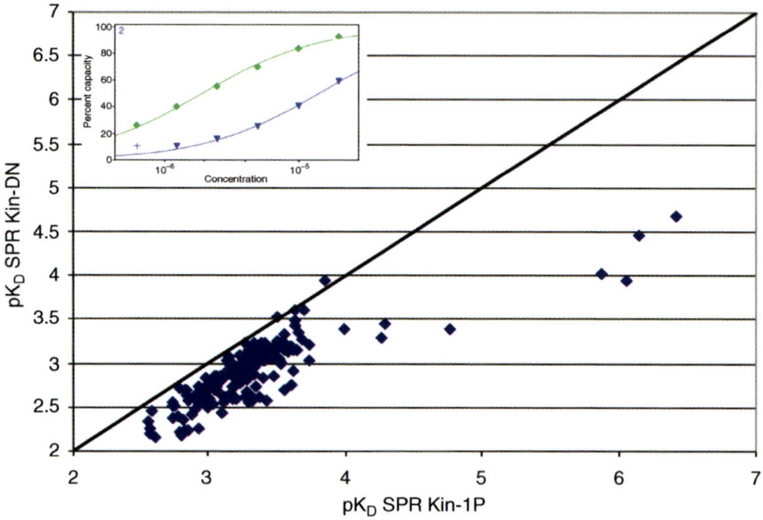


**Figure 6.3** Comparison of SPR binding on Kin-1P versus Kin-DN. Note the systematic weakening of the point cloud below the 1 : 1 line (solid black). Inset shows the dose–response curve for a control compound binding the Kin-1P (green) versus the Kin-DN (blue) showing a ~7-fold weakening in affinity.

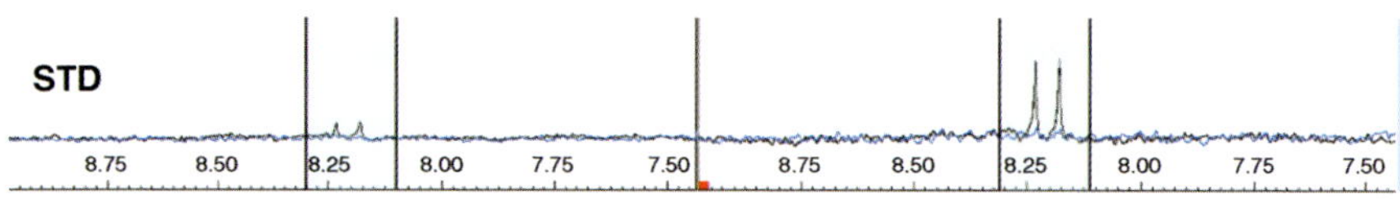


**Figure 6.4** STD binding of adenine to Kin-DN (left) versus Kin-TA. Note the substantial improvement in signal intensity. Black: adenine alone. Blue: adenine + 30 μM staurosporine.

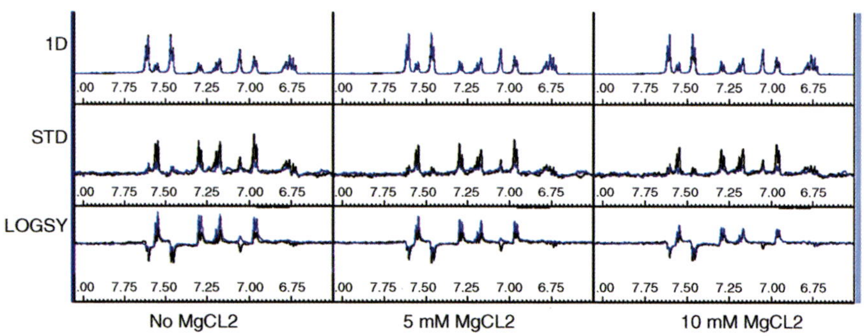


**Figure 6.5** 1D, STD, and water-LOGSY NMR spectra of fragment mixtures binding Kin-TA in the absence (left, pH 8.0) or presence of 5 mM (middle, pH 7.5) or 10 mM $MgCl_2$. Magnesium suppresses nonspecific binding and reduces artifactual LOGSY spectra, while the STD spectra are unaffected. Blue: same as black but after addition of 30 μM staurosporine.

experiments may not be immediately mechanistically understood, the result was clear that there were behaviors in the system we did not expect and we should pause the screen and evaluate the assay components more thoroughly.

Given the number of disconnects between NMR, SPR, and enzymology, an additional biophysical technique and protein construct (Kin1-TA) were introduced to further characterize the differences between proteins and experimental conditions. Kin1-TA has a substitution that results in higher expression, but the substitution is more conservative and remote from the ATP site than the Kin1-DN substitution. ITC (isothermal titration calorimetry) was used to assess a small number of tool compounds to several protein preparations. When a control compound was tested on Kin1-DN, very little heat was evolved and a binding affinity could not be determined. This was despite the same protein preparation giving consistent and clean binding and inhibition in SPR and enzyme inhibition assays, respectively. However, the titrations of the same compound on Kin1-1P (not shown) or Kin1-TA (Figure 6.6) showed identical results and a larger amount of heat displayed. Given the enthalpy profile disconnect between the Kin1-DN protein and the other two forms, we decided to abandon the Kin1-DN protein for further screening and characterization. Changing protein solved the odd behavior with the LOGSY experiment as well as significantly increased the signal to noise for positive controls in the STD experiment. However, nonspecific binding and high hit rates continued to be an issue. The inclusion of ITC at this juncture was critical as it obviated a change in the binding mechanism that neither SPR nor

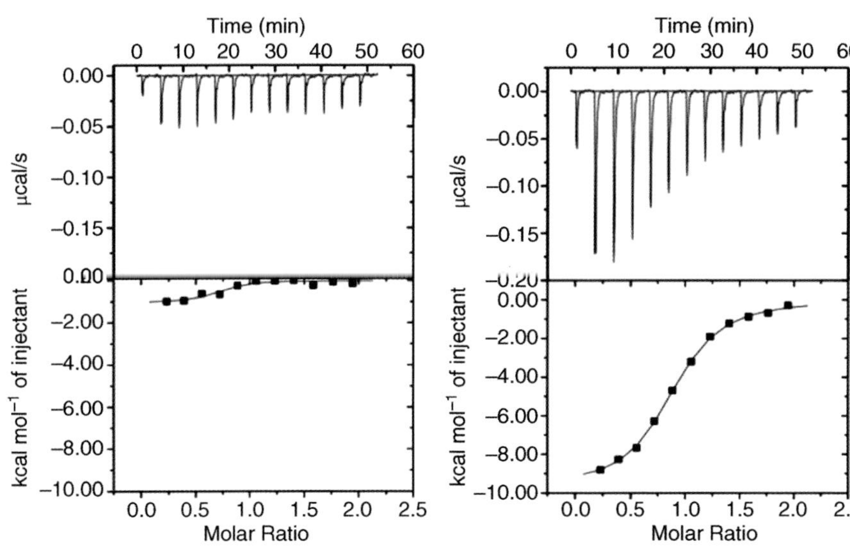


**Figure 6.6** Titration of a control compound onto Kin1-DN (a) or Kin1-TA (b) proteins. Despite showing potent binding to both of these protein batches by SPR, the thermodynamics of the interaction appear to be altered with little heat being evolved by the Kin1-DN mutation.

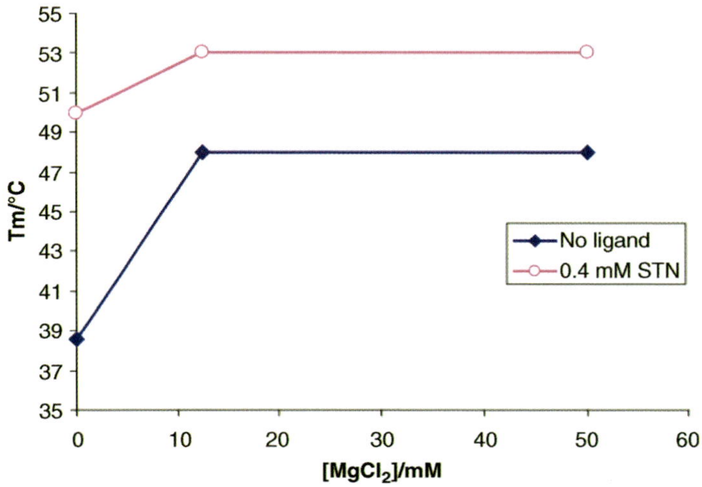


**Figure 6.7** Observed thermal melt temperature versus magnesium concentration. Even with the addition of 400 uM staurosporine (pink), the Kin1-TA protein continues to show an increase in thermal stability with magnesium.

enzyme reactions could observe. Changes in thermodynamic parameters that leave affinity and binding kinetics untouched have been observed before [17] but indicate a change in the system that could have significant ramifications for understanding binding and, at worse, drive potentially unproductive diversions in medicinal chemistry pursuing subtle, but real, artifacts.

During assay review, it was noted that the SPR and biochemical assays used 10 mM $MgCl_2$ as part of the buffer, whereas the initial NMR and crystallization experiments used none. Although this is rarely an issue with kinases when not using ATP as a substrate, we investigated the potential effects of magnesium in more detail. We added the biophysical technique of protein thermal melt experiments to assess the effects of magnesium on protein stability. The Kin-TA protein showed a rise in protein stability from ~38 C to 48 C by increasing $MgCl_2$ from zero to 10 mM (Figure 6.7). Interestingly, a rise in stability of ~4 degrees was noted when the protein was prebound by saturating amounts of staurosporine, indicating a stabilizing role for magnesium even when the protein is prebound to a potent ligand ($K_D \sim 1$ nM for Kin1-TA). 1D NMR spectra show the titration of a small set of protein peaks upon titration of $MgCl_2$ (not shown). Taken together, these biophysical data indicate that magnesium is important to the structure of Kin1 and that its elimination destabilizes the protein and leads to an increase in nonspecific binding. While this is not observed for many kinases, it indicates the importance of systematically evaluating all buffer components before elimination, as well as coordinating the assay development and conditions across all experiment types. The NMR screen of the Vernalis fragment library performed with Kin1-TA protein and 10 mM $MgCl_2$ yielded hit rates similar to the SPR screen of the Genentech library. In addition, under these conditions the NMR confirmation of SPR hits from the Genentech library, and the SPR confirmation of NMR hits from the Vernalis library, showed ~99% concordance indicating few false positives for either technique. Working together, both techniques contributed to advancing fragment hits to late-stage chemical series.

## 6.3 Application of Orthogonal Biophysical Methods to Identify and Overcome an Unusual Ligand: Protein Interaction

The peptidyl-prolyl *cis-trans* isomerase Pin1 is a small protein that catalyzes the *cis-trans* isomerization of pSer/pThr-Pro peptide bonds [18]. Pin1 has been implicated as having a role in a wide range of biological processes, and was the molecular target in a number of campaigns to develop anticancer therapeutics [19,20]. In one of these, fragment-based hit generation was used to identify and develop a series of a-benzimidazolyl-substituted amino acids. Although these compounds inhibited Pin1 activity with an $IC_{50}$ below 100 nM, they were inactive on cells. However, the process by which these compounds were identified provides an example of how synergy between orthogonal biophysical methods can be used to identify and overcome potential issues in the development of high-affinity ligands.

The initial screen of a curated library of approximately 1200 well-characterized fragments [21] was performed using 1D ligand observed NMR methods. Three independent NMR experiments were performed on each sample (STD, water-LOGSY, and $T_2$-filtered) in order to minimize the potential for false positives or false negatives arising from experimental artifacts in any of these experiments [22]. A high-affinity competitor was then added in order to displace any potential ligands, and a decrease in binding signal (i.e., a decrease in the signal in the STD experiment, a shift toward negative signal in the water-LOGSY experiment, and an increase in the signal in the $T_2$-filtered experiment) was interpreted as being consistent with the decrease in the bound population of the ligand and a concurrent increase in the free population of the ligand. Displacement by a potent competitor also provides an indirect indication of binding to the active site of interest.

In this screening campaign, the fragment library was screened twice using different competitor molecules, the D-peptide inhibitor **1** [23] and the phenylalaninol phosphate **2** [24] (Figure 6.8). Given the low tractability of the target site, this redundancy was deemed necessary in order to minimize the risk that any putative ligands identified by the screen were false positives. Compounds that showed binding and displacement by both competitors in all three of the ligand-observed NMR binding experiments were further

**Compound 1**

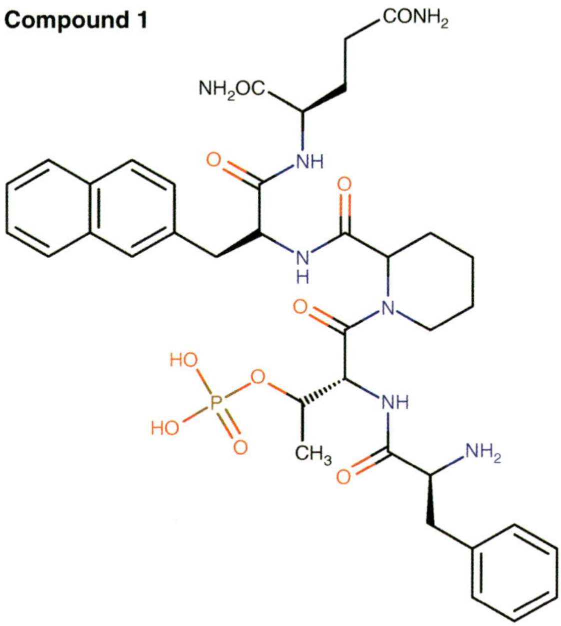

**Compound 2**

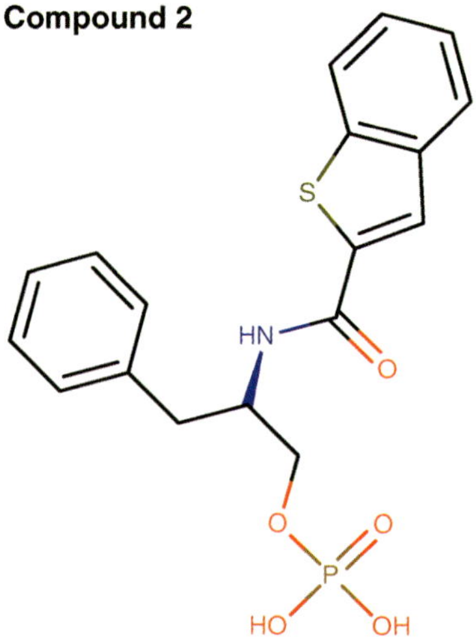

**Compound3**

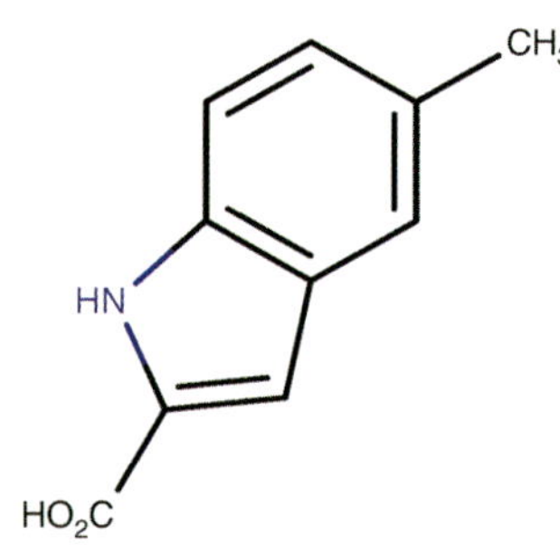

**Figure 6.8** Compounds involved in the FBLD campaign against Pin1. Compounds **1** and **2** are potent literature ligands for Pin1 [23,24]; compound **3** was identified as a hit from the Vernalis fragment library [21].

characterized by orthogonal biophysical methods before incorporation into a medicinal chemistry campaign. Compounds that did not show this behavior were also characterized, but as a lower priority; the validation rate of this set of compounds was substantially lower than that of the set of compounds that showed idealized behavior in each of the experiments.

In total, nine preliminary fragment hits were identified, a hit rate of 0.75%, and were studied further using a range of biophysical techniques. This typically consisted of protein observed NMR ($^{15}$N-$^{1}$H HSQC) titrations, SPR, ITC, and crystallography. Five fragment hits were validated as ligands across this suite of experiments, a final hit rate of 0.4%, and these formed the basis for the medicinal chemistry effort within the project.

One fragment hit in particular, compound **3**, was identified as being of particular interest, based on its chemical tractability, binding mode, and relatively high potency in a coupled functional PPIase activity assay with a $K_i$ of 25 μM (LE 0.48). In order to fully validate this fragment, further biophysical characterization was performed.

A crystal structure of compound **3** was obtained using a readily crystallizable mutant of Pin1, Pin1 R14A, and the indole moiety of compound **3** was observed to fit snugly into a hydrophobic pocket close to the active site (Figure 6.9); the R14A mutation is distal to the ligand binding site. The indole NH forms a hydrogen bond to the side chain of C113, while the carboxylic acid forms interactions with K63 and an extensive water network in the phosphate binding pocket of the active site.

Protein-observed NMR experiments were performed, using 2D $^{15}$N-$^{1}$H HSQC spectra at constant protein concentration and titrating compound **3** to a solubility-limited concentration of 2.5 mM. The observed chemical shift perturbations were consistent with the crystal structure, localized around the hydrophobic pocket-forming part of the active site. However, analysis of the magnitude of the chemical shift perturbations caused by titration of compound **4** gave a $K_D$ value of ~ 280 μM (LE 0.37), significantly higher than the $K_i$ obtained from the coupled functional assay.

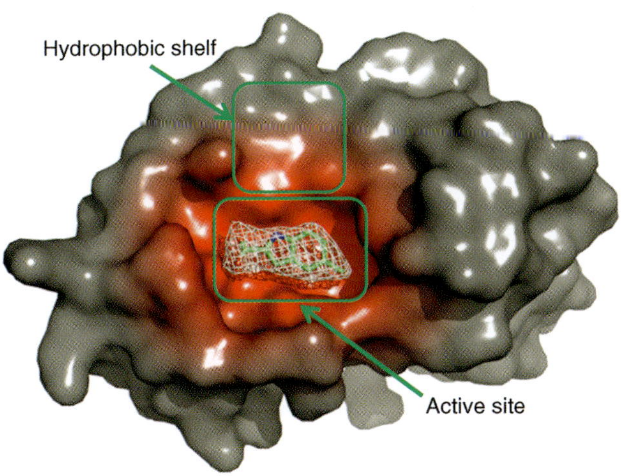


**Figure 6.9** The crystal structure of compound **3** bound to Pin1 R14A. Electron density is shown for the ligand. The active site and "hydrophobic shelf" targeted by evolved compounds are highlighted.

SPR experiments were performed using double-His tagged PPIase domain [25]; the double His tag was found to significantly improve immobilization to a Ni-NTA chip and to reduce the degree of drift in the baseline during the SPR experiments. While most of the fragment series were observed to give conventional responses in the SPR experiments, compound **3** was observed to bind anomalously; at a concentration of 400 μM, the stoichiometric ratio was estimated to be 2 : 1 ligand:protein. This stoichiometric ratio increased up to the limiting solubility of the compound, and no robust $K_D$ value could be determined.

The discrepancy between the relatively high activity of compound **3** in the functional assay and the lower affinity observed in the protein-observed NMR experiments, the well-defined binding mode observed in both the NMR and the crystallographic experiments and the anomalous behavior identified in the SPR experiments could not immediately be reconciled. Since relatively few validated hits had been obtained from the fragment screen against Pin1, and given the chemical tractability of compound **3**, there was interest in characterizing the system further in order to test whether the compound was suitable as a start point for a medicinal chemistry program to inhibit Pin1 activity.

A clue as to the physical processes underlying the observed behavior of compound **3** came from crystallographic studies on a series of related compounds. In these crystal structures, electron density was frequently observed showing a second molecule stacking on top of the molecule bound in the primary binding site (Figure 6.10). These compounds also showed superstoichiometric binding in the SPR experiments, but perturbed only a single region of the protein in $^{15}N$-$^{1}H$ HSQC NMR experiments.

A consistent feature of the compounds that formed this stacking interaction was the formation, on binding of the first molecule into the active site, of an extended planar region of hydrophobic surface. Additional molecules then

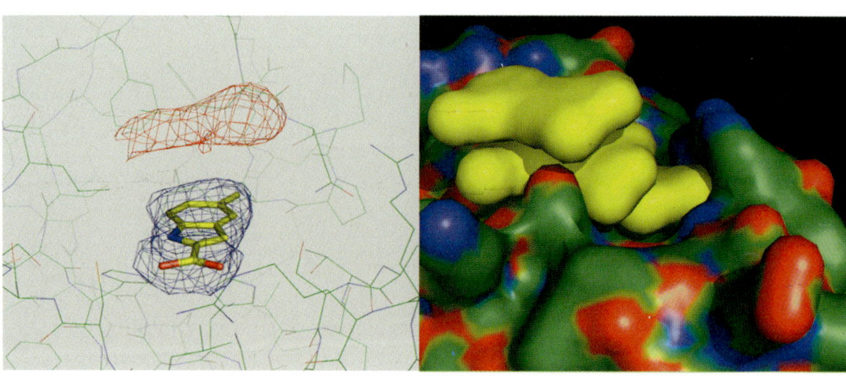

(a) (b)

**Figure 6.10** (a) Crystal structure of compound **3**, showing weak electron density consistent with a second ligand molecule binding on top of the first molecule. (b) Crystal structure of a related molecule, where clear stacking of two additional molecules on top of the primary site was observed.

appear to interact with this extended hydrophobic surface, at least in the crystal lattice, and data from other biophysical techniques (in particular NMR and SPR) were consistent with this model.

In order to test this hypothesis, and in an attempt to overcome these issues, a series of compounds was synthesized where groups were added in order to break the planar hydrophobic surface formed by the binding of the first molecule to the primary binding site. As these molecules were extended further to the "hydrophobic shelf" region, the observed overbinding in the SPR experiments decreased dramatically and affinities obtained from a variety of methods converged and gave consistent values. The series eventually gave rise to compounds with double-digit nM affinity and cellular activity, and which importantly bound with 1 : 1 stoichiometry and had binding affinities SAR that were consistent with the observed binding mode.

While the link between the phenomena observed across a range of biophysical techniques and the physical mechanism underlying these phenomena was not proven, and could not be in the context of an active research program in an industrial environment, it is clear that a hypothesis could be formed based on the observed data that guided the medicinal chemistry program in a direction that gave rise to useful and efficacious molecules. This interplay between biophysical techniques, and the guidance it gives to medicinal chemistry programs, is key to overcoming some of the more challenging and obscure issues that can arise during the hits-to-leads phase of a drug discovery program.

## 6.4 Direct Comparison of Orthogonal Screening Methods Against a Well-Characterized Protein System

In a recent study, a number of biophysical screening techniques were compared directly against each other using a carefully curated library on a well-known tractable protein target [26]. While this was carried out primarily in order to test the robustness of a new primary screening technique (WAC), it gives an excellent data set with which to compare the pros and cons of various screening methodologies and to examine the benefits and potential hazards of using orthogonal techniques to identify and validate weak binding ligands.

A library of 111 compounds was screened against N-terminal ATPase domain of Hsp90, the "workhorse" of fragment-based lead discovery. This library was a subset of the Vernalis fragment library [21,27], which has been carefully designed and extensively curated in order to ensure diversity, solubility, and stability in both aqueous and DMSO solutions. This fragment library was screened to identify weakly binding ligands using surface plasmon resonance (SPR), thermal shift assay (TSA; thermal melts), a robust fluorescence polarization (FP) probe displacement assay, weak affinity chromatography, and a consensus of three ligand observed NMR experiments (STD, water-LOGSY, and CPMG) (Figure 6.11). A subset of hits from these techniques was then taken forward

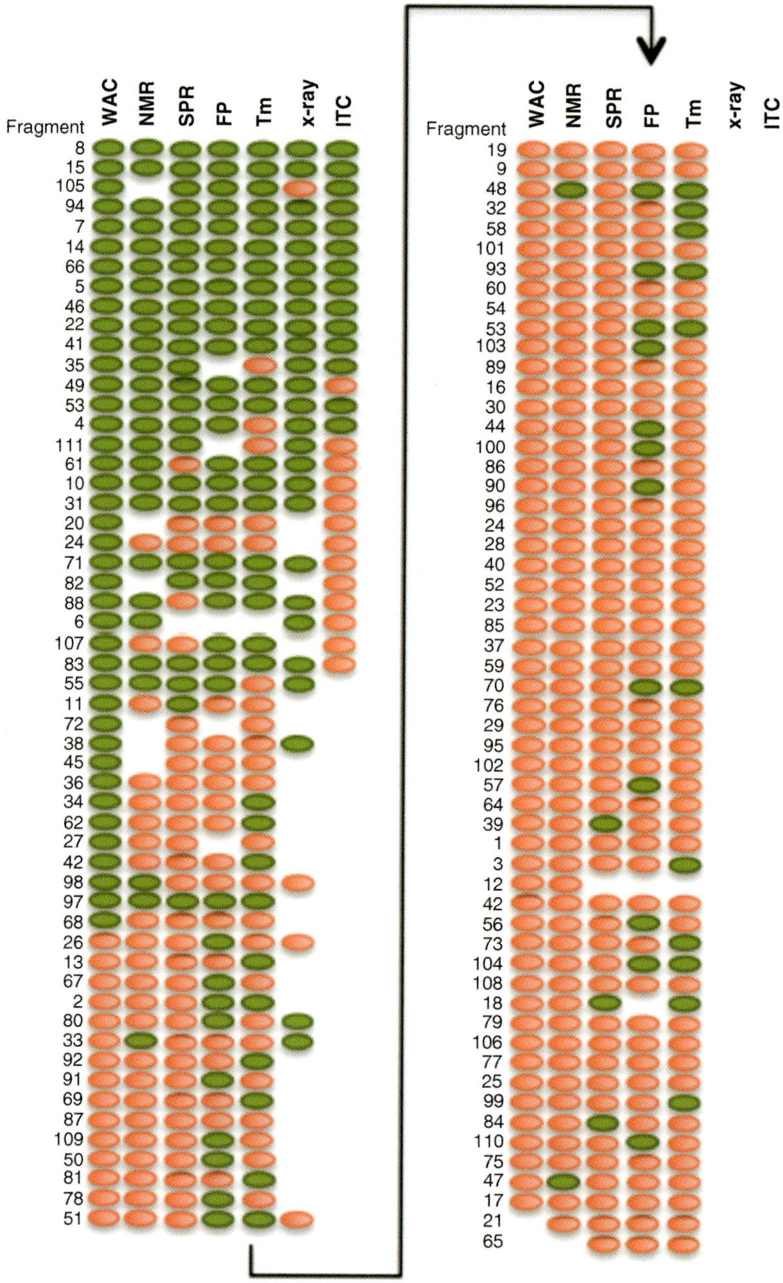


**Figure 6.11** Screening of 111 fragments by WAC, NMR, SPR, FP, and Tm shift. Selected fragments were also assessed by ITC. Fragments are sorted in decreasing affinity order according to WAC. Hits are denoted as green, nonhits as red, and missing cells indicate that no data is available. For X-ray crystallography, the green shows the fragments for which structures were attempted and determined.

into lower throughput techniques for "validation" of the binding observed in the primary technique, including isothermal titration calorimetry and X-ray crystallography. It should be noted that Hsp90 is very much an ideal situation for studying protein–fragment interactions and that the correlation rates for other less tractable proteins are frequently considerably lower. However, in our and others' experience, similar patterns are still observed.

When considering the correlation between techniques, it is important to recognize that there is no "right" answer. No method for detecting binding is 100% reliable, and all are prone to artifacts resulting from their specific methods of detection and interpretation. Since no absolute result is available, comparing correlation rates between the methods is necessarily rather complicated, and is best done by considering pairs or sets of techniques. Empirically, it has been widely observed that the more direct an observation, and the closer it is to the solution-phase conditions of a ligand–protein interaction, the more it has proved to be robust and reliable at identifying "real" weak binding ligands. This has resulted in NMR and SPR being widely regarded as the gold standards of fragment screening, with high-concentration biochemical screening being rather more artifact prone and concurrently less reliable. X-ray crystallography, while yielding high-resolution data about the structure of the ligand–protein complex, is highly dependent on the crystal form and conditions used and so its reliability and relevance to interactions in solution is also rather variable.

The correlation between the various approaches used in the WAC correlation study is shown in Table 6.1. The precise number of compounds screened in each technique varies slightly, as experimental factors (such as data quality, compound availability, and spectral overlap) sometimes prevented data from being acquired on every compound for each technique. However, several clear trends are observed. Of the primary screening techniques, SPR and NMR show a high degree of cross-correlation (90% agreement across 102 common compounds tested), with WAC having a lower correlation with these techniques (83–88% across 103–107 compounds). The thermal shift assay shows rather low correlation with

**Table 6.1** Correlation between various fragment screening methods.

| | WAC | NMR | SPR | ITC | X-ray | FP | Tm shift |
|---|---|---|---|---|---|---|---|
| WAC | — | 88% ($n = 103$) | 83% ($n = 107$) | 52% ($n = 27$) | 83% ($n = 30$) | 66% ($n = 109$) | 71% ($n = 107$) |
| NMR | | — | 90% ($n = 102$) | 63% ($n = 24$) | 93% ($n = 28$) | 74% ($n = 104$) | 74% ($n = 102$) |
| SPR | | | — | 73% ($n = 26$) | 76% ($n = 29$) | 73% ($n = 109$) | 73% ($n = 109$) |
| ITC | | | | — | 57% ($n = 23$) | 63% ($n = 27$) | 58% ($n = 26$) |
| X-ray | | | | | — | 73% ($n = 30$) | 66% ($n = 29$) |
| FP | | | | | | — | 73% ($n = 109$) |
| Tm shift | | | | | | | — |

WAC, SPR, and NMR (71–73% across 102–107 compounds), while the high-concentration biochemical assay (FP) shows a similar correlation degree of correlation with NMR and SPR (73–74% across 104–109 compounds), but rather less with WAC (66%, 109 compounds).

Each screening method was carried out under conditions optimal for that technique; thus, while the signal/noise in each experiment was optimal, subtle differences in conditions may be responsible for the observation of binding or nonbinding of certain compounds by different methods. However, each technique also has a certain degree of experimental noise, both false positives and false negatives, and this will inevitably result in different sets of hits from different methods. Assays with the highest signal/noise should give the highest correlation, and this appears to be the case for ligand-observed NMR and SPR. WAC has a reasonable correlation with both NMR and SPR. Thermal shift and FP assays, both of which are known to be error prone, do indeed give a lower correlation with the more robust methods.

Subsets of compounds, mostly hits in the primary assays, were then taken forward and studied by lower throughput validation assays, X-ray crystallography and ITC. Although the lower throughput of these techniques is apparent from the number of compounds studied (23–30), the data obtained from these techniques is vital in confirming the validity of the preliminary hits. Similarly, these techniques are often relied upon to routinely generate data to guide medicinal chemistry efforts, so reliability must be thoroughly established.

In general, the degree of correlation of the primary screening methods with ITC was lower than almost all other techniques used (52–73% across 23–27 compounds screened). From examination of the raw experimental data, this appears to be due to the stringent requirements of ITC in terms of compound solubility and behavior resulting in a rather high false negative rate; compounds that were consistently identified as valid ligands by other methods frequently failed to generate useable data in the ITC experiments.

In contrast to ITC, hits taken forward into X-ray were successful in generating crystal structures; this is particularly noticeable with hits from the ligand-observed NMR screening, where 93% of the 28 hits taken forward yielded structures. WAC, FP, and SPR hits showed a lower validation rate (83% and 76% from 29–30 hits), while hits from the thermal shift assay showed only a 66% success rate (30 hits) in X-ray crystallography; for this system, this is substantially low and reflects the higher error rate generally observed in identifying hits from thermal shift assays.

However, it is important to recognize that the X-ray structures obtained were not the result from a single soak or cocrystallization experiment. Success or failure in crystallization can depend on many factors, ranging from the easily modified (such as ligand concentration, soak duration, and temperature) to more dramatic changes (such as mutation of the protein construct). In our experience, success in crystallization frequently requires up to six attempts, and may well require identification of alternative crystal forms (Figure 6.12).

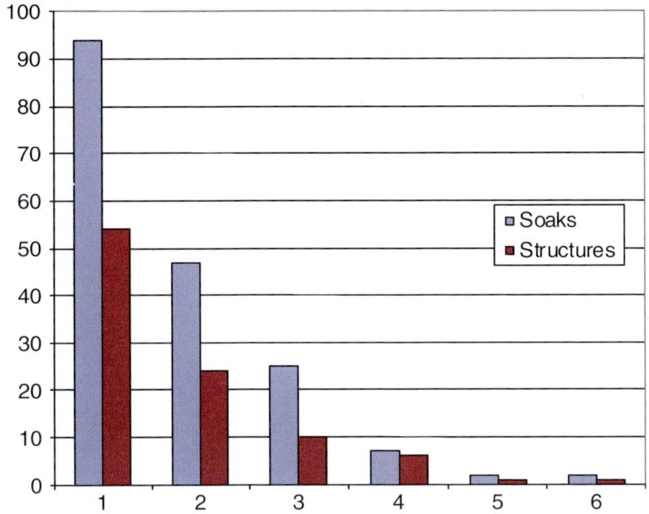


**Figure 6.12** The number of soaks required to obtain a crystal structure of a protein–fragment complex. These data represent the sum of four targets from among different target classes. If a compound has been identified as a validated ligand but fails to give a crystal structure, it is typically taken forward and repeated under different conditions. If no crystal structure is obtained after ~6 different soaking attempts, alternative crystal forms or mutation of the protein may be required, or move on to other fragments or hit-expansion compounds depending on evolving project knowledge.

## 6.5 Conclusions

Since the output from the primary fragment screen will go on to form the foundations for much subsequent research, and the application of a great deal of time and resources hinges on the quality of this data, it is vital that as high a level of experimental confidence as possible is achieved on hits identified and validated by this screen. As discussed in this chapter, orthogonal biophysical techniques have the capability to identify issues with the screening conditions, with the target itself, and with the hits identified by the screen. These issues may not always be immediately apparent, though, and can initially manifest themselves as discrepancies between well-understood techniques, "unusual" behavior of protein or compounds or other anomalous observations. While these types of behavior can occur even with well-behaved and tractable systems, they can be a useful flag that more detailed study is required and that the experimental data may not be as simple to interpret as initially appeared to be the case.

When combining the output from multiple experimental assay techniques, whether they are biochemical, biophysical, or functional, it is vital to remember that simple dogmatic reliance on agreement between orthogonal techniques is not the best approach as this can result in simply propagating the signal/noise ratio of the least sensitive experiment across the combined data set. Rather, it is important to understand the limitations and restrictions of each method used, and to consider that the summed output merely increases the probability and confidence that an observed hit will turn out to be a validated ligand. All techniques will have an inherent rate of false positives and false negatives, and the power of using orthogonal methods is to combine the output in such a way as to minimize these error rates rather than maximize them. Through such efforts, investigators are likely to garner insights that can give them an edge in understanding how to drug a system, as well as to continue to increase confidence and adoption of fragment-based approaches.

## References


1 Davis, M.J. and Schlessinger, J. (2012) The genesis of Zelboraf: targeting mutant B-Raf in melanoma. *The Journal of Cell Biology*, **199**, 15–19.

2 Kutchukian, P.S., Wassermann, A.M., Lindvall, M.K., Wright, S.K., Ottl, J., Jacob, J., Scheufler, C., Marzinzik, A., Brooijmans, N., and Glick, M. (2015) Large scale meta-analysis of fragment-based screening campaigns: privileged fragments and complementary technologies. *Journal of Biomolecular Screening*, **20**, 588–96.

3 Albert, J.S., Blomberg, N., Breeze, A.L., Brown, A.J., Burrows, J.N., Edwards, P.D., Folmer, R.H., Geschwindner, S., Griffen, E.J., Kenny, P.W., Nowak, T., Olsson, L.L., Sanganee, H., and Shapiro, A.B. (2007) An integrated approach to fragment-based lead generation: philosophy, strategy and case studies from AstraZeneca's drug discovery programmes. *Current Topics in Medicinal Chemistry*, **7**, 1600–1629.

4 Congreve, M., Rich, R.L., Myszka, D.G., Figaroa, F., Siegal, G., and Marshall, F.H. (2011) Fragment screening of stabilized G-protein-coupled receptors using biophysical methods. *Methods in Enzymology*, **493**, 115–136.

5 Perspicace, S., Banner, D., Benz, J., Muller, F., Schlatter, D., and Huber, W. (2009) Fragment-based screening using surface plasmon resonance technology. *Journal of Biomolecular Screening*, **14**, 337–349.

6 Dalvit, C. (2009) NMR methods in fragment screening: theory and a comparison with other biophysical techniques. *Drug Discovery Today*, **14**, 1051–1057.

7 Danielson, U.H. (2009) Fragment library screening and lead characterization using SPR biosensors. *Current Topics in Medicinal Chemistry*, **9**, 1725–1735.

8 Davis, B.J. and Erlanson, D.A. (2013) Learning from our mistakes: the 'unknown knowns' in fragment screening. *Bioorganic and Medicinal Chemistry Letters*, **23**, 2844–2852.

9 Edwards, P.D., Albert, J.S., Sylvester, M., Aharony, D., Andisik, D., Callaghan, O., Campbell, J.B., Carr, R.A., Chessari, G., Congreve, M., Frederickson, M., Folmer, R.H., Geschwindner, S., Koether, G., Kolmodin, K., Krumrine, J., Mauger, R.C., Murray, C.W., Olsson, L.L., Patel, S., Spear, N., and Tian, G. (2007) Application of fragment-based lead generation to the discovery of novel, cyclic amidine beta-secretase inhibitors with nanomolar potency, cellular activity, and high ligand efficiency. *Journal of Medicinal Chemistry*, **50**, 5912–5925.

10 Erlanson, D.A. (2012) Introduction to fragment-based drug discovery. *Topics in Current Chemistry*, **317**, 1–32.

11 Giannetti, A.M. (2011) From experimental design to validated hits a comprehensive walk-through of fragment lead identification using surface plasmon resonance. *Methods in Enzymology*, **493**, 169–218.

12 Hennig, M., Ruf, A., and Huber, W. (2012) Combining biophysical screening and

X-ray crystallography for fragment-based drug discovery. *Topics in Current Chemistry*, **317**, 115–143.

13 Jhoti, H., Cleasby, A., Verdonk, M., and Williams, G. (2007) Fragment-based screening using X-ray crystallography and NMR spectroscopy. *Current Opinion in Chemical Biology*, **11**, 485–493.

14 Pollack, S.J., Beyer, K.S., Lock, C., Muller, I., Sheppard, D., Lipkin, M., Hardick, D., Blurton, P., Leonard, P.M., Hubbard, P.A., Todd, D., Richardson, C.M., Ahrens, T., Baader, M., Hafenbradl, D.O., Hilyard, K., and Burli, R.W. (2011) A comparative study of fragment screening methods on the p38alpha kinase: new methods, new insights. *Journal of Computer-Aided Molecular Design*, **25**, 677–687.

15 Scott, D.E., Coyne, A.G., Hudson, S.A., and Abell, C. (2012) Fragment-based approaches in drug discovery and chemical biology. *Biochemistry*, **51**, 4990–5003.

16 Talamas, F.X., Ao-Ieong, G., Brameld, K.A., Chin, E., de Vicente, J., Dunn, J.P., Ghate, M., Giannetti, A.M., Harris, S.F., Labadie, S.S., Leveque, V., Li, J., Lui, A.S., McCaleb, K.L., Najera, I., Schoenfeld, R.C., Wang, B., and Wong, A. (2013) *De novo* fragment design: a medicinal chemistry approach to fragment-based lead generation. *Journal of Medicinal Chemistry*, **56**, 3115–3119.

17 Papalia, G.A., Giannetti, A.M., Arora, N., and Myszka, D.G. (2008) Thermodynamic characterization of pyrazole and azaindole derivatives binding to p38 mitogen-activated protein kinase using Biacore T100 technology and van't Hoff analysis. *Analytical Biochemistry*, **383**, 255–264.

18 Lu, K.P. and Zhou, X.Z. (2007) The prolyl isomerase PIN1: a pivotal new twist in phosphorylation signalling and disease. *Nature Reviews Molecular Cell Biology*, **8**, 904–916.

19 Potter, A.J., Ray, S., Gueritz, L., Nunns, C.L., Bryant, C.J., Scrace, S.F., Matassova, N., Baker, L., Dokurno, P., Robinson, D.A., Surgenor, A.E., Davis, B., Murray, J.B., Richardson, C.M., and Moore, J.D. (2010) Structure-guided design of alpha-amino acid-derived Pin1 inhibitors. *Bioorganic and Medicinal Chemistry Letters*, **20**, 586–590.

20 Potter, A., Oldfield, V., Nunns, C., Fromont, C., Ray, S., Northfield, C.J., Bryant, C.J., Scrace, S.F., Robinson, D., Matossova, N., Baker, L., Dokurno, P., Surgenor, A.E., Davis, B., Richardson, C. M., Murray, J.B., and Moore, J.D. (2010) Discovery of cell-active phenyl-imidazole Pin1 inhibitors by structure-guided fragment evolution. *Bioorganic and Medicinal Chemistry Letters*, **20**, 6483–6488.

21 Baurin, N., Aboul-Ela, F., Barril, X., Davis, B., Drysdale, M., Dymock, B., Finch, H., Fromont, C., Richardson, C., Simmonite, H., and Hubbard, R.E. (2004) Design and characterization of libraries of molecular fragments for use in NMR screening against protein targets. *Journal of Chemical Information and Computer Sciences*, **44**, 2157–2166.

22 Davis, B. (2013) Screening protein–small molecule interactions by NMR. *Methods in Molecular Biology (Clifton, NJ)*, **1008**, 389–413.

23 Zhang, Y., Daum, S., Wildemann, D., Zhou, X.Z., Verdecia, M.A., Bowman, M.E., Lucke, C., Hunter, T., Lu, K.P., Fischer, G., and Noel, J.P. (2007) Structural basis for high-affinity peptide inhibition of human Pin1. *ACS Chemical Biology*, **2**, 320–328.

24 Dong, L., Marakovits, J., Hou, X., Guo, C., Greasley, S., Dagostino, E., Ferre, R., Johnson, M.C., Kraynov, E., Thomson, J., Pathak, V., and Murray, B.W. (2010) Structure-based design of novel human Pin1 inhibitors (II). *Bioorganic and Medicinal Chemistry Letters*, **20**, 2210–2214.

25 Massey, A.J., Williamson, D.S., Browne, H., Murray, J.B., Dokurno, P., Shaw, T., Macias, A.T., Daniels, Z., Geoffroy, S., Dopson, M., Lavan, P., Matassova, N., Francis, G.L., Graham, C.J., Parsons, R., Wang, Y., Padfield, A., Comer, M., Drysdale, M.J., and Wood, M. (2010) A novel, small molecule inhibitor of Hsc70/Hsp70 potentiates Hsp90 inhibitor induced apoptosis in HCT116 colon carcinoma cells. *Cancer Chemotherapy and Pharmacology*, **66**, 535–545.

26 Meiby, E., Simmonite, H., le Strat, L., Davis, B., Matassova, N., Moore, J.D.,

Mrosek, M., Murray, J., Hubbard, R.E., and Ohlson, S. (2013) Fragment screening by weak affinity chromatography: comparison with established techniques for screening against HSP90. *Analytical Chemistry*, **85**, 6756–6766.

27 Chen, I.J. and Hubbard, R.E. (2009) Lessons for fragment library design: analysis of output from multiple screening campaigns. *Journal of Computer-Aided Molecular Design*, **23**, 603–620.

# 7
# Differential Scanning Fluorimetry as Part of a Biophysical Screening Cascade

*Duncan E. Scott, Christina Spry, and Chris Abell*

## 7.1
## Introduction

Differential scanning fluorimetry (DSF) or thermal shift (TS) has emerged in recent years as a high-throughput method for screening compounds [1,2]. At its heart is the idea that ligands can bind to and stabilize a protein, causing the protein to unfold at a higher temperature than otherwise. The difference between the "melting" temperature of a protein in the presence and absence of a ligand is termed the "thermal shift." In practice, an environmentally sensitive dye is mixed with a protein of interest and the solution is slowly heated while the fluorescence of the dye is monitored. As the protein unfolds, the dye binds to the exposed hydrophobic core of the protein and the fluorescence of the dye is substantially increased. In this way, the protein unfolding transition can be observed and the thermal melting temperature of the protein, the temperature corresponding to the half-way transition point, is determined. The technology has proven amenable to miniaturization into 96- and 384-well formats [2] and has found particular application within fragment screening cascades.

In this chapter, we discuss the relationship between thermal shifts and the affinity of ligands and other thermodynamic parameters. Large thermal shifts (>10 K) are theoretically possible for potent compounds under certain conditions; however, fragments are typically weakly binding molecules ($K_D \sim 0.1$–1 mM) that give thermal shifts of the order 0.5–2.0 K. We discuss the theory behind the DSF experiment and explain why thermal shifts are larger for some proteins than others, even for ligands with the same affinity at the melting temperature of the apo protein. We also discuss how the thermodynamics of ligand binding to protein changes with temperature and the implications of this for the design of experiments, and highlight some of the intrinsic limitations of the technique. Negative thermal shifts are commonly observed in experiments and we review a series of equilibrium models that can explain these observations. In the second half of this chapter, we discuss literature examples of the application of DSF to fragment-based drug design, making comparisons where we can to other biophysical screening techniques.

*Fragment-based Drug Discovery: Lessons and Outlook*, First Edition. Edited by Daniel A. Erlanson and Wolfgang Jahnke.

## 7.2
## Theory

### 7.2.1
### Equilbria are Temperature Dependent

At a given temperature, the Gibbs free energy of any reversible process and the thermodynamic components of enthalpy and entropy are related by Equation 7.1.

$$\Delta G = \Delta H - T\Delta S. \tag{7.1}$$

$\Delta G$ is the Gibbs free energy of the process, $T$ is the temperature, and $\Delta H$ and $\Delta S$ are the enthalpy and entropy components, respectively. This equation indicates that Gibbs free energy is temperature dependent and that this dependence on temperature is proportional to the entropy of the reversible process. Hence, equilibria are temperature dependent. For reversible equilibria such as protein unfolding, ligand binding, or acid–base equilibria, there is an intrinsic role of temperature that is sometimes forgotten when discussing biophysical processes and screens performed at different temperatures. The equilibrium constant of a reversible process is only a constant at a given temperature and varies according to Equation 7.2.

$$\Delta G = -RT\ln(K). \tag{7.2}$$

$K$ is the equilibrium constant of a process (equivalent to $K_A$ or $1/K_D$ for bimolecular interactions) and $R$ is the universal gas constant. In fact, the picture is more complicated than this; both $\Delta H$ and $\Delta S$ in turn vary with temperature. Equations 7.3 and 7.4 describe mathematically how $\Delta H$ and $\Delta S$ at a given temperature $T$ can be calculated from knowledge of $\Delta H$ and $\Delta S$ at a reference temperature $T_0$ ($\Delta H_0$ and $\Delta S_0$, respectively), and a new parameter, the change in "heat capacity" ($\Delta C_p$).

$$\Delta H(T) = \Delta H_0 + \Delta C_p(T - T_0). \tag{7.3}$$

$$\Delta S(T) = \Delta S_0 + \Delta C_p\ln(T/T_0). \tag{7.4}$$

$\Delta C_p$ is perhaps the least intuitive of the thermodynamic parameters to understand. Mathematically it can be seen that a positive value of $\Delta C_p$ yields a more positive enthalpy and entropy for a process where the temperature ($T$) exceeds the reference temperature ($T_0$). Alternatively, $\Delta C_p$ can be thought of as the change in the microstates available in a system. In the case of protein unfolding, a compact ordered structure opens to produce a flexible, disordered species. This corresponds to an increase in the number of accessible microstates for the system and hence $\Delta C_p$ for protein unfolding tends to be large and positive. As discussed later, this has important consequences for protein stability. Conversely, a ligand binding to a folded protein will bury surface area and is associated with a small and negative $\Delta C_p$. For such processes where the $\Delta C_p$ is relatively small, the effect of $\Delta C_p$ can be ignored over small temperature changes

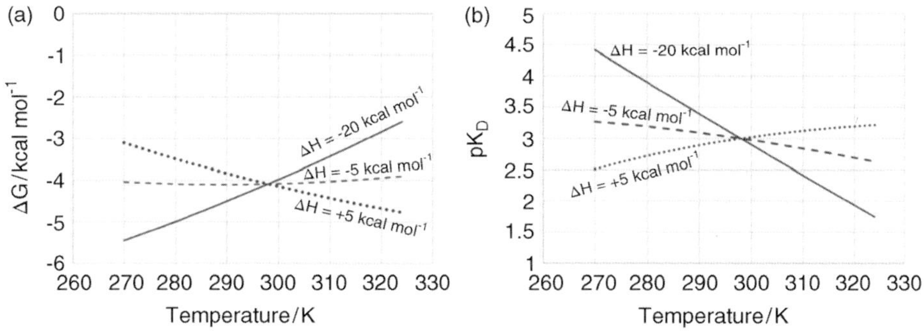


**Figure 7.1** (a) Calculated change in $\Delta G$ for ligand–protein binding and (b) the corresponding change in $pK_D$, for three ligands, as a function of temperature. Each ligand has a $K_D$ of 1 mM at 298 K and $\Delta H$ (at 298 K) as indicated. Note that the small but nonzero $\Delta C_p$ ($-0.1$ kcal mol$^{-1}$ K$^{-1}$) causes only a small deviation from linearity for ligand–protein binding equilibria and can mostly be ignored. Plots created using Equations 7.1–7.4.

and $\Delta G$ approximated by Equation 7.1. Therefore, small increases in temperature and a favorable $\Delta S$ (positive) will lead to a more favorable (more negative) Gibbs free energy. Conversely, an unfavorable entropy change (negative) will lead to a less favorable Gibbs free energy (more positive). To state this another way, the affinity of compounds that are entropically driven actually *increases* as temperature increases, while the affinity of enthalpically driven binders *decreases* as temperature increases. Of course, if the temperature is reduced, as is typical in ligand-observed NMR experiments, the reverse is true – the affinity of enthalpically driven binders increases.

In Figure 7.1, the change in free energy and equilibrium constant for ligand binding is shown as a function of temperature for three theoretical ligands with a $K_D$ (at 298 K) of 1 mM, which is typical for a fragment. Ligand binding at 298 K is chosen to be either strongly enthalpic ($\Delta H = -20$ kcal mol$^{-1}$), weakly enthalpic ($\Delta H = -5$ kcal mol$^{-1}$), or entropically driven ($\Delta H = +5$ kcal mol$^{-1}$). The change in $\Delta G$ with increasing temperature is largely linear for all the curves due to the relatively small $\Delta C_p$ and is most marked for the strongly enthalpic binder (solid curve). For this ligand, the change in $\Delta G$ observed when the temperature is increased from 280 K (a typical temperature for a ligand-observed NMR screen) to 315 K (a temperature at which many proteins unfold), corresponds to a 60-fold change in $K_D$, in the direction of poorer binding. On the other hand, the affinity of the entropically driven binder (dotted curve) increases with the same temperature increase, with a threefold improvement in binding affinity. The $\Delta G$ of the weakly enthalpic binder is essentially unchanged, but the $K_D$ increases by threefold (a consequence of $K_D$ being temperature dependent at a constant $\Delta G$). From these simple cases, it can be seen that one of the caveats of DSF experiments is the bias toward entropically driven binders, with the possibility of missing enthalpically driven binders. It is, however, worth noting that an enthalpy of $-20$ kcal mol$^{-1}$ is rare for a fragment, with $-5$ kcal mol$^{-1}$ being more common [3].

### 7.2.2 Thermodynamics of Protein Unfolding

Protein unfolding can be approximated to a reversible two-state equilibrium with equilibrium constant $K_U$ between the folded native state (N) and the unfolded state (U) represented by Equation 7.5.

$$N \overset{K_U}{\rightleftharpoons} U. \quad (7.5)$$

For mathematical convenience, it is typical to use the apo unfolding temperature as the reference temperature ($T_0$) in Equations 7.3–7.4. We will, therefore, refer to the apo unfolding temperature as $T_0$ throughout this chapter, which casts all thermal shifts relative to it. By combining Equations 7.1–7.4 with a knowledge of $\Delta H$, $\Delta S$, and $\Delta C_p$ of the unfolding process at a given temperature, it is possible to fully describe the function of $\Delta G(T)$ and hence derive the apo thermal melting temperature ($T_0$), the "half-way point" temperature at which there are equal proportions of folded and unfolded species ($\Delta G(T_0) = 0$, $K(T_0) = 1$). There is an implicit assumption here that protein unfolding can be well described by a two-state model and indeed that it is reversible. Monomeric proteins will most likely give a single-transition unfolding curve, as will some multimeric proteins. In reality, unfolded protein has a tendency to aggregate over time through association of the exposed hydrophobic cores, but on a short enough timescale it has been argued that the unfolding process is at least pseudo-reversible and can be treated as a fully reversible system [4]. The justification for this is both empirical and computational. It has been observed that data from irreversible thermal protein denaturations fit well to van't Hoff plots, for example, denaturation of *Escherichia coli* aspartate transcarbamoylase [5] and the core protein of *lac* repressor [6]. In addition, simulations of thermal denaturation that include an irreversible kinetic deactivation step generate data that are very similar to that from a fully reversible model [6].

Robertson and Murphy [7] have tabulated values of $T_0$, $\Delta H$, and $\Delta C_p$ for a collection of proteins. Typical values are $341 \pm 16$ K, $104 \pm 58$ kcal mol$^{-1}$, and $1.8 \pm 1.2$ kcal mol$^{-1}$ K$^{-1}$ for $T_0$, $\Delta H$, and $\Delta C_p$, respectively [7]. How $\Delta G$ changes with temperature for three different proteins is shown in Figure 7.2. The parabolic-like relationship between $\Delta G$ and temperature contrasts with the relatively linear change in the $\Delta G$ for ligand binding as a function of temperature. This departure from linearity is due to the large $\Delta C_p$ associated with protein unfolding. A large and positive $\Delta C_p$ accounts for the fact that the function of $\Delta G(T)$ is indeed a curve and therefore that a maximum protein stability exists [7,8]. The exact shape of the curve depends upon the melting temperature $T_0$ and the $\Delta H$ of unfolding. A further interesting consequence of a curving $\Delta G(T)$ profile is that while a thermal melting temperature exists at a temperature greater than the maximum stability temperature, there also exists another denaturing temperature below this temperature, representing the phenomenon of "cold denaturation."

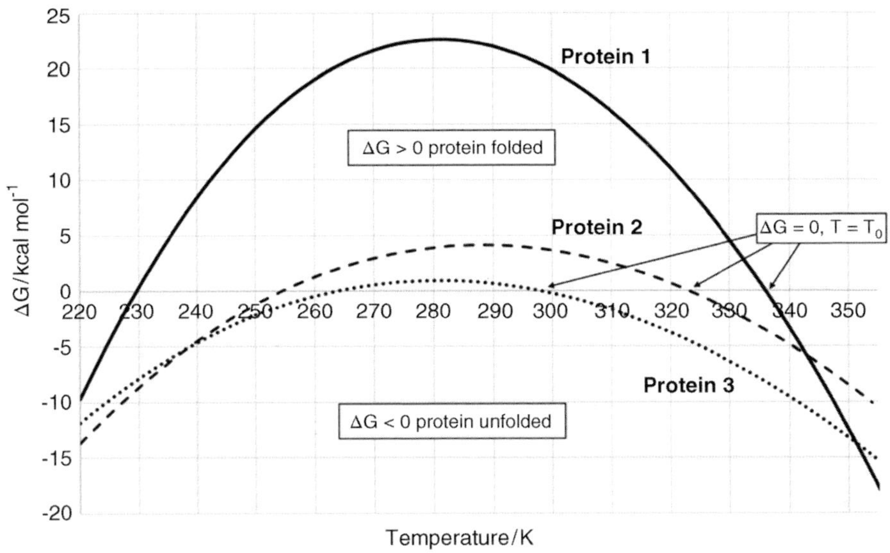


**Figure 7.2** Theoretical protein denaturation curves are shown for three proteins with differing values of $\Delta H$ of unfolding and $\Delta C_p$. Plots created using Equations 7.1–7.4. Protein 1: $T_0 = 336$ K, $\Delta H$ (336 K) = 270 kcal mol$^{-1}$, $\Delta C_p$ (336 K) = 4.5 kcal mol$^{-1}$ K$^{-1}$ (solid); protein 2: $T_0 = 323$ K, $\Delta H$ (323 K) = 75 kcal mol$^{-1}$, $\Delta C_p$ (323 K) = 2.0 kcal mol$^{-1}$ K$^{-1}$ (dashed); protein 3: $T_0 = 298$ K, $\Delta H$ (298 K) = 32 kcal mol$^{-1}$, $\Delta C_p$ (298 K) = 1.8 kcal mol$^{-1}$ K$^{-1}$ (dotted).

### 7.2.3
### Exact Mathematical Solutions to Ligand-Induced Thermal Shifts

When a ligand is added to an apo protein in a DSF experiment, both the unfolding equilibrium and the equilibrium of the ligand binding to the native folded protein must be considered (Equation 7.6). By virtue of binding to the folded protein, a ligand shifts the unfolding equilibrium to produce a different protein melting temperature ($T_m$). The position of the unfolding equilibrium of a protein in the presence of a ligand can be calculated exactly by considering how the equilibria change as a function of temperature according to Equation 7.7, which incorporates Equations 7.1–7.4 and the simple equilibrium model represented by Equation 7.6 [9].

$$N.L \overset{K_D}{\rightleftharpoons} L + N \overset{K_U}{\rightleftharpoons} U, \quad K_D = \frac{[N][L]}{[N.L]} \quad K_U = \frac{[U]}{[N]}. \tag{7.6}$$

$$K_D(T_m) = \frac{[L]_{T_m}}{\exp\left\{\frac{-\Delta H(T_0)}{R}\left(\frac{1}{T_m} - \frac{1}{T_0}\right) + \frac{\Delta C_p}{R}\left[\frac{T_0}{T_m} - 1 + \ln\left(\frac{T_m}{T_0}\right)\right]\right\} - 1}. \tag{7.7}$$

$K_D$ is the equilibrium constant of ligand binding and $K_U$ is the equilibrium constant of unfolding. $T_0$ is the thermal melting temperature of the apo protein and $T_m$ is the melting temperature in the presence of the ligand. $K_D(T_m)$ is the equilibrium constant of binding for the ligand at temperature $T_m$ and $[L]_{Tm}$ is the free ligand concentration at $T_m$ needed to affect this thermal shift. $\Delta H(T_0)$ is the enthalpy change associated with unfolding the protein at temperature $T_0$ and $\Delta C_p$ is the change in heat capacity associated with protein unfolding. To understand the implication of Equation 7.7, consider two ligands at different concentrations that give rise to the same thermal shift. In this scenario, the exponential term in the denominator is the same in both cases and therefore, the ligand that is at a higher concentration must have a higher $K_D$ (weaker binder) than the ligand at lower concentration, which must have a lower $K_D$ (stronger binder). This is in line with our intuitive expectations that a lower concentration of stronger binding ligands is required to generate the same thermal shift as a higher concentration of weaker binding ligands. Interestingly, it is possible to show that Equation 7.7 has no physical solution for $K_D(T_m)$ if $T_m$ is less than $T_0$, that is, for the case where a ligand produces the so-called negative thermal shift. The important consequence of this is that negative thermal shifts, which are commonly observed in experiments, cannot be explained using this model. More complex models that adequately describe negative thermal shifts are discussed later. It should be noted that the $-1$ term of the denominator of Equation 7.7 is sometimes intentionally omitted when the equation is expressed with $K_A$ rather than $K_D$ as the subject and the exponential term is much greater than 1 [1,2,9]. A further common approximation of the equation involves replacement of the $[L]_{Tm}$ term (the free ligand concentration at $T_m$) with the total ligand concentration $[L]_{\text{Total}}$, as in Equation 7.8.

$$K_D(T_m) \approx \exp\left\{\frac{\Delta H(T_0)}{R}\left(\frac{1}{T_m}-\frac{1}{T_0}\right)-\frac{\Delta C_p}{R}\left[\frac{T_0}{T_m}-1+\ln\left(\frac{T_m}{T_0}\right)\right]\right\}[L]_{\text{Total}}. \tag{7.8}$$

Equation 7.8 provides a good approximation of the equilibrium constant for ligand binding at $T_m$ when the ligand concentration far exceeds the protein concentration, as is the case in a typical DSF experiment in which the fragment concentration is between 1 and 10 mM, and the protein concentration is between 1 and 10 μM. As discussed in a previous section, binding constants change with temperature. Equation 7.9 can be used to convert the equilibrium constant of ligand binding at $T_m$ to the corresponding equilibrium constant at room temperature (or any other temperature).

$$K_D(T) = K_D(T_m) \cdot \exp\left\{\frac{\Delta H(T)}{R}\left(\frac{1}{T}-\frac{1}{T_m}\right)-\frac{\Delta C_{pL}}{R}\left[1-\frac{T}{T_m}+\ln\left(\frac{T}{T_m}\right)\right]\right\}. \tag{7.9}$$

In Equation 7.9, $K_D(T)$ is the equilibrium constant at temperature $T$, $\Delta H(T)$ is the enthalpy of ligand binding at temperature $T$ and $\Delta C_{pL}$ is the change in heat capacity upon ligand binding. As the last term is typically small it is sometimes ignored and $K_D(T)$ approximated using Equation 7.10 [2].

$$K_D(T) \approx K_D(T_m) \cdot \exp\left\{\frac{\Delta H(T)}{R}\left(\frac{1}{T}-\frac{1}{T_m}\right)\right\}. \tag{7.10}$$

It can be seen from Equation 7.10 that when $K_D$ (at $T_m$) of an entropically driven ligand ($\Delta H(T) > 0$) is converted to the corresponding $K_D$ at a lower temperature $T$, an increase in $K_D$ is observed. By contrast, $K_D$ (at $T_m$) of an enthalpically driven ligand ($\Delta H(T) < 0$) will decrease. In other words, the affinity of entropically driven ligands increases with increasing temperature, while the affinity of enthalpically driven ligands increases with decreasing temperature. Equation 7.10 recapitulates a well-known result from thermodynamics – the van't Hoff equation. Before the advent of accurate calorimetry techniques, the van't Hoff equation was used to deduce the thermodynamics of binding interactions from changes in ligand binding constant as a function of temperature. This has found particular application in studying ligand–receptor interactions [10], for example, small molecules binding to IL-2 [11].

### 7.2.4 Ligand Binding and Protein Unfolding Thermodynamics Contribute to the Magnitude of Thermal Shifts

A number of factors can affect the size of thermal shifts. One factor is certainly the affinity of the ligand *at the melting temperature*; however, properties of the target protein also come into play. Figure 7.3a shows how the $pK_D$

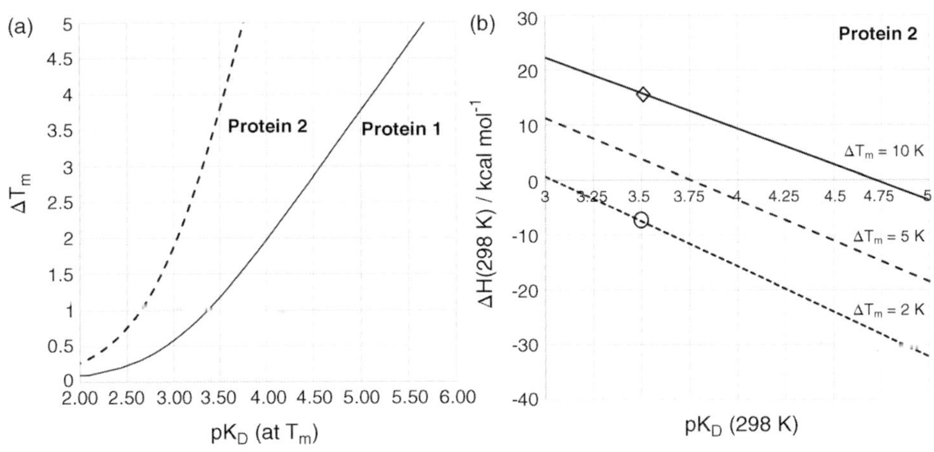


**Figure 7.3** (a) Graph showing the simulated effect of ligand $pK_D$ at $T_m$ on $\Delta T_m$ observed for two proteins with different unfolding thermodynamics. The total ligand concentration is 1 mM and is assumed to be essentially equivalent to $[L]_{Tm}$. The unfolding thermodynamics of the two proteins are as follows: protein 1, $T_0 = 336$ K, $\Delta H$ (336 K) = 270 kcal mol$^{-1}$, $\Delta C_p$ (336 K) = 4.5 kcal mol$^{-1}$ K$^{-1}$; protein 2, $T_0 = 323$ K, $\Delta H$ (323 K) = 75 kcal mol$^{-1}$, $\Delta C_p$ (323 K) = 2.0 kcal mol$^{-1}$ K$^{-1}$. (b) Graph showing the relationship between the ligand $pK_D$ at 298 K and ligand $\Delta H$ of binding at 298 K and the effect on $\Delta T_m$ for protein 2. The plotted lines correspond to constant $\Delta T_m$ values of 2 K (dotted), 5 K (dashed), or 10 K (solid). The diamond and circle indicate a theoretical pair of compounds with the same $pK_D$ at 298 K that are discussed in the text. Plots created using Equation 7.7 (assuming $[L]_{Tm} = [L]_{Total}$ = 1 mM) and 7.10, which ignores $\Delta C_p$ of ligand binding.

value (at $T_m$) of a ligand correlates with the observed thermal shifts ($\Delta T_m = T_m - T_0$) for two different proteins. Above a threshold value of p$K_D$, an approximately linear relationship between $\Delta T_m$ and p$K_D$ is observed for both of the proteins. Nonetheless, it can be seen that the relationship between $\Delta T_m$ and p$K_D$ is different for each protein. This is because the unfolding properties ($T_0$, $\Delta H(T_0)$ and $\Delta C_p(T_0)$) of the three simulated proteins differ. These three quantities, which fully define the thermodynamics of protein unfolding, are typically determined by differential scanning calorimetry [7]. It is for this reason that some proteins may be more amenable to DSF screens than others. As discussed in Section 7.2.3, the degree to which the affinity of a compound changes with temperature depends upon $\Delta H$ of ligand binding at 298 K. As shown in Figure 7.3b, for a given protein it is a combination of $\Delta H$ (at 298 K) and $K_D$ (at 298 K) of ligand binding that determines the magnitude of $\Delta T_m$; three lines of constant values of $\Delta T_m$ are plotted. The downward trend of all the lines reflects the fact that more enthalpically driven binders must bind with higher affinity at 298 K to compensate for the decrease in affinity with increasing temperature. For example, consider a pair of compounds that bind a target protein with a p$K_D$ (298 K) of 3.5 ($K_D = 300$ μM); in this system, if the enthalpy is positive (+16 kcal mol$^{-1}$, diamond Figure 7.3b) or negative (−7.6 kcal mol$^{-1}$, circle Figure 7.3b), a $\Delta T_m$ of either 10 or 2 K is observed, respectively.

A useful feature of DSF screens is that observed $\Delta T_m$ values can correlate with affinity for a given protein; however, it is important to be aware that this is not always the case. This is, as discussed above, because $\Delta H$ of protein unfolding, $\Delta H$ of ligand binding, and ligand $K_D$ all contribute to the observed thermal shift. Only within a series of compounds with similar enthalpies will $\Delta T_m$ increase directly with affinity. In a study by Lo *et al.* [1], the binding of ligands to BACE1 (beta-secretase 1) was investigated using both DSF and isothermal titration calorimetry (ITC). The authors of this study found a good correlation between the results obtained using the two techniques, both in terms of rank order of affinity and absolute affinity. As the authors point out, this set of ligands represents a special case where $\Delta H$ does not vary much between compounds. In another study, Fedorov *et al.* [12] used DSF to screen 156 kinase inhibitors against 60 human kinases. In an attempt to validate $\Delta T_m$ as an indicator of affinity, the authors investigated the correlation between the measured $\Delta T_m$ and $IC_{50}$ values determined for the compounds against two targets, PIM1 and GSK3ß. They observed a reasonable correlation, again probably due to similar binding enthalpies for the compounds. Clearly, the advantage of running a DSF-based screen as compared to an ITC-based screen, for example, is the vastly higher throughput and fewer resources required. Although some compounds may be missed, this becomes increasingly unlikely as the affinity improves. It is worth bearing in mind that potential sources for affinity not correlating with $\Delta T_m$ values are (i) large variations in enthalpies within a series, (ii) binding of ligands to the unfolded state of the protein, and perhaps to a lesser extent, (iii) $\Delta C_p$ of ligand binding being nonnegligible.

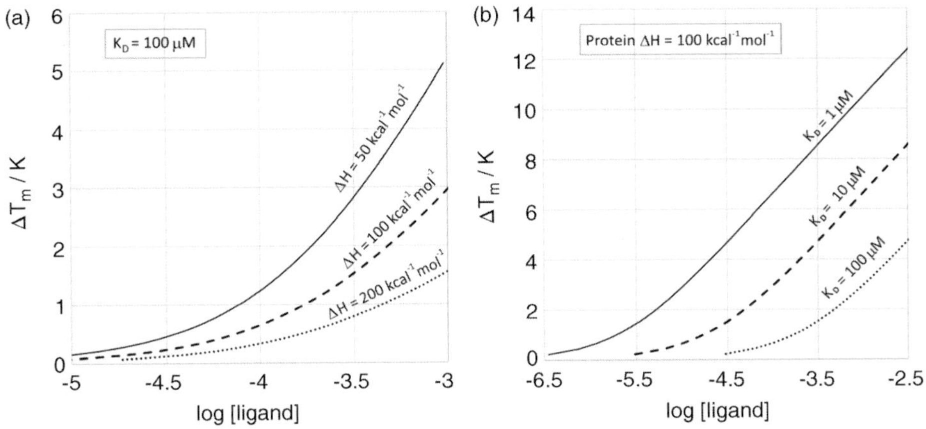


**Figure 7.4** Simulations showing the effect of $\Delta H$ of unfolding (a) and the effect of ligand $K_D$ (b) on $\Delta T_m$. Unless otherwise indicated, $\Delta H$ (unfolding) at 323 K = 100 kcal mol$^{-1}$, $\Delta C_p$ (unfolding) = 2 kcal mol$^{-1}$ K$^{-1}$, $T_0$ = 323 K, [protein] = 1 μM, $K_D$ of the ligand is 100 μM at 298 K, $\Delta C_p = -0.2$ kcal mol$^{-1}$ K$^{-1}$, and $\Delta H = -5$ kcal mol$^{-1}$. The exact concentration of free ligand at $T_m$ has been calculated for these plots. Plots created using Equations 7.7, 7.9 and 7.11.

### 7.2.5
### Ligand Concentration and the Magnitude of Thermal Shifts

Ligand concentration plays an important role in determining the size of thermal shifts. Matulis *et al.* [13] have studied the theoretical basis for ligand concentration-dependent changes in $T_m$, as measured by DSF analysis, using carbonic anhydrase, a well-characterized protein for which inhibitory ligands have been identified. Figure 7.4 demonstrates how the enthalpy of unfolding, ligand concentration, and ligand affinity affect the size of thermal shifts, and summarizes some of the key findings of Matulis *et al.* The total ligand concentrations ($[L]_{\text{Total}}$) needed to induce the thermal shifts plotted in Figure 7.4 were calculated using Equation 7.11. Equation 7.11 is a correction of a typographical error in Equation 13 of the Matulis paper [14,15].

$$[L]_{\text{Total}} = (K_U(T_m) - 1)\left(\frac{[P]_{\text{Total}}}{2K_U(T_m)} + K_D\right). \tag{7.11}$$

$K_U(T_m)$ is the unfolding equilibrium constant (Equation 7.6) at $T_m$, $[P]_{\text{Total}}$ is the total protein concentration and $K_D$ is the ligand binding constant at $T_m$. In Figure 7.4a, simulated titrations are shown for three proteins with the same $T_0$ and $\Delta C_p$, but different enthalpies of unfolding. It shows that at a given concentration of a single ligand proteins with a lower $\Delta H$ of unfolding give larger $\Delta T_m$ values and that $\Delta T_m$ continues to increase with ligand concentration. In theory, the protein stabilization can increase indefinitely with rising ligand concentration, as this will pull the unfolding equilibrium further toward the folded liganded species. In practice, however, nonspecific effects, solubility limits, and binding of the ligand to the unfolded form of the protein are likely to limit the thermal shift observed. Simulated titrations of three different ligands against a single protein are shown in Figure 7.4b. Under this particular set of conditions, the size of $\Delta T_\mathrm{m}$ directly correlates with the affinity of the ligand at a given concentration. When the affinity is sufficiently high, the titrations become more sigmoidal (not shown). Under these circumstances, it is possible to calculate binding stoichiometry [13].

### 7.2.6
### Models of Protein Unfolding Equilibria and Ligand Binding

Waldron and Murphy [16] have reported explicit simulations of the thermal unfolding process in the presence of ligands. They consider the unfolding equilibria and the effect of ligands binding to the folded state (model A), the unfolded state (model B), and binding to both states (model C) (Figure 7.5). Model A is the same model discussed previously and also shown in Equation 7.6. Although in general the ligands they considered bind with much higher affinity than is usual for fragments, the results provide some important insights into the interpretations of thermal shifts.

Simulations based on model A, in which ligands bind only to the native state, showed that two compounds that bind a protein (with a $T_0$ of 330 K) with the same $K_\mathrm{D}$ (1 µM) at 298 K, but with differing $\Delta H$ and $\Delta S$ contributions, can produce very different thermal shifts (5 K or 37 K). In practice, thermal shifts as high as 37 K are rarely observed, presumably due to ligand binding to the unfolded state (model C) becoming significant as the ligand concentration is increased. Figure 7.6 shows how the Gibbs free energy for unfolding of the apo protein used in the simulations changes as a function of temperature, becoming zero at $T_0$ of the protein (330 K). The change in the Gibbs free energy as a function of temperature for the binding of both of the ligands used in the model A-based simulations, and the effect of the ligands on the unfolding of the protein, according to the model A-based simulations, are also shown in Figure 7.6. It is possible to calculate the concentration of free ligand $[L]$ at temperature $T$ by Equation 7.12 [9].

$$\begin{aligned} b &= K_\mathrm{D}(1 + K_\mathrm{U}) + [P]_\mathrm{Total} - [L]_\mathrm{Total}. \\ c &= -[L]_\mathrm{Total} K_\mathrm{D}(1 + K_\mathrm{U}). \\ [L] &= \frac{-b + \sqrt{b^2 - 4c}}{2}. \end{aligned} \tag{7.12}$$

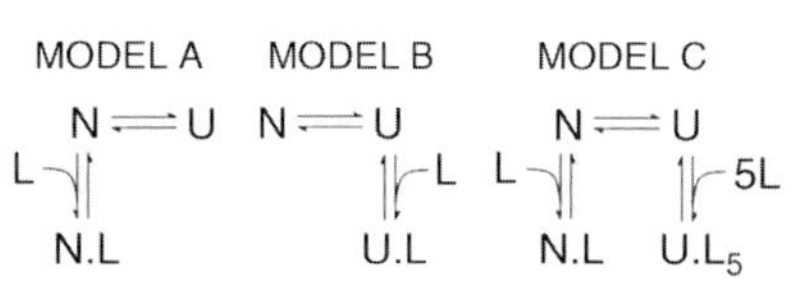


**Figure 7.5** Relevant equilibrium models considered by Waldron and Murphy [16]. Ligand *L* binds to the native folded protein (*N*) but not to the unfolded state (model A), *L* binds to the unfolded state only (model B), or *L* binds to the folded state and nonspecifically multiple times (five times) to the unfolded state (model C).

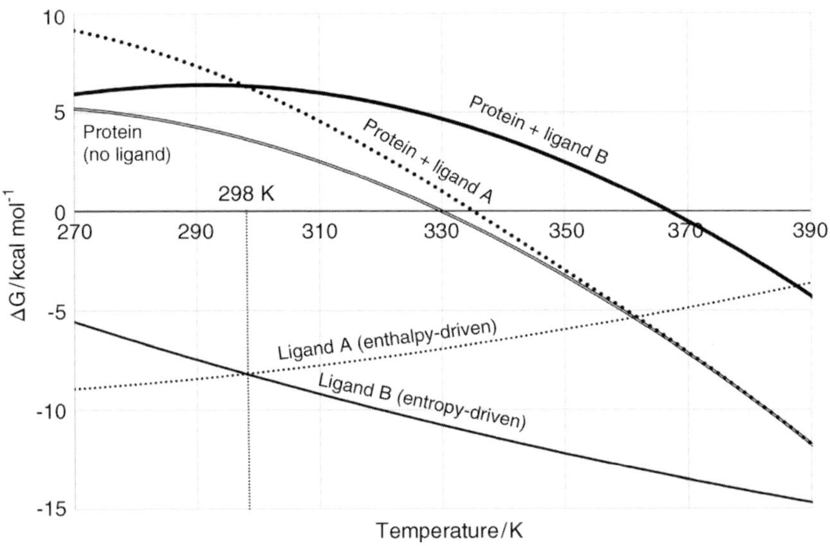


**Figure 7.6** Simulations demonstrating how $\Delta G(T)$ varies for a protein, for two ligands (A and B) and for the same protein in the presence of the two ligands according to model A (Figure 7.5). The figure shows $\Delta G(T)$ for unfolding of the protein in the absence of any ligand (double line), in the presence of ligand A (an enthalpic binder, large dots, $\Delta T_m = 5$ K), and in the presence of ligand B (an entropic binder, thick solid line, $\Delta T_m = 37$ K). The change in $\Delta G$ for binding of ligands A and B is also shown. The simulation conditions are as follows: $T_0 = 330$ K, $\Delta H$ (330 K) = 48 kcal mol$^{-1}$, $\Delta C_p = 0.60$ kcal mol$^{-1}$ K$^{-1}$, [protein] = 10 μM. Both ligands have a $K_D$ of 1 μM at 298 K, and a $\Delta C_p$ of −0.11 kcal mol$^{-1}$ K$^{-1}$. [Ligand] = 100 μM. For ligand A, $\Delta H = -18$ kcal mol$^{-1}$ and for ligand B, $\Delta H = +18$ kcal mol$^{-1}$. Plots created using Equations 7.1–7.4, 7.12 and 7.13.

$[P]_{\text{Total}}$ is the total concentration of protein in the experiment. By calculating $K_U$, $K_D$, and $[L]$ at a given temperature, it is possible to work out the equilibrium constant for the unfolding equilibrium defined in Equation 7.13 [9].

$$K_{eq} = \frac{[U]}{[N] + [N \cdot L]} = \frac{K_D K_U}{K_D + [L]}. \tag{7.13}$$

From this analysis, it is possible to rationalize the observations – the entropically driven ligand increases its affinity as the temperature rises, at 330 K it is 72 nM. Conversely, the enthalpically driven ligand's affinity decreases as the temperature rises (from 1 μM at 298 K to 26 μM at 330 K). A compound that binds more tightly to the native protein will act so as to pull the equilibrium in model A toward the folded complex, away from the unfolded species, and hence cause a larger $\Delta T_m$. Simulations based on model B, in which compounds only bind the unfolded state, showed that two compounds that bind a protein with the same $K_D$ (10 μM) at 298 K but with different thermodynamics of binding produce very different thermal shifts (zero and −20 K). This result can be understood in the context of the changes in equilibrium constants as a function of temperature.

The compounds giving the most negative thermal shifts correspond to the most entropically driven ligands. Again, this is readily reasoned by virtue that entropically driven binders increase their affinity for the unfolded state as the temperature increases, and as binding to the unfolded state increases, the equilibrium in model B is pulled toward the unfolded complex and the protein is destabilized.

Waldron and Murphy also presented simulations based on model C (Figure 7.5). In model C, a ligand can bind not only to the native folded protein but also non-specifically to the unfolded protein, up to five times. This more complex model is probably a more accurate model of the reality of what happens in a DSF experiment, where both positive and negative thermal shifts are frequently observed. In this model, the thermodynamics of binding to the native state and to the unfolded state interplay to produce a thermal shift. Waldron and Murphy present a particular case of binding to both the native and the unfolded state (model C), but by changing the nature of the binding to the native state either a positive or negative thermal shift can result. In summary, although positive thermal shifts provide evidence of binding to a protein, a compound that binds to the native state of the protein can also yield no thermal shift or a negative thermal shift if binding to the native state is masked by additional binding to the unfolded protein.

### 7.2.7 Negative Thermal Shifts and General Confusions

As discussed in the previous section, a model in which a ligand binds to the native state only cannot explain negative thermal shifts (model A, Figure 7.5). Despite this, there are some spurious claims in the literature that negative thermal shifts are indicative of specific binding to the native state. More complex models than model A are needed to describe negative thermal shifts [16]. The most general model discussed by Waldron and Murphy is model C (Figure 7.5), in which a ligand can potentially bind to both the folded and the unfolded protein forms.

The relative affinities of a ligand for the folded and unfolded species at any given temperature determines the $\Delta T_m$. Negative thermal shifts are evidence of a ligand binding to the unfolded state of the protein, but this does not preclude the possibility that the ligand binds to the folded protein and that this has been masked. It is our experience that there is confusion in the literature concerning the interpretation of negative thermal shifts. Occasionally, an argument is made that negative thermal shifts are not a consequence of binding to the unfolded form but instead are a result of binding to a less stable form of the protein ($N^*$). Implicitly, a model such as that depicted in Figure 7.7a is being proposed.

It is sometimes argued that protein conformers ($N^*$) that are less stable or "destabilized" relative to the native protein (N) can bind the ligand to form a complex $N^* \cdot L$ and because $N^*$ is less stable than $N$, it is further concluded that a negative thermal shift is to be expected. Perhaps superficially appealing, this reasoning is inconsistent with the equilibria model that is implied. If $N^* \cdot L$ were less stable than $N + L$, L would not bind. To put this the other way, any perturbation that prevents the concentration of U increasing (i.e., by increasing the

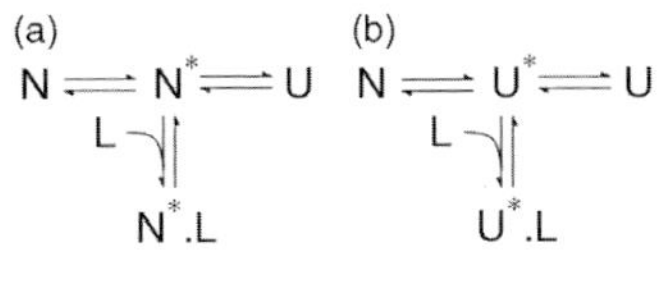


**Figure 7.7** (a) Model of ligand binding to a "destabilized" folded protein $N^*$ that is sometimes used to explain negative thermal shifts. (b) Proposed corrected model of ligand binding to a partially unfolded protein $U^*$ species.

population of $N^* \cdot L$) can only serve to delay unfolding to higher temperatures, not prematurely cause it. In fact, it depends how $N^* \cdot L$ interacts with the fluorescent dye in the experiment. As the DSF experiment works through detection of dye binding, this must be considered when thermal shifts are interpreted. If the fluorescent dye binds $N^* \cdot L$, then in the scope of the experiment it is detected as an unfolded or pseudo-unfolded species rather than a pseudo-folded species. Therefore, increasing the population of this species will cause a negative thermal shift. In general, ligands that bind solely to any folded forms of the protein will cause a positive thermal shift, while those that bind solely to unfolded forms will cause a negative thermal shift. The subtlety here is that "folded forms of the protein" are defined as protein structures that do not bind the fluorescent dye and "unfolded forms" are protein structures that bind the dye. Therefore, negative thermal shifts should be interpreted as the result of stabilization of a form of the protein that also binds the fluorescent dye. This could be the unfolded form or perhaps a pseudo-unfolded form $U^*$ in which a significant hydrophobic surface is exposed such that the dye is simultaneously bound (model represented in Figure 7.7b). Note this latter case is a development of the initial destabilization argument, but properly clarified.

### 7.2.8 Lessons Learnt from Theoretical Analysis of DSF

The thermodynamics of individual proteins impact on the size of the thermal shifts caused by ligands. Proteins with lower unfolding enthalpies are expected to generate greater thermal shifts for a given ligand $K_D$, ligand concentration, and protein $T_0$. Some proteins will, therefore, respond better in DSF experiments than others because of this fundamental property. The higher the affinity of the ligand at $T_m$, the greater the thermal shift observed for a given protein. Affinity changes with temperature, mostly as a function of binding enthalpy, introducing a mismatch between room-temperature $K_D$ and $K_D$ at $T_m$. The affinity of enthalpic binders decreases with temperature and therefore such compounds may be missed entirely at high $T_0$. As $T_0$ of the apo protein is increased, the affinity of enthalpically driven binders is decreased, and the bias toward identifying entropically driven binders is enhanced. In some DSF programs, screening conditions that yield the highest possible $T_0$ for the apo protein are sometimes selected. However, this is, in fact, to the detriment of identifying enthalpically driven binders, something that is generally considered favorable in drug development [17]. Of course, the protein needs to be stable enough to work with and to not denature during assay set up, for example, when mixed with DMSO, but selecting conditions that yield a lower $T_0$ should improve the chances of finding enthalpically driven fragments. Using conditions where the $\Delta H$ of unfolding is lowered should have the advantage of yielding larger thermal shifts for all hits. Finding conditions that maximize $\Delta T_m$ with a known ligand is a good strategy for establishing suitable screening conditions, in the case that the known ligand binds enthalpically. Prior to performing a fragment screen, our own practice is to measure the melting temperature of the target protein in the presence and absence of the known ligand using a variety of buffers, pH values, and salt concentrations. In our own experience, we have observed "shielding" effects from salt in thermal shift experiments; we often see the $T_0$ increase as the salt concentration is increased and in turn the relative thermal shift of a native ligand is diminished, possibly due to reduced ligand affinity with increased $T_0$. In a study of 221 different proteins, Vedadi *et al.* [18] also observed that $T_0$ commonly increases with increasing salt concentration.

## 7.3 Practical Considerations for Applying DSF in Fragment-Based Approaches

Although DSF is operationally simple, careful assay design is important for maximizing hit rates and minimizing false positive identification. It is important that under the selected experimental conditions, an unambiguous melt curve is obtained for the apo protein; a clear sigmoidal transition such as those shown in Figure 7.8 should be observed. Initially, the fluorescence intensity should be

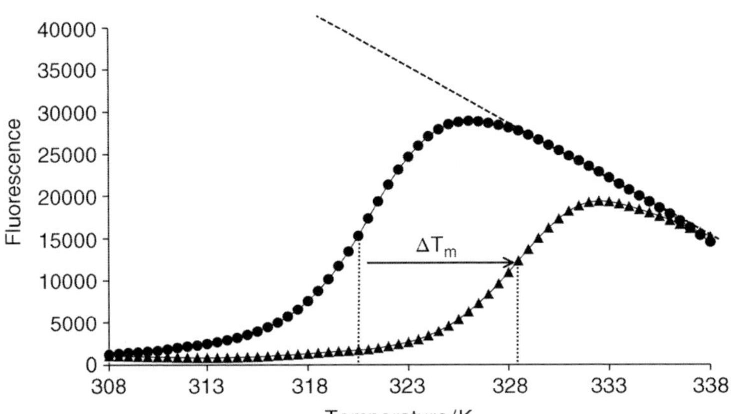


**Figure 7.8** Raw DSF data showing unfolding of a protein in the absence (circles) and presence (triangles) of a ligand. The difference in the melting temperature in the presence and absence of the ligand is indicated by the vertical dotted lines and denoted as $\Delta T_m$. The decline in fluorescence after the transition peak is highlighted by the extrapolated dashed line.

relatively low; however, as the temperature is increased there should be a sharp and significant increase. Typically, it is observed that after unfolding there is a decrease in fluorescence. This has been attributed to protein aggregation [19]; however Kranz and Schalk-Hihi [20] have suggested it to be due to a linear temperature dependence of the dye's fluorescence. In addition to optimizing the shape of the curve, it is also important, as discussed in Section 7.2.8, to identify conditions under which $T_0$ for the protein is as low as practically possible so as to ensure that fragments that bind enthalpically are not missed.

A number of different fluorescent dyes can be used for monitoring protein unfolding. SYPRO orange [19,21] is the most popular, in part because it is associated with a high signal-to-noise ratio. Additionally, the optimal wavelength at which the dye is excited (close to 500 nm) is higher than some of the alternatives, which means that the number of small molecules interfering with the fluorescence readout of protein unfolding (e.g., because of fluorescence quenching) is reduced [19]. The latter is particularly relevant in fragment screens where, by necessity, high ligand concentrations (typically 1–10 mM) are tested. It is recommended that initially a range of dye and protein concentrations are tested. The optimal dye concentration is that which gives the maximum ratio between the intensity of the fluorescent signal of the dye when bound to unfolded protein and the background fluorescence (from the free dye and dye bound to the folded protein). Where SYPRO orange is used in screens, typically the commercially available stock solution (supplied as a 5000× concentrate) is diluted 500–5000-fold. In order to minimize protein consumption, typically the lowest concentration of protein that provides a sufficient fluorescence increase upon protein unfolding (commonly between 1 and 10 μM) is used.

Once suitable dye and protein concentrations are chosen, the optimal buffer, pH, and ionic strength for fragment screening should be determined. In selecting buffers, it is advantageous to avoid those where the $pK_A$ is highly temperature dependent, for example, Tris. Additionally, it is advantageous to consider the conditions that will be used for orthogonal hit validation techniques and, if practical, use similar or matching buffers, pH, and ionic strength. This provides the best chance that true hits will be validated by independent methods.

Melting temperatures are commonly determined from the raw fluorescence data generated in a thermal shift experiment by fitting the data to a thermodynamic model (Equation 7.14), the Boltzmann equation (Equation 7.15), or higher order polynomial equations [14]. As discussed by Schulz, Landstrom, and Hubbard [14], various typographical errors in the thermodynamic model equations have appeared in the literature. We believe Equation 7.14 to be correct.

$$\gamma(T) = \gamma_{\mathrm{F},T_\mathrm{m}} + m_\mathrm{F}(T - T_\mathrm{m}) + \frac{\gamma_{\mathrm{U},T_\mathrm{m}} - \gamma_{\mathrm{F},T_m} + (m_\mathrm{U} - m_\mathrm{F})(T - T_\mathrm{m})}{1 + \exp\left\{\frac{\Delta H(T_0) + \Delta C_\mathrm{p}(T - T_0) - T(\Delta S(T_0) + \Delta C_\mathrm{p}\ln(T/T_0))}{RT}\right\}}. \tag{7.14}$$

$$\gamma(T) = \min + \frac{\max - \min}{1 + \exp\left\{\frac{T_\mathrm{m} - T}{a}\right\}}. \tag{7.15}$$

In Equation 7.14, $\gamma(T)$ is the observed fluorescence at temperature $T$, $\gamma_{F,Tm}$ and $\gamma_{U,Tm}$ are the fluorescence levels before and after the unfolding transition with the slopes $m_F$ and $m_U$, respectively. $\Delta H(T_0)$ and $\Delta S(T_0)$ are the enthalpy and entropy of protein unfolding at reference temperature $T_0$. $\Delta C_p$ is the change in heat capacity for unfolding. In Equation 7.15, min and max are the minimum and maximum fluorescence values before and after the transition and $a$ is the Hill slope.

As fragments are inherently weak binders, small thermal shifts (0.5–2 K) are typically associated with fragment binding. For this reason, there is often only a narrow margin by which to distinguish between hits and nonhits, and even the choice of the method used for data analysis can affect which compounds are classed as hits [14]. Commonly, the threshold for a hit is set to a thermal shift two or three times the standard deviation of the control measurements. In some cases, where hit rates are low, researchers may reduce this threshold and take the least destabilizing fragments as a potentially enriched set. It is worth bearing in mind that some destabilizing fragments may also bind the native folded state and therefore be of interest. The solvent in which fragments are solubilized (typically DMSO) can affect the melting temperature of the protein and as such it is important that all thermal shifts are measured relative to a solvent-matched control. Typically, fragment concentrations between 1 and 10 mM are screened. In theory higher fragment concentrations should yield higher thermal shifts and in turn higher hit rates; however, in reality higher fragment concentrations will also increase the probability of compounds precipitating and causing optical interference. For initial thermal shift fragment hits, retesting from fresh fragment solutions is good practice.

## 7.4 Application of DSF to Fragment-Based Drug Discovery

DSF is now commonly used in fragment screening campaigns [22], which in large part is due to the ease with which the technique can be implemented (specialist equipment and training are not required) and because screens can be performed rapidly and at a low cost. As highlighted in examples discussed below, DSF is most often adopted early on in fragment screening cascades, as a hit-finding mechanism.

### 7.4.1 DSF as a Primary Enrichment Technique

DSF is traditionally used in fragment-based approaches as a means to enrich for binders and to remove problem compounds such as denaturants (that destabilize the protein), and aggregators and compounds with poor solubility (that often yield complex melt curves). This enrichment step is subsequently followed by validation using orthogonal, lower throughput techniques that can often provide

**Table 7.1** Summary of results from fragment screening cascades involving an initial DSF-enrichment step.

| Target | DSF hit rate (%) | Follow-up screen | Confirmation rate (%) | References |
|---|---|---|---|---|
| Pantothenate synthetase | 3.1 | Ligand-observed NMR | 44 | Hung *et al.* [25]<br>Silvestre *et al.* [27] |
| CYP121 | 9.9 | Ligand-observed NMR | 46 | Hudson *et al.* [24] |
| Rad51-BRCA2 | 3.2 | Ligand-observed NMR | NR[a)] | Scott *et al.* [29] |
| EthR | 6.9 | SPR-based functional assay | 52 | Surade *et al.* [28] |
| NS3 helicase | 7.2 | X-ray crystallography | 0 | Coutard *et al.* [30] |
| NS5 mRNA methyltransferase | 6.4 | X-ray crystallography | 22 | Coutard *et al.* [30] |

a) Not reported.

additional information (e.g., on the mode of binding). In our own fragment screening campaigns, we commonly start with a DSF-based enrichment and follow this with a ligand-observed NMR validation step [23]. As highlighted in the following examples, DSF has been applied as an enrichment tool in a number of fragment-based projects focused on a variety of targets including classical enzyme targets [24–27] as well as targets considered more challenging (e.g., transcription factors [28] and protein–protein interactions [29]).

The results of two fragment screening cascades targeting enzymes from *Mycobacterium tuberculosis* (pantothenate synthetase [25,27] and cytochrome P450 121 or CYP121 [24]) are summarized in Table 7.1. In both screens, DSF was used as the frontline hit-finding mechanism and ligand-observed NMR (a combination of saturation transfer difference (STD) and water–ligand observed via gradient spectroscopy (water-LOGSY) NMR) was used to confirm binding of DSF hits to the target protein. In the pantothenate synthetase screen, 39 fragments from a library of 1250 were found to stabilize the protein by >0.5 K (a threshold 2.5 times the standard deviation of the control measurements) when tested at a concentration of 10 mM, corresponding to a hit rate of 3.1%. In the ligand-observed NMR validation step, evidence of binding was obtained for 17 of the 39 DSF hits; 16 were shown to compete with the natural substrate ATP and/or the pantoyladenylate reaction intermediate for binding. The hit confirmation rate of 44% provides evidence of successful enrichment by DSF and that the technique has served its purpose; by comparison, a water-LOGSY NMR screen of a pool of randomly selected fragments against pantothenate synthetase yielded just one hit (2% hit rate) [27].

The hit rate from the CYP121 DSF screen was higher than that from the pantothenate synthetase DSF screen. Sixty-six fragments from a library of 665 stabilized CYP121 by >0.8 K (a threshold set at eight times the standard deviation of the control measurements) when tested at a concentration of 5 mM (9.9% hit rate). A selection of the 56 most chemically diverse DSF hits were progressed to ligand-observed NMR screening and 46% of these were observed both to bind and to compete with the enzyme's substrate cyclodityrosine.

In both the pantothenate synthetase and the CYP121 screening cascades, some or all of the NMR-confirmed hits were further evaluated by ITC and X-ray crystallography. $K_D$ values were determined by direct ITC for 14 of the 17 pantothenate synthetase hits, with the compound causing the greatest stabilization ($\Delta T_m$ of 3.0 K) also showing the highest affinity ($K_D$ value of 0.48 mM). Cocrystal structures were obtained (by crystal soaking) for eight fragments and revealed that the fragments were found in three different regions of the binding site. With just one exception, the fragments that failed to give cocrystal structures were among those hits giving rise to thermal shifts only just above the DSF hit threshold ($\Delta T_m$ of 0.5–0.7 K) and included the three fragments not validated by ITC. One of the fragment hits validated both by ITC and X-ray crystallography has been utilized in successful fragment linking strategies to produce inhibitors of pantothenate synthetase binding with low to submicromolar affinity [25,31]. Interestingly, however, the second fragment that formed the basis of these linking strategies (as well as a successful fragment growing strategy [25]) did not stabilize pantothenate synthetase ($\Delta T_m$ of 0 K) under the conditions of the screen and was instead identified in the aforementioned water-LOGSY NMR screen of a random pool of compounds. This reinforces that DSF is sensitive to false negatives and hence that quality fragment hits can be missed if DSF is used as a frontline screening technique. Nonetheless, the DSF-based screen did yield eight validated hits.

In the CYP121 screen, eight of the most soluble NMR-validated fragment hits were taken forward into crystal soaking experiments and cocrystal structures were obtained for four. $K_D$ values were determined for each of these hits by direct ITC ($K_D$ values of 0.4–1.7 mM). The crystal structures confirmed binding in the CYP121 active site and revealed possibilities for fragment merging and linking. Two fragments were successfully linked, and a third fragment that bound in two overlapping poses became the basis for a successful merging strategy. Elaborated compounds with $K_D$ values of 28 μM and 40 μM were generated [24,32].

Analogous to the screens described above, a study aimed at identifying fragments that could form the basis for inhibitors of the interaction between the tumor suppressor BRCA2 and the recombination enzyme RAD51 also made use of DSF as a frontline screening technology and ligand-observed NMR to validate DSF hits (Table 7.1) [29]. In this study, when a library of 1249 fragments was screened (each at 5 mM) against a surrogate RAD51 (a modified archeal orthologue), 3.2% of the fragments were found to stabilize the protein by >1 K (a hit threshold twice the standard deviation of the control measurements), 60% were essentially without effect ($-0.5\,K < \Delta T_m < 0.5\,K$), 22% actually destabilized the protein by >0.5 K, and 8% gave rise to complex thermal melt curves (attributed to optical interference, compound insolubility, or aggregation). 5-Methoxyindole and 5-hydroxyindole, the two well-behaved fragments that induced the greatest stabilization (1.5 and 2 K, respectively), were subsequently pursued. STD-NMR and ITC provided confirmation of binding to the model RAD51 ($K_D$ values of 2.1 mM as measured by direct ITC) and the results of competition experiments (STD-NMR and ITC) were consistent with binding to the shallow site that accommodates a key binding motif of BRC4. This was confirmed by X-ray crystallography. Through subsequent STD-NMR experiments, initially using 5-hydroxyindole as a reporter ligand, fragments that bind to the same site with slightly improved affinity and the potential for elaboration were identified.

In a fragment screen targeting the *M. tuberculosis* transcription factor EthR, a DSF-based enrichment was performed ahead of surface-plasmon resonance (SPR)-based validation of the DSF hits (Table 7.1). In this study, 86 fragments from a library of 1250 were found to stabilize the protein by more than 1 K when tested at a concentration of 10 mM (6.9% hit rate) [28]. Notably, 22 of the fragments actually stabilized the protein by more than 3 K, and five by more than 5 K. These thermal shifts are much higher than the 0.5–2 K shifts more typically observed for fragment hits, which is likely related to the lipophilic nature of the allosteric binding site of the protein. Among the 86 fragment hits, 45 (52%) showed greater than 10% inhibition when tested at a concentration of 0.5 mM in an SPR-based functional assay in which the interaction between EthR and DNA was measured (overall hit rate of 3.6%). By comparison, only one fragment with inhibitory activity in the SPR assay was identified from a selection of 45 of the fragments that did not stabilize the protein. Hence, DSF proved effective as an enrichment technique and gave few false negatives. $IC_{50}$ values were subsequently determined for a selection of the hits with novel scaffolds, and four were soaked into EthR crystals in order to elucidate the binding modes. All four fragments were observed in the binding site of the allosteric regulator hexadecyl octanoate and three were actually found to bind twice. One of these three fragments, a fragment that stabilized the protein by 5.5 K and inhibited the activity of the protein with an $IC_{50}$ value of 280 μM, formed the basis of subsequent chemical elaboration efforts.

In fragment screening campaigns, targeting two different Dengue virus proteins that form part of the flaviviral replication machinery (NS3 helicase and NS5 mRNA methyltransferase), a library of 500 fragments was initially screened by DSF and hits were progressed to X-ray crystallography for hit confirmation and determination of binding modes [30]. In this study, the authors found that the effectiveness of the DSF-based enrichment differed dramatically for the two targets (Table 7.1). The DSF-based screens yielded 36 and 32 fragment hits that reproducibly stabilized the helicase and methyltransferase, respectively, by >0.5 K (a hit threshold based on twice the standard deviation of control measurements). This corresponded to hit rates of 7.2% for the helicase and 6.4% for the methyltransferase. All hits were subsequently individually soaked into helicase or methyltransferase crystals. Surprisingly, none of the crystal structures of the fragment-soaked helicase crystals showed bound fragments. The methyltransferase soaking experiments were, however, more successful, with cocrystal structures obtained for 7 of the 32 DSF hits (22% visualization rate). By comparison, an X-ray crystallography-based fragment screen against another RNA virus target that did not involve an initial DSF enrichment step showed a hit-rate of only 2% [33]. Almost one-third of the helicase fragment hits inhibited the ATPase activity of the protein by more than 25% at a concentration of 2 mM, which was consistent with at least some of the DSF hits being true hits. The authors of the study hypothesized that the inability to obtain hit confirmation by X-ray crystallography is due to fragment binding requiring a conformational change that cannot occur in the crystals. Whether a higher confirmation rate (perhaps after a second biophysical confirmation step) could be obtained by performing cocrystallization experiments would certainly be interesting to see. The seven confirmed methyltransferase hits were observed to bind to four distinct binding sites on the protein, three of which were novel. Interestingly, however, cocrystal structures were not obtained by soaking (or cocrystallization) either for the methyltransferase fragment hit that stabilized the protein most effectively ($\Delta T_m = 4.1$ K) or for three of the four fragment hits showing >30% inhibitory activity in methyltransferase activity assays, indicating that these parameters per se were not effective predictors of crystallographic success. The former helps to warn against pursuing fragments only on the basis of high thermal shifts.

Certainly, as for any technique, DSF should be used with an awareness of the potential drawbacks and pitfalls. A recent fragment-based study by Larsson *et al.* [26] targeting the poly-ADP-ribosylating enzyme tankyrase provides a good example of how DSF can be used carefully and successfully to not only identify quality fragment hits but also guide fragment elaboration. In this study, DSF formed the basis of the primary fragment screen. A library of 500 rule-of-three compliant fragments was screened against the catalytic domain of human tankyrase (isoform 2) initially at a single concentration (1 mM). Fragments that increased the melting temperature of the protein were subsequently tested at a range of concentrations (typically between 5 μM and 4 mM) and only those fragments that caused a dose-dependent stabilization were classified as hits. Tankyrase 2 produced a biphasic melt curve (Figure 7.9a) rather than the more typical sigmoidal curve, which somewhat complicated the analysis of the

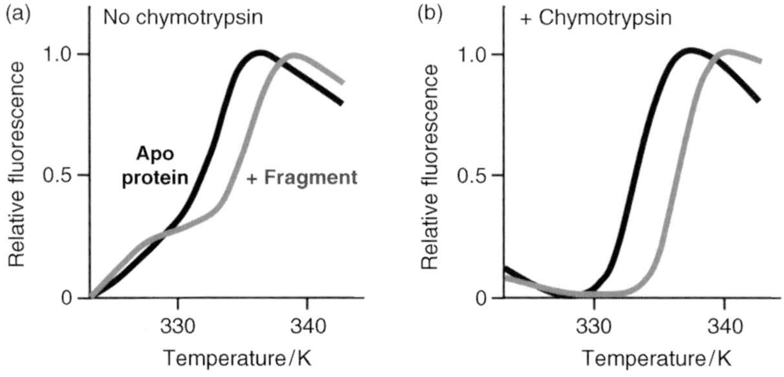


**Figure 7.9** Tankyrase 2 melt curves in the presence and absence of a stabilizing fragment hit, before and after *in situ* chymotrypsination. (Adapted from Larsson *et al.* [26].)

screening data, particularly in the case of weakly stabilizing fragments. However, the authors found that known inhibitors stabilized the most stable transition and therefore focused on this. Additionally, they were even able to remove the less stable transition and simplify the melt curves by performing an *in situ* digestion with chymotrypsin (Figure 7.9b). *In situ* digestion with chymotrypsin had previously been shown to promote crystallization of tankyrase 2, which was presumed to be because it served to remove less-ordered contaminants. Cocrystal structures were obtained (by crystal soaking) for two fragments that stabilized tankyrase 2 in a dose-dependent manner and by more than 1 K at a concentration of 1 mM, and revealed that the compounds bind in the binding site of the natural substrate nicotinamide. Both fragment hits were additionally shown to inhibit tankyrase 2 in a poly-ADP-ribosylation-based activity assay ($IC_{50}$ values of 12 and 4 μM) and one became the focus of a fragment growing strategy. Elaborated compounds were assessed by DSF and in the activity assay, and for a selection of compounds $K_D$ values were determined in direct SPR experiments. Notably, the $\Delta T_m$, $IC_{50}$, and $K_D$ values determined throughout were observed to correlate well. The elaboration efforts resulted in high-affinity inhibitors of tankyrase that, as shown by X-ray crystallography, maintain the binding pose of the parent fragment. The best in the series stabilizes the protein by 10.1 K at a concentration of 125 μM (a concentration at which the initial fragment hits have no effect), inhibits its poly-ADP-ribosylation-based activity with an $IC_{50}$ value on 9 nM, and binds with a $K_D$ value of 16 nM.

### 7.4.2
### DSF Compared with Other Hit Identification Techniques

Attempts to address the question of how well the results of DSF fragment screens correlate with those of fragment screens performed utilizing orthogonal screening technologies have been made in a number of independent studies involving different target proteins. Although it is difficult to draw general conclusions, in part due to the use of nonidentical conditions for different techniques and in part due to the scarcity of complete data sets, we highlight some of the key findings from the comparative studies in the literature.

Basse *et al.* [34] reported the results of two parallel screens of a library of 1895 fragments using either DSF or water-LOGSY NMR as an initial filter to identify fragments binding to a surface pocket of the Y220C mutant of the tumor suppressor p53. The hits from each screen were then taken forward into a protein $^{15}N/^{1}H$ heteronuclear single-quantum coherence (HSQC) NMR validation stage (Figure 7.10). Although the water-LOGSY screen was more sensitive to hits than the DSF screen, producing 70 and 17 validated hits, respectively, it did not produce higher quality hits, that is, both the DSF hits and the water-LOGSY NMR screen hits showed similar validation rates in the subsequent HSQC experiments (approximately 35%).

Interestingly, only three structures overlapped between the two sets of validated hits, presumably reflecting differences in the assays. One might speculate

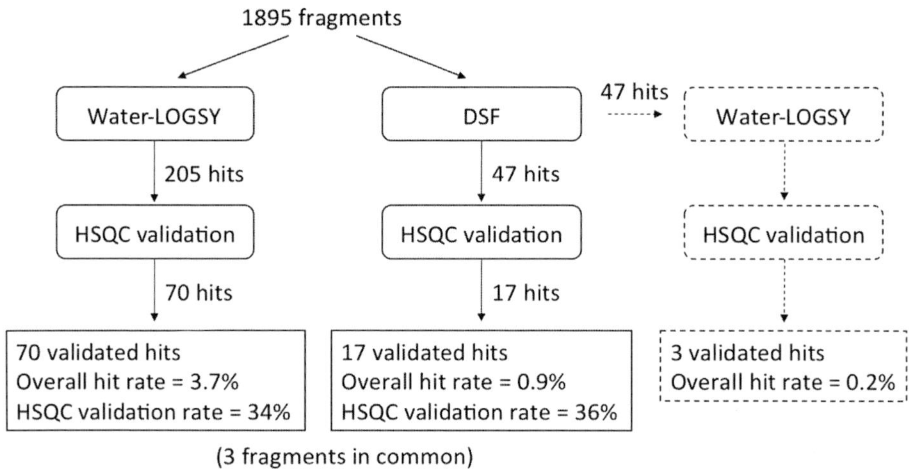


**Figure 7.10** Results of fragment screens carried out against the Y220C mutant of p 53 [34]. The dashed lines indicate a hypothetical screening pipeline for which the results can be inferred from the number of overlapping validated fragments from the other two screening cascades.

that while DSF biases toward entropic binders, ligand-observed NMR assays, which are often performed at lower temperatures, bias toward enthalpic binders. DSF screening is repeatedly purported to be an enrichment process, and is at least as good as water-LOGSY in this respect. However, it would be interesting to know what the HSQC validation rate is for compounds that were negative in the DSF assay, in order to confirm the enrichment hypothesis. If there were no enrichment of a pool of hits in either the DSF or the water-LOGSY assays, one would expect the same validation rate at the HSQC stage (i.e., a random selection of compounds from the library in both cases). Indeed, this is what is observed, but it seems unlikely that neither technique would offer a selection bias for binders versus nonbinders and that the intrinsic hit rate would be as high as 35%. DSF assays remain vastly more efficient in terms of time, protein consumption, and labor than ligand-observed NMR spectroscopy and in this study at least DSF delivers a very similar enrichment rate to water-LOGSY NMR. There is a concern though that there are only three structures that survived the DSF, water-LOGSY, and HSQC validation processes, as it potentially has ramifications for designing screening pipelines. If the authors had taken their DSF hits through water-LOGSY validation and finally HSQC (dashed lines, Figure 7.10), they would have identified only 3 hits rather than the 17 identified by going straight from DSF to HSQC.

To compare DSF, NMR, and SPR, Hubbard and Murray [35] took 80 fragments from their fragment library at Vernalis and screened these against HSP90 (a well-characterized target) using the three different techniques in parallel. In this study, similar hit rates were obtained for the different techniques. As shown in Figure 7.11, 19 fragment hits were identified using DSF (24% hit rate), 17 fragments showed binding by all three NMR techniques used (STD, water-LOGSY,

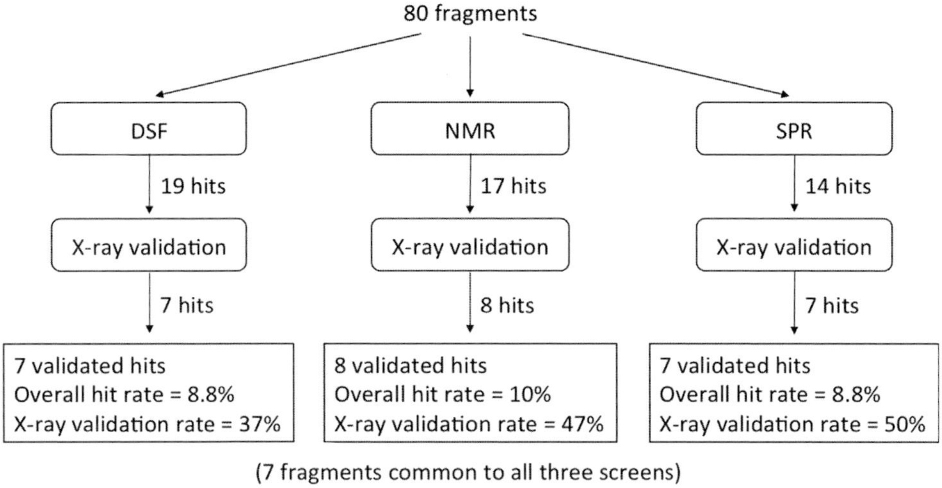


**Figure 7.11** Results of fragment screens carried out against HSP90, as reported by Hubbard and Murray [35].

and Carr–Purcell–Meiboom–Gill (CPMG) NMR; 21% hit rate), and 14 fragments were identified as binders in direct SPR binding experiments (18% hit rate). The more comparable hit rates observed in the primary DSF and NMR screens here (relative to the p53 study described above) may in part reflect the use of three complementary NMR techniques that would reduce the number of NMR false positives. Importantly, each of the screens yielded seven (DSF and SPR) or eight (NMR) fragments that subsequently produced cocrystal structures on the first attempt of crystal soaking, indicating that the quality of hits was also on par. Seven fragments were identified as hits by all three techniques and only one of these failed to yield a cocrystal structure in soaking/cocrystallization experiments. While all the initial SPR hits were also identified as hits in the DSF and/or NMR screens, both DSF and NMR yielded unique hits (seven and two, respectively). The seven unique DSF hits, however, were not validated by X-ray crystallography.

How DSF and SPR perform as tools for validating NMR positive hits was further investigated in this study by following up on hits resulting from NMR screens carried out against a number of different (mostly enzyme) targets using the same in-house library of 1200 compounds. A low confirmation rate (between 8 and 23%) was encountered when DSF was used to validate the NMR hits identified against three enzyme targets (two of which were kinases). The confirmation rate observed for a different yet overlapping set of five targets was substantially higher when SPR (direct binding assays) was used as the follow-up technique (83–100%). For one of the targets, however, they found DSF to serve as an excellent predictor of crystallographic success, with 92% of the NMR hits that yielded crystal structures on the first attempt of crystal soaking having been confirmed as hits by DSF. At the other end of the spectrum, however, DSF predicted only 21% of the NMR hits for a second target that gave rise to crystal structures on the first attempt. The observed target-dependent variability in the performance of DSF as a validation technique, which could result from variability in target thermodynamics, the value of $T_0$ under the experimental conditions, and/or a discrepancy between the conditions used for the DSF screens and those used for NMR and X-ray crystallography, led these researchers to favor other techniques over DSF as frontline hit identification/confirmation technologies.

In an attempt to evaluate a new fragment screening technique known as weak affinity chromatography (WAC), Meiby *et al.* [36] used this technique in parallel with a range of other screening methods (including DSF) to test a library of 111 fragments against the ATPase domain of HSP90 (one of the targets also used by Hubbard and Murray to compare screening techniques). Based on a comparison between which fragments were identified as hits and nonhits, good correlation was observed between the results obtained by DSF and those obtained by WAC (71%), NMR (74%; a combination of STD, water-LOGSY, and relaxation edited experiments – a fragment was considered a hit only when binding and competition with a known binder was observed in all three experiments), SPR (73%; ADP displacement experiments), and fluorescence polarization (FP, 73%) (Figure 7.12). However, higher correlation was observed between NMR and SPR (90%), WAC and NMR (88%), and WAC and SPR (83%). Notably, DSF showed a number of false negatives, with four fragments identified as hits by WAC, NMR, and SPR failing to be detected by DSF, and a further two fragments identified as hits by two of these three alternate screening techniques failing to be detected by DSF. Although FP yielded the highest number of false positives (20 FP hits were not identified as hits by WAC, NMR, or SPR), DSF was not far behind (17 DSF hits were not identified by WAC, NMR, or SPR). Twenty-seven of the highest ranking fragment hits identified by WAC were selected for affinity measurements by ITC and for 14 of these $K_D$ values for HSP90 could be obtained.

| | NMR | SPR | ITC | X-ray | FP | DSF |
|---|---|---|---|---|---|---|
| WAC | 88% (n = 103) | 83% (n = 107) | 52% (n = 27) | 83% (n = 30) | 66% (n = 109) | 71% (n = 107) |
| NMR | | 90% (n = 102) | 63% (n = 24) | 93% (n = 28) | 74% (n = 104) | 74% (n = 102) |
| SPR | | | 73% (n = 26) | 76% (n = 29) | 73% (n = 109) | 73% (n = 109) |
| ITC | | | | 57% (n = 23) | 63% (n = 27) | 58% (n = 26) |
| X-ray | | | | | 73% (n = 30) | 66% (n = 29) |
| FP | | | | | | 73% (n = 109) |

50 – 59% | 60 – 69% | 70 – 79% | 80–89% | ≥ 90%

**Figure 7.12** Correlation observed between the results of fragment screens carried out against HSP90 using different screening techniques [36]. "*n*" represents the number of compounds on which each correlation is based. (Adapted from Meiby *et al.* [36].)

Of these 14 fragment hits, only 1 had not been identified as a hit in the DSF screen. As a predictor of crystallography success, however, DSF performed less well. Thirty-two fragments were taken forward to X-ray crystallography and soaked into HSP90 crystals; cocrystal structures were obtained for 26 of these fragments. Although 72% of the fragments for which crystal structures were obtained were also identified as hits by DSF, seven fragments were not, and when compared with WAC, NMR, SPR, and FP, DSF was the least effective predictor of crystallography success (Figure 7.12).

The authors highlight that they chose to use the optimal conditions for each technique rather than aiming for the conditions that would allow the best direct comparison between techniques. Hence, this may in part contribute to some of the observed discrepancies between the screening results obtained using DSF and the results obtained using other techniques. One notable difference between the DSF screen and the WAC, NMR, SPR, and FP screens performed is that the screen did not discriminate between fragments binding to the active site and fragments binding elsewhere to the protein, either type of fragment would be classified as a hit. The WAC, NMR, SPR, and FP assays all incorporated competition with a known active site binder into the criteria for a hit, and hence should yield only active site binding fragments. This, at least in part, may explain the number of seemingly false positive hits identified.

In a screening campaign targeting the protein kinase MEK1, Amaning *et al.* [37] compared DSF and SPR as techniques for prioritizing fragment hits from a virtual screen for follow-up by X-ray crystallography. As in the HSP90 study of Meiby *et al.* [36] described above, the ATP binding region of the protein was the target. At the onset of the campaign against this oncology target, a biochemical assay had not been established, and the combination of a virtual fragment screen with biophysical validation was seen as a means by which they could jump-start their drug discovery program. A virtual screen of an in-house library of 10 000 fragments was performed using the published MEK1 crystal structure, and the top 196 scoring fragments (~2%) were screened in parallel by DSF and SPR (Figure 7.13). Based on a low thermal shift cutoff of 0.2 K, 44 virtual screening hits were identified as hits by DSF (22% hit rate). Thirty fragments were identified as hits by SPR (15% hit rate, direct binding experiments); however, only twelve (40%) of these were also identified as hits by DSF. The DSF hits with thermal shifts greater than or equal to 1.2 K (seven fragments, 4%) were subsequently progressed to crystal soaking experiments and five (71%) yielded cocrystal structures. Notably only two of the five DSF hits for which crystal structures were obtained were identified as hits by SPR, revealing the SPR assay to be sensitive to false negatives. The authors of this study echo the hypothesis that because DSF hits stabilize the target to thermal denaturation, they will have a higher propensity to yield cocrystal structures [18]. Certainly a high proportion of the hits giving rise to thermal shifts greater than 1.2 K yielded crystal structures; however, how the SPR positive hits giving rise to little or no thermal shift behaved in crystal soaking experiments was not reported and hence whether DSF is a better predictor of crystallographic success than SPR could not

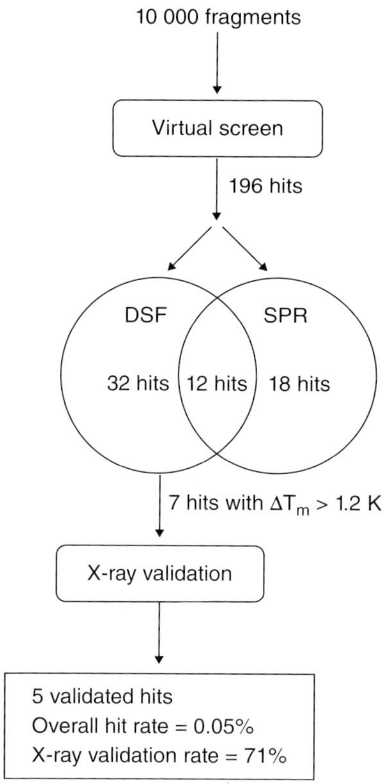


**Figure 7.13** Results of fragment screens carried out against MEK1 as reported by Amaning *et al.* [37].

be concluded for this target. Interestingly, the crystal soaking experiments were performed at 277 K, and hence the affinity of the fragments under the DSF and crystallography experiments might differ substantially.

DSF was also recently compared with a high concentration AlphaScreen™ kinase assay as a hit-finding technology in a fragment screen carried out against checkpoint kinase 2 (CHK2), a serine/threonine kinase [38]. When a library of 1869 fragments was screened in an AlphaScreen™ kinase assay, 45 fragments were observed to inhibit the phosphorylation of the substrate peptide by more than 50% (2.4% hit rate) (Figure 7.14) when tested at a single concentration (300 μM). Fragments found to interfere with the readout of the AlphaScreen™ were subsequently removed from this hit list, yielding a refined hit list of 20 fragments (1.1% hit rate). When the same library was screened by DSF, 63 fragments were found to increase the melting temperature of the protein above the hit threshold (3.4% hit rate). The thermal shifts measured for the hits ranged between 0.9 and 7.0 K. Fourteen fragment hits were common between the two screens; however, two were among those discarded due to interference in the

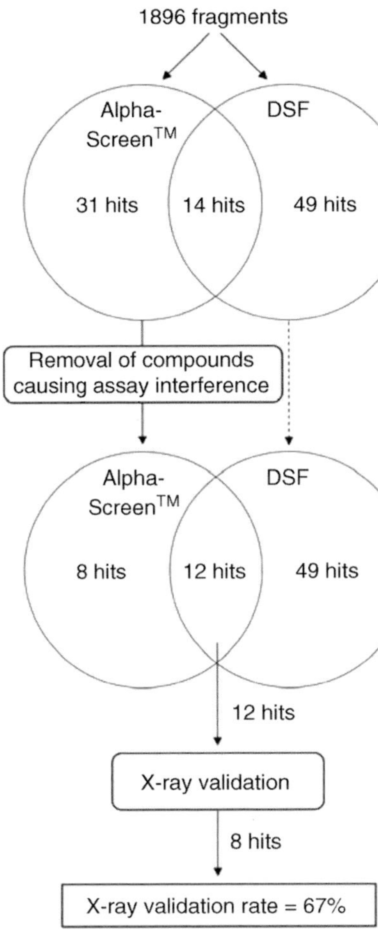


**Figure 7.14** Results of fragment screen carried out against checkpoint kinase 2 as reported by Silva-Santisteban *et al.* [38].

AlphaScreen™. Interestingly, good correlation was observed between concentrations causing 50% inhibition ($IC_{50}$ values) and thermal shifts determined for the twelve fragment hits common between the two screens ($R^2 = 0.9225$), with the compound with the lowest $IC_{50}$ value (2.7 μM) also causing the biggest thermal shift (7 K). Structures were obtained (by cocrystallization) for eight of the fragments in complex with the kinase domain of CHK2. All were observed to bind to the hinge region of the ATP binding site.

The set of compounds identified as hits in the AlphaScreen™ but not in the DSF screen, contained mostly fragments that were subsequently found to interfere with the readout of the AlphaScreen™ and/or fragments with limited solubility. Of the 49 fragments identified as hits by DSF but not in the AlphaScreen™ assay, 13 were found to have limited solubility. It is possible that at least some of the remainder bind yet do not affect the activity of the protein or at least not at the concentration used in the AlphaScreen™ (which was nearly seven times lower than that used in the DSF assay). Interestingly, three of the five fragments giving rise to the highest thermal shifts were found in this set of compounds. Although all were confirmed to be soluble, the $IC_{50}$ values measured for these fragments were much higher than those determined for the fragments also identified as hits in the AlphaScreen™. Only one of these three fragments (a fragment with an $IC_{50}$ value of 228 µM, $\mathbf{\Delta} T_m = 4.0$ K) yielded a cocrystal structure. Like the AlphaScreen™ positive hits, the fragment was observed to bind to the hinge region of the ATP binding site. In this study, DSF proved to be effective at enriching for checkpoint 2 kinase inhibitors as well as removing compounds that interfered with the AlphaScreen™. Furthermore, at least among the AlphaScreen™ positive fragment hits, the measured thermal shifts were predictive of inhibitory activity.

### 7.4.3 Pursuing Destabilizing Fragment Hits

In a typical fragment screen, a number of fragments that destabilize the target are identified. These fragments are commonly discarded. In a recent fragment-based project aimed at targeting *M. tuberculosis* 7,8-diaminopelargonic acid synthase (BioA), the aminotransferase that catalyzes the second step in biotin biosynthesis, destabilizing fragments were pursued with interesting outcomes [39]. In this study, DSF was used to screen a library of ~1000 fragments against BioA (a homodimer that utilizes the cofactor pyridoxal-5′-phosphate, PLP, Figure 7.15a), and compounds that stabilized or destabilized holo (PLP-bound) BioA by more than 2 K (9 and 12 compounds, respectively) were progressed to X-ray crystallography-based validation experiments. A structure was obtained (by cocrystallization) for just one hit in complex with the protein – a fragment that destabilizes the protein by 7 K (fragment **1**, Figure 7.15a). Holo BioA was observed to accommodate the compound in one of the two available active sites (in the position typically occupied by the natural substrate *S*-adenosyl methionine, SAM), without undergoing significant conformational change (Figure 7.15b). Binding of this compound to BioA in solution was independently confirmed using STD-NMR.

Commercially available analogs of fragment hit **1** were tested, and a hydrazine-bearing analog that destabilizes holo BioA by 18 K was identified (fragment **2**, Figure 7.15a). In both soaking and cocrystallization experiments, this compound was observed to change the color of holo BioA from yellow to red, indicative of a chemical reaction taking place. A crystal structure of BioA in complex with this compound (obtained by cocrystallization) revealed that the hydrazine moiety of the fragment reacts with the aldehyde of PLP to form a stable covalent quinonoid adduct (compound **3**, Figure 7.15a); the bond that covalently links PLP to a lysine residue (Lys283) of BioA is as a consequence broken and two local conformational changes occur in the active site (Figure 7.15c). Dialysis experiments

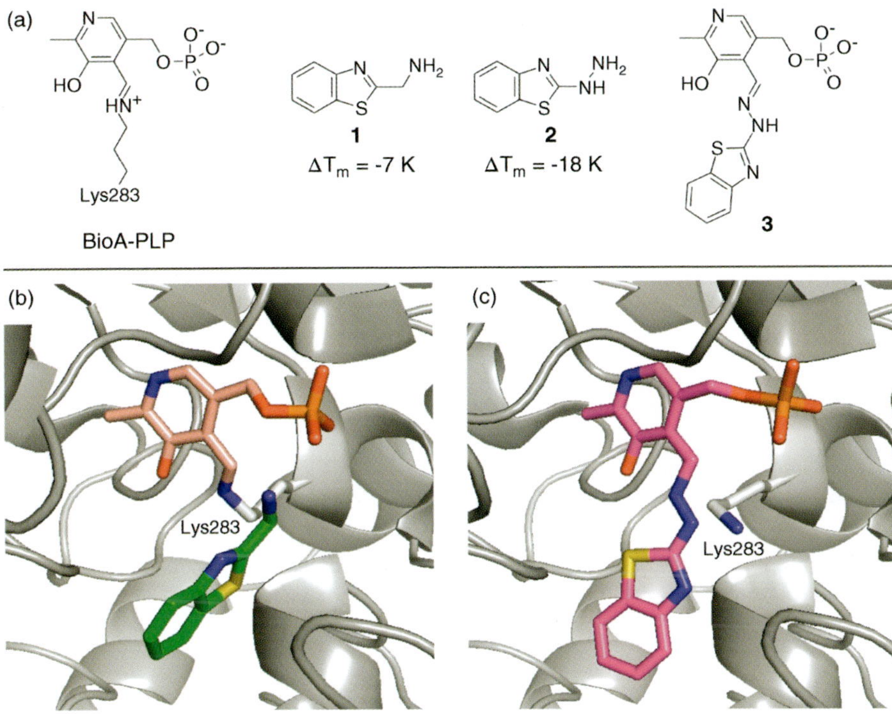


**Figure 7.15** Binding of destabilizing fragment hits to BioA. (a) Chemical structures of PLP, the original fragment hit (**1**), the hydrazine analog of fragment **1** (**2**), and the quinonoid adduct formed between hydrazine **2** and PLP (**3**). (b and c) X-ray crystal structures showing binding of fragment **1** (green) to BioA with PLP (pink) bound (b, PDB ID: 4CXR) and the quinoid adduct (magenta) formed upon reaction of hydrazine fragment **2** with PLP (c, PDB ID: 4MQP). The side chain of Lys283 is shown. Atoms are colored as follows: nitrogen, blue; oxygen, red; phosphorus, orange; sulfur, yellow.

revealed that adduct formation is reversible, and kinetic analyses were consistent with the compound acting as a reversible competitive inhibitor of BioA with respect to the natural substrate SAM (inhibition constant ($K_i$) value of 10 μM) and an uncompetitive inhibitor with respect to a second substrate, 2-keto-8-aminopelargonic acid or KAPA ($K_i$ value of 85 μM). Two hydrazide-bearing analogs tested (one of which was the antituberculosis agent isoniazid) were also observed (by X-ray crystallography) to form reversible adducts with holo BioA and to reduce the protein's melting temperature by more than 8 K; however, neither showed an inhibitory effect. This was hypothesized to be because the resultant adducts are less stable than that formed by the hydrazine.

How the initial fragment hit and the hydrazine and hydrazide analogs are able to so effectively destabilize holo BioA is poorly understood; however, the most likely explanation is that they stabilize the unfolded state of the protein (refer to Section 7.2.7). This study certainly highlights that destabilizing fragments can be effective inhibitors and that if the dogma of pursuing only stabilizing hits is strictly adhered to, valuable hits may be discarded. The authors of this paper also suggest that destabilizing compounds per se may be of value; they suggest that they could be used to destabilize proteins *in vivo*, which in turn may result in the target protein being more rapidly degraded and in some cases depleted. This hypothesis certainly requires in cell investigations on a target-by-target basis.

### 7.4.4
### Lessons Learnt from Literature Examples of DSF in Fragment-Based Drug Discovery

Using DSF, a fragment library can be screened rapidly and at a low cost. Nonetheless, some researchers have concerns over the use of DSF as a frontline fragment screening technique, largely because of the potential for false negatives and to a lesser extent false positives, as well as target-dependent variability [22]. In the literature examples discussed herein, fragments that significantly increase the melting temperature of all targets were identified; however, varying hit rates were encountered (from about 3% to greater than 20%). As discussed earlier in this chapter, thermal shifts are affected both by the inherent properties of the protein and by the screening conditions, and this in turn contributes to the variability in hit rates observed.

Evidence that DSF-based screens facilitate successful enrichment of fragment hits has been presented. Subsequent confirmation rates varied greatly, which is likely to be because of inherent differences between the conditions of orthogonal screens performed. In one extreme presented, no DSF hits were confirmed in a subsequent X-ray crystallography screen; however, in an analogous screening cascade against another target, 22% of hits were confirmed, and in other studies where secondary validation was performed by NMR or SPR, confirmation rates were above 40%. Possible reasons for high attrition rates (or "false positives") include large temperature discrepancies between screens, differences in the availability of relevant target conformations, and, where validation involves competition with known binders, binding to additional sites on the protein. In the studies discussed, there were examples of confirmed hits identified by orthologous screening technologies (some of which were even validated by X-ray crystallography) that failed to be detected by DSF (false negatives). Although it is unlikely this can be avoided entirely, careful optimization of assay conditions to minimize $T_0$ of the apo protein should help to reduce the numbers of false negatives and maximize hit rates. Nonetheless, at least for some targets, the capacity to screen more compounds is likely to offset higher rates of false negatives.

Although examples in which thermal shifts correlate with affinity are found in the literature, it is important to be aware that this correlation is typically observed only for similar compounds (with similar binding enthalpies) within a series, and it is not necessarily advantageous to pursue fragments only on the basis of large thermal shifts. With new literature examples of compounds that produce negative thermal shifts being validated as binders of the native target by orthogonal techniques, interest in compounds producing negative thermal shifts has increased. As compounds can interact with both the unfolded and the folded states of the protein, a negative thermal shift does not preclude binding to the native state. However, to the best of our knowledge, there is as yet no evidence that negative thermal shifts per se are indicative of binding to the native state of a protein.

## 7.5 Concluding Remarks

On first inspection, DSF is conceptually a simple technique, but there is some subtlety when one considers how the interplay of the ligand binding thermodynamics and the protein unfolding thermodynamics give rise to a thermal shift for a ligand. For a given melting temperature and $\Delta C_p$ of a protein, proteins with a lower enthalpy of unfolding are stabilized to a greater degree by a given ligand and give larger thermal shifts. There is an inherent bias toward identifying entropically driven binders in these experiments, which is due to the way in which binding constants for different ligands change with temperature. Optimizing conditions to minimize $T_0$ may help to mediate against this entropy-driven bias, as should attempts to maximize $\Delta T_m$ of a known enthalpically driven ligand. Despite this bias, there is some evidence of correlation between $\Delta T_m$ and $K_D$ within a series of compounds for a given protein target. Thermal shift experiments are further complicated by the possibility of ligand binding to the unfolded state, which gives rise to negative thermal shifts. A combination of native state binding and denatured state binding can give rise to either positive or negative thermal shifts. Therefore, there may be cases where a negative shift is observed but the compound binds the folded state and is still of interest.

The often-missing correlation between room temperature $K_D$ values and observed thermal shifts is, in part, due to the temperature dependence of $K_D$ values. Room temperature alternatives to thermal denaturation, such as guanidinium titrations, do exist. Such experiments have the advantage of bringing the experiment to a constant temperature so that the ligand affinity does not change. However, multiple data points are required to obtain a complete titration, reducing throughput [40]. In the future, it will be interesting to see if the use of DSF can be extended to studying compounds that disrupt dimers or multimers, with the expectation that interpretation of the results will be more complicated.

Compared to other screening techniques (e.g., NMR, SPR, or X-ray crystallography), DSF remains a relatively high-throughput technique that is simple and quick to execute. Despite the caveats previously discussed, a reproducible positive shift in a DSF screen is indicative of protein binding. We have discussed a number of studies in which DSF has worked as an enrichment tool that justify its use in frontline screens against many target proteins.

## Acknowledgments

We would like to thank Walter Ward for helpful discussions and advice.

## References


1 Lo, M.C., Aulabaugh, A., Jin, G., Cowling, R., Bard, J., Malamas, M., and Ellestad, G. (2004) Evaluation of fluorescence-based thermal shift assays for hit identification in drug discovery. *Analytical Biochemistry*, **332**, 153–159.

2 Pantoliano, M.W., Petrella, E.C., Kwasnoski, J.D., Lobanov, V.S., Myslik, J., Graf, E., Carver, T., Asel, E., Springer, B.A., Lane, P., and Salemme, F.R. (2001) High-density miniaturized thermal shift assays as a general strategy for drug discovery. *Journal of Biomolecular Screening*, **6**, 429–440.

3 Torres, F.E., Recht, M.I., Coyle, J.E., Bruce, R.H., and Williams, G. (2010) Higher throughput calorimetry: opportunities, approaches and challenges. *Current Opinion in Structural Biology*, **20**, 598–605.

4 Sturtevant, J.M. (1987) Biochemical applications of differential scanning calorimetry. *Annual Review of Physical Chemistry*, **38**, 463–488.

5 Edge, V., Allewell, N.M., and Sturtevant, J. M. (1985) High-resolution differential scanning calorimetric analysis of the subunits of *Escherichia coli* aspartate transcarbamoylase. *Biochemistry*, **24**, 5899–5906.

6 Manly, S.P., Matthews, K.S., and Sturtevant, J.M. (1985) Thermal denaturation of the core protein of lac repressor. *Biochemistry*, **24**, 3842–3846.

7 Robertson, A.D. and Murphy, K.P. (1997) Protein structure and the energetics of protein stability. *Chemical Reviews*, **97**, 1251–1268.

8 Brandts, J.F. (1964) The thermodynamics of protein denaturation. I. The denaturation of chymotrypsinogen. *Journal of the American Chemical Society*, **86**, 4291–4301.

9 Brandts, J.F. and Lin, L.N. (1990) Study of strong to ultratight protein interactions using differential scanning calorimetry. *Biochemistry*, **29**, 6927–6940.

10 Li, J.G., Raffa, R.B., Cheung, P., Tzeng, T.B., and Liu-Chen, L.Y. (1998) Apparent thermodynamic parameters of ligand binding to the cloned rat μ-opioid receptor. *European Journal of Pharmacology*, **354**, 227–237.

11 Arkin, M.R., Randal, M., DeLano, W.L., Hyde, J., Luong, T.N., Oslob, J.D., Raphael, D.R., Taylor, L., Wang, J., McDowell, R.S., Wells, J.A., and Braisted, A.C. (2003) Binding of small molecules to an adaptive protein–protein interface. *Proceedings of the National Academy of Sciences of the United States of America*, **100**, 1603–1608.

12 Fedorov, O., Marsden, B., Pogacic, V., Rellos, P., Muller, S., Bullock, A.N., Schwaller, J., Sundstrom, M., and Knapp, S. (2007) A systematic interaction map of validated kinase inhibitors with Ser/Thr kinases. *Proceedings of the National Academy of Sciences of the United States of America*, **104**, 20523–20528.

13 Matulis, D., Kranz, J.K., Salemme, F.R., and Todd, M.J. (2005) Thermodynamic stability of carbonic anhydrase: measurements of binding affinity and stoichiometry using ThermoFluor. *Biochemistry*, **44**, 5258–5266.

14 Schulz, M.N., Landstrom, J., and Hubbard, R.E. (2013) MTSA: a Matlab program to fit thermal shift data. *Analytical Biochemistry*, **433**, 43–47.

15 Zubriene, A., Matuliene, J., Baranauskiene, L., Jachno, J., Torresan, J., Michailoviene, V., Cimmperman, P., and Matulis, D. (2009) Measurement of nanomolar dissociation constants by titration calorimetry and thermal shift assay: radicicol binding to Hsp90 and ethoxzolamide binding to CAII. *The International Journal of Molecular Science*, **10**, 2662–2680.

16 Waldron, T.T. and Murphy, K.P. (2003) Stabilization of proteins by ligand binding: application to drug screening and determination of unfolding energetics. *Biochemistry*, **42**, 5058–5064.

17 Scott, A.D., Phillips, C., Alex, A., Flocco, M., Bent, A., Randall, A., O'Brien, R., Damian, L., and Jones, L.H. (2009) Thermodynamic optimisation in drug discovery: a case study using carbonic anhydrase inhibitors. *ChemMedChem*, **4**, 1985–1989.

18 Vedadi, M., Niesen, F.H., Allali-Hassani, A., Fedorov, O.Y., Finerty, P.J. Jr., Wasney, G.A., Yeung, R., Arrowsmith, C., Ball, L.J., Berglund, H., Hui, R., Marsden, B.D., Nordlund, P., Sundstrom, M., Weigelt, J., and Edwards, A.M. (2006) Chemical screening methods to identify ligands that promote protein stability, protein crystallization, and structure determination. *Proceedings of the National Academy of Sciences of the United States of America*, **103**, 15835–15840.

19 Niesen, F.H., Berglund, H., and Vedadi, M. (2007) The use of differential scanning fluorimetry to detect ligand interactions that promote protein stability. *Nature Protocols*, **2**, 2212–2221.

20 Kranz, J.K. and Schalk-Hihi, C. (2011) Protein thermal shifts to identify low molecular weight fragments. *Methods in Enzymology*, **493**, 277–298.

21 Steinberg, T.H., Jones, L.J., Haugland, R.P., and Singer, V.L. (1996) SYPRO orange and SYPRO red protein gel stains: one-step fluorescent staining of denaturing gels for detection of nanogram levels of protein. *Analytical Biochemistry*, **239**, 223–237.

22 Davis, B.J. and Erlanson, D.A. (2013) Learning from our mistakes: the 'unknown knowns' in fragment screening. *Bioorganic and Medicinal Chemistry Letters*, **23**, 2844–2852.

23 Mashalidis, E.H., Sledz, P., Lang, S., and Abell, C. (2013) A three-stage biophysical screening cascade for fragment-based drug discovery. *Nature Protocols*, **8**, 2309–2324.

24 Hudson, S.A., McLean, K.J., Surade, S., Yang, Y.Q., Leys, D., Ciulli, A., Munro, A.W., and Abell, C. (2012) Application of fragment screening and merging to the discovery of inhibitors of the *Mycobacterium tuberculosis* cytochrome P450 CYP121. *Angewandte Chemie*, **51**, 9311–9316.

25 Hung, A.W., Silvestre, H.L., Wen, S., Ciulli, A., Blundell, T.L., and Abell, C. (2009) Application of fragment growing and fragment linking to the discovery of inhibitors of *Mycobacterium tuberculosis* pantothenate synthetase. *Angewandte Chemie*, **48**, 8452–8456.

26 Larsson, E.A., Jansson, A., Ng, F.M., Then, S.W., Panicker, R., Liu, B., Sangthongpitag, K., Pendharkar, V., Tai, S.J., Hill, J., Dan, C., Ho, S.Y., Cheong, W.W., Poulsen, A., Blanchard, S., Lin, G.R., Alam, J., Keller, T.H., and Nordlund, P. (2013) Fragment-based ligand design of novel potent inhibitors of tankyrases. *Journal of Medicinal Chemistry*, **56**, 4497–4508.

27 Silvestre, H.L., Blundell, T.L., Abell, C., and Ciulli, A. (2013) Integrated biophysical approach to fragment screening and validation for fragment-based lead discovery. *Proceedings of the National Academy of Sciences of the United States of America*, **110**, 12984–12989.

28 Surade, S., Ty, N., Hengrung, N., Lechartier, B., Cole, S.T., Abell, C., and Blundell, T.L. (2014) A structure-guided fragment-based approach for the discovery of allosteric inhibitors targeting the lipophilic binding site of transcription factor EthR. *The Biochemical Journal*, **458**, 387–394.

29 Scott, D.E., Ehebauer, M.T., Pukala, T., Marsh, M., Blundell, T.L., Venkitaraman, A.R., Abell, C., and Hyvonen, M. (2013) Using a fragment-based approach to target protein–protein interactions. *ChemBioChem*, **14**, 332–342.

30 Coutard, B., Decroly, E., Li, C., Sharff, A., Lescar, J., Bricogne, G., and Barral, K. (2014) Assessment of Dengue virus helicase and methyltransferase as targets for fragment-based drug discovery. *Antiviral Research*, **106**, 61–70.

31 Sledz, P., Silvestre, H.L., Hung, A.W., Ciulli, A., Blundell, T.L., and Abell, C. (2010) Optimization of the interligand Overhauser effect for fragment linking: application to inhibitor discovery against *Mycobacterium tuberculosis* pantothenate synthetase. *Journal of the American Chemical Society*, **132**, 4544–4545.

32 Hudson, S.A., Surade, S., Coyne, A.G., McLean, K.J., Leys, D., Munro, A.W., and Abell, C. (2013) Overcoming the limitations of fragment merging: rescuing a strained merged fragment series targeting *Mycobacterium tuberculosis* CYP121. *ChemMedChem*, **8**, 1451–1456.

33 Antonysamy, S.S., Aubol, B., Blaney, J., Browner, M.F., Giannetti, A.M., Harris, S.F., Hebert, N., Hendle, J., Hopkins, S.,

Jefferson, E., Kissinger, C., Leveque, V., Marciano, D., McGee, E., Najera, I., Nolan, B., Tomimoto, M., Torres, E., and Wright, T. (2008) Fragment-based discovery of hepatitis C virus NS5b RNA polymerase inhibitors. *Bioorganic and Medicinal Chemistry Letters*, **18**, 2990–2995.

34 Basse, N., Kaar, J.L., Settanni, G., Joerger, A.C., Rutherford, T.J., and Fersht, A.R. (2010) Toward the rational design of p53-stabilizing drugs: probing the surface of the oncogenic Y220C mutant. *Chemistry and Biology*, **17**, 46–56.

35 Hubbard, R.E. and Murray, J.B. (2011) Experiences in fragment-based lead discovery. *Methods in Enzymology*, **493**, 509–531.

36 Meiby, E., Simmonite, H., le Strat, L., Davis, B., Matassova, N., Moore, J.D., Mrosek, M., Murray, J., Hubbard, R.E., and Ohlson, S. (2013) Fragment screening by weak affinity chromatography: comparison with established techniques for screening against HSP90. *Analytical Chemistry*, **85**, 6756–6766.

37 Amaning, K., Lowinski, M., Vallee, F., Steier, V., Marcireau, C., Ugolini, A., Delorme, C., Foucalt, F., McCort, G., Derimay, N., Andouche, C., Vougier, S., Llopart, S., Halland, N., and Rak, A. (2013) The use of virtual screening and differential scanning fluorimetry for the rapid identification of fragments active against MEK1. *Bioorganic and Medicinal Chemistry Letters*, **23**, 3620–3626.

38 Silva-Santisteban, M.C., Westwood, I.M., Boxall, K., Brown, N., Peacock, S., McAndrew, C., Barrie, E., Richards, M., Mirza, A., Oliver, A.W., Burke, R., Hoelder, S., Jones, K., Aherne, G.W., Blagg, J., Collins, I., Garrett, M.D., and van Montfort, R.L. (2013) Fragment-based screening maps inhibitor interactions in the ATP-binding site of checkpoint kinase 2. *PLoS ONE*, **8**, e65689.

39 Dai, R., Wilson, D.J., Geders, T.W., Aldrich, C.C., and Finzel, B.C. (2014) Inhibition of Mycobacterium tuberculosis transaminase BioA by aryl hydrazines and hydrazides. *ChemBioChem*, **15**, 575–586.

40 Mahendrarajah, K., Dalby, P.A., Wilkinson, B., Jackson, S.E., and Main, E.R. (2011) A high-throughput fluorescence chemical denaturation assay as a general screen for protein–ligand binding. *Analytical Biochemistry*, **411**, 155–157.

# 8
# Emerging Technologies for Fragment Screening

*Sten Ohlson and Minh-Dao Duong-Thi*

## 8.1
## Introduction

Fragment-based drug discovery (FBDD) has recently established itself as a viable technology in the drug discovery tool box for discovering, developing, and characterizing drug molecule binding to various targets as can be judged by the number of reviews [1–9] and edited books [10–12] in the area. FBDD, as well as virtual screening and structure-based design, should be regarded as complementary approaches in addition to high-throughput screening (HTS) [13–15], especially in cases where HTS has not provided a successful outcome (about 50–60% of drug discovery campaigns [13]). The truth of the matter is that basically no drug discovery procedure can rely on only one method for hit and lead identification as they all have their inherent pros and cons and they have to be integrated with each other to enhance the chances for success in a drug discovery program. Many pharma companies adopt in parallel both HTS and FBDD for the same targets to achieve a higher rate of promising lead candidates. FBDD has now come a long way since the pioneering work by Shuker *et al.* [16] in 1996 and the technology has been successfully used by many pharmaceutical companies and to some extent by academia.

FBDD is also used under other acronyms including FBLD; fragment-based ligand or lead discovery and FBLG; fragment-based lead generation. However, FBDD is the most commonly used term. Striking evidence that FBDD has proven itself as a valuable tool is the fact that more than 25 drug candidates are now in clinical trials based on FBDD [9]. Only a few years ago, one fragment-discovered drug (vemurafenib) was approved by the regulatory authorities for the treatment of metastatic melanoma [17,18].

One of the major challenges in fragment screening is to detect and characterize the weak binding occurring (μM–mM) between the target and the interacting ligands. Therefore, it is critical to have robust and sensitive methods that can detect this fragile binding between target and fragment. Fragments typically bind, if they bind, with affinity in the range from μM to mM. There are two major consequences with weak binders. First, the amount of coverage on the

*Fragment-based Drug Discovery: Lessons and Outlook*, First Edition. Edited by Daniel A. Erlanson and Wolfgang Jahnke.

target is much less at weaker affinities. For example, for nM binders you only need nM concentrations to reach a state with 50% bound at equilibrium whereas a million times higher concentration is required (mM) for mM binders. This necessitates screening approaches to be at high concentrations of the fragment and the consumption of materials can, therefore, be considerable. Second, the kinetics is much faster for weak binders especially in the off-rates where the receptor–ligand pair can dissociate within a fraction of a second. The weak interaction is much more dynamic and the equilibrium can be changed almost momentarily upon changes in the ligand concentration. Furthermore, the specificity is also of a major concern when exploring weak receptor–ligand interactions as there are possibilities for numerous competing weak binding events in the system under study. For example, a protein target can present numerous weak binding spots for interaction especially in the mM range of affinities that are actually outside of the active site of the protein. Another issue is that a fragment is a small molecule usually defined as having a molecular weight (MW) less than 300 Da, which can cause problems with sensitivity in certain fragment screening approaches.

Fortunately, there are a number of methods available that can readily detect and give information on the binding event in terms of affinity, kinetics, stoichiometry, and specificity. It is now a common procedure that fragment hits should be confirmed by multiple orthogonal approaches in order to have confidence in them when progressing toward lead candidates. No single technique can cover all aspects of binding for all types of targets. A specific important area, which presents special challenges for fragment screening techniques, is membrane proteins as they represent 60% of all drug targets [19]. In addition, protein–protein interaction targets are generally considered to be challenging.

As of today, the most established technologies in FBDD in the pharma industry and to some extent in academia are NMR, SPR, and X-ray crystallography. They are all presented in Chapter 6 of this book. In addition, bioassays in particular for enzyme targets also play a major role in fragment screening. All these methods carry their inherent pros and cons and to some extent they complement each other. They all involve specific features in terms of capacity to detect binding, sensitivity, throughput, interference of nonspecific binding, and robustness. Hence, they should all be included in the toolbox of every user seriously considering fragment screening as a tool for drug discovery.

However, there are a number of emerging technologies that hold promise to remedy some of the deficiencies of the more established methods. These novel approaches are not currently in widespread usage because they are not that well known to the FBDD community and they have not yet been validated toward a number of diverse targets. It remains to be seen how suitable these technologies will be for fragment screening. In this chapter, we will cover some new approaches as new tools for fragment screening including weak affinity chromatography (WAC), mass spectroscopy (MS), and microscale thermophoresis (MST).

## 8.2
## Emerging Technologies

### 8.2.1
### Weak Affinity Chromatography

#### 8.2.1.1 Introduction

Since its birth in a modern format in the 1960s, affinity chromatography [20], also called affinity purification, has developed into a powerful purification procedure where a molecule of interest can be isolated to full purity and homogeneity in one step from a crude matrix such as blood plasma or cell culture supernatant. It is not unusual that purification factors of 1000-fold can be achieved in this process. There are numerous applications of this technique; two illustrative examples are the isolation of antibodies with the bacterial $F_c$ protein A and of recombinant proteins equipped with poly His-tags using metal-chelate chromatography. Affinity chromatography is based on the specific binding between molecules such as a target (e.g., a protein) interacting with various ligands. The protein in this case is immobilized to a solid support such as a carbohydrate-based material. When a crude sample is passed through the immobilized protein usually in a column format, the interacting molecules are specifically bound to the immobilized protein whereas all other substances are washed out by the flow of solvent through the affinity unit. After thorough washing, the bound ligands are eluted by a drastic change in the physiochemical conditions such as pH or ionic strength resulting in a pure sample of the desired molecules.

In the 1970s, affinity chromatography was further developed into a high-performance mode where the technique was integrated into an HPLC format [21,22]. By doing so, high-speed separations with maximum resolution and specificity were obtained and the benefits of the HPLC platform were fully utilized. In this way, the merits of both techniques of affinity chromatography and HPLC were exploited. This technique was called HPLAC, high-performance liquid affinity chromatography, or HPAC, high-performance affinity chromatography. HPLC support matrices were here used such as derivatized porous microparticulate silica, highly cross-linked polysaccharides, and synthetic polymer monoliths. HP(L)AC has now found many, mainly analytical, applications especially to examine biomolecular interactions (see a recent review: [23]) and for diagnostic assays to quickly quantify biomolecules in crude media [24,25].

Affinity chromatography is typically a high-affinity technique where molecules of interest are basically adsorbed by the immobilized target such as a protein under high affinity ($K_D < \mu M$). It can be argued that if this type of an affinity separation is really a chromatographic approach, it should rather be considered as an extraction procedure based on the strong interactions between target and ligand. Chromatographic procedures in general are mostly characterized as differential retention of various analytes interacting differently with a matrix resulting in a chromatogram where they are separated due to their inherent partition to the matrix. However, by performing affinity chromatography under weak or

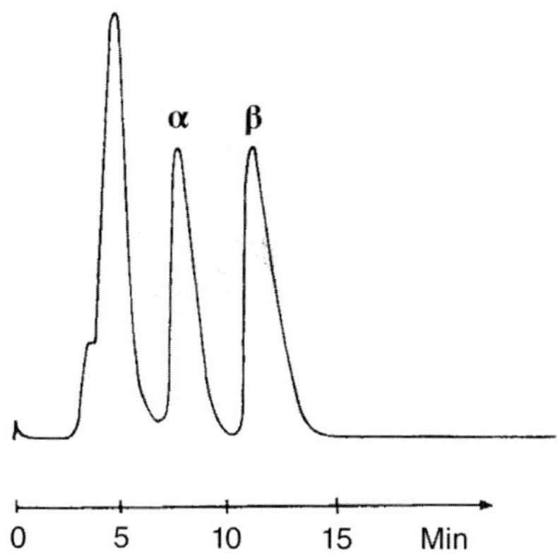


**Figure 8.1** Weak affinity chromatography by immobilized monoclonal antibodies of the anomeric forms of α- and β-*p*-nitrophenyl glycosides of maltose in the presence of bovine serum [27]. The first peak at 5 min represents serum components not having any affinity to target, whereas the separated carbohydrates bind to the monoclonal antibodies in the range of $K_D = 10^{-3}$–$10^{-4}$ M. The separation was performed isocratically in phosphate buffer and the analytes were detected at 300 nm.

low-affinity conditions ($K_D > \mu M$; usually in the range of mM to μM of $K_D$), it should be possible to separate and differentiate between analytes according to their differing affinities for the target protein and resulting in chromatograms where each molecule is easily recognized by its respective retention. This opportunity to run affinity chromatography in a weak affinity mode was more rationally recognized during the 1980s and was consequently coined as weak affinity chromatography [26–29]. The key to the realization of WAC was the implementation of high loads of immobilized targets to achieve appreciable retention of analytes in the chromatographic system.

An early example of the technology is shown in Figure 8.1, where isocratic separation of two anomeric forms of a tagged (*p*-nitrophenyl) maltose with immobilized monoclonal antibodies was achieved in the presence of serum [27]. The maltose forms were separated in the sub-mM range of affinity where they interacted weakly but differently with the immobilized monoclonal IgG antibody. It is evident that the weak affinity binding was sufficient to separate the carbohydrates from the major bulk of serum components. A consequence of WAC is that it has the potential to resolve similar molecules only differing slightly in affinities to target even in the presence of a crude sample. Furthermore, separations can be performed isocratically or with minor gradients under mild conditions such as in a physiological buffer as no harsh elution procedure is required. If properly designed, WAC can detect and quantify very weak interactions; an example is the separation of carbohydrates with lectin affinity chromatography in the 10 mM range of $K_D$ [30].

WAC and related techniques have been applied in a number of areas for analytical applications for separation and characterization of biomolecules. This includes chiral separations [31–33] and the evaluation of protein–analyte interactions [34–36]. WAC has also been used to study overloaded conditions for preparative chromatography [37]. WAC has been applied successfully to study carbohydrates with immobilized lectins [38–41]. The binding of drugs to plasma proteins such as albumin and $\alpha_1$-acid glycoprotein can also be easily assessed using WAC/HP(L)AC [42,43].

#### 8.2.1.2 Theory

The theory and basic concepts behind interactive liquid chromatography was originally developed in the 1960s and further refined in the following decades for analytical affinity chromatography and finally adopted and applied to WAC in the 1990s [44–48]. By performing WAC, the characteristics in terms of both affinity and kinetics can be directly assessed. In this chapter, we will focus on the evaluation of binding constants with WAC but the interested reader in determining kinetic constants is advised to refer to the literature [43,47].

To simplify the theory, we are considering a 1 : 1 interaction of a small molecule ligand (such as a fragment) (L) to a receptor (R) according to:

$$\mathrm{R} + \mathrm{L} \underset{k_{\mathrm{off}}}{\overset{k_{\mathrm{on}}}{\rightleftharpoons}} \mathrm{RL}.$$

$$K_{\mathrm{A}} = \frac{k_{\mathrm{on}}}{k_{\mathrm{off}}} = \frac{[\mathrm{RL}]}{[\mathrm{R}][\mathrm{L}]} = \frac{1}{K_{\mathrm{D}}}. \tag{8.1}$$

$K_{\mathrm{A}}$ and $K_{\mathrm{D}}$ are association constant ($\mathrm{M}^{-1}$) and dissociation constant (M), respectively; $k_{\mathrm{on}}$ and $k_{\mathrm{off}}$ are the association ($\mathrm{M}^{-1}\mathrm{s}^{-1}$) and dissociation ($\mathrm{s}^{-1}$) rate constants, respectively. The total concentration of the receptor $[\mathrm{R}]_{\mathrm{o}}$ is the sum of the free receptor concentration and receptor in complex with ligand (Equation 8.2) and from this the concentration of RL can be deduced as follows:

$$[\mathrm{R}]_{\mathrm{o}} = [\mathrm{R}] + [\mathrm{RL}] \quad \text{or} \quad [\mathrm{R}] = [\mathrm{R}]_{\mathrm{o}} - [\mathrm{RL}]. \tag{8.2}$$

Integrating Equation 8.2 into Equation 8.1 gives

$$\frac{[\mathrm{RL}]}{([\mathrm{R}]_{\mathrm{o}} - [\mathrm{RL}]) \times [\mathrm{L}]} = \frac{1}{K_{\mathrm{D}}} \quad \text{or} \quad [\mathrm{RL}] \times K_{\mathrm{D}} = ([\mathrm{R}]_{\mathrm{o}} - [\mathrm{RL}]) \times [\mathrm{L}],$$

which can be rearranged as follows:

$$[\mathrm{RL}] \times K_{\mathrm{D}} + [\mathrm{RL}] \times [\mathrm{L}] = [\mathrm{R}]_{\mathrm{o}} \times [\mathrm{L}],$$

which can be rewritten as follows:

$$[\mathrm{RL}] = \frac{[\mathrm{R}]_{\mathrm{o}} \times [\mathrm{L}]}{K_{\mathrm{D}} + [\mathrm{L}]}. \tag{8.3}$$

Equation 8.3 is the well-known Langmuir adsorption equation or isotherm. Equation 8.3 shows, for example, that when $[\mathrm{L}] = K_{\mathrm{D}}$, then 50% of all receptor sites are bound with ligand. When $[\mathrm{L}] \ll K_{\mathrm{D}}$, then

$$[\mathrm{RL}] = [\mathrm{R}]_{\mathrm{o}} \times [\mathrm{L}]/K_{\mathrm{D}}, \tag{8.4}$$

which is the linear adsorption isotherm at low concentration of L. Equation 8.4 is considered to be valid for most practical situations when performing WAC.

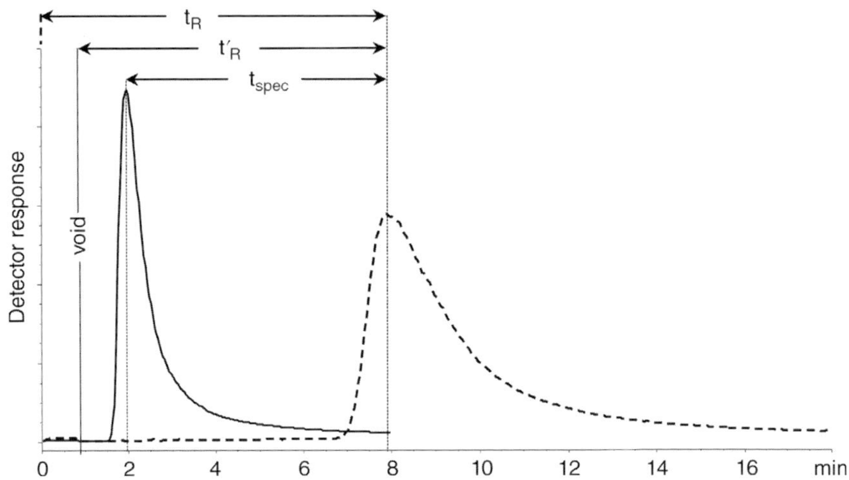


**Figure 8.2** Retention time ($t_R$), adjusted retention time ($t'_R$) and specific retention time ($t_{spec}$) from a chromatogram on an active (dashed-lined chromatogram) and an inhibited target (solid-lined chromatogram) [50].

In chromatographic theory, the capacity (retention) factor ($k'$) is defined as the ratio of the ligand in the complexed state to the ligand in the free state at equilibrium and can be written as follows:

$$k' = \frac{[\mathrm{RL}]V_s}{[\mathrm{L}]V_M}, \tag{8.5}$$

where $V_M$ is the volume of mobile phase and $V_S$ is the volume of stationary phase. Equation 8.4 can then be integrated into Equation 8.5 leading to

$$k' = \frac{V_R - V_M}{V_M} = \frac{[\mathrm{R}]_o \times V_s}{K_D \times V_M} = \frac{R_{tot}}{K_D \times V_M}, \tag{8.6}$$

where $V_R$ is the elution volume of analyte and $R_{tot}$ is the number of moles of active sites of target ($R_{tot} = [\mathrm{R}]_o \times V_s$). As can be seen from Figure 8.2, $t_R$ is the retention time for a ligand ($V_R = F \times t_R$, where $F$ is the flow rate) and the void time ($t_R - t'_R$; $V_M = F \times (t_R - t'_R)$. As can be seen from Equation 8.6, retention is dependent on the affinity and concentration or amount of active target sites. $R_{tot}$ can be determined by frontal affinity chromatography where the WAC column is saturated with varying concentrations of ligand and a Langmuir adsorption isotherm is established [49] or by WAC using a compound with a documented known affinity for the target. In cases where $R_{tot}$ is not determined, a good approximation of the active amount of target is to use 50% of total immobilized target. In order to achieve appreciable retention at weak affinities, high amounts of active target are required; that is, if you need to see a mM binder, you need to have a mM density of active target. Again, it is important to state that operation at low concentrations of ligand is required for Equation 8.3 to be valid. At high concentrations of ligand, retention times will be less than predicted by Equation 8.6.

#### 8.2.1.3 Fragment Screening

Fragment screening by WAC is practical and straightforward to perform. After immobilization of target (μg to mg) either *in situ* on a preactivated column or in a batch followed by packing of the column, the target column is set up on a standard HPLC/MS platform. The fragment library is run isocratically (with no changes of mobile phase) as singletons or preferably in mixtures of 50–100 fragments automatically where the total library of thousands of fragments is evaluated within a day. Specificity of binding can be determined by, for example, active-site blockage of target if inhibitor is available. Mass spectrometry is used to identify the fragments with UV detection as a possible complement. Affinity and kinetics are determined directly from chromatograms without running any concentration series of ligand. The precision and robustness of the system is high and if the target is not highly unstable the affinity unit can be used for thousands of analyses rendering the consumption of target per fragment screening campaign very low. In many cases, the protein stability is enhanced by immobilization contributing to its high stability. As WAC for fragment screening can be applied on any commercial LC/MS platform, it is widely accessible for routine usage.

##### Matrices for WAC in Fragment Screening

WAC is an HPLC-based procedure and requires matrices with high mechanical stability such as cross-linked organic polymers or silica supports. The preferred support for fragment screening by WAC is microparticulate, porous, and spherical silicas (5–10 μm in diameter and 100–300 Å in porosity). A high surface area is recommended (typically 100 $m^2/g$ silica) so the support can accommodate 0.1–1 mM of target. Silica materials cannot be used as such because of the presence of high nonspecific binding to the silica surface. To minimize these effects, the silica is derivatized into diol groups (two vicinal hydroxyl groups) that minimizes any nonspecific binding to the silica surface (Figure 8.3). Diols are excellent functionalities for protein immobilization through oxidation of aldehyde groups on the support. Primary amino groups of the protein (mainly lysine residues) react with the aldehydes to form a reversible Schiff base that is locked into a permanent covalent bond by reductive amination with the mild reducing agent sodium cyanoborohydride (Figure 8.3). These bonds are stable with a preserved positive charge structure and this immobilization procedure may stabilize the protein considerably. Remarkably, some proteins bound in this way to modified silica have been shown to retain >90% of their activity after 12 months of storage at fridge temperatures. This presents the user with a very high level of convenience compared to most other methods of studying ligand binding to active proteins. In our laboratories, we have so far successfully immobilized approximately 20 proteins and used them for WAC.

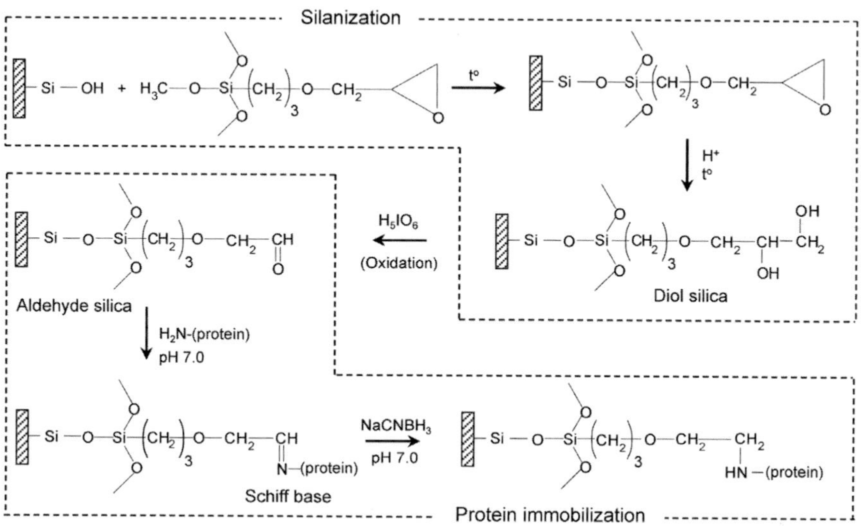


**Figure 8.3** Reaction scheme for silanization and immobilization of a protein target on the silica surface [50].

**WAC Platforms for Fragment Screening**

Fragment screening with WAC can be performed on any standard HPLC platform that is equipped with MS detection. The MS unit is typically a single quadrupole detector with electrospray injection. The fragments are followed by their individual masses either as singles or in mixtures. Single ions (SIM mode) can be monitored to increase sensitivity.

UV detection can also be used to follow the overall analysis, but this is not really necessary. Mobile phase such as volatile water-based phase, for example, ammonium acetate, is delivered isocratically and high flow rates can be used to reduce analysis time. Sensitivity can be increased by postcolumn addition of organic solvents such as acetonitrile and methanol, but usually water-based volatile phases give sufficient sensitivity. Optimization is carried out by adjusting parameters such as concentration and pH of the mobile phase, capillary and fragmentor voltage, nebulizer pressure, and flow and temperature of drying gas. Fragment screening is temperature controlled and the whole operation is fully automatic where the samples are delivered from a sample rack such as a 96-well plate. The immobilized targets are typically run as capillary/microcolumns of various sizes.

**Characteristics of WAC for Fragment Screening**

**Throughput** WAC is an efficient high-throughput technique where a whole library can be screened within a day. For example, it was demonstrated that a library of 3500 fragments can be screened within 24 h [51]. This high speed is achievable because the fragments are screened in mixtures of up to 100 compounds. In each mixture, each fragment should be unique in its MW so the MS

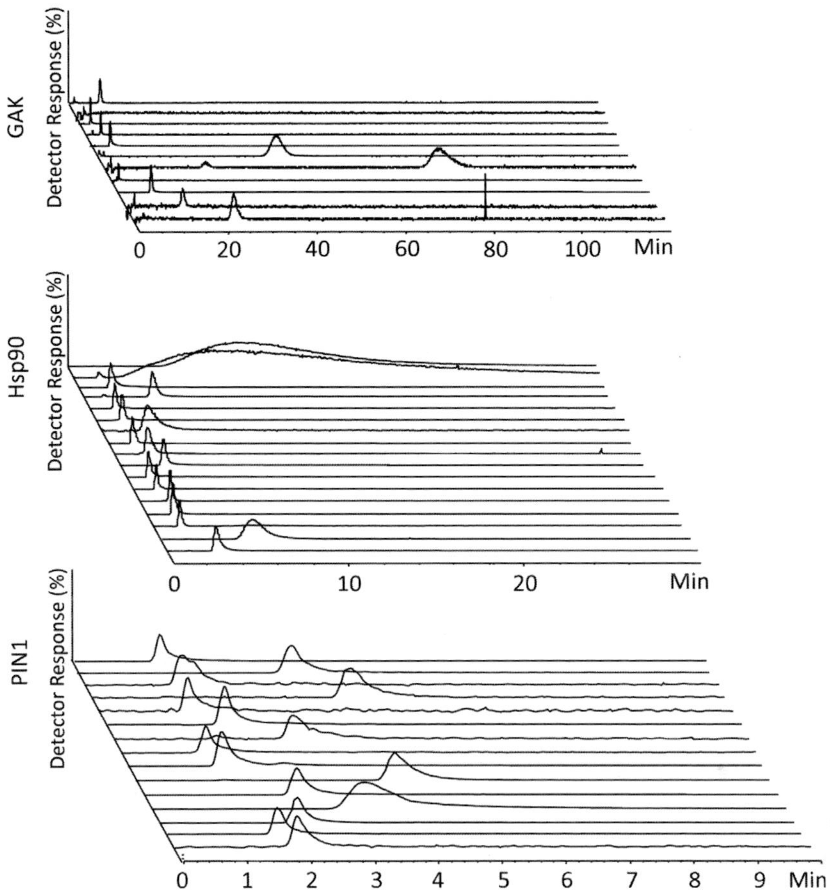


**Figure 8.4** Extracted ion chromatograms of fragments in mixtures from screening on three different drug targets [53].

detector can follow each fragment individually. If the fragments have the same affinity for the target and hence retention time, they can still be distinguished by the MS detector. According to our experience, in a screening campaign 5–10% of fragments are hard to detect due to poor ionizing that in turn is dependent on the quality of the library and choice of MS detector. Speed can further be improved by parallelization of column operation. A demonstration of this is the multichannel cartridge where 24 columns were run in parallel [52]. In Figure 8.4, an example is given of fragment screening of different mixtures to three different targets [53].

**Affinity** An important feature of WAC for fragment screening is that only one, low concentration ($\leq$0.1 mM) of fragment is needed to obtain the affinity without running a whole concentration range of fragments to establish a Langmuir adsorption isotherm. The detectable affinity during isocratic running ranges from 10 mM (or even higher) to 1 μM. Tighter binders will be severely retarded and elute impractically late. In this case, they can be eluted by gradient elution by pH, temperatures, and/or small amount of organic solvents. In the rare situation of a tight binder (<1 μM in affinity) in a fragment screening campaign, it can be indirectly seen by its absence in the operational chromatogram or identified by proper elution.

**Specificity/Selectivity** The binding of a fragment to a covalently attached protein target can occur via many types of interactions to various sites of the protein or to the surrounding environment where the target is immobilized. Nonspecific/nonselective binding is defined as interactions to sites other than the desired active sites of the target. As fragment screening is characterized by weak binding events between the target active site and the fragments (with affinities in the range 10 mM–1 μM), there are many opportunities for nonspecific weak binding to other sites outside the active site of the target. This is especially true for very weak binders such as when $K_D > 0.5$ mM, the contribution from nonspecific binding could be substantial (>50%) [54]. This situation is, of course, not unique to WAC and is also prevalent for other fragment screening methods such as SPR and ligand-based NMR. To control the amount of off-site interactions leading to false positives, there are different ways to approach this problem. The most efficient way is to study the target with a modified active site or to use blockers of the active site of the target. Both covalent and noncovalent high-affinity inhibitors must be available toward the target as tool compounds. As for noncovalent blockers, they are continuously added to the mobile phase and are hence constantly available at an appropriate concentration to completely block the active site. The net specific affinity is given by Equation 8.7 below (see Figure 8.2):

$$K_{D\ \mathrm{specific}} = \frac{R_{\mathrm{tot}}}{F \times t_{\mathrm{spec}}}. \tag{8.7}$$

Using a blocked active site of the target is probably the most relevant control of specific binding even though the blocker in itself can cause nonspecific binding and/or will not be able to inhibit the target completely. Apparent nonspecific binding to other off-sites on the target can still be of value as it can represent possible allosteric binders. When inhibitors are not available, other controls for nonspecific binding can be used such as other irrelevant protein targets and/or references with no bound target. However, these reference systems are not ideal and give only some guidance on the presence of nonspecific interactions.

**Consumption of Target and Ligand (Fragment)** The effective amount of the target protein is dependent on the format of the column and its stability during operation and storage. By miniaturization of the separation unit into a capillary column, consumption of the target is reduced to the μg range, whereas microcolumns require low mg range. With a chip format of the immobilized target, consumption of protein can be reduced even further into the ng range.

By immobilization, some target proteins may be stabilized as these units have been used repeatedly for hundreds or even thousands of analyses of fragment mixtures over long periods of time [54–56]. By using internal standards, the operational stability of the WAC columns can be constantly monitored. As detection of the fragments in a screening session is achieved by MS and no isotherms are required to estimate affinity, the consumption of each individual fragment is minimal (in the nanogram range).

#### Applications

As WAC is a new technology for fragment screening in drug discovery and was introduced as such only a few years back [55], the number of applications is still limited. Nevertheless, it has been applied in several published studies as well as toward proprietary targets in the pharma industry. In published reports, WAC has been used for fragment screening toward proteases (thrombin and trypsin) [51,55,56], kinases (cyclin G-associated kinase, GAK) [57], isomerases (peptidyl-propyl *cis-trans* isomerase NIMA-interacting 1, PIN1) [53], and chaperones (heat shock protein 90, HSP90) [54].

An illustrative example (Figure 8.5) of using WAC for fragment screening is the analysis of 568 fragments binding to the target thrombin from a library including two reference compounds (3-aminobenzamidine, 3-ABA; 4-aminobenzamidine, 4-ABA) [51].

Figure 8.5 is an example of a primary screen where 30 hits were defined that were above the threshold value. On secondary screening, only one hit out of the 30 hits showed specific binding to the target ($K_D = 124\,\mu M$) when evaluated on a blocked thrombin column with a covalent inhibitor [51]. The whole analysis of this selected fragment library took less than 4 h.

Another key advantage of WAC for fragment screening is its ability to distinguish stereoisomers according to their target affinity. Figure 8.6 illustrates WAC's potential to characterize a racemic mixture where the enantiomers show different affinities to the target that in this case is thrombin [56]. These two

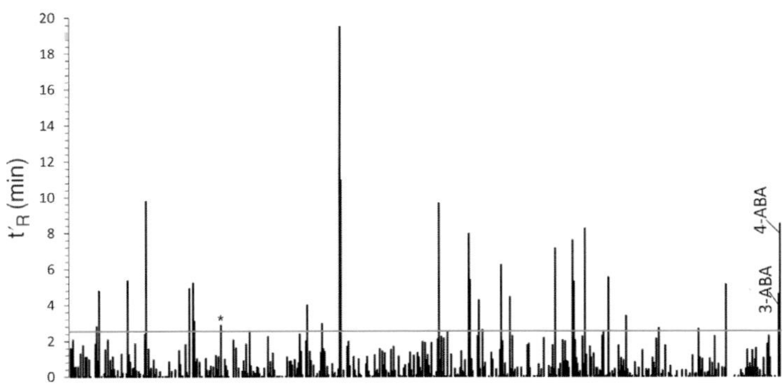


**Figure 8.5** Adjusted retention times ($t'_R$) of 568 fragments including two reference binders. The horizontal line shows the cutoff value for potential hits. All fragments were run in 11 mixtures of each 35–65 compounds. (Adapted from Ref. [51].)

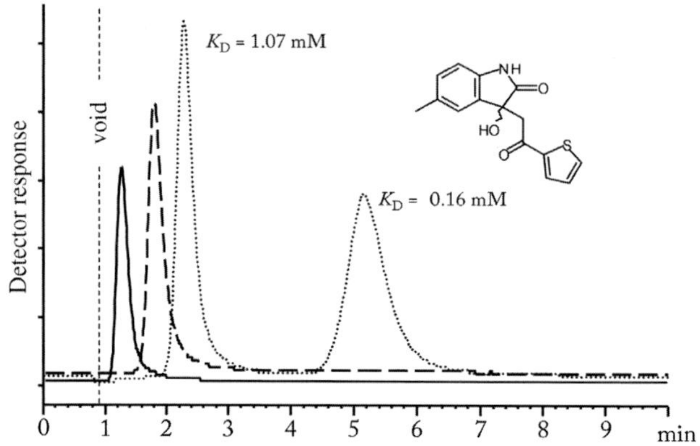


**Figure 8.6** Overlaid chromatograms of the enantiomers of a fragment on a diol silica column (solid line), a covalently inhibited thrombin column (dashed line) and an active thrombin column (dotted line) [50].

enantiomers bind with a sixfold difference to the target. It is interesting, as shown in Figure 8.6, that when the active site is blocked both isomers show the same remaining binding affinity, which represents the nonspecific binding predominantly toward the target protein surface. The separation capacity of WAC makes it useful for screening of stereoisomeric libraries that can be valuable in selection of good hits for lead development.

#### Validation of WAC for Fragment Screening

Validation of WAC versus other quantitative techniques such as NMR, SPR, and X-ray crystallography is essential to establish WAC as a tool in academia and industry. In a thorough study of binding of fragments to HSP90, WAC was validated toward a number of techniques including NMR, SPR, and X-ray crystallography [54]. WAC correlated well in terms of set discrimination values with all three methods: NMR (88%; $n = 103$), SPR (83%; $n = 107$), and X-ray crystallography (83%; $n = 30$). However, the optimized conditions for screening with the respective techniques were somewhat different leading to discrepancies when correlating the absolute values of $K_D$ of the various techniques. This can be due to a number of factors such as variations of experimental conditions during screening of, for example, buffer composition, pH, and temperature as well as different designs to estimate nonspecific binding. It was evident in this study of HSP90 that variations between the methods are most pronounced when binding of fragment to target is very weak ($K_D > 0.5$ mM).

Another study also showed good correlation of the absolute values of $K_D$ when comparing WAC with SPR [58]. The correlation was highly significant ($R^2 = 0.995$; $P < 0.0001$) at a slope near to one. In this case, similar experimental conditions were adopted for both methods, which could partly explain the excellent correlation. However in this study, there were a number of fragments that could not produce reliable results with SPR as opposed to WAC.

## 8.2.2
## Mass Spectrometry

### 8.2.2.1 Introduction

Mass spectrometry can be used directly to detect fragment binding to a protein target. This detector arrangement has yet to find widespread use in fragment screening. However, with the introduction of WAC, MS is expected to be used for more fragment screening applications.

An early approach to use MS was introduced by Erlanson and co-workers [59,60], called "tethering," where thiol-containing fragments are covalently linked to preferably cysteine residues closest to the binding site of the fragment in the target. These are then readily detected by MS (see also Chapter 11).

Another approach using MS that seems attractive for fragment screening is native MS: electrospray ionization MS (ESI-MS). The applications in this area are growing as evidenced by some recent reviews [61,62]. This technique measures directly the complex between fragment and target as the MS can easily detect the mass difference between target and the complex. A key to the usage of the method is that the fragile complex is maintained during the transfer from liquid to gas phase. This can be a challenge for very weak interactions and especially when there are hydrophobic effects between fragment and target. Extensive sample handling is also required to achieve a successful screening with ESI-MS and background effects from DMSO have to be elucidated. As with WAC using MS to increase throughput, mixtures can be analyzed simultaneously as long as the molecular weights of the fragments are different where the mass of the fragments serves as a bar-code for each fragment.

An example of an ESI mass spectrum with a cocktail of ligands is given in Figure 8.7 where the target, carbonic anhydrase II, is mixed with five small ligands [61]. The signals from the protein and the individual ligands are clearly

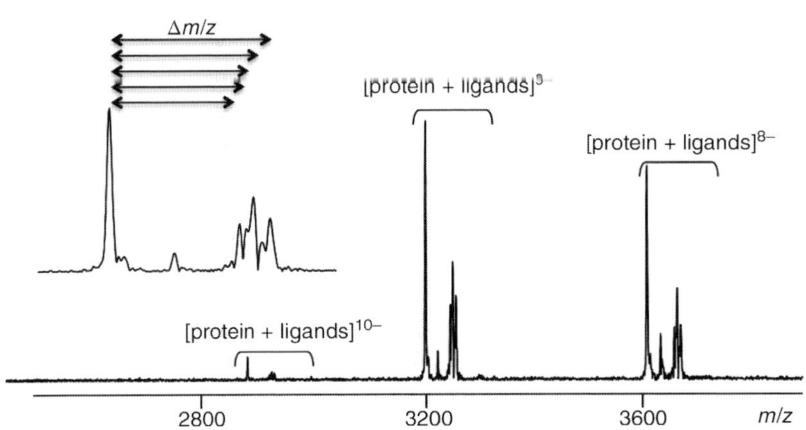


**Figure 8.7** ESI-MS spectrum of carbonic anhydrase II under native state conditions with five binding ligands in a mixture (Reproduced with permission from [61].)

seen. The MW of each ligand can be calculated by the relationship: $MW_{Ligand} = \Delta m/z \times z$, where $MW_{Ligand}$ is the molecular weight of ligand, $\Delta m/z$ is the difference in mass-to-charge ratio, $m$ is the molecular mass, and $z$ is the charge.

Native ESI-MS with time-of-flight (TOF) or TOF hybrid instruments for fragment screening is characterized by its high sensitivity of both target and fragments, large dynamic range of affinities ($K_D$s of μM to mM), and high throughput by multiplexing of a thousand fragments per day. In addition, it can give information on the stoichiometry of binding. However, there is concern of the survival of the complex in the gas phase and of introducing nonspecific binding events during the electrospray process leading to false positives. As with WAC, there are currently no routine approaches to study screening of membrane proteins as targets, although research on this is on its way.

#### 8.2.2.2 **Theory**

In a native state ESI-MS experiment (see Figure 8.7), the detector response at equilibrium of R and RL is recorded. The peak heights of R and RL can be converted into concentrations by assuming they both give the same linear response and that the sum of [R] and [RL] is equal to the total receptor concentration $[R]_o$. The initial ligand concentration $[L]_o$ is equal to the sum of free concentration of ligand [L] and complexed ligand [RL]. The concentrations of [RL], [R], and [L] are then put into Equation 8.1 to give $K_A$ or $K_D$.

#### 8.2.2.3 **Applications**

Currently, there are only a few applications using native state mass spectroscopy [62–65]. In the following two examples, the use of direct ESI-MS is given as representative of this technique.

Screening of a fragment library against HSP90 was performed on an automated chip-based nano unit coupled to an ESI-TOF mass spectrometer [62]. This configuration has significantly increased throughput for fragment screening and reduced need for protein target. In this case, with HSP90 as the target, the consumption was approximately 2 μg per fragment. In this screen from a 350-member library, a hit rate of 12.5% was found where the hits were subsequently analyzed by X-ray crystallography. In an alternative approach, *in silico* primary screening was carried out from a 2200-member library yielding 60 fragment hits that were then followed up by native MS as a secondary screening method. Many of these initial virtual hits showed binding in MS and were further characterized by X-ray crystallography. The fragment hits showed in both screens a typical binding in the μM range.

In another extensive nano-ESI-MS study by Maple *et al.* [63], a model system, using hen egg white lysozyme (HEWL) binding to four known ligands with a $K_D$ range from low μM to mM, was used to evaluate the parameters for screening with native MS. Automation and unattended screening were introduced by a robotic nanospray source to increase throughput. It was important to carefully optimize the conditions for MS to allow successful detection of the fragile complexes of target–ligand. Proper sample preparation and setting of mass spectrometer conditions (temperature, nebulizer pressure, and voltages) were essential to achieve transmission and detection of the noncovalent weak complexes in the MS instrument. An increased spectrometer nebulizer pressure was crucial to be able to detect and measure very weak binding such as of glucose ($K_D = 5$ mM) to HEWL that could not be detected by either NMR or isothermal titration calorimetry (ITC). However, it is not clear if this binding is nonspecific and could be explained by binding to other scaffolds of the protein as well. The correlation was satisfactory with NMR and ITC for other binders to HEWL. Screening was then performed on the optimized system of a selected 157-member library (MW range 240–500 Da) against the antiapoptotic Bcl-$x_L$ protein. By running this screen at one low ligand to protein ratio (2 : 1) to save analysis time and minimize complications due to solubility of ligands, it was possible to do a broad ranking of the apparent affinities of ligand to target. From this library of 157 compounds, 66% were found at $K_D < 150\,\mu$M, 18% with $K_D$ in the range of 150–300 μM, and 2% >300 μM, while 9% were nonbinding and 4% failed to spray. The high hit rate may be caused by many false positives due to nonspecific interactions both in the liquid and in the gas phase during the electrospray process. In addition, the quality of the ranking is due to the concentration of ligand used in the assay where, for example, high ratios of ligand to target (L:R) as well as extensive modifications of the experimental protocol are required to observe any weak interactions in the mM range. The whole screen consumed less than 1 mg of target and less than 50 μg of ligand (at a 2 : 1 L:R ratio) at a speed of 100 s per compound. However, a considerable amount of time was needed to implement the screen in terms of instrument setup and sample preparation. The correlation was convincing concerning validation of native MS versus NMR and ITC, although many discrepancies occurred especially at weak affinities.

## 8.2.3 Microscale Thermophoresis

### 8.2.3.1 Introduction

MST was recently introduced as a novel technology to study and quantify biomolecular interactions [66,67]. Although the method has been applied in various drug discovery programs [68–70], its application as a screening approach is in its infancy and currently there are no reports of primary fragment screening using MST in the literature. Nonetheless, the unique principle of MST with its low sample consumption and robustness make the technique an attractive approach for FBDD.

MST is a fluorescence-based technique that measures the directed movement of molecules under a temperature gradient. In theory, MST works with all kinds of molecules but normally macromolecules such as protein and DNA are used. The thermophoretic movement is influenced by a number of factors including size, charge, hydration shell, and the conformation of the macromolecules, all of which can be altered upon ligand binding. In a typical MST experiment (Figure 8.8), the macromolecules are monitored by fluorescence from intrinsic

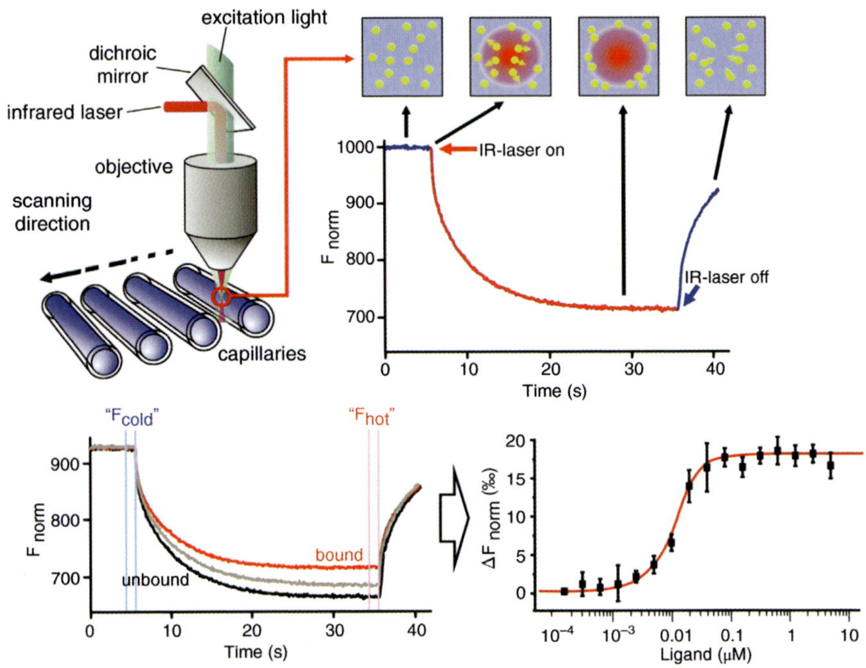


**Figure 8.8** MST setup and experiment. Top left: schematic presentation of optics and capillaries. Top right: typical signal of an MST experiment. Lower left: typical change in signals on binding of a ligand. Lower right: binding curve where signal change is plotted versus concentration of ligand. (Adapted from Ref. [71].)

or extrinsic fluorophores in the target molecule. Sample is loaded into a glass capillary that is rapidly heated by a focused beam of infrared laser to create a temperature gradient in a microscopic area of the glass capillary. Subsequently, the target macromolecules at the heated spot move along the temperature gradient, thus changing the fluorescence intensity of the region. Since the movement is affected by ligand binding, incubating the macromolecules with different concentrations of a ligand can provide a binding curve to calculate ligand–macromolecule affinity.

MST possesses many attractive features for characterization of ligand–protein binding that could be valuable for FBDD. First, MST covers a broad range of affinities ($K_D$ from pM to mM) [68,71]. However, in most applications it has been used mostly to detect rather tight interactions (from pM to low μM). In order for the technology to be useful for fragment screening in which affinities are weak (μM to mM), MST has to work with high concentrations of ligand, which raise demands on solubility and stability of ligand. Second, MST is also a robust technology where it can perform measurements in complex matrices such as serum or cell lysate. Third, the most significant feature of MST is probably its ability to measure the binding in the native state of the target without any immobilization or labeling. However, in some cases labeling with fluorophores is required to achieve sufficient sensitivity. Fourth, another attractive feature of MST for fragment screening is the low consumption of target protein due to high sensitivity of fluorescence signal and miniaturization of thermophoresis. On the other hand, consumption of ligand is rather high but could be reduced when capillary-free approaches are used [72]. As to throughput, the efficiency of MST has been recently improved by employing automated equipment, but it is still far behind other fragment screening approaches such as SPR and WAC [71].

#### 8.2.3.2 **Theory**

The thermophoretic movement of molecules in solution is dependent on the thermal diffusion coefficient or thermophoretic mobility ($D_T$) and mass diffusion coefficient ($D$). The quotient of these coefficients forms the Soret coefficient ($S_T$):

$$S_T = D_T/D. \tag{8.8}$$

It can be shown [73] that the normalized fluorescence of the sample ($F_{norm}$) is as an approximation dependent on the temperature jump (T-jump; $\partial F/\partial T$); the change of fluorescence yield due to the temperature increase ($\Delta T$):

$$F_{norm} = F_{hot}/F_{cold} = 1 + (\partial F/\partial T - S_T)\Delta T, \tag{8.9}$$

where $F_{cold}$ is fluorescence before IR-laser heating and $F_{hot}$ is after heating.

If we define $X$ as the fraction of target bound to ligand, $F_{norm(unbound)}$ as the normalized fluorescence of free target and $F_{norm(bound)}$ as the normalized fluorescence of target–ligand complexes, then total normalized fluorescence can be calculated [67,73] as follows:

$$F_{norm} = (1 - X)F_{norm(unbound)} + X_{Fnorm(bound)}. \tag{8.10}$$

From the differences in $F_{norm(bound)}$ and $F_{norm(unbound)}$, we can calculate $X$ and subsequently the $K_D$ value by plotting $X$ against ligand concentration to build a binding curve. An alternative is to directly plot the change in $F_{norm}$ ($\Delta F_{norm}$) against ligand concentrations [71].

#### 8.2.3.3 **Applications**

The number of publications on MST in characterizing biomolecular interactions is increasing as can be seen by recent reviews [71,73]. However, the impact of MST on FBDD remains to be established. In a comparative study on the p38α kinase [68], MST was used to confirm binding of a few fragments to the target and the results correlated well with SPR.

## 8.3 Conclusions

It is evident as presented in this book that there has been a rapid development in drug discovery of technologies for fragment screening since the early pioneering works in the 1990s [16]. Established technologies including NMR, X-ray crystallography, and SPR play an important role and they are now used routinely as both primary and secondary fragment screening methods. These methods have now been complemented by other approaches to fragment screening of which we have presented WAC, MS, and MST in this chapter of emerging technologies. It is clear that all techniques for fragment screening are not equally well adapted to all targets and they all possess both pros and cons. Probably the most efficient way to perform fragment screening on a wider scale is to rely on several orthogonal methods working in concert. This will allow the user to evaluate the different aspects of fragment screening and to conclude more firmly the results of a screening campaign. Furthermore, it will give the investigator more choices of qualified hits to work from when developing useful leads. Many pharma companies employ multiple techniques for fragment screening.

In Table 8.1, we have attempted to compare different screening methodologies with emphasis on the new methods versus the established. We realize that it is very hard to do a "fair" comparison as the technologies are judged differently depending on the preferences of individual researchers and laboratories. In addition, the individual characteristics of the methods can vary significantly depending on the optimization of the screening operation. Nevertheless, our ambition in Table 8.1 has been to give some kind of guidance of a few key performance indicators as it is of value to try to compare head-to-head the different methods that are available. To conclude, the main take-home message is that there are now available novel technology platforms, as presented in this chapter, which

**Table 8.1** Comparison of fragment screening methods.

| Parameter | NMR | X-ray | SPR | WAC | Native MS | MST |
|---|---|---|---|---|---|---|
| Throughput[a)] | Medium | Low | High | High | Low–medium | Medium |
| Consumption of target[b)] | Medium–high | Medium–high | Low | Low | Low–medium | Low–medium |
| Target immobilization | No | No | Yes | Yes | No | No |
| Label free | Yes | Yes | Yes | Yes | Yes | Yes/no |
| Consumption of ligand (fragment)[c)] | Medium | Medium–high | Medium–high | Low | Medium–high | Medium–high |
| Affinity data | Yes | No | Yes | Yes | Yes | Yes |
| Kinetic data | No | No | Yes | Yes | No | No |
| Information rich[d)] | High | High | Low–medium | Low–medium | Low–medium | Low–medium |
| False positives | Low | Low | Medium | Low | Low–medium | NA |

a) Low = <100/week, medium = 100–1000/week, and high = >1000/week.
b) Low = <1 mg, medium = 1–10 mg, and high = >10 mg.
c) Low = <1 ng, medium = 1 ng – 1 μg, and high > 1 μg.
d) Structure–activity relationships.
NA = non-applicable. No information available.

may complement in a beneficial way current practices to conduct fragment screening.

## Acknowledgments

The authors are grateful to Brian Dymock and Mats Wikström for their valuable comments and suggestions on the manuscript. Financial support from Nanyang Technological University (NTU), Singapore, is also gratefully acknowledged.

## References


1 Joseph-McCarthy, D., Campbell, A.J., Kern, G., and Moustakas, D. (2014) Fragment-based lead discovery and design. *Journal of Chemical Information and Modeling*, **54**, 693–704.

2 Davis, B.J. and Erlanson, D.A. (2013) Learning from our mistakes: the 'unknown knowns' in fragment screening. *Bioorganic and Medicinal Chemistry Letters*, **23**, 2844–2852.

3 Larsson, A., Jansson, A., Åberg, A., and Nordlund, P. (2011) Efficiency of hit generation and structural characterization in fragment-based ligand discovery. *Current Opinion in Chemical Biology*, **15**, 482–488.

4 Leach, A.R. and Hann, M.M. (2011) Molecular complexity and fragment-based drug discovery: ten years on. *Current Opinion in Chemical Biology*, **15**, 489–496.

5 Sun, C.H., Petros, A.M., and Hajduk, P.J. (2011) Fragment-based lead discovery: challenges and opportunities. *Journal of Computer-Aided Molecular Design*, **25**, 607–610.

6 Coyne, A.G., Scott, D.E., and Abell, C. (2010) Drugging challenging targets using fragment-based approaches. *Current Opinion in Chemical Biology*, **14**, 299–307.

7 Murray, C.W. and Blundell, T.L. (2010) Structural biology in fragment-based drug design. *Current Opinion in Structural Biology*, **20**, 497–507.

8 Schulz, M.N. and Hubbard, R.E. (2009) Recent progress in fragment-based lead discovery. *Current Opinion in Pharmacology*, **9**, 615–621.

9 Congreve, M., Chessari, G., Tisi, D., and Woodhead, A.J. (2008) Recent developments in fragment-based drug discovery. *Journal of Medicinal Chemistry*, **51**, 3661–3680.

10 Erlanson, D.A. and Jahnke, W. (2006) The concept of fragment-based drug discovery, in *Fragment-based Approaches in Drug Discovery* (eds W. Jahnke and D.A. Erlanson), Wiley-VCH Gerlag GmbH, Weinheim.

11 Hubbard, R.E. and Murray, J.B. (2011) Experiences in fragment-based lead discovery, in *Fragment-based Drug Design: Tools, Practical Approaches, and Examples* (ed. L.C. Kuo), Elsevier Academic Press Inc., San Diego.

12 Cherry, M. and Mitchell, T. (2008) Introduction to fragment-based drug discovery, in *Fragment-based Drug Discovery: A Practical Approach* (eds E.R. Zartler and M.J. Shapiro), John Wiley & Sons, Ltd, Chichester.

13 Macarron, R., Banks, M.N., Bojanic, D., Burns, D.J., Cirovic, D.A., Garyantes, T., Green, D.V.S., Hertzberg, R.P., Janzen, W.P., Paslay, J.W., Schopfer, U., and Sittampalam, G.S. (2011) Impact of high-throughput screening in biomedical research. *Nature Reviews. Drug Discovery*, **10**, 188–195.

14 Macarrón, R. and Hertzberg, R.P. (2011) Design and implementation of high throughput screening assays. *Molecular Biotechnology*, **47**, 270–285.

15 van Dongen, M., Weigelt, J., Uppenberg, J., Schultz, J., and Wikstrom, M. (2002) Structure-based screening and design in

drug discovery. *Drug Discovery Today*, **7**, 471–478.

16 Shuker, S.B., Hajduk, P.J., Meadows, R.P., and Fesik, S.W. (1996) Discovering high-affinity ligands for proteins: SAR by NMR. *Science*, **274**, 1531–1534.

17 Tsai, J., Lee, J.T., Wang, W., Zhang, J., Cho, H., Mamo, S., Bremer, R., Gillette, S., Kong, J., Haass, N.K., Sproesser, K., Li, L., Smalley, K.S.M., Fong, D., Zhu, Y.L., Marimuthu, A., Nguyen, H., Lam, B., Liu, J., Cheung, I., Rice, J., Suzuki, Y., Luu, C., Settachatgul, C., Shellooe, R., Cantwell, J., Kim, S.H., Schlessinger, J., Zhang, K.Y.J., West, B.L., Powell, B., Habets, G., Zhang, C., Ibrahim, P.N., Hirth, P., Artis, D.R., Herlyn, M., and Bollag, G. (2008) Discovery of a selective inhibitor of oncogenic B-Raf kinase with potent antimelanoma activity. *Proceedings of the National Academy of Sciences of the United States of America*, **105**, 3041–3046.

18 Bollag, G., Hirth, P., Tsai, J., Zhang, J.Z., Ibrahim, P.N., Cho, H.N., Spevak, W., Zhang, C., Zhang, Y., Habets, G., Burton, E., Wong, B., Tsang, G., West, B.L., Powell, B., Shellooe, R., Marimuthu, A., Nguyen, H., Zhang, K.Y.J., Artis, D.R., Schlessinger, J., Su, F., Higgins, B., Iyer, R., D'Andrea, K., Koehler, A., Stumm, M., Lin, P.S., Lee, R.J., Grippo, J., Puzanov, I., Kim, K.B., Ribas, A., McArthur, G.A., Sosman, J.A., Chapman, P.B., Flaherty, K.T., Xu, X.W., Nathanson, K.L., and Nolop, K. (2010) Clinical efficacy of a RAF inhibitor needs broad target blockade in BRAF-mutant melanoma. *Nature*, **467**, 596–599.

19 Zheng, C.J., Han, L.Y., Yap, C.W., Xie, B., and Chen, Y.Z. (2006) Progress and problems in the exploration of therapeutic targets. *Drug Discovery Today*, **11**, 412–420.

20 Cuatrecasas, P., Wilchek, M., and Anfinsen, C.B. (1968) Selective enzyme purification by affinity chromatography. *Proceedings of the National Academy of Sciences of the United States of America*, **61**, 636–643.

21 Ohlson, S., Hansson, L., Larsson, P.O., and Mosbach, K. (1978) High performance liquid affinity chromatography (HPLAC) and its application to the separation of enzymes and antigens. *FEBS Letters*, **93**, 5–9.

22 Ohlson, S., Hansson, L., Glad, M., Mosbach, K., and Larsson, P.O. (1989) High-performance liquid affinity-chromatography: a new tool in biotechnology. *Trends in Biotechnology*, **7**, 179–186.

23 Zheng, X.W., Li, Z., Beeram, S., Podariu, M., Matsuda, R., Pfaunmiller, E.L., White, C.I., Carter, N., and Hage, D.S. (2014) Analysis of biomolecular interactions using affinity microcolumns: a review. *Journal of Chromatography B*, **968**, 49–63.

24 Ohlson, S., Gudmundsson, B.M., Wikström, P., and Larsson, P.O. (1988) High-performance liquid affinity-chromatography: rapid immunoanalysis of transferrin in serum. *Clinical Chemistry*, **34**, 2039–2043.

25 Ohlson, S. and Wieslander, J. (1987) High-performance liquid affinity chromatographic-separation of mouse monoclonal-antibodies with protein A silica. *Journal of Chromatography*, **397**, 207–212.

26 Ohlson, S., Lundblad, A., and Zopf, D. (1988) Novel approach to affinity chromatography using weak monoclonal antibodies. *Analytical Biochemistry*, **169**, 204–208.

27 Zopf, D. and Ohlson, S. (1990) Weak-affinity chromatography. *Nature*, **346**, 87–88.

28 Strandh, M., Andersson, H., and Ohlson, S. (2000) Weak affinity chromatography, in *Methods in Molecular Biology* (eds P. Bailon, G.K. Ehrlich, W.J. Fung, and W. Berthold), Humana Press, Inc., Totowa.

29 Stevenson, R. (1990) Weak affinity chromatography: a new tool for immunoassays. *American Biotechnology Laboratory*, **6** (8), 8.

30 Leickt, L., Bergström, M., Zopf, D., and Ohlson, S. (1997) Bioaffinity chromatography in the 10mM range of $K_d$. *Analytical Biochemistry*, **253**, 135–136.

31 Hofstetter, O., Lindstrom, H., and Hofstetter, H. (2002) Direct resolution of enantiomers in high-performance immunoaffinity chromatography under isocratic conditions. *Analytical Chemistry*, **74**, 2119–2125.

32 Ravelet, C., Michaud, M., Ravel, A., Grosset, C., Villet, A., and Peyrin, E.

(2004) Streptavidin chiral stationary phase for the separation of adenosine enantiomers. *Journal of Chromatography A*, **1036**, 155–160.

33 Matsunaga, H. and Haginaka, J. (2014) Separation of enantiomers on chiral stationary phase based on chicken alpha (1)-acid glycoprotein: effect of silica particle diameters on column performance. *Journal of Chromatography A*, **1363**, 96–100.

34 Moaddel, R., Calleri, E., Massolini, G., Frazier, C.R., and Wainer, I.W. (2007) The synthesis and initial characterization of an immobilized purinergic receptor (P2Y(1)) liquid chromatography stationary phase for online screening. *Analytical Biochemistry*, **364**, 216–218.

35 Bergström, M., Liu, S., Kiick Kristi, L., and Ohlson, S. (2009) Cholera toxin inhibitors studied with high-performance liquid affinity chromatography: a robust method to evaluate receptor–ligand interactions. *Chemical Biology and Drug Design*, **73**, 132–141.

36 Yoo, M.J., Smith, Q.R., and Hage, D.S. (2009) Studies of imipramine binding to human serum albumin by high-performance affinity chromatography. *Journal of Chromatography B*, **877**, 1149–1154.

37 Hubble, J. (2001) Competition and intrinsic displacement effects in overloaded weak affinity chromatography. *Separation Science and Technology*, **36**, 541–554.

38 Ma, Z., Skorobogatko, Y., and Vosseller, K. (2013) Tandem lectin weak affinity chromatography for glycoprotein enrichment, in *Mass Spectrometry of Glycoproteins: Methods and Protocols* (eds J.J. Kohler and S.M. Patrie), Humana Press, New York.

39 Vosseller, K., Trinidad, J.C., Chalkley, R.J., Specht, C.G., Thalhammer, A., Lynn, A.J., Snedecor, J.O., Guan, S.H., Medzihradszky, K.F., Maltby, D.A., Schoepfer, R., and Burlingame, A.L. (2006) O-linked *N*-acetylglucosamine proteomics of postsynaptic density preparations using lectin weak affinity chromatography and mass spectrometry. *Molecular and Cellular Proteomics*, **5**, 923–934.

40 Sato, C. (2014) *Frontal Affinity Chromatography: Practice of Weak Interaction Analysis Between Lectins and Fluorescently Labeled Oligosaccharides* (ed. J. Hirabayashi), Lectins Springer New York, New York.

41 van Leeuwen, S.S., Schoemaker, R.J.W., Timmer, C., Kamerling, J.P., and Dijkhuizen, L. (2012) Use of *Wisteria floribunda* agglutinin affinity chromatography in the structural analysis of the bovine lactoferrin N-linked glycosylation. *Biochimica et Biophysica Acta*, **1820**, 1444–1455.

42 Vuignier, K., Guillarme, D., Veuthey, J.L., Carrupt, P.A., and Schappler, J. (2013) High performance affinity chromatography (HPAC) as a high-throughput screening tool in drug discovery to study drug–plasma protein interactions. *Journal of Pharmaceutical and Biomedical Analysis*, **74**, 205–212.

43 Talbert, A.M., Tranter, G.E., Holmes, E., and Francis, P.L. (2002) Determination of drug–plasma protein binding kinetics and equilibria by chromatographic profiling: exemplification of the method using L-tryptophan and albumin. *Analytical Chemistry*, **74**, 446–452.

44 Dunn, B.M. and Chaiken, I.M. (1974) Quantitative affinity chromatography. Determination of binding constants by elution with competitive inhibitors. *Proceedings of the National Academy of Sciences of the United States of America*, **71**, 2382–2385.

45 Kubín, M. (1964) Beitrag zur theorie der chromatographie. *Collection of Czechoslovak Chemical Communication*, **30**, 1104–1118.

46 Kučera, E. (1965) Contribution to the theory of chromatography linear non-equilibrium elution chromatography. *Journal of Chromatography*, **19**, 237–248.

47 Wikström, M. and Ohlson, S. (1992) Computer-simulation of weak affinity-chromatography. *Journal of Chromatography*, **597**, 83–92.

48 Leickt, L., Månsson, A., and Ohlson, S. (2001) Prediction of affinity and kinetics in biomolecular interactions by affinity chromatography. *Analytical Biochemistry*, **291**, 102–108.

49 Kasai, K.I., Oda, Y., Nishikata, M., and Ishii, S.I. (1986) Frontal affinity chromatography: theory for its application to studies on specific interactions of biomolecules. *Journal of Chromatography B*, **376**, 33–47.

50 Duong-Thi, M.D. (2013) *Introducing Weak Affinity Chromatography to Drug Discovery with Focus on Fragment Screening*, Linnaeus University, Växjö, Sweden.

51 Duong-Thi, M.D., Bergström, M., Fex, T., Isaksson, R., and Ohlson, S. (2013) High-throughput fragment screening by affinity LC–MS. *Journal of Biomolecular Screening*, **18**, 160–171.

52 Ohlson, S., Duong-Thi, M.D., Bergström, M., Fex, T., Hansson, L., Pedersen, L., Guazotti, S., and Isaksson, R. (2010) Toward high-throughput drug screening on a chip-based parallel affinity separation platform. *Journal of Separation Science*, **33**, 2575–2581.

53 Meiby, E. (2013) *Progress of Weak Affinity Chromatography as a Tool in Drug Development*, Linnaeus University, Växjö, Sweden.

54 Meiby, E., Simmonite, H., le Strat, L., Davis, B., Matassova, N., Moore, J.D., Mrosek, M., Murray, J., Hubbard, R.E., and Ohlson, S. (2013) Fragment screening by weak affinity chromatography: comparison with established techniques for screening against HSP90. *Analytical Chemistry*, **85**, 6756–6766.

55 Duong-Thi, M.D., Meiby, E., Bergström, M., Fex, T., Isaksson, R., and Ohlson, S. (2011) Weak affinity chromatography as a new approach for fragment screening in drug discovery. *Analytical Biochemistry*, **414**, 138–146.

56 Duong-Thi, M.D., Bergström, M., Fex, T., Svensson, S., Ohlson, S., and Isaksson, R. (2013) Weak affinity chromatography for evaluation of stereoisomers in early drug discovery. *Journal of Biomolecular Screening*, **18**, 748–755.

57 Meiby, E., Knapp, S., Elkins, J.M., and Ohlson, S. (2012) Fragment screening of cyclin G-associated kinase by weak affinity chromatography. *Analytical and Bioanalytical Chemistry*, **404**, 2417–2425.

58 Duong-Thi, M.D., Bergström, G., Mandenius, C.F., Bergström, M., Fex, T., and Ohlson, S. (2014) Comparison of weak affinity chromatography and surface plasmon resonance in determining affinity of small molecules. *Analytical Biochemistry*, **461**, 57–59.

59 Erlanson, D.A. and Hansen, S.K. (2004) Making drugs on proteins: site-directed ligand discovery for fragment-based lead assembly. *Current Opinion in Chemical Biology*, **8**, 399–406.

60 Erlanson, D.A., Braisted, A.C., Raphael, D.R., Randal, M., Stroud, R.M., Gordon, E. M., and Wells, J.A. (2000) Site-directed ligand discovery. *Proceedings of the National Academy of Sciences of the United States of America*, **97**, 9367–9372.

61 Poulsen, S.A. (2013) Fragment screening by native state mass spectrometry. *Australian Journal of Chemistry*, **66**, 1495–1501.

62 Hannah, V.V., Atmanene, C., Zeyer, D., Van Dorsselaer, A., and Sanglier-Cianferani, S. (2010) Native MS: an 'ESI' way to support structure- and fragment-based drug discovery. *Future Medicinal Chemistry*, **2**, 35–50.

63 Maple, H.J., Garlish, R.A., Rigau-Roca, L., Porter, J., Whitcombe, I., Prosser, C.E., Kennedy, J., Henry, A.J., Taylor, R.J., Crump, M.P., and Crosby, J. (2012) Automated protein–ligand interaction screening by mass spectrometry. *Journal of Medicinal Chemistry*, **55**, 837–851.

64 Drinkwater, N., Vu, H., Lovell, K.M., Criscione, K.R., Collins, B.M., Prisinzano, T.E., Poulsen, S.A., McLeish, M.J., Grunewald, G.L., and Martin, J.L. (2010) Fragment-based screening by X-ray crystallography, MS and isothermal titration calorimetry to identify PNMT (phenylethanolamine *N*-methyltransferase) inhibitors. *The Biochemical Journal*, **431**, 51–61.

65 Moore, C.D., Wu, H.H., Bolaños, B., Bergqvist, S., Brooun, A., Pauly, T., and Nowlin, D. (2009) Structural and biophysical characterization of XIAP BIR3 G306E mutant: insights in protein dynamics and application for fragment-based drug design. *Chemical Biology and Drug Design*, **74**, 212–223.

66 Wienken, C.J., Baaske, P., Rothbauer, U., Braun, D., and Duhr, S. (2010) Protein-

binding assays in biological liquids using microscale thermophoresis. *Nature Communications*, **1**. doi:10.1038/ncomms1093.

67 Baaske, P., Wienken, C.J., Reineck, P., Duhr, S., and Braun, D. (2010) Optical thermophoresis for quantifying the buffer dependence of aptamer binding. *Angewandte Chemie International Edition in English*, **49**, 2238–2241.

68 Pollack, S.J., Beyer, K.S., Lock, C., Muller, I., Sheppard, D., Lipkin, M., Hardick, D., Blurton, P., Leonard, P.M., Hubbard, P.A., Todd, D., Richardson, C.M., Ahrens, T., Baader, M., Hafenbradl, D.O., Hilyard, K., and Burli, R.W. (2011) A comparative study of fragment screening methods on the p38 alpha kinase: new methods, new insights. *Journal of Computer-Aided Molecular Design*, **25**, 677–687.

69 Seidel, S.A., Wienken, C.J., Geissler, S., Jerabek-Willemsen, M., Duhr, S., Reiter, A., Trauner, D., Braun, D., and Baaske, P. (2012) Label-free microscale thermophoresis discriminates sites and affinity of protein–ligand binding. *Angewandte Chemie International Edition in English*, **51**, 10656–10659.

70 Fang, Z., Simard, J.R., Plenker, D., Nguyen, H.D., Phan, T., Wolle, P., Baumeister, S., and Rauh, D. (2014) Discovery of inter-domain stabilizers: a novel assay system for allosteric Akt inhibitors. *ACS Chemical Biology*, **10** (1), 279–288.

71 Jerabek-Willemsen, M., André, T., Wanner, R., Roth, H.M., Duhr, S., Baaske, P., and Breitsprecher, D. (2014) MicroScale thermophoresis: interaction analysis and beyond. *Journal of Molecular Structure*, **1077**, 101–113.

72 Seidel, S.A.I., Markwardt, N.A., Lanzmich, S.A., and Braun, D. (2014) Thermophoresis in nanoliter droplets to quantify aptamer binding. *Angewandte Chemie International Edition in English*, **53**, 7948–7951.

73 Jerabek-Willemsen, M., Wienken, C.J., Braun, D., Baaske, P., and Duhr, S. (2011) Molecular interaction studies using microscale thermophoresis. *Assay and Drug Development Technologies*, **9**, 342–353.

# 9
# Computational Methods to Support Fragment-based Drug Discovery

*Laurie E. Grove, Sandor Vajda, and Dima Kozakov*

## 9.1
## Computational Aspects of FBDD

Computational FBDD has the potential to provide complementary information to the experimental FBDD process throughout the entire course of drug discovery. Computational methods can be used to detect fragment binding sites, assess binding site druggability, build compact but diverse fragment libraries, screen fragment libraries using molecular docking, and improve ligand binding affinity and selectivity through fragment extension. Computational FBDD methods lack many of the challenges encountered in experimental techniques [1]. Because fragments tend to bind in the millimolar to micromolar range, they require sensitive but time-intensive screening techniques, including protein crystallography and NMR spectroscopy. Furthermore, there are costs associated with equipment and infrastructure. Computational methods can be performed with relatively little cost and often on a much faster timescale. Furthermore, molecular databases have grown drastically and contain millions of members, thus creating the potential for large fragment libraries [2]. Virtual methods may provide the only means to efficiently screen such large fragment libraries. In addition, computational methods can provide information complementary to experimental approaches.

The current methods available to FBDD are not without their own challenges and limitations. With the exception of the development of fragment screening libraries, all methods discussed in this chapter require knowledge of the protein three-dimensional structure. Additionally, many computational methods used in FBDD grew out of techniques used in more general drug discovery methods. In some cases, the application of these methods to FBDD has been without problems; in others, most notably, fragment docking, modifications are required to properly address the unique energetics and dynamics of fragment binding. Due to the inherently weaker binding affinities of fragments within a binding site, new methods of scoring and binding site identification have been warranted. More generally, results obtained by computational methods are predictions that

*Fragment-based Drug Discovery: Lessons and Outlook*, First Edition. Edited by Daniel A. Erlanson and Wolfgang Jahnke.

have to be validated experimentally. Specific challenges and prospects are discussed in this chapter.

## 9.2
## Detection of Ligand Binding Sites and Binding Hot Spots

One of the initial areas that can be aided by computational methods is in the detection and description of the ligand binding sites for a given protein target. As discussed in later sections, knowledge of the binding site size, shape, and chemical functionalities can be used throughout the FBDD process, steering fragment library design, assessing druggability of a binding site, and extending core fragments.

Computational binding site detection is most obviously useful in cases where there may be little to no structural data available for the protein–ligand complex, which is usually the case when trying to target novel allosteric sites. However, these methods can also be useful in cases for verifying binding sites as in some cases either protein–ligand structures are not accurate or they contain a low-affinity ligand that does not provide complete information on the binding site [3,4]. These problems are seen in particular in cases of membrane-bound proteins, binding sites of protein–protein interactions (PPI), or highly flexible targets. For instance, the much studied G-protein-coupled receptors (GCPRs) comprise a huge family of potential targets, but these proteins are challenging [5]. For other protein targets, pockets may be relatively well-studied due to a large number of available protein–ligand structures as well as other data, including binding energetics. However, even in these cases, computational binding site detection can still be used to identify secondary binding sites.

Perhaps, two of the biggest challenges encountered in ligand binding site detection are the treatment of protein flexibility and the identification of binding sites such as those encountered in PPI targets. Currently, the methods for accounting for protein flexibility are computationally expensive; thus, they are not ideal for large screening studies. In terms of PPI targets, binding sites involved in PPIs tend to be shallow and diffuse, substantially differing from the deeper, well-formed pockets of traditional protein targets. More specifically, compared to protein–ligand and protein–protein targets, traditional protein targets tend to have one or two large pockets that are used by an inhibitor [6]. Conversely, for PPIs affinity is achieved by summing over many weak interactions [7]. Thus, many of the purely geometry-based methods are not well suited for these types of pockets and more robust methods are required.

Protein binding site detection largely uses one of the four methods: (1) geometry- based, (2) energy-based, (3) evolutionary-based, and (4) a combination of methods [8]. Within these categories, the method can be grid based or some other method. These methods are organized into Table 9.1. While a detailed description of each method is beyond the scope of this chapter, examples of the different types of methods as well as their usefulness and limitations with regard to binding site detection are discussed.

**Table 9.1** Binding site detection methods: geometry-based, energy-based, evolutionary, and combination methods.

| Geometry-based methods | Energy-based methods | Evolutionary-based methods | Combined |
|---|---|---|---|
| Grid-based | GRID (1985) | Rate4Site (2002) | LIGSITE$^{CSC}$ (2006) |
| DOCK (1982) | MCSS (1993) | ConSurf (2003) | SURFNET-Consurf (2006) |
| Pocket (1992) | QSiteFinder (2005) | FindSite (2008) | PocketFinder (2005) |
| VOIDOO (1994) | PocketFinder (2005) | 3dLIGANDsite (2009) | SiteMap (2009) |
| LIGSITE (1997) | BindingResponse (2007) | DrugSite (2013) | ConCavity (2009) |
| SCREEN (2006) | Method by Morita (2008) | | MetaPocket (2011) |
| PocketPicker (2007) | SITEHound (2009) | | |
| PocketDepth(2008) | FTSite (2011) | | |
| VICE (2009) | | | |
| ghecom (2010) | | | |
| Roll (POCASA) (2010) | | | |
| PLBSave (2013) | | | |
| LISE (2013) | | | |
| DoGSite (2010) | | | |
| Sphere filling | | | |
| SURFNET (1995) | | | |
| PASS (2000) | | | |
| Alpha shape | | | |
| APROPOEs(1996) | | | |
| CAST (CASTp) (1998) | | | |
| Fpocket (2009) | | | |

### 9.2.1 Geometry-based Methods

In geometry-based methods, the algorithm searches for pockets by measuring the size or other properties of cavities on a protein surface. Generally, the largest pockets in terms of volume are identified as the ligand binding site, but shape and buriedness descriptors can also be used. Geometry-based methods are computationally inexpensive to run and can thus be used to screen large numbers of structures. For targets that have moderately to well-defined ligand binding sites, these methods can be relatively successful. Some of the earlier geometry-based methods include Pocket [9], VOIDOO [10], SURFNET [11], APROPOS [12], LIGSITE [13], and CAST [14]. These methods all provide examples of the three different strategies used in geometry- based algorithms: grid based, sphere-filling based, and alpha shape based.

In the grid-based method, the protein structure is projected onto a three-dimensional grid and the grid points are classified based upon their positions relative to the protein surface. The nonprotein grid points are then clustered and assigned to pockets using definitions specific to the method. The earlier methods include Pocket [9] and LIGSITE [13]. While grid-based methods can be simple and require no *a priori* ligand binding information, they are orientation specific. Additionally, to help identify pockets that are not the typical, concave pockets more recent methods have considered various definitions of binding site pockets besides volume.

Many of the newer geometry-based methods are based on DOCK, which uses a sphere-filling technique on a grid to assess grid points and hence detect pockets based on a variety of properties [15]. In this manner, the grid-based technique can be used to measure more than just pocket volume. For instance, PocketPicker uses probes of different sizes to assess the pocket type, depth, and accessibility, thereby calculating a measure of "buriedness" [16]. Ghecom is similar in method but strives to detect both deep and shallow pockets by using several different probes sizes [17]. Like Ghecom, VICE was designed to target shallow binding regions, in addition to traditional pockets, using a grid-based approach and by defining a degree of buriedness [18]. DoGSite uses a difference of Gaussian filter to detect potential pockets [19]. It then splits these pockets into subpockets. The DoGSite server outputs the size, shape, and chemical features of the calculated pockets. In a different approach, ROLL and SCREEN both use a rolling sphere approach to define grid points as either within a cavity or outside of a cavity [20,21]. By using probes of different sizes, ROLL can categorize pockets and cavities based on depth and shape descriptors.

In the second type of geometry-based method, a sphere-based approach is used. In this case, the protein surface is layered with spheres of different radii. The final pocket sites are defined as those that are rich with spheres. This method has the advantage over grid-based approaches in being orientation independent. SURFNET is an early example of this type of program [11]. In SURFNET, probe spheres are placed midway between pairs of atoms on the protein. If a probe overlaps any neighboring atoms, its radius is reduced until no overlap occurs. The final set of probes defines the pockets and cavities. PASS is a similar method in which only probes with low solvent exposure are retained [22].

In the last category are methods based on the alpha shape representation of the protein. For these methods, proteins are triangulated with a series of Delaunay tetrahedra. The alpha shape is then derived by removing tetrahedra that are outside of the protein. The alpha shapes are grouped by letting smaller shapes flow toward their neighboring larger shapes. Pockets are defined as collections of empty shapes. APROPOS [12] and CAST [14] are two of the first of this type and a more recent addition is FPocket, which uses alpha spheres [23]. FPocket has several advantages, including computational efficiency and the ability to directly identify atoms forming the pocket.

VICE, DoGSite, Fpocket, PocketPicker, LIGSITE, LSite, DSite, CAST, PASS, and SURFNET were all tested on a set of 48 targets. These targets included a diverse range of binding site types, including grooves, PPI sites, and sites with flexible loops [18]. For its 48 targets, VICE identified the ligand binding site as the top ranked site for 83% of unbound and 85% of bound structures, making

VICE the top performing geometry-based method to date. The scores of DoG-Site and Fpocket were close to those of VICE.

### 9.2.2
### Energy-based Methods

Pure geometric methods are straightforward and computationally efficient to run, but do not take into account one of the primary factors motivating ligand binding in the first place – energetics. In fact, targeting multiple small pockets that are capable of contributing to the ligand binding free energy may be a key strategy to develop drugs for PPI targets [7]. In energy-based methods, an interaction energy is calculated between the protein and a probe or chemical group.

The first two energy-based methods were GRID and MCSS (multiple copy simultaneous search) [24,25]. These methods calculate an interaction energy between small probes, such as water, the methyl group and the hydroxyl group, and the protein. More recent versions of these two methods provide different probe clustering steps and additional probe types. QSiteFinder, the method by Morita, Nakamura, and Shimizu, and PocketFinder all identify sites that interact favorably with methyl probes; energetically favorable sites are clustered with nearby sites and ranked based on the total of interaction energies [26–28]. The method by Morita, Nakamura, and Shimizu has been found to perform well for unbound proteins. SITEHOUND offers the use of two types of probes, a carbon probe for drug-like binding site detection and a phosphate probe for phosphorylated ligands [29]. Binding response uses the sphere-based method of DOCK as well as clustering [30]. Specifically, the binding response is a measure of a match between a binding site and a set of probe compounds, in terms of binding energies and agreement in geometries.

The last type of energy-based method uses energy functions to detect hot spots. These methods are modeled after the experimental approaches of SAR by NMR and the multiple solvent crystal structures (MSCS) methods [31,32]. In these methods, regions that contribute to the free energy of ligand binding are those that also bind a large number of diverse small organic probe molecules [33]. FTSite [34] is based on the FTMap method [35], which uses a library of 16 probes to identify regions that can bind multiple probe molecules. Of the energy-based methods, FTSite uses the most detailed energy function, implemented using the rigid body FFT correlation approach. FTSite showed an accuracy of 94% and 97% for predicting the top ranked site as the ligand binding site for the LIGSITE$^{CSC}$ and QSiteFinder data set.

Energy-based methods have the potential to explicitly account for some degree of flexibility, unlike geometry-based methods. If there are multiple structures available, then all structures within the ensemble can be analyzed as is done in FTProd [36]. In other cases, MD simulations were used to generate an ensemble and that ensemble was mapped using a separate server, such as FTMap [37–39]. The ensembles can also be used as input for geometry-based methods, as has been done in modified versions of PASS and MDPocket [40]. In particular, using this method, PASS was able to identify known ligand binding sites for three well-studied PPI targets. Other groups have used MD simulations to directly model probe molecule interactions with the protein target [41,42]. For example, Guvench and MacKerell used a computational mapping strategy similar to that of FTSite but used MD simulations to identify probe molecule binding positions [43]. In this respect, protein flexibility is directly accounted for. Kulp *et al.* have used Monte Carlo sampling to generate ensembles of protein–ligand complexes to identify low-energy binding regions [44]. In their method, they considered high-affinity binding positions for eight organic probes as well as water to account for solvation.

### 9.2.3 Evolutionary and Structure-based Methods

There are a number of computational options that compare an input protein sequence to compiled data for known systems. For instance, Rate4Site compares the evolution of surface residues among homologous proteins to detect regions of slowly evolving residues. These regions are often functionally or structurally important [45]. ConSurf follows a similar procedure in searching for highly conserved regions among a set of proteins with a high degree of sequence conservation [46]. However, global fold is more strongly conserved than protein sequence [47], hence algorithms that consider three-dimensional descriptions of binding sites have been developed. Methods such as FINDSITE, DrugSite, and 3DLigandSite are based on the observation that certain protein folds bind substrates at similar locations, despite the low degree of sequence similarity [48–50].

### 9.2.4 Combination Methods

By combining different search techniques, a number of new methods have been developed that are quite well-performing. LIGSITE$^{CSC}$ combines the LIGSITE algorithm with knowledge of the degree of conservation of involved surface residues [51]. LIGSITE$^{CSC}$ outperforms pure geometry-based methods: CAST, PASS, LIGSITE, and SURFNET, but the more recent VICE and Fpocket have published higher success rates. The SURFNET-ConSurf method is similar to LIGSITE$^{CSC}$ – combining the geometry-based algorithm of SURFNET to detect pockets and then using ConSurf to filter pockets based on evolutionary considerations [52]. Both Concavity and SiteMap start with a grid creation [53,54]. Concavity then uses mathematical morphology to determine well-shaped pockets, as in ghecom. SiteMap combines aspects of GRID and energy considerations, including hydrophobic and hydrophilic contributions to the pocket.

MetaPocket 2.0 combines the results of eight different high-performing binding site predictor algorithms: LIGSITE$^{CSC}$, PASS, SURFNET, QSiteFinder, Fpocket, ghecom, ConCavity, and ROLL. MetaPocket 2.0 outperformed all of these methods, achieving a success rate of 61% for correctly identifying the ligand binding site as the top ranked site for a set of 198 drug–target complexes [55]. But MetaPocket 2.0 had difficulties for targets with flat or small binding sites, and binding sites at the interface of two domains.

## 9.3 Assessment of Druggability

Once the binding site has been determined, it can be assessed for druggability, that is, whether a protein binding site is amenable to binding drug-like molecules. Typically, drug-like molecules are those molecules falling within Lipinski's "rule of five" description [56]. For traditional enzyme binding sites, the largest cavity is often the active site [57] and these sites are often druggable using drug-like molecules. These sites are characterized by regions of high structural stability that are used to form intermolecular interactions, as well as regions of flexibility that are used for induced fit to accommodate ligands [58]. However, as the targets become more and more diverse, simply relying on a cavity shape to determine druggability is unreliable. As mentioned previously, interfaces involved in PPIs have larger, less concave binding sites, which may exhibit mixed flexibility. Hence, current computational methods to assess druggability need to account for more than just size and shape of a cavity.

A number of methods have been developed to assess whether a binding site is druggable. The most common and generally easiest method in terms of computational expense is to classify the target by whether it belongs to a family that is known to be druggable [59]. This approach may work well for some well-studied families, such as GPCRs [60], but fail for other families, despite sequence similarity. Furthermore, there have certainly been targets that traditionally have been believed to be undruggable, for which novel drug molecules have been found [61].

In a different approach, Hajduk, Huth, and Fesik provide what can be considered as an experimental description of a druggability score. They have shown that there is a high correlation between hit rates from NMR-based fragment screening and druggability [33]. Observations from this study as well as others have found that the hit rate has a positive correlation with pocket hydrophobicity, pocket size, and buriedness, while factors such as polar contacts and charged residues have a negative impact on hit rate [62]. Ultimately, high hit rates were found to be the result of a number of different positive contributions. Many of the current computational algorithms to determine druggability thus use descriptors related to hydrophobicity, size, and buriedness to determine druggability and use only structural information as input. To develop the algorithms, training sets are heavily relied upon to extract pocket parameters that correlate with known druggability of a target, and these pocket properties are then selected to create a druggability score. Methods within this category are generally computationally quite efficient. The first method of this type, MAPPOD, was developed by Cheng *et al.* to approximate $K_d$ for drug binding by primarily considering the fraction of nonpolar solvent accessible surface area relative to the curvature of the cavity [63].

Their method correctly predicted $K_d$ values of greater than 100 nM for the four undruggable targets in their set of 27 total targets. However, the method incorrectly predicted five targets to be undruggable and this error provides an example of a problem frequently encountered in such methods. The five targets incorrectly predicted to be undruggable are known to have ligands; however, these ligands contain at least one heavily polar group, and these five targets are indeed druggable using prodrugs or active transport mechanisms. Hence, the exclusion of polarity terms presents a problem with the underlying theory of their method. In fact, throughout these studies, druggability really meant ligandability, whereas prodrug and active transport refer to permeability that should not influence ligandability, but still may determine if an approved drug exists for the target. A number of other groups have developed approaches similar to the method of Cheng *et al.* using different metrics based on their selected training sets. Halgren includes a term to account for the negative effect of general hydrophilicity in his calculation of a druggability score, Dscore, in SiteMap [54]. With this implementation, SiteMap can better differentiate between undruggable and difficult targets. Similarly, Schmidtke and Barril have introduced a method for Fpocket to consider the local hydrophobic density of the binding site in addition to the polarity and hydrophobicity of amino acids within the pocket [64], recognizing that polar groups can contribute to binding, as in hydrogen bonding. Fpocket performs similar to SiteMap, though it operates on the timescale of seconds, whereas SiteMap requires a few minutes. By analyzing approximately 290 000 pockets, Sheridan *et al.* devised a drug-like density (DLID) score [65]. Their analysis revealed that there is a correlation between the logarithm of the volume, buriedness, and hydrophobicity. Their method primarily stumbled in detecting pockets that are very small or shallow; however, the assignments made using DLID show a correlation to the assignments made by Cheng using their set of targets.

Krasowski *et al.* developed DrugPred that is based on 16 descriptors for polarity, size, and compactness and 40 descriptors for amino acid composition [66]. This method was trained using 115 structures that purposely exclude covalent drugs and prodrugs. By excluding these groups, DrugPred is much more focused and outperforms the methods listed above, with the exception of Cheng's method. However, this method was developed to identify pockets that are druggable to traditional small molecule drug targets. Perola, Herman, and Weiss performed another exhaustive analysis, considering seven measures [67]. Five parameters were found to be important for predicting druggability: volume, depth, enclosure, % of charged residues, and hydrophobicity. One key difference between this method and others is that while methods such as SiteMap and Fpocket were trained on specific targets, Perola, Herman, and Weiss did not focus their training set. In DoGScorer, the DoGSite method is used to detect binding sites and calculate the pocket volume and surface. From here, support vector machine learning is employed to analyze the input structure using a selected subset of descriptors [68].

One of the downfalls of the methods discussed above is that they are not based on first principles, and instead rely on training sets, which are necessarily biased toward targets for which there already exists information on druggability. Targets are labeled as either druggable or undruggable; however, a key problem with this method is that it is impossible to absolutely prove a target to be undruggable. Furthermore, the pool of undruggable targets is relatively small, since there is very little information publicly available for failed targets. Several methods have been developed that do not rely on such training sets. These methods are based on the observation that protein binding sites can bind multiple, diverse small organic probe molecules – a direct observation of Hajduk, Huth, and Fesik [33]. As discussed in the previous section, these methods can be used to detect binding sites, but they can also be used to assess the drug affinity of a binding site.

In one such method, binding site energetics ($K_d$ values) are calculated from final trajectories of MD simulations using varying concentrations of isopropanol (2–100%) [41]. The choice of isopropanol was made to model both hydrophobic and polar interactions. Tested on five targets, this method performed well; however, as it involves MD simulations, the method is computationally intensive. In another method, Brenke *et al.* have defined druggability scores that can be applied to the output from their FTMap method [35]. The FTMap method uses 16 probe molecules and a detailed energy function to detect locations of high binding affinity, the so-called hot spots. Using a set of 150 protein–ligand complexes, three conditions of druggable hot spots were devised: strength, compactness, and size. Using these three measures, binding sites can be categorized as druggable, borderline druggable, and undruggable. Furthermore, the type of molecule by which the binding site is druggable is also identified as a small molecule, peptide, or macrocycle. This assessment escapes the limitations set for traditional drug-like molecules, which may not have a binding affinity for certain proteins, which are nevertheless druggable by nondrug-like molecules.

## 9.4 Generation of Fragment Libraries

Fragment libraries need not be large and in fact unnecessarily large libraries require too many resources without any guarantee that the results will be better than those obtained using a small library. In fact, properties such as chemical diversity, key physiochemical properties, and fragment availability are more important in delineating a good library from a weaker library [69]. Since FBDD is a combinatorial approach, in terms of covering the chemical space, libraries of 1000–3000 fragments can replace much larger compound libraries. However, the search for small molecule fragments can be a lengthy process and many compounds and libraries are proprietary. Using computational screening methods, fragment screening libraries can be developed efficiently based on large chemical libraries.

In the generation of fragment libraries, diversity is important; however, molecules should possess properties that are known to be relevant to drug-like molecules. In this way, computational methods are very useful for both generating fragments from large chemical libraries and filtering those fragments. To increase diversity in fragments, large libraries are often used as starting points to computationally extract fragments based on a set of parameters. These parameters commonly include ease of synthesis of the resulting fragment, size of the fragment, and complexity of the fragment. Both before and after this step, the compounds can be filtered, generally using any or all of the following considerations [70]. First, the compound should be "rule of three" compliant [71]. That is, the fragment should have a molecular weight of less than 300, the calculated octanol–water partition coefficient (cLog*P*) should be less than or equal to 3, and the number of hydrogen bond donors and acceptors should each be less than or equal to 3. Second, the compound cannot contain undesirable moieties that impart toxicity or particularly acute reactivity. Furthermore, it must be organic, soluble, and synthetically feasible.

Besides these methods, computational techniques can also be employed in clustering steps. Most frequently, clustering steps are used to generate fragment pools for fragment cocktail screening. In fragment cocktail screening, diverse pools of 4–10 diverse compounds are screened in batches using either crystallograhy or NMR techniques. This method can significantly reduce the experimental screening time [72,73].

For fragment library generation, there are three possible fragment scaffold sources as categorized by Boyd and Kloe [70]. For the first type, known drugs are used to identify the so-called privileged scaffolds that occur within different drug classes. Natural compounds are a second source for scaffolds. Natural compounds have the benefit of immense diversity, though many compounds have still not been cataloged. Furthermore, many compounds within this category likely meet the required drug-like properties. Last are novel scaffolds, which are best reached by computational means. By generating a filter of desired physiochemical, drug-like, and structural properties, large libraries can be searched for scaffolds that meet certain criteria. Of course, one of the challenges within this area is the extensive amount of proprietary information. Chemical libraries and filtering methods are often not available publicly.

### 9.4.1 Known Drugs

Methods that use known drugs or biologically active molecules as their basis for fragment library development have several advantages. Synthetic routes are known for the molecules and the molecules already have drug-like features. In many of these methods, catalogs of known drugs are used to select a group of fragments with particular properties. These fragments and properties are then used to steer a search within larger chemical libraries.

Several of the earliest methods for computational fragment design used libraries of biologically active compounds as a starting point for fragment generation. The SHAPES library [74–76], designed at Vertex Pharmaceuticals, used the

Comprehensive Medicinal Chemistry (CMC) database to decompose compounds into categories – rings, linkers, and side chains (approximately 5000 entries used) – thus identifying the structural features found in known drugs. This method generated 41 unique scaffolds, to which side chains were added. The resulting compounds were filtered based on solubility and synthetic simplicity, yielding 132 compounds. An advantage of this library is that the entire SHAPES library can be screened for a target in several hours using NMR techniques. The RECAP method cleaves molecules within a database so as to generate searchable building blocks with nonredundant functionalities [77]. Eleven bond types were selected as cleavable, based on their ability to be formed using common chemical reactions (starting from 35 000 molecules). The fragment library developed using the RECAP method as applied to the World Drug Index yielded one of the first publicly available fragment databases, DAYLIGHT THOR, which is searchable.

Some of the more recent library designs have used larger databases or multiple databases for fragment design, hoping to increase diversity. Baurin *et al.* have used multiple libraries of known drug-like compounds to filter and screen fragments in their SeeDs method [78]; the primary goal was to create a diverse library of fragments that are likely to be good hits for a wide variety of targets. This version (VER2004) has been updated to include or remove fragments, but keeping it at the same size to ensure quick screening [79]. The decomposition and identification of molecules (DAIM) method decomposes molecules and assigns fingerprints to describe key structural features of each fragment [80]. Thus, the generated fragments can easily be compared and sorted within large libraries. Comparison of the fragments generated using this method on a library of known protein–ligand complexes versus the large ZINC database (1.85 million unique compounds at publication [81]) showed a high degree of similarity in the identified fragments, though some key scaffolds were missing from the ZINC database. However, this result indicates that this method can identify and select fragments known to be important for binding affinity even when applied to non-drug-related libraries. Gianti and Sartori used the commercially available Pipeline Pilot to identify privileged fragments within the commercial databases of drug-like compounds: CMC, the Symyx Drug Data Report (MDDR), and the drug-like subset of the ZINC database [82]. Fragmentation and filtering steps included the rule-of-three considerations in addition to shape similarity relative to known biologically active compounds. By including compounds from the virtual screening library, ZINC, the authors were hoping to identify novel fragments. The generated fragments were then used to extract commercially available compounds from the ACD.

### 9.4.2
### Natural Compounds

An advantage of using natural product databases is the wide variety of known structures that have been minimally explored in terms of their applicability to drug discovery. Compounds within natural products databases tend to have more stereocenters and greater complexity in general. Recently, Over *et al.* have used the Dictionary of Natural Products (DNPs) to identify fragments and filter fragments within Pipeline Pilot using a method similar to the SHAPES method of identifying rings, linkers, and side chains [83]. The fragments identified using this method and filtration were found to have only small overlap in chemical properties with fragments from the ZINC and SynLib databases and thus showing promise as novel fragments.

In a different approach, the "fragments of life" (FOL) database was created to maximize the ligand efficiency of the fragments within the library by including molecules that have most likely evolved in conjunction with proteins [84]. As such, this database includes soluble metabolites, natural products, and biaryl molecules, similar to those motifs found in proteins. Fragments were selected manually from metabolic pathways, as well as from the ACD and the Maybridge Fragment Library to ensure diversity of fragments. The authors also illustrate the use of Monte Carlo optimization, as implemented by Accelrys, to generate diverse collections of fragments for use in fragment cocktail screening.

### 9.4.3 Novel Scaffolds

As the concern grows over identifying unique and patentable drug scaffolds, searching larger databases has become common. Furthermore, using generic databases to build fragment libraries remains the best strategy for fragment libraries to be used for targets of which little is known – diversity of fragments is very important. Blomberg *et al.* describe a number of techniques used to generate the generic fragment screening library (GFSL05) and the generic NMR library (APGNMR07) [85]. The first, GFSL05, was developed to be widely applied to many targets. A primary consideration of this library was to make the screening process fast and flexible. On the other hand, the APGNMR07 library was designed with NMR screening in mind, though it too is designed to be generic with respect to targets. In both cases, the fragments were then clustered using Monte Carlo protocols to form diverse pools for fragment cocktail screening, again the aim being to produce a fragment library of maximum versatility. Schulz *et al.* also experimented with a variety of filtering and clustering steps to develop a maximally diverse library [86]. In this method, fragments were pooled from the ZINC database, but focused on commercially available compounds.

On the other hand, screening libraries should not be too large, and it has been argued that many compounds within these large libraries are not biologically relevant. In this respect, known drug or natural product databases are often used as references to extract fragments from the larger libraries. Perhaps, one of the largest sources of fragments used to date was the GDB-17 database, containing approximately 166 billion organic molecules of up to 17 atoms [87]. In this case, Ruddigkeit, Blum, and Reymond used molecular graph theory to provide a structure description of possible fragments from the GDB-17 database [88].

## 9.5 Docking Fragments and Scoring

Fragment screening typically employs NMR assays or other techniques. In computational screening, fragments are docked and scored. The best poses can then be grown to take advantage of nearby binding pockets. The fragment poses from docking experiments can also be used during fragment linking and fragment joining methods. Docking presents a number of benefits over traditional fragment screening: it is low cost, fast, and escapes any solubility concerns. It also allows one to screen compounds that have not been synthesized. There has also been some evidence that docking can allow fragments to be prioritized within a library [89,90].

### 9.5.1 Challenges of Fragment Docking

Despite its potential benefits, fragment docking presents a number of challenges. Indeed, the assumption is that fragments are more difficult to dock than drug-like molecules because fragments are weak binders and thus it is far from simple to differentiate between native and other low-energy poses. Furthermore, fragments are more likely to occupy binding sites on a protein surface that can accommodate low molecular weight compounds with limited specificity. The smaller size of fragments relative to the volume of the binding pocket may result in incorrect binding modes. However, Verdonk *et al.* have demonstrated that a key indicator of docking performance for both fragment and drug-like molecule docking is ligand efficiency [91] and there was no significant difference in docking performance for those two groups. The main cause of failure for the docking of fragments does remain to be scoring though, whereas the main cause of failure for the drug-like molecules was insufficient sampling. In fact, the use of a robust, physics-based rescoring function (AMBER, in this case) provided better results for fragment docking.

Of the challenges presented in fragment docking, the most important are scoring and receptor flexibility. In terms of scoring, three types of scoring functions can be chosen from: physics based, empirical, and knowledge based [92]. Physics-based scoring functions are capable of accounting for electrostatics, van der Waals, and desolvation energies. The molecular mechanics generalized Born surface area (MM-GBSA) method is one in particular that has been tested for fragments. A more rigorous treatment of solvation and electrostatics using MD or Monte Carlo methods can be used to improve scoring [93]; however, such calculations can become quite computationally intensive. Essentially, all methods ignore the entropy change of ligand binding. While computationally expensive, physics-based functions have the advantage of not being biased toward any particular type of compound, which is a problem with empirical scoring functions. Furthermore, they have been found to perform well in terms of identifying correct binding poses [88], but they come at a higher computation cost [94].

Empirical functions use weighted energy terms to estimate the binding affinity. The weights are determined by fitting the binding affinity data of a training set of protein–ligand complexes. Since this binding data typically consists of ligands that are known drug-like molecules, this method works well with drug screening for well- known targets, but is not optimized for fragments. In particular, fragments generally have fewer hydrogen bond donors and acceptors than drug-like compounds; hence enthalpic parameterization terms associated with desolvation and protein–ligand interactions may be quite different [95]. Similarly, entropic terms are likely underestimated for fragments that may have several binding modes. However, empirical scoring functions can be quite computationally efficient and are hence frequently used; examples include GlideScore and ChemScore. A couple of fragment-based profiles for protein–fragment interactions have also been developed [96,97], as well as the fragment-specific method, LUDI, which uses an empirical scoring function to dock small and fairly rigid fragments [98,99].

Knowledge-based scoring functions employ energy potentials derived from experimentally determined atomic structures. In this case, pair-wise potentials are obtained from the frequency of certain atom pairs in a database. While most knowledge-based potentials are trained on protein–ligand pairs and not protein–fragment pairs, these methods tend to be robust as the training set can be quite large. Furthermore, these methods are comparable to empirical methods in terms of speed. A fragment-specific method has recently been reported and is capable of identifying fragments that correspond to the bound ligand with a success rate of 74% [1].

The second significant problem encountered in fragment docking is flexibility. In terms of flexibility, fragments themselves tend to have few if any degrees of freedom; however, receptor flexibility can be quite an important consideration. Because fragments are smaller, they are more likely to bind to proteins in different conformational states; thus, it is optimal to account for receptor flexibility when docking fragments, though far from feasible for most large fragment libraries. A number of solutions to this problem exist – though all increase computational time. An ensemble of conformations can be generated using MD simulations and individual conformations can be tested. Another approach is to model different regions of a receptor as flexible independent of other regions and in this way creates different snapshots that can be independently docked. The FlexX method provides a fast approach for the treatment of ligand and receptor flexibility, and has been shown to successfully identify fragment inhibitors from a fragment library of over 10 000 members [100].

### 9.5.2 Examples of Fragment Docking

Of the commercially available programs, GOLD and Glide are perhaps the most frequently used. Both programs provide a range of options in terms of the types of available scoring functions as well as pre- and postprocessing options. Indeed, in one study GOLD and Glide were found to perform similarly well, predicting the correct pose for fragments 70% of the time, relative to 60% and 40% for Surflex and FlexX, respectively [101]. DOCK is also widely used and is freely available for academic institutions. In terms of which scoring functions work best for fragments, the results are somewhat mixed. Studies by Shoichet and coworkers have shown that the use of the physics-based MM-GBSA technique as implemented in DOCK was able to predict fragment poses in agreement with X-ray crystallographic data [89,102]. The combined quantum mechanics/molecular mechanics (QM/MM) method has also been tested specifically in terms of its ability to account for polarizibility; however, the intensive computational requirements of this technique significantly limit its applicability. It was tested in the rescoring of GOLD-derived protein–fragment poses for a series of kinase-based fragments using the density functional theory (DFT) hydride functional B3LYP [103]. Based on the calculated energies, the correct fragment pose was selected 77% of the time relative to using GOLD (44% enrichment).

The GlideScore empirical-based scoring function has also been evaluated in several studies. Sandor and coworkers found that for Glide, the default Glide SP (standard precision) protocol combined with expanded sampling showed the best performance for fragment docking [95]. In this study, no advantage was found by using the Glide XP (extra precision) protocol, though the Glide XP protocol had been previously tested on a set of 12 targets and showed good performance by predicting the correct pose for each ligand within an RMSD of 1 Å [104]. Compared to GlideScore, the use of physics-based scoring techniques was assumed to be preferable as these techniques are not biased toward drug-like molecules. However, Glide XP performed better than Glide SP in combination with MM-GBSA rescoring [105], even though MM-GBSA rescoring has been previously found to increase the docking accuracy in the docking of drug-like molecules. The use of the combination scoring method, Emodel, performed equally as well as the default implementation of Glide SP. Emodel combines GlideScore with the nonbonded interaction energy between the ligand and the protein.

In a somewhat different application of docking, Loving, Salam, and Sherman used Glide XP to dock fragments and then identify key pharmocophore features using an energetic ranking of the docking results [104]. After generating these features, they can be used in a number of ways: to identify key features of molecular recognition for a particular binding site, for database screening, and for identifying new ligands from a database. This method was shown to be successful at finding both diverse compounds and known actives from a database. However, the study was performed using ligand-bound rather than ligand-free protein structures, which potentially reduces the generality of the results.

The Caflisch and coworker have contributed a number of docking resources specifically aimed at fragment docking [106]. Their program SEED (solvation energy for exhaustive docking) uses a vector-based approach for docking apolar fragments that considers the desolvation energy of the receptor and the van der Waals interaction energy between the protein and the fragment [107].

The LIE (linear interaction energy) model can be used to rank the docking poses and the binding energy can be calculated using MD or Monte Carlo methods. This group has also developed a docking method for flexible fragments, FFLD (fragment-based flexible ligand docking) [108].

## 9.6 Expansion of Fragments

There are two primary concerns in fragment expansion: (1) where to expand to within the binding site and (2) how to expand a core fragment or link neighboring fragments. Some of the binding site detection algorithms are also used for determining secondary binding sites. For instance, FTMap has shown that the primary hot spot corresponds to the core fragment, while nearby secondary hot spots are used for the binding of extensions to yield high-affinity ligands [109].

Fragment expansion is by necessity addressed using computational methods and is often approached as a combinatorial problem. Hence, the pool of molecules generated from fragment expansion can quickly exceed a number that is reasonable to test experimentally. However, computational methods provide a way to efficiently search a large chemical space for molecules meeting specified criteria. To reduce the number of generated molecules, fragment expansion is often guided using defined pharmacophore constraints, docking scores, or other properties. Thus, the testing of inappropriate compounds can be avoided. However, fragment expansion presents a number of challenges. First, due to their smaller size, fragments often bind in a number of low-energy conformations. Depending on the binding mode, the expansion of a fragment may lead to a molecule that cannot bind in a low-energy conformation. Furthermore, it may be difficult or impossible to link two nearby fragments due to steric constraints or proximity. In both cases, by allowing for slight displacements of the fragment within the binding site, possible solutions may be found. Finally, fragments generated *in silico* are not always generated with synthetic feasibility in mind. There are several techniques that can be used for fragment expansion. Ligand growing is perhaps the simplest method in which functional groups or smaller fragments are added to a single fragment, sometimes in a combinatorial style. At each step, the new fragment is assessed. Generally, the fragments within a range of energies are retained to account for the fact that intermediate fragments may be slightly higher in energy than the initial or final molecules. In other cases, a molecular skeleton is generated and chemical moieties are added to it to construct a molecule. The merging method is similar: two nearby fragments are joined, but no new atoms are added to the system. In these methods, ligand affinity was found to be roughly linear with respect to molecular weight [110].

In fragment linking techniques, known moderate affinity fragments in neighboring pockets are linked using one or more atoms. In this manner, fragments with known binding affinity are used to form a larger molecule, in theory simplifying the search process and minimizing the number of molecules that need to be generated. In this manner, there is an assumption that the binding affinity of fragments is additive [111]. However, as there can be entropic losses due to the loss of a mostly rigid-body fragment, binding affinity can decrease.

Since there are often multiple low-energy fragment positions within a binding site, it may be difficult to design a linker that accommodates the lowest energy configuration of each fragment. Or, spatially, the location of neighboring fragments may not allow for a sterically favorable linking configuration. These problems can be addressed by allowing the fragments some amount of translational freedom during the linking process. Conversely, the RMSD between the fragment's original position and its new position in the expanded molecule can be monitored during the expansion process to ensure that the original high-affinity intermolecular interactions are maintained, as in S4MPLE [112] or CONFIRM [113]. In CAVEAT, linkers are selected so as to maintain the original three-dimensional orientations of the fragments and to position key functional groups in specific orientations [114].

In any fragment expansion method, the combinatorial possibilities can quickly become too numerous to be realistically tested. Hence, some method of assessment is used throughout the expansion process to guide the design. A fitness function of user- defined properties can help this effort as is used by genetic algorithms in LigBuilder [115] and LEA3D [116]. FlexNovo uses fragments from a large fragment space for fragment growth while considering desired physiochemical and geometric properties [117]. In the BREED method, fragments are grown by decomposing known ligands of the target and then merging the fragments together to form new molecules [118]. In this case, because the functional groups present are already known to bind to the target, there is a high likelihood that high-affinity ligands will be generated. However, the novelty of such ligands may not be great. GANDI uses structure-based properties of known inhibitors in addition to a force field energy to join predocked fragments and new fragments [119].

More computationally expensive techniques such as MD or Monte Carlo simulations are often used in growing approaches and similar structural sampling methods. In some cases, fragments are randomly dispersed and connections are randomly made between fragments; thus leading to a new series of molecules that are then scored [120]. In CONCERTS, one of the first MD-based *de novo* design approaches, fragments are allowed to sample low-energy positions and larger molecules are formed when it is possible for nearby fragments to form a bond. Bonds are broken and formed throughout the process with the overall goal of creating low-energy molecules [121]. To avoid the possibility of a combinatorial explosion due to the large number of new molecules, restrictions are typically applied such as steric constraints and electrostatic considerations. An example is the SPROUT method, which generates a molecular skeleton using molecular graphing and then assigns three-dimensional structural elements to build the molecule [122]. In the second method, MCDNLG (Monte Carlo *de novo* ligand generator), the so-called "super molecule," composed of densely and randomly packed molecules in the binding site, is annealed to generate a chemically stable molecule [123]. A similar method is described for dynamic ligand design (DLD) [124].

Another problem with molecules made via fragment expansion is their lack of validated synthetic feasibility and as such several methods have been developed to address this challenge. SYNOPSIS designs molecules based on the possible synthetic routes, starting from a database of available molecules [125]. The more recent version of SkelGen uses a Monte Carlo-based growing algorithm that favors synthetically feasible molecules [126].

## 9.7 Outlook

From the time when computational methods were first used in drug discovery in the late 1980s to the current era, the computing power available to researchers has increased to such a point that entire fragment libraries can be screened efficiently. One of the primary advantages and potentials provided by computational methods is the vast resources that can be saved in terms of time and experimental equipment. However, all computational techniques are a balance of efficiency and accuracy. What works for one target may fail for another, and hence a certain amount of trial and error is required when using computational resources. However, as has been shown by the increasing number of high-affinity inhibitors developed using computational FBDD resources [127,128], the continued research and development in this area has led to results. Again, we emphasize that by definition all computational results are predictions, and although they can be very useful, they have to be validated experimentally.

## References


1 Tang, G.W. and Altman, R.B. (2014) Knowledge-based fragment binding prediction. *PLoS Computational Biology*, **10**, e1003 589.

2 Bohacek, R.S., McMartin, C., and Guida, W.C. (1996) The art and practice of structure-based drug design: a molecular modeling perspective. *Medicinal Research Reviews*, **16**, 3–50.

3 Davis, A.M., St-Gallay, S.A., and Kleywegt, G.J. (2008) Limitations and lessons in the use of X-ray structural information in drug design. *Drug Discovery Today*, **13**, 831–841.

4 Reynolds, C.H. (2014) Protein–ligand cocrystal structures: we can do better. *ACS Medicinal Chemistry Letters*, **5**, 727–729.

5 Hubbard, R.E. (2011) Structure-based drug discovery and protein targets in the CNS. *Neuropharmacology*, **60**, 7–23.

6 Fuller, J.C., Burgoyne, N.J., and Jackson, R.M. (2009) Predicting druggable binding sites at the protein–protein interface. *Drug Discovery Today*, **14**, 155–161.

7 Clackson, T. and Wells, J. (1995) A hot spot of binding energy in a hormone–receptor interface. *Science*, **267**, 383–386.

8 Henrich, S., Salo-Ahen, O.M.H., Huang, B., Rippmann, F.F., Cruciani, G., and Wade, R.C. (2010) Computational approaches to identifying and characterizing protein binding sites for

ligand design. *Journal of Molecular Recognition*, **23**, 209–219.
9 Levitt, D.G. and Banaszak, L.J. (1992) Pocket: a computer graphics method for identifying and displaying protein cavities and their surrounding amino acids. *Journal of Molecular Graphics*, **10**, 229–234.
10 Kleywegt, G.J. and Jones, T.A. (1994) Detection, delineation, measurement and display of cavities in macromolecular structures. *Acta Crystallographica Section D Biological Crystallography*, **50**, 178–185.
11 Laskowski, R.A. (1995) SURFNET: a program for visualizing molecular surfaces, cavities, and intermolecular interactions. *Journal of Molecular Graphics*, **5**, 323–330.
12 Peters, K.P., Fauck, J., and Frömmel, C. (1996) The automatic search for ligand binding sites in proteins of known three-dimensional structure using only geometric criteria. *Journal of Molecular Biology*, **256**, 201–213.
13 Hendlich, M., Rippmann, F., and Barnickel, G. (1997) LIGSITE: automatic and efficient detection of potential small molecule-binding sites in proteins. *Journal of Molecular Graphics and Modelling*, **15**, 359–363.
14 Liang, J., Woodward, C., and Edelsbrunner, H. (1998) Anatomy of protein pockets and cavities: measurement of binding site geometry and implications for ligand design. *Protein Science*, **7**, 1884–1897.
15 Kuntz, I.D., Blaney, J.M., Oatley, S.J., Langridge, R., and Ferrin, T.E. (1982) A geometric approach to macromolecule–ligand interactions. *Journal of Molecular Biology*, **161**, 269–288.
16 Weisel, M., Proscak, E., and Schneider, G. (2007) PocketPicker: analysis of ligand binding-sites with shape descriptors. *Chemistry Central Journal*, **1**.
17 Kawabata, T. (2010) Detection of multiscale pockets on protein surfaces using mathematical morphology. *Proteins: Structure, Function, and Bioinformatics*, **78**, 1195–1211.
18 Tripathi, A. and Kellogg, G.E. (2010) A novel and efficient tool for locating and characterizing protein cavities and binding sites. *Proteins: Structure, Function, and Bioinformatics*, **78**, 825–842.
19 Volkamer, A., Griewel, A., Grombacher, T., and Rarey, M. (2010) Analyzing the topology of active sites: on the prediction of pockets and subpockets. *Journal of Chemical Information and Modeling*, **50**, 2041–2052.
20 Yu, J., Zhou, Y., Tanaka, I., and Yao, M. (2010) Roll: a new algorithm for the detection of protein pockets and cavities with a rolling probe sphere. *Bioinformatics*, **26**, 46–52.
21 Nayal, M. and Honig, B. (2006) On the nature of cavities on protein surfaces: application to the identification of drug-binding sites. *Proteins: Structure, Function, and Bioinformatics*, **63**, 892–906.
22 Brady, G.P. and Stouten, P.F. (2000) Fast prediction and visualization of protein binding pockets with pass. *Journal of Computer-Aided Molecular Design*, **14**, 383–401.
23 Guilloux, V.L., Schmidtke, P., and Tuffrey, P. (2009) Fpocket: an open source platform for ligand pocket detection. *BMC Bioinformatics*, **10**, 168.
24 Goodford, P.J. (1985) A computational procedure for determining energetically favorable binding sites on biologically important macromolecules. *Journal of Medicinal Chemistry*, **28**, 849–857.
25 Caflisch, A., Miranker, A., and Karplus, M. (1993) Multiple copy simultaneous search and construction of ligands in binding sites: application to inhibitors of HIV-1 aspartic proteinase. *Journal of Medicinal Chemistry*, **36**, 2142–2167.
26 Laurie, A.T.R. and Jackson, R.M. (2005) Q-SiteFinder: an energy-based method for the prediction of protein–ligand binding sites. *Bioinformatics*, **21**, 1908–1916.
27 Morita, M., Nakamura, S., and Shimizu, K. (2008) Highly accurate method for ligand-binding site prediction in unbound state (apo) protein structures. *Proteins: Structure, Function, and Bioinformatics*, **73**, 468–479.
28 An, J., Totrov, M., and Abagyan, R. (2005) Pocketome via comprehensive

identification and classification of ligand binding envelopes. *Molecular and Cellular Proteomics*, **4**, 752–761.

29 Hernandez, M., Ghersi, D., and Sanchez, R. (2009) SITEHOUND-web: A server for ligand binding site identification in protein structures. *Nucleic Acids Research*, **37**, W413–W416.

30 Zhong, S. and MacKerell, A.D. (2007) Binding response: a descriptor for selecting ligand binding site on protein surfaces. *Journal of Chemical Information and Modeling*, **47**, 2303–2315.

31 Liepinsh, E. and Otting, G. (1997) Organic solvents identify specific ligand binding sites on protein surfaces. *Nature Biotechnology*, **15**, 264–268.

32 Mattos, C., Bellamacina, C.R., Peisach, E., Pereira, A., Vitkup, D., Petsko, G.A., and Ringe, D. (2006) Multiple solvent crystal structures: probing binding sites, plasticity and hydration. *Journal of Molecular Biology*, **357**, 1471–1482.

33 Hajduk, P.J., Huth, J.R., and Fesik, S.W. (2005) Druggability indices for protein targets derived from NMR-based screening data. *Journal of Medicinal Chemistry*, **48**, 2518–2525.

34 Ngan, C.H., Hall, D., Zerbe, B., Grove, L. E., Kozakov, D., and Vajda, S. (2011) FTSite: high accuracy detection of ligand binding sites on unbound protein structures. *Bioinformatics*, **28**, 286–287.

35 Brenke, R., Kozakov, D., Chuang, G.Y., Beglov, D., Hall, D., Landon, M.R., Mattos, C., and Vajda, S. (2009) Fragment-based identification of druggable 'hot spots' of proteins using Fourier domain correlation techniques. *Bioinformatics*, **25**, 621–627.

36 Votapka, L. and Amaro, R.E. (2013) Multistructural hot spot characterization with FTProd. *Bioinformatics*, **29**, 393–394.

37 Ivetac, A. and Andrew McCammon, J. (2010) Mapping the druggable allosteric space of G-protein coupled receptors: a fragment-based molecular dynamics approach. *Chemical Biology and Drug Design*, **76**, 201–217.

38 Ortiz-Sanchez, J.M., Nichols, S.E., Sayyah, J., Brown, J.H., McCammon, J.A., and Grant, B.J. (2012) Identification of potential small molecule binding pockets on Rho family GTPases. *PLoS ONE*, **7**, e40 809.

39 Miao, Y., Nichols, S.E., and McCammon, J.A. (2014) Mapping of allosteric druggable sites in activation-associated conformers of the M2 muscarinic receptor. *Chemical Biology and Drug Design*, **83**, 237–246.

40 Schmidtke, P., Bidon-Chanal, A., Luque, F.J., and Barril, X. (2011) MDpocket: open-source cavity detection and characterization on molecular dynamics trajectories. *Bioinformatics*, **27**, 3276–3285.

41 Seco, J., Luque, F.J., and Barril, X. (2009) Binding site detection and druggability index from first principles. *Journal of Medicinal Chemistry*, **52**, 2363–2371.

42 Lexa, K.W. and Carlson, H.A. (2011) Full protein flexibility is essential for proper hot-spot mapping. *Journal of the American Chemical Society*, **133**, 200–202.

43 Guvench, O. and MacKerell, A.D. Jr. (2009) Computational fragment-based binding site identification by ligand competitive saturation. *PLoS Computational Biology*, **5**, e1000 435.

44 Kulp, J.L., Kulp, J.L., Pompliano, D.L., and Guarnieri, F. (2011) Diverse fragment clustering and water exclusion identify protein hot spots. *Journal of the American Chemical Society*, **133**, 10 740–10 743.

45 Pupko, T., Bell, R.E., Mayrose, I., Glaser, F., and Ben-Tal, N. (2002) Rate4Site: an algorithmic tool for the identification of functional regions in proteins by surface mapping of evolutionary determinants within their homologues. *Bioinformatics*, **18**, S71–S77.

46 Landau, M., Mayrose, I., Rosenberg, Y., Glaser, F., Martz, E., Pupko, T., and Ben-Tal, N. (2005) ConSurf 2005: the projection of evolutionary conservation scores of residues on protein structures. *Nucleic Acids Research*, **33**, W299–W302.

47 Chothia, C. and Lesk, A.M. (1986) The relation between the divergence of sequence and structure in proteins. *EMBO Journal*, **5**, 823–826.

48 Brylinski, M. and Skolnick, J. (2008) A threading-based method (FINDSITE) for ligand-binding site prediction and functional annotation. *Proceedings of the National Academy of Sciences of the United States of America*, **105**, 129–134.

49 Van Voorst, J.R. and Finzel, B.C. (2013) Searching for likeness in a database of macromolecular complexes. *Journal of Chemical Information and Modeling*, **53**, 2634–2647.

50 Wass, M.N. and Sternberg, M.J.E. (2009) Prediction of ligand binding sites using homologous structures and conservation at CASP8. *Proteins: Structure, Function, and Bioinformatics*, **77**, 147–151.

51 Huang, B. and Schroeder, M. (2006) LIGSITE$^{csc}$: predicting ligand binding sites using the Connolly surface and degree of conservation. *BMC Structural Biology*, **6**, 19.

52 Glaser, F., Morris, R.J., Najmanovich, R.J., Laskowski, R.A., and Thornton, J.M. (2006) A method for localizing ligand binding pockets in protein structures. *Proteins: Structure, Function, and Bioinformatics*, **62**, 479–488.

53 Capra, J.A., Laskowski, R.A., Thornton, J. M., Singh, M., and Funkhouser, T.A. (2009) Predicting protein ligand binding sites by combining evolutionary sequence conservation and 3D structure. *PLoS Computational Biology*, **5**, e1000 585.

54 Halgren, T.A. (2009) Identifying and characterizing binding sites and assessing druggability. *Journal of Chemical Information and Modeling*, **49**, 377–389.

55 Zhang, Z., Li, Y., Lin, B., Schroeder, M., and Huang, B. (2011) Identification of cavities on protein surface using multiple computational approaches for drug binding site prediction. *Bioinformatics*, **27**, 2083–2088.

56 Lipinski, C.A., Lombardo, F., Dominy, B. W., and Feeney, P.J. (2001) Experimental and computational approaches to estimate solubility and permeability in drug discovery and development settings. *Advanced Drug Delivery Reviews*, **46**, 3–26.

57 Laskowski, R.A., Luscombe, N.M., Swindells, M.B., and Thornton, J.M. (1996) Protein clefts in molecular recognition and function. *Protein Science*, **5**, 2438–2452.

58 Lugue, I. and Freire, E. (2000) Structural stability of binding sites: consequences for binding affinity and allosteric effects. *Proteins: Structure Function, and Genetics*, **41**, 63–71.

59 Fauman, E.B., Rai, B.K., and Huang, E.S. (2011) Structure-based druggability assessment: identifying suitable targets for small molecule therapeutics. *Current Opinion in Chemical Biology*, **15**, 463–468.

60 Kuhn, M., Campillos, M., González, P., Jensen, L.J., and Bork, P. (2008) Large-scale prediction of drug–target relationships. *FEBS Letters*, **582**, 1283–1290.

61 Hopkins, A.L. and Groom, C.R. (2002) The druggable genome. *Nature Reviews Drug Discovery*, **1**, 727–730.

62 Nisius, B., Sha, F., and Gohlke, H. (2012) Structure-based computational analysis of protein binding sites for function and druggability prediction. *Journal of Biotechnology*, **159**, 123–134.

63 Cheng, A.C., Coleman, R.G., Smyth, K.T., Cao, Q., Soulard, P., Caffrey, D.R., Salzberg, A.C., and Huang, E.S. (2007) Structure-based maximal affinity model predicts small-molecule druggability. *Nature Biotechnology*, **25**, 71–75.

64 Schmidtke, P. and Barril, X. (2010) Understanding and predicting druggability. A high-throughput method for detection of drug binding sites. *Journal of Medicinal Chemistry*, **53**, 5858–5867.

65 Sheridan, R.P., Maiorov, V.N., Holloway, M.K., Cornell, W.D., and Gao, Y.D. (2010) Drug-like density: a method of quantifying the "bindability" of a protein target based on a very large set of pockets and drug-like ligands from the protein data bank. *Journal of Chemical Information and Modeling*, **50**, 2029–2040.

66 Krasowski, A., Muthas, D., Sarkar, A., Schmitt, S., and Brenk, R. (2011) DrugPred: a structure-based approach to predict protein druggability developed using an extensive nonredundant data set. *Journal of Chemical Information and Modeling*, **51**, 2829–2842.

67 Perola, E., Herman, L., and Weiss, J. (2012) Development of a rule-based method for the assessment of protein druggability. *Journal of Chemical Information and Modeling*, **52**, 1027–1038.

68 Volkamer, A., Kuhn, D., Grombacher, T., Rippmann, F., and Rarey, M. (2012) Combining global and local measures for structure-based druggability predictions. *Journal of Chemical Information and Modeling*, **52**, 360–372.

69 Huth, J.R., Sun, C., Sauer, D.R., and Hajduk, P.J. (2005) Utilization of NMR-derived fragment leads in drug design. *Methods in Enzymology*, **394**, 549–571.

70 Boyd, S.M. and de Kloe, G.E. (2010) Fragment library design: efficiently hunting drugs in chemical space. *Drug Discovery Today: Technologies*, **7**, e173–e180.

71 Congreve, M., Carr, R., Murray, C.W., and Jhoti, H. (2003) A 'rule of three' for fragment-based lead discovery. *Drug Discovery Today*, **8**, 876–877.

72 Nienaber, V.L., Richardson, P.L., Klighofer, V., Bouska, J.J., Giranda, V.L., and Greer, J. (2000) Discovering novel ligands for macromolecules using X-ray crystallographic screening. *Nature Biotechnology*, **18**, 1105–1108.

73 Bosch, J., Robien, M.A., Mehlin, C., Boni, E., Riechers, A., Buckner, F.S., Van Voorhis, W.C., Myler, P.J., Worthey, E.A., DeTitta, G., Luft, J.R., Lauricella, A., Gulde, S., Anderson, L.A., Kalyuzhniy, O., Neely, H.M., Ross, J., Earnest, T.N., Soltis, M., Schoenfeld, L., Zucker, F., Merritt, E. A., Fan, E., Verlinde, C.L.M.J., and Hol, W.G.J. (2006) Using fragment cocktail crystallography to assist inhibitor design of *Trypanosoma brucei* nucleoside 2-deoxyribosyltransferase. *Journal of Medicinal Chemistry*, **49** 5939–5946.

74 Bemis, G.W. and Murcko, M.A. (1996) The properties of known drugs. 1. Molecular frameworks. *Journal of Medicinal Chemistry*, **39**, 2887–2893.

75 Bemis, G.W. and Murcko, M.A. (1999) Properties of known drugs. 2. Side chains. *Journal of Medicinal Chemistry*, **42**, 5095–5099.

76 Fejzo, J., Lepre, C.A., Peng, J.W., Bemis, G.W., Ajay, Murcko, M.A., and Moore, J. M. (1999) The SHAPES strategy: an NMR-based approach for lead generation in drug discovery. *Chemistry and Biology*, **6** (10), 755–769.

77 Lewell, X.Q., Judd, D.B., Watson, S.P., and Hann, M.M. (1998) RECAP – retrosynthetic combinatorial analysis procedure: a powerful new technique for identifying privileged molecular fragments with useful applications in combinatorial chemistry. *Journal of Chemical Information and Computer Sciences*, **38**, 511–522.

78 Baurin, N., Aboul-Ela, F., Barril, X., Davis, B., Drysdale, M., Dymock, B., Finch, H., Fromont, C., Richardson, C., Simmonite, H., and Hubbard, R.E. (2004) Design and characterization of libraries of molecular fragments for use in NMR screening against protein targets. *Journal of Chemical Information and Computer Sciences*, **44**, 2157–2166.

79 Chen, I.J. and Hubbard, R. (2009) Lessons for fragment library design: analysis of output from multiple screening campaigns. *Journal of Computer-Aided Molecular Design*, **23**, 603–620.

80 Kolb, P. and Caflisch, A. (2006) Automatic and efficient decomposition of two-dimensional structures of small molecules for fragment-based high-throughput docking. *Journal of Medicinal Chemistry*, **49**, 7384–7392.

81 Irwin, J.J. and Shoichet, B.K. (2005) ZINC: a free database of commercially available compounds for virtual screening. *Journal of Chemical Information and Modeling*, **45**, 177–182.

82 Gianti, E. and Sartori, L. (2008) Identification and selection of "privileged fragments" suitable for primary screening. *Journal of Chemical Information and Modeling*, **48**, 2129–2139.

83 Over, B., Wetzel, S., Grütter, C., Nakai, Y., Renner, S., Rauh, D., and Waldmann, H. (2013) Natural-product-derived fragments for fragment-based ligand discovery. *Nature Chemistry*, **5**, 21–28.

84 Davies, D.R., Mamat, B., Magnusson, O. T., Christensen, J., Haraldsson, M.H., Mishra, R., Pease, B., Hansen, E., Singh, J., Zembower, D., Kim, H., Kiselyov, A.S.,

Burgin, A.B., Gurney, M.E., and Stewart, L.J. (2009) Discovery of leukotriene A4 hydrolase inhibitors using metabolomics biased fragment crystallography. *Journal of Medicinal Chemistry*, **52**, 4694–4715.

85 Blomberg, N., Cosgrove, D., Kenny, P., and Kolmodin, K. (2009) Design of compound libraries for fragment screening. *Journal of Computer-Aided Molecular Design*, **23**, 513–525.

86 Schulz, M., Landström, J., Bright, K., and Hubbard, R. (2011) Design of a fragment library that maximally represents available chemical space. *Journal of Computer-Aided Molecular Design*, **25**, 611–620.

87 Ruddigkeit, L., van Deursen, R., Blum, L.C., and Reymond, J.L. (2012) Enumeration of 166 billion organic small molecules in the chemical universe database GDB-17. *Journal of Chemical Information and Modeling*, **52**, 2864–2875.

88 Ruddigkeit, L., Blum, L.C., and Reymond, J.L. (2013) Visualization and virtual screening of the chemical universe database GDB-17. *Journal of Chemical Information and Modeling*, **53**, 56–65.

89 Chen, Y. and Shoichet, B.K. (2009) Molecular docking and ligand specificity in fragment-based inhibitor discovery. *Nature Chemical Biology*, **5**, 358–364.

90 Teotico, D.G., Babaoglu, K., Rocklin, G.J., Ferreira, R.S., Giannetti, A.M., and Shoichet, B.K. (2009) Docking for fragment inhibitors of AmpC β-lactamase. *Proceedings of the National Academy of Sciences of the United States of America*, **106**, 7455–7460.

91 Verdonk, M.L., Giangreco, I., Hall, R.J., Korb, O., Mortenson, P.N., and Murray, C.W. (2011) Docking performance of fragments and druglike compounds. *Journal of Medicinal Chemistry*, **54**, 5422–5431.

92 Chen, Y. and Pohlhaus, D.T. (2010) *In silico* docking and scoring of fragments. *Drug Discovery Today: Technologies*, **7**, e149–e156.

93 Clark, M., Guarnieri, F., Shkurko, I., and Wiseman, J. (2006) Grand canonical Monte Carlo simulation of ligand–protein binding. *Journal of Chemical Information and Modeling*, **46**, 231–242.

94 Loving, K., Alberts, I.L., and Sherman, W. (2010) Computational approaches for fragment-based and *de novo* design. *Current Topics in Medicinal Chemistry*, **10**, 14–32.

95 Sándor, M., Kiss, R., and Keserű, G.M. (2010) Virtual fragment docking by Glide: a validation study on 190 protein–fragment complexes. *Journal of Chemical Information and Modeling*, **50**, 1165–1172.

96 Chan, A.E., Laskowski, R.A., and Selwood, D.L. (2010) Chemical fragments that hydrogen bond to Asp, Glu, Arg, and His side chains in protein binding sites. *Journal of Medicinal Chemistry*, **53**, 3086–3094.

97 Wang, L., Xie, Z., Wipf, P., and Xie, X.Q. (2011) Residue preference mapping of ligand fragments in the protein data bank. *Journal of Chemical Information and Modeling*, **51**, 807–815.

98 Böhm, H.J. (1992) The computer program LUDI: a new method for the de novo design of enzyme inhibitors. *Journal of Computer-Aided Molecular Design*, **6**, 61–78.

99 Grädler, U., Gerber, H.D., Goodenough-Lashua, D.M., Garcia, G.A., Finer, R., Reuter, K., Stubbs, M., and Klebe, G. (2001) A new target for Shigellosis: rational design and crystallographic studies of inhibitors of tRNA-guanine transglycosylase. *Journal of Molecular Biology*, **306**, 455–467.

100 Rummey, C., Nordhoff, S., Thiemann, M., and Metz, G. (2006) *In silico* fragment-based discovery of DPP-IV{S1} pocket binders. *Bioorganic and Medicinal Chemistry Letters*, **16**, 1405–1409.

101 Marcou, G. and Rognan, D. (2007) Optimizing fragment and scaffold docking by use of molecular interaction fingerprints. *Journal of Chemical Information and Modeling*, **47**, 195–207.

102 Huang, N., Shoichet, B.K., and Irwin, J.J. (2006) Benchmarking sets for molecular docking. *Journal of Medicinal Chemistry*, **49**, 6789–6801.

103 Gleeson, M.P. and Gleeson, D. (2009) QM/MM as a tool in fragment based drug discovery: a cross-docking, rescoring study of kinase inhibitors.

*Journal of Chemical Information and Modeling*, **49**, 1437–1448.

104 Loving, K., Salam, N., and Sherman, W. (2009) Energetic analysis of fragment docking and application to structure-based pharmacophore hypothesis generation. *Journal of Computer-Aided Molecular Design*, **23**, 541–554.

105 Kawatkar, S., Wang, H., Czerminski, R., and Joseph-McCarthy, D. (2009) Virtual fragment screening: an exploration of various docking and scoring protocols for fragments using glide. *Journal of Computer-Aided Molecular Design*, **23**, 527–539.

106 Huang, D. and Caflisch, A. (2010) Library screening by fragment-based docking. *Journal of Molecular Recognition*, **23**, 183–193.

107 Majeux, N., Scarsi, M., Apostolakis, J., Ehrhardt, C., and Caflisch, A. (1999) Exhaustive docking of molecular fragments with electrostatic solvation. *Proteins: Structure, Function, and Bioinformatics*, **37**, 88–105.

108 Budin, N., Majeux, N., and Caflisch, A. (2001) Fragment-based flexible ligand docking by evolutionary optimization. *Biological Chemistry*, **382**, 1365–1372.

109 Hall, D.R., Ngan, C.H., Zerbe, B.S., Kozakov, D., and Vajda, S. (2012) Hot spot analysis for driving the development of hits into leads in fragment-based drug discovery. *Journal of Chemical Information and Modeling*, **52**, 199–209.

110 Hajduk, P.J. (2006) Fragment-based drug design: how big is too big? *Journal of Medicinal Chemistry*, **49** (24), 6972–6979.

111 Murray, C.W. and Verdonk, M.L. (2002) The consequences of translational and rotational entropy lost by small molecules on binding to proteins. *Journal of Computer-Aided Molecular Design*, **16** (10), 741–753.

112 Hoffer, L., Renaud, J.P., and Horvath, D. (2013) *In silico* fragment-based drug discovery: setup and validation of a fragment-to-lead computational protocol using S4MPLE. *Journal of Chemical Information and Modeling*, **53**, 836–851.

113 Thompson, D., Aldrin Denny, R., Nilakantan, R., Humblet, C., Joseph-McCarthy, D., and Feyfant, E. (2008) CONFIRM: connecting fragments found in receptor molecules. *Journal of Computer-Aided Molecular Design*, **22**, 761–772.

114 Lauri, G. and Bartlett, P. (1994) CAVEAT: a program to facilitate the design of organic molecules. *Journal of Computer-Aided Molecular Design*, **8**, 51–66.

115 Wang, R., Gao, Y., and Lai, L. (2000) LigBuilder: a multi-purpose program for structure-based drug design. *Molecular Modeling Annual*, **6**, 498–516.

116 Douguet, D., Munier-Lehmann, H., Labesse, G., and Pochet, S. (2005) LEA3D: a computer-aided ligand design for structure-based drug design. *Journal of Medicinal Chemistry*, **48**, 2457–2468.

117 Degen, J. and Rarey, M. (2006) FlexNovo: structure-based searching in large fragment spaces. *ChemMedChem*, **1**, 854–868.

118 Pierce, A.C., Rao, G., and Bemis, G.W. (2004) BREED: generating novel inhibitors through hybridization of known ligands. Application to CDK2, p38, and HIV protease. *Journal of Medicinal Chemistry*, **47**, 2768–2775.

119 Dey, F. and Caflisch, A. (2008) Fragment-based *de novo* ligand design by multiobjective evolutionary optimization. *Journal of Chemical Information and Modeling*, **48**, 679–690.

120 Liu, H., Duan, Z., Luo, Q., and Shi., Y. (1999) Structure-based ligand design by dynamically assembling molecular building blocks at binding site. *Proteins: Structure, Function, and Bioinformatics*, **36**, 462–470.

121 Pearlman, D.A. and Murcko, M.A. (1996) CONCERTS: dynamic connection of fragments as an approach to *de novo* ligand design. *Journal of Medicinal Chemistry*, **39**, 1651–1663.

122 Mata, P., Gillet, V.J., Johnson, A.P., Lampreia, J., Myatt, G.J., Sike, S., and Stebbings, A.L. (1995) SPROUT: 3D structure generation using templates. *Journal of Chemical Information and Computer Sciences*, **35**, 479–493.

123 Gelhaar, D.K., Moerder, K.E., Zichi, D., Sherman, C.J., Ogden, R.C., and Freer, S.

T. (1995) *De novo* design of enzyme inhibitors by Monte Carlo ligand generation. *Journal of Medicinal Chemistry*, **38**, 466–472.

124 Miranker, A. and Karplus, M. (1995) An automated method for dynamic ligand design. *Proteins: Structure, Function, and Bioinformatics*, **23**, 472–490.

125 Vinkers, H.M., de Jonge, M.R., Dayaert, F. F.D., Heeres, J., Koymans, L.M.H., van Lenthe, J.H., Lewi, P.J., Timmerman, H., Van Aken, K., and Janssen, P.A.J. (2003) SYNOPSIS: SYNthesize and OPtimize System In Silico. *Journal of Medicinal Chemistry*, **46**, 2765–2773.

126 Dean, P.M., Firth-Clark, S., Harris, W., Kirton, S.B., and Todorov, N.P. (2006) SkelGen: a general tool for structure-based *de novo* ligand design. *Expert Opinion on Drug Discovery*, **1**, 179–189.

127 Congreve, M., Chessari, G., Tisi, D., and Woodhead, A.J. (2008) Recent developments in fragment-based drug discovery. *Journal of Medicinal Chemistry*, **51**, 3661–3680.

128 Vnagrevelinghe, E. and Rüdisser, S. (2007) Computational approaches for fragment optimization. *Current Computer-Aided Drug Design*, **3**, 69–83.

# 10
# Making FBDD Work in Academia

*Stacie L. Bulfer, Frantz L. Jean-Francois, and Michelle R. Arkin*

## 10.1
## Introduction

Fragment-based drug discovery (FBDD) is widely practiced in the pharmaceutical industry, and has also been increasingly adopted in academia. A FBDD approach takes advantage of biophysical and structural methods, including nuclear magnetic resonance (NMR) [1–3], differential scanning fluorimetry (DSF) [4,5], surface plasmon resonance (SPR) [6], disulfide trapping (also called tethering) [7,8], X-ray crystallography [9,10], and computational modeling [11,12], to detect binding of low molecular weight (MW) compounds or "fragments" (MW < 280 Da). Because chemical space is immensely smaller for fragments compared to a drug-sized compound of 500 Da ($10^9$ versus $10^{60}$) [13,14], fragment screens can use smaller libraries (1000–30 000 compounds) to efficiently cover this space. Fragments also tend to home to adaptive regions within a protein, for example, protein–protein interfaces and allosteric sites, which are often hard to target using typical high-throughput screen (HTS) [15–17]. Although fragments typically bind with weaker affinities compared to HTS compounds (μM–mM versus nM–μM), fragments tend to use a large fraction of their surface to interact with the protein target, resulting in ligand-efficient hits. Efforts to optimize these hits, by linking two fragments, merging overlapping scaffolds, or growing out fragments to lead-sized compounds, can lead to highly potent and selective ligands. Most fragment optimization programs utilize structure-guided design; hence, biophysics, structural biology, and chemistry must be well integrated.

As a small molecule discovery strategy, FBDD has potential benefits for academia. Conceptually, fragment screening has proven useful for exploring fundamental questions in molecular recognition and allostery; FBDD has also yielded chemical starting points for heretofore "undruggable" target classes. Practically, unlike HTS, which requires a large screening library (100 000+ compounds), fragment screening can be accomplished with a less costly collection of ~1000+ fragments and does not require extensive liquid-handling robotics. Many methods of FBDD screening use instrumentation (e.g., NMR spectrometer, RT-PCR machine, and biosensor instrument) already in place at most academic institutions.

*Fragment-based Drug Discovery: Lessons and Outlook*, First Edition. Edited by Daniel A. Erlanson and Wolfgang Jahnke.

At the same time, interest in FBDD has increased as academic HTS has matured. Between 2004 and 2010, at least 32 new academic or nonprofit small molecule drug discovery centers were established in the United States, but only two reported doing FBDD [18]. With the increasing success of FBDD, fragment libraries are now appearing at more of these centers. There are also more academic drug discovery centers – the Academic Drug Discovery Consortium (ADDC) lists 108 centers among its membership. To get a sense of the academic fragment screening landscape, we posted a brief survey on the Practical Fragments Blog and also sent a mail to the members of the ADDC. Practical Fragments is a go-to site for discussing the breakthroughs, challenges, and trends in FBDD; the ADDC brings together the community of academics, industry, and funding organizations interested in academic drug discovery and public/private partnerships. In total, 82 individuals responded (34 responses from the Practical Fragments Blog and 48 from the ADDC community), representing at least 70 labs from 59 academic institutions. Of all the respondents, 68% are practicing FBDD and 13% expressed interest in doing so (Figure 10.1); clearly, academic FBDD is becoming more prevalent, and successful case studies are likely to increase the likelihood of further growth.

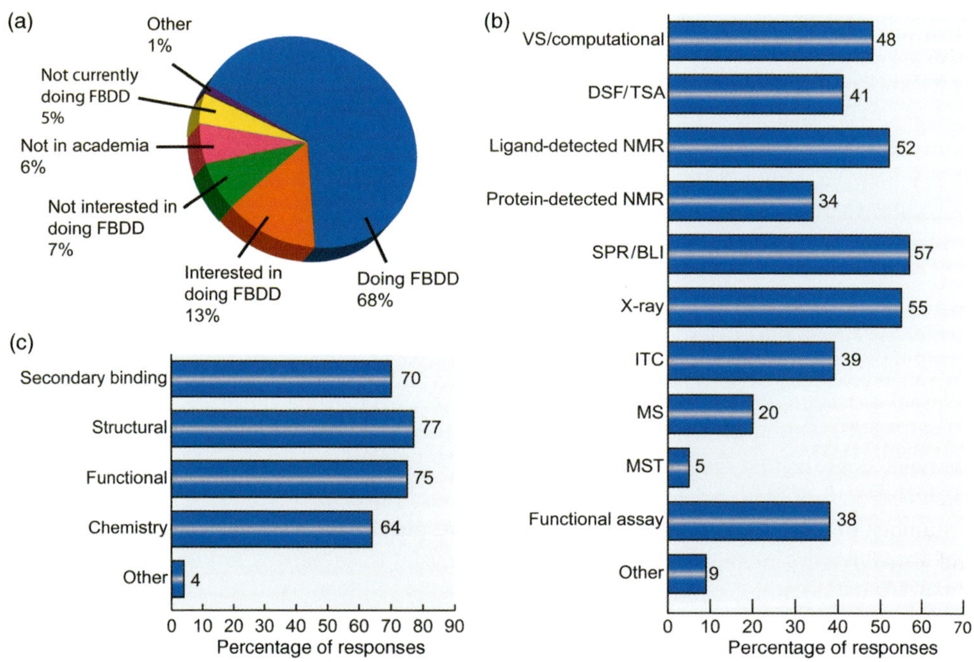


**Figure 10.1** FBDD in academia survey results. (a) A number of academic labs are doing FBDD (56 responses from 42 institutions) or are interested in doing FBDD (11 responses from at least 9 institutions). (b) Methods used by academic labs doing FBDD. (c) How academic labs doing FBDD follow up on a fragment hit.

## 10.2
## How Academic and Industry Drug Discovery Efforts Differ

Small molecule discovery efforts in academia and industry differ in some key respects that also inform the differences in how FBDD strategies are prosecuted. These key differences include how targets are selected, which technologies are used, the structure of project teams, and the scope and mechanisms of funding.

In academics, small molecule projects may be primarily focused on chemical biology (using small molecules to elucidate biological pathways), on drug discovery, or on uncovering fundamental principles of molecular recognition [18,19]. Chemical biology and drug discovery projects are most often initiated because the target shows an important effect on a biological pathway, and targets are not generally discounted due to their perceived undruggability. Hence, the small molecule discovery project often centers on transcription factors, protein–protein interactions, and as-yet-undrugged enzymes that have not yielded to traditional, target-based HTS. This biological focus is reflected in the high percentage of phenotypic, cell-based screens run in academia. Fragment screening is also playing an increasingly large role due to its reputation for finding initial compounds for these challenging target classes. Furthermore, many laboratories are focused on developing a fundamental understanding of protein dynamics and molecular recognition – principles that are then applied to drug discovery efforts in both academia and pharmaceutical companies. The structure-focused nature of fragment discovery is well suited, and increasingly appreciated, for addressing these fundamental questions.

Several pharmaceutical companies engaged in FBDD point out the importance of using multiple technologies, either sequentially or in parallel, to prosecute a fragment discovery program [20,21]. Thus, a pharmaceutical program might use computational screening, HTS, several methods of FBDD, and mining of the patent literature for their high-value targets. They invest in a range of technologies, including instrumentation and experienced scientists. An average sized academic lab, on the other hand, is more likely to develop deep expertise on a handful of technologies. Methodology-oriented groups are, therefore, in a strong position to innovate in that technology but need to reach across laboratories to access a broad suite of technological expertise. For instance, there are several NMR-based methods to discover and characterize fragments, and there continue to be innovative approaches, such as using spin labels or ever-larger protein complexes [22]. A number of NMR-focused academic groups have purchased fragment libraries and run FBDD screens; given the expense of running high-field NMR instruments, a 1000-member fragment library seems an incrementally small cost.

Project team structure is generally very different in academia compared to industry. Ideally, a pharmaceutical project team includes several labs with functional-area expertise but all have the same goal – the success of the company, which will come from the successful execution of their drug discovery program. An interdisciplinary project team in academia, on the other hand, is likely a voluntary affiliation of independent laboratories that have their own cultures, reward systems, and goals (of which the collaboration is but one). Leadership of such a team requires understanding these various goals and cultures, and finding common ground. Additionally, one of the major missions of an academic institute is to educate future scientists; thus, academics invest time and resources in training, perhaps in a variety of methodologies. The goals of education and innovation are potentially at odds with "critical path" driven drug discovery research. However, FBDD, with its emphasis on novel targets, technology development, and structure-guided design can be very appealing to an academic project structure.

The multimillion-dollar costs and multiyear time horizons for running a state-of-the-art drug discovery effort are prohibitive for most academic grant mechanisms. NIH and disease foundations have sought to provide some translationally oriented grants, with many focused on screening or preclinical biology. Academics' interest in poorly understood diseases, the strong support of rare disease foundations, and the inability to compete dollar-for-dollar with pharmaceutical companies are all factors directing academic drug discovery, and FBDD in particular, toward niche targets, rare diseases, and a more focused selection of discovery technologies.

## 10.3
## The Making of a Good Academic FBDD Project

Our group, the small molecule discovery center (SMDC) at the University of California, San Francisco, is a multidisciplinary "collaboratory" focused on drug discovery and chemical biology. We run an open-access HTS facility, engage in multiple fragment screening technologies, and have expertise in biochemistry, structural biology, medicinal chemistry, and drug mechanism. FBDD and HTS are clearly complementary approaches; thus, when we can, we try to use both simultaneously to identify small molecule probes and leads for certain targets. However, as mentioned above, many academic project teams do not have the bandwidth to use both HTS and FBDD when screening a target. So, when is using FBDD a good choice? A comparison of an HTS workflow and a FBDD workflow reveals parallel tactics for discovering leads (Figure 10.2), but there are a few key differences that may make a project more tractable for FBDD versus HTS [23,24].

One major difference between HTS and FBDD is that FBDD commonly uses biophysical screening techniques instead of a functional activity assay. This is advantageous if your target does not have biochemical activity, for example, the inactive form of a kinase [25,26], a proprotease [27,28], or a potential allosteric site, where binding might or might not lead to functional modulation. Additionally, FBDD has often been used as a salvage effort to develop drugs for challenging targets including BACE1, Pin1, HSP90, and protein–protein interactions (PPI) [17,29–34], which failed to generate chemical matter using traditional HTS. Because academic groups tend to work on nontraditional drug discovery targets [18,19], for certain targets it may be advantageous to use FBDD from the start, rather than using it as a last resort.

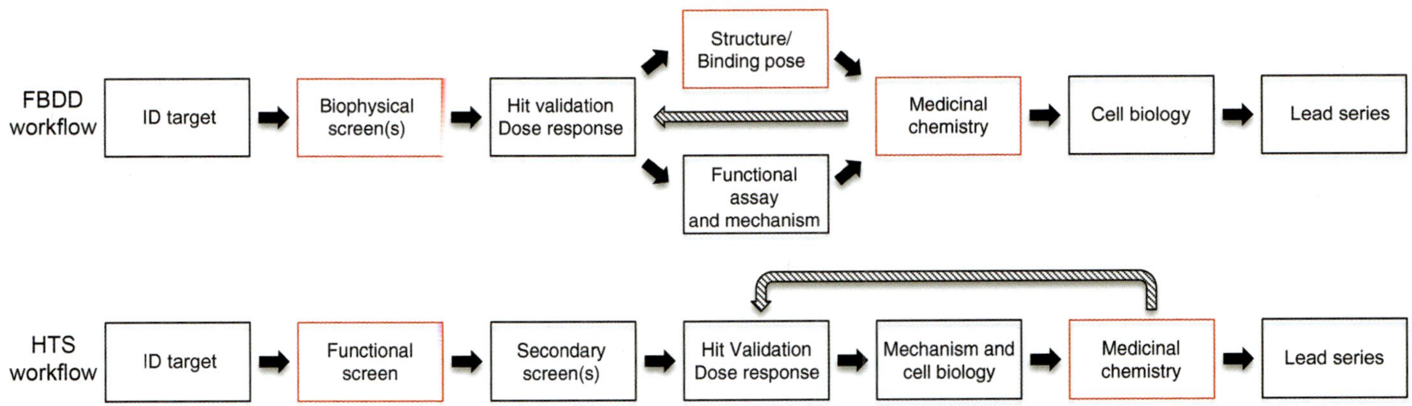


**Figure 10.2** Workflow for a FLBD approach and for a HTS approach. Major differences between the two workflows, including the use of biophysical screens, the importance of protein-fragment structural information, and integration of chemistry earlier in an FBDD project compared to HTS, are highlighted with red boxes.

In contrast to an HTS approach, which relies on rapid assays and high chemical diversity, FBDD relies heavily on structure-based design [23,24]. Once a fragment hit has been identified, a cycle of fragment elaboration involving structural studies, activity studies, secondary binding assays, and medicinal chemistry ensues (Figure 10.2). When asked how academics follow-up on fragment screens, responders reported that they used on average 2.8 approaches, with the vast majority (77%) reporting that they use structural studies (Figure 10.1). Having structural information for your target (X-ray, NMR, or docking model) is fundamental for prioritizing hits and optimizing their binding affinity through medicinal chemistry. FBDD efforts on targets that lack a crystal or NMR structure may prove challenging; by the same token, developing technologies to optimize fragments without crystallography could be a fertile and important area for academic research, as this is the cutting edge for FBDD.

FBDD could also be a first-choice screening approach when the goal is to test the importance of a specific site on a protein. Often, an HTS effort is geared to find inhibitors of a protein, and is less concerned with the details of how the inhibitor works. By contrast, site-directed methods, like protein-detected NMR, crystallography, or tethering, can be used to find compounds that bind at specific locations. These methods are better suited for primary screening of fragment libraries (~1000 compounds) than for HTS libraries (~100 000 compounds). We have worked on a number of projects where we used fragment screens, particularly tethering (discussed in Section 10.4 and Chapter 11) [7,8], to find compounds that preferentially bind to one protein conformation, such as a protein in a specific signaling state.

The typically weak binding affinity for fragment hits requires that medicinal chemistry comes into play earlier in the project than it does for HTS, where cell active compounds can be identified during the primary screen. With the combined efforts of structural biology, computation, and parallel chemical synthesis, potent compounds can be efficiently evolved from fragment hits. Having medicinal chemistry expertise, along with efficient and clear communication between the structural biology and chemistry groups, is vital in the FBDD process. Similar to the importance of structural studies in fragment validation, the majority (64%) of survey responders stated that they used chemical synthesis as a follow-up technique for their FBDD projects (Figure 10.1). The observation that 36% of laboratories do not claim to do medicinal chemistry, however, suggests that many FBDD groups work with outside chemistry collaborators, are focused primarily on screening as a service or are most interested in fundamental aspects of protein/compound interactions.

## 10.4
## FBDD Techniques Currently Used in Academia

The implementation of FBDD in drug discovery requires the use of biophysical techniques capable of detecting weak affinity interactions. As with HTS, all

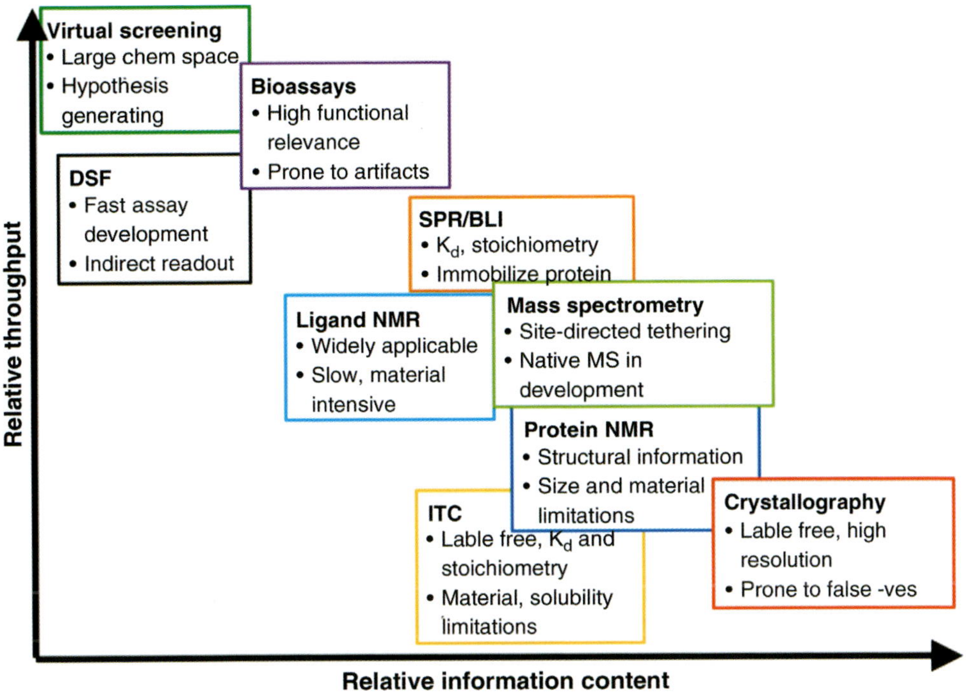


**Figure 10.3** Comparison of the different screening techniques. Advantages and drawbacks for the main techniques used for FBDD. The techniques are represented on a graph based on their relative throughput and the relevance of the information provided. Some techniques can significantly move vertically depending on the type of equipment used or the state of the project (Crystallography, protein detected NMR).

potential hits must be validated by orthogonal methods to confirm binding and stoichiometry and to prioritize chemical scaffolds for further development. When asked, "What techniques do you use to do FBDD?" academics doing FBDD responded that they use an average of 4–5 different methods to evaluate compounds throughout the screening campaigns. Surface plasmon resonance (SPR), X-ray crystallography, and ligand-detected NMR topped the list with >50% of responders stating that they use these methods (Figure 10.1). The most common biophysical methodologies are summarized here, listing their potential benefits and drawbacks (Figure 10.3) and noting examples from academic labs.

### 10.4.1 Nuclear Magnetic Resonance

NMR is one of the most widely used and versatile techniques to study protein–ligand interactions [1,35]. Fifty-two percent of academics responded that they use ligand-detected NMR, while 34% use protein-detected NMR routinely during their FBDD efforts. NMR not only can be used as a primary screen to detect compound/protein binding but also offers the advantage of providing structural information on both the ligand binding epitope (ligand-detected NMR) and the binding pocket of the receptor (protein-detected NMR). In both cases, the interaction is studied at a molecular level, in a nondestructive manner in solution. Disadvantages to NMR are the reliance on expensive instrumentation and expert analysis. The choice between ligand-detected NMR and protein-detected NMR depends on the size of the target and $K_D$ of the complex [36,37]. NMR experiments can be used to detect $K_D$s in the range of $10^{-3}$–$10^{-8}$ M.

Ligand-based NMR screening is more commonly used than protein-detected NMR mainly because it does not require isotope labeling, uses lower concentrations of protein, and does not require assignment of the protein resonances. One common method is saturation transfer difference (STD) NMR, which detects magnetization that is transferred from the protein to the ligand [38]. Compounds that bind to the protein will show reduced proton signals due to this transfer. Which chemical groups on the compound interact with the protein can be determined through a ligand-detected Group Epitope Mapping experiment [38]. Since the compounds are visualized through these methods, they can be screened in pools to reduce the amount of protein needed and to increase throughput. Academic labs have also applied more complex ligand-based NMR, such as interligand NOE measurements, to FBDD [39].

In protein-detected NMR, chemical shift changes within the protein are monitored as the ligand is added. This technology requires $^{15}N$ and/or $^{13}C$ labeling of the protein and heteronuclear multidimensional experiments [40,41]. A spectrum of the labeled protein is acquired and the chemical shifts of the labeled nuclei are assigned to specific residues. As ligand binds to a particular site, the local environment and the chemical shifts in that region are affected. $^{15}N$-$^{1}H$ HSQC has been used extensively for FBDD by Fesik and coworkers at Vanderbilt University [42]. Similar methods have been developed to determine SAR and ligand orientation in the presence of several analogues. Recently, Isabelle Krimm and colleagues at the University of Lyon illustrated the potential and limitations of NMR in predicting the binding poses of analogous fragments using peroxiredoxin 5 as a model [43,44]. There have been important recent advances in protein isotope labeling techniques that allow the study of larger proteins (ClpC and proteasome) [45]. However, NMR studies with such large proteins are very time-consuming and may be better suited to secondary assays than to primary fragment screens. Typically, protein-detected NMR FBDD primary screens are limited to 30–40 kDa proteins.

### 10.4.2
### X-Ray Crystallography

Macromolecular X-ray crystallography is an important and powerful technique in FBDD [9]. Fifty-five percent of academics doing FBDD use X-ray crystallography, primarily to follow up on fragment hits (Figure 10.1). X-ray crystallography is the gold standard for gaining detailed analysis of protein–ligand interaction, thereby allowing chemists to make structure-based design decisions to improve affinity.

The starting point for X-ray crystallography is the preparation of suitable crystals, which can be achieved by cocrystallization or soaking [46]. In cocrystallization, the ligand is added to the protein prior to the crystal formation. Soaking is achieved by adding the fragment directly onto preexisting crystal. The latter strategy is usually favored since it significantly increases the speed at which fragments can be screened; however, the ligand sometime disrupts the crystal lattice. It is now possible to determine the structure of many protein–ligand complexes for a particular protein and to collect hundreds of data sets at a time. These data sets can be collected and analyzed in a few days, which allow some academic drug discovery programs, including Cambridge University, to use X-ray crystallography as the primary screening technique [47].

X-ray crystallography has a very low rate of false positives, can detect very low-affinity binders, and defines the fragment binding site in detail [48]. Unfortunately, the requirements for obtaining stable cocrystals also gives rise to high numbers of false negatives. Although X-ray crystallography has been a central technique in FBDD, its utility for a given target is dictated by how readily the target crystallizes and the ability to generate ~10–50 mg of highly pure protein. It is also important to note that crystallography does not provide binding affinities; a different biophysical or biochemical method is used to measure structure–activity relationships.

### 10.4.3
### Surface Plasmon Resonance/Biolayer Interferometry

SPR and biolayer interferometry (BLI) play an important role in academic screening. Nearly, 60% of academics involved in FBDD that responded to our survey use SPR/BLI as screening techniques (Figure 10.1). In a typical SPR/BLI assay, the protein target is immobilized to a surface and the fragment is injected over the surface. If the fragment binds the protein, an increase in SPR signal (response units, RU) or BLI signal (interference) is observed. Dose–response analysis of fragment hits can then be completed to determine the equilibrium dissociation constant ($K_D$) and the binding stoichiometry. Since nonstoichiometric binding could suggest that the compound is a binding artifact [49], this data is critical for prioritizing fragments for follow-up, for example, crystallography.

SPR and BLI are popular choices for fragment screening because they are fast, efficient, cost-effective, and quantitative [50]. Moreover, SPR biosensor assays can be developed for a wide range of proteins. Transmembrane proteins, such as G-protein-coupled receptors, are particularly challenging SPR targets that have been successfully tackled by academic labs [51,52].

However, there are caveats to using SPR. SPR and BLI require significant technical expertise, and SPR/BLI assays can also miss very weak binders (mM) that would be identified by NMR techniques. Finally, although SPR competition assays can identify if a fragment is competitive with a substrate or known inhibitor [53], SPR data will not provide detailed information describing the binding site of the fragments. Protein-detected NMR and X-ray crystallography are, therefore, common secondary steps after SPR screening.

### 10.4.4 Differential Scanning Fluorimetry

DSF (also called thermal shift assays, TSA) is a rapid and inexpensive screening method to identify low molecular weight ligands that bind and stabilize purified proteins. This makes it a method of choice for more than 40% of academic laboratories surveyed. The temperature at which a protein unfolds is measured by an increase in the fluorescence of a dye with affinity for hydrophobic parts of the protein, which are exposed as the protein unfolds [54]. A simple comparison of the protein melting temperature with and without fragments identifies fragments that bind and stabilize the protein. However, what about fragments that destabilize a protein? In a recent paper by the Finzel lab (University of Minnesota) on the *M. tuberculosis* protein BioA, several fragments caused a downshift in the melting temperature but were determined to be true binders by NMR and X-ray [55]. The indirect nature of the DSF readout (melting temperature) may not be suitable for all targets, and although fast and convenient, it can generate a large number of false positive and negatives.

### 10.4.5 Isothermal Titration Calorimetry

Isothermal titration calorimetry (ITC) is used to determine the thermodynamic parameters of interactions in solution. Compound and protein are mixed together, and the change in heat is precisely measured. From the heat released, one can determine the enthalpy change ($\Delta H$) and Gibbs energy change ($\Delta G$) of binding (and thus the binding affinity, $K_a$). Moreover, the binding stoichiometry ($n$) of the interaction between the two or more molecules in solution can be calculated. In practice, fragment interactions may be too weak for detection. Based on our survey, just under 40% of academic drug discovery centers use ITC to identify hits. Most instruments measure a few interactions per day and utilize large amounts of protein, and so ITC is used in conjunction with higher throughput methods, as recently reviewed by Ciulli and coworkers [56]. Ladbury, Klebe, and Freire from the University of Texas recently highlighted ITC's potential and the development of higher throughput instruments [57].

### 10.4.6 Virtual Screening

Virtual screening (VS) computationally predicts the binding of large compound libraries to a protein target. VS allows prioritization of the >10 million commercially available compounds to a manageable number that can be purchased and tested, or can be used to design focused libraries. As the accuracy of the method has increased, virtual screening has become an integral part of the drug discovery process. Indeed, about 40% of the academic centers utilize VS for FBDD. Screens can be ligand based (pharmacophore modeling) or target based (focused on the protein binding site). These are hypothesis-generating methods that are followed up by experimental testing.

In the ligand-based/pharmacophore approach, a set of structurally diverse ligands that bind to a receptor is used to predict the key shapes and molecular features that are associated with binding. Candidate ligands are then compared to the pharmacophore model to determine whether they are compatible with it and therefore likely to bind [58].

The target-based approach, also called structure-based virtual screening, involves docking candidate ligands into a protein target, followed by applying a scoring function to estimate the likelihood that the ligand will bind to the protein with high affinity. Academic labs have played key roles in developing FBDD tools for VS; for example, beta lactamase has been thoroughly evaluated by docking methods, leading to the first potent, noncovalent inhibitors of this enzyme [59,60]. Another target-based approach called FTMAP, developed at Boston University, uses virtual fragments to identify potential small molecule binding sites (hot spots) on a protein surface [61]. FTMAP uses high concentrations of small fragments to globally sample the protein surface; patches where fragments tend to pool are ranked as potential fragment binding sites.

### 10.4.7 Mass Spectrometry

Mass spectrometry (MS) is not as routinely used as NMR, SPR, and X-ray crystallography, but offers some significant advantages. In fact, fewer than 20% of academics polled use MS as part of their drug discovery efforts. Here we present two main MS screening techniques: native MS and site-directed disulfide trapping (tethering). The main drawback for both techniques is the high sensitivity to buffer conditions.

#### 10.4.7.1 Native MS

Native MS is a highly sensitive and accurate screening technique. Under careful conditions (soft ionization, ESI) protein–small molecule complexes can be observed [62]. The process requires very small amounts of protein and does not require immobilization or conjugation. However, the required conditions (lack of detergent) can lead to false positives due to aggregation. The use of native MS is very appealing, and researchers continue to develop methods that ionize proteins without losing weak, noncovalent complexes. In 2006, Steven Hofstadler summarized in a review the potential of the various ESI-MS techniques [63]. This methodology is currently used for FBDD by Ronald Quinn and colleagues at the Griffith University, Australia [64–66]. MS can also be used to detect ligands following affinity separation [67].

#### 10.4.7.2 Site-Directed Disulfide Trapping (Tethering)

Tethering involves reacting a native or engineered cysteine near a binding site with a library of thiol-containing fragments under partially reducing conditions. Fragments that can form the covalent interaction (disulfide) and have intrinsic affinity for the binding pocket will be stabilized and detected by mass spectrometry or by plate-based functional assays [8,68]. This approach has also proven successful using reversible electrophilic fragments and irreversible covalent fragments [69,70]. As a drawback, tethering and related technologies require a significant chemistry effort to build the thiol reactive fragment library [71,72].

The site-directed nature of disulfide trapping makes it is auspiciously suited for identifying compounds that bind to allosteric sites (Chapter 11). Using engineered cysteine residues, Sadowsky *et al.* identified fragments that bind to an allosteric site on the kinase PDK1 [16]. Fragments bound to the same site could allosterically activate or inhibit kinase activity, depending on how they interacted with the regulatory C-helix. Tethering has the great advantage of targeting predetermined sites on a protein; on the other hand, the disulfide tether must be removed for drug discovery applications if the fragment binds to an engineered residue. By contrast, tethering to native cysteines allows a rapid transition to cell-based probes and leads; Ostrem *et al.* recently developed covalent fragments that bind to an oncogenic mutant (G12C) in Ras protein and allosterically inhibit Ras activity [73].

### 10.4.8 High-Concentration Bioassays

Although fragments can form high-quality interactions and have good ligand efficiency, their small size means that they usually have weak affinity. Regular bioassays have to be reconfigured to detect binding in the range of 100 μM–1 mM [74]. High-concentration bioassays provide a rapid and quantitative method for detection, which makes them an attractive method of academic institutions. Forty per cent of academic centers mentioned the use of bioassays in their fragment-screening pipeline. High-concentration bioassays can be successfully used for primary screening, as shown by the van Montfort lab at the Institute of Cancer Research and the Fast lab at the University of Texas, Austin [75,76]. At the SMDC, we sometimes screen the fragment library at a high concentration (ca. 100–150 μM) during routine HTS, particularly for targets where historical success with HTS has been low, for example, with protein–protein interactions.

However, many high-concentration fragment bioassays prove unsuitable, with false positives as a result of compound aggregation at the concentration of ligands used, interference with assay end point (e.g., optical interference fluorescence, quenching, toxicity, etc.), disruption of the protein by unfolding, disruption of protein complexes or precipitation. Additionally, false negatives can occur due to lack of effective solubility of compounds in assay buffers. While in theory any bioassay could be used for fragment-based screening, they often lack the sensitivity to identify weakly binding fragments.

## 10.5
## Project Structures for Doing FBDD in Academia

How labs organize their FBDD efforts can vary widely across academia. Successful structures can include a single lab, collaboration among academic labs, or working with an academic drug discovery facility. In addition to all-academic project structures, the SMDC also engages in FBDD projects as part of a government-funded academic consortium and as partners in industry–academic collaborations. Both of these project structures aim to use the strengths of academia (innovation, target expertise, and interest in high-risk or neglected diseases) in combination with additional resources (financial and technical expertise) provided by the collaboration to facilitate the advancement of a molecule into clinical development. The following subsections will describe recently published work involving FBDD that highlights these two types of project structures.

### 10.5.1
### Targeting p97: A Chemical Biology Consortium Project

The Chemical Biology Consortium (CBC) is part of the National Cancer Institute's (NCI's) Experimental Therapeutics (NExT) program that aims to facilitate the discovery and development of novel targets and agents to treat cancer. Applicants submit proposals, often focused on high-risk or underrepresented targets, to gain access to a network of 12 academic research centers. These centers have expertise in chemical biology, cellular oncology, medicinal chemistry, screening, and structural biology and collaborate with each other and the NCI to meet the goals of the project. The p97 project was brought to the CBC by Ray Deshaies (Caltech) and stemmed from his lab's collaboration with the Molecular Libraries Probes Center Network (MLPCN), which identified potent and selective cell active inhibitors to p97 [77–81]. Within the CBC, researchers at the University of Pittsburg, University of Minnesota, UCLA, UCSF, and SRI International contributed to this project.

The protein p97 belongs to the AAA+ (ATPase associated with diverse cellular activities) family of ATPases and functions to separate proteins from protein complexes in cellular processes ranging from DNA repair to membrane fusion and the ubiquitin–protein degradation pathways [82–84]. Because p97 has an essential role in protein homeostasis, but acts only on a subset of protein degradation pathways, p97 is a potential target for novel cancer therapies that may have fewer toxicities than proteasome inhibitors [85].

Structurally, p97 functions as a hexameric protein with each monomer consisting of three major domains: the D1 and D2 ATPase domains, and an N-terminal domain. The D2 domain is responsible for the majority of the ATPase activity of full-length p97 [86,87] and has been the target for the majority of p97 inhibitors [77,81,88,89]. In contrast, the activity of the D1 domain has been debated, with recent findings indicating that it is also an active ATPase that can be inhibited independent of D2 [78]; however, specific D1 inhibitors have yet to be developed. A variety of p97 cofactor proteins bind to the N domain through a common protein–protein interaction site and direct p97 to specific cellular functions [83,90–92]. In addition, the position of the N domain is highly dynamic and intricately connected to the ATPase activity of the D1 and D2 domains [78,93,94]. Allosteric modulation of these protein dynamics could inhibit overall p97 activity or could preferentially alter the function of specific adapter proteins.

In an effort to identify active site and allosteric inhibitors of p97, the CBC team developed a multifaceted approach involving both HTS and FBDD. To target full-length p97, the team developed a biochemical HTS assay and screened ~300 000 compounds. In parallel, we targeted the N and D1 domains using FBDD approaches. We specifically chose to use FBDD to target these domains because (1) we wanted to use a biophysical readout to target the adapter-binding site on the N domain and (2) we wanted to differentiate the inhibition of the D1 domain from the inhibition of the D2 domain, which is the primary readout of a functional activity assay.

Screening a library of ~2500 fragments, a combination of SPR and STD-NMR identified and confirmed that several fragments bound specifically to a 97 construct consisting of only the N and D1 domains (Chimenti M. *et al* in press). The binding poses of these fragments were further characterized by SPR competition assays and 2D [$^{13}$C,$^{1}$H] -methyl-TROSY NMR (M-TROSY) methods. M-TROSY NMR is suitable for large proteins (the ND1 hexamer has a molecular weight of 320 kDa) and was used because obtaining high-diffracting, reproducible crystals of the ND1 protein proved difficult. The SPR ADP competition and NMR structural studies indicated that some fragments bound at the nucleotide binding site, whereas others bound elsewhere (Figure 10.4). M-TROSY data correlated well with SPR competition studies and molecular docking; structural information from M-TROSY, along with computational modeling, will be useful in guiding structure-based chemical synthesis to improve the potency of these fragments. Additional, ongoing efforts on p97 are focused on inhibitors identified through HTS and protein-detected NMR fragment screens of the N domain.

Developing both a functional HTS screen and FBDD biophysical assays has been extremely beneficial for the project. Each screening approach targeted different domains of p97, enabling the team to identify a diverse set of chemical matter that binds to p97 at several distinct sites. In addition, having validated SPR and NMR FBDD assays at the start of the project saved time and resources downstream when they were used to determining the binding mechanism of HTS hits and optimized leads.

### 10.5.2
### Targeting Caspase-6: An Academic–Industry Partnership

The drug discovery industry has faced declining R&D productivity due to high costs and high project attrition rates. To better capitalize on the surge in academic drug discovery, the industry has embraced sponsored research, private–

(a)

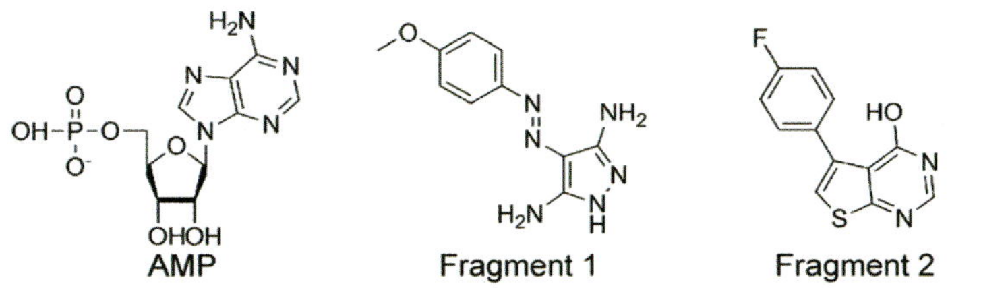


| | AMP | Fragment 1 | Fragment 2 |
|---|---|---|---|
| % STD | 75 | 77 | 85 |
| $K_D$ (μM) | 94 | 220 | 180 |
| % SPR response with ADP | 16 | 20 | 96 |

(b)

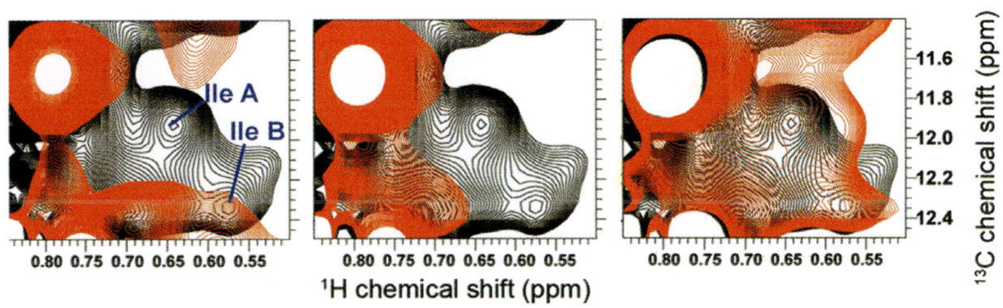


(c)

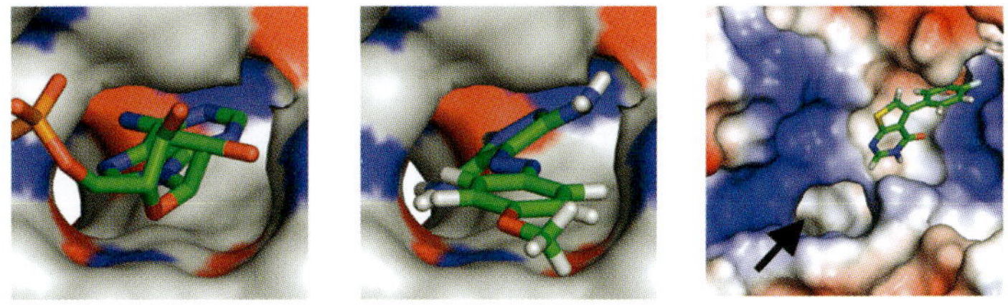

**Figure 10.4** SPR competition assays and M-TROSY NMR experiments identify the binding pose of p97 ND1 fragments. (a) NMR and SPR results for AMP and two fragments that bind to the ND1 domains of p97. (b) The isoleucine region from the 2D M-TROSY spectra of p97 ND1 (black) plus AMP or fragment (red). (c) Docking poses for AMP and the two fragments. The arrow indicates the nucleotide binding sites. Fragment 1 displays both a reduction in SPR response in the presence of ADP and similar M-TROSY chemical shifts to AMP indicating that it binds at the nucleotide binding site. In contrast, fragment 2 binds elsewhere with the ND1 domains of p97.

public centers, "precompetitive" collaborations, and academic–industry partnerships [95,96]. The SMDC interacts with industry in various ways ranging from fee-for-service structures to academic–industry collaborations [97]. Establishing public–private partnerships caters to the nonredundant strengths of the individual parties and provides mutual benefit by derisking projects [98,99]. Despite the benefits of academic–industry collaborations, there are still barriers (cultural differences, intellectual property, etc.) to setting up of these partnerships [100].

These barriers are reflected in the fact that only 6% of funding to academic drug discovery centers comes from these types of relationships [18,19]. For the caspase-6 project, the underlining scientific interest motivated both institutions (UCSF and Genentech) to overcome these obstacles and initiate a sponsored collaboration [27].

Caspases belong to the cysteine-aspartate family of proteases and have a critical role in cellular processes including apoptosis, inflammation, cell proliferation, and differentiation, and these have also been implicated in disease states such as cancer and neurodegenerative disorders [101,102]. In particular, Genentech and the SMDC were targeting caspase-6 for treatment of Huntington's disease [103] and to evaluate the link between caspase-6 and Alzheimer's disease [104]. Caspase-6, an executioner caspase, is expressed as a dimeric procaspase (zymogen) that requires cleavage between its large and small subunit to become active [105]. Despite many years of effort, specific caspases have proven difficult to drug, since the primary recognition features (a cysteine nucleophile and aspartic acid binding pocket) are common for all caspases.

A multidisciplinary approach involving both HTS and FBDD was used to identify inhibitors to mature and zymogen caspase-6. HTS, followed by extensive hit validation and chemical optimization, yielded a tight binding, uncompetitive inhibitor that recognized both the substrate peptide sequence and the fluorometric dye used in the screening assay [106]. In ongoing work, the SMDC also ran a tethering-based screen to identify compounds that bound to cysteine residues outside of the catalytic site. From this screen, several series of compounds are being optimized for potency and selectivity for caspase-6 over other cysteine proteases. Tethering has been a powerful method for discovering inhibitors of caspases, also uncovering a novel allosteric site at the dimer interface in caspases 1, 3, and 7.

Finally, the Genentech team developed an FBDD approach to target the procaspase-6 in the hopes of stabilizing the zymogen, and thereby preventing its activation. Because the caspase-6 zymogen does not have biochemical activity, FBDD was the optimal choice in targeting this form of the enzyme. SPR was used to screen a ~2300 fragment library against both the active caspase-6 and the zymogen. A total of 84 hits ($K_D$s = 3 μM–2.3 mM) were identified as specific binders to the zymogen [28]. X-ray crystallography revealed that fragments bound at the zymogen dimer interface. Most fragments formed π–π stacking interactions between symmetrical tyrosines within the interface, with one asymmetric fragment extending into an adjacent hydrophobic pocket. Two structure-based design strategies focused on (1) fragment merging or (2) elaboration of the asymmetric fragment to bind both halves of the symmetric pro-caspase-6 dimer; remarkably, both strategies led to ~500 nM ligands (Figure 10.5). These leads significantly stabilized the zymogen (5–7 °C) and represented the first class of noncovalent, sub-μM affinity compounds that bound the dimer interface of a caspase zymogen.

The collaborative Genentech/UCSF project team was able to aggregate strategies and technologies to develop three unique series targeting caspase-6. As with

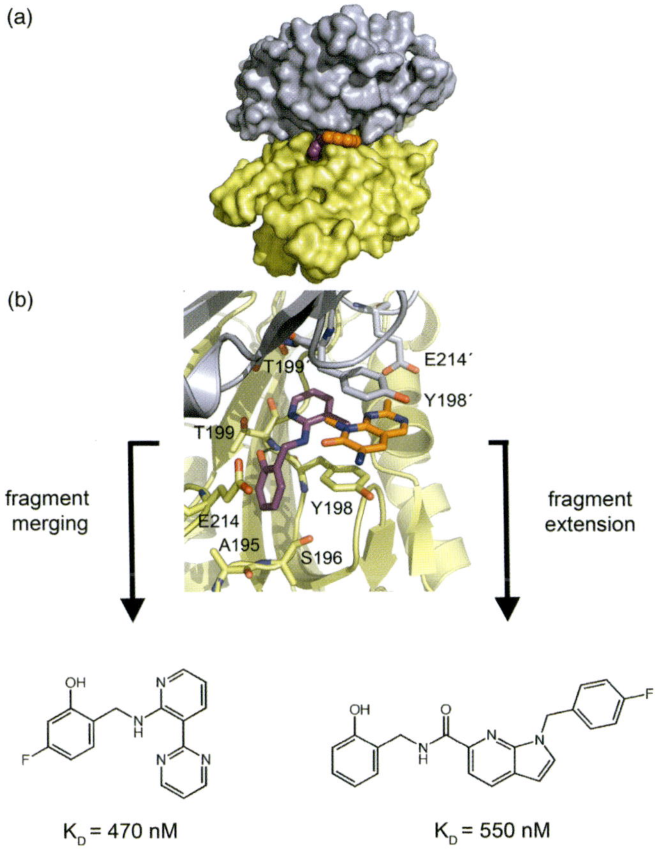


**Figure 10.5** Caspase-6 fragment merging and fragment elaboration. (a) Fragment screens identified binders (orange and violet spheres) at the dimeric interface of the Caspase-6 zymogen (blue and yellow monomers). (b) Close-up view of two classes of fragments binding at the dimer interface. Fragment merging and symmetry-based fragment extension strategies were used to develop potent nM-range fragments.

p97, these complementary approaches allowed rapid validation of compounds by orthogonal methods. Throughout, regular communication, reassessment of the project, and alignment to common goals were critical for making FBDD work in this academic–industry partnership.

## 10.6 Conclusions and Perspectives

As academic drug discovery matures and diversifies, FBDD is becoming more commonly practiced throughout academia. An increase in academic FBDD will have a positive effect on the discipline, as challenging targets, technical innovation, and fundamental questions fuel development of fragment-based technologies. Academic FBDD is not without challenges, however. The interdisciplinary nature of FBDD requires project teams and organizational structures that are relatively uncommon in university settings. Though academic HTS has gotten more sophisticated in assay design and in rooting out artifacts, the literature is still plagued by "frequent hitters" and otherwise fishy compounds that are inappropriate as chemical tools and drug leads [107]. The academic community must build on this awareness of nondrug-like mechanisms and apply this awareness to FBDD. The opportunity is that the increasing use of structural and biophysical tools for fragment discovery will in turn make these methods more common in validation of HTS hits. All in all, the mainstreaming of FBDD in the pharmaceutical industry has led to novel drugs and has reinvigorated efforts to tackle so-far undrugged targets. Similarly, this attitude will stimulate academics to develop new biological probes and drug leads, a deeper understanding of molecular recognition, and new technologies for FBDD.

## References


1 Shuker, S.B., Hajduk, P.J., Meadows, R.P., and Fesik, S.W. (1996) Discovering high-affinity ligands for proteins: SAR by NMR. *Science*, **274**, 1531–1534.

2 Klages, J., Coles, M., and Kessler, H. (2007) NMR-based screening: a powerful tool in fragment-based drug discovery. *Analyst*, **132**, 693–705.

3 Lepre, C.A. (2011) Practical aspects of NMR-based fragment screening. *Methods in Enzymology*, **493**, 219–239.

4 Vedadi, M., Niesen, F.H., Allali-Hassani, A., Fedorov, O.Y., Finerty, P.J. Jr., Wasney, G.A., Yeung, R., Arrowsmith, C., Ball, L.J., Berglund, H., Hui, R., Marsden, B.D., Nordlund, P., Sundstrom, M., Weigelt, J., and Edwards, A.M. (2006) Chemical screening methods to identify ligands that promote protein stability, protein crystallization, and structure determination. *Proceedings of the National Academy of Sciences of the United States of America*, **103**, 15835–15840.

5 Mashalidis, E.H., Sledz, P., Lang, S., and Abell, C. (2013) A three-stage biophysical screening cascade for fragment-based drug discovery. *Nature Protocols*, **8**, 2309–2324.

6 Giannetti, A.M. (2011) From experimental design to validated hits: a comprehensive walk-through of fragment lead identification using surface plasmon resonance. *Methods in Enzymology*, **493**, 169–218.

7 Erlanson, D.A., Braisted, A.C., Raphael, D.R., Randal, M., Stroud, R.M., Gordon, E.M., and Wells, J.A. (2000) Site-directed ligand discovery. *Proceedings of the National Academy of Sciences of the United States of America*, **97**, 9367–9372.

8 Erlanson, D.A., Wells, J.A., and Braisted, A. C. (2004) Tethering: fragment-based drug discovery. *Annual Review of Biophysics and Biomolecular Structure*, **33**, 199–223.

9 Davies, T.G. and Tickle, I.J. (2012) Fragment screening using X-ray crystallography. *Topics in Current Chemistry*, **317**, 33–59.

10 Jhoti, H., Cleasby, A., Verdonk, M., and Williams, G. (2007) Fragment-based screening using X-ray crystallography and NMR spectroscopy. *Current Opinion in Chemical Biology*, **11**, 485–493.

11 Desjarlais, R.L. (2011) Using computational techniques in fragment-based drug discovery. *Methods in Enzymology*, **493**, 137–155.

12 Barelier, S., Eidam, O., Fish, I., Hollander, J., Figaroa, F., Nachane, R., Irwin, J.J., Shoichet, B.K., and Siegal, G. (2014) Increasing chemical space coverage by combining empirical and computational fragment screens. *ACS Chemical Biology*, **9**, 1528–1535.

13 Bohacek, R.S., McMartin, C., and Guida, W.C. (1996) The art and practice of structure-based drug design: a molecular modeling perspective. *Medicinal Research Reviews*, **16**, 3–50.

14 Blum, L.C. and Reymond, J.L. (2009) 970 million druglike small molecules for virtual screening in the chemical universe database GDB-13. *Journal of the American Chemical Society*, **131**, 8732–8733.

15 Coyne, A.G., Scott, D.E., and Abell, C. (2010) Drugging challenging targets using fragment-based approaches. *Current Opinion in Chemical Biology*, **14**, 299–307.

16 Sadowsky, J.D., Burlingame, M.A., Wolan, D.W., McClendon, C.L., Jacobson, M.P., and Wells, J.A. (2011) Turning a protein kinase on or off from a single allosteric site via disulfide trapping. *Proceedings of the National Academy of Sciences of the United States of America*, **108**, 6056–6061.

17 Wilson, C.G. and Arkin, M.R. (2013) Probing structural adaptivity at PPI interfaces with small molecules. *Drug Discovery Today: Technologies*, **10**, e501–508.

18 Frye, S., Crosby, M., Edwards, T., and Juliano, R. (2011) US academic drug discovery. *Nature Reviews Drug Discovery*, **10**, 409–410.

19 Tralau-Stewart, C., Low, C.M., and Marlin, N. (2014) UK academic drug discovery. *Nature Reviews Drug Discovery*, **13**, 15–16.

20 Carr, R.A., Congreve, M., Murray, C.W., and Rees, D.C. (2005) Fragment-based drug discovery: leads by design. *Drug Discovery Today*, **10**, 987–992.

21 Hennig, M., Ruf, A., and Huber, W. (2012) Combining biophysical screening and X-ray crystallography for fragment-based drug discovery. *Topics in Current Chemistry*, **317**, 115–143.

22 Barrett, P.J., Chen, J., Cho, M.K., Kim, J.H., Lu, Z., Mathew, S., Peng, D., Song, Y., Van Horn, W.D., Zhuang, T., Sonnichsen, F.D., and Sanders, C.R. (2013) The quiet renaissance of protein nuclear magnetic resonance. *Biochemistry*, **52**, 1303–1320.

23 Rees, D.C., Congreve, M., Murray, C.W., and Carr, R. (2004) Fragment-based drug discovery. *Nature Reviews Drug Discovery*, **3**, 660–672.

24 Murray, C.W. and Rees, D.C. (2009) The rise of fragment-based drug discovery. *Nature Chemistry*, **1**, 187–192.

25 Moffett, K., Konteatis, Z., Nguyen, D., Shetty, R., Ludington, J., Fujimoto, T., Lee, K.J., Chai, X., Namboodiri, H., Karpusas, M., Dorsey, B., Guarnieri, F., Bukhtiyarova, M., Springman, E., and Michelotti, E. (2011) Discovery of a novel class of non-ATP site DFG-out state p38 inhibitors utilizing computationally assisted virtual fragment-based drug design (vFBDD). *Bioorganic and Medicinal Chemistry Letters*, **21**, 7155–7165.

26 Gill, A.L., Frederickson, M., Cleasby, A., Woodhead, S.J., Carr, M.G., Woodhead, A.J., Walker, M.T., Congreve, M.S., Devine, L.A., Tisi, D., O'Reilly, M., Seavers, L.C., Davis, D.J., Curry, J., Anthony, R., Padova, A., Murray, C.W., Carr, R.A., and Jhoti, H. (2005) Identification of novel p38alpha MAP kinase inhibitors using fragment-based lead generation. *Journal of Medicinal Chemistry*, **48**, 414–426.

27 Flygare, J.A. and Arkin, M.R. (2014) Inhibiting caspase-6 activation and catalytic activity for neurodegenerative diseases. *Current Topics in Medicinal Chemistry*, **14**, 319–325.

28 Murray, J., Giannetti, A.M., Steffek, M., Gibbons, P., Hearn, B.R., Cohen, F., Tam, C., Pozniak, C., Bravo, B., Lewcock, J., Jaishankar, P., Ly, C.Q., Zhao, X., Tang, Y., Chugha, P., Arkin, M.R., Flygare, J., and Renslo, A.R. (2014) Tailoring small molecules for an allosteric site on procaspase-6. *ChemMedChem*, **9**, 73–77, 72.

29 Arkin, M.R., Randal, M., DeLano, W.L., Hyde, J., Luong, T.N., Oslob, J.D.,

Raphael, D.R., Taylor, L., Wang, J., McDowell, R.S., Wells, J.A., and Braisted, A.C. (2003) Binding of small molecules to an adaptive protein–protein interface. *Proceedings of the National Academy of Sciences of the United States of America*, **100**, 1603–1608.

30 Wilson, C.G. and Arkin, M.R. (2011) Small-molecule inhibitors of IL-2/IL-2R: lessons learned and applied. *Current Topics in Microbiology and Immunology*, **348**, 25–59.

31 Brough, P.A., Aherne, W., Barril, X., Borgognoni, J., Boxall, K., Cansfield, J.E., Cheung, K.M., Collins, I., Davies, N.G., Drysdale, M.J., Dymock, B., Eccles, S.A., Finch, H., Fink, A., Hayes, A., Howes, R., Hubbard, R.E., James, K., Jordan, A.M., Lockie, A., Martins, V., Massey, A., Matthews, T.P., McDonald, E., Northfield, C.J., Pearl, L.H., Prodromou, C., Ray, S., Raynaud, F.I., Roughley, S.D., Sharp, S.Y., Surgenor, A., Walmsley, D.L., Webb, P., Wood, M., Workman, P., and Wright, L. (2008) 4,5-Diarylisoxazole Hsp90 chaperone inhibitors: potential therapeutic agents for the treatment of cancer. *Journal of Medicinal Chemistry*, **51** 196–218.

32 Potter, A., Oldfield, V., Nunns, C., Fromont, C., Ray, S., Northfield, C.J., Bryant, C.J., Scrace, S.F., Robinson, D., Matossova, N., Baker, L., Dokurno, P., Surgenor, A.E., Davis, B., Richardson, C.M., Murray, J.B., and Moore, J.D. (2010) Discovery of cell-active phenyl-imidazole Pin1 inhibitors by structure-guided fragment evolution. *Bioorganic and Medicinal Chemistry Letters*, **20**, 6483–6488.

33 Wang, Y.S., Strickland, C., Voigt, J.H., Kennedy, M.E., Beyer, B.M., Senior, M. M., Smith, E.M., Nechuta, T.L., Madison, V.S., Czarniecki, M., McKittrick, B.A., Stamford, A.W., Parker, E.M., Hunter, J.C., Greenlee, W.J., and Wyss, D.F. (2010) Application of fragment-based NMR screening, X-ray crystallography, structure-based design, and focused chemical library design to identify novel microM leads for the development of nM BACE-1 (beta-site APP cleaving enzyme 1) inhibitors. *Journal of Medicinal Chemistry*, **53**, 942–950.

34 Baker, M. (2013) Fragment-based drug discovery grows up. *Nature Reviews Drug Discovery*, **12**, 5–7.

35 Vanwetswinkel, S., Heetebrij, R.J., van Duynhoven, J., Hollander, J.G., Filippov, D.V., Hajduk, P.J., and Siegal, G. (2005) TINS, target immobilized NMR screening: an efficient and sensitive method for ligand discovery. *Chemistry and Biology*, **12**, 207–216.

36 Carlomagno, T. (2005) Ligand–target interactions: what can we learn from NMR? *Annual Review of Biophysics and Biomolecular Structure*, **34**, 245–266.

37 Ludwig, C. and Guenther, U.L. (2009) Ligand based NMR methods for drug discovery. *Frontiers in Bioscience*, **14**, 4565–4574.

38 Mayer, M. and Meyer, B. (2001) Group epitope mapping by saturation transfer difference NMR to identify segments of a ligand in direct contact with a protein receptor. *Journal of the American Chemical Society*, **123**, 6108–6117.

39 Becattini, B., Culmsee, C., Leone, M., Zhai, D., Zhang, X., Crowell, K.J., Rega, M.F., Landshamer, S., Reed, J.C., Plesnila, N., and Pellecchia, M. (2006) Structure–activity relationships by interligand NOE-based design and synthesis of antiapoptotic compounds targeting Bid. *Proceedings of the National Academy of Sciences of the United States of America*, **103**, 12602–12606.

40 Fesik, S.W. (1993) Nmr structure-based drug design. *Journal of Biomolecular NMR*, **3**, 261–269.

41 Homans, S.W. (2004) NMR spectroscopy tools for structure-aided drug design. *Angewandte Chemie International Edition*, **43**, 290–300.

42 Harner, M.J., Frank, A.O., and Fesik, S.W. (2013) Fragment-based drug discovery using NMR spectroscopy. *Journal of Biomolecular NMR*, **56**, 65–75.

43 Aguirre, C., ten Brink, T., Guichou, J.F., Cala, O., and Krimm, I. (2014) Comparing binding modes of analogous fragments using NMR in fragment-based drug design: application to PRDX5. *PLoS One*, **9**, e102300.

44 Barelier, S., Linard, D., Pons, J., Clippe, A., Knoops, B., Lancelin, J.M., and

Krimm, I. (2010) Discovery of fragment molecules that bind the human peroxiredoxin 5 active site. *PLoS One*, **5**, e9744.

45 Religa, T.L., Ruschak, A.M., Rosenzweig, R., and Kay, L.E. (2011) Site-directed methyl group labeling as an NMR probe of structure and dynamics in supramolecular protein systems: applications to the proteasome and to the ClpP protease. *Journal of the American Chemical Society*, **133**, 9063–9068.

46 Hassell, A.M., An, G., Bledsoe, R.K., Bynum, J.M., Carter, H.L. III, Deng, S.-J. J., Gampe, R.T., Grisard, T.E., Madauss, K.P., Nolte, R.T., Rocque, W.J., Wang, L., Weaver, K.L., Williams, S.P., Wisely, G.B., Xu, R., and Shewchuk, L.M. (2007) Crystallization of protein–ligand complexes. *Acta Crystallographica Section D: Biological Crystallography*, **63**, 72–79.

47 Blundell, T.L., Jhoti, H., and Abell, C. (2002) High-throughput crystallography for drug discovery in drug design. *Nature Reviews Drug Discovery*, **1**, 45–54.

48 Williams, S.P. and Sigler, P.B. (1998) Atomic structure of progesterone complexed with its receptor. *Nature*, **393**, 392–396.

49 Giannetti, A.M., Koch, B.D., and Browner, M.F. (2008) Surface plasmon resonance based assay for the detection and characterization of promiscuous inhibitors. *Journal of Medicinal Chemistry*, **51**, 574–580.

50 Neumann, T., Junker, H.D., Schmidt, K., and Sekul, R. (2007) SPR-based fragment screening: advantages and applications. *Current Topics in Medicinal Chemistry*, **7**, 1630–1642.

51 Rich, R.L., Errey, J., Marshall, F., and Myszka, D.G. (2011) Biacore analysis with stabilized G-protein-coupled receptors. *Analytical Biochemistry*, **409**, 267–272.

52 Aristotelous, T., Ahn, S., Shukla, A.K., Gawron, S., Sassano, M.F., Kahsai, A.W., Wingler, L.M., Zhu, X., Tripathi-Shukla, P., Huang, X.P., Riley, J., Besnard, J., Read, K.D., Roth, B.L., Gilbert, I.H., Hopkins, A.L., Lefkowitz, R.J., and Navratilova, I. (2013) Discovery of beta2 adrenergic receptor ligands using biosensor fragment screening of tagged wild-type receptor. *ACS Medicinal Chemistry Letters*, **4**, 1005–1010.

53 Perspicace, S., Banner, D., Benz, J., Muller, F., Schlatter, D., and Huber, W. (2009) Fragment-based screening using surface plasmon resonance technology. *Journal of Biomolecular Screening*, **14**, 337–349.

54 Niesen, F.H., Berglund, H., and Vedadi, M. (2007) The use of differential scanning fluorimetry to detect ligand interactions that promote protein stability. *Nature Protocols*, **2**, 2212–2221.

55 Dai, R., Wilson, D.J., Geders, T.W., Aldrich, C.C., and Finzel, B.C. (2014) Inhibition of *Mycobacterium tuberculosis* transaminase BioA by aryl hydrazines and hydrazides. *ChemBioChem*, **15**, 575–586.

56 Silvestre, H.L., Blundell, T.L., Abell, C., and Ciulli, A. (2013) Integrated biophysical approach to fragment screening and validation for fragment-based drug discovery. *Proceedings of the National Academy of Sciences of the United States of America*, **110**, 12984–12989.

57 Ladbury, J.E., Klebe, G., and Freire, E. (2010) Adding calorimetric data to decision making in drug discovery: a hot tip. *Nature Reviews Drug Discovery*, **9**, 23–27.

58 Wolber, G., Seidel, T., Bendix, F., and Langer, T. (2008) Molecule–pharmacophore superpositioning and pattern matching in computational drug design. *Drug Discovery Today*, **13**, 23–29.

59 Chen, Y. and Shoichet, B.K. (2009) Molecular docking and ligand specificity in fragment-based inhibitor discovery. *Nature Chemical Biology*, **5**, 358–364.

60 Nichols, D.A., Jaishankar, P., Larson, W., Smith, E., Liu, G., Beyrouthy, R., Bonnet, R., Renslo, A.R., and Chen, Y. (2012) Structure-based design of potent and ligand-efficient inhibitors of CTX-M class A beta-lactamase. *Journal of Medicinal Chemistry*, **55**, 2163–2172.

61 Ngan, C.H., Bohnuud, T., Mottarella, S.E., Beglov, D., Villar, E.A., Hall, D.R., Kozakov, D., and Vajda, S. (2012)

FTMAP: extended protein mapping with user-selected probe molecules. *Nucleic Acids Research*, **40**, W271–W275.

62 Vivat Hannah, V., Atmanene, C., Zeyer, D., Van Dorsselaer, A., and Sanglier-Cianferani, S. (2010) Native MS: an 'ESI' way to support structure- and fragment-based drug discovery. *Future Medicinal Chemistry*, **2**, 35–50.

63 Hofstadler, S.A. and Sannes-Lowery, K.A. (2006) Applications of ESI-MS in drug discovery: interrogation of noncovalent complexes. *Nature Reviews Drug Discovery*, **5**, 585–595.

64 Maresca, A., Temperini, C., Vu, H., Pham, N.B., Poulsen, S.A., Scozzafava, A., Quinn, R.J., and Supuran, C.T. (2009) Non-zinc mediated inhibition of carbonic anhydrases: coumarins are a new class of suicide inhibitors. *Journal of the American Chemical Society*, **131**, 3057–3062.

65 Vu, H., Pham, N.B., and Quinn, R.J. (2008) Direct screening of natural product extracts using mass spectrometry. *Journal of Biomolecular Screening*, **13**, 265–275.

66 Vu, H., Roullier, C., Campitelli, M., Trenholme, K.R., Gardiner, D.L., Andrews, K.T., Skinner-Adams, T., Crowther, G.J., Van Voorhis, W.C., and Quinn, R.J. (2013) Plasmodium gametocyte inhibition identified from a natural-product-based fragment library. *ACS Chemical Biology*, **8**, 2654–2659.

67 O'Connell, T.N., Ramsay, J., Rieth, S.F., Shapiro, M.J., and Stroh, J.G. (2014) Solution-based indirect affinity selection mass spectrometry: a general tool for high-throughput screening of pharmaceutical compound libraries. *Analytical Chemistry*, **86**, 7413–7420.

68 Lodge, J.M., Rettenmaier, T.J., Wells, J.A., Pomerantz, W.C., and Mapp, A.K. (2014) FP tethering: a screening technique to rapidly identify compounds that disrupt protein–protein interactions. *MedChemComm*, **5**, 370–375.

69 Kathman, S.G., Xu, Z., and Statsyuk, A.V. (2014) A fragment-based method to discover irreversible covalent inhibitors of cysteine proteases. *Journal of Medicinal Chemistry*, **57**, 4969–4974.

70 Miller, R.M., Paavilainen, V.O., Krishnan, S., Serafimova, I.M., and Taunton, J. (2013) Electrophilic fragment-based design of reversible covalent kinase inhibitors. *Journal of the American Chemical Society*, **135**, 5298–5301.

71 Burlingame, M.A., Tom, C.T., and Renslo, A.R. (2011) Simple one-pot synthesis of disulfide fragments for use in disulfide-exchange screening. *ACS Combinatorial Science*, **13**, 205–208.

72 Turner, D.M., Tom, C.T., and Renslo, A.R. (2014) Simple plate-based, parallel synthesis of disulfide fragments using the CuAAC click reaction. *ACS Combinatorial Science*, **16**, 661–664.

73 Ostrem, J.M., Peters, U., Sos, M.L., Wells, J.A., and Shokat, K.M. (2013) K-Ras (G12C) inhibitors allosterically control GTP affinity and effector interactions. *Nature*, **503**, 548–551.

74 Wolkenberg, S.E., Zhao, Z., Mulhearn, J.J., Harrison, S.T., Sanders, J.M., Cato, M.J., Jovanovska, A., Panigel, J., Cook, S.P., Henze, D.A., Kane, S.A., Hartman, G.D., and Barrow, J.C. (2011) High concentration electrophysiology-based fragment screen: discovery of novel acid-sensing ion channel 3 (ASIC3) inhibitors. *Bioorganic and Medicinal Chemistry Letters*, **21**, 2646–2649.

75 Linsky, T.W. and Fast, W. (2012) Discovery of structurally-diverse inhibitor scaffolds by high-throughput screening of a fragment library with dimethylarginine dimethylaminohydrolase. *Bioorganic and Medicinal Chemistry*, **20**, 5550–5558.

76 Silva-Santisteban, M.C., Westwood, I.M., Boxall, K., Brown, N., Peacock, S., McAndrew, C., Barrie, E., Richards, M., Mirza, A., Oliver, A.W., Burke, R., Hoelder, S., Jones, K., Aherne, G.W., Blagg, J., Collins, I., Garrett, M.D., and van Montfort, R.L. (2013) Fragment-based screening maps inhibitor interactions in the ATP-binding site of checkpoint kinase 2. *PLoS One*, **8**, e65689.

77 Chou, T.F., Brown, S.J., Minond, D., Nordin, B.E., Li, K., Jones, A.C., Chase, P., Porubsky, P.R., Stoltz, B.M., Schoenen, F. J., Patricelli, M.P., Hodder, P., Rosen, H., and Deshaies, R.J. (2011) Reversible inhibitor of p97, DBeQ, impairs both

ubiquitin-dependent and autophagic protein clearance pathways. *Proceedings of the National Academy of Sciences of the United States of America*, **108**, 4834–4839.

78 Chou, T.F., Bulfer, S.L., Weihl, C.C., Li, K., Lis, L.G., Walters, M.A., Schoenen, F. J., Lin, H.J., Deshaies, R.J., and Arkin, M. R. (2014) Specific inhibition of p97/VCP ATPase and kinetic analysis demonstrate interaction between D1 and D2 ATPase domains. *Journal of Molecular Biology*, **426**, 2886–2899.

79 Chou, T.F. and Deshaies, R.J. (2011) Development of p97 AAA ATPase inhibitors. *Autophagy*, **7**, 1091–1092.

80 Chou, T.F. and Deshaies, R.J. (2011) Quantitative cell-based protein degradation assays to identify and classify drugs that target the ubiquitin-proteasome system. *Journal of Biological Chemistry*, **286**, 16546–16554.

81 Chou, T.F., Li, K., Frankowski, K.J., Schoenen, F.J., and Deshaies, R.J. (2013) Structure–activity relationship study reveals ML240 and ML241 as potent and selective inhibitors of p97 ATPase. *ChemMedChem*, **8**, 297–312.

82 Bug, M. and Meyer, H. (2012) Expanding into new markets: VCP/p97 in endocytosis and autophagy. *Journal of Structural Biology*, **179**, 78–82.

83 Meyer, H., Bug, M., and Bremer, S. (2012) Emerging functions of the VCP/p97 AAA-ATPase in the ubiquitin system. *Nature Cell Biology*, **14**, 117–123.

84 Dantuma, N.P., Acs, K., and Luijsterburg, M.S. (2014) Should I stay or should I go: VCP/p97-mediated chromatin extraction in the DNA damage response. *Experimental Cell Research*, **329**, 9–17.

85 Deshaies, R.J. (2014) Proteotoxic crisis, the ubiquitin-proteasome system, and cancer therapy. *BMC Biology*, **12**, 94.

86 Briggs, L.C., Baldwin, G.S., Miyata, N., Kondo, H., Zhang, X., and Freemont, P. S. (2008) Analysis of nucleotide binding to P97 reveals the properties of a tandem AAA hexameric ATPase. *Journal of Biological Chemistry*, **283**, 13745–13752.

87 Song, C., Wang, Q., and Li, C.C. (2003) ATPase activity of p97-valosin-containing protein (VCP). D2 mediates the major enzyme activity, and D1 contributes to the heat-induced activity. *Journal of Biological Chemistry*, **278**, 3648–3655.

88 Magnaghi, P., D'Alessio, R., Valsasina, B., Avanzi, N., Rizzi, S., Asa, D., Gasparri, F., Cozzi, L., Cucchi, U., Orrenius, C., Polucci, P., Ballinari, D., Perrera, C., Leone, A., Cervi, G., Casale, E., Xiao, Y., Wong, C., Anderson, D.J., Galvani, A., Donati, D., O'Brien, T., Jackson, P.K., and Isacchi, A. (2013) Covalent and allosteric inhibitors of the ATPase VCP/p97 induce cancer cell death. *Nature Chemical Biology*, **9** 548–556.

89 Polucci, P., Magnaghi, P., Angiolini, M., Asa, D., Avanzi, N., Badari, A., Bertrand, J., Casale, E., Cauteruccio, S., Cirla, A., Cozzi, L., Galvani, A., Jackson, P.K., Liu, Y., Magnuson, S., Malgesini, B., Nuvoloni, S., Orrenius, C., Sirtori, F.R., Riceputi, L., Rizzi, S., Trucchi, B., O'Brien, T., Isacchi, A., Donati, D., and D'Alessio, R. (2013) Alkylsulfanyl-1,2,4-triazoles, a new class of allosteric valosine containing protein inhibitors. Synthesis and structure–activity relationships. *Journal of Medicinal Chemistry*, **56** 437–450.

90 Madsen, L., Seeger, M., Semple, C.A., and Hartmann-Petersen, R. (2009) New ATPase regulators: p97 goes to the PUB. *International Journal of Biochemistry and Cell Biology*, **41**, 2380–2388.

91 Schuberth, C. and Buchberger, A. (2008) UBX domain proteins: major regulators of the AAA ATPase Cdc48/p97. *Cellular and Molecular Life Sciences*, **65**, 2360–2371.

92 Yeung, H.O., Kloppsteck, P., Niwa, H., Isaacson, R.L., Matthews, S., Zhang, X., and Freemont, P.S. (2008) Insights into adaptor binding to the AAA protein p97. *Biochemical Society Transactions*, **36**, 62–67.

93 Chia, W.S., Chia, D.X., Rao, F., Bar Nun, S., and Geifman Shochat, S. (2012) ATP binding to p97/VCP D1 domain regulates selective recruitment of adaptors to its proximal N-domain. *PLoS One*, **7**, e50490.

94 Niwa, H., Ewens, C.A., Tsang, C., Yeung, H.O., Zhang, X., and Freemont, P.S. (2012) The role of the N-domain in the ATPase activity of the mammalian AAA ATPase p97/VCP. *Journal of Biological Chemistry*, **287**, 8561–8570.

95 Khanna, I. (2012) Drug discovery in pharmaceutical industry: productivity challenges and trends. *Drug Discovery Today*, **17**, 1088–1102.

96 Tralau-Stewart, C.J., Wyatt, C.A., Kleyn, D.E., and Ayad, A. (2009) Drug discovery: new models for industry–academic partnerships. *Drug Discovery Today*, **14**, 95–101.

97 Arkin, M.R., Ang, K.K., Chen, S., Davies, J., Merron, C., Tang, Y., Wilson, C.G., and Renslo, A.R. (2014) UCSF small molecule discovery center: innovation, collaboration and chemical biology in the Bay Area. *Combinatorial Chemistry and High Throughput Screening*, **17**, 333–342.

98 Huryn, D.M. (2013) Drug discovery in an academic setting: playing to the strengths. *ACS Medicinal Chemistry Letters*, **4**, 313–315.

99 Huryn, D.M., Resnick, L.O., and Wipf, P. (2013) Contributions of academic laboratories to the discovery and development of chemical biology tools. *Journal of Medicinal Chemistry*, **56**, 7161–7176.

100 Bruneel, J., D'Este, P., and Salter, A. (2010) Investigating the factors that diminish the barriers to university–industry collaboration. *Research Policy*, **39**, 858–868.

101 Li, J. and Yuan, J. (2008) Caspases in apoptosis and beyond. *Oncogene*, **27**, 6194–6206.

102 McIlwain, D.R., Berger, T., and Mak, T.W. (2013) Caspase functions in cell death and disease. *Cold Spring Harbor Perspectives in Biology*, **5**, a008656.

103 Graham, R.K., Deng, Y., Carroll, J., Vaid, K., Cowan, C., Pouladi, M.A., Metzler, M., Bissada, N., Wang, L., Faull, R.L., Gray, M., Yang, X.W., Raymond, L.A., and Hayden, M.R. (2010) Cleavage at the 586 amino acid caspase-6 site in mutant huntingtin influences caspase-6 activation *in vivo*. *Journal of Neuroscience*, **30**, 15019–15029.

104 LeBlanc, A.C. (2013) Caspase-6 as a novel early target in the treatment of Alzheimer's disease. *European Journal of Neuroscience*, **37**, 2005–2018.

105 Crawford, E.D. and Wells, J.A. (2011) Caspase substrates and cellular remodeling. *Annual Review of Biochemistry*, **80**, 1055–1087.

106 Heise, C.E., Murray, J., Augustyn, K.E., Bravo, B., Chugha, P., Cohen, F., Giannetti, A.M., Gibbons, P., Hannoush, R.N., Hearn, B.R., Jaishankar, P., Ly, C.Q., Shah, K., Stanger, K., Steffek, M., Tang, Y., Zhao, X., Lewcock, J.W., Renslo, A.R., Flygare, J., and Arkin, M.R. (2012) Mechanistic and structural understanding of uncompetitive inhibitors of caspase-6. *PLoS One*, **7**, e50864.

107 Baell, J. and Walters, M.A. (2014) Chemistry: chemical con artists foil drug discovery. *Nature*, **513**, 481–483.

# 11
# Site-Directed Fragment Discovery for Allostery

*T. Justin Rettenmaier, Sean A. Hudson, and James A. Wells*

## 11.1
## Introduction

Protein allostery is one of the most prevalent and intriguing mechanisms for regulating protein function [1]. Allostery is mediated through conformational changes where binding at one site remotely affects binding at another. Allosteric ligands can turn proteins on or off, and can even do so by binding at the same allosteric site. For example, the binding of ligands to allosteric sites in G-protein-coupled receptors (GPCRs), nuclear receptors, and enzymes can remotely control their ability to interact with other proteins or substrates, which can result in a wide range of functional outcomes.

Allosteric sites have been of increasing interest in the drug discovery community. Such sites are often less conserved than enzyme active sites that require strict retention and orientation of catalytic and substrate binding sites. Thus, allosteric sites can afford lower hurdles for selectivity, and have the option to either activate or inhibit the target protein. Also, active sites can be challenging when they bind highly charged or large diffuse substrates that are not easily mimicked by drug-like small molecules. Indeed, there are now a number of small molecules developed for allosteric sites in GPCRs, kinases, viral proteins, and many other target classes [2,3].

However, allosteric sites can be challenging for drug discovery especially when there are no known natural allosteric ligands for the target of interest. Many allosteric interactions involve protein–protein interfaces, and though functional, these sites provide little chemical matter to guide the discovery of drug-like small molecules. In addition, allosteric sites are challenging to annotate, since to do so requires a known ligand or protein to which the target binds and is functionally modulated by. Indeed chemical screening campaigns often reveal new allosteric sites, "orphan sites," to which no natural effector is known [4]. Protein dynamics and bioinformatics can reveal new cavities not evident in native protein, which may also be opportunities for small molecule drug discovery [5,6], but it is challenging to screen for molecules to bind specifically to these sites.

*Fragment-based Drug Discovery: Lessons and Outlook*, First Edition. Edited by Daniel A. Erlanson and Wolfgang Jahnke.

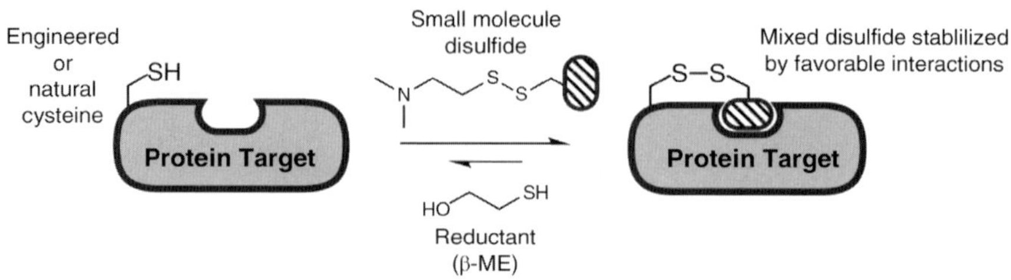


**Figure 11.1** Tethering: site-directed ligand discovery.

We have developed a site-directed drug discovery technology, called tethering [7]. In this approach, a cysteine residue is introduced near a known or putative binding site (Figure 11.1). The thiol-containing protein is then mixed with a small molecule fragment (MW 200–300) that contains a disulfide capped with dimethyl-cysteamine [8]. Each disulfide fragment is allowed to undergo disulfide exchange with the protein in the presence of a redox buffer, typically containing β-mercaptoethanol (β-ME). If the compound binds noncovalently in the vicinity of the cysteine residue, it will preferentially form the mixed protein–fragment disulfide, which is detected by mass spectrometry. This technology was developed at Sunesis Pharmaceuticals and has been applied to over 40 protein targets, with on average 4 cysteine mutants for each target [9]. Many fragments identified from tethering have been advanced to free-standing compounds that target proteases [10,11], kinases [12], and protein–protein interfaces [13].

Tethering is ideally suited for discovering and validating allosteric sites. The site-directed aspect allows one to interrogate and discover ligands that engage known or suspected surface sites of interest. Moreover, tethering can be used to identify "orphan" allosteric sites either from adventitious natural cysteines or engineered ones placed near cavities of interest. Here, we describe multiple examples of both. First, we will describe work on caspases-1, -3, -5, and -7 where tethering discovered a natural cysteine at the dimer interface that bound specific tethered compounds that drives each protease to the inactive zymogen-like state. We then review work on K-Ras where an oncogenic cysteine near the active site allowed discovery of tethered inhibitors. Next, we describe the use of tethering to stabilize a dynamic domain of the transcription factor CBP/p300, allowing crystallization and solving of its first X-ray structure as well as generating inhibitors and activators of transcription. Finally, we focus on two systems for which putative peptide binding sites were known, the kinase PDK1 and the GPCR C5a receptor. In both of these cases, there was a modeled peptide allosteric binding site for which we found tethered compounds, some of which were allosteric activators and others allosteric inhibitors. These studies reveal the power of the tethering technology to discover small molecule fragments and to identify and mechanistically characterize allosteric sites.

## 11.2 Caspases

Caspases are a family of dimeric cysteine proteases that cleave after the Asp residue in a short tetrapeptide (often XXXD) motif [14]. Like many proteases, the caspases are expressed as catalytically inactive zymogens that must be proteolytically activated by an upstream protease. Caspases can be subdivided into two main classes: the inflammatory caspases (-1, -2, -4, -5) and the apoptotic caspases (-3, -6, -7, -8, -9). The apoptotic caspases deconstruct the cell during programmed cell death. In response to apoptotic stimuli, the initiator caspases (-8, -9) are activated and proceed to cleave and activate the executioner caspases (-3, -6, -7). Through the coordinated cleavage of hundreds of different proteins, the executioner caspases trigger many of the hallmarks of apoptosis including membrane blebbing and chromatin laddering.

Caspases represent an important class of drug targets for stroke, ischemia, cancer, and inflammatory diseases. However, despite their biological significance, no caspase-directed therapies are currently available. The active sites of caspases have been difficult to target with drug-like compounds owing to their preference for negatively charged substrates. Recently, several successful tethering campaigns have identified novel allosteric inhibitors of caspases.

### 11.2.1 Tethered Allosteric Inhibitors of Executioner Caspases-3 and -7

In 2004, Hardy *et al.* of Sunesis Pharmaceuticals reported the first tethering screen against caspases [15]. A library of 10 000 disulfide fragments was screened against endogenous cysteines of wild-type active caspase-3. Active caspase-3 contains five surface-exposed cysteines in addition to the catalytic cysteine. Two compounds were discovered, named FICA and DICA, which completely inhibited caspase-3 activity following 100% covalent modification of a single cysteine (Figure 11.2a). MS analysis revealed that the modified cysteine was not the catalytic cysteine but a cysteine present in the small subunit. Remarkably, this allosteric cysteine was also present in the executioner caspase-7, and FICA and DICA also quantitatively inhibited active caspase-7. Labeling of the allosteric cysteine with FICA/DICA was mutually exclusive with labeling of the active site cysteine with the peptide Z-DEVD-fluoromethylketone, suggesting that the binding of FICA/DICA induced a large conformational change that was incompatible with substrate binding.

An X-ray crystal structure of FICA- or DICA-modified caspase-7 showed the allosteric inhibitors conjugated to Cys290 in a central deep cavity at the caspase dimer interface (Figure 11.2b and c). Despite being 14 Å from the active site, this site functionally couples to the active site and prevents substrate binding. FICA forms two intermolecular hydrogen bonds with itself across the dimer interface, and each FICA molecule displaces Y223. When comparing with the structure of active caspase-7 bound by substrate-mimic DEVD, we see that DEVD forces

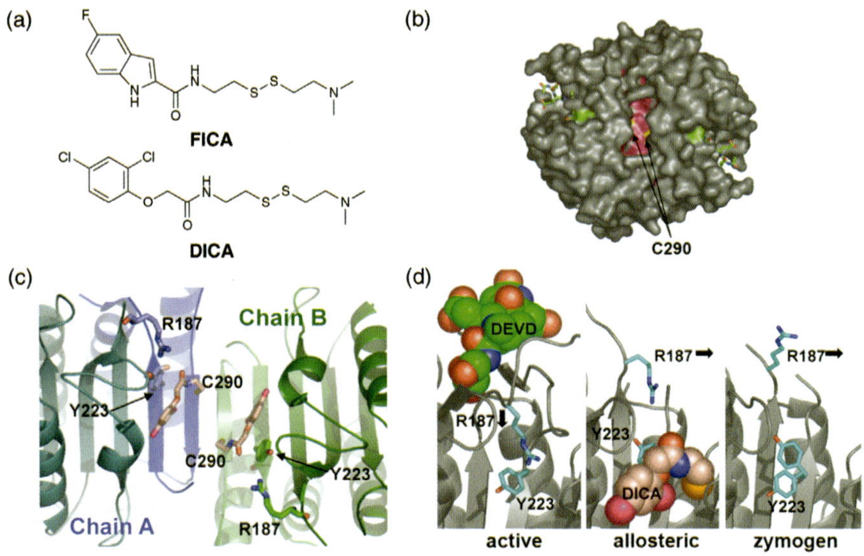


**Figure 11.2** (a) Chemical structure of FICA and DICA. (b) Active caspase-7 dimer central cavity (red) with C290 labeled. The active site is colored green. (c) Crystal structure of caspase-7 bound by DICA. (d) Conformations of residues Y223 and R187 for active caspase-7 bound to substrate-mimic DEVD (left), caspase-7 bound to allosteric inhibitor DICA (middle), or zymogen caspase-7 (right).

Y223 and adjacent R187 down (Figure 11.2d). FICA and DICA force Y223 and subsequently R187 into the up position by steric exclusion. In the procaspse-7 zymogen structure, R187 is also up and Y223 is free to assume either the up or the down conformation. Hence, the FICA and DICA allosteric inhibitors appear to trap a zymogen-like conformation. The inhibitors also induce conformational changes particularly in substrate binding region loops into zymogen-like conformations (not illustrated). Subsequent mutational analysis further evaluated an allosteric "switch" that connects the allosteric and active sites [16,17].

### 11.2.2
### Tethering Inflammatory Caspase-1

To probe the generality of this dimer interface site, Scheer, Romanowski, and Wells performed a tethering screen against human inflammatory caspase-1 [18]. Caspase-1 is primarily responsible for processing and activating the inflammatory cytokine interleukin 1β. Despite sharing only a 20–30% amino acid sequence identity with the apoptotic caspases, caspase-1 also contains a cavity at the dimer interface with a cysteine (Cys331) in a position close to those in caspase-3 and -7.

A library of ~8000 disulfide fragments was screened by mass spectrometry and then an *in vitro* enzymatic assay, which led to the identification of the

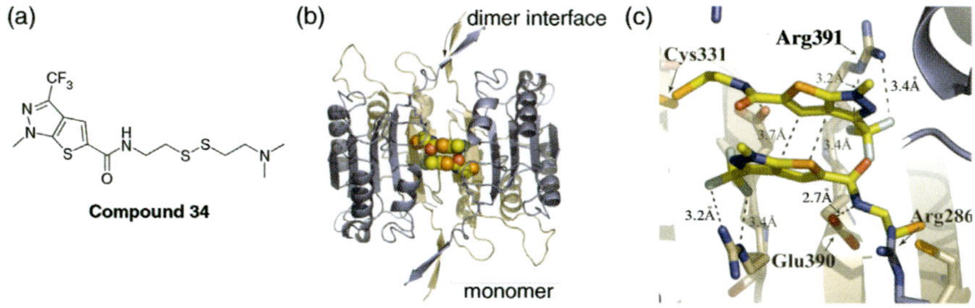


**Figure 11.3** (a) Chemical structure of compound **34**. (b) Crystal structure of caspase-1 bound to compound **34** at the dimer interface. (c) Close-up view of the ligand binding site.

thienopyrazole compound **34** (Figure 11.3a). The caspase-1 mutant C331A was neither labeled nor inhibited by compound **34**. A crystal structure of caspase-1 in complex with compound **34** shows two molecules of the allosteric inhibitor trapped to Cys331 in the central cavity at the dimer interface (Figure 11.3b and c). The compound binding site is 15 Å away from the active site, yet these sites appear to be functionally coupled much like the FICA/DICA binding site in caspases-3 and -7. Binding of compound **34** causes a large conformational change wherein Arg286-Glu390 (located next to the catalytic cysteine) break a salt bridge and shift >12 Å to interact with the compound. A comparison of the global structures of caspase-1 in the allosterically inhibited or apo states revealed that compound **34** traps an inactive conformation that preexists in the ligand-free enzyme. Subsequent mutational analysis revealed a clear allosteric circuit that connected the allosteric and active sites similar to that described for caspase-7 [19].

### 11.2.3
### Tethered Allosteric Inhibitors of Caspase-5

More recently, Gao and colleagues extended tethering to target caspase-5 [20]. Caspase-5 is also a inflammatory caspase, like caspase-1, but its specific cellular roles remain unclear. Selective inhibitors would be helpful to dissect the role of caspase-5 in the innate immune response. From a screen of ~15 000 disulfide fragments, the naphthyl-thiazole compound **8** was the top hit (Figure 11.4). Compound **8** inhibited 50% of caspase-5 activity at a concentration of 1 μM. Labeling of caspase-5 by compound **8** could be reduced to 50% by 2.3 mM β-ME, which indicated high conjugation strength as this value was 23-fold more stringent than that used during the primary screen. Compound **8** selectively labeled the allosteric cysteine on the subunit at the caspase-5 dimer interface, but did not label or inhibit caspase-1. This selectivity is remarkable considering the high degree of sequence conservation between these inflammatory caspases (59% identity and 71% similarity in the catalytic subunits).

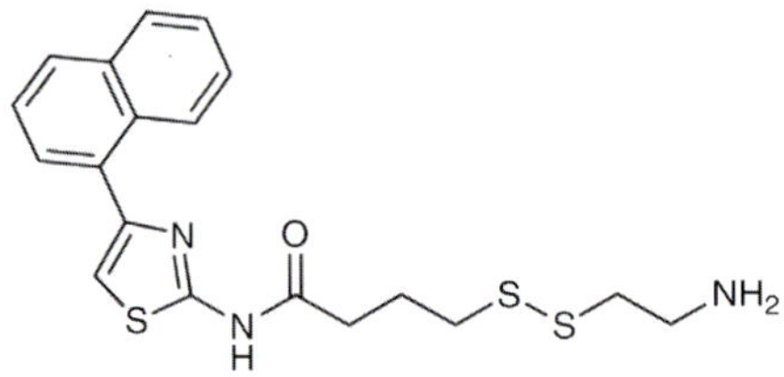

**Compound 8**

**Figure 11.4** Chemical structure of compound **8**.

### 11.2.4
### General Allosteric Regulation at the Caspase Dimer Interface

The success of tethering against the allosteric cavity located at the dimer interface of caspases-1, -3, -5, and -7 suggests that this regulatory site is common between inflammatory and apoptotic caspases. This possibility inspired mutational studies to further define the residues that functionally couple the allosteric and active sites in both caspase-1 and caspase-7 [16,19]. The deadly nature of caspases to cells has necessitated the evolution of many levels of regulation to ensure their activity only at the appropriate times. Thus, it is tempting to speculate that the central cavity may serve as a regulatory site for an as yet undiscovered effector molecule that acts as an additional level of native regulation. This allosteric cavity featuring a natural cysteine appears across most active caspases (Figure 11.5). The sequence diversity at the allosteric site across the caspases may enable the future development of selective small-molecules inhibitors for each family member.

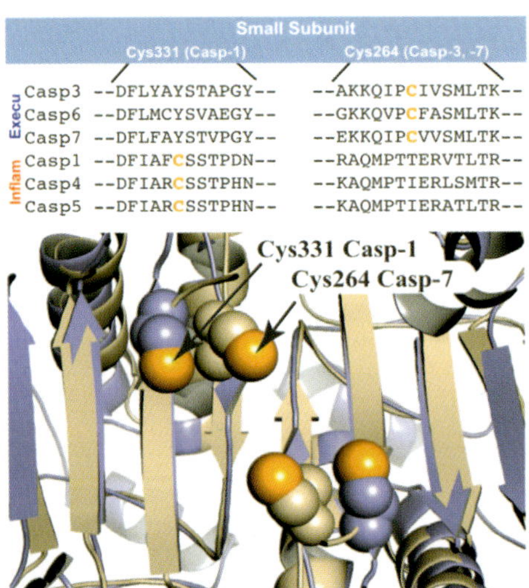


**Figure 11.5** Conservation of a cysteine across caspases in the allosteric cavity discovered by tethering.

### 11.2.5
### Using Disulfide Fragments as "Chemi-Locks" to Generate Conformation-Specific Antibodies

The dynamic conformational states of caspase-1 (active, allosterically inhibited, and ligand free) present potential antigens that could be used to find antibodies that selectively trap each conformation (Figure 11.6). Conformation-specific antibodies would have utility in probing the equilibrium distribution of caspase-1 conformational states both *in vitro* and in cells. Gao and colleagues from UCSF and Genentech used covalent ligands as "chemi-locks" to homogeneously trap caspase-1 into one of the two alternative conformations [21]. The covalent active-site inhibitor (Ac-YVAD-chloromethylketone) locked the on-state, and the disulfide fragment **34** (Figure 11.3a) was used to lock the off-state. These conformation-trapped forms of the protease were then used as antigens in alternating rounds of selection and antiselection for antibody fragments (Fabs) displayed on phage (Figure 11.6). After affinity maturation, two high-affinity Fabs were isolated with $K_d$ values ranging from 2 to 5 nM, and each bound to their cognate conformer 20- to 500-fold more tightly than the opposite conformational state.

The Fabs were able to activate ($Fab_{on}$, $EC_{50} = 11.7$ nM) or inhibit ($Fab_{off}$, $IC_{50} = 920$ nM) caspase-1 in an *in vitro* proteolysis assay. When converted into full-length human IgGs, both bound to ligand-free caspase-1 like their Fab counterparts. $IgG_{off}$ inhibited caspase-1-mediated processing of interleukin-1β in monocyte (THP-1) cell lysates. The $IgG_{on}$ monoclonal antibody was used to probe and localize active conformers of caspase-1 in THP-1 cells, showing that activated caspase-1 is concentrated in a central structure in the cytosol, similar to what has been described as the pyroptosome. In all, the approach provides a general strategy for producing multiple conformation-specific antibodies to proteins.

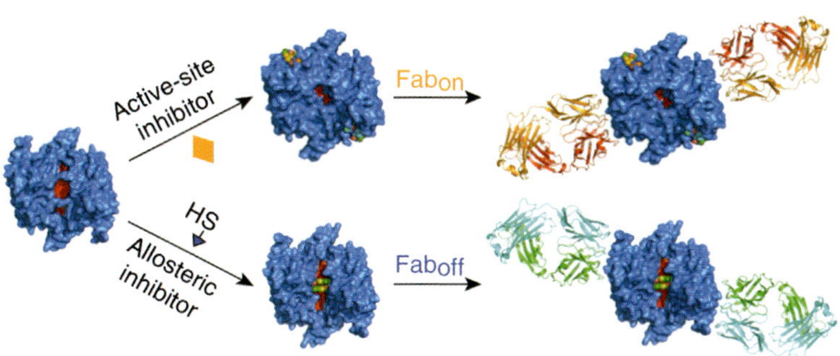


**Figure 11.6** A covalent peptide substrate and the allosteric tethering inhibitor were used to lock caspase-1 into different conformations for antibody generation via phage display. The caspase–antibody complexes shown on the right are models.

## 11.3 Tethering K-Ras(G12C)

One-third of all tumors harbor mutations in Ras genes, but GTPases encoded by these mutant genes have eluded targeting by therapeutic agents [22]. Ras serves as an important switch for cell signaling, critical to cell cycle progression [23]. Ras cycles between the GTP-bound active and the GDP-bound inactive signaling states. Oncogenic mutations to K-Ras such as G12C impair its ability to hydrolyze GTP to GDP. With diminished GTPase activity, the GTP-bound active state of Ras accumulates, which leads to unregulated signaling and ultimately tumorigenesis.

Drug discovery efforts have long attempted to halt unregulated Ras, but have failed to identify potent and selective small molecules. GTP and GDP are present in the cell at millimolar concentrations and bind to Ras with picomolar affinity, which renders the active site of Ras virtually intractable for inhibitor development. Drug development thus turned to inactivating RAS indirectly through the development of farnesyltransferase inhibitors, but these drugs had limited efficacy in clinical trials [24]. Attention has recently focused on inhibitors of the prenyl binding protein PDEδ, as correct subcellular localization and signaling by farnesylated K-Ras is regulated by binding with PDEδ [25].

Ostrem *et al.* [26] recently reported a new tethering approach specifically targeting the K-Ras(G12C) mutant, which is found in a substantial proportion of lung cancers. A library of 480 disulfide fragments was screened for binding the oncogenic cysteine mutant K-Ras(G12C) in the GDP-bound inactive state. Disulfide fragment 6H05 gave the greatest degree of modification (94%; Figure 11.7a). Compound 6H05 did not label wild-type K-Ras, which contains three endogenous cysteine residues. The labeling of K-Ras (G12C) by 6H05 was not diminished by high-concentration 1 mM GDP, which suggested that the compound bound to a site that did not overlap with GDP. However, preloading of K-Ras(G12C) with GTP significantly impaired the binding of 6H05.

A small number of analogs of 6H05 were made to increase its affinity and the top analog **6H05-2** was 4.2-fold more potent (Figure 11.7a). A 1.3 Å cocrystal structure of **6H05-2** bound to K-Ras(G12C) in the GDP state revealed that **6H05-2** does not bind in the nucleotide pocket, but rather extends from Cys12 into a previously unobserved binding pocket composed largely of switch-II, called the switch-II pocket (Figure 11.7b). The switch-I and switch-II loops intimately interact with the nucleotide binding pocket of K-Ras and mediate interactions with a multitude of downstream effectors in a GTP-dependent manner. The switch-II pocket is not visible in other structures of K-Ras, and thus it is probably highly dynamic when GDP is bound. In the GTP-bound active state conformation of RAS, residues from switch-II entirely fill the switch-II pocket. Thus, compound **6H05-2** binds to a cryptic allosteric pocket and displaces switch-II but has little effect on switch-I (Figure 11.7c).

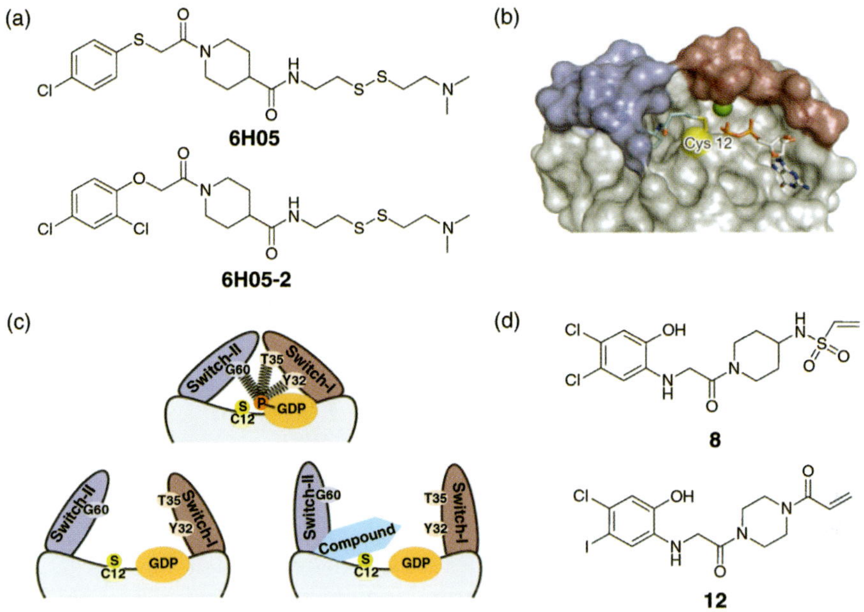


**Figure 11.7** (a) Chemical structures of compound 6H05 and analog 6H05-2. (b) Cocrystal structure K-Ras(G12C) labeled by 6H05-2 (cyan) and bound to GDP (gray) and $Ca^{2+}$ (green). (c) Structural basis of K-Ras(G12C) modulation by disulfide fragments. On GTP hydrolysis (top to bottom, left), removal of the γ-phosphate leads to relaxation of the spring-loaded active Ras-GTP state. Drug binding (bottom right) moves switch-II even further away and interferes with GTP binding. (d) Chemical structures of the optimized electrophiles **8** and **12**.

Since disulfide fragments are susceptible to reduction in cells, the tethering hits were converted to irreversible cysteine reactive electrophiles, including acrylamides and vinyl sulfonamides. The fragments were chemically optimized through the generation of 100 analogs, resulting in electrophiles **8** and **12** that efficiently labeled K-Ras(G12C) at 10 μM (Figure 11.7d). Treatment with electrophile **12** modified K-Ras(G12C) but not BSA or wild-type K-Ras, indicating the electrophile did not promiscuously label cysteines but selectively reacted with K-Ras(G12C). The covalent bond between **12** and G12C was essential for binding, as the naked fragment without its electrophile did not prevent labeling of K-Ras(G12C) by **12**.

The authors first assessed the functional consequences of labeling K-Ras (G12C) using an *in vitro* nucleotide binding assay. Labeling K-Ras(G12C) with **8** or **12** decreased its affinity for GTP relative to GDP by nearly fourfold. Based on this result, they expected that treating K-Ras with these compounds would lead to the accumulation of its GDP-loaded inactive state. Moreover, the conformational rearrangements to K-Ras were predicted to be incompatible with binding to downstream effectors. Indeed, treatment of K-Ras(G12C) mutant lung cancer cells with **12** diminished the interaction between K-Ras and its binding partners B-RAF and C-RAF. Excitingly, treatment of a panel of lung cancer cell lines with **12** led to apoptosis only in those cells annotated to harbor the K-Ras(G12C) mutation. In all, the lead electrophile compound needs additional affinity optimization and evaluation *in vivo*, but Ostrem *et al.* provide a good starting point for new drug discovery efforts targeting K-Ras(G12C) and its cryptic allosteric switch-II pocket.

## 11.4 The Master Transcriptional Coactivator CREB Binding Protein

Small molecules that directly modulate transcriptional regulators promise direct control of gene expression by bypassing upstream regulatory pathways. However, with the exception of steroid hormone receptors and histone modifying enzymes, most transcriptional regulators have eluded modulation by small molecules due to the chemical intractability of their highly charged DNA-binding sites and extended protein–protein interfaces. CREB-binding protein (CBP), also known as p300, is a master transcriptional coactivator that serves as a transcriptional hub in all metazoans. CBP interacts with numerous transcriptional activators using several different domains. The KIX domain of CBP is a 90-residue three-helix bundle that interacts with at least 10 different transcriptional activators at two distinct binding sites (Figure 11.8a). Interestingly, the two binding sites exhibit positive allosteric coupling where the binding of a transcriptional activator at one site enhances the binding of a second activator at the other site. Unfortunately, the dynamic nature of the KIX domain has complicated in-depth analyses of its structure and the allosteric coupling between its binding sites. Using Tethering, Mapp and colleagues have identified fragments that act as pharmacological chaperones to stabilize the protein for structure/function studies. Excitingly, many of these fragments inhibit the binding of transcriptional activators to one or both binding sites on the KIX domain of CBP.

### 11.4.1 Tethering to Find Stabilizers of the KIX Domain of CBP

In 2013, Wang *et al.* reported a tethering screen against the KIX domain of CBP with the aim of identifying fragments to stabilize KIX and facilitate crystallographic analysis [27]. The authors introduced a cysteine mutation at L664 of the KIX domain, which lies adjacent to the known binding site for the transcriptional activation domain of mixed lineage leukemia protein (MLL) (Figure 11.8a). A screen of 480 disulfide fragments yielded two hits, **1–10** and **2–64**, which strongly tethered to the KIX domain (Figure 11.8b). These fragments increased the melting temperature of KIX by 15–18 °C, protected 40 and

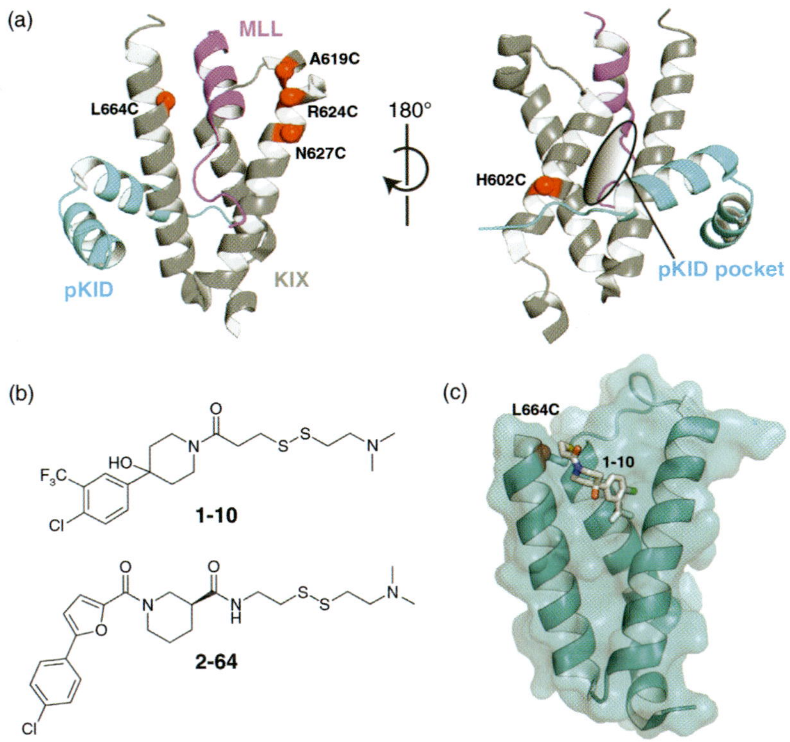


**Figure 11.8** (a) The two binding surfaces on opposite faces of the KIX domain (gray) are shown bound to MLL in magenta and pKID in cyan [40]. Tethering has been conducted at all positions marked with red spheres. (b) Chemical structures of top hits from a tethering screen against KIX L664C. (c) Crystal structure of KIX L664C in complex with disulfide fragment **1–10**.

55% of amides from hydrogen–deuterium exchange, and led to a fivefold and 35-fold increased resistance to degradation by the protease thermolysin. Lastly, both fragments bound to KIX competitively with MLL. Altogether, these results suggested that these covalent small molecules dramatically stabilized this highly dynamic protein by occupying the MLL binding site. Ultimately, labeling KIX with disulfide fragment **1–10** enabled its crystallization and the determination of the first KIX crystal structure at 2.0 Å resolution (Figure 11.8c). This study revealed the utility of disulfide fragments as small molecule stabilizers for enabling the structural characterization of highly dynamic proteins.

### 11.4.2
### Dissecting the Allosteric Coupling between Binding Sites on KIX

Wang *et al.* next sought to understand the mechanism of allosteric communication between the two binding sites on KIX [28]. In principle, the allosteric coupling between KIXs two binding sites could be explained by changes in the rate of formation of a ternary complex or the rate of its dissociation. To discriminate between these possibilities, the authors performed stopped flow binding experiments to measure the rates of binding ($k_{on}$) and dissociation ($k_{off}$) for a fluorescent peptide (pKID) when mixed with various liganded states of KIX. These experiments required prior formation of a complex between KIX and a ligand at the MLL binding site, so that binding at the pKID binding site could be observed. Using the covalent tool compound **1–10**, KIX was quantitatively labeled and trapped into a state of positive allosteric coupling (modification of N627C) or a state of negative allosteric coupling (modification of L664C). Surprisingly, the kinetic binding experiments revealed that the basis of allosteric coupling between KIXs two binding sites could be explained solely by modulation of the dissociation rate ($k_{off}$) of the ternary complex. That is, positive allosteric modulators stabilized the ternary complex and resulted in a slower off-rate, whereas negative allosteric modulators had the opposite effect. These findings may extend to understanding binding cooperativity in the context of other protein–protein interactions that involve intrinsically disordered domains like the transcriptional activation domains of MLL and pKID studied here. Given the ease of identifying tethering fragments, mechanistic studies of this type could be initiated in many other systems.

### 11.4.3
### Rapid Identification of pKID-Competitive Fragments for KIX

Despite tethering against cysteine residues placed directly adjacent to the MLL binding site in the KIX domain (Figure 11.8a), Mapp and colleagues remarked that a large fraction of fragments discovered from these screens did not efficiently disrupt the interaction between KIX and MLL. Thus, in order to select only those fragments that are effective PPI inhibitors, tethering was adapted to a competitive binding assay where the disruption of the PPI would be monitored by fluorescence polarization (Figure 11.9) [29]. To create displaceable binding probes, the authors conjugated fluorophores to the peptide ligands MLL, pKID, and c-Myb. As a proof of concept, they showed that this new approach, termed FP tethering, could correctly identify known MLL-competitive fragments against KIX N627C. Next, they screened 960 disulfide fragments against KIX H602C to find ligands that could displace pKID and c-Myb on the opposite face of the protein. The screen was performed at low, medium, and high stringency (0.2 mM, 1 mM, and 5 mM β-ME) to expedite the identification of the most potent fragments all at once. The authors selected the top 10 of 63 hits (6.6% hit rate) to validate by the traditional mass spectrometry readout. Nine of the ten hits were confirmed, including the extremely efficient ligand **3D4** (β-$ME_{50}$ >5 mM). Surprisingly, **3D4** impeded the binding not only of pKID (6.5-fold) and c-Myb (9-fold) but also of MLL (9-fold), which binds on the opposite face of the KIX domain. The authors are further pursuing covalent analogues of **3D4** for use in studying CBP/p300 biology.

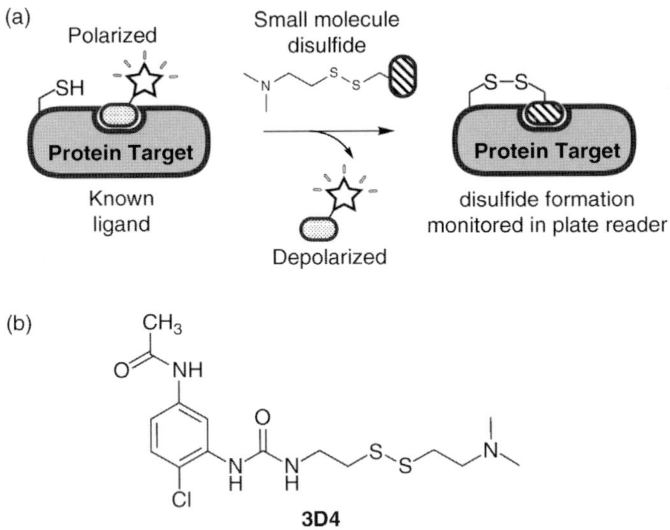


**Figure 11.9** (a) Premise of FP tethering. Disulfide formation is detected by monitoring displacement of a known ligand that has been appended with a fluorophore. (b) Chemical structure of the disulfide fragment 3D4, which labels KIX H602C and binds to the pKID pocket.

## 11.5 Tethering Against the PIF Pocket of Phosphoinositide-Dependent Kinase 1 (PDK1)

The critical role of protein kinases as regulators of cell state has fueled enormous interest over the past 15 years to develop kinase inhibitors as chemical probes and therapeutics. Nearly all kinase inhibitors target the highly conserved ATP binding pocket, and as a consequence these molecules usually inhibit many kinases simultaneously. These off-target effects diminish the value of chemical tools and can cause undesirable side effects in patients. To address these limitations, significant effort has focused on targeting allosteric sites distinct from the ATP binding pocket. Several allosteric kinase inhibitors are now in clinical trials, and many more are being developed [3], but the discovery of chemically tractable allosteric sites remains a major bottleneck in the development of new compounds.

The protein kinase PDK1 is a member of the AGC family of serine/threonine kinases, which collectively regulate aspects of cell growth, proliferation, metabolism, and programmed cell death [30]. PDK1 controls the activation of at least 23 other AGC kinases, including the oncogenic kinases AKT and S6Ks. PDK1 has, therefore, been targeted by numerous drug discovery programs, but potent and highly selective inhibitors have had limited anticancer efficacy [31]. In order to overcome the limitations of PDK1 inhibitors, recent efforts have focused on targeting the PDK1 interacting fragment (PIF) pocket of PDK1, which is an allosteric peptide-docking site that PDK1 uses to recruit its substrates [32].

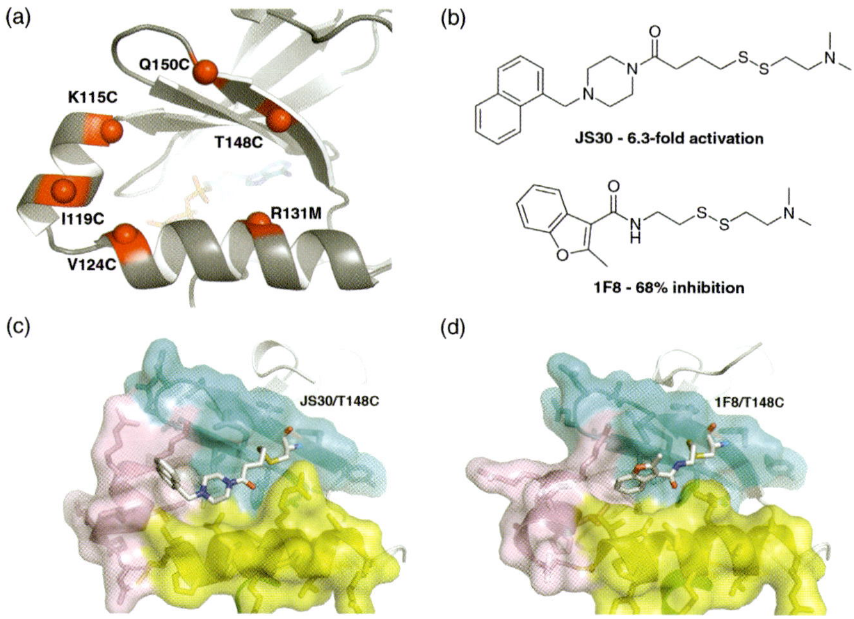


**Figure 11.10** (a) Residues lining the PIF pocket of PDK1 that were mutated to cysteine are shown as red spheres. (b) Chemical structures of the best activator JS30 and the best inhibitor 1F8. (c, d) Crystal structures of PDK1 T148C in complex with the activator JS30 or the inhibitor 1F8. The αB helix, αC helix, and β4/5 strands that form the PIF pocket are colored pink, yellow, and cyan, respectively.

Sadowsky *et al.* used tethering to probe the structure, chemical tractability, and functional plasticity of the PIF pocket of PDK1 [33]. We first mapped the interaction between the PIF pocket and a known peptide ligand PIFtide through peptide disulfide trapping experiments. Six residues lining the PIF pocket were individually mutated to cysteine (Figure 11.10a) and the efficiency of disulfide formation was measured for each mutant across a panel of cysteine-mutant PIFtides. Next, the cross-linking frequencies were converted to distance restraints and were used to create a structural model that suggested how hot spot residues within PIFtide engage the PIF pocket of PDK1. This structural model closely mimics a recent crystal structure of the PDK1-PIFtide complex, validating the disulfide cross-linking approach [34].

Next, we explored whether disulfide fragments could be identified that mimic the PIFtide and what range of effects such fragments would exert on PDK1's catalytic activity. A library of 480 disulfide fragments was screened against each of the 6 cysteine mutants of PDK1 under stringent conditions (10 mM β-ME). The hit rate varied widely from 1.3% for I119C to 13.8% for K115C. Remarkably, these disulfide fragments revealed a large degree of functional plasticity at the PIF pocket: both activators and inhibitors could be found at every cysteine position. The widest range of functional effects was observed for T148C spanning a 19-fold difference in catalytic activity (32–630%; Figure 11.10b). To understand the structural basis of these allosteric effects, we solved crystal structures of PDK1 T148C in complex with activating or inhibiting fragments. The best activator **JS30** (Figure 11.10c) enforced a closed conformation of the αC helix and correctly positioned residues key to catalysis. In contrast, the inhibitory fragment **1F8** held the αC helix in an open conformation in which the ATP binding pocket is unliganded and catalytic residues are out of register (Figure 11.10d). These studies revealed that different ligands binding to the same allosteric site can trap a range of conformations between full on- and off-states. Moreover, this work established a number of chemical starting points for the further development of allosteric PDK1 inhibitors that might overcome the limitations of existing active-site inhibitors.

## 11.6 Tethering Against GPCRs: Complement 5A Receptor

G-protein-coupled receptors are transmembrane receptors that sense a multitude of extracellular signals including hormones, neurotransmitters, odors, and even photons. Given their chemical tractability and central roles in numerous cellular processes, GPCRs are the target of nearly 40% of all approved drugs [35]. Despite the intense focus of the pharmaceutical industry on GPCRs, there is still much we have to learn about this protein family. For example, knowing that two drugs bind to the same site on a GPCR tells us nothing about whether each drug activates or inhibits the receptor. Moreover, we have very limited structural information for the vast majority of GPCRs, due in part to a lack of stabilizing ligands. New technologies to rapidly identify GPCR ligands could provide tool compounds that enable the mechanistic and structural studies needed to address these questions.

In 2005, Buck and colleagues reported two studies that extended the scope of tethering from beyond soluble proteins to targeting a GPCR, the complement 5a receptor (C5aR) [36,37]. C5aR binds to a 74-amino acid protein called C5a, which is a highly inflammatory molecule that mediates the pathology of sepsis and several autoimmune disorders. Medicinal chemistry efforts in the early 1990s showed that 6-mer peptide analogs derived from the C-terminus of C5a can bind to C5aR and either activate or inhibit the receptor [38]. In the first study, the first three amino acids of these 6-mer peptides were removed, which destroyed their ability to bind. However, when the truncated peptides were appended with a cysteine residue they could tether to specific cysteines introduced into the C5aR and signal (Figure 11.11a and b). Much like PDK1 it was possible to activate or inhibit the receptor by conjugating different disulfide ligands to the same cysteine.

In a follow-up study, Buck and Wells used tethering to identify nonpeptidic fragments that agonize and antagonize C5aR. Since mass spectrometry was not suitable to monitor disulfide formation with this transmembrane receptor, the authors instead adapted a radioligand displacement assay to identify fragments

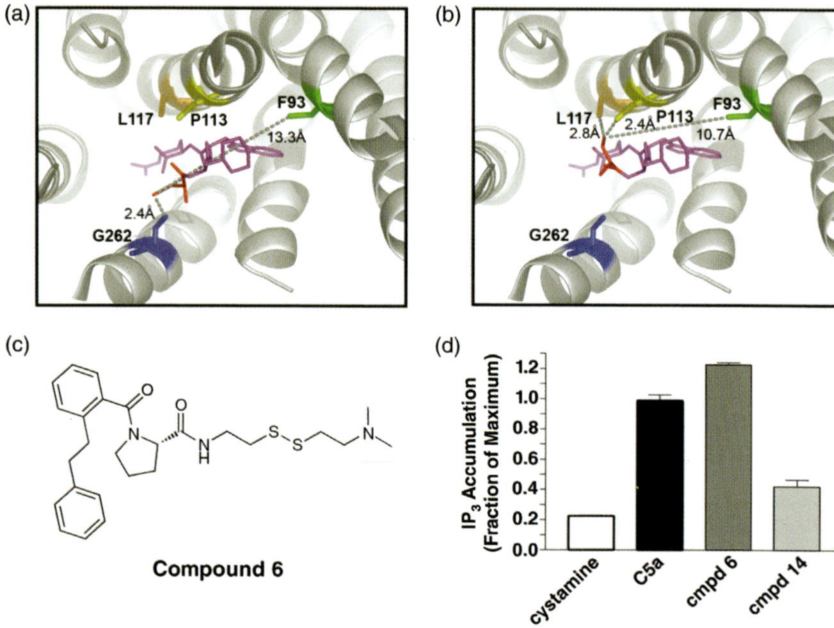


**Figure 11.11** (a, b) Placement of cysteine mutations within the C5aR peptide binding pocket. Thiol-to-thiol distances between C5aR and a peptide ligand are shown in dotted gray lines. F93 is predicted to be >10 Å away from the ideal disulfide distance of 2.4 Å. (c) Chemical structure of agonist compound **6**. (d) Compound **6** activates C5aR G262C more than C5a does, but its enantiomer compound **14** is much less active.

that bind competitively with C5a. To prepare the four C5aR mutants for tethering, the authors expressed each mutant in COS-7 cells and prepared membranes by hypotonic lysis and centrifugation through a sucrose cushion. Next, they screened 10 000 fragments in pools of 10 to identify fragments that diminished the binding of $^{125}$I-C5a by >15%. The hit rates ranged from 0.2 to 0.7% for the three cysteine mutants that were predicted to be within range for disulfide formation. No hits were identified for the control mutation F93C, which was predicted to be 10–15 Å from the ligand binding site (Figure 11.11a and b). Remarkably, the disulfide ligands could label C5aR mutants expressed on the surface of living cells and activate or inhibit C5aR as read out by $IP_3$ accumulation. Importantly, the effects of the disulfide fragments were stereospecific, as compound **14**, the enantiomer of the active compound **6**, had minimal effect on C5aR signaling (Figure 11.11c and d).

Overall, these studies demonstrated that tethering is suitable for rapidly identifying ligands for membrane proteins. This approach could be particularly powerful for generating chemical tools that enable the mechanistic study of the basis of agonism versus antagonism or covalent ligands that stabilize receptors to enable structural characterization. Perhaps most excitingly, this approach could be exploited to explore the function of the nearly 150 "orphan" GPCRs that have no known ligands.

## 11.7
## Conclusions and Future Directions

These studies exemplify the power of tethering to discover ligands to known and orphaned allosteric sites. The caspases show that disulfide fragments can be found for natural cysteines that lock proteins in transient off-conformations even when the protein is predominantly in the on-state. The K-Ras case similarly used a natural cysteine to discover disulfide fragments that bind to a new site not evident in previous structures. The case of KIX shows that engineered cysteines can be used to discover disulfide fragments that stabilize a dynamic protein to enable crystallization and structural analysis. The case of PDK1 shows that tethering against a known allosteric site can reveal a much wider range of possible functional effects than were appreciated from studying natural ligands. Finally, the case of the C5a receptor shows that tethering can target GPCRs and potentially other membrane proteins to identify both agonists and antagonists.

Tethering is amenable to many screening paradigms. Most of the cases described here used mass spectrometry to detect the mixed disulfide between a purified protein and a disulfide fragment. This mass spectrometry approach can be applied to single compounds or pools of 5–10 compounds provided they differ in mass sufficiently to be easily resolved. Importantly, however, mass spectrometry is not required to perform tethering. The cases of KIX and C5aR show how functional assays like fluorescence polarization and radioligand displacement can be used to detect disulfide fragments with the desired functional effect. In an unpublished work, we have also performed tethering screens using enzymatic assays on GTPases and proteases. More broadly, tethering could be combined with any functional or biophysical assay that is compatible with disulfide-containing compounds and β-mercaptoethanol, increasing the accessibility of this powerful technology.

Tethering has unique advantages over other fragment screening approaches that use X-ray, NMR, or SPR. While fragments often bind with high ligand efficiency, their low affinities for the target protein can make it challenging to assay their functional impact or to saturate the binding site for structural analysis. Without knowing the binding mode or functional impact of a fragment, efforts to derivatize the compounds or understand the function of the putative allosteric site are extremely challenging. In contrast, the covalent nature of the disulfide fragments used in tethering allows one to readily generate homogeneous protein–fragment conjugates for structural and functional analysis. While the use of a covalent bond during the tethering screen reduces the degrees of freedom for binding, the fragments discovered by tethering have comparable noncovalent binding affinities to fragments discovered by other means (0.1–5 mM). In addition, it is simple to derivatize disulfide fragments and monitor their SAR by determining the 50% conjugation point in the presence of increasing β-ME (often shifting from 0.5 to >10 mM) or increasing compound concentration (typically 1–50 μM in 0.5 mM β-ME). Although one can readily advance the optimized disulfide fragment in noncovalent mode just like other fragment approaches (IL-2 [39] and caspase-3 [10]), it is also possible to convert the disulfide into a covalent electrophile for intracellular work as was shown for K-Ras.

In addition to its utility as a drug discovery technique for challenging targets, tethering is also a very powerful approach for studying the mechanisms of allostery. In virtually every example, we could trap proteins in active or inactive states and determine their structures. We could also use these "chemi-locked" proteins to develop conformation-specific antibodies that were useful for tracking enzyme activation in cells. Allostery is a fascinating property of protein dynamics and tethering is ideally suited to generate useful fragments for understanding and exploiting it for drug discovery.

## References


1 Kuriyan, J. and Eisenberg, D. (2007) The origin of protein interactions and allostery in colocalization. *Nature*, **450** (7172), 983–990.

2 Wootten, D., Christopoulos, A., and Sexton, P.M. (2013) Emerging paradigms in GPCR allostery: implications for drug discovery. *Nature Reviews Drug Discovery*, **12** (8), 630–644.

3 Fang, Z., Grütter, C., and Rauh, D. (2013) Strategies for the selective regulation of kinases with allosteric modulators: exploiting exclusive structural features. *ACS Chemical Biology*, **8** (1), 58–70.

4 Hardy, J.A. and Wells, J.A. (2004) Searching for new allosteric sites in enzymes. *Current Opinion in Structural Biology*, **14** (6), 706–715.

5 Kozakov, D. *et al.* (2011) Structural conservation of druggable hot spots in protein–protein interfaces. *Proceedings of the National Academy of Sciences of the United States of America*, **108** (33), 13528–13533.

6 Bowman, G.R. and Geissler, P.L. (2012) Equilibrium fluctuations of a single folded protein reveal a multitude of potential cryptic allosteric sites. *Proceedings of the National Academy of Sciences of the United States of America*, **109** (29), 11681–11686.

7 Erlanson, D.A. *et al.* (2000) Site-directed ligand discovery. *Proceedings of the National Academy of Sciences of the United States of America*, **97** (17), 9367–9372.

8 Burlingame, M.A., Tom, C.T., and Renslo, A.R. (2011) Simple one-pot synthesis of disulfide fragments for use in disulfide-exchange screening. *ACS Combinatorial Science*, **13** (3), 205–208.

9 Erlanson, D.A., Wells, J.A., and Braisted, A.C. (2004) Tethering: fragment-based drug discovery. *Annual Review of Biophysics and Biomolecular Structure*, **33**, 199–223.

10 Choong, I.C. *et al.* (2002) Identification of potent and selective small-molecule inhibitors of caspase-3 through the use of extended tethering and structure-based drug design. *Journal of Medicinal Chemistry*, **45** (23), 5005–5022.

11 Yang, W. *et al.* (2009) Fragment-based discovery of nonpeptidic BACE-1 inhibitors using tethering. *Biochemistry*, **48** (21), 4488–4496.

12 Erlanson, D.A. *et al.* (2011) Discovery of a potent and highly selective PDK1 inhibitor via fragment-based drug discovery.

*Bioorganic and Medicinal Chemistry Letters*, **21** (10), 3078–3083.

13 Arkin, M.R. *et al.* (2003) Binding of small molecules to an adaptive protein–protein interface. *Proceedings of the National Academy of Sciences of the United States of America*, **100** (4), 1603–1608.

14 Tait, S.W. and Green, D.R. (2010) Mitochondria and cell death: outer membrane permeabilization and beyond. *Nature Reviews Molecular Cell Biology*, **11** (9), 621–632.

15 Hardy, J.A., Lam, J., Nguyen, J.T., O'Brien, T., and Wells, J.A. (2004) Discovery of an allosteric site in the caspases. *Proceedings of the National Academy of Sciences of the United States of America*, **101** (34), 12461–12466.

16 Hardy, J.A. and Wells, J.A. (2009) Dissecting an allosteric switch in caspase-7 using chemical and mutational probes. *The Journal of Biological Chemistry*, **284** (38), 26063–26069.

17 Cade, C., Swartz, P., MacKenzie, S.H., and Clark, A.C. (2014) Modifying caspase-3 activity by altering allosteric networks. *Biochemistry*, **53** (48), 7582–7595.

18 Scheer, J.M., Romanowski, M.J., and Wells, J.A. (2006) A common allosteric site and mechanism in caspases. *Proceedings of the National Academy of Sciences of the United States of America*, **103** (20), 7595–7600.

19 Datta, D., Scheer, J.M., Romanowski, M.J., and Wells, J.A. (2008) An allosteric circuit in caspase-1. *Journal of Molecular Biology*, **381** (5), 1157–1167.

20 Gao, J. and Wells, J.A. (2012) Identification of specific tethered inhibitors for caspase-5. *Chemical Biology and Drug Design*, **79** (2), 209–215.

21 Gao, J., Sidhu, S.S., and Wells, J.A. (2009) Two-state selection of conformation-specific antibodies. *Proceedings of the National Academy of Sciences of the United States of America*, **106** (9), 3071–3076.

22 Bollag, G. and Zhang, C. (2013) Drug discovery: pocket of opportunity. *Nature*, **503** (7477), 475–476.

23 Baines, A.T., Xu, D., and Der, C.J. (2011) Inhibition of Ras for cancer treatment: the search continues. *Future Medicinal Chemistry*, **3** (14), 1787–1808.

24 Rowinsky, E.K. (2006) Lately, it occurs to me what a long, strange trip it's been for the farnesyltransferase inhibitors. *Journal of Clinical Oncology: Official Journal of the American Society of Clinical Oncology*, **24** (19), 2981–2984.

25 Zimmermann, G. *et al.* (2013) Small molecule inhibition of the KRAS-PDEdelta interaction impairs oncogenic KRAS signalling. *Nature*, **497** (7451), 638–642.

26 Ostrem, J.M., Peters, U., Sos, M.L., Wells, J.A., and Shokat, K.M. (2013) K-Ras (G12C) inhibitors allosterically control GTP affinity and effector interactions. *Nature*, **503** (7477), 548–551.

27 Wang, N. *et al.* (2013) Ordering a dynamic protein via a small-molecule stabilizer. *Journal of the American Chemical Society*, **135** (9), 3363–3366.

28 Wang, N., Lodge, J.M., Fierke, C.A., and Mapp, A.K. (2014) Dissecting allosteric effects of activator–coactivator complexes using a covalent small molecule ligand. *Proceedings of the National Academy of Sciences of the United States of America*, **111** (33), 12061–12066.

29 Lodge, J.M., Rettenmaier, T.J., Wells, J.A., Pomerantz, W.C., and Mapp, A.K. (2014) FP tethering: a screening technique to rapidly identify compounds that disrupt protein–protein interactions. *Medicinal Chemistry Communications*, **5**, 370–375.

30 Pearce, L.R., Komander, D., and Alessi, D.R. (2010) The nuts and bolts of AGC protein kinases. *Nature Reviews Molecular Cell Biology*, **11** (1), 9–22.

31 Medina, J.R. (2013) Selective 3-phosphoinositide-dependent kinase-1 (PDK1) inhibitors: dissecting the function and pharmacology of PDK1. *Journal of Medicinal Chemistry*, **56** (7), 2726–2737.

32 Arencibia, J.M., Pastor-Flores, D., Bauer, A.F., Schulze, J.O., and Biondi, R.M. (2013) AGC protein kinases: From structural mechanism of regulation to allosteric drug development for the treatment of human diseases. *Biochimica et Biophysica Acta*, **1834** (7), 1302–1321.

33 Sadowsky, J.D. *et al.* (2011) Turning a protein kinase on or off from a single allosteric site via disulfide trapping.

*Proceedings of the National Academy of Sciences of the United States of America*, **108** (15), 6056–6061.

34 Rettenmaier, T.J. *et al.* (2014) A small-molecule mimic of a peptide docking motif inhibits the protein kinase PDK1. *Proceedings of the National Academy of Sciences of the United States of America*, **111** (52), 18590–18595.

35 Rask-Andersen, M., Almen, M.S., and Schioth, H.B. (2011) Trends in the exploitation of novel drug targets. *Nature Reviews Drug Discovery*, **10** (8), 579–590.

36 Buck, E., Bourne, H., and Wells, J.A. (2005) Site-specific disulfide capture of agonist and antagonist peptides on the C5a receptor. *The Journal of Biological Chemistry*, **280** (6), 4009–4012.

37 Buck, E. and Wells, J.A. (2005) Disulfide trapping to localize small-molecule agonists and antagonists for a G protein-coupled receptor. *Proceedings of the National Academy of Sciences of the United States of America*, **102** (8), 2719–2724.

38 Konteatis, Z.D. *et al.* (1994) Development of C5a receptor antagonists. Differential loss of functional responses. *Journal of Immunology*, **153** (9), 4200–4205.

39 Braisted, A.C. *et al.* (2003) Discovery of a potent small molecule IL-2 inhibitor through fragment assembly. *Journal of the American Chemical Society*, **125** (13), 3714–3715.

40 Bruschweiler, S., Konrat, R., and Tollinger, M. (2013) Allosteric communication in the KIX domain proceeds through dynamic repacking of the hydrophobic core. *ACS Chemical Biology*, **8** (7), 1600–1610.

# 12
# Fragment Screening in Complex Systems

*Miles Congreve and John A. Christopher*

## 12.1
## Introduction

Fragment-based drug discovery (FBDD) has been a maturing and evolving field since its inception in the late 1990s. At that time, Bemis and Murcko introduced the idea of molecular drug frameworks as a concept for drug discovery and Fesik and Hajduk demonstrated that small low-affinity molecules could be experimentally identified and then optimized (by linking together) in their "SAR by NMR" approach [1,2]. The intuitive simplicity and elegance of this strategy inspired a number of companies pioneering FBDD to spring up shortly afterward, pursuing this new approach using various methods to screen fragment libraries at high concentrations. Notably Astex Technology, Sunesis, Structural Genomics (SGX), and Plexxikon were born. In 2011, the first drug derived from FBDD reached the marketplace, vemurafenib, a B-Raf kinase inhibitor [3]. During the last 15 years, FBDD has been widely adopted within pharmaceutical organizations as a complementary approach to high-throughput screening (HTS) and is very popular in academia, due mainly to the fact that only a small library of compounds is required and so avoiding the need for an extensive HTS infrastructure for screening. Multiple success stories have been published and many molecules are being progressed in clinical trials [4–8]. Some of the significant targets for FBDD are discussed elsewhere in this book.

The many examples of FBDD we see in the literature largely deal with targets that can be relatively easily crystallized in complex with ligands, including weakly active fragment molecules. There is then a clear path forward using structure-based drug design (SBDD) to optimize one or more of the fragment hits – although this is not trivial as often potency needs to progress from mM to nM, that is, over six orders of magnitude, never an easy task. As the field of FBDD continues to evolve, attention is starting to turn to protein targets that are less straightforward to work with. Generally, these proteins are difficult to crystallize and require cocrystallization approaches that are less likely to yield fragment structures than soaking techniques that can be used with more tractable, soluble protein systems. These targets may be membrane associated, multidomain

*Fragment-based Drug Discovery: Lessons and Outlook*, First Edition. Edited by Daniel A. Erlanson and Wolfgang Jahnke.

complexes, conformationally highly flexible, or poorly expressed. This chapter explores the potential of FBDD for targets of this nature and that we are terming "complex systems". The key question that is the subject of this chapter is: with all of the effort and focus on FBDD in the last 15 years, what can we say we have learned about fragment screening that is relevant to difficult or complex protein targets? In addition to addressing the question just posed, there are some associated considerations we will try to take account of. Is FBDD worth considering if there is no X-ray structure available? Without the aid of a clear and precise view of the interactions the fragment is forming and the adjacent pockets that one can target, surely this makes optimization from such a poorly active starting point extremely challenging? When should we consider FBDD and when may it not be considered appropriate? The last aspect we shall try to address is: are there any general learnings from the targets addressed so far by FBDD in terms of the optimization process itself that might be applied to more complex examples or, indeed, is every target different?

## 12.2
## Fragment Screening and Detection of Fragment Hits

Upon contemplating a fragment screening campaign, the first area to focus on is the screening protocol itself and the subsequent screening cascade. Any FBDD project requires a robust primary screening approach, preferably by a direct binding (biophysical) technique. If the assays available have not been applied to screening at high concentrations, have multiple components, or are in a cellular setting, the fragment screening process is likely to fail due to high levels of nonspecific binding and false positives. In addition to the primary screening method, at least one orthogonal assay, preferably another biophysical approach, is required for hit confirmation. If hits cannot be readily confirmed by X-ray crystallography structure solution, it becomes critically important to have approaches that build good confidence in the specificity of the fragment hits being detected. If it is not possible to configure a screening cascade where weakly binding hits can be confidently validated FBDD should not be considered.

The screening concentration and hence the sensitivity of the detection method is critical to the success of a fragment screen. Figure 12.1 plots the relationship for a given molecule between HAC (heavy atom count, the number of nonhydrogen atoms) and affinity ($pIC_{50}$). LE (ligand efficiency) is defined as the binding energy of a molecule for its target divided by its HAC, that is, $LE = \Delta G/HAC$ or $LE = 1.4(-\log IC_{50})/HAC$ [9]. In the figure, the lines of constant $LE = 0.3$ and $LE = 0.6$ have been added. A compound with $LE = 0.3$ is generally considered to have the minimum LE required to be optimized to a drug that obeys Lipinski's rules [10]. In practice, an LE value of 0.6 is infrequently exceeded, and if so is rarely maintained during optimization, and thus is a meaningful upper limit of what might be achieved for highly tractable drug targets. The area between the $LE = 0.3$ and 0.6 lines is therefore a useful one to consider as "ligand efficient

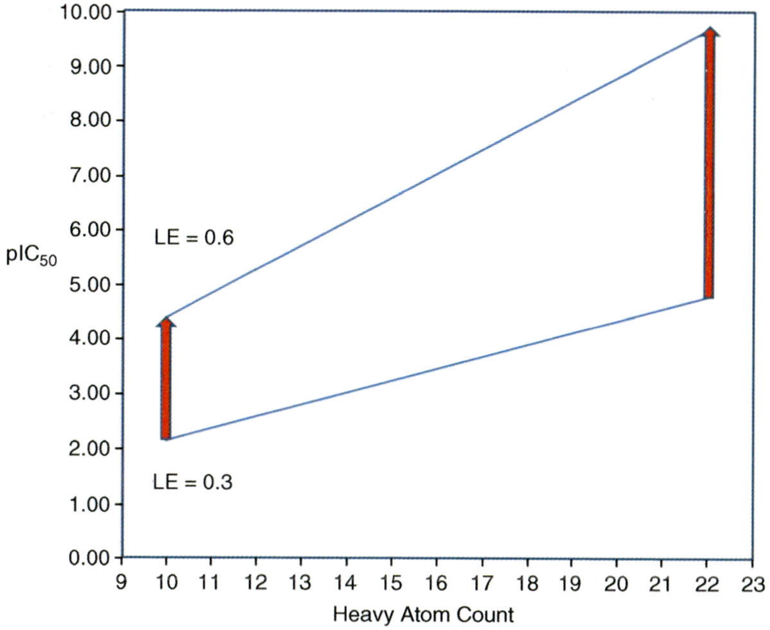


**Figure 12.1** The relationship between heavy atom count (number of non-H atoms) and affinity ($pIC_{50}$) defines an area of protein–ligand interaction space where ligand-efficient fragments and lead compounds will generally reside.

space" for fragments and lead compounds. Notice that the two lines of constant LE 0.3 and 0.6 are not parallel, that is, affinity varies less for LE = 0.3 molecules as they are optimized than it does for compounds with LE = 0.6. The use of LE has recently been both questioned and subsequently defended in the literature [11–14]. It is the view of these authors that LE is a useful guide when considering which hits to optimize and when tracking the optimization process itself, particularly in the earlier stages of a drug discovery project. The progress of different hit series can be, for example, plotted on the graph shown in Figure 12.1 and this may be a useful way of considering how well a series is performing. However, it is also our view that the derived efficiency metrics should not be used as strict criteria within medicinal chemistry projects (and that they were never intended to be used in this way).

Another application of this analysis is when considering which molecules to screen and at what concentration. It is important to understand that the tractability of a given drug discovery target should be a driver of the choice of screening method used and the nature, size, and composition of the library being screened. If there are preexisting ligands for the target of interest, the LE values can be calculated and a preliminary assessment of the tractability of the target to a small molecule approach can be considered. Hypothetically, we can consider three scenarios with reference to Figure 12.1.

1) Hits and leads known with LE values in the range 0.5 or higher. The target is highly tractable. Screening even small fragments at 100 μM is likely to allow detection of hits, and larger fragments should be detectable at concentrations often used in HTS campaigns.
2) Hits and leads known with LE values of around LE = 0.4. Good target tractability. Small fragments will be detectable at mM concentrations and larger fragments at 100 μM.
3) Hits and leads known with LE values of around LE = 0.3. Acceptable tractability. Very small fragments will not be detectable by all but the most sensitive methods (NMR, crystal soaking, and X-ray structure determination). Larger fragments should be detectable at mM concentrations.

This simple analysis helps to guide us to the appropriate screening technology for the target in question and also may suggest if a fragment library is, therefore, appropriate, for example, if highly sensitive methods of fragment detection are not available. It is perhaps useful to now consider some examples of different screening systems for more challenging targets to see how the field is developing in practice.

In a number of cases below, results from our own laboratories at Heptares Therapeutics are described for G protein-coupled receptor (GPCR) targets. Adaptation of biophysical and some other techniques to membrane proteins is hampered by the poor expression levels and low stability of these proteins. For biophysical methods, membrane proteins must also be solubilized away from the cell membrane while maintaining their functional state, which is also often challenging. Our approach to overcome these limitations is to introduce a small number of mutations in positions that do not affect the binding site characteristics of the receptor but dramatically increase thermostability and therefore the yield of folded pure protein in solution with detergent. These modified GPCRs are known as StaR® proteins [15].

### 12.2.1 Fragment Screening Using NMR Techniques

NMR screening has been the most widely used biophysical method in more complex protein systems to date. Target-immobilized NMR screening (TINS) is a technique that has been used to detect weak interactions between proteins and small molecules and until recently had typically been used with soluble proteins such as kinases and proteases [16]. In the first demonstration of GPCR fragment screening by NMR, the TINS methodology was applied to a thermostabilized (StaR®) version of $\beta_1$AR, detergent solubilized in decyl maltoside (DM) and immobilized onto resin [17]. A small library of 579 fragments was screened, yielding a number of weak hits, a subset of which was confirmed to bind to the receptor in an orthogonal radioligand binding assay. The TINS technique has also been applied to fragment screening of the *Escherichia coli* inner membrane protein DsbB [18]. Two solubilization approaches were compared, first in the detergent DPC (*n*-dodecyl phosphocholine) and second in nanodiscs, which consist of a lipid bilayer

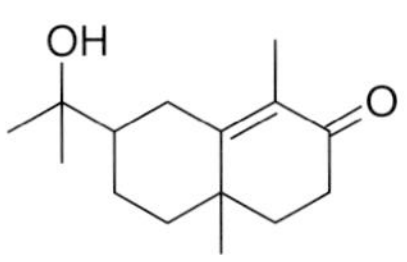

**1** $IC_{50}$ 7 μM (LE 0.41)

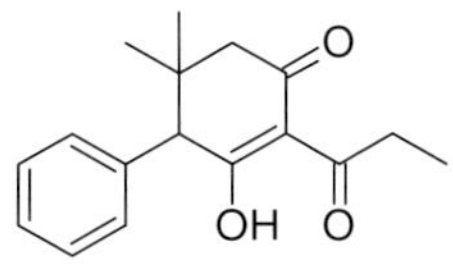

**2** $IC_{50}$ 100 μM (LE 0.27)

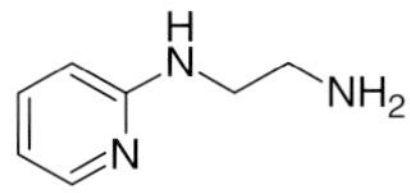

**3** $K_D$ 70 μM (LE 0.57)

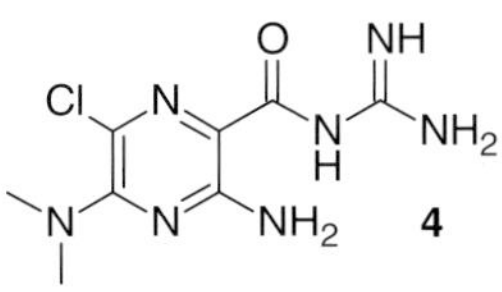

**4**

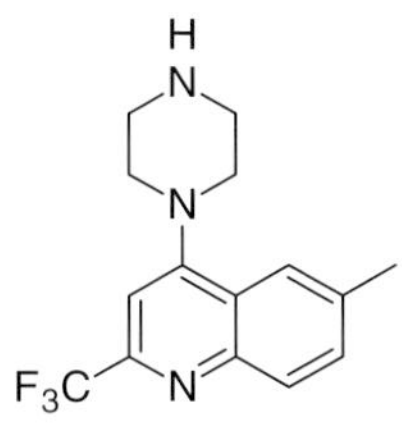

**5** $\beta_2$AR p$K_D$ = 7.8 (LE 0.51)

**Figure 12.2** Hits from NMR screening of *E. coli* inner membrane protein DsbB (**1** and **2**), *E. coli* A-site RNA (**3**) and HBV ε RNA (**4**), and from SPR screening of $\beta_2$AR (**5**).

surrounded by an amphiphilic α-helical membrane scaffold protein. Nanodiscs are better mimics of the native cell membrane environment than detergent micelles, and in the TINS fragment screen of DsbB, the nanodisc approach yielded a better signal-to-noise ratio than detergent solubilization. After hit validation in an enzymatic assay, the eight highest affinity fragment hits, including **1** and **2**, Figure 12.2, were further characterized by 2D NMR experiments, which indicated that fragments with differing modes of action had been identified.

NMR screening has also been applied to the fragment discovery of binders to RNA, an area that has been recently reviewed [19]. An early example of NMR-based screening of *E. coli* A-site RNA from Fesik and colleagues was published in 2003 and described the discovery of fragment-sized, submillimolar affinity, benzimidazole, 2-aminoquinoline, and 2-aminopyridine (e.g., **3**, Figure 12.2) hits [20]. Hits were identified using a simple target-observed protocol where changes to the 1D NMR spectrum of the RNA in the imino proton region of RNA are characteristic of ligand binding. More recently, $^{15}$N-labeling, which aids in defining the footprint of ligand binding, has been used in a fragment screen of the encapsidation signal (ε) portion of RNA from the hepatitis B virus (HBV) [19]. A screen using TROSY (transverse relaxation-optimized spectroscopy) experiments identified dimethylamiloride **4** (affinity data not reported) as a binder to the ε stem region of $^{15}$N-labeled HBV ε RNA. RNA screening has also been undertaken using techniques based on mass spectroscopy, as detailed later in Section 12.4.2 on fragment linking [21].

### 12.2.2 Fragment Screening Using Surface Plasmon Resonance

In a similar way to NMR approaches, surface plasmon resonance (SPR) screening of membrane proteins is often accompanied by the twin challenges of low expression and poor stability of the target receptor when isolated from the cell membrane. GPCR SPR screening was demonstrated for the first time in 2011 by immobilization of an $A_{2A}$ StaR® protein, C-terminally tagged with a His-10 sequence, on to a nickel (Ni-NTA) chip [17,22]. Once immobilized, the captured GPCR displayed sufficient stability to be used for proof-of-concept fragment screening and kinetic analysis of the binding characteristics of several known $A_{2A}$ antagonists. Following this successful demonstration, SPR screening has been used for several further GPCR fragment hit identification campaigns, including the $\beta_1$-adrenergic receptor ($\beta_1$AR) StaR® which is detailed further in Section 12.5.1. The prior thermostabilization of the GPCR target before SPR capture and screening is often necessary to provide a biosensor surface suitably stable for the demands of screening significant numbers of compounds. Thermostabilization also results in a higher percentage of correctly folded protein on the chip, particularly for less stable GPCRs, yielding higher sensitivity. However, exceptions to this are reports of the capture of wild-type chemokine receptor CCR5 and the subsequent detection of binding of probe molecules, described by Navratilova, Besnard, and Hopkins [23] and subsequently extended to screening of a set of 656 fragments against the immobilized wild-type $\beta_2$AR [24,25]. In the second study, five hits (p$K_D$ 4.7–7.8), including quinoline **5** that was the highest affinity (Figure 12.2), were identified from a single concentration screen and subsequently confirmed in dose–response experiments. It is worth noting that the use of SPR for membrane proteins is nontrivial and requires experience and expertise to carry out successfully. There are a number of issues related to nonspecific binding, nonstoichiometric binding, and protein leaching from the chip. Importantly, many hit compounds will exhibit some components of nonspecific binding as well as their specific interaction with a druggable binding site and can be easily missed if the appropriate control experiments are not carried out or if the fragment screening is carried out at too high a concentration (when nonspecific binding will dominate, often at concentrations above 50 μM). Additionally, it is critical to hone assay parameters such as temperature and flow rate to optimize specific versus nonspecific binding. As a community of scientists working on FBDD, sharing of best practice is very important and helps researchers develop the required know-how and confidence in using a particular technique. SPR fragment screening is certainly one such case where sharing of experiences is very helpful to maximize the utility of the approach for a broader set of targets.

### 12.2.3
### Fragment Screening Using Capillary Electrophoresis

Capillary electrophoresis (CE) is a sensitive biophysical screening technique that has been used with several soluble protein targets, for example, heat shock protein 90, to identify low affinity, low MW ligands in competition assays with a UV active or fluorescent probe ligand [26]. The approach has been validated for use in fragment screening with stabilized GPCRs, where among other hits, the fragment-sized antagonist caffeine could be detected specifically binding to a stabilized form (StaR®) of the adenosine $A_{2A}$ receptor [27]. This again shows that if the GPCR protein can be produced in a well-behaved form in solution it can be interrogated in a similar way to soluble proteins, suggesting protein quality is just as important as the nature of the drug target itself. Unlike NMR and SPR fragment screening techniques, the CE approach will only detect fragments binding at the same site as the probe ligand, which has advantages and disadvantages (see Section 12.3).

### 12.2.4
### Fragment Screening Using Radioligand and Fluorescence-Based Binding Assays

The dogma has been that with more challenging proteins biochemical screening at high concentration is fraught with difficulty due to the propensity for detection of false positives, dissuading the use of this approach in favor of direct binding biophysical techniques such as the ones described above. However, competition binding with radioligands is a well-established screening paradigm for a broad range of proteins in HTS and targeted screening campaigns, and the approach is now beginning to be used, for example, for GPCR fragment hit identification [28,29]. de Esch and coworkers have screened HEK293 cell membranes expressing the histamine $H_4$ receptor in a radioligand binding displacement assay with [$^3$H]histamine, where a fragment library of 1010 molecules screened at 10 μM yielded 56 hits (6%) with the ability to displace >50% of the radioligand from the receptor [30]. Pyridine **6** was the highest affinity, selective, hit identified (Figure 12.3). Additional radioligand binding fragment screening efforts directed at the $\beta_2$AR, histamine $H_1$, and histamine $H_3$ receptors have also been described by the same authors [31] and representative hit molecules **7–9** are shown in Figure 12.3 (affinity data were not reported). The histamine $H_1$ and histamine $H_3$ screening examples elegantly combined the use of radioligand binding as the screening methodology with the use of virtual fragment screening based on available $H_1$ X-ray crystallographic information [32,33]. A similar approach has been recently published by Keserű and coworkers where virtual fragment screening, using an X-ray structure and homology model of the dopamine $D_3$ and histamine $H_4$ receptors, respectively, was complemented by docking into multiple protein conformations (ensemble docking) of each receptor [34]. From a library of 12 905 fragments, approximately 50 compounds from both methodologies for each receptor were progressed to radioligand binding assays, yielding 16–32% hit rates (defined as >20% inhibition at 10 μM). The highest affinity examples from each screen (**10**, **11**) are shown in Figure 12.3. In the above cases, the targets are actually known to be relatively tractable, which is typically reflected in high ligand efficiencies of the identified molecules and the low screening concentrations used, minimizing the risk of detecting high numbers of nonspecific hits (see Figure 12.1 and the earlier discussion). By contrast, an AstraZeneca high-concentration screen of the melanocortin 4 receptor (MC4), a challenging GPCR target that to date has largely yielded large and

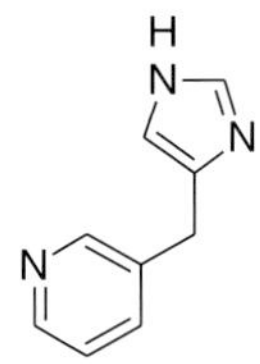

**6** $H_4$ p$K_i$ = 7.0

(LE 0.80)

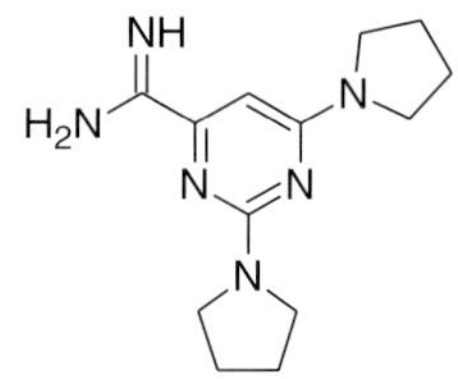

**7** $\beta_2$ AR hit

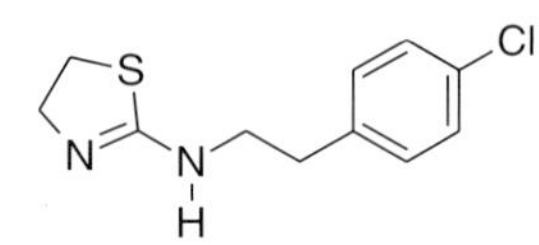

**8** $H_1$ hit

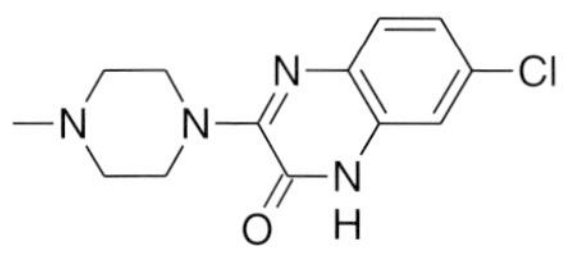

**9** $H_3$ hit

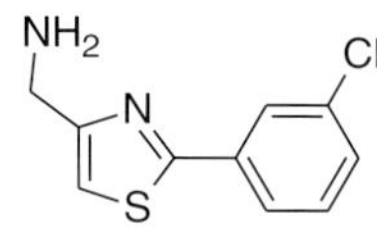

**10** $D_3$ p$K_i$ = 6.8

(LE 0.66)

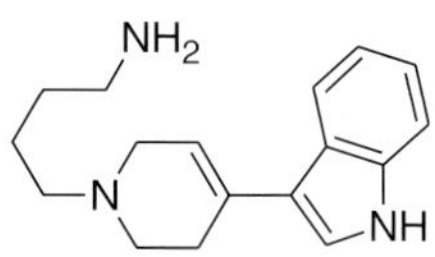

**11** $H_4$ p$K_i$ = 5.1

(LE 0.35)

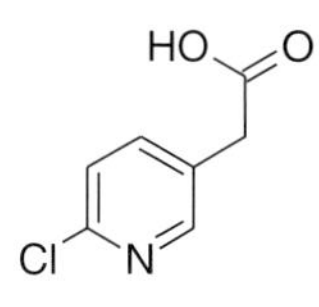

**12** MC4 p$IC_{50}$ = 4.9

(LE 0.45)

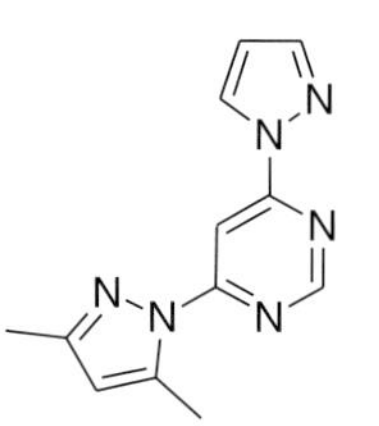

**13** mGlu$_5$ p$K_i$ = 5.2

(LE 0.39)

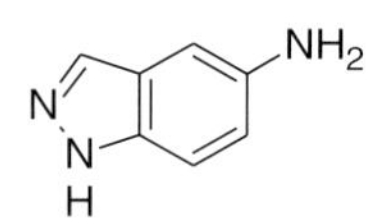

**14** $A_3$ p$K_i$ = 5.5

(LE 0.75)

**Figure 12.3** GPCR fragment hits from radioligand and fluorescence-based binding assays.

lipophilic molecules [35], identified fragments that typically had low to moderate ligand efficiency (LE 0.2–0.3). An exception was the acidic fragment **12** that has very high LE (Figure 12.3) [36].

In our laboratories, we have used StaR® proteins (described earlier) in several high-concentration radioligand binding fragment screening campaigns, where the use of the thermostabilized GPCR constructs has a number of distinct advantages over wild-type GPCRs. First, the stabilized proteins tolerate higher concentrations of DMSO in the assay buffer, enabling screening of fragment collections at high concentrations while ensuring the potential ligands remain in solution. Second, it has been frequently observed that the expression of StaR® proteins is higher than wild-type, making it easier to secure sufficient reagent stocks for screening campaigns. Finally, a specific conformation is usually selected during the stabilization process [15], allowing identification of fragments that show biased binding toward agonists or antagonists. This is particularly relevant to detection of agonists as fragments are perhaps more likely to bind to the more stable antagonist conformation rather than the higher energy agonist conformation when screening wild-type protein [37]. In a recent example of GPCR radioligand binding fragment screening, metabotropic glutamate receptor 5 ($mGlu_5$) was thermostabilized and assayed against a general fragment screening collection and also a bespoke class C focused fragment set that had been assembled for hit identification campaigns against this important family of GPCRs. The screen targeted negative allosteric modulators (NAMs) of the receptor. Both portions of the screening effort yielded tractable hits (e.g., pyrimidine **13**, Figure 12.3) with a combined hit rate of 6% (>30% radioligand displacement at 300 μM) and several series were progressed, initially by synthetic enumeration and purchase of commercial near-neighbor analogs (see Section 12.6).

In contrast to using membrane preparations for high-concentration radioligand binding assays, or stabilized, isolated GPCRs in detergent solubilized biophysical techniques, Hill and coworkers have used a fluorescence-based binding assay in living cells for fragment screening of the adenosine $A_3$ receptor [38]. The technique centers on a high-affinity fluorescent $A_3$ antagonist with a slow off-rate at the receptor, derived from the known adenosine receptor xanthine amine congener (XAC). A high-content screening procedure using a confocal imaging plate reader to detect competitive binding was validated with a range of known literature antagonists, which returned affinities in good accordance with their reported functional potencies. Subsequent extension to screening of 248 commercial fragment molecules at 1 mM yielded 38 hits, with follow-up dose–response screening yielding compounds with affinities ranging from $pK_i$ 4.0 to $pK_i$ 6.4 (**14**, Figure 12.3). A potential advantage of this whole cell fluorescence approach to fragment screening is that, in contrast to biophysical assays using purified detergent-solubilized receptors or biochemical membrane-based approaches, the cell membrane is maintained under physiological condition. Additionally, intracellular signaling proteins will be present which can allosterically influence the binding of molecules to GPCRs [39]. This striking example challenges the thinking that cell-based assays and fragment-based screening do not mix, suggesting there may be potential for other screening paradigms in which competition of a fragment for a selective ligand could be used in a cell-based format.

### 12.2.5
### Ion Channel Fragment Screening

Ion channels are a highly important class of membrane associated targets and are pore-forming proteins that allow the flow of ions across plasma membranes or membranes of intracellular organelles. Taken together, ligand-gated and voltage-gated ion channels represent the second largest protein target class for existing drug molecules after GPCRs [40,41]. In line with their importance, and the advent of new, faster, screening techniques for ion channel hit identification such as high-throughput electrophysiology [42], fragment screening techniques for ion channels are beginning to emerge. Wolkenberg and coworkers at Merck reported the first example of high-concentration ion-channel patch clamp screening in 2011, screening a set of 1280 compounds against acid-sensing ion channel 3 (ASIC3) at 1 mM (see also Sections 12.3 and 12.4.1). Lummis and colleagues have described an information-rich biochemical method for fragment screening that uses a voltage-sensitive dye to probe the function of the ion-channel, exemplified by screening the ligand-gated target 5-$HT_3$ [43]. After dye-loading, cells are loaded into a fluorescence plate reader, baseline fluorescence is read, followed by addition of fragment test compound and changes in fluorescence are monitored. Analysis of the fluorescence change due to test compound compared with the response due to 5-HT allowed hits to be categorized as agonists or antagonists. From a screen of 1116 fragments, 72 hits were identified and confirmed by radioligand–ligand binding competition experiments with the 5-$HT_3$ antagonist [$^3$H]granisetron, and at least two hit series have been optimized further to high-affinity analogs (see Section 12.5.1). This new direction in fragment screening into the field of ion channels could become an important application of FBDD; ion channels are modulated by very small ligands with high LE in many cases making them potentially ideal targets for fragment libraries. The limitation, until very recently, was the robustness and throughput of the assays to study this important class of targets; however, these recent reports suggest that these issues may soon be resolved.

## 12.3 Validating Fragment Hits

In the previous section, the thought processes and technical considerations required to commit to a fragment screen for a protein target were discussed and some practical examples and results were presented. Here, it is assumed that a fragment screen has been carried out and that a number of fragment hits have been identified. To validate one or more fragment hits requires not only the confirmation of their binding in at least one orthogonal assay but also evidence that the hit can be optimized by synthesis and testing of analogs. To drive this early chemical optimization of fragment hits, SBDD should ideally be employed and the crystal structure of the target will have already been solved. Unfortunately, for many complex systems and particularly multidomain or membrane-associated targets, this is not the case.

If no crystal structure exists for the target protein, there may still be options for rational optimization of fragment hits. If there is at least one structure known of a protein in the same family as the target, then it may be possible to construct a homology model. For many years, for example, the structure of bacteriorhodopsin was used as a template for homology modeling for GPCRs to attempt to identify the binding site of ligands using site-directed mutagenesis (SDM), with some success [44,45]. With a suitable homology model, SBDD can then be attempted, developing ideas for compound optimization using the putative binding pose and refining the model as data emerges. In an analogous manner, if known ligands can be docked into the homology model and overlaid with each fragment hit, this can quickly generate structure-based hypotheses for new target compounds. Once synthesized, if active, these hits will help confirm the proposed binding mode of the chemical series and improve confidence in the homology model. If unsuccessful, other poses for the fragment hits must be considered and further iterations attempted, or the optimization abandoned.

Complementary ligand-based approaches may also be possible if sufficient structure–activity relationship (SAR) knowledge is available for one or more chemical series [45,46]. If a ligand-based pharmacophore is known or can be generated, then fragment hits can be considered in this context and ideas for their optimization developed. Overall, data from homology modeling and/or a pharmacophore can be extremely valuable to give confidence in early fragment hits and drive their optimization. In particular, if there are common motifs in known ligands, it may be possible to incorporate these moieties into a number of the fragments and rapidly identify more active hit compounds by making "hybrid" target molecules.

In parallel with the development of structure- and ligand-based hypotheses for optimization, it is often fruitful or at least informative to empirically purchase a selection of near neighbours of each fragment hit in order to generate simple SAR. This can help to prioritize which fragments to commit to further chemistry efforts, especially if a more potent hit can be readily identified. It is important to demonstrate, as early as possible, that one or more very simple analogs, for example, containing halogen substitution, retain activity. Alternatively, if the original fragment hit is a singleton after screening of appropriate close analogs, it may be prudent to stop further efforts.

Two examples of hit validation, relevant to our discussion of complex protein systems, are outlined here.

An intriguing use of *in silico* methods to validate and prioritize the output from a biophysical GPCR fragment screen has been recently reported in the context of TINS NMR screening of the adenosine $A_{2A}$ receptor [47]. NMR screening of a set of 500 fragments against thermostabilized $A_{2A}$ [48] was complemented by parallel prospective docking of the same molecules into an X-ray crystal structure of the receptor, ranking the fragments by their predicted affinities. The NMR screen yielded 94 primary hits that were triaged using a radioligand binding assay, resulting in 5 orthosteric compounds with well-behaved dose–response curves, including the highest affinity and LE hit **15** (Figure 12.4), which in

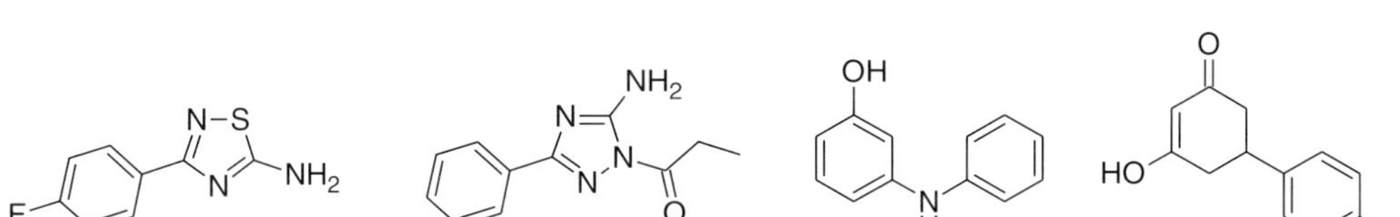


**15** $A_{2A}$ $K_i$ 14 μM (LE 0.53) **16** $A_{2A}$ $K_i$ 6.4 μM (LE 0.46) **17** ($A_{2A}$ NAM) **18** ($A_{2A}$ PAM)

**Figure 12.4** $A_{2A}$ orthosteric TINS NMR (**15**) and *in silico* screening (**16**) hits; negative (**17**) and positive (**18**) allosteric modulators of $A_{2A}$.

common with the other hits was unselective against the related $A_1$ receptor. The parallel docking approach was able to identify four of these five ligands within the top 5% of the ranked structures, suggesting that the parallel use of biophysical and *in silico* approaches, at least for a system such as $A_{2A}$ where X-ray structural information is available, is a viable way to improve the efficiency of fragment screening and aid in the prioritization and validation of primary hits. In an extension to the docking approach, 328 000 commercially available fragments were docked into an $A_{2A}$ X-ray structure and ranked, with 22 of the top scoring molecules selected for radioligand binding assay. Fourteen hits were confirmed in this screen, with affinities in the range of 2–240 μM, of which **16** is one of the highest efficiency examples. Evaluation of primary hits from the NMR screen also resulted in the identification of allosteric binders to $A_{2A}$, characterized by their differential effects on the dissociation rates of the compounds of $A_1$ and $A_{2A}$ radioligands [48]. Three hits significantly decreased the off-rate of an orthosteric radioligand for each receptor and four significantly increased off-rate, suggesting that both positive and negative allosteric modulators (PAMs and NAMs) had been identified, examples of each are shown in Figure 12.4 (**17**, **18**. Affinity data were not reported). Unlike competition experiments, the TINS NMR, and indeed SPR, approaches to fragment screening in complex systems are direct binding methods, meaning that hit fragments could be binding anywhere on the protein, not just on the same site as a labeled probe such as a radioligand. As in the example above, this opens up the possibility for hit discovery at allosteric sites, but it must be acknowledged that in the absence of structural insights or prior detailed knowledge of allosteric site SAR trends, it will be highly challenging to progress compounds in anything but an empirical manner and this may be an arduous and frustrating task.

In an application of near-neighbor screening workers at Merck applied the approach to 12 structurally distinct fragment hit series that had been identified from the high-concentration ASIC3 screen detailed in Section 12.2 [49]. Similarity searching of internal and external collections, capped in each case by the molecular weight of the parent hits, was used to validate the hits, and in several cases this was achieved with concomitant increases in potency and LE, for example, **19** to **20**, Figure 12.5. This example clearly illustrates how confidence can be quickly gained in the quality of a hit and of how progressable it is before a significant synthetic chemistry commitment is made.

**19** ASIC3 $IC_{50}$ 763 μM (LE 0.33)

**20** ASIC3 $IC_{50}$ 130 μM (LE 0.41)

**Figure 12.5** Elaboration of ASIC3 hit **19** to **20** through near-neighbor screening.

## 12.4
## Fragment to Hit

Development of a fragment to a hit series with lead-like size and complexity is a key part of the FBDD process. Indeed, it might be argued if this phase is successful then the remaining optimization to a drug can be seen as conventional medicinal chemistry. As we consider hit identification for more complex protein targets, using only models of the molecule's binding mode to the target, we must be prepared to spend longer on this step than if X-ray structures were available. As discussed in the previous section, the SAR that will have been mapped out so far may be largely empirical, even if homology or pharmacophore models have been constructed and are being tested.

The majority of successful FBDD campaigns described in the literature use a strategy called fragment evolution [50–52]. Typically, a fragment hit will be built up atom by atom to better fit the binding site described by its ligand–protein X-ray structure. This works particularly well where the drug binding site is relatively small and not largely exposed to bulk solvent. Reasonable confidence in the design of new target molecules for synthesis can be gained by consideration of the molecule's binding mode, justifying even quite complex synthetic chemistry efforts. However, in a scenario where there are several hypotheses of how the molecule may be binding, based on homology models (perhaps supported by site-directed mutagenesis), it may not be prudent to design targets that require multiple synthetic steps. A better approach may be to investigate several vectors around the fragment with simple substitutions in order to establish activity and synthetic tractability and then to explore a broader range of modifications at the most promising positions. These targets will still be informed by the molecular modeling, but will be designed both to explore the structure-guided SAR and to test the accuracy and value of the model itself.

If the binding site of the target is predicted to be larger or more solvent exposed, other approaches may need to be considered in addition to attempting fragment evolution. In this scenario, simple analogs of the initial fragment hits may not have increased affinity and the SAR may appear rather "flat." It is then helpful to consider if the fragment hits are actually an expression of only a small part of a sophisticated pharmacophore for the binding site, which requires a much larger molecule for nanomolar potency in order to satisfy multiple "hot spots" for ligand binding. To test these ideas, much larger substituent changes might be considered to attempt to fill the entire site. Alternatively, fragment linking, another well-precedented fragment optimization strategy, could be attempted [50–52]. The fragment hits identified could potentially be linked together or hybridized, testing if they occupy different parts of a larger site, and one or more novel fragments could be hybridized with larger and much more potent ligands described in the literature for the target. Overall, it is important to create and test multiple models of how the fragments and any known larger ligands are binding to the target and test each hypothesis with several compound design ideas. This is not easy and may be time and resource intensive. At this stage, building confidence in the ability to progress one or more fragments to a hit series and in the usefulness of the computational models being used is important. Considerations of the physicochemical properties, lipophilicity, and other drug-like parameters might be seen as less important in the short-term. However, it is always helpful to keep in mind that the next stage of the process is to produce a high-quality lead series and that we should not "lose control" of the molecular properties of our hits.

### 12.4.1
### Fragment Evolution

An initial fragment is optimized by adding functionality to bind to adjacent regions of the active site.

A demonstration of the fragment evolution technique is provided by the ASIC3 example from Section 12.3. After near-neighbor screening had been completed around the initial fragments, a synthetic strategy of fragment growing was employed, which in the absence of structural or homology information was purely empirical [49]. One aspect of the strategy comprehensively scanned all synthetically substitutable vectors, sampling various geometries with $sp^3$, $sp^2$, and sp-hybridized substituents and employing C, N, and O linker atoms. The second approach used an iterative library technique to synthesize 475 analogs, which enabled progression of tetrahydroisoquinoline **21** (Figure 12.6) to the less ligand-efficient, but higher potency, potential tool compound **22**. Another fragment evolution example is provided by a group from AstraZeneca who identified moderately efficient piperazine **23** as an inhibitor of the T-type ion channel $Ca_v3.2$ (Figure 12.6) [53]. Encouraged by the presence of SAR from near-neighbor fragments and with the opportunity to hybridize **23** with literature molecules, initial optimization yielded micromolar hit **24** after the synthesis of

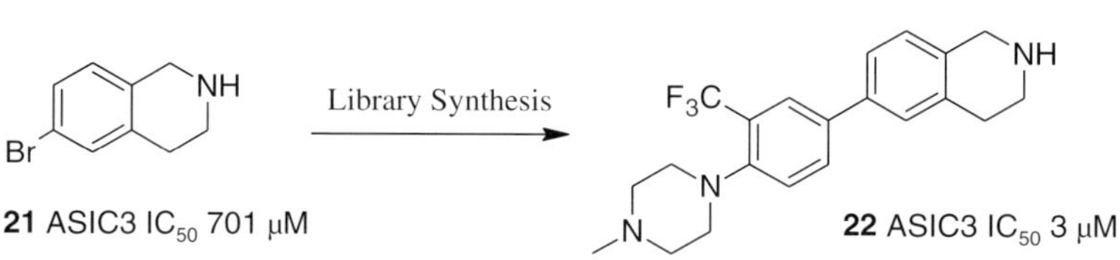


**21** ASIC3 $IC_{50}$ 701 μM (LE 0.39)

**22** ASIC3 $IC_{50}$ 3 μM (LE 0.28)

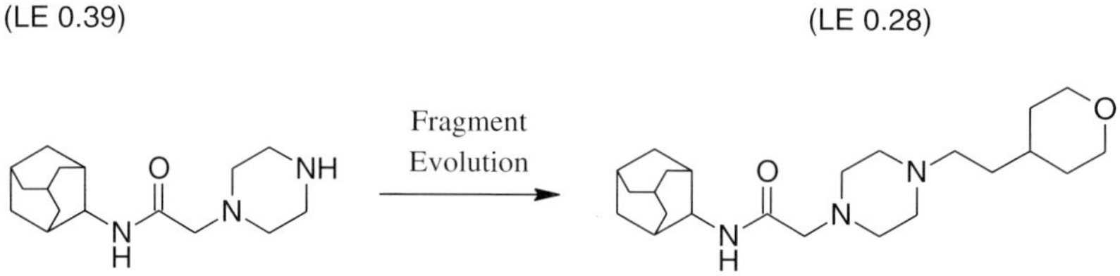


**23** $Ca_v3.2$ $IC_{50}$ 70 μM (LE 0.28)

**24** $Ca_v3.2$ $IC_{50}$ 2 μM (LE 0.28)

**Figure 12.6** ASIC3 and $Ca_v3.2$ fragment to hit evolution examples.

approximately 30 analogs, which maintained the ligand efficiency of the starting fragment and had suitable rat PK parameters to be used in an *in vivo* PK/PD model. It should be noted that in both of the above examples the evolution process is largely empirical and not guided by homology model or X-ray crystallographic information. While this can be successful in a complex system, especially when SAR information from near neighbors and alternative scaffolds can be incorporated, it should be acknowledged that the challenges of fragment progression are higher than if a structure-based strategy can be employed.

### 12.4.2
### Fragment Linking

Two (or more) fragments, which bind to proximal parts of the active site, are joined together to give a larger, higher affinity binding molecule.

Swayze *et al.* have demonstrated the fragment linking approach, using a mass spectrometry screen to identify ligands for the 1061 region of bacterial 23S rRNA [21]. A D-amino acid class of hits including **25** was identified, where the positive charge was crucial for RNA binding, together with quinoxalin-2,3-dione **26**, Figure 12.7. MS competition experiments between **25** and the carboxylic acid derivative of **26** supported the hypothesis that the different chemotypes bound to distinct regions of the target RNA. Distance constraints suggesting that the furan portion of **25** needed to be approximately three atoms away from the carboxyl function of the **26** series were derived and subsequently linked compounds such as **27** were synthesized with significantly higher affinities than the parent hits ($K_D > 100\,\mu M$).

The second RNA fragment linking example is provided by Tisné and coworkers who used an NMR fragment screen to identify binders to $tRNA_3^{Lys}$ [19]. In the primary screen, perturbations to the tRNA imino protons in 1D NMR experiments were monitored, allowing the identification of diaminocyclopentanol **28** and aniline **29** (Figure 12.7) which both bound with affinities estimated to be in the millimolar range. TROSY NMR experiments using $^{15}$N-labeled RNA indicated that the molecules bound to adjacent sites on the target with a degree of overlap, providing confidence to pursue a fragment linking strategy. Several families of compounds were prepared with different linkers, and examples of triazole- or triazine-based compounds, **30** and **31**, respectively, demonstrated significantly enhanced affinities in the micromolar region.

## 12.5
## Fragment to Lead Approaches

In this section, we will examine some further examples of optimization of fragment hits, building on the previous discussion. Here, we will focus on examples that have given lead compounds (rather than hits as above) and help to demonstrate very clearly that fragment-based approaches can be applied successfully

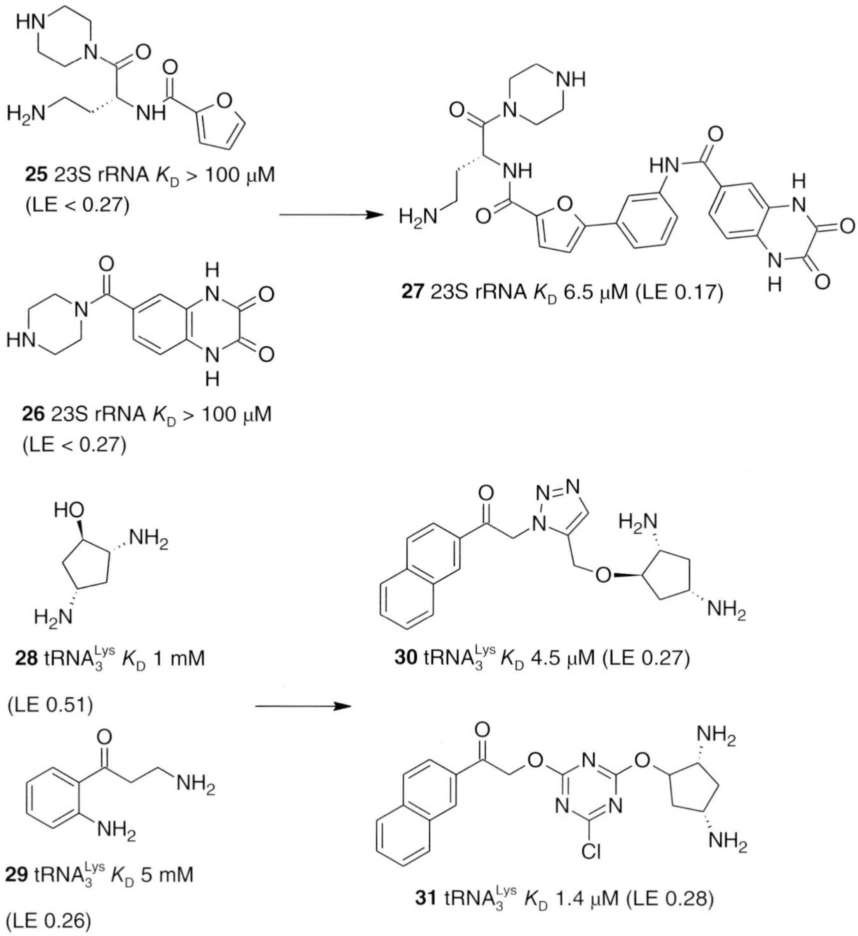


**Figure 12.7** Fragment linking of hits for bacterial 23S rRNA and $tRNA_3^{Lys}$.

even for targets where the protein is difficult to work with and SBDD is challenging in practice. The two main optimization strategies employed, as before, are fragment evolution and fragment linking and the examples are categorized in this way.

### 12.5.1
### Fragment Evolution

An initial fragment is optimized by adding functionality to bind to adjacent regions of the active site.

In our laboratories, we have optimized moderate affinity $\beta_1$-adrenergic receptor ($\beta_1$AR) fragments to lead molecules using structure-based design techniques [54]. Using an SPR screen with thermostabilized $\beta_1$AR, fragment hits

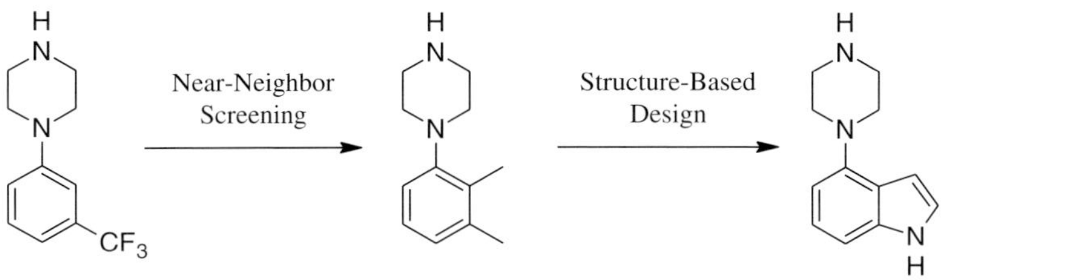


**32** $\beta_1$AR p$K_D$ 4.8 (LE 0.41)

**33** $\beta_1$AR p$K_i$ 6.3 (LE 0.61)

**34** $\beta_1$AR p$K_i$ 7.2 (LE 0.65)

**Figure 12.8** $\beta_1$AR structure-based fragment to hit to lead progression.

including **32** (Figure 12.8) were identified (Section 12.2.1), and their activity confirmed in an orthogonal radioligand binding assay. Guided by docking of molecules into a crystal structure of $\beta_1$AR (with ligand removed), a number of near neighbors of the initial fragments were purchased and screened, validating the chemotype and yielding hit molecules such as **33** with improved affinity and ligand efficiency. The docking and selection process was guided by a thorough druggability analysis [55,56] of the ligand binding site that assisted in the appropriate placement of fragments, which can often present multiple plausible docked binding modes. With a validated fragment series in hand, we were able to overlay hits with liganded X-ray structures from both $\beta_1$AR and the closely related $\beta_2$AR and use this information to prospectively design compounds with the anticipation that further advantageous interactions with residues in the binding site would be made. Concomitantly, the ligand overlays allowed us to identify an appropriate region to introduce polarity into the designed molecules to reduce the lipophilicity of the series. The strategy yielded lead compounds such as indole **34**, with high affinity, ligand efficiency, and solubility, and examples were subsequently crystallized in to $\beta_1$AR, confirming the binding mode anticipated from the docking and forming a platform for further structure-based optimization.

A ligand-based pharmacophoric approach to fragment evolution has been used by de Esch and colleagues to optimize 5-HT$_3$ quinazoline fragment hit **35** (Figure 12.9), identified from the voltage-sensitive dye-based fragment screen detailed earlier (Section 12.2.5) [57]. A largely empirical exploration of SAR, within the context of an established 5-HT$_3$ pharmacophore model, led to high-affinity piperazine **36** with excellent selectivity against related receptors. A 5-HT$_3$ binding site homology model was constructed from an X-ray crystal structure of acetylcholine binding protein, which serves as a surrogate for the extracellular domain of ligand-gated ion channels, and the observed quinazoline SAR could be retrospectively rationalized in the context of the model. A key future challenge in this area will be to use the validated homology model prospectively in structure-based design approaches to the target. Separately, SAR around a second hit from the 5-HT$_3$ screen, quinoxaline **37** (Figure 12.9), has been reported [58]. Quinoxaline **37** was identified to have more than 80-fold selectivity in a radioligand binding assay between the 5HT$_3$A and the 5-HT$_3$AB

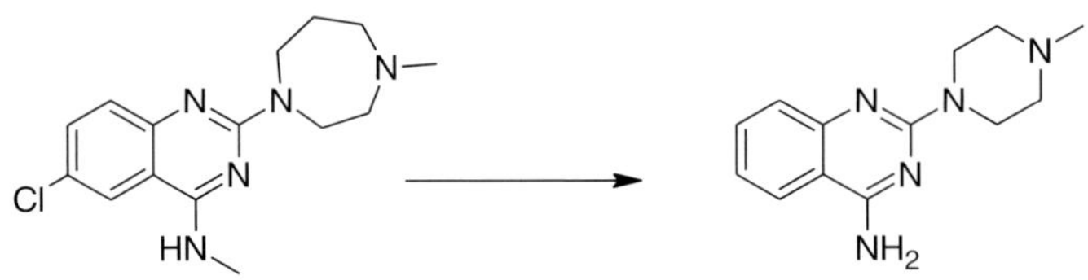

**35** 5-HT$_3$ p$K_i$ 7.7 (LE 0.50)

**36** 5-HT$_3$ p$K_i$ 10.3 (LE 0.78)

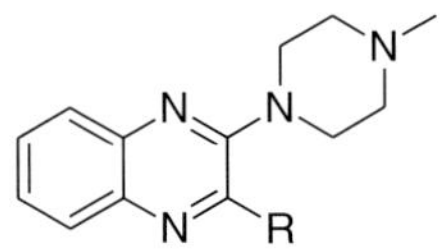

**37 (R=Cl)** 5-HT$_3$A p$K_i$ 9.8 (LE 0.74), 5-HT$_3$AB p$K_i$ 7.9

**38 (R=H)** 5-HT$_3$A p$K_i$ 8.2, 5-HT$_3$AB p$K_i$ 9.1 (LE 0.73)

**Figure 12.9** 5-HT$_3$ fragment screening hit **35** and subsequent high-affinity analog **36**; subtype selective hit **37** and close analog **38**.

receptor subtypes, and quinoxalines with reversed selectivity profiles, albeit only approximately eightfold at best (**38**) were identified through synthesis of close analogs.

### 12.5.2
### Fragment Linking

Two (or more) fragments, which bind to proximal parts of the active site, are joined together to give a larger, higher affinity binding molecule.

In a recent example of an *in silico* FBDD approach, Keserű and coworkers have reported a sequential docking methodology for the prediction of starting points for fragment linking [59], in the context of the dopamine $D_3$ and $D_2$ receptors that have been described as having primary and secondary binding sites [60]. The primary, or orthosteric, binding site that binds dopamine has high sequence homology between $D_3$ and $D_2$, making the discovery of selective ligands that only occupy this area challenging. An adjacent secondary binding site that is significantly less conserved between dopamine receptors can potentially be exploited by the so-called bitopic compounds to yield enhanced selectivity. In the linking study, two focused libraries of fragments were selected based on prior knowledge of the pharmacophores of the receptor primary and secondary binding sites, and the first of these libraries (196 fragments containing an aryl or heteroaryl portion substituted appended with a basic amine in an aliphatic ring) was docked into the primary site of the $D_3$ receptor X-ray crystal structure. Using the criterion of an interaction with a key acidic residue in the binding site, 145 of the fragments were classed as binders and ranked by docking score. The top 15 were docked into the analogous site of a ($D_3$ X-ray structure derived) $D_2$ receptor homology model, and as expected due to the highly conserved nature of this area of the dopamine receptors, also achieved high docking scores. The 145 well-docked fragments were merged with the apo $D_3$ structure and new grids constructed

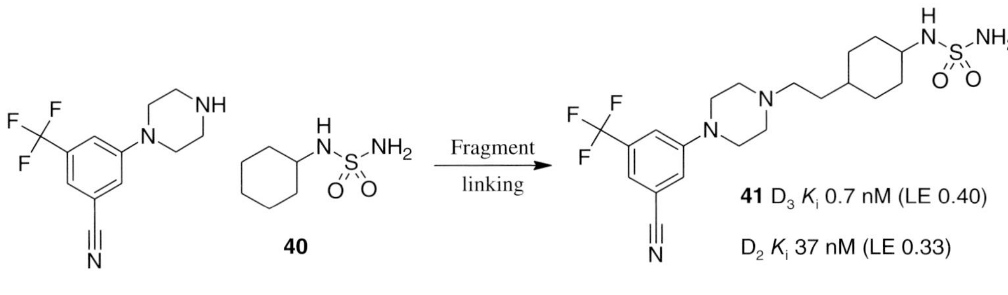


**Figure 12.10** Primary (**39**) and secondary (**40**) site dopamine receptor binders and linked, $D_3$-selective compound **41**.

including the ligands, into which the second library (consisting of 266 cyclohexyl or piperidine ring containing fragments) was docked and ranked by mean docking score across all 145 grids. A similar process was repeated in $D_2$ where the top ranking primary site fragment was merged with the apo homology model and the secondary site library docked and ranked. After visual inspection of the docking poses of the top ranking secondary site fragments, three were selected for linking to the top ranking fragment from the primary site screen, resulting in linked compounds with subnanomolar $D_3$ affinities (e.g., **41**, Figure 12.10. The affinities of the individual fragments **39** and **40** were not given). Piperazine **41** demonstrated more than 50-fold selectivity for $D_3$ over $D_2$, which could be rationalized by an extra H-bonding interaction not formed in the latter receptor.

## 12.6
## Perspective and Conclusions

In this chapter, we have examined FBDD in situations where the protein targets are challenging to work with particularly from the perspective of generating X-ray structural information. We have assessed when fragment screening is likely to be useful, focusing on tractability and its relationship to the screening paradigms available, giving examples of the various types of biophysical and biochemical screens across a range of targets. A number of strategies for validating and optimizing fragment hits have also been outlined. Overall, the growing body of literature reviewed here appears to support the notion that fragment screening and optimization should be considered for complex targets, particularly where the druggability (to small molecules) of the targets is good.

In Figure 12.11, the fragment hits illustrated above have been plotted onto the same graph depicted in Figure 12.1. Only the initial fragments have been included, not optimized hits derived from them, and the key illustrates which targets and the compound numbers that are included. Examination of this plot suggests that neither does there appear to be a very strong bias toward highly

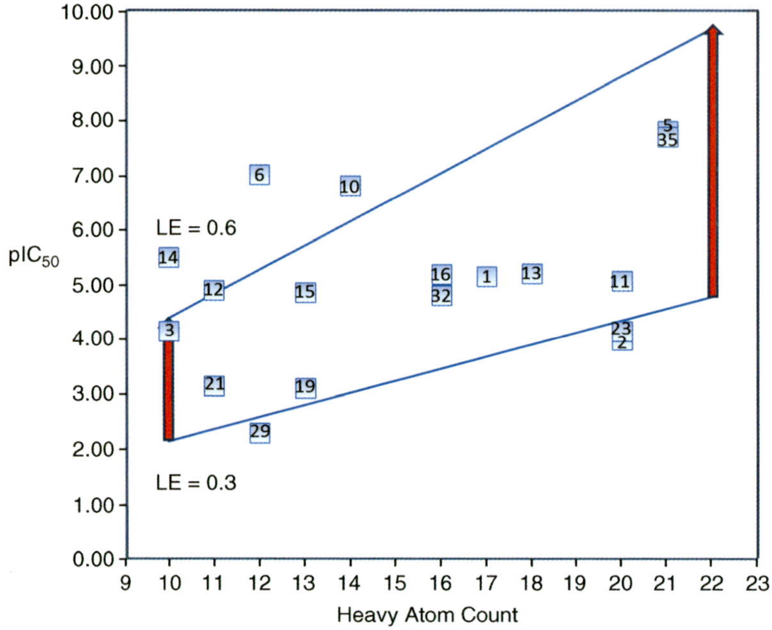


**Figure 12.11** Hits plotted showing $pIC_{50}$ versus number of heavy atoms in the ligand-efficient fragment space from Figure 12.1. Hits are as follows: **1** and **2** DsbB, **3** *E. coli* A-site RNA, **5** $\beta_2$AR, **6** $H_4$, **10** $D_3$, **11** $H_4$, **12** MC4, **13** $mGlu_5$, **14** $A_3$, **15** and **16** $A_{2A}$, **19** and **21** ASIC3, **23** $Ca_v3.2$, **29** $tRNA_3^{Lys}$, **32** $\beta_1$AR, **35** 5-$HT_3$.

tractable targets nor is there a requirement for relatively high potency for fragments to be detectable. It is notable, however, that there are few fragment hits being detected in the millimolar range with hits generally being high to low micromolar. This is an obvious limitation and suggests further development of biophysical screening methodology is required for low tractability targets that are also challenging proteins to work with.

One example from our own studies is the efficient optimization from initial fragment hit **13** (briefly outlined in Section 12.2.4) to a preclinical candidate molecule, driven initially by molecular modeling and docking into an $mGlu_5$ homology model. Parallel efforts in structural biology led to the crystallization of the receptor with the clinical-stage negative allosteric modulator mavoglurant [61], and subsequently multiple further crystal structures of literature and in-house ligands (unpublished results), opening up the potential for $mGlu_5$ SBDD [62]. This fragment to candidate process took only one year and illustrates the potential of FBDD for GPCRs at least where the target has good druggability. In this case, the target was an allosteric binding site and this also highlights another opportunity when using FBDD. A clear advantage of a biophysical approach is that neither the identification of ligands is limited to competition for a specific site nor does the ligand have to be sufficiently potent to invoke a functional consequence on the function of the protein to be detected. The mGluR family is a good example of this – the orthosteric ligands are small polar acids at odds with the properties usually required for a CNS target and for which receptor selectivity has been extremely challenging to engineer. However, allosteric ligands for multiple members of the family have been discovered in the CNS drug space and with high receptor selectivity. The example noted above is the first FBDD case study for a class C GPCR illustrating that this approach is very promising for this subclass of targets, perhaps allowing the identification of a broader range of chemotypes.

In the early days of FBDD, there was much skepticism that hits detected with very weak binding affinity were "nonspecific" and that even with an X-ray structure the binding mode was not going to be predictable or maintained as compounds were optimized. Today, the general consensus is that fragments can be rationally optimized using SBDD, even if starting with very low affinity. In the area of complex proteins and especially membrane proteins, many of the same concerns pervade the discussion of FBDD as they did for enzymes 15 years ago. Are the hits nonspecific? Will hits maintain their binding pose or simply move around to totally new binding modes as they are optimized? Overall, there are many similarities between ligand recognition for soluble and membrane-bound proteins, but the lack of structural information has so far held back FBDD for membrane proteins [55]. The lessons learned from many years of FBDD with soluble proteins is making progress more feasible in the absence of structural information, while the recent advances in obtaining structures of membrane proteins suggest successful FBDD with membrane proteins will become more common. The data in Figure 12.11 supports the detection of fragment hits for a broad range of complex protein targets. The work presented above and in our own laboratories on GPCRs suggests to these authors that FBDD will track the same trajectory as it did for enzymes, as our ability to screen and crystallize these more challenging targets develops. It is our conclusion that in 10–15 years from now FBDD will often be successfully applied to GPCRs, other membrane proteins, and more classes of complex targets.

## Acknowledgments

The authors would like to thank Fiona Marshall, Rob Cooke, and Ali Jazayeri for their help in proofreading and finalizing this chapter.

## References

1 Bemis, G.W. and Murcko, M.A. (1996) The properties of known drugs. 1. Molecular frameworks. *Journal of Medicinal Chemistry*, **39**, 2887–2893.

2 Shuker, S.B., Hajduk, P.J., Meadows, R.P., and Fesik, S.W. (1996) Discovering high-affinity ligands for proteins: SAR by NMR. *Science*, **274**, 1531–1534.

3 Bollag, G., Tsai, J., Zhang, J., Zhang, C., Ibrahim, P., Nolop, K., and Hirth, P. (2012) Vemurafenib: the first drug approved for *BRAF*-mutant cancer. *Nature Reviews Drug Discovery*, **11**, 873–886.

4 Jhoti, H., Williams, G., Rees, D.C., and Murray, C.W. (2013) The 'rule of three' for fragment-based drug discovery: where are we now? *Nature Reviews Drug Discovery*, **12**, 644–645.

5 Hajduk, P.J. and Greer, J. (2007) A decade of fragment-based drug design: strategic advances and lessons learned. *Nature Reviews Drug Discovery*, **6**, 211–219.

6 Alex, A.A. and Flocco, M.M. (2007) Fragment-based drug discovery: what has it achieved so far? *Current Topics in Medicinal Chemistry*, **7**, 1544–1567.

7 Stamford, A. and Strickland, C. (2013) Inhibitors of BACE for treating Alzheimer's disease: a fragment-based drug discovery story. *Current Opinion in Chemical Biology*, **17**, 320–328.

8 Woodhead, A.J., Angove, H., Carr, M.G., Chessari, G., Congreve, M., Coyle, J.E., Cosme, J., Graham, B., Day, P.J., Downham, R., Fazal, L., Feltell, R., Figueroa, E., Frederickson, M., Lewis, J., McMenamin, R., Murray, C.W., O'Brien, M.A., Parra, L., Patel, S., Phillips, T., Rees, D.C., Rich, S., Smith, D.-M., Trewartha, G., Vinkovic, M., Williams, B., and Woolford, A.J.-A. (2010) Discovery of (2,4-Dihydroxy-5-isopropylphenyl)-[5-(4-methylpiperazin-1-ylmethyl)-1,3-dihydroisoindol-2-l]methanone (AT13387), a novel inhibitor of the molecular chaperone Hsp90 by fragment based drug design. *Journal of Medicinal Chemistry*, **53**, 5956–5969.

9 Hopkins, A.L., Groom, C.R., and Alex, A. (2004) Ligand efficiency: a useful metric for lead selection. *Drug Discovery Today*, **9**, 430–431.

10 Lipinski, C. (2000) Drug-like properties and the causes of poor solubility and poor permeability. *Journal of Pharmacological and Toxicological Methods*, **44**, 235–249.

11 Shultz, M.D. (2013) Setting expectations in molecular optimizations: Strengths and limitations of commonly used composite parameters. *Bioorganic and Medicinal Chemistry Letters*, **23**, 5980–5991.

12 Shultz, M.D. (2014) Improving the plausibility of success with inefficient metrics. *ACS Medicinal Chemistry Letters*, **5**, 2–5.

13 Hopkins, A.L., Keserü, G.M., Leeson, P.D., Rees, D.C., and Reynolds, C.H. (2014) The role of ligand efficiency metrics in drug discovery. *Nature Reviews Drug Discovery*, **13**, 105–121.

14 Murray, C.W., Erlanson, D.A., Hopkins, A.L., Keserü, G.M., Leeson, P.D., Rees, D.C., Reynolds, C.H., and Richmond, N.J. (2014) Validity of ligand efficiency metrics. *ACS Medicinal Chemistry Letters*, **5**, 616–618.

15 Robertson, N., Jazayeri, A., Errey, J., Baig, A., Hurrell, E., Zhukov, A., Langmead, C.J., Weir, M., and Marshall, F.H. (2011) The properties of thermostabilised G protein-coupled receptors (StaRs) and their use in drug discovery. *Neuropharmacology*, **60**, 36–44.

16 Vanwetswinkel, S., Heetebrij, R.J., van Duynhoven, J., Hollander, J.G., Filippov, D.V., Hajduk, P.J., and Siegal, G. (2005) TINS, target immobilized NMR screening: an efficient and sensitive method for ligand discovery. *Chemistry and Biology*, **12**, 207–216.

17 Congreve, M., Rich, R.L., Myszka, D.G., Figaroa, F., Siegal, G., and Marshall, F.H. (2011) Fragment screening of stabilized G-protein-coupled receptors using biophysical methods, in *Methods in Enzymology*, vol. **493** (ed. L.C. Kuo), Academic Press, Burlington, MA, U.S.

18 Fruh, V., Zhou, Y., Chen, D., Loch, C., Ab, E., Grinkova, Y.N., Verheij, H., Sligar, S.G., Bushweller, J.H., and Siegal, G. (2010) Application of fragment-based drug discovery to membrane proteins: identification of ligands of the integral membrane enzyme DsbB. *Chemistry and Biology*, **17**, 881–891.

19 Moumné, R., Catala, M., Larue, V., Micouin, L., and Tisné, C. (2012) Fragment-based design of small RNA binders: promising developments and contribution of NMR. *Biochimie*, **94**, 1607–1619.

20 Yu, L., Oost, T.K., Schkeryantz, J.M., Yang, J., Janowick, D., and Fesik, S.W. (2003) Discovery of aminoglycoside mimetics by NMR-based screening of *Escherichia coli*

A-site RNA. *Journal of the American Chemical Society*, **125**, 4444–4450.

21 Swayze, E.E., Jefferson, E.A., Sannes-Lowery, K.A., Blyn, L.B., Risen, L.M., Arakawa, S., Osgood, S.A., Hofstadler, S.A., and Griffey, R.H. (2002) SAR by MS: a ligand based technique for drug lead discovery against structured RNA targets. *Journal of Medicinal Chemistry*, **45**, 3816–3819.

22 Rich, R.L., Errey, J., Marshall, F., and Myszka, D.G. (2011) Biacore analysis with stabilized G-protein-coupled receptors. *Analytical Biochemistry*, **409**, 267–272.

23 Navratilova, I., Besnard, J., and Hopkins, A.L. (2011) Screening for GPCR ligands using surface plasmon resonance. *ACS Medicinal Chemistry Letters*, **2**, 549–554.

24 Aristotelous, T., Ahn, S., Shukla, A.K., Gawron, S., Sassano, M.F., Kahsai, A.W., Wingler, L.M., Zhu, X., Tripathi-Shukla, P., Huang, X.-P., Riley, J., Besnard, J., Read, K.D., Roth, B.L., Gilbert, I.H., Hopkins, A.L., Lefkowitz, R.J., and Navratilova, I. (2013) Discovery of $\beta 2$ adrenergic receptor ligands using biosensor fragment screening of tagged wild-type receptor. *ACS Medicinal Chemistry Letters*, **4**, 1005–1010.

25 Shepherd, C.A., Hopkins, A.L., and Navratilova, I. (2014) Fragment screening by SPR and advanced application to GPCRs. *Progress in Biophysics and Molecular Biology*, **116**, 113–123.

26 Austin, C., Pettit, S.N., Magnolo, S.K., Sanvoisin, J., Chen, W., Wood, S.P., Freeman, L.D., Pengelly, R.J., and Hughes, D.E. (2012) Fragment screening using capillary electrophoresis (CEfrag) for hit identification of heat shock protein 90 ATPase inhibitors. *Journal of Biomolecular Screening*, **17**, 868–876.

27 Andrews, S.P. and Tehan, B. (2013) Stabilised G protein-coupled receptors in structure-based drug design: a case study with adenosine $A_{2A}$ receptor. *Medicinal Chemistry Communications*, **4**, 52–67.

28 Murray, C.W., Verdonk, M.L., and Rees, D.C. (2012) Experiences in fragment-based drug discovery. *Trends in Pharmacological Sciences*, **33**, 224–232.

29 Erlanson, D.A. (2012) Introduction to fragment-based drug discovery. *Topics in Current Chemistry*, **317**, 1–32.

30 Verheij, M.H.P., de Graaf, C., de Kloe, G.E., Nijmeijer, S., Vischer, H.F., Smits, R.A., Zuiderveld, O.P., Hulscher, S., Silvestri, L., Thompson, A.J., van Muijlwijk-Koezen, J.E., Lummis, S.C.R., Leurs, R., and de Esch, I.J.P. (2011) Fragment library screening reveals remarkable similarities between the G protein-coupled receptor histamine $H_4$ and the ion channel serotonin 5-$HT_{3A}$. *Bioorganic and Medicinal Chemistry Letters*, **21**, 5460–5464.

31 de Graaf, C., Vischer, H.F., de Kloe, G.E., Kooistra, A.J., Nijmeijer, S., Kuijer, M., Verheij, M.H.P., England, P.J., van Muijlwijk-Koezen, J.E., Leurs, R., and de Esch, I.J.P. (2013) Small and colorful stones make beautiful mosaics: fragment-based chemogenomics. *Drug Discovery Today*, **18**, 323–330.

32 de Graaf, C., Kooistra, A.J., Vischer, H.F., Katritch, V., Kuijer, M., Shiroishi, M., Iwata, S., Shimamura, T., Stevens, R.C., de Esch, I.J.P., and Leurs, R. (2011) Crystal structure-based virtual screening for fragment-like ligands of the human histamine $H_1$ receptor. *Journal of Medicinal Chemistry*, **54**, 8195–8206.

33 Sirci, F., Istyastono, E.P., Vischer, H.F., Kooistra, A.J., Nijmeijer, S., Kuijer, M., Wijtmans, M., Mannhold, R., Leurs, R., de Esch, I.J.P., and de Graaf, C. (2012) Virtual fragment screening: discovery of histamine $H_3$ receptor ligands using ligand-based and protein-based molecular fingerprints. *Journal of Chemical Information and Modeling*, **52**, 3308–3324.

34 Vass, M., Schmidt, E., Horti, F., and Keserű, G.M. (2014) Virtual fragment screening on GPCRs: a case study on dopamine D3 and histamine H4 receptors. *European Journal of Medical Chemistry*, **77**, 38–46.

35 Ujjainwalla, F. and Sebhat, I.K. (2007) Small molecule ligands of the human melanocortin-4 receptor. *Current Topics in Medicinal Chemistry*, **7**, 1068–1084.

36 Albert, J.S., Blomberg, N., Breeze, A.L., Brown, A.J.H., Burrows, J.N., Edwards, P.D., Folmer, R.H.A., Geschwindner, S., Griffen, E.J., Kenny, P.W., Nowak, T.,

Olsson, L.-L., Sanganee, H., and Shapiro, A.B. (2007) An integrated approach to fragment-based lead generation: philosophy, strategy and case studies from AstraZeneca's drug discovery programmes. *Current Topics in Medicinal Chemistry*, 7, 1600–1629.

37 Bennett, K.A., Tehan, B., Lebon, G., Tate, C.G., Weir, M., Marshall, F.H., and Langmead, C.J. (2013) Pharmacology and structure of isolated conformations of the adenosine $A_{2A}$ receptor define ligand efficacy. *Molecular Pharmacology*, **83**, 949–958.

38 Stoddart, L.A., Vernall, A.J., Denman, J.L., Briddon, S.J., Kellam, B., and Hill, S.J. (2012) Fragment screening at adenosine-$A_3$ receptors in living cells using a fluorescence-based binding assay. *Chemistry and Biology*, **19**, 1105–1115.

39 Kenakin, T.P. (2012) Biased signalling and allosteric machines: new vistas and challenges for drug discovery. *British Journal of Pharmacology*, **165**, 1659–1669.

40 Overington, J.P., Al-Lazikani, B., and Hopkins, A.L. (2006) How many drug targets are there? *Nature*, **5**, 993–996.

41 Bagal, S.K., Brown, A.D., Cox, P.J., Omoto, K., Owen, R.M., Pryde, D.C., Sidders, B., Skerratt, S.E., Stevens, E.B., Storer, R.I., and Swain, N.A. (2013) Ion channels as therapeutic targets: a drug discovery perspective. *Journal of Medicinal Chemistry*, **56**, 593–624.

42 Dunlop, J., Bowlby, M., Peri, R., Vasilyev, D., and Arias, R. (2008) High-throughput electrophysiology: an emerging paradigm for ion-channel screening and physiology. *Nature Reviews Drug Discovery*, **7**, 358–368.

43 Thompson, A.J., Verheij, M.H.P., Leurs, R., De Esch, I.J.P., and Lummis, S.C.R. (2010) An efficient and information-rich biochemical method design for fragment library screening on ion channels. *Biotechniques*, **49**, 822–829.

44 Blaney, F.E., Capelli, A.-M., and Tedesco, G. (2006) 7TM models in structure-based drug design, in *Methods and Principles in Medicinal Chemistry, Ligand Design for G Protein-coupled Receptors* (ed. D. Rognan), Wiley-VCH Verlag GmbH, Weinheim.

45 Barton, N., Blaney, F.E., Garland, S., Tehan, B., and Wall, I. (2007) Seven transmembrane G protein-coupled receptors: insights for drug design from structure and modeling, in *Comprehensive Medicinal Chemistry II* (eds J.B. Taylor and D.J. Triggle), Elsevier, Oxford.

46 Clark, D.E. and Higgs, C. (2006) Ligand-based rational design: virtual screening, in *Methods and Principles in Medicinal Chemistry, Ligand Design for G Protein-coupled Receptors* (ed. D. Rognan), Wiley-VCH Verlag GmbH, Weinheim.

47 Chen, D., Ranganathan, A., IJzerman, A.P., Siegal, G., and Carlsson, J. (2013) Complementarity between *in silico* and biophysical screening approaches in fragment-based lead discovery against the $A_{2A}$ adenosine receptor. *Journal of Chemical Information and Modeling*, **53**, 2701–2714.

48 Chen, D., Errey, J.C., Heitman, L.H., Marshall, F.H., IJzerman, A.P., and Siegal, G. (2012) Fragment screening of GPCRs using biophysical methods: identification of ligands of the adenosine $A_{2A}$ receptor with novel biological activity. *ACS Chemical Biology*, **7**, 2064–2073.

49 Wolkenberg, S.E., Zhao, Z., Mulhearn, J.J., Harrison, S.T., Sanders, J.M., Cato, M.J., Jovanovska, A., Panigel, J., Cook, S.P., Henze, D.A., Kane, S.A., Hartman, G.D., and Barrow, J.C. (2011) High concentration electrophysiology-based fragment screen: Discovery of novel acid-sensing ion channel 3 (ASIC3) inhibitors. *Bioorganic and Medicinal Chemistry Letters*, **21**, 2646–2649.

50 Congreve, M., Chessari, G., Tisi, D., and Woodhead, A.J. (2008) Recent developments in fragment-based drug discovery. *Journal of Medicinal Chemistry*, **51**, 3661–3680.

51 de Kloe, G.E., Bailey, D., Leurs, R., and de Esch, I.J. (2009) Transforming fragments into candidates: small becomes big in medicinal chemistry. *Drug Discovery Today*, **14**, 630–646.

52 Rees, D.C., Congreve, M., Murray, C.W., and Carr, R. (2004) Fragment-based lead discovery. *Nature Reviews Drug Discovery*, **3**, 660–672.

53 Giordanetto, F., Knerr, L., Selmi, N., Llinàs, A., Lindqvist, A., Wang, W.-D., Ståhlberg, P., Thorstensson, F., Ullah, V.,

Nilsson, K., O'Mahony, G., Högberg, Å., Lindhardt, E., Åstrand, A., and Duker, G. (2011) Discovery of *N*-(1-adamantyl)-2-(4-alkylpiperazin-1-yl)acetamide derivatives as T-type calcium channel ($Ca_v3.2$) inhibitors. *Bioorganic and Medicinal Chemistry Letters*, **21**, 5557–5561.

54 Christopher, J.A., Brown, J., Doré, A.S., Errey, J.C., Koglin, M., Marshall, F.H., Myszka, D.G., Rich, R.L., Tate, C.G., Tehan, B., Warne, T., and Congreve, M. (2013) Biophysical fragment screening of the $\beta_1$-adrenergic receptor: identification of high affinity arylpiperazine leads using structure-based drug design. *Journal of Medicinal Chemistry*, **56**, 3446–3455.

55 Mason, J.S., Bortolato, A., Congreve, M., and Marshall, F.H. (2012) New insights from structural biology into the druggability of G protein-coupled receptors. *Trends in Pharmacological Sciences*, **33**, 249–260.

56 Congreve, M., Dias, J.M., and Marshall, F.H. (2014) Structure-based drug design for G protein-coupled receptors. *Progress in Medicinal Chemistry*, **53**, 1–63.

57 Verheij, M.H.P., Thompson, A.J., van Muijlwijk-Koezen, J.E., Lummis, S.C.R., Leurs, R., and de Esch, I.J.P. (2012) Design, synthesis, and structure–activity relationships of highly potent 5-$HT_3$ receptor ligands. *Journal of Medicinal Chemistry*, **55**, 8603–8614.

58 Thompson, A.J., Verheij, M.H.P., van Muijlwijk-Koezen, J.E., Lummis, S.C.R., Leurs, R., and de Esch, I.J.P. (2013) Structure–activity relationships of quinoxaline-based 5-$HT_3$A and 5-$HT_3$AB receptor-selective ligands. *ChemMedChem*, **8**, 946–955.

59 Vass, M., Ágai-Csongor, E., Horti, F., and Keserű, G.M. (2014) Multiple fragment docking and linking in primary and secondary pockets of dopamine receptors. *ACS Medicinal Chemistry Letters*, **5**, 1010–1014.

60 Newman, A.H., Beuming, T., Banala, A.K., Donthamsetti, P., Pongetti, K., LaBounty, A., Levy, B., Cao, J., Michino, M., Luedtke, R.R., Javitch, J.A., and Shi, L. (2012) Molecular determinants of selectivity and efficacy at the dopamine D3 receptor. *Journal of Medicinal Chemistry*, **55**, 6689–6699.

61 Doré, A.S., Okrasa, K., Patel, J.C., Serrano-Vega, M., Bennett, K., Cooke, R.M., Errey, J.C., Jazayeri, A., Khan, S., Tehan, B., Weir, M., Wiggin, G.R., and Marshall, F.H. (2014) Structure of class C GPCR metabotropic glutamate receptor 5 transmembrane domain. *Nature*, **511**, 557–562.

62 Christopher, J.A., Aves, S.J., Bennett, K.A., Congreve, M., Jazayeri, A., Khan, S., Patel, J.C., Tehan, B., Wiggin, G., and Marshall, F.H. (2014) Poster presentation, 247th ACS National Meeting, Dallas, Texas, U.S.

# 13 Protein-Templated Fragment Ligation Methods: Emerging Technologies in Fragment-Based Drug Discovery

*Mike Jaegle, Eric Nawrotzky, Ee Lin Wong, Christoph Arkona, and Jörg Rademann*

## 13.1 Introduction: Challenges and Visions in Fragment-Based Drug Discovery

Considerable progress has been achieved in the field of fragment-based drug discovery over the past decade both in the development of biophysical fragment detection methods and in the conversion of fragment hits into potent drug-like molecules. As described in other chapters of this book, numerous success stories leading to clinical candidates and possibly to clinically admitted molecules have been reported so far. Despite these impressive achievements, however, two major challenges remain inherent to the development of drug-like molecules from fragments and threaten the success of many fragment-based drug discovery projects: the detection problem (1) and the linkage problem (2) (Figure 13.1).

The detection problem arises from the observation that molecules with a molecular weight below 250 g $mol^{-1}$ (or Da) often bind to proteins with dissociation constants in the millimolar range making the detection of ligands for a defined binding site an enormously difficult task. The low affinity of primary fragment hits can require the chemical combination of at least two hit fragments to obtain a strongly binding fragment combination. The linkage problem arises as two fragments have to be identified with adjacent, nonoverlapping binding sites; the initial fragments both have to contain a reactive site or functionality that enables the covalent linking of fragment while significantly increasing the free binding energy of the fragment combination. As the number of chemical variations is high, one can expect that numerous chemical linkages and fragment combinations have to be tested before finding a successful solution. In most FBDD projects without using protein-templated fragment ligation, the ligation problem has been tackled by starting from a single fragment hit, which was derivatized systematically in order to find improved fragment combinations.

*Fragment-based Drug Discovery: Lessons and Outlook*, First Edition. Edited by Daniel A. Erlanson and Wolfgang Jahnke.

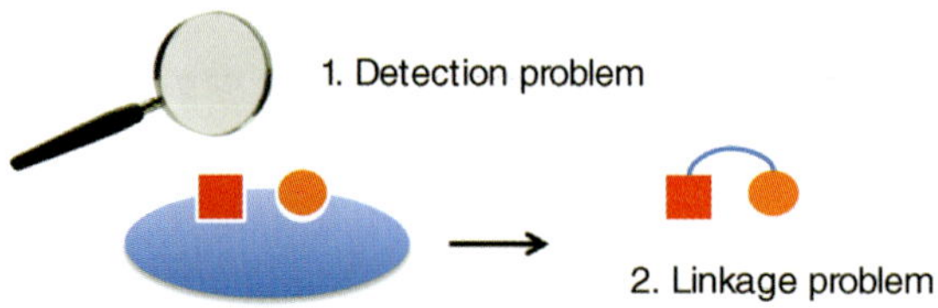


**Figure 13.1** The two main challenges of fragment-based drug discovery: the site-specific detection of millimolar protein ligands (1) and the connection of two fragments to a more potent fragment combination (2).

## 13.2 Target-Guided Fragment Ligation: Concepts and Definitions

This chapter will cover alternative approaches to FBDD, which tackle both the detection problem and the linkage problem, using the target protein itself as a template for the assembly and detection of fragment combinations (Figure 13.2, Table 13.1). All target-guided strategies are based on the fact that binding of fragments to proteins may induce chemical reactivity, which can be exploited for the detection and ligation of fragments to fragment combinations. Thus, in principle these strategies employ the protein as a catalyst or a template for the selection and formation of improved protein ligands – often in a single step – and thereby directly realize the seminal concept of the combination of single fragments to higher affinity ligands described by Jencks *et al.* in 1981 [1]. Since the early days of considering fragment binding to proteins, several different approaches of target-guided ligand discovery have been conceived and implemented that here will be employed for the structuring of this chapter (Table 13.1).

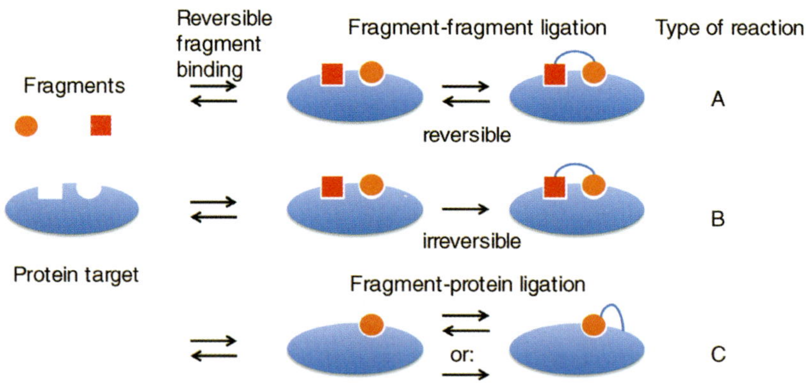


**Figure 13.2** Target-guided fragment ligation methods for ligand discovery. Reversible binding of small molecule fragments to protein binding sites can induce the formation of a covalent chemical bond between fragments through a reversible (a) or irreversible (b) chemical reaction. Alternatively, bound fragments can undergo bond-forming reactions with the protein template itself.

**Table 13.1** Types of protein-guided fragment ligation reactions.

| Entry | Reaction partners | Type of reaction | Reaction mode | Section |
|---|---|---|---|---|
| 1 | Fragment–fragment | A | Reversible | 3 |
| 2 | Fragment–fragment | B | Irreversible | 4 |
| 3 | Fragment-protein | C | Reversible | 5 |
| 4 | Fragment–protein | C | Irreversible | 5 |

First of all, target-guided ligation reactions can be differentiated in reversible reactions in equilibrium (see Section 13.3) and irreversible reactions (see Section 13.4), though the distinction between both groups of reactions is not strict and several intermediate cases are found. Reversible ligation reactions often yield products that are unstable, precluding their isolation, whereas products of irreversible ligations are mostly stable, but there are also examples for reversibly formed products being stable after isolation. As another category, fragment ligations on proteins can occur between binding fragments, but there are several applications of reactions between a bound fragment and the protein itself, reversibly or irreversibly. As target-guided methods in FBDD have been referenced extensively in earlier review articles [2–8], we will focus here entirely on contributions since 2006.

## 13.3 Reversible Fragment Ligation

### 13.3.1 Dynamic Reversible Fragment Ligation Strategies

Reversible ligation reactions can be expected to have comparably low activation barriers, a situation facilitating the formation of the thermodynamically most favored fragment combination – both in the presence of the protein target and without. The ideal equilibrium reaction for fragment ligation should be fast, thereby allowing for the rapid formation of the ligation product and with little or no time-dependent effects. It should be noted, however, that the thermodynamically most favored fragment combination formed in the presence of the target protein does not necessarily have to be the best binding fragment combination product (Figure 13.3). Especially if the reactivity of fragments is high and the free reaction energy of the fragment ligation reaction ($\Delta G_1$) is already strongly negative without the presence of the protein, the thermodynamically most favored fragment combination ($\Delta G_1 + \Delta G_2 = \text{min.}$) might not be the best binding one. To solve this serious conceptual problem, various detection strategies have been applied, which do not identify the most stable ligation product but instead the biologically most active one (see Section 13.3.3).

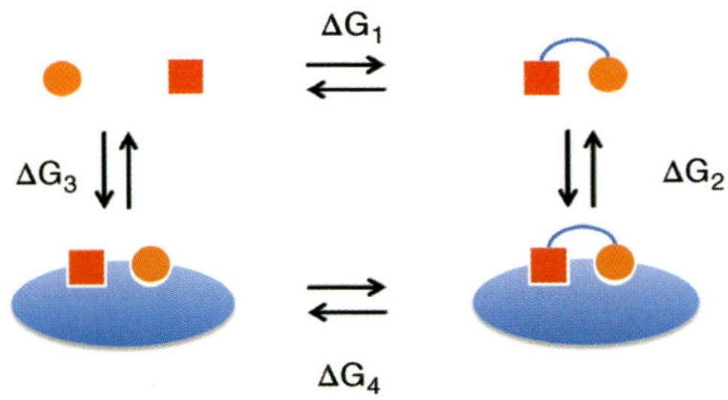


**Figure 13.3** Dynamic fragment ligation reactions can be dissected into fragment ligation in solution ($\Delta G_1$) or at the protein template ($\Delta G_4$) and into binding of single fragments ($\Delta G_3$) or binding of the linked fragments ($\Delta G_2$). According to Hess's law, the free energy of the fragment ligation and binding amounts to $\Delta G_1 + \Delta G_2 = \Delta G_3 + \Delta G_4$.

A general feature of reversible fragment ligation strategies with a low activation barrier is that they do not form stable fragment combinations, which can be isolated, stored, and analyzed by spectroscopic methods or used directly in follow-up experiments, for example, in a cellular assay. Consequently, unstable ligation products have to be converted into noncleavable fragment combinations with some uncertainty that the resulting irreversibly fragment combination has the same bioactivity as the reversible combination used as a starting point.

If instead reversible fragment ligations are applied that enable the isolation of the fragment combination products, usually the equilibration reaction is slow and the formation of fragment combinations becomes time dependent. Here, detection does not require chemical fixation; however, for slow equilibrating reactions, the thermodynamically most-favored fragment combination formed in the presence of the protein template does not have to be the best binding fragment combination.

### 13.3.2
### Chemical Reactions Used in Dynamic Fragment Ligations

Several different reversible fragment ligation reactions have been employed for the detection and development of fragment-based bioactive molecules over recent years (Table 13.2). Older work has already been referenced extensively in earlier review articles [2–7,9], thus here the focus is entirely on contributions since the year 2006.

The largest group of work employing reversible ligation reactions for FBDD uses nucleophilic addition reactions to carbonyl electrophiles. These reactions include the formation of imines, hemi-acetals, hemi-aminals, hemi-thioacetals, hydrazones, and oximes from aldehydes and – much less common – from ketones. Addition products of nucleophiles and aldehydes/ketones vary strongly with respect to the stability of formed products. While aliphatic imines and hemi-(thio)-acetals usually cannot be isolated and have to be converted into chemically stable products, hydrazones, oximes, and to a certain degree also

**Table 13.2** Reversible fragment ligation reactions employed in FBDD: (a) imines, (b) hemiaminals, (c) hemiacetals, (d) acetals, (e) hemithioacetals, (f) nitroaldols, (g) hydrazones, (h) oximes, (i) disulfide exchange, (j) thioester exchange, (k) boronates, (l) alkene metathesis, (m) alkyne metathesis, (n) hydrogen bond exchange.

| | | References |
|---|---|---|
| (a) |  | [5,10–16] |
| (b) |  | [17,18] |
| (c) |  | [16,19] |
| (d) | 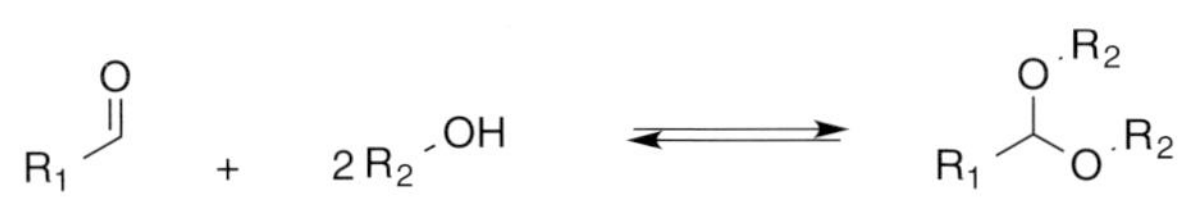 | [20,21] |
| (e) |  | [16,22,23] |
| (f) | 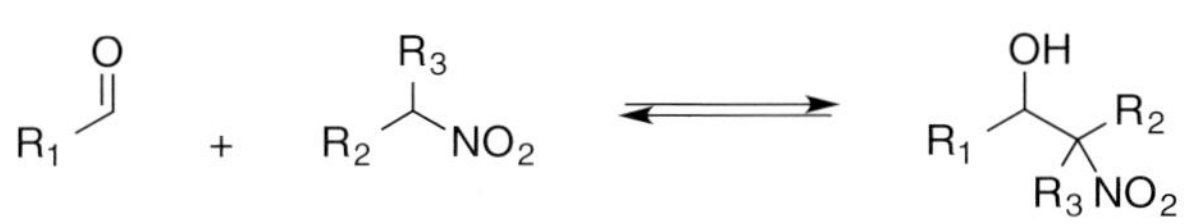 | [24–28] |
| (g) | 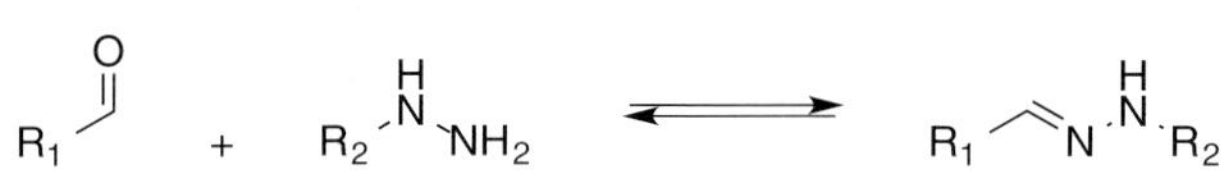 | [29–36] |
| (h) | 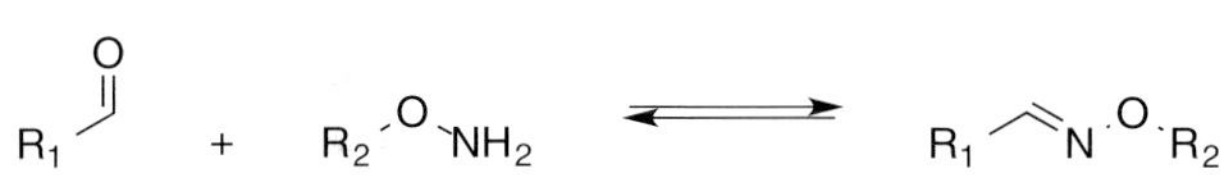 | [37,38] |
| (i) | 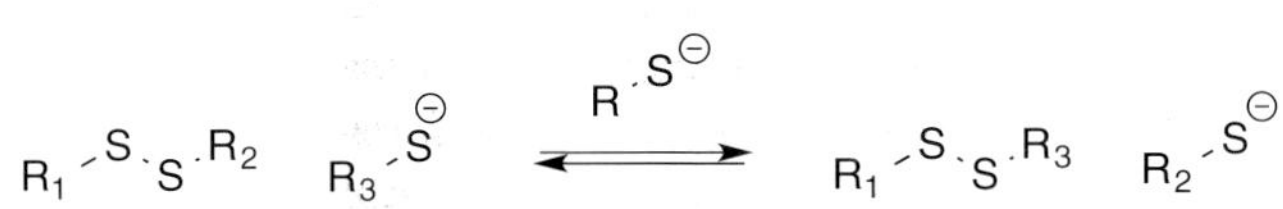 | [39–45] |

(*continued*)

**Table 13.2** (*Continued*)

| | References |
|---|---|
| (j) 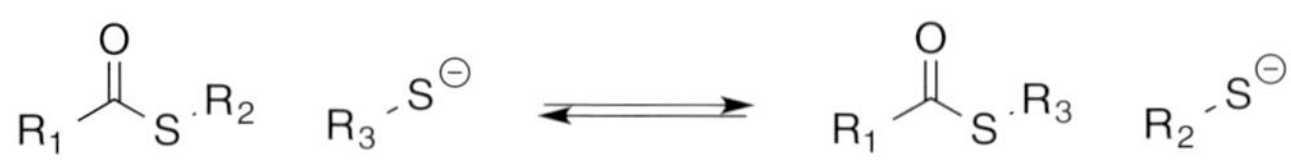 | [46–51] |
| (k) 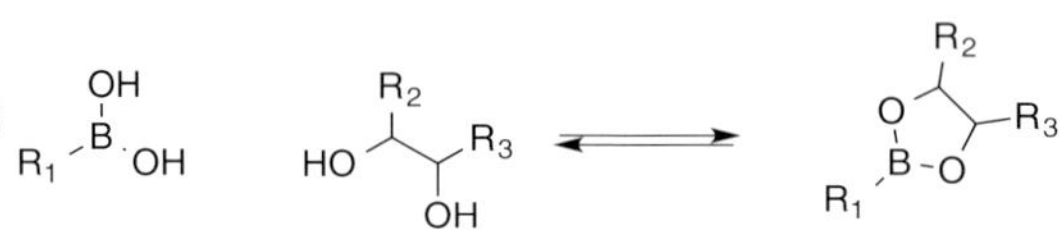 | [52,53] |
| (l) 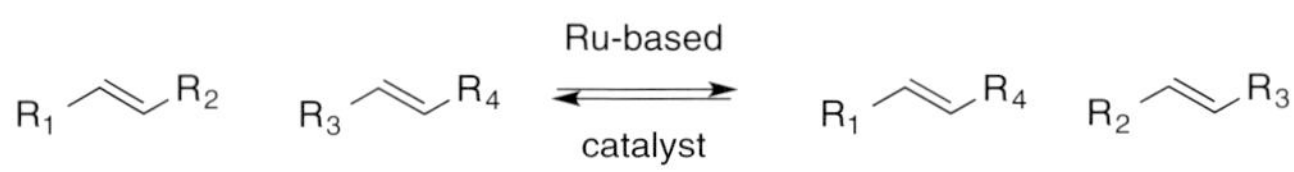 | [54–56] |
| (m) 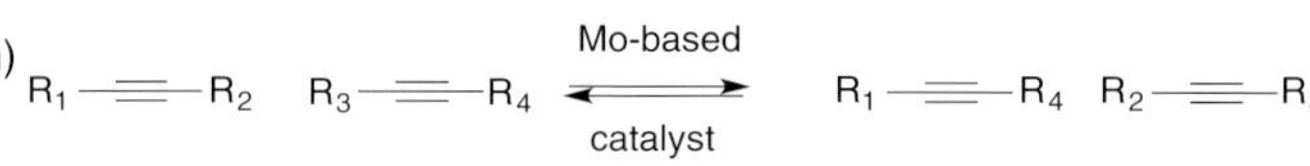 | [57] |
| (o) 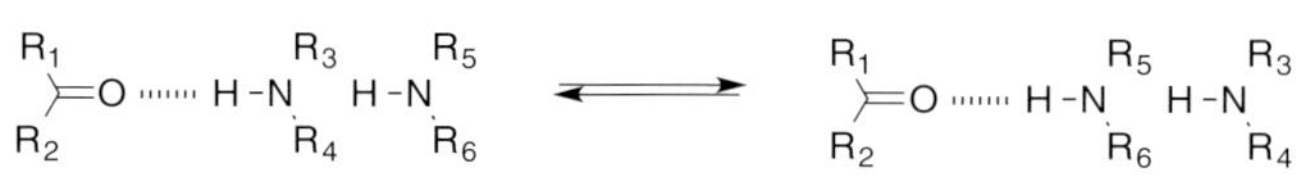 | [58–60] |

imines of aromatic amines constitute chemically stable entities by themselves, which can be isolated without a subsequent fixation reaction. Considering the rate of nucleophile exchange, concise differences can be observed for the various electrophile–nucleophile combinations. In principle, additions of amines, alcohols, and thiols to aldehydes and ketones proceed rapidly in water in accordance with a low activation barrier. For amines, the product equilibrium depends strongly on the $pK_a$ values of the nucleophile and on the pH of the aqueous phase [10]. Reactivity of the carbonyl electrophile is tuned by the electron-withdrawing effect of substituents: electron-deprived aliphatic aldehydes become hydrated in water reducing their reactivity with a second nucleophile. As the activation barrier is considerably higher in the formation of more stable ligation products such as hydrazones and oximes, these reactions can be very slow requiring acceleration, for example, with aniline as a catalyst. Therefore, ligations yielding hydrazones or oximes are at the boundary of irreversible reactions, but are nevertheless covered in this paragraph due to their close similarity to the other nucleophile additions to carbonyl groups. Over the recent years, reversible reactions involving aldehydes as electrophiles have been further extended and now include CC-bond forming reactions such as variants of aldol reactions, the nitroaldol reaction.

Equilibria between thiols and disulfides are a second broadly used class of reactions in dynamic ligations. Equilibration of these reactions is slow compared to that of nucleophilic additions of amines, alcohols, or thiols to carbonyl groups and requires electron transfer between the reduced thiols and the oxidized disulfide products. Usually, the equilibrium reaction proceeds with an excess of thiols with oxygen present in the reaction. Ligation reactions of thioesters with thiols have also been described especially well during recent years. Another reaction type useful for reversible ligation has been the exchange reaction of boronic acids and borane esters with diols. Finally, metathesis reactions have been proposed as reversible ligation reactions, for example, between olefins or alkynes. Strictly speaking, H-bonding assembly cannot be considered as covalent ligation reactions. However, this type of exchange reaction has recently found application in fragment binding and exchange.

### 13.3.3
### Detection Strategies in Dynamic Fragment Ligations

As described above, the challenge in dynamic fragment ligations is the reliable identification of the best binding/most biologically active fragment ligation products from high numbers of potential fragment combinations. Several methods to fulfill this task have been developed and applied. These methods can be distinguished as ligand-directed detection methods, which directly identify fragment combinations, and protein-directed detection methods, which scrutinize the bioactivity or binding site occupation of the protein target (Figure 13.4).

Ligand-directed detection methods have been mostly employed to investigate reversible ligations with low activation barriers. One prominent approach to identify best-binding combinations has been the introduction of dynamic combinatorial libraries (DCL), which have been investigated in the presence of the protein target. DCLs are defined as mixtures of fragments that can undergo fragment ligation, for example, a number of $x$ aldehydes and $y$ amines, potentially forming $x$ times $y$ fragment combination products. The idea is now to identify

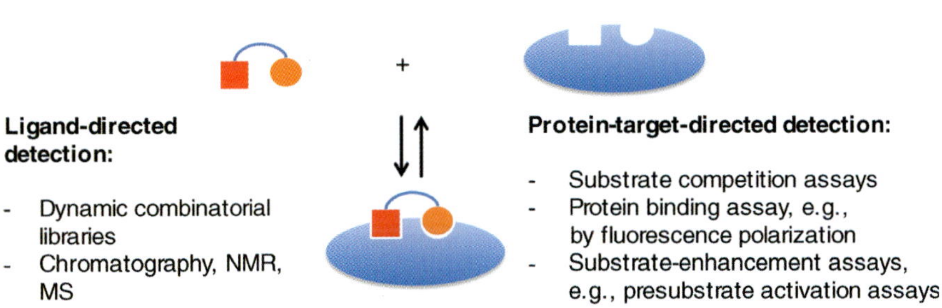


**Figure 13.4** Methods for the detection of protein binding and bioactive fragment combinations.

the best binding combination by adding an equimolar amount or stoichiometric excess of the target protein. Under these conditions, the fragment combinations compete for the binding site and the protein favors formation of the best binding fragment combination and protects it from further conversion, thereby shifting the equilibrium to the preferred product. This equilibrium shift can be observed by NMR or by direct MS analysis. In most examples, however, the preferred ligation product was converted into a chemically stable species, for example, by reduction of the unstable imines to stable amine products. The approach requires the use of a roughly equimolar amount or even stoichiometric excess of the protein target in order to assure formation of quantifiable amounts of ligation product that subsequently is identified by HPLC, HPLC-MS, or $^{1}$H-NMR spectroscopy. Identification of the best binding fragment combination is assured only if the free energy of binding of the fragment combination is significantly high compared to its free energy of formation and if the ligation product is stable enough to enable the fixation reaction. Due to these limitations of the detection, the use of DCL has been mostly reported for small proof-of-concept studies often with libraries forming already known potent fragment combinations.

Similar detection protocols were followed if chemically stable ligation products such as hydrazones were formed. Dynamic combinatorial libraries or dynamic covalent libraries with higher activation barriers do not require chemical fixation of the formed ligation products. Still, even for stable ligation products the free energy of binding has to be significantly higher than the free reaction energy in order to enable the enrichment of a best binding combination in the product. At present, complex fragment mixtures for ligation purposes are used less intensively, possibly due to the availability of alternative detection protocols with less experimental uncertainty. Fragment mixtures are, however, still broadly in use for fragment screening by biophysical detection such as NMR and X-ray crystallography.

An interesting alternative to the ligand-directed detection methods is offered by the protein target-directed detection of bioactive fragment combinations (see Figure 13.3). The big advantage of this approach is that it records directly the bioactivity of the protein in the presence of a given fragment combination. So far, several bioassays have been adapted to bioactivity-guided fragment ligations including substrate competition and substrate enhancement assays. In substrate competition assays, a fluorogenic or chromogenic substrate or a FRET substrate is replaced by the fragment combination, whereas in the substrate enhancement assay, affinity of the substrate itself is increased by reversible ligation resulting in a more rapid turnover of the substrate. Recently, the detection of fragments has been extended to the ligation of presubstrates [16]; these are nonsubstrate molecules that are turned into enzyme substrates through a ligation reaction. Also, binding assays have been adapted to fragment ligation screening using, for example, fluorescence polarization (FP) or radioisotopes for detection. Again fragment ligation can be observed as fragment enhancement, if the fluorophore is coupled to a ligating fragment, or in a competitive mode, in which the binding fragment combination is detected by replacing the fluorescent probe from the protein binding pocket. Bioassay-based fragment detection methods have the strength to be conducted in high-throughput formats using established screening infrastructure including microtiter plates, plate readers, and pipetting robots. Therefore, single fragment combinations can be tested and usually no complex fragment mixtures are used. As a result, the observed bioactivity can be attributed to a pair of reactive fragments without requiring deconvolution.

### 13.3.4 Applications of Dynamic Fragment Ligations in FBDD

While the seminal studies on dynamic combinatorial chemistry had impressively demonstrated that protein targets added in excess could shift the equilibrium reaction to the thermodynamically favored reaction products, they still lacked the integration into functional biological assays and therefore had limited impact on the development of novel, fragment-based protein ligands [2,61]. During recent years, however, several advanced strategies of dynamic fragment ligation techniques have been developed, which are represented in the following as selected highlights.

The direct integration of dynamic fragment ligation with an enzyme assay has been demonstrated first for the SARS coronavirus main protease (SARS CoV $M^{pro}$, Figure 13.5) [11]. SARS CoV $M^{pro}$ is a virus-encoded cysteine protease

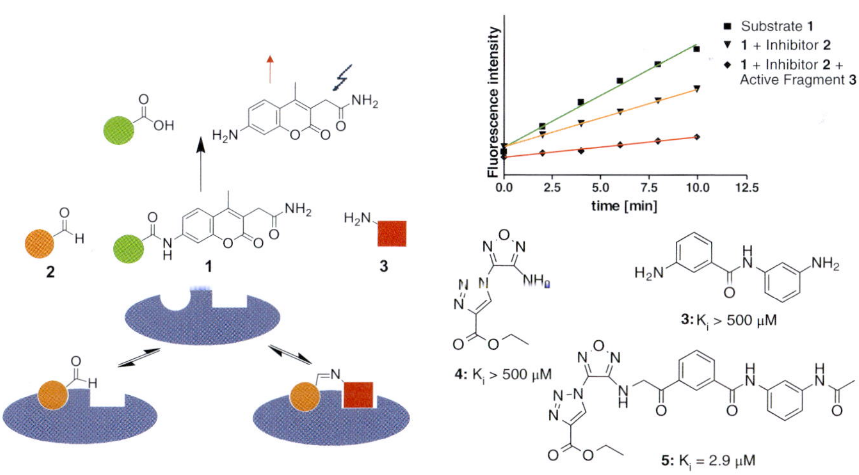


**Figure 13.5** Concept of dynamic ligation screening illustrated by the development of nonpeptidic SARS-CoV $M^{pro}$ inhibitor [11]. Substrate **1** (green) competes with peptide aldehyde inhibitor **2** (orange) targeting the S1 pocket of the SARS-CoV main protease (blue). An active amine fragment **3** (red) binds to the S1′ pocket, leading to increased inhibition via dynamic ligation in the active protein site. Based on the active amine fragment **3** (red), an analogous electrophilic aldehyde fragment that also binds to S1′ was synthesized and used for DLS of the S1 pocket, yielding active fragment **4**, and the combination of low-affinity fragments **3** and **4** yielded nonpeptidic inhibitor **5**.

essential for the replication of the virus inside the infected host cells; inhibitors targeting $M^{pro}$, therefore, are relevant as potential antivirals. Due to the fact that proteases possess defined binding pockets for the side chains of peptide, it was possible to develop a peptide aldehyde inhibitor positioning an electrophilic aldehyde precisely at the active site of the protease. Dynamic ligation screening (DLS) for nonpeptidic fragment inhibitors has been performed, exploiting the reversible reaction of the aldehyde and one amine per microtiter plate well forming a peptidyl imine (Figure 13.5). Bioactive fragment combinations were detected in a substrate competition assay employing a fluorogenic tetrapeptidyl-7-amino-4-methyl coumarinyl-2-acetamide (AMCA) as substrate. Partial inhibition by the peptide aldehyde was amplified considerably by S1′ binding amine fragments and the peptide aldehyde enables the site-specific, sensitized detection of low-affinity fragments with mM $K_I$. The best fragment hit was converted into the corresponding aldehyde fragment and assayed again for enhanced inhibition against an amine library, providing a secondary hit fragment for the adjacent S1-pocket of the enzyme's binding site. Only two iterations of dynamic ligation screening starting from a peptide inhibitor resulted in the selection of two millimolar small molecule fragments and the covalent combination of these fragments by reductive amination revealed an entirely nonpeptidic inhibitor of SARS CoV $M^{pro}$ with a $K_I$ value of 2.9 μM.

The nature of protein-templated fragment ligation was further scrutinized by extending dynamic ligation screening from an enzymatic assay toward a general protein binding assay employing fluorescence polarization for detection [12]. Attachment of carboxy-fluorescein as a fluorophore to a protein binding fragment A enables a clear distinction between the cooperative binding of secondary fragment B from additive binding or competitive binding (Figure 13.6). While additive binding of B does not influence the binding of A leaving the fluorescence polarization unchanged ($FP_{AB} = FP_A$), competitive binding of B results in the reduced binding of A ($FP_{AB} < FP_A$) and cooperative binding in the enhanced binding of A ($FP_{AB} > FP_A$). The assay was established for caspase 3, a protein playing an essential role in the execution phase of cell apoptosis, thus making it a potential drug target in traumatic brain injury, amyotrophic lateral sclerosis, Alzheimer's and Parkinson's disease. A total of 7397 fragments including 4019 nucleophilic and primary amines from the ChemBioNet library have been tested in 384-well microtiter plates. We observed no change for most of the fragments tested, 78 fragments led to FP signal reductions (negative cooperative), and 176 fragments evoked a 20% stronger FP (positive cooperative) signal than the control. Twenty-one fragments were confirmed as competitive inhibitors of caspase-3 with $K_i < 10$ μM. The best cooperatively binding fragment was linked covalently to fragment A by reductive amination resulting in a potent inhibitor of caspase-3 with a $K_I$ value of 80 pM.

Dynamic ligation assays were further implemented for secondary site screening of four closely related protein tyrosine phosphatases (PTP): human PTP1B, PTPN7, PTPN12 (= SHP2), and mycobacterial MPTPA [14]. Generally, the development of specific PTP inhibitors is considered to be a major challenge

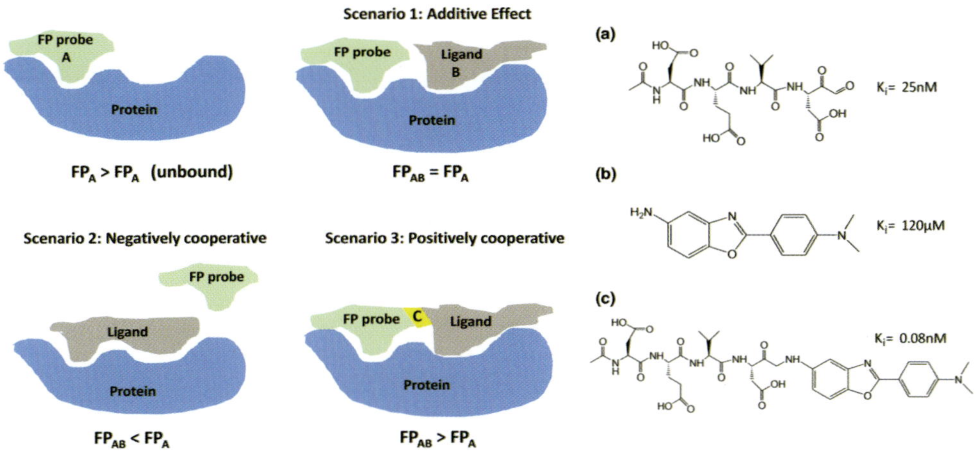


**Figure 13.6** Selective detection of cooperatively binding fragments in a dynamic ligation screening [12]. Fluorescence polarization was used to determine the binding of fragment A. Three different scenarios (S1-3) of fragment interaction can be distinguished: Additive binding of fragment B does not affect FP (S1), competitive or negatively cooperative binding leads to decreased FP (S2), positively cooperative binding amplifies FP (S3). Coupling of the nonfluorescent derivative of A with positively cooperative fragment B furnished a picomolar inhibitor of the target protein, the protease caspase-3.

raising doubts on the druggability of this physiologically relevant class of enzymes. Using 4-formylphenyl phosphate as an electrophilic PTP substrate, specific secondary site binding fragments could be selected for each of the PTP from a library of only 110 primary amines. Specific secondary site binders were detected by the amplified release of phosphate from the phenyl phosphate fragment as determined in a Malachite green assay. This assay format was denominated as a “dynamic substrate enhancement”: active fragments form a ligation product with a lowered $K_M$ value, corresponding to the higher stability of the enzyme–substrate complex. Replacement of the phenyl phosphate substrate by a noncleavable phosphotyrosine mimetic and covalently linking it to the selected amines resulted in inhibitors of *Mycobacterium tuberculosis* protein tyrosine phosphatase A (MPTPA) with no activity for the three other protein tyrosine phosphatases, suggesting that second-site targeting of PTP is sufficient to develop selective PTP inhibitors.

Dynamic ligation screening has further been extended toward aspartic proteases, this time employing a *F*luorescence *R*esonance *E*nergy *T*ransfer (FRET) assay [15]. Peptide aldehydes were used as directing probes to identify the active-site binding fragments of the aspartic protease β-secretase (BACE-1). This enzyme was known as a main culprit in aggregation of amyloid-β-peptide, which is a major pathological hallmark of Alzheimer’s disease (AD). Thus, efforts have been made to discover BACE-1 inhibitors for potential therapeutic treatment for AD. Instead of reversible hemithioacetal formation, aldehyde hydrate was

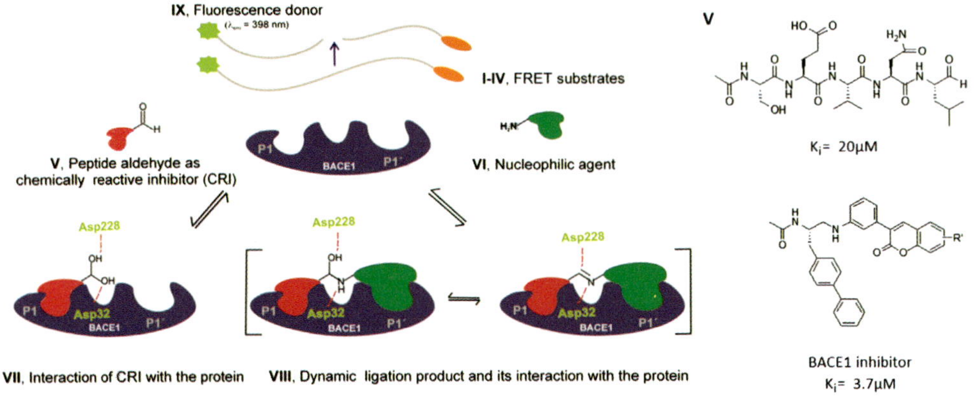


**Figure 13.7** The concept of DLS applied to the identification of BACE-1 inhibitors [15].

formed to bind the catalytic aspartic acid dyad through hydrogen bond formation. Then, dynamic ligation of the peptide aldehyde with an amine nucleophile yields a reversible imine product. The peptide aldehyde, which serves as a chemically reactive inhibitor (CRI), revealed 3-(3-aminophenyl)-2*H*-chromen-2-one to be a competitive BACE-1 inhibitor. The identified 3-(aminophenyl) coumarin fragment was used as a starting point for hit optimization and a low micromolar ($K_I = 3.7\,\mu M$) BACE-1 inhibitor was developed [15] (Figure 13.7).

Although most initial work on dynamic ligation was conducted with aldehydes and amines forming imines as ligation products, there have been numerous attempts to extend the scope of dynamic ligation to other nucleophiles including alcohols [16,19], thiols [16,22,23], and even C-nucleophiles like nitroalkanes [24–28]. The detection of hemiacetals and hemithioacetals in aqueous solution, however, is severely impeded by the instability of these ligation products. Several strategies have been applied to overcome this problem. Ramström and colleagues recently investigated an elegant approach: the protein-templated formation of hemithioacetals in aqueous medium by using proton saturation transfer difference NMR spectroscopy ($^1$H-STD-NMR) Figure 13.8 [23]. For a proof-of-concept study, β-galactosidase was selected as a target protein catalyzing the hydrolysis of *O*-β-galactosides and containing no cysteine residues in the active site. A mixture of five thiols and two aldehydes was employed resulting in the potential formation of 10 hemithioacetals. $^1$H-STD NMR spectroscopy is a method exploiting the selective transfer of proton magnetization from the protein to reversibly bound ligands and was used here to identify binding fragments from this dynamic combinatorial library (DCL). It was demonstrated that 1-β-mercapto-D-galactose binds more strongly to the target enzyme than the four other thiols, while both aldehydes displayed strong signals in the ST difference spectrum. Surprisingly, binding of the two aldehydes could not be suppressed by the addition of substrate, suggesting additional allosteric binding sites for these fragments. Significant inhibition of substrate hydrolysis could be confirmed for

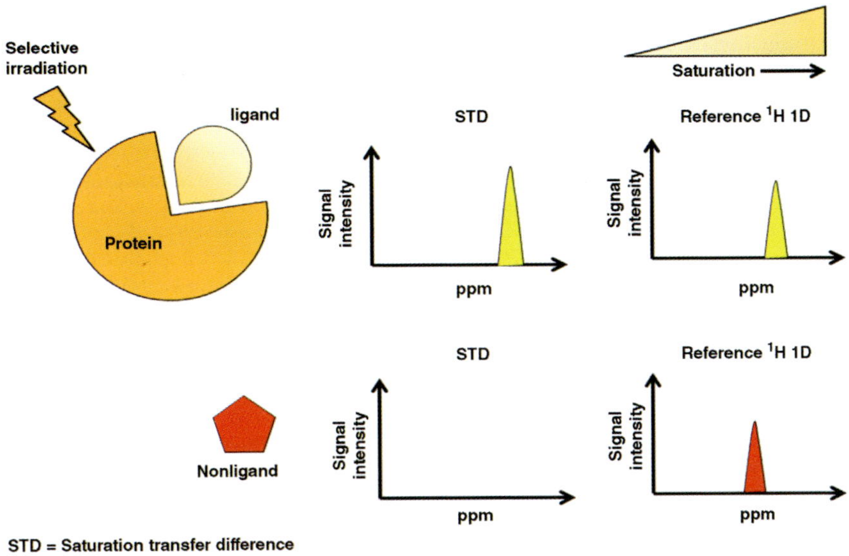


**Figure 13.8** Application of $^1$H-STD NMR spectroscopy for characterizing protein-bound fragments from dynamic combinatorial library [23].

one thiol-aldehyde combination, strongly suggesting the formation of hemithioacetals as active inhibitors.

Reversibly formed, unstable ligation products such as hemiacetals and hemithioacetals can also be detected directly through biochemical assays as recently demonstrated in the "presubstrate activation" concept [16]. In this approach, protein binding fragments are exploited to induce enzymatic reactions on the protein surface via dynamic fragment ligation. "Presubstrates" are completely inactive ligation probes containing a fluorophore like amino-methylcoumarin (AMC) and an aldehyde as an electrophilic reactive site. Protein binding fragments containing a suitable nucleophile can react covalently with the presubstrate, position it into the active site of the enzyme, and subsequently induce the enzymatic hydrolysis of the protein–fragment–presubstrate (P-F-PS) complex. In the initial work, three electrophilic presubstrates were tested with 14 protein binding fragments and 2 serine proteases, trypsin and factor Xa (FXa), a pharmacological target within the blood coagulation cascade. It could be proven that the increase in fluorescence is directly correlated with the hydrolytic cleavage of the presubstrates and the release of 7-amino-methyl coumarin. The rate of presubstrate cleavage ($V$) increased with growing fragment concentration and reached saturation at maximal conversion rate ($V_{max}$). The stability of the protein–fragment–presubsubstrate complex ($K_M*$ value) was determined, providing a quantitative description of the templating effect exerted by the target protein quantitatively for each fragment–presubstrate combination. Thus, the approach enables the site-specific identification of protein binding fragments, the functional characterization of enzymatic sites, and the quantitative analysis of protein-templated ligation reactions. It could be demonstrated that amines, alcohols, and thiols can activate presubstrates through reversible ligation, significantly increasing the scope of dynamic ligation reactions, while irreversible Michael ligations were observed only with thiol nucleophiles.

Contrary to the dynamic ligations of amines, alcohols, and thiols, the ligation products of hydrazines and hydroxylamines are stable enough for product isolation and thus have been the focus of numerous contributions in the past decade. Initially, the requirement of an acidic pH was a major drawback of acylhdrazone formation as this limited the use of this reaction for protein-templated ligation reactions [24,62]. In 2010, the group of Greaney reported that reversible formation of hydrazones can be achieved in the presence of target proteins at almost neutral pH using aniline as a nucleophilic catalyst [31]. In consequence, the time for dynamic combinatorial library to reach equilibrium was drastically shortened from 5 days to only 6 h at pH 6.2 in the presence of aniline. The authors recruited two recombinant isozymes from the gluthatione *S*-transferase (GST) enzyme family and exposed them to a preequilibrated DCL comprising one aromatic aldehyde and ten hydrazides in the presence of 10 mM aniline [23] (Figure 13.8). The aldehyde was tagged with glutathione (GSH) in order to address the GSH binding site or the proteins; thus, the acylhydrazide moiety was expected to target an adjacent hydrophobic binding site (H-site). Formation of acylhydrazones was analyzed using LC–MS, showing that different acylhydrazone products were amplified by each isozyme; for each isozyme, a compound with a low micromolar $IC_{50}$ was identified. Based on the initial findings, a bivalent DCC approach was further developed to screen for potential inhibitors using four different GST isozymes (mGSTM1-1, hGST1-1, SjGST, and mGSTA4-4) as target proteins [36]. Taking advantage of the fact that a majority of GST isozymes form homodimers, a selective nanomolar inhibitor was discovered for mGSTM1-1 isozymes using this approach.

In 2013, the group of Wanner introduced an efficient way to screen for γ-aminobutyric acid transporter 1 (GAT1) inhibitors using a competitive MS-based binding assay and a pseudo-static dynamic combinatorial library [33,34]. The pseudo-static DCL was formed in the presence of target protein mGAT1 with an excess of hydrazine. For the proof of concept, the authors screened a focused hydrazone library, which contained structures that resembled mGAT1 inhibitors. This led to the generation of 36 hydrazones with 1 selected hydrazine and 36 aldehydes (Figure 13.9). Deconvolution experiments were carried out using nine one-dimensional sublibraries containing only one hydrazine and four aldehydes. The competitive MS-based binding assay was then performed with a previously reported native marker and further verified using [$^{3}$H]-GABA uptake assay in mGAT1-expressing HEK293 cells. A low nanomolar range mGAT1 inhibitor was discovered and shown to contain a 2′-4′dichloro biphenyl moiety. The method introduced is highly sensitive and avoids the use of radioisotopes.

The same strategy was adopted for static oxime libraries to discover new potent mGAT1 inhibitors in a subsequent publication in 2015 [37]. The library

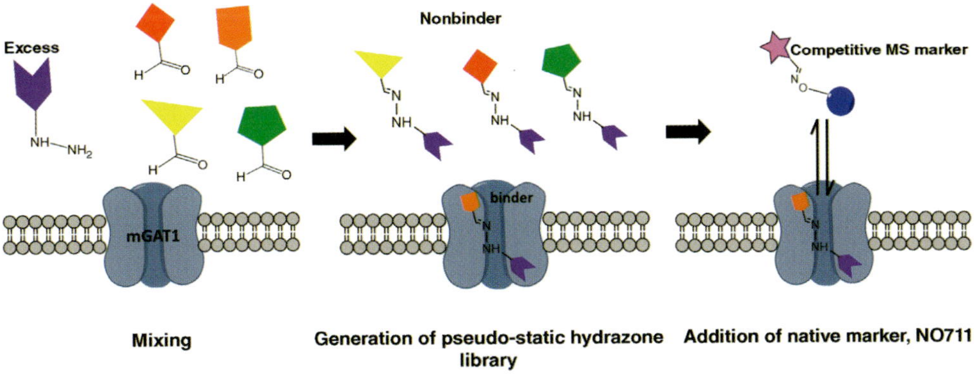


**Figure 13.9** Principal steps in the discovery process of mGAT1 inhibitors starting from pseudostatic hydrazone library formation to competitive MS binding upon addition of native marker, NO711.

was generated by reaction of guvacine derivatives with various sets of four aldehydes, and the most potent mGAT inhibitor was identified with a p$K_i$ value ≥8.

Another recent application of acylhydrazone libraries was recently published by the groups of Hirsch and Klebe, describing the combination of *de novo* structure-based design (SBD) with dynamic combinatorial chemistry (DCC) for hit identification and optimization [63]. Endothiapepsin, an aspartic protease, plays role in numerous diseases from hypertension to malaria. As the protein is stable at a slightly acidic pH of 4.5, aniline can be omitted in the equilibration reaction. Two acylhydrazones were designed by referring to two crystal structures of endothiapepsin, which display two alternate ligand binding modes, in which ligands are either bound directly to the catalytic dyad of this aspartic protease or form a hydrogen bond network including the catalytic water molecule. The ligand design was validated by construction of a dynamic library of five hydrazides and five aldehydes. Next, saturation-transfer difference NMR spectroscopy ($^1$H-STD NMR) was used to screen for the best component associations that bound to endothiapepsin. This method allows the direct characterization of protein–ligand interactions in solution and is thus a powerful analytical tool for DCLs. Hit validation was carried out using fluorescence-based enzyme activity assays and two potent, low micromolar range inhibitors were found. Cocrystallization of the two inhibitors with endothiapepsin confirmed the hydrogen bond formation from *in silico* prediction. This combination of methods was proposed to provide a less risky alternative for *de novo* design and to be beneficial at the early stage of drug discovery.

Beyond condensation reactions of carbonyl electrophiles and (hetero)-nucleophiles, reversible disulfide formation has been recognized as a powerful method for fragment combination. An impressive application of dynamic disulfide exchange was recently demonstrated by the group of Miller [64]. These authors work on myotonic dystrophy type 1 (DM1), the most common form of muscular

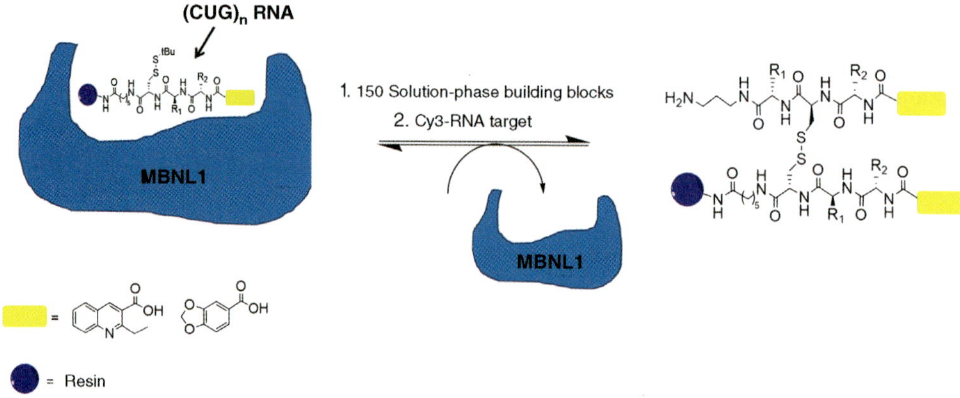


**Figure 13.10** Preparation of disulfide-bridged tripeptide library directed against binding of MBNL1 to (CUG) repeat RNA.

dystrophy in adults. This disease is characterized by dramatically expanded CUG repeats in heterogeneous nuclear (hn) RNA species that lead to a pathologically enhanced sequestering of RNA-binding proteins, for example, the splicing regulator MBNL1 (muscleblind-like splicing regulator 1) [64] (Figure 13.10). Therefore, molecules that inhibit the interaction of RNA-based CUG repeats with MBNL1 may be a treatment option. In searching for such 'interaction inhibitors,' the authors used a library of 150 cysteine-containing tripeptide derivatives covalently attached to resins. This resin-bound tripeptide library was screened against an identical set of solution-phase tripeptides allowing for the oxidative generation of disulfide bonds between two tripeptides. This ligation screening was performed in the presence of an artificial RNA target resembling the principal conditions found in myotonic dystrophy type 1: CCG-$(CUG)_{10}$-CCG RNA.

For detection, this RNA was 5′-labeled with the fluorescent dye cyanine-3 (Cy3-RNA). The first readout of successful resin-bound tripeptide combinations was done by fluorescence microscopy as a result of Cy3-RNA binding to the resin beads and resulted in a maximum number of 10 possible disulfides from which the best ones displayed low μM $K_i$ values for the inhibition of MBNL1 binding to artificial (CUG) repeat RNA. In addition, enzyme fragment complementation (EFC) assay was employed to evaluate the ability of selected ligands in disrupting the interaction of the CUG repeat RNA with the protein MBNL1. A luminescent product was detected only when β-Gal enzyme acceptor bound to the prolabel (PL)-MBNL1 enzyme donor. Thus, any luminescence correlates with the amount of (CUG) repeat bound MBNL1 that allows $K_i$ value determination of lead compounds. The inventive idea in this work in particular is the target-driven generation of a library of disulfide-linked tripeptides in covalent attachment to a resin that facilitates the readout of active peptide dimers.

In 2011, Claridge and colleagues reported that the inhibition of nucleophilic enzymes by boronic acids was greatly improved by addition of specific diols/sugar [53]. For a proof-of-concept study, serine protease α-chymotrypsin (αCT) was used as a model enzyme. Two methods, namely, $^{11}$B-NMR and $^{1}$H-water-LOGSY, were used to observe the formation of ternary complexes of enzyme, boronic acid, and sugar. $^{11}$B-NMR efficiently distinguished free ligands from the protein-bound boron species without the interference from protein background signal, and $^{1}$H-water LOGSY spectroscopy was used to provide higher sensitivity with small amounts of proteins.

The initial boronic acid fragments were selected based on known serine protease inhibitors, **1a-e** ($IC_{50}$ of 5 μM for **1a** and 14 μM for **1b**) while D-fructose (**2a**) and L-fructose (**2b**) were chosen as the diol fragments for dynamic ligation. The pH of the ligation experiment was maintained at 5.8, given that the $pK_a$ value of most boronic acids is around 7–9. Without the protein α-CT, both boronic acids **1a** and **1b** are equally active in boronate ester formation with diol **2a**. In the presence of protein, however, the ternary complex composed of α-CT, **1a**, and **2a** was predominantly formed over α-CT-**1b**-**2a** (Figure 13.11). In this pioneering study, serine protease α-chymotrypsin served as a template for reversible boronate ester formation, showing potential use of this reaction for protein-templated DCC. However, better analytical techniques were required to analyze the labile ternary complexes formed.

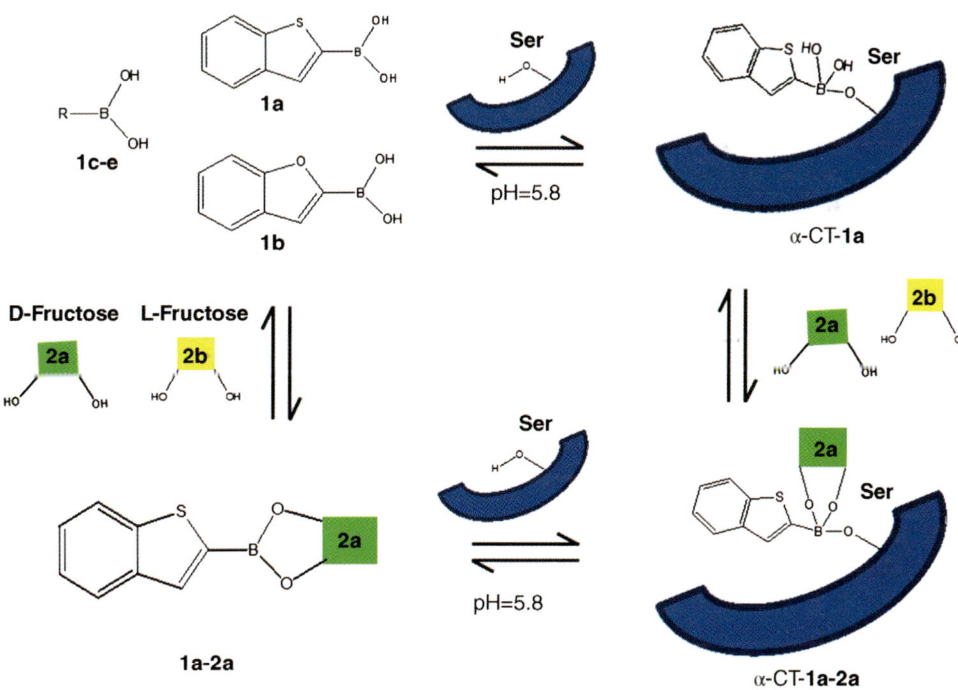


**Figure 13.11** Schematic representation of the generation of boronate ester-based dynamic combinatorial libraries using α-chymotrypsin as an enzyme model with the presence of building block (**1a–e**, **2a**, and **2b**).

Thus, the preliminary findings further encouraged the groups of Schofield and Claridge to apply dynamic combinatorial mass spectrometry (DCMS) of nondenatured proteins for the identification of reversibly formed boronate ester as potential inhibitors of 2-oxoglutarate (2-OG) oxygenases [65]. The prolyl hydroxylase domain isoform 2 (PHD2) was taken as a model enzyme for the screening of a dynamic combinatorial library. In brief, PHD2 is a $Fe^{II}$ containing 2-OG oxygenase, which helps to regulate the human hypoxic response, making it an excellent drug target for treatment of anemia and ischemia-related diseases.

In the experiments, a heterocyclic strongly iron-binding boronic acid fragment was incubated with several pools of diol ligands and subsequently analyzed by nondenaturing ESI-MS. Best binding fragment combinations were identified by the shift in protein mass, and converted into stable, boron-free inhibitors using Suzuki cross-coupling reactions, resulting in nanomolar inhibitors of the enzyme. The results of the fragment ligation assays were further validated in another study using an $^{1}$H-NMR-based method [66]. In short, for this assay the enzyme was deactivated by substituting Fe(II) ions with catalytically inactive Zn (II) ions. The cosubstrate, 2-oxo glutarate, served in this assay as a reporter ligand and its displacement upon binding of reversibly formed competitive ligand was monitored by $^{1}$H-NMR spectroscopy and was able to confirm the initial screening of PHD2/2OG inhibitors by validating hits discovered by DCMS screening reported in their previous studies.

Dynamic fragment ligation screening in the presence of the target can efficiently steer the underlying equilibrium of covalent-reversible reactions toward selective amplification of the strongest target binders. On the other hand, the best reaction conditions of the chemical fragments are not always compatible with the biochemical properties of the target. Most importantly, the following two points have to be considered when setting up an assay for dynamic fragment ligation: (i) are the chemical fragments soluble/stable in aqueous medium at a pH value that helps keep the target biologically intact and (ii) are unwanted side reactions with the functional groups of the target (mostly) excluded?

In cases where dynamic library generation in the presence of the target might be problematic, dynamic fragment ligation can still be applied in the form of preequilibrated fragment ligation. An example of this was shown in a publication from Poulsen and Bornaghi who searched for small molecule inhibitors of carboanhydrase II utilizing alkene cross metathesis as the reversible reaction [56]. Carboanhydrase II catalyzes the interconversion of $CO_2$ and $HCO_3^-$ and has been linked to the pathology of glaucoma and epilepsy. In the mentioned work, the application of a preequilibrated fragment ligation (i.e., in the absence of the target carboanhydrase II) was necessary as the Grubbs catalyst employed for cross metathesis, a ruthenium (II) carbene complex, is not sufficiently soluble in water. Following equilibrium incubation, the samples were serially diluted with water and read out for carboanhydrase II inhibition in a fluorescence competition assay. Some of the selected products revealed $K_I$ values in the low nanomolar range.

## 13.4
## Irreversible Fragment Ligation

### 13.4.1
### Irreversible Fragment Ligation Strategies: Pros and Cons

The major advantage of irreversible fragment ligations is that they can directly form the active protein ligand without the necessity to transform or stabilize the protein binding fragment combination chemically. Ideally, two fragments, which do not react with each other in the absence of protein, bind to adjacent protein binding sites, thereby lowering the energy of activation and stabilizing the transition state of the ligation reaction. In this mechanism, the protein catalyzes the formation of its ligands, though progression of the catalytic cycle might be limited by product inhibition, if the protein binding sites are blocked by the fragment combination. As irreversible fragment ligations usually require activation energy, they can be slow on the timescale of protein assays.

Considering the proposed mechanism of target-guided reactions, it can be expected that reactions should be favored for fragment ligations in which the starting fragments, the transition state of the reaction, and the formed fragment combination bind in a similar orientation to the protein. These conditions seem to be met by many reaction types and accordingly numerous reactions have been applied to irreversible fragment ligation (Table 13.3).

Among these reactions, pericyclic reactions have been employed most frequently. Classical Diels-Alder cycloadditions have been used in older studies on DNA-templated reactions, which usually involve higher binding affinities and thus can be operated also with slower reactions, whereas in recent years dipolar cycloadditions became quite popular for irreversible target-guided synthesis. Several studies have applied the dipolar cycloadditon of an azide dipole with a terminal alkyne residue for the formation of target-guided synthesis of protein ligands, and recently the reaction of sulfonyl azides with thioacid fragments has been implemented for the purpose and termed the "sulfo-click" reaction, furnishing sulfonamides. Alternative reactions used for irreversible ligations beyond cycloadditions include nucleophilic reactions at $sp^3$ carbon centers, Michael additions, and acylations. However, all these alternatives have been employed only occasionally compared to the dominating click reactions. As a result, the diversity of fragment linkages obtained by irreversible ligations is rather limited and does not represent the linkages most often represented in drug molecules.

### 13.4.2
### Detection in Irreversible Fragment Ligation

Due to the stability of the products, detection of irreversibly formed fragment ligations can be conducted with standard analytical methods such as HPLC–MS or NMR. In many cases, however, the amount of formed product and thus its

**Table 13.3** Irreversible fragment ligation reactions employed in FBDD: (a) sulfides, (b) acylsulfonamides, (c) triazoles, (d) sulfides, (e) disulfide exchange, (f) acyltransfer, and (g) nucleophilic substitution.

| **Irreversible covalent bond formation** | **References** |
|---|---|
| (a)  | [67] |
| (b) 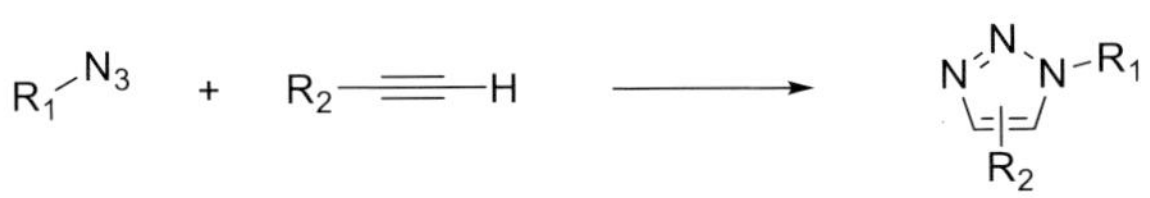 | [68–73] |
| (c)  | [74–76] |
| (d) 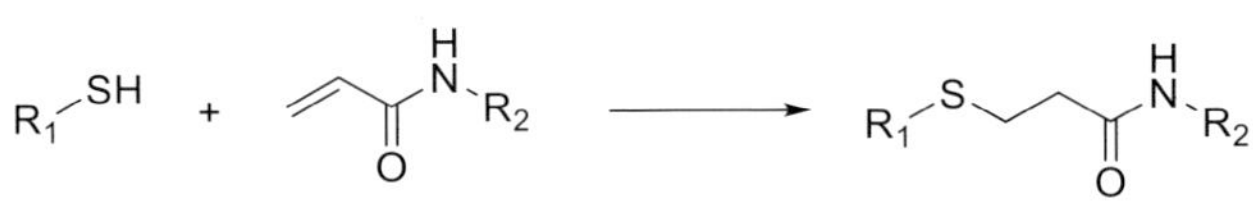 | [16,77] |
| Ligations involving covalent reactions with the target | |
| (e) 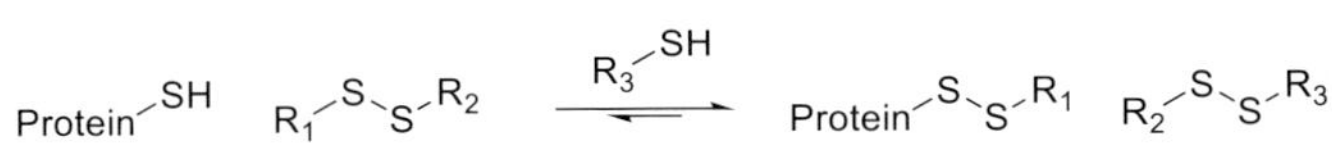 | [78–81] |
| (f) 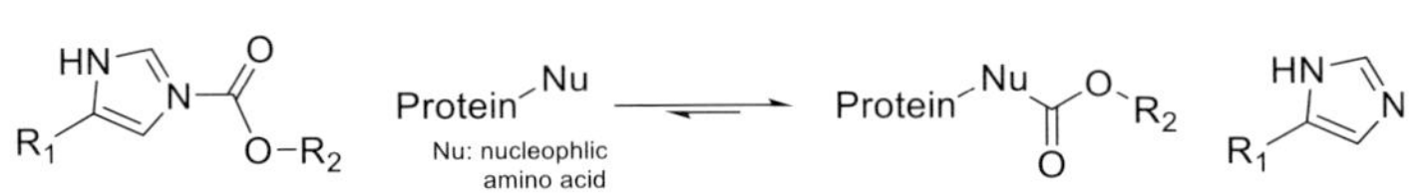 | [82] |
| (g)  | [83] |

detection is limited by the product inhibition that is inherent to the method. As a consequence, the identification of ligation products by NMR is feasible only with large amounts of protein, whereas advanced MS techniques are sensitive enough to detect even small traces of the ligation product formed.

Bioassay-based detection of best binding fragments could be advantageous as well, though this mode of detection seems to be neglected in the area of irreversible fragment ligation so far. Possibly, the establishment of such methods would stimulate the field and enable the identification of ligation products from more diverse chemistries, overcoming the current focus on the click chemistry.

### 13.4.3
### Applications of Irreversible Fragment Ligations in FBDD

An early proof-of-concept study on kinetic target-guided synthesis was conducted by the Sharpless group, introducing the *in situ* click reaction as an irreversible ligation reaction templated by the target protein [84]. In a follow-up paper from 2005, the method was again applied to acetylcholinesterase (AChE), a target in the treatment of Alzheimer's desease [71]. The goal of the project was to identify and synthesize a fragment combination associating both with the active center and with the peripheral binding site (Figure 13.12). As a starting point, the well-known AChE inhibitor tacrine was used, which binds to the active center of the enzyme. Tacrine was modified by introducing an azide functionality. A compound collection of 23 terminal acetylenes was composed of peripheral site binding fragments and tested to detect the formation of the 1,2,3-triazole formed by dipolar cycloaddition. First, LC–MS results confirmed the templating effect of the enzyme for protein binding fragments. Exclusively in the presence of the enzyme, 1,5-disubstituted (syn) triazoles were formed, showing little or no background reaction. After having confirmed the reaction with single-fragment pairs, the authors turned their attention to multicomponent reactions using mixtures of up to 10 acetylene compounds simultaneously. Indeed, the protein template induced formation of highly potent fragment

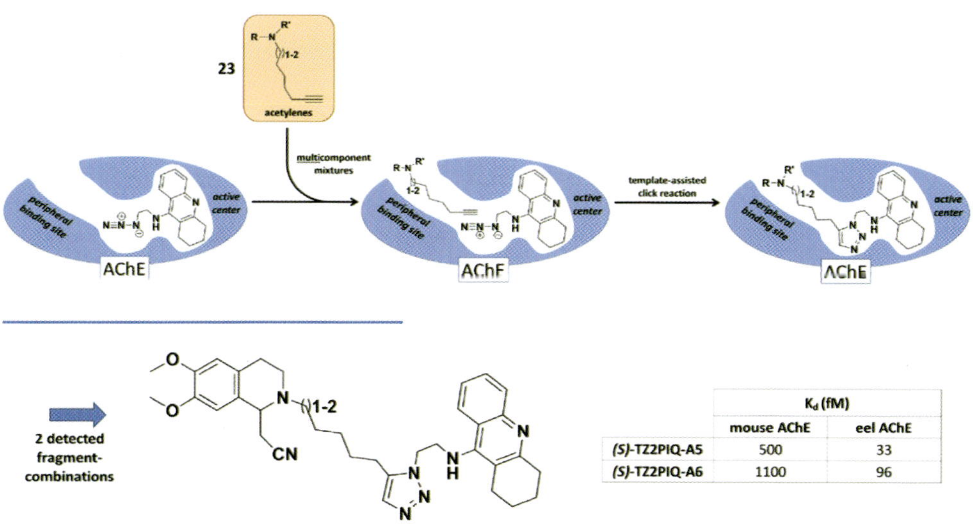


| | $K_d$ (fM) | |
|---|---|---|
| | mouse AChE | eel AChE |
| *(S)*-TZ2PIQ-A5 | 500 | 33 |
| *(S)*-TZ2PIQ-A6 | 1100 | 96 |

**Figure 13.12** *In situ* click chemistry screening for acetylcholine-esterase inhibitors. Step 1: known active center binding tacrine azide serves as anchoring molecule. Addition of up to 10 acetylene compounds per mixture. Step 2: assembly of the best fitting acetylene compound. Due to the close proximity of the reactive groups, triazole-formation becomes possible, templated by the target protein. Via LC/MS–SIM measurements, two fragment combinations could be detected.

combinations, displaying dissociation constants in the low picomolar or even femtomolar range.

Azide-alkyne ligations have been applied to other protein targets as well. For example, formation of a known nanomolar inhibitor of HIV protease containing an antisubstituted 1,2,3-triazole was accelerated in the presence of the enzyme [85], although the reaction was considerably slower than reported for the pico/femtomolar AChE inhibitors, and this time a clear background reaction to the syn-triazole was observed. Protein-template azide-alkyne "Click" reactions have also been successfully applied to the Abl tyrosine kinase [69]. The Fukase group has demonstrated the templated synthesis of functional mimetics of the Grb2-SH2 domain, which inhibited the growth of cancer cells *in vitro* [68]. In order to obtain the best inhibitor through the templated reaction of azide and alkyne fragments, it is very crucial to exclude even tiny amounts of copper ions from the reaction. The groups of Miyata and Finn have reported the *in situ* synthesis of inhibitors of the histone deacetylase 8 (HDAC-8) from an alkyne fragment bearing a hydroxamate and an azide fragment [86]. Surprisingly, the formed ligation product was far from the best inhibitor and comprised a 1,4-disubstituted antitriazole instead of the more active syn-triazole. A detailed scrutiny of these results revealed that the reaction contained traces of Cu (I) ions, which were bound by the metal binding site of HDAC-8 and were sufficient to catalyze the reaction to the less favored cycloaddition product.

Beyond azide-alkyne ligations, other dipolar cycloadditions have been investigated, too. Reactions of thioacids with sulfonyl azide fragments, which provide acylsulfonamides via a cyclic intermediate, constitute especially successful examples and have been denominated as sulfo-click reactions. In a 2011 paper, the Manetsch group demonstrated protein-templated, irreversible ligation reactions with an especially challenging target, the protein–protein interaction domain Bcl-$X_L$ [76]. The interaction of Bcl-$X_L$ with BH3 peptide is a crucial step in the regulation of apoptosis, the programmed cell death. Inhibitors of this interaction might be potent anticancer agents. In a screen for new modulators, Bcl-$X_L$ protein was used as a template for the reaction of a sulfonyl azide and thioacid fragments via sulfo-click reaction (Figure 13.13).

Out of 81 possible fragment-combinations, only 4 combinations were detected via LC/MS-SIM measurements of which all showed biological activity. To validate the concept that Bcl-$X_L$ serves as a template for the sulfo-click reaction, several control experiments were conducted, all confirming that the best inhibitors have been generated by a template-assisted reaction induced by the protein target (Figure 13.14).

The results proved the template function of Bcl-$X_L$, therefore, making the sulfo-click reaction an interesting new reaction type for kinetic target-guided lead discovery.

Another reaction that might find broader application in protein template fragment ligation is the thio-Michael addition. Following an initial communication by Shi and Greaney [50], in 2014 several further applications have been

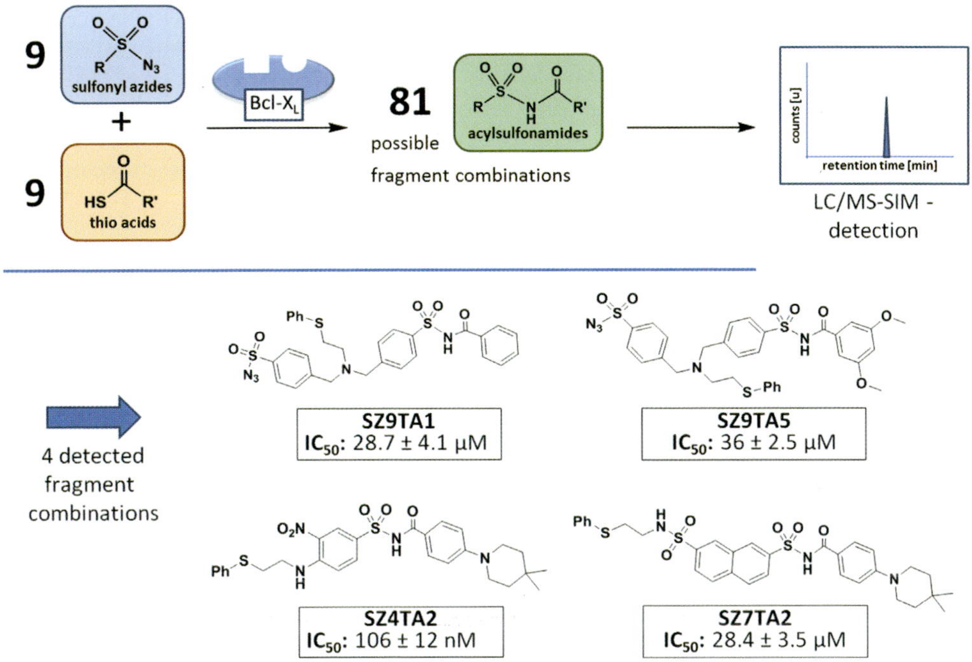


**Figure 13.13** Sulfo-click reaction for the discovery of proapoptotic Bcl-X$_L$ ligands.

reported [16,77]. The group of Pierre-Yves Renard [77] demonstrated the thio-Michael addition with acetylcholinesterase serving as template (Figure 13.15). To avoid the preparation of highly reactive thiols, leading to various possible side products, the authors used acetylated precursors. The screening becomes possible by taking advantage of both the enzyme's natural hydrolase activity and its templating effect.

For the proof of concept, one active site binding fragment was derivatized with an acrylamide serving as the potential Michael acceptor, while the second fragment known to bind to the peripheral binding site was functionalized with *S*-acetyl thioester. The *S*-acetyl group serves as a temporary protection of the reactive thiol, which is, however, rapidly cleaved by the esterase activity of the enzyme, a serine protease. The goal of the experiment was to investigate the optimal length of the linker between both binding fragments. Therefore, two small libraries were prepared, containing both fragments with chain lengths of two–four carbon atoms, leading to nine possible fragment combinations. The screening resulted in the detection of two active fragment combinations, which showed a 14–57-fold increase in inhibitory activity compared to the fragments alone. Thus, the thio-Michael addition is suitable for protein-templated fragment ligation and constitutes a valuable addition to the still limited pool of reactions being useful for this purpose.

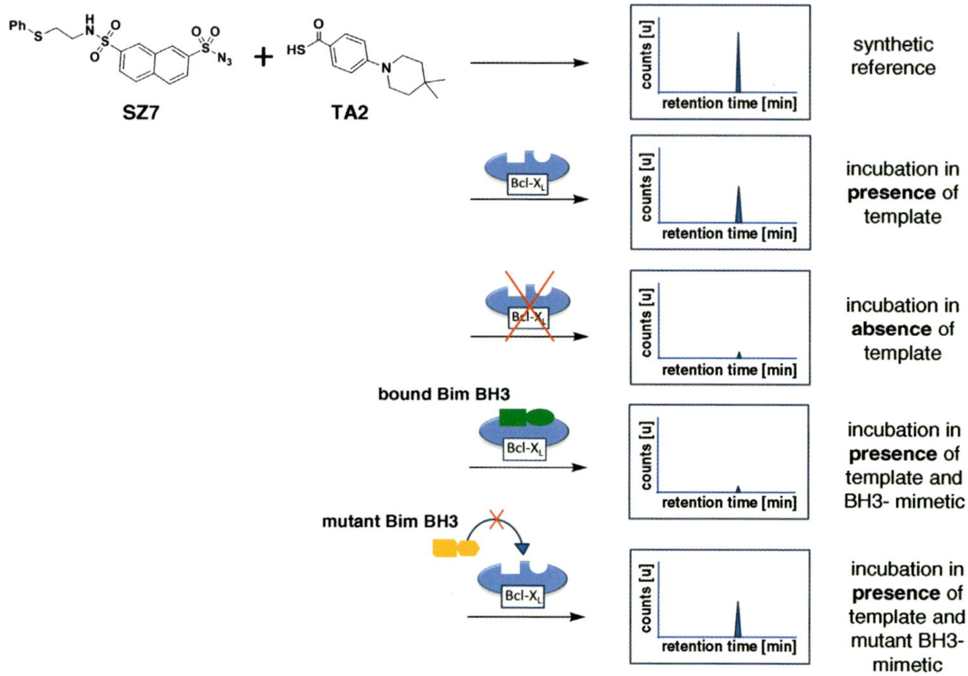


**Figure 13.14** Control experiments with fragments SZ7 and TA2 from top. (1) Reference compound from chemical synthesis; (2) templated reaction: fragments incubated in the presence of target protein form product; (3) background reaction: fragment incubated without protein template form no product; (4) Bim BH3, a native ligand of the protein template inhibits fragment binding to Bcl-X$_L$, no product formation; (5) mutant Bim BH3 has low affinity to Bcl-X$_L$, fragment ligation product formed.

## 13.5
## Fragment Ligations Involving Covalent Reactions with Proteins

Fragment ligations between a reactive fragment and the protein itself are to be covered in this article as they also constitute examples for protein-templated reactions and thus might be applicable to FBDD as well. The recent years have witnessed further applications of the tethering approach for site-directed ligand discovery [79]. This method exploits the reversible disulfide exchange reaction on the protein surface between a natural or an engineered cysteine residue and a disulfide-containing small molecule fragment in solution. Further developments by the Erlanson group using the so-called "extenders" made this approach even more applicable for fragment combination reactions [80]. In 2008, the group conducted an inhibitor screen targeting the Aurora kinase (Figure 13.16) [81]. The first step was to introduce a cysteine residue near the ATP binding site.

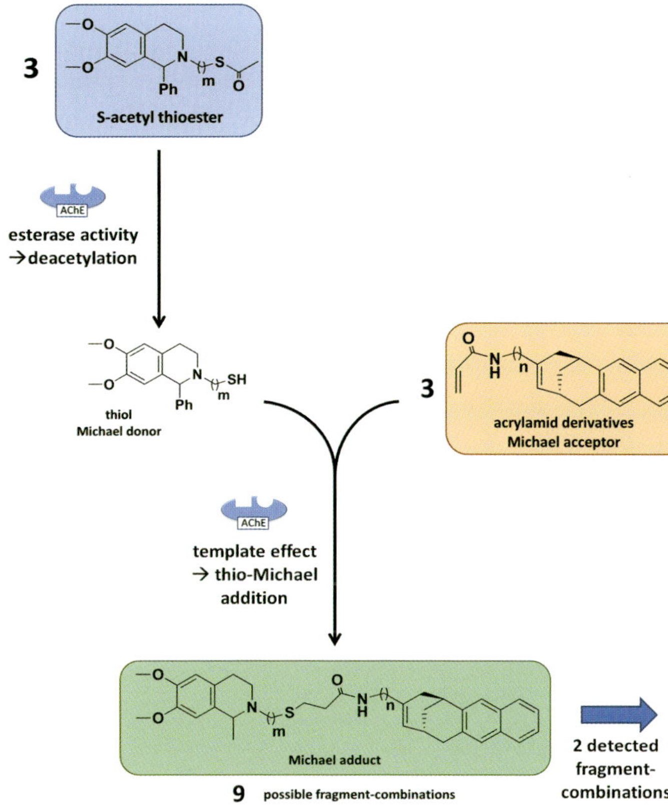


| | n | m | $IC_{50}$ (nM) |
|---|---|---|---|
| 1 | 3 | 2 | 0.9 ± 0.1 |
| 2 | 3 | 4 | 4.0 ± 0.2 |

**Figure 13.15** Thio-Michael ligations resulting in the formation of inhibitors of acetyl choline esterase (AChE).

Next, an extender was synthesized containing two disulfide-containing appendages and a diaminopyrimidine group, which is known to bind at the purine binding site. One appendage targets the introduced cysteine residue, the other one is able to bind the compound. If the compound associates with the adaptive region of the protein, with the extender in close proximity, two stabilized disulfide bonds are formed. This stabilized complex can be detected via MS measurements looking for modification of the protein's mass. In their experiments, a library containing roughly 4500 disulfide-containing fragments was screened in pools of 10 compounds, resulting in the identification of several fragment combinations. A few modifications of the corresponding fragments led to the formation of a stable μM inhibitor. A similar approach has been successfully applied to another kinase, 3-phosphoinositide-dependent protein kinase-1 (PDK1), attaching the extender through an irreversible alkylation reaction [87].

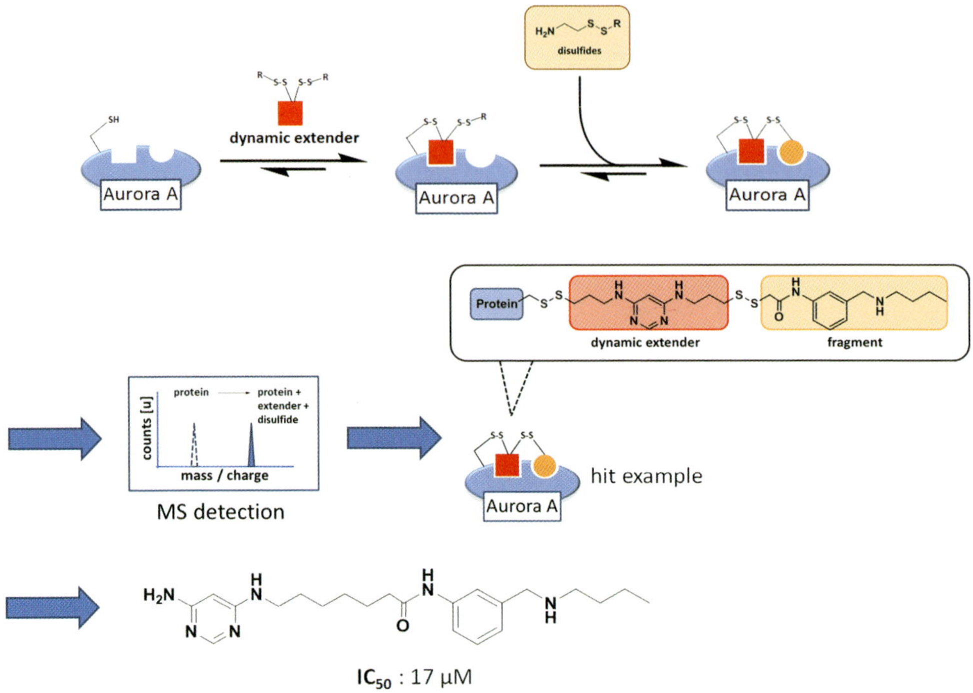


**Figure 13.16** Tethering approach using a dynamic extender has been applied to the screening for fragments binding to Aurora kinase [81].

Several examples of specific and irreversible protein modifications have been reported in recent years. These protein modifications are considered as protein-templated fragment ligations, as a fragment is recognized by a protein target thereby inducing the irreversible reaction. The Hamachi group initially used phenylsulfonates as reactive leaving groups for the ligand-directed alkylation of living cells and even animals. The aim of their work was to develop a method to label membrane-bound proteins on live cells without prior incorporation of unique reactive moieties into the protein via genetic engineering. They developed labeling reagents for the folate receptor, but slow rate and labeling efficiency prevented a broader use of this method [83]. A promising new method has been published by the Hamachi group in 2012, called "ligand-directed acyl imidazole chemistry" (Figure 13.17). In order to provide labeling reagents with increased reaction rate and labeling efficiency, an acyl imidazole-containing labeling reagent was synthesized [82]. This reagent contains a probe (e.g., fluorescein), an alkyloxyacyl imidazole moiety, and methotrexate (MTX), as an affinity ligand for the folate receptor. With this affinity ligand attached, selective binding of the reagent to the target protein is possible, followed by an acyl transfer reaction from the reagent to a natural amino acid on the protein. The probe

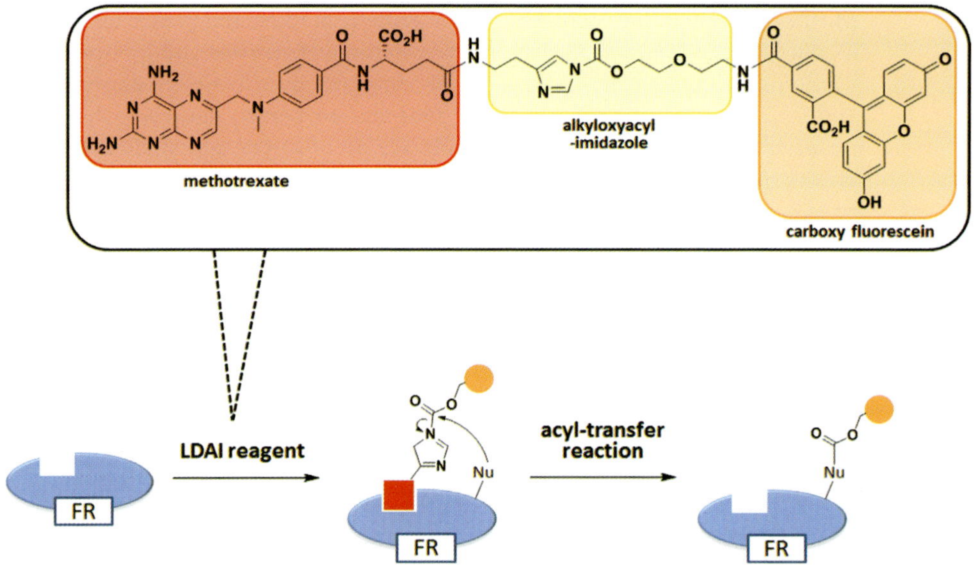


**Figure 13.17** Protein labeling via ligand-directed acyl imidazole chemistry.

is now bound to the protein and the ligand is released, leading to a labeled protein with a vacant binding pocket. Since the natural ligand binding ability of the protein is maintained, this method can be considered as a helpful tool in the future for drug target identification and drug screening via combination reactions even in living cell systems. Moreover, one might expect that the protein-templated labeling chemistry will also find useful applications for template ligand ligation reactions.

## 13.6 Conclusions and Future Outlook: How Far did We Get and What will be Possible?

Without a doubt, the formation and discovery of protein ligands through protein-templated fragment ligations is a tantalizing concept. The method is more elegant and potentially more efficient, rapid, and direct than any preceeding approaches. However, it remains to be scrutinized to what extent this strategy can enable discoveries that exceed the potential of competing methods. In other words, the two ultimate questions seem to be: will fragment ligation provide new protein ligands that could not be identified by classical screening methods and concepts of fragment-based drug discovery? And: can fragment ligation identify ligands with higher efficiency, saving time, human and material resources, and finally money compared to classical methods?

We will focus here on a brief discussion of the first question. Considering published results, there are several examples that answer this question with a yes.

Most advanced is the identification of protein binding fragments for defined binding sites of enzymes as demonstrated both by reversible and by irreversible ligation methods as demonstrated, for example, by the works illustrated in Figures 13.6, 13.7, 13.12, and 13.13. One has to realize, however, that most research projects conducted in this area are proof-of-concept studies. Often, a known protein ligand and/or drug is rebuilt through a fragment ligation reaction or fragments binding to sites that have been defined already. While such proof-of-concept studies are valuable to establish new methodology, they are only the first step in demonstrating the real potential of the concept. The real challenge will consist of going one step further and developing new potent ligands and finally new drugs from a fragment ligation hit. In our opinion, it is not clear yet whether this will be possible. However, considering the growing success of FBDD as a whole and the enormous impact single-fragment hits can have on the creation of new drugs, we expect to see examples of fragment ligation-based drugs in the future.

## References


1 Jencks, W.P. (1981) On the attribution and additivity of binding energies. *Proceedings of the National Academy of Sciences of the United States of America*, **78**, 4046–4050.

2 Corbett, P.T., Leclaire, J., Vial, L., West, K.R., Wietor, J.-L., Sanders, J.K.M., and Otto, S. (2006) Dynamic combinatorial chemistry. *Chemical Reviews*, **106**, 3652–3711.

3 Lehn, J.-M. (2007) From supramolecular chemistry towards constitutional dynamic chemistry and adaptive chemistry. *Chemical Society Reviews*, **36**, 151–160.

4 Ladame, S. (2008) Dynamic combinatorial chemistry: on the road to fulfilling the promise. *Organic and Biomolecular Chemistry*, **6**, 219–226.

5 Schmidt, M.F. and Rademann, J. (2009) Dynamic template-assisted strategies in fragment-based drug discovery. *Trends in Biotechnology*, **27**, 512–521.

6 Jin, Y., Yu, C., Denman, R.J., and Zhang, W. (2013) Recent advances in dynamic covalent chemistry. *Chemical Society Reviews*, **42**, 6634–6654.

7 Mondal, M. and Hirsch, A.K. (2015) Dynamic combinatorial chemistry: a tool to facilitate the identification of inhibitors for protein targets. *Chemical Society Reviews*, **44**, 2455–2488.

8 Ulrich, S. and Dumy, P. (2014) Probing secondary interactions in biomolecular recognition by dynamic combinatorial chemistry. *Chemical Communications*, **50**, 5810.

9 Moulin, E., Cormos, G., and Giuseppone, N. (2012) Dynamic combinatorial chemistry as a tool for the design of functional materials and devices. *Chemical Society Reviews*, **41**, 1031–1049.

10 Godoy-Alcántar, C., Yatsimirsky, A.K., and Lehn, J.M. (2005) Structure–stability correlations for imine formation in aqueous solution. *Journal of Physical Organic Chemistry*, **18**, 979–985.

11 Schmidt, M.F., Isidro-Llobet, A., Lisurek, M., El-Dahshan, A., Tan, J., Hilgenfeld, R., and Rademann, J. (2008) Sensitized detection of inhibitory fragments and iterative development of non-peptidic protease inhibitors by dynamic ligation screening. *Angewandte Chemie International Edition*, **47**, 3275–3278.

12 Schmidt, M.F., El-Dahshan, A., Keller, S., and Rademann, J. (2009) Selective identification of cooperatively binding fragments in a high-throughput ligation assay enables development of a picomolar caspase-3 inhibitor. *Angewandte Chemie International Edition*, **48**, 6346–6349, S6346/1–S6346/13.

13 Herrmann, A. (2009) Dynamic mixtures and combinatorial libraries: imines as probes for molecular evolution at the interface between chemistry and biology. *Organic and Biomolecular Chemistry*, **7**, 3195–3204.
14 Schmidt, M.F., Groves, M.R., and Rademann, J. (2011) Dynamic substrate enhancement for the identification of specific, second-site-binding fragments targeting a set of protein tyrosine phosphatases. *ChemBioChem*, **12**, 2640–2646.
15 Fernández-Bachiller, M.I., Lisurek, M., Horatscheck, A., and Rademann, J. (2011) Alzheimer's disease: Identification and development of BACE-1 binding fragments and inhibitors by Dynamic Ligation Screening (DLS), *ChemMedChem*, **8**, 1041–1056.
16 Burda, E. and Rademann, J. (2014) Catalytic activation of pre-substrates via dynamic fragment assembly on protein templates. *Nature Communications*, **5** (5170), 1–9.
17 Miljanic, O.S. (2011) Reversible covalent chemistry: alcohols in the mix. *Nature Chemistry*, **3**, 909–910.
18 You, L., Long, S.R., Lynch, V.M., and Anslyn, E.V. (2011) Dynamic multicomponent hemiaminal assembly. *Chemistry: A European Journal*, **17**, 11017–11023, S11017/1–S11017/14.
19 You, L., Berman, J.S., and Anslyn, E.V. (2011) Dynamic multi-component covalent assembly for the reversible binding of secondary alcohols and chirality sensing. *Nature Chemistry*, **3**, 943–948.
20 Berkovich-Berger, D. and Gabriel Lemcoff, N. (2008) Facile acetal dynamic combinatorial library. *Chemical Communications (Cambridge, England)*, 1686–1688.
21 Berrocal, J.A., Cacciapaglia, R., and Di Stefano, S. (2011) A well-behaved dynamic library of cyclophane formaldehyde acetals incorporating diphenylmethane units. *Organic and Biomolecular Chemistry*, **9**, 8190–8194.
22 Caraballo, R., Sakulsombat, M., and Ramström, O. (2010) Towards dynamic drug design: identification and optimization of β-galactosidase inhibitors from a dynamic hemithioacetal system. *ChemBioChem*, **11**, 1600–1606.
23 Caraballo, R., Dong, H., Ribeiro, J.P., Jiménez-Barbero, J., and Ramström, O. (2010) Direct STD NMR identification of β-galactosidase inhibitors from a virtual dynamic hemithioacetal system. *Angewandte Chemie*, **122**, 599–603.
24 Beeren, S.R. and Sanders, J.K.M. (2010) *History and Principles of Dynamic Combinatorial Chemistry*, Wiley-VCH Verlag GmbH, Weinheim, pp. 1–22.
25 Angelin, M., Larsson, R., Vongvilai, P., Sakulsombat, M., and Ramström, O. (2010) *Dynamic Combinatorial Resolution*, John Wiley & Sons, Inc., Hoboken, pp. 169–200.
26 Ramström, O. (2007) Dynamic combinatorial resolution. *Latvijas Kimijas Zurnals*, **4**, 354–360.
27 Angelin, M., Fischer, A., and Ramstroem, O. (2008) Crystallization-induced secondary selection from a tandem driven dynamic combinatorial resolution process. *Journal of Organic Chemistry*, **73**, 3593–3595.
28 Vongvilai, P., Angelin, M., Larsson, R., and Ramstrom, O. (2007) Dynamic combinatorial resolution: direct asymmetric lipase-mediated screening of a dynamic nitroaldol library. *Angewandte Chemie International Edition in English*, **46**, 948–950.
29 Dirksen, A., Dirksen, S., Hackeng, T.M., and Dawson, P.E. (2006) Nucleophilic catalysis of hydrazone formation and transimination: implications for dynamic covalent chemistry. *Journal of the American Chemical Society*, **128**, 15602–15603.
30 Kalia, J. and Raines, R.T. (2008) Hydrolytic stability of hydrazones and oximes. *Angewandte Chemie International Edition in English*, **47**, 7523–7526.
31 Bhat, V.T., Caniard, A.M., Luksch, T., Brenk, R., Campopiano, D.J., and Greaney, M.F. (2010) Nucleophilic catalysis of acylhydrazone equilibration for protein-directed dynamic covalent chemistry. *Nature Chemistry*, **2**, 490–497.
32 Simpson, M.G., Pittelkow, M., Watson, S.P., and Sanders, J.K. (2010) Dynamic combinatorial chemistry with hydrazones: libraries incorporating heterocyclic and

steroidal motifs. *Organic and Biomolecular Chemistry*, **8**, 1181–1187.

33 Sindelar, M., Lutz, T.A., Petrera, M., and Wanner, K.T. (2013) Focused pseudostatic hydrazone libraries screened by mass spectrometry binding assay: optimizing affinities toward gamma-aminobutyric acid transporter 1. *Journal of Medicinal Chemistry*, **56**, 1323–1340.

34 Sindelar, M. and Wanner, K.T. (2012) Library screening by means of mass spectrometry (MS) binding assays: exemplarily demonstrated for a pseudostatic library addressing γ-aminobutyric acid (GABA) transporter 1 (GAT1). *ChemMedChem*, **7**, 1678–1690.

35 Larsen, D., Pittelkow, M., Karmakar, S., and Kool, E.T. (2015) New organocatalyst scaffolds with high activity in promoting hydrazone and oxime formation at neutral pH. *Organic Letters*, **17**, 274–277.

36 Clipson, A.J., Bhat, V.T., McNae, I., Caniard, A.M., Campopiano, D.J., and Greaney, M.F. (2012) Bivalent enzyme inhibitors discovered using dynamic covalent chemistry. *Chemistry: A European Journal*, **18**, 10562–10570.

37 Kern, F.T. and Wanner, K.T. (2015) Generation and screening of oxime libraries addressing the neuronal GABA transporter GAT1. *ChemMedChem*, **10**, 396–410.

38 Ulrich, S., Boturyn, D., Marra, A., Renaudet, O., and Dumy, P. (2014) Oxime ligation: a chemoselective click-type reaction for accessing multifunctional biomolecular constructs. *Chemistry: A European Journal*, **20**, 34–41.

39 Black, S.P., Sanders, J.K.M., and Stefankiewicz, A.R. (2014) Disulfide exchange: exposing supramolecular reactivity through dynamic covalent chemistry. *Chemical Society Reviews*, **43**, 1861–1872.

40 Leclaire, J., Vial, L., Otto, S., and Sanders, J.K. (2005) Expanding diversity in dynamic combinatorial libraries: simultaneous exchange of disulfide and thioester linkages. *Chemical Communications (Cambridge, England)*, 1959–1961.

41 Rasmussen, B., Sorensen, A., Gotfredsen, H., and Pittelkow, M. (2014) Dynamic combinatorial chemistry with diselenides and disulfides in water. *Chemical Communications (Cambridge, England)*, **50**, 3716–3718.

42 Rodriguez-Docampo, Z. and Otto, S. (2008) Orthogonal or simultaneous use of disulfide and hydrazone exchange in dynamic covalent chemistry in aqueous solution. *Chemical Communications (Cambridge, England)*, 5301–5303.

43 Zhu, Z., Zhang, F., and Liu, G. (2010) Dynamic covalent chemistry of disulfides offers a highly efficient synthesis of diverse benzofused nitrogen–sulfur heterocycles in one pot. *Journal of Combinatorial Chemistry*, **12**, 531–540.

44 Brisig, B., Sanders, J.K.M., and Otto, S. (2003) Selection and amplification of a catalyst from a dynamic combinatorial library. *Angewandte Chemie International Edition*, **42**, 1270–1273.

45 Atcher, J. and Alfonso, I. (2013) The effect of DMSO in the aqueous thiol-disulfide dynamic covalent chemistry of model pseudopeptides. *RSC Advances*, **3**, 25605–25608.

46 Angelin, M., Larsson, R., Vongvilai, P., and Ramström, O. (2010) Introducing dynamic combinatorial chemistry: probing the substrate selectivity of acetylcholinesterase. *Journal of Chemical Education*, **87**, 1248–1251.

47 Ghosh, S., Ingerman, L.A., Frye, A.G., Lee, S.J., Gagné, M.R., and Waters, M.L. (2010) Dynamic cyclic thiodepsipeptide libraries from thiol-thioester. *Organic Letters*, **12**, 1860–1863.

48 Larsson, R., Pei, Z., and Ramstrom, O. (2004) Catalytic self-screening of cholinesterase substrates from a dynamic combinatorial thioester library. *Angewandte Chemie International Edition in English*, **43**, 3716–3718.

49 Larsson, R. and Ramström, O. (2006) Dynamic combinatorial thiolester libraries for efficient catalytic self-screening of hydrolase substrates. *European Journal of Organic Chemistry*, **2006**, 285–291.

50 Shi, B. and Greaney, M.F. (2005) Reversible Michael addition of thiols as a new tool for dynamic combinatorial chemistry. *Chemical Communications (Cambridge, England)*, 886–888.

51 Sutton, L.R., Donaubauer, W.A., Hampel, F., and Hirsch, A. (2004) Tris(thioacetals) from benzene hexathiol: towards covalent self-assembly. *Chemical Communications (Cambridge, England)*, 1758–1759.

52 Demetriades, M., Leung, I.K., Chowdhury, R., Chan, M.C., McDonough, M.A., Yeoh, K.K., Tian, Y.M., Claridge, T.D., Ratcliffe, P.J., Woon, E.C., and Schofield, C.J. (2012) Dynamic combinatorial chemistry employing boronic acids/boronate esters leads to potent oxygenase inhibitors. *Angewandte Chemie International Edition in English*, **51**, 6672–6675.

53 Leung, I.K.H., Brown, T. Jr., Schofield, C.J., and Claridge, T.D.W. (2011) An approach to enzyme inhibition employing reversible boronate ester formation. *Medicinal Chemistry Communications*, **2**, 390.

54 Hunter, L., Condie, G.C., and Harding, M.M. (2010) Reversible aqueous metathesis reactions for potential application in dynamic combinatorial chemistry. *Tetrahedron Letters*, **51**, 5064–5067.

55 Edwards, P. (2007) Fragment-based drug discovery of carbonic anhydrase II inhibitors by dynamic combinatorial chemistry. *Drug Discovery Today*, **12**, 497–498.

56 Poulsen, S.-A. and Bornaghi, L.F. (2006) Fragment-based drug discovery of carbonic anhydrase II inhibitors by dynamic combinatorial chemistry utilizing alkene cross metathesis. *Bioorganic and Medicinal Chemistry*, **14**, 3275–3284.

57 Zhang, W. and Moore, J.S. (2005) Reaction pathways leading to arylene ethynylene macrocycles via alkyne metathesis. *Journal of the American Chemical Society*, **127**, 11863–11870.

58 ten Cate, A.T., Dankers, P.Y.W., Sijbesma, R.P., and Meijer., E.W. (2005) Disulfide exchange in hydrogen-bonded cyclic assemblies: stereochemical self-selection by double dynamic chemistry. *Journal of Organic Chemistry*, **70**, 5799–5803.

59 Kerckhoffs, J.M., Mateos-Timoneda, M.A., Reinhoudt, D.N., and Crego-Calama, M. (2007) Dynamic combinatorial libraries based on hydrogen-bonded molecular boxes. *Chemistry*, **13**, 2377–2385.

60 Mansfeld, F.M., Au-Yeung, H., Sanders, J.K.M., and Otto, S. (2010) Dynamic combinatorial chemistry at the phospholipid bilayer interface. *Journal of Systems Chemistry*, **1**, 12.

61 Huc, I. and Lehn, J.-M. (1997) Virtual combinatorial libraries: dynamic generation of molecular and supramolecular diversity by self-assembly. *Proceedings of the National Academy of Sciences*, **94**, 2106–2110.

62 Sanders, J.K. (2001) Adventures in molecular recognition. The ins and outs of templating. *ChemInform*, **72**, 2265–2274.

63 Mondal, M., Radeva, N., Köster, H., Park, A., Potamitis, C., Zervou, M., Klebe, G., and Hirsch, A.K.H. (2014) Structure-based design of inhibitors of the aspartic protease endothiapepsin by exploiting dynamic combinatorial chemistry. *Angewandte Chemie International Edition*, **53**, 3259–3263.

64 Gareiss, P.C., Sobczak, K., McNaughton, B.R., Palde, P.B., Thornton, C.A., and Miller, B.L. (2008) Dynamic combinatorial selection of molecules capable of inhibiting the (CUG) repeat RNA–MBNL1 interaction *in vitro*: discovery of lead compounds targeting myotonic dystrophy (DM1). *Journal of the American Chemical Society*, **130**, 16254–16261.

65 Demetriades, M., Leung, I.K.H., Chowdhury, R., Chan, M.C., McDonough, M.A., Yeoh, K.K., Tian, Y.-M., Claridge, T.D.W., Ratcliffe, P.J., Woon, E.C.Y., and Schofield, C.J. (2012) Dynamic combinatorial chemistry employing boronic acids/boronate esters leads to potent oxygenase inhibitors. *Angewandte Chemie International Edition*, **51**, 6672–6675.

66 Leung, I.K.H., Demetriades, M., Hardy, A.P., Lejeune, C., Smart, T.J., Szöllössi, A., Kawamura, A., Schofield, C.J., and Claridge, T.D.W. (2013) Reporter ligand NMR screening method for 2-oxoglutarate oxygenase inhibitors. *Journal of Medicinal Chemistry*, **56**, 547–555.

67 Nguyen, R. and Huc, I. (2001) Using an enzyme's active site to template inhibitors. *Angewandte Chemie International Edition*, **40**, 1774–1776.

68 Tanaka, K., Shirotsuki, S., Iwata, T., Kageyama, C., Tahara, T., Nozaki, S., Siwu,

E.R., Tamura, S., Douke, S., Murakami, N., Onoe, H., Watanabe, Y., and Fukase, K. (2012) Template-assisted and self-activating clicked peptide as a synthetic mimic of the SH2 domain. *ACS Chemical Biology*, **7**, 637–645.

69 Peruzzotti, C., Borrelli, S., Ventura, M., Pantano, R., Fumagalli, G., Christodoulou, M.S., Monticelli, D., Luzzani, M., Fallacara, A.L., Tintori, C., Botta, M., and Passarella, D. (2013) Probing the binding site of abl tyrosine kinase using *in situ* click chemistry. *ACS Medicinal Chemistry Letters*, **4**, 274–277.

70 Suzuki, T., Ota, Y., Kasuya, Y., Mutsuga, M., Kawamura, Y., Tsumoto, H., Nakagawa, H., Finn, M.G., and Miyata, N. (2010) An unexpected example of protein-templated click chemistry. *Angewandte Chemie International Edition*, **49**, 6817–6820.

71 Krasiński, A., Radić, Z., Manetsch, R., Raushel, J., Taylor, P., Sharpless, K.B., and Kolb, H.C. (2005) *In situ* selection of lead compounds by click chemistry: target-guided optimization of acetylcholinesterase inhibitors. *Journal of the American Chemical Society*, **127**, 6686–6692.

72 Kolb, H.C. and Sharpless, K.B. (2003) The growing impact of click chemistry on drug discovery. *Drug Discovery Today*, **8**, 1128–1137.

73 Roman, M. and Sharpless, K.B. (2006) *In situ* click chemistry: a powerful means for lead discovery. *Expert Opinion in Drug Discovery*, **1**, 525–538.

74 Rijkers, D.T., Merkx, R., Yim, C.B., Brouwer, A.J., and Liskamp, R.M. (2010) 'Sulfo-click' for ligation as well as for site-specific conjugation with peptides, fluorophores, and metal chelators. *Journal of Peptide Science*, **16**, 1–5.

75 Hu, X., Sun, J., Wang, H.-G., and Manetsch, R. (2008) Bcl-XL-templated assembly of its own protein–protein interaction modulator from fragments decorated with thio acids and sulfonyl azides. *Journal of the American Chemical Society*, **130**, 13820–13821.

76 Kulkarni, S.S., Hu, X., Doi, K., Wang, H.G., and Manetsch, R. (2011) Screening of protein–protein interaction modulators via sulfo-click kinetic target-guided synthesis. *ACS Chemical Biology*, **6**, 724–732.

77 Oueis, E., Nachon, F., Sabot, C., and Renard, P.-Y. (2014) First enzymatic hydrolysis/thio-Michael addition cascade route to synthesis of AChE inhibitors. *Chemical Communications*, **50**, 2043–2045.

78 Lodge, J.M., Rettenmaier, T.J., Wells, J.A., Pomerantz, W.C., and Mapp, A.K. (2014) FP tethering: a screening technique to rapidly identify compounds that disrupt protein–protein interactions. *Medicinal Chemistry Communications*, **5**, 370–375.

79 Erlanson, D.A., Braisted, A.C., Raphael, D.R., Randal, M., Stroud, R.M., Gordon, E.M., and Wells, J.A. (2000) Site-directed ligand discovery. *Proceedings of the National Academy of Sciences*, **97**, 9367–9372.

80 Erlanson, D.A., Lam, J.W., Wiesmann, C., Luong, T.N., Simmons, R.L., DeLano, W.L., Choong, I.C., Burdett, M.T., Flanagan, W.M., Lee, D., Gordon, E.M., and O'Brien, T. (2003) *In situ* assembly of enzyme inhibitors using extended tethering. *Nature Biotechnology*, **21**, 308–314.

81 Cancilla, M.T., He, M.M., Viswanathan, N., Simmons, R.L., Taylor, M., Fung, A.D., Cao, K., and Erlanson, D.A. (2008) Discovery of an Aurora kinase inhibitor through site-specific dynamic combinatorial chemistry. *Bioorganic and Medicinal Chemistry Letters*, **18**, 3978–3981.

82 Fujishima, S.H., Yasui, R., Miki, T., Ojida, A., and Hamachi, I. (2012) Ligand-directed acyl imidazole chemistry for labeling of membrane-bound proteins on live cells. *Journal of the American Chemical Society*, **134**, 3961–3964.

83 Tsukiji, S., Miyagawa, M., Takaoka, Y., Tamura, T., and Hamachi, I. (2009) Ligand-directed tosyl chemistry for protein labeling *in vivo*. *Nature Chemical Biology*, **5**, 341–343.

84 Lewis, W.G., Green, L.G., Grynszpan, F., Radić, Z., Carlier, P.R., Taylor, P., Finn, M.G., and Sharpless, K.B. (2002) Click chemistry *in situ*: acetylcholinesterase as a reaction vessel for the selective assembly of a femtomolar inhibitor from an array of

building blocks. *Angewandte Chemie*, **114**, 1095–1099.

85 Whiting, M., Muldoon, J., Lin, Y.-C., Silverman, S.M., Lindstrom, W., Olson, A.J., Kolb, H.C., Finn, M.G., Sharpless, K.B., Elder, J.H., and Fokin, V.V. (2006) Inhibitors of HIV-1 protease by using *in situ* click chemistry. *Angewandte Chemie International Edition*, **45**, 1435–1439.

86 Suzuki, T., Ota, Y., Kasuya, Y., Mutsuga, M., Kawamura, Y., Tsumoto, H., Nakagawa, H., Finn, M.G., and Miyata, N. (2010) An unexpected example of protein-templated click chemistry. *Angewandte Chemie International Edition in English*, **49**, 6817–6820.

87 Erlanson, D.A., Arndt, J.W., Cancilla, M.T., Cao, K., Elling, R.A., English, N., Friedman, J., Hansen, S.K., Hession, C., Joseph, I., Kumaravel, G., Lee, W.C., Lind, K.E., McDowell, R.S., Miatkowski, K., Nguyen, C., Nguyen, T.B., Park, S., Pathan, N., Penny, D.M., Romanowski, M.J., Scott, D., Silvian, L., Simmons, R.L., Tangonan, B.T., Yang, W., and Sun, L. (2011) Discovery of a potent and highly selective PDK1 inhibitor via fragment-based drug discovery. *Bioorganic and Medicinal Chemistry Letters*, **21** 3078–3083.

# Part III
# Successes from Fragment-based Drug Discovery

# 14 BACE Inhibitors

*Daniel F. Wyss, Jared N. Cumming, Corey O. Strickland, and Andrew W. Stamford*

## 14.1 Introduction

β-Secretase, widely known as β-site amyloid precursor protein cleaving enzyme 1 (BACE1), initiates production of the toxic amyloid-β (Aβ) peptides that are believed to play a crucial early part in Alzheimer's disease (AD) pathogenesis. AD is a devastating neurodegenerative disease characterized by the accumulation of two hallmark structures in the brain – amyloid plaques and neurofibrillary tangles. Amyloid plaques are extracellular deposits of insoluble fibrils of Aβ peptides that are primarily 40 and 42 residues in length. Neurofibrillary tangles are intracellular aggregates of aberrantly hyperphosphorylated forms of tau, a microtubule-associated protein. The amyloid hypothesis postulates that accumulation of Aβ plaques and soluble neurotoxic Aβ oligomers in the brain is a critical, required early step that leads to formation of neurofibrillary tangles, neuroinflammation, synaptic loss, neuronal cell death, and ultimately dementia [1–3]. If this hypothesis is correct, then preventing accumulation of Aβ in the brain, either through reduced production (e.g., BACE1 inhibition) or through increased clearance (e.g., Aβ-antibodies [4,5]), should be an effective approach to slow or halt progression of AD when given early enough in the disease process [3,6].

Aβ peptides are formed as minor products (5–10%) of the metabolism of amyloid precursor protein (APP) via two consecutive cleavages – first by BACE1, in competition with the major pathway (90–95%) of nonamyloidogenic processing of APP by α-secretase, followed by γ-secretase (see Figure 14.1 in [6]). The active site of BACE1 is within the acidic environment of the endosomes, with a pH optimum correspondingly in the 4.5–5.5 range, and is relatively shallow and hydrophilic. BACE1 is mainly expressed in the central nervous system (CNS). A closely related enzyme, BACE2, is predominantly expressed in the periphery and does not function as a β-secretase

*Fragment-based Drug Discovery: Lessons and Outlook*, First Edition. Edited by Daniel A. Erlanson and Wolfgang Jahnke.

(a)

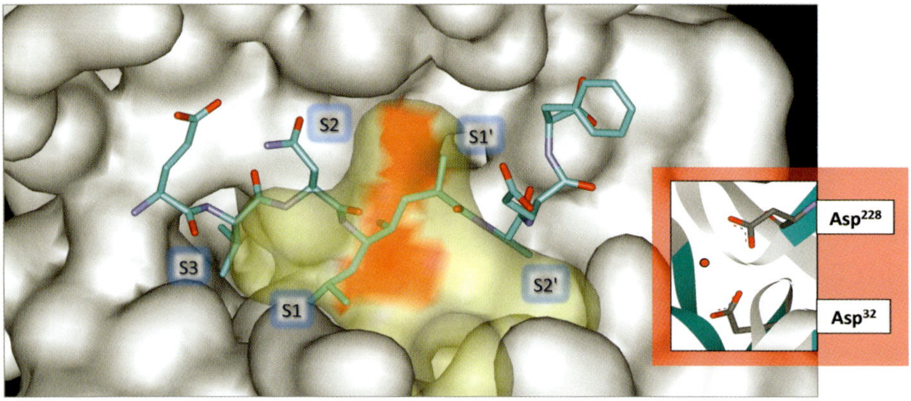


**Figure 14.1** (a) BACE1 structural characteristics and binding mode of an early-generation peptidomimetic BACE1 inhibitor (OM99-2) showing its relatively large size and the extended BACE1 active site (PDB code 1FKN) [17]. The overall folding of BACE1 is typical for an aspartic acid protease, consisting of N- and C-terminal lobes with the substrate binding site located in a crevice between the two lobes. A flexible hairpin loop, called the flap (yellow transparent surface), partially covers the active site of BACE1 and can adopt many different conformations as a result of inhibitor binding. In the center of the active site are the two aspartic acid residues Asp32 and Asp228 (orange and inset with a bound water molecule at the catalytic center) that are involved in the enzymatic reaction.

(b)

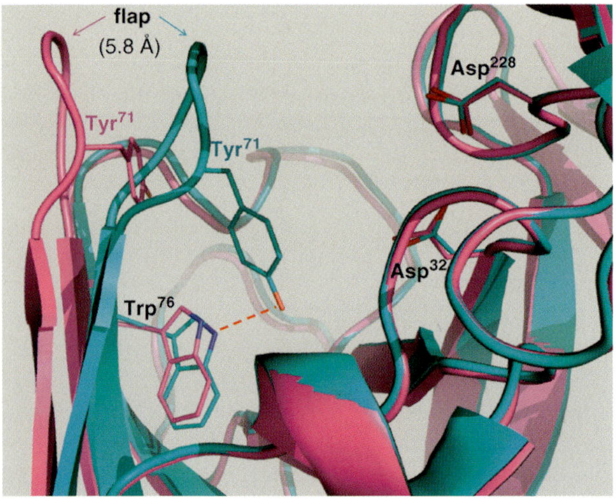


**Figure 14.1** (b) Close-up view of the active site of superimposed crystal structures of flap-open (PDB code 2OHU, magenta) [28] and flap-closed (PDB code 1W51, cyan) [29] BACE1. Ligands have been omitted for clarity. Due to the conformation of Tyr71 in the flap-open 2OHU structure (magenta), the hydrogen bond with Trp76 is broken and Tyr71 no longer occupies the position between the S1 and S2′ subsites. The changing position of the flap relative to the catalytic dyad provides a means for the substrate and the ligands to diffuse into and out of the active site.

(c)

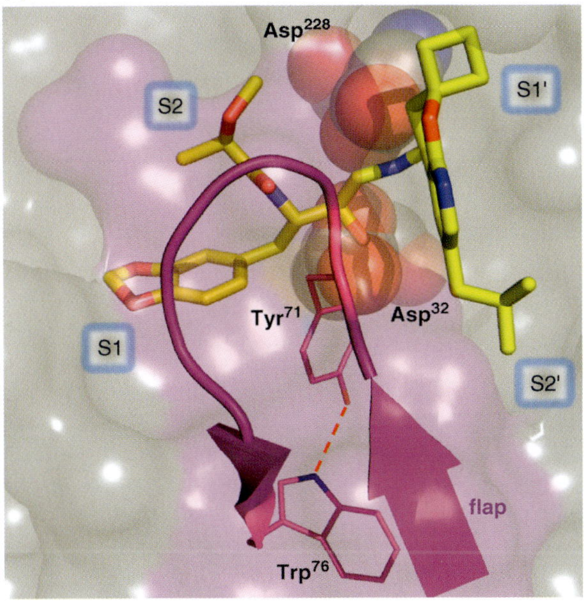


**Figure 14.1** (c) Bound conformation of an evolved hydroxyethylamine containing peptidomimetic inhibitor (yellow) bound with BACE1 (PDB code 4DH6). This compound demonstrated robust reduction of central β-amyloid in a preclinical rat animal model [27]. BACE1 is in the flap-closed conformation (purple surface) where the Tyr71 side chain hydroxyl is hydrogen-bonded to the NH of the Trp76 side chain that physically separates the S1 and S2′ subsites. The catalytic aspartyl dyad is shown as spheres and the proximal subsites are labeled.

*in vivo* [3,6]. BACE1 belongs to the pepsin family under the aspartyl protease superfamily and has sequence identity not only with BACE2 (52%) but also with cathepsin D (29%), cathepsin E (27%), pepsin (27%), and renin (24%) [7].

The amyloid hypothesis is supported by a large body of genetic, biochemical, and histopathological evidence (see [3,6] for recent reviews). Since its discovery in 1999 [8–12], inhibition of BACE1 has been viewed as the best small molecule approach to test this hypothesis, underscored by initial observations that BACE1 knockout (KO) mice do not produce Aβ and show a relatively benign overt phenotype [13]. Furthermore, the well-studied Swedish mutation of APP causes early onset familial AD by specifically enhancing BACE1-dependent cleavage by as much as 100-fold [14,15]. More recently, identification of a naturally occurring variant of APP (the "Icelandic mutation") with a reduced propensity to proteolysis by BACE1 and association with a decreased risk of dementia further solidified the role of this enzyme in the etiology of AD [16]. Although BACE1 has emerged as a prime therapeutic target for AD, the recent identification of multiple new substrates and reports of its involvement in a much wider range of physiological processes within and outside the CNS have raised concerns about potential mechanism-based side effects of drugs targeting this enzyme [3,6]. While the science around BACE1 continues to evolve, the bulk of the data, including both heterozygous KO and chronic small molecule treatment across multiple species, suggests that long-term dosing with a sufficient therapeutic window should be possible with a potent, selective, brain-penetrant small molecule inhibitor.

Localization of BACE1 in the CNS, however, has posed a significant challenge in identifying low molecular weight (MW) compounds with the aforementioned profile. Initial drug design efforts focused on noncleavable peptide-based transition state analogs in which the residues of APP around the site of BACE1 cleavage served as a template [10,17,18]. The first wave of published BACE1 crystal structures featured molecules that interacted with the catalytic dyad using classical aspartyl protease transition state mimics, for example, hydroxyethylene (HE) [17,18] and hydroxyethylamine (HEA) isosteres [19,20] (see [21] for a recent review). These peptidomimetic structures bound to the flap-closed form of the enzyme and filled most of the pockets that the substrate occupies around the active site (Figure 14.1). While potent *in vitro*, these molecules had poor ligand efficiencies (LE) [22] and high polar surface areas (PSA) due to their high MW and peptidic nature. These properties contributed to deficient profiles common to many peptidomimetics, including poor oral bioavailability and low blood–brain barrier penetration driven by susceptibility to permeability glycoprotein (P-gp) efflux. In fact, only a handful of literature examples in this class of inhibitors exhibited brain permeability and measurable central pharmacodynamic (PD) effects *in vivo* (e.g., see examples in [23–27]).

Beyond the peptidomimetics, it has proven challenging to develop nonpeptidic "second-generation" BACE1 inhibitors that exhibit brain Aβ lowering with a dose and safety profile suitable for clinical progression. Fragment-based approaches have provided novel and structurally diverse chemotypes that, together with extensive SAR exploration using *in vitro* and *in vivo* model systems, have afforded a wealth of structure- and property-based knowledge over the past decade, significantly advancing our understanding of what makes a "good" BACE inhibitor [30,31]. This chapter will discuss the major role FBDD played in the development of such chemotypes, including the nonplanar cyclic amidines, a chemotype that ultimately enabled researchers to develop selective BACE inhibitors that are more drug-like in terms of MW, Clog*P*, and PSA, that have lower susceptibility to P-gp efflux, and most importantly that penetrate the CNS and effectively reduce Aβ levels, some of which progressed into clinical development [32–41].

## 14.2
## FBDD Efforts on BACE1

### 14.2.1
### Fragment Hit Identification, Validation, and Expansion

Diverse approaches to FBDD have been employed across many groups in the discovery of novel nonpeptidic low MW BACE1 inhibitors (Table 14.1). These efforts ranged from extensive fragment screening campaigns that tested hundreds to thousands of compounds from diverse libraries to more knowledge-based (e.g., structure- and property-based) campaigns that tested focused libraries of tens to hundreds of fragments. As can be seen from the left two columns in Table 14.1, a combination of both biophysical/biochemical screening and confirmation methods were used in early efforts to discover diverse pharmacophores that bind to the BACE1 active site. In the majority of cases, an X-ray crystal structure of BACE1 with a fragment or fragment-derived analog was ultimately obtained to validate those hits and subsequently to serve as a basis to guide challenging and diverse hit expansion campaigns. While some of these efforts were successful, many others failed to produce legitimate leads against this challenging drug target. Nevertheless, the wealth of new chemical matter that was discovered and the associated structural information that was generated through fragment-based approaches turned out to be invaluable for BACE1 drug discovery. Among the nonpeptidic scaffolds identified through this process, a number of amidine- or guanidine-containing structures, both cyclic and acyclic, were discovered early on and, through X-ray crystallography, were shown to form complex and highly productive hydrogen-bonding networks with the catalytic aspartic acid dyad of BACE1 (Table 14.1).

As can be seen from the middle column in Table 14.1, multiple approaches were taken for initial elaboration of the fragment hits, generally using either "SAR by catalog" and/or parallel synthesis. These efforts were aimed both at SAR development for occupancy of neighboring subsites to increase affinity in a ligand-efficient manner and at improvement of PK and other properties. This process was complex for BACE1 given the many challenges for drug discovery that were mentioned in the introduction. A combination of biophysical and biochemical screening methods were used, supported steadily with X-ray crystal structures, to measure progress with the weakly binding low MW compounds (Table 14.1, right-hand columns). The amidine- and guanidine-based hits ultimately provided, after significant design and exploration, an excellent platform for optimization by providing what would later prove to be favorable substitution vectors that led to a series of potent BACE1 inhibitors with high ligand efficiency.

### 14.2.2
### Fragment Optimization

The fragment-derived hits in Table 14.1 were optimized using a combination of structure-based design and biochemical functional assays to develop SAR and

**Table 14.1** Summary of BACE1 inhibitors derived from various FBDD approaches.

| Ref. | Fragment-based approach | Representative validated fragment hit | Fragment hit to lead expansion strategy | Representative evolved hit/lead class | Evolved hit/lead binding mode characteristics |
|---|---|---|---|---|---|
| [28,42] | • 347 Drug fragments and 65 virtual screening (VS) hits selected for X-ray soaking<br>• Access to S3 and S2' from S1 based on binding mode of 2,3-diaminopyridines | [20HM] 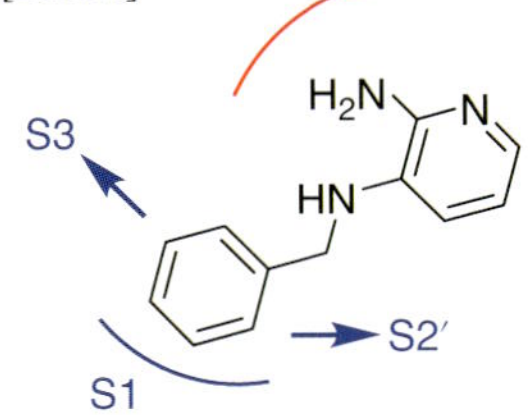  BACE1 $IC_{50}$ = 310 μM<br>LE = 0.32 | • Biaryl substitution to better occupy S1–S3 hydrophobic pocket<br>• Extensions at *ortho*-position of S1 phenyl explore hydrophobic S2' subsite under the flap<br>• Enzyme assay and X-ray for hit validation | 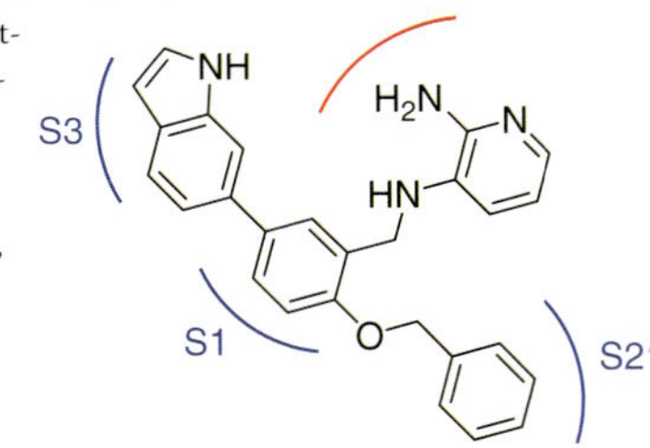  BACE1 $IC_{50}$ = 0.69 μM<br>LE = 0.26 | • *2,3-Diaminopyridine* H-bonds to catalytic aspartates and *N*-benzyl occupies S1<br>• Phenyl-indolyl group sits in S1–S3 and benzyloxy at *ortho*-position extends toward S2′ |
| [43,44] | • Competition water-LOGSY NMR screen of 2000 fragment library<br>• Competition SPR and NMR to test analogs, followed by endo-thiapepsin X-ray hit validation<br>• Isocytosine scaffold hopping and fragment growing toward S1–S3 | 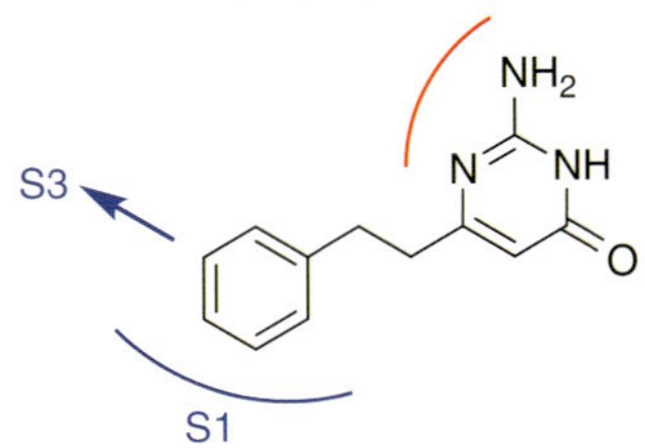  ISA SPR $K_d$ = 660 μM<br>LE = 0.27 | • 6-Substituted isocytosine and dihydro-isocytosine analogs to optimize occupancy of S1/S3 subsites<br>• Enzyme assay and BACE1 X-ray for hit validation | 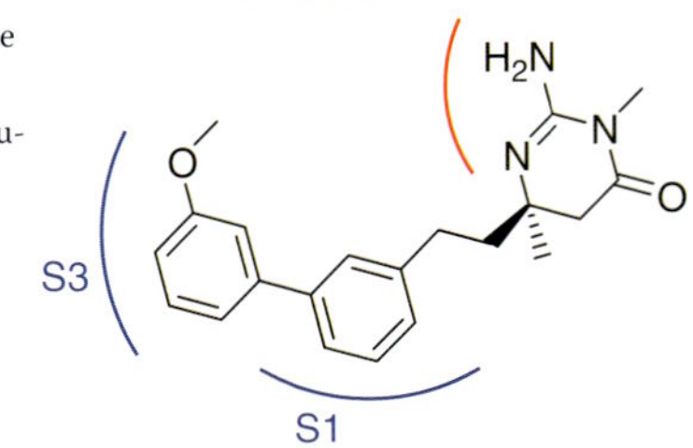  BACE1 $IC_{50}$ = 0.08 μM<br>LE = 0.37 | • Dihydroisocytosine core tilted toward prime side while maintaining key interactions with catalytic aspartates<br>• Less favorable pseudo-axial conformation stabilized by quaternary center on core<br>• Biphenyl deeper in S1–S3 with 3′-methoxy H-bound to hydroxyl group of Ser229 displacing water molecule deep in S3 |

| | | | | | |
|---|---|---|---|---|---|
| [45] | • VS and SPR to select 48 fragments for X-ray soaking<br>• Binding mode of tyramine fragment hit as starting point for design of low MW inhibitors | [3BRA] 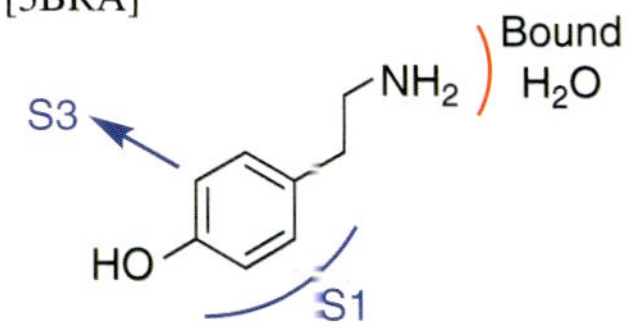  SPR $F_d$ = 2000 μM<br>LE = 0.37 | • Fragment growing toward S3 through substructure searches of Roche compound collection<br>• SPR and X-ray soaking for analog hit validation | 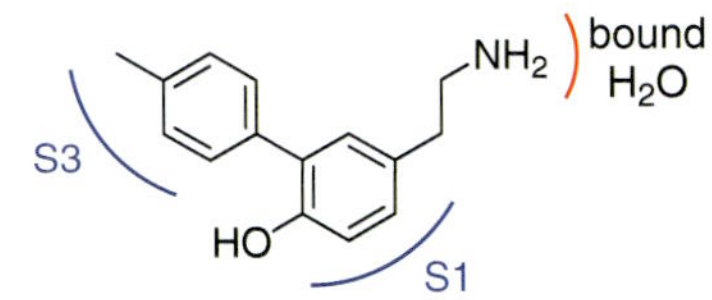  SPR $K_d$ = 60 μM<br>LE = 0.34 | • Binds to S1 and basic amine H-bonds to catalytic water<br>• *Ortho*-phenol extensions reach toward S3 |
| [46] | • Tethering technology to screen 15 000 monophores<br>• X-ray hit validation of aminobenzyl-piperidines (ABP) and 3-carboxamide-benzylpiperazines (CBP)<br>• Conversion of tethered ligands to noncovalent BACE1 inhibitors | [2ZJH] 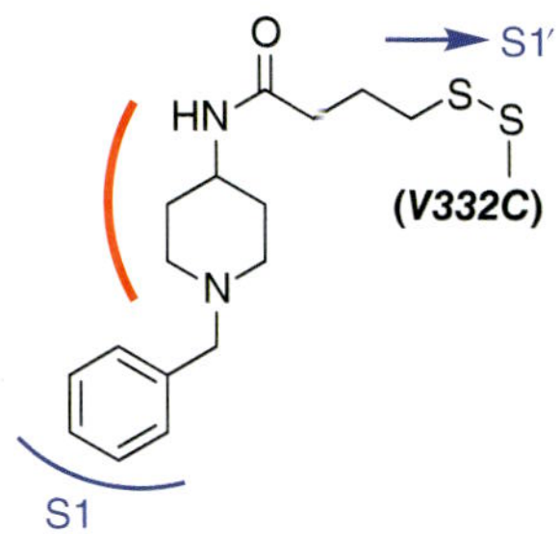  | • SAR and structural studies of affinity-improved tethered analogs to fill S1 subsite<br>• Follow-up by replacing thiol-containing linker with substituents that pack against S1′ subsite<br>• FL-substrate cleavage assay and X-ray for hit validation | [2ZJN] 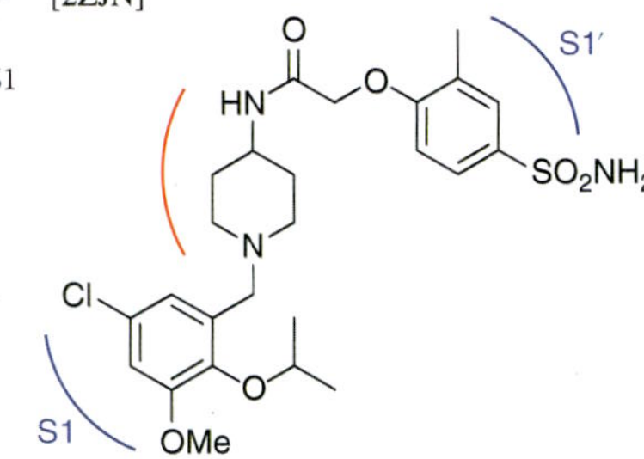  BACE1 $K_i$ = 74 μM<br>LE = 0.16 | • Aminopiperidine core H-bonds to catalytic aspartates<br>• Benzyl binds to hydrophobic S1 subsite<br>• 3-Sulfonamide phenoxy substituents directed toward the S1′ subsite causing BACE1 conformational change |

(*continued*)

Table 14.1 (Continued)

| Ref. | Fragment-based approach | Representative validated fragment hit | Fragment hit to lead expansion strategy | Representative evolved hit/lead class | Evolved hit/lead binding mode characteristics |
|---|---|---|---|---|---|
| [47,48] | • HCS of 20 000 fragment library in enzyme primary assay and hit triaging by counterassay<br>• Hit confirmation with a secondary BACE1 assay and competition SPR to select hits for X-ray validation<br>• Structure-based growing of 2-aminobenzimidazoles into S3 and S2′ | [3MSJ] 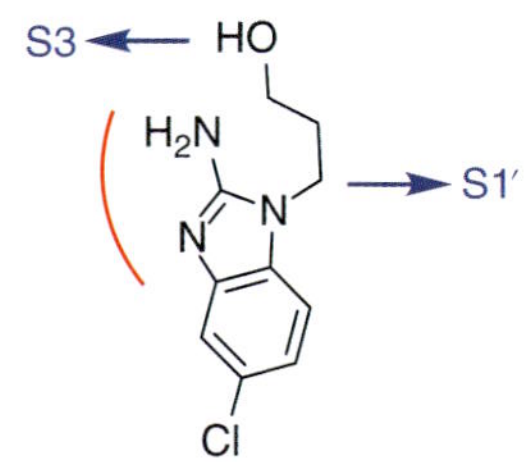  BACE1 $IC_{50}$ = 770 μM LE = 0.28 | • Structure overlay with related 2-amino-3,4-dihydroquinazoline [56] suggested α-branched analogs to grow toward S3 and prime side, respectively<br>• Enzyme assay and X-ray for hit validation<br>• Series deprioritized due to hERG off-target activity | [3MSL] 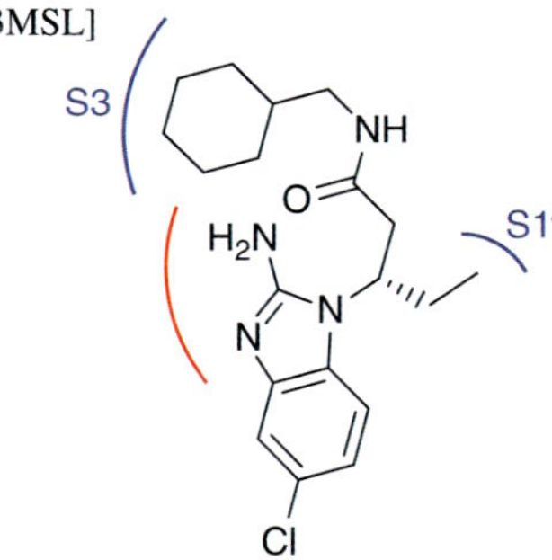  BACE1 $IC_{50}$ = 7 μM LE = 0.28 | • 2-Aminoimidazole and hydroxyl H-bond to catalytic aspartates<br>• Aryl unit forms face-to-edge π-interaction with flap Tyr71<br>• 5-Chloro sits in a pocket between Tyr71 and Trp76<br>• α-Branched ethyl occupies prime side and cyclohexylmethyl reaches S3 subsite |
| [49,50] | • Target-detected NMR to screen 10 000 fragment library and heterocyclic isothiourea isosteres yielding several X-ray structures<br>• *Novel cyclic acylguanidine core designs* as starting points for structure- and property-based fragment growing to explore S1-S3 and S2′ | [3KMX] 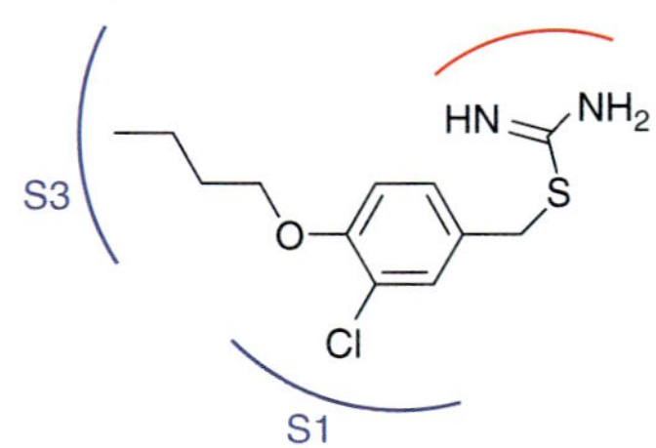  NMR $K_d$ = 15 μM LE = 0.39 | • Structure-guided design of heterocyclic isothiourea isosteres led to discovery of 2-amino-pyridines, imino-hydantoins and iminopyrimidinones<br>• Structure- and property-based SAR exploration of S1-S3 and S2′<br>• Target-detected NMR, enzyme assay and X-ray for hit validation | [4DJU] 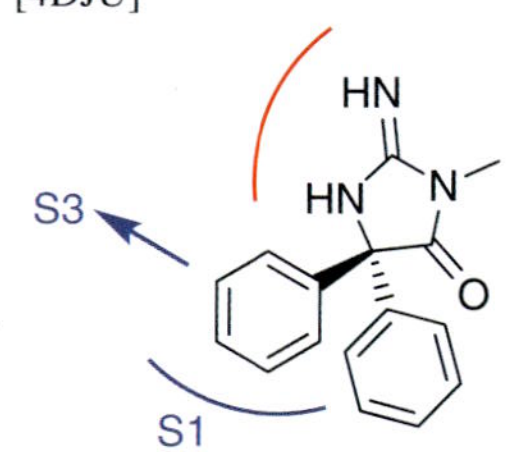  BACE1 $K_i$ = 4 μM LE = 0.37 | • *Non-planar cyclic acylguanidine* motif in the imino-hydantoin head group H-bonds to catalytic aspartates<br>• One of the phenyl groups is placed into S1<br>• Other phenyl group is placed into F′ near S2′ which is occupied by Tyr71 in the "closed" flap conformation |

| | | | | | |
|---|---|---|---|---|---|
| [51] | • SPR screen of 4000 fragment library<br>• [19]F-NMR functional assay for hit validation<br>• 6-Substituted 2-aminoquinolines yield X–ray fragment structures<br>• Structure- and property-based fragment growing to explore S1 and S1′–S2′ | [3RTH]<br>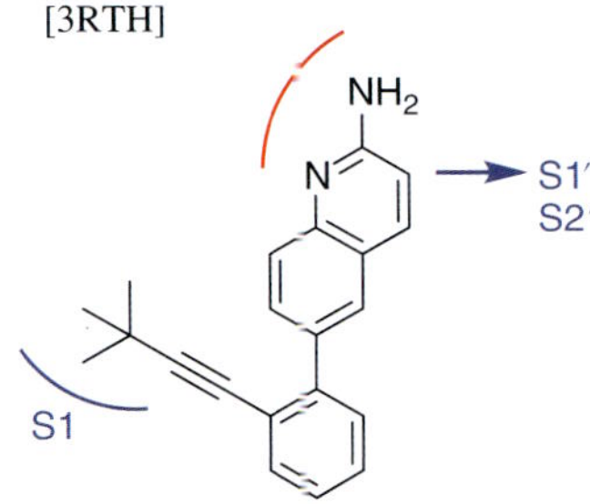<br><br>SPR $K_c$ = 6.1 μM<br>LE = 0.31 | • Structure- and property-based SAR exploration of S1′ and S2′ using 3,6-substituted 2-amino-quinolines provided more ligand efficient potency than toward S1<br>• Enzyme assay and X-ray for hit validation | [3RTN]<br>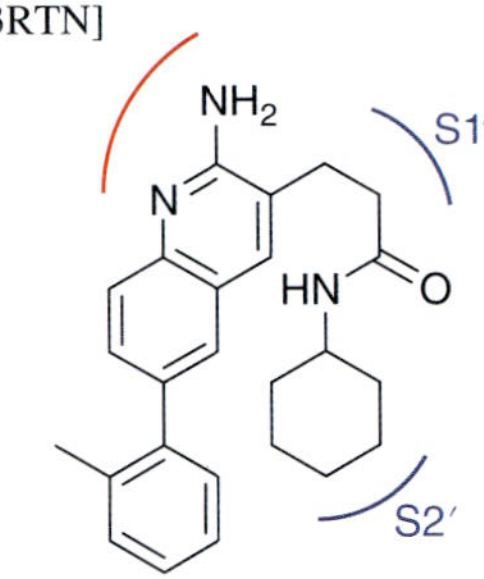<br><br>BACE1 $IC_{50}$ = 0.074 μM<br>LE = 0.34 | • 3,6-Substituted 2-amino-quinoline core H-bonds to catalytic aspartates<br>• 6-Tolyl group changes flap conformation and makes face-edge interaction with Phe108<br>• 3-amide chain oriented toward S1′ side and N-cyclohexyl occupies S2′ |
| [38] | • HCS of 8000 fragment library in enzyme assay identified 2-aminobenzo-thiazine and 2-amino-thiadiazine hits<br>• X-ray structure (not disclosed) guided exploration of S1–S3 | 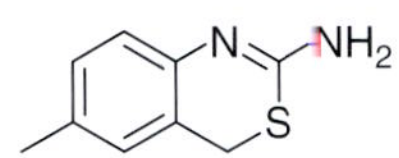<br><br>ELISA $IC_{50}$ = 790 μM<br>LE = 0.35<br>(Two copies reported in active site, one bound to catalytic dyad and one across S1-S3) | • Deplanarization of rigid bicyclic scaffold for optimal fragment growth into S1–S3 followed by property-based optimization<br>• Enzyme assay and X-ray for hit validation | 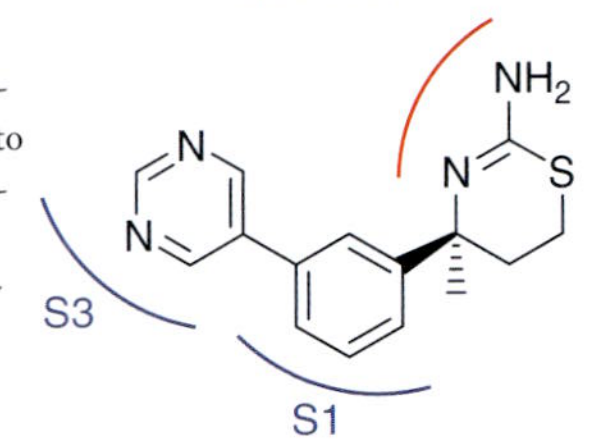<br><br>BACE1 $IC_{50}$ = 0.24 μM<br>LE = 0.41 | • 2-Amino-dihydrothiazine core H-bonds to catalytic aspartates<br>• Difluoro-phenyl placed into S1<br>• Pyrimidine extended into S3 |
| [52] | • 340 member diverse subset of proprietary fragment library for X-ray soaking<br>• Fragment growing toward S1′/S2′ based on unique binding mode of spiropyrrolidine | [3UDH]<br>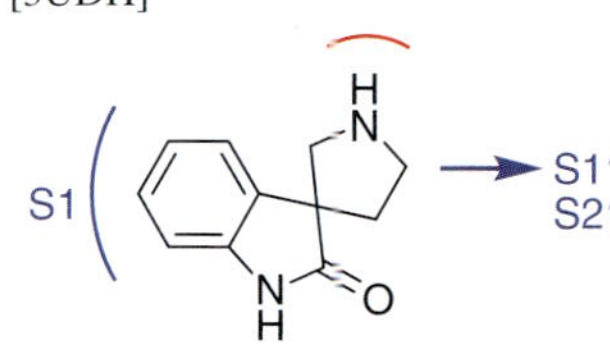<br><br>NMR $IC_{50}$ = 1100 μM<br>LE = 0.29 | • Hit assessment by competition NMR and biolayer interferometry, and functional NMR activity assay<br>• Structure- and property-based SAR exploration of S1′ and S2′ | 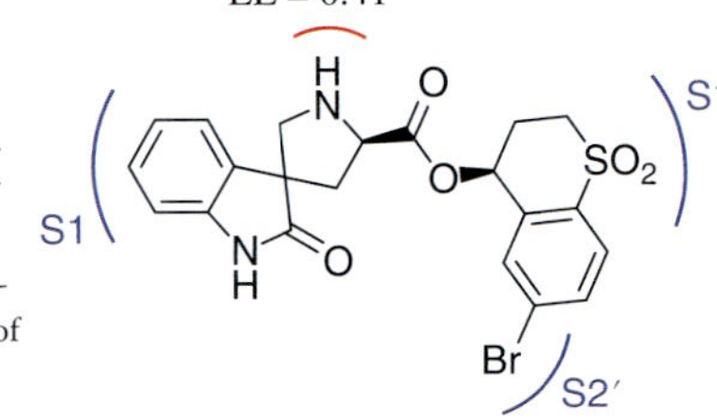<br><br>NMR $IC_{50}$ = 1 μM<br>LE = 0.27 | • Spirocyclic pyrrolidine NH H-bonds to catalytic aspartates and rotates CO of Asp228 by ~90°<br>• Oxindole sits in S1 with its NH H-bonding to the flap and no vectors to extend toward S3<br>• Several spirocyclic pyrrolidine extensions occupy S1′/S2′ |

(*continued*)

Table 14.1 (*Continued*)

| Ref. | Fragment-based approach | Representative validated fragment hit | Fragment hit to lead expansion strategy | Representative evolved hit/lead class | Evolved hit/lead binding mode characteristics |
|---|---|---|---|---|---|
| [53] | • VS, NMR and HCS culminated in discovery of an *acyclic acylguanidine* hit for library synthesis SAR exploration<br>• X-ray structure-guided exploration of S1-S3 | 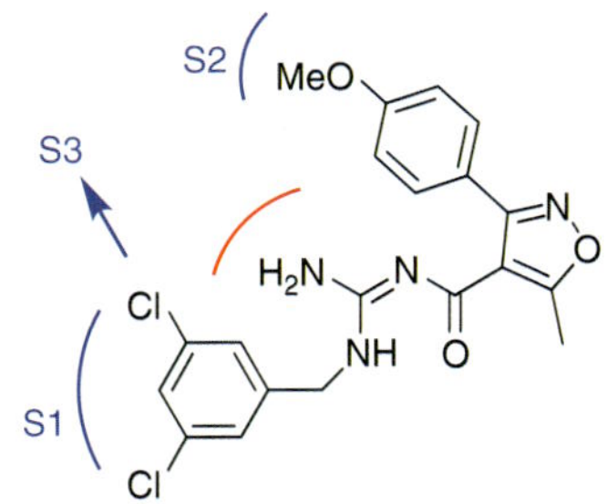<br>BACE1 $K_i = 0.67\ \mu M$<br>LE = 0.29 | • Structure-assisted library synthesis to explore optimal occupancy into S1-S3<br>• Enzyme assay and X-ray for hit validation | [4FSE]<br>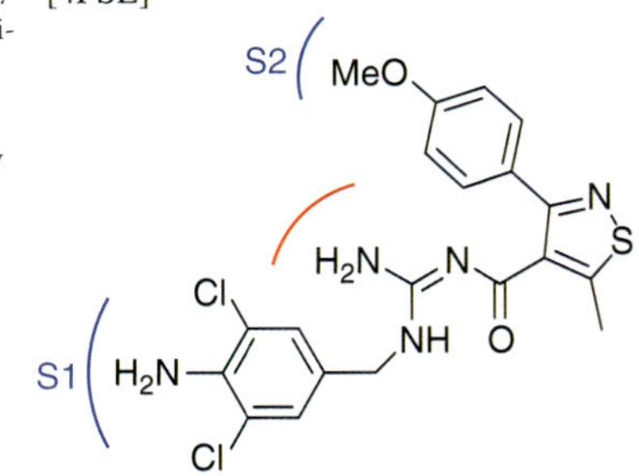<br>BACE1 $K_i = 0.04\ \mu M$<br>LE = 0.34 | • Exocyclic amine of *acyclic acylguanidine* motif H-bonds to aspartates<br>• Methoxyphenyl extends into S2<br>• 3,5-Dichlorobenzyl occupies S1 and 4-amino-group H-bonds to backbone carbonyl of Phe108<br>• Flap Gln73 side chain H-bonds to isoxazole carbonyl oxygen |
| [54,55] | • HTS enzyme assay and X-ray cocrystallization of aminothiazine fragment hit<br>• Property-based head group replacement and structure-based fragment growing into S3<br>• Med. chem. optimization | [4BEK]<br>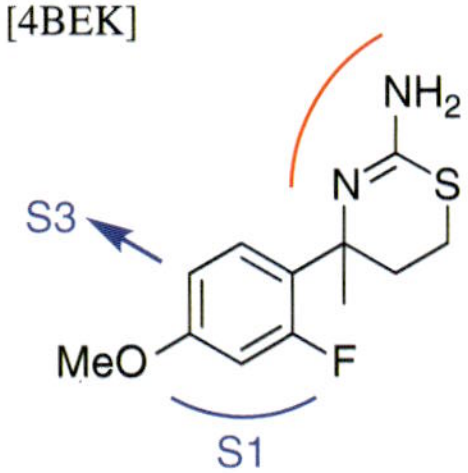<br>BACE1 $IC_{50} = 41.2\ \mu M$<br>LE = 0.37 | • Head group scan to improve PK and extending toward S3<br>• Extensive fluorine scan of 1,3-oxazines to improve PK profile<br>• Enzyme assay and X-ray for hit validation | 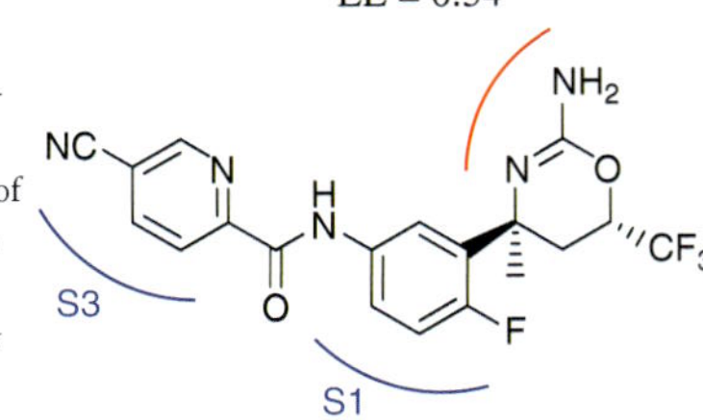<br>BACE1 $IC_{50} = 0.012\ \mu M$<br>LE = 0.36 | • Nonplanar cyclic amidine motifs in thiazine/oxazine head groups H-bond to catalytic aspartates and place phenyl into S1<br>• Amide NH H-bonds to CO of Gly291 and pyridyl *N*<br>• *p*-Cyano projects deep into S3 displacing bound water |

For each structure (as applicable), the interaction with the catalytic dyad (or bound water where specified) is shown by the red arc, subsite interactions (or potential interactions) are shown in blue, and the Protein Data Bank (PDB) ID for the X-ray structure with BACE1 is indicated in brackets.

generate bona fide leads. These efforts eventually led several groups to discover nonplanar cyclic amidines that can all be described with general structure **1** (Figure 14.2a), which would later be regarded as a key minimal pharmacophore [57] and ultimately yield a large variety of novel, potent BACE1 inhibitors that possess drug-like properties. In addition to forming a complex hydrogen-bonding network with the catalytic aspartyl dyad of BACE1, this framework provided opportunities to explore SAR in both S1 and S2′ subsites through direct substitution of the core (Figure 14.2b). Furthermore, extensions at the *meta*-position of the S1 phenyl group provided a favorable trajectory to access the

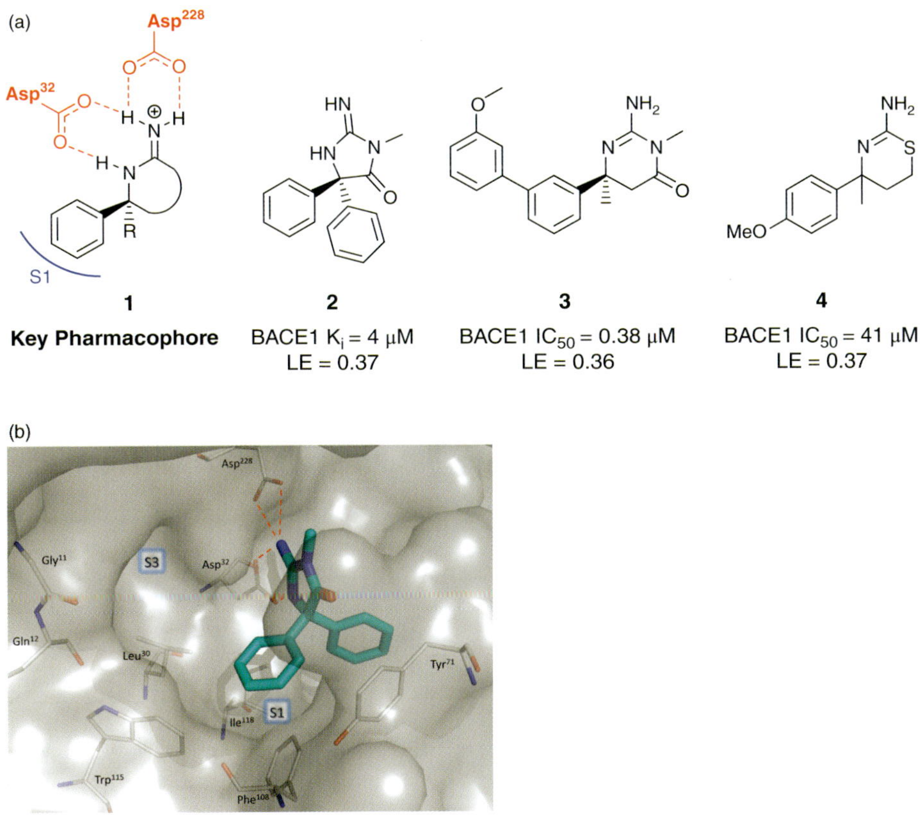


**Figure 14.2** (a) Nonplanar cyclic amidine key pharmacophore **1** and derived BACE1 inhibitor leads discovered by diverse FBDD approaches. (b) Crystal structure of **2** bound to BACE1 (PDB code 4DJU) [58]. (c) Crystal structure of **3** bound to BACE1 (PDB code 2VA6) [44]. The amidine portion of the key pharmacophore **1** present in hits **2–4** hydrogen bonds (dashed red lines) to the catalytic dyad, Asp32 and Asp228, while the nonplanar $sp^3$ carbon displays the pendant phenyl moiety into the hydrophobic S1 pocket to enable extensions toward S3 and allowing additional access to the prime side.

(c)

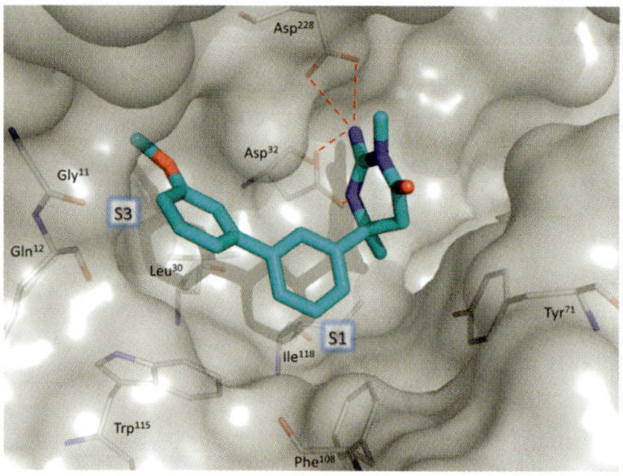


**Figure 14.2** *(Continued)*

adjacent S3 pocket (Figures 14.2c), offering the potential for ligand-efficient inhibitors for this challenging drug target.

Relative to cores that share minimal pharmacophore **1**, the other structures in Table 14.1 offered less direct access to the subpockets around the catalytic dyad, generally leading to less efficient ligand designs. For example, vectors from the planar cyclic amidines [28,42,47,48,51], the acyclic acylguanidines [53], and the 4-aminobenzylpiperidine [46] were perpendicular to the S1 subsite due to the binding pose required to achieve a productive binding to the catalytic aspartic acids. With no direct access to S1, many optimization strategies have used less ligand efficient access to S1 and/or have required occupation of alternate subsites including S1′, S2, and S2′ to gain affinity. The spiropyrrolidine fragment core has been plagued by similar limitations since the phenyl group that occupies S1 does so with a suboptimal trajectory into S3 [52]. These cores have all faced similar challenges linked to inferior PK properties, P-gp issues, poor brain penetration, or unacceptable off-target activities that have been difficult to mitigate during fragment optimization.

### 14.2.3
### From a Key Pharmacophore to Clinical Candidates

The first iminoheterocycles exemplifying pharmacophore **1**, for example, iminohydantoin **2** and iminopyrimidinone **3**, were reported in the patent literature in 2005 [59]. These and related structures were independently discovered using FBDD by scientists at Schering-Plough (**2**, **3**) [50,58,60–62] and AstraZeneca (**3**) [44] as well as through traditional high-throughput screening (HTS) methods by the Wyeth group (**2**) [63,64]. Beginning in 2007 [65–67], a number of groups disclosed dihydrothiazine-based structures (e.g., **4**). Both Lilly [38] and Roche [54] identified low MW dihydrothiazine fragments through high-concentration biochemical screening, whereas Shionogi [65] and Eisai [66] arrived at their structures through undisclosed methods. X-ray crystal structures of a number of initial isothiourea leads with BACE1 have recently been published, including compound **4** from Roche [54].

Researchers at Schering-Plough discovered iminoheterocycles **2** and **3** via a design effort based on isothiourea fragment hit **5** (Figure 14.3a and Table 14.1), which they had identified through screening by NMR and validated with X-ray crystallography [49,50]. The array of hydrogen bond donor–acceptor interactions formed between the amidine portion of **5** and the catalytic dyad of BACE1 was at the time unprecedented for aspartyl protease inhibitors. Recognizing the potential liabilities of an isothiourea – susceptibility to oxidative degradation, chemical reactivity, and toxicity – they designed cyclic acylguanidine scaffolds **6**

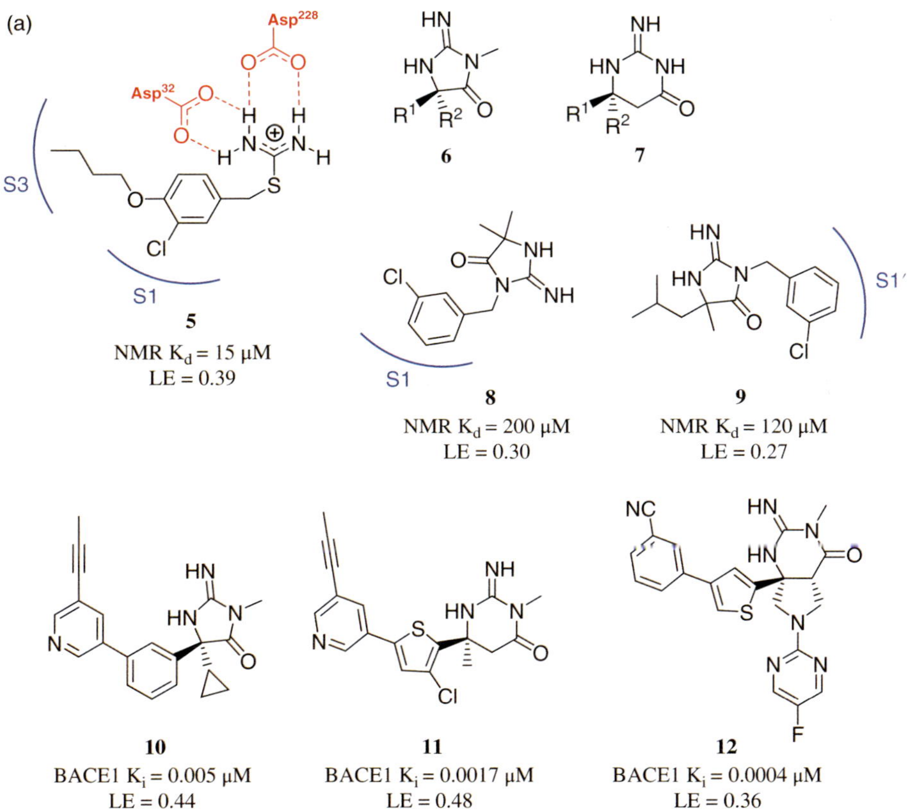


**Figure 14.3** (a) Optimization of NMR isothiourea fragment hit **5** from Schering-Plough [49,50,58,60–62]. Merck has progressed the most advanced lead from this effort into two phase 3 clinical trials. (b) Optimization of NMR pyrimidinone fragment hit **13** to an early clinical compound from AstraZenenca [34,37,43,44,68]. (c) Optimization of HTS dihydrothiazine fragment hit **21** to first two clinical candidates from Eli Lilly [38,39].

(b)

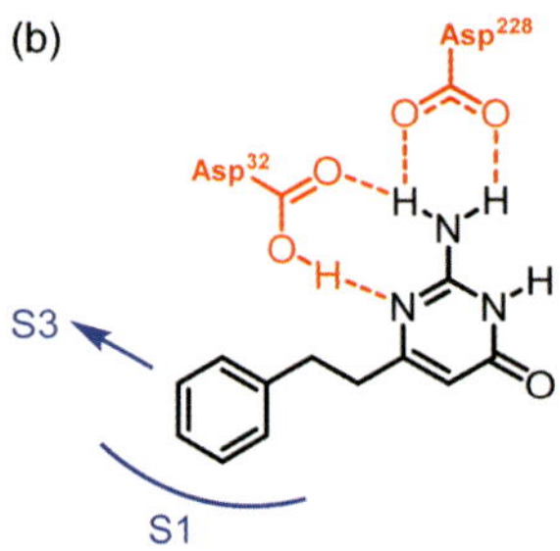


**13**
ISA SPR $K_d$ = 660 μM
LE = 0.27

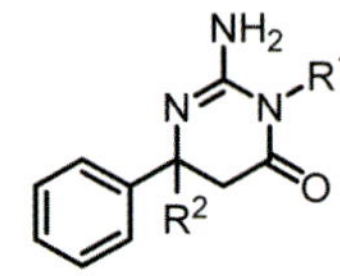

**14**: $R^1 = R^2 = H$
**15**: $R^1 = H, R^2 = Me$
BACE1 $IC_{50}$ = 190 μM
LE = 0.34
**16**: $R^1 = R^2 = Me$;
BACE1 $IC_{50}$ = 6 μM
LE = 0.45

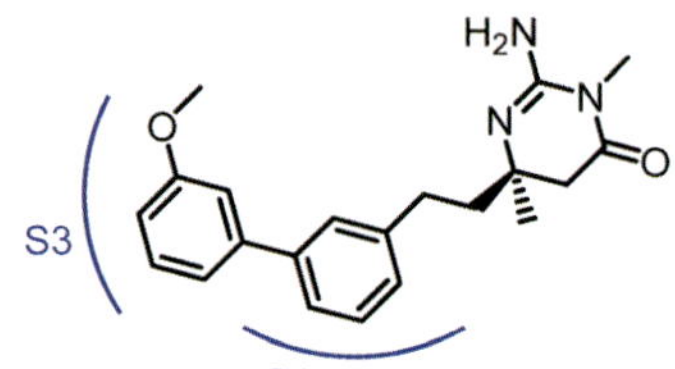


**17**
BACE1 $IC_{50}$ = 0.08 μM
LE = 0.37

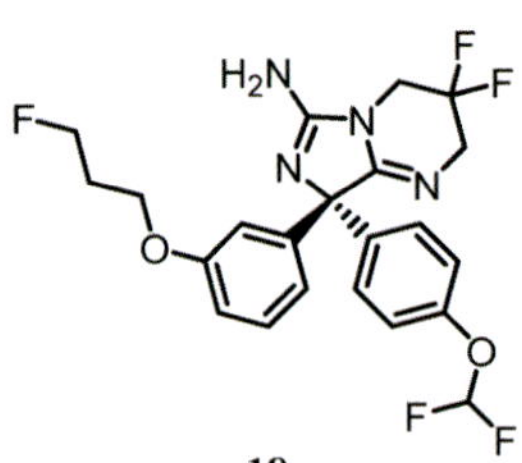

**18**
BACE1 $IC_{50}$ = 0.032 μM
LE = 0.31

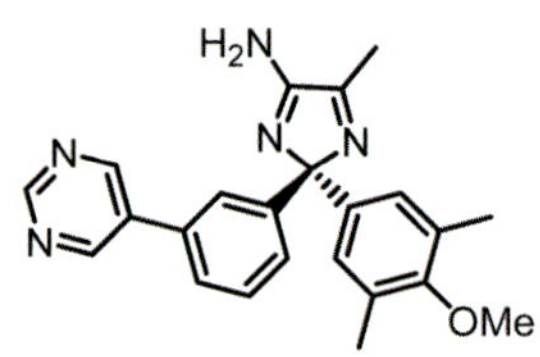

**19**
BACE1 $IC_{50}$ = 0.024 μM
LE = 0.36

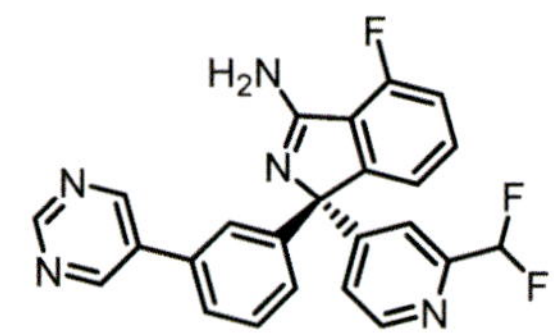

**AZD-3839 (20)**
BACE1 $K_i$ = 0.026 μM
LE = 0.32

(c)

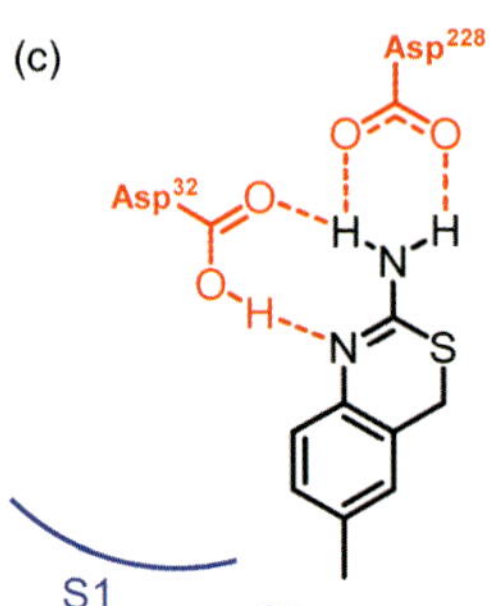


**21**
ELISA $IC_{50}$ = 790 μM
LE = 0.35

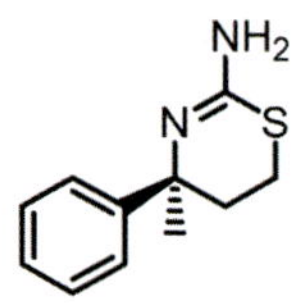

**22**
BACE1 $IC_{50}$ = 22 μM
LE = 0.45

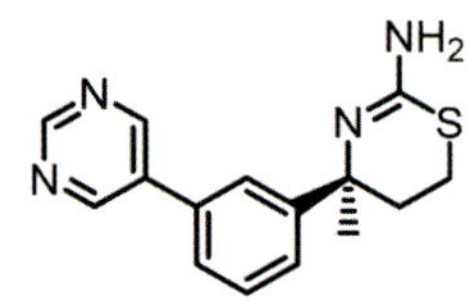

**23**
BACE1 $IC_{50}$ = 0.92 μM
LE = 0.41

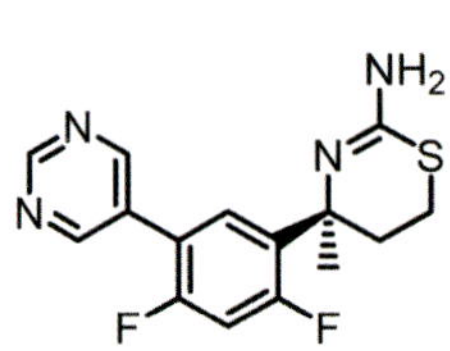

**LY-2811376 (24)**
BACE1 $IC_{50}$ = 0.24 μM
LE = 0.41

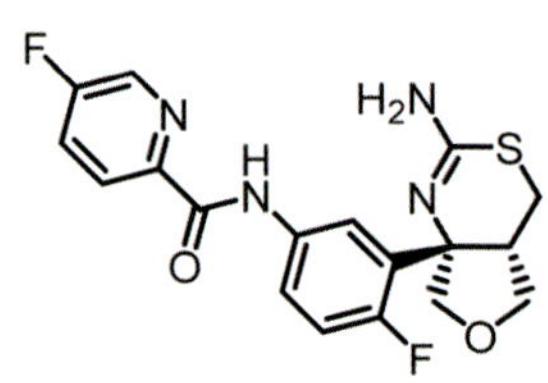

**LY-2886721 (25)**
BACE1 $IC_{50}$ = 0.020 μM
LE = 0.39

**Figure 14.3** *(Continued)*

and **7** using the detailed structural information provided by the X-ray cocrystal structure of isothiourea **5** in conjunction with *in silico* modeling. The concept of these chemotypes was to preserve the key interactions with the catalytic aspartates using a minimal number of hydrogen bond donors while tuning basicity and having fewer rotable bonds. In addition to these characteristics, which were intended to favor CNS penetration, the group also incorporated an $sp^3$ center on the cores to enable ligand-efficient access to both the prime and the nonprime sides of the active site.

Initially, both N3-substituted (e.g., **6**) and regioisomeric N1-substituted iminohydantoins were examined, but only the N3-substituted analogs showed binding to BACE1 by NMR, likely due to the nonbasic nature of the N1-substituted analogs (calculated $pK_a$ values for N3- and N1-substitution are ~7 and ~4, respectively). As SAR developed within the N3-substituted series, minor differences in the substitution pattern, for example, C5-gem-dimethyl analog **8** versus C5-methyl-isobutyl compound **9**, unexpectedly caused the core binding mode to flip by 180°, projecting the benzyl group into either S1′ or S1 as evidenced by X-ray crystallography [50,60]. Taken together, these early findings emphasized both the importance of the protonation state of the core in maintaining the productive hydrogen bonding interactions with the catalytic dyad and the key role that X-ray crystallography played in understanding emerging, often complex SAR.

Through continued structure-enabled optimization and a focus on drug-like properties, both iminohydantoin **2** and iminopyrimidinone **3** were eventually identified as ligand-efficient leads that demonstrated a minimal shift in potency in the cellular assay and reasonable rat PK [50,60,61]. It should be noted that the ring size of the cyclic acylguanidine core (i.e., five-membered iminohydantoins **2, 6,** and **10** versus six-membered iminopyrimidinones **3, 7,** and **11**) had a profound impact on both the physicochemical properties and the binding conformation of these structures, which therefore required very different strategies for optimization [61]. This highlights the fact that, even starting with a minimal pharmacophore like **1**, a significant degree of optimization is required to find advanced leads. A major driver in subsequent development of both series was the use of an *in vivo* rat pharmacodynamic assay early in the design cycle, which allowed differentiation of compounds based on their ability to lower both brain and cerebrospinal fluid (CSF) Aβ levels. This integrated approach ultimately resulted in the discovery of orally bioavailable, brain-penetrant inhibitors **10–12** that afforded reductions of central Aβ40 levels in rats following single oral doses [58,61,62]. Additional extensive SAR development and optimization of properties produced the most clinically advanced BACE inhibitor to date, **MK-8931** (verubecestat) [69]. This compound dramatically lowered CSF Aβ levels both in healthy volunteers [70] and in AD patients [33] in phase 1 clinical trials. Based on positive feedback from a planned interim safety analysis by their Data Monitoring Committee [71], MK-8931 initiated two phase 3 trials in 2013, one in mild-to-moderate AD patients [72] and one in patients with prodromal AD [73].

AstraZeneca arrived at iminopyrimidinone **3** from a very different starting point, in their case from isocytosine fragment hit **13** (Figure 14.3b) [43,44].

Deplanarization and removal of rotatable bonds from this hit gave C6-phenyl iminopyrimidinone **14**. X-ray crystallography of this compound with BACE1 showed that the $sp^3$ nature of the C5 and C6 centers enabled the C6-phenyl group to occupy the S1 subpocket, albeit via an energetically disfavored pseudoaxial conformation. Quaternization of the C6 center through introduction of a methyl group afforded compound **15** and imparted modest potency gains through stabilization of this conformation. Finally, derivatization of N3 with a methyl group based on the SAR from the planar isocytosine series gave core **16** with significant and ligand-efficient potency gains, and subsequent elaboration of the S1 phenyl to an S1–S3 biaryl ultimately produced **3**.

Guided by their X-ray data, they further modified **3** to more fully occupy the S1–S3 subsites. To that end, reintroduction of the rotatable bonds present in their initial hit **13** afforded analog **17**. The methoxybiphenyl portion of this compound did indeed sit deeper in the S1–S3 pockets with the iminopyrimidinone core tilted toward the prime side while maintaining productive contacts with the catalytic aspartates. This modification afforded a fourfold boost in BACE1 potency with no loss in ligand efficiency.

AstraZeneca subsequently pursued scaffold hopping in designing their core-modified inhibitors **18–20** [34,37,68], all of which retained key pharmacophore **1**. Interestingly, the Wyeth group independently discovered a bicyclic amidine scaffold similar to **18** through their own HTS campaign and optimized to very similar chemical space [74], highlighting the sometimes convergent nature of efforts in this arena. Compound **20** potently lowered central Aβ levels in preclinical species [35], and it was advanced into phase 1 clinical trials as **AZD-3839** [36,75] but was ultimately halted. AstraZeneca has since partnered with Eli Lilly to develop **AZD-3293/LY-3314814**, which recently entered phase 2/3 clinical trials [76].

In their dihydrothiazine work, scientists at Eli Lilly also pursued a strategy of deplanarization, in this case of screening hit **21** [38] to give core **22** similar to lead **4** reported by Roche [54]. Compound **22** was optimized along similar lines as iminopyrimidinone **16**, building into the S3 subsite by elaborating the S1 phenyl into a *meta*-biaryl. With introduction of an S3 pyrimidinyl group to afford **23**, they identified a ligand efficient submicromolar BACE1 inhibitor lead that, with property-based optimization, ultimately led to identification of their first clinical compound **LY-2811376** (**24**). This compound afforded significant reductions in CSF Aβ levels in healthy volunteers following single doses, but its development was discontinued due to preclinical toxicity findings unrelated to inhibition of BACE1 [38]. Lilly subsequently turned to a related tetrahydrofuran-fused dihydrothiazine core, also investigated extensively by Eisai [66], from which they identified their second clinical candidate, **LY-2886721** (**25**) [39]. Development of this compound was stopped in phase 2 due to abnormal liver function, which they reported as unrelated to inhibition of BACE1 [77].

In addition to the case studies described above, a number of other companies are or have been in phase 1 clinical trials with BACE1 inhibitors, including Eisai/Biogen (**E2609**) [78], Shionogi/Janssen (**JNJ-54861911**) [79], Boeringher-

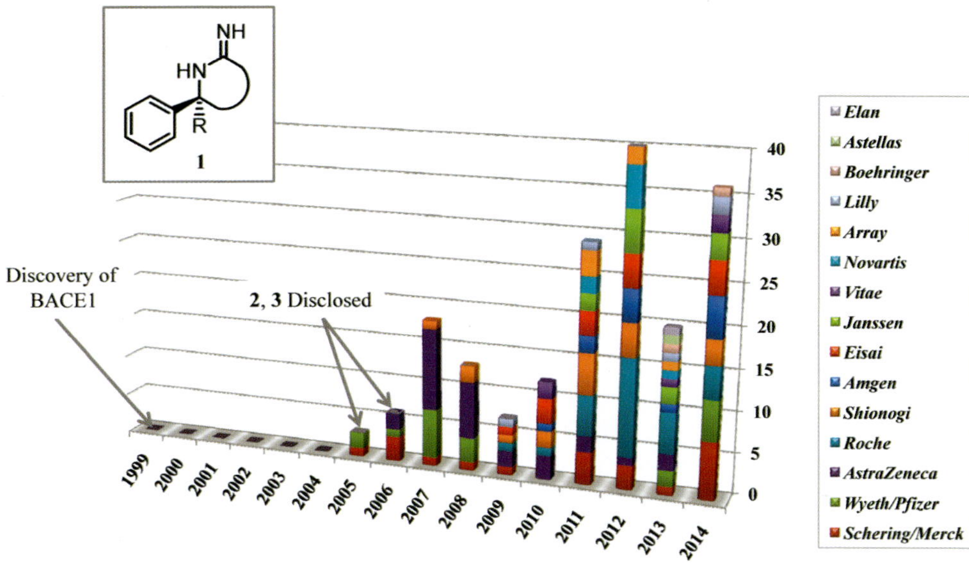


**Figure 14.4** Survey of small molecule BACE1 inihibitor patent applications disclosing structures that contain key pharmacophore **1**. For each year, the bars are colored by company. After an inital period with relatively few applications and limited companies working in this space, the number of applications and diversity of companies significantly increased beginning in 2011.

Ingelheim/Vitae (**BI1181181**, discontinued) [80], and Roche (**RG7129**, discontinued) [81]. While the structures of these compounds have not been disclosed, a review of the patent literature shows that, despite a wide array of structural differences, most if not all of the possible candidates contain the minimal pharmacophore **1** or a close variant thereof. More broadly, a recent review by Oehlrich *et al.* elegantly summarized the extensive diversity of cyclic nonplanar amidines that have been explored across the field (see Figure 30 in Ref. [40]). Looking at BACE1 inhibitor patent applications over time, Figure 14.4 represents a survey of those applications that have employed key pharmacophore **1**. Following its first disclosure in 2005 [59], there was an incubation period before a significant increase in prevalence is seen beginning in 2011 as more researchers recognized the value of this structural motif, particularly given the challenges of finding tractable, brain-penetrant chemical matter when pursuing other chemotypes. It is striking to note that from the first iminoheterocycle disclosures, it took 8 years to start the first phase 3 clinical trials with a BACE1 inhibitor in 2013, reflective of the extensive development and optimization of inhibitors that was still required to identify chemical matter suitable for long-term trials.

Finally, while a number of different hits exemplifying pharmacophore **1** were identified though FDBB and served as launching points for numerous BACE1 inhibitor discovery programs, this amidine moiety has created its own drug discovery challenges. As was found early on by the Schering-Plough group, the amidine must be protonated for productive interactions with the catalytic dyad, and this basic center contributed to hERG affinity and P-gp susceptibility of the molecules. Several companies have published on their approaches to these issues (e.g [7,40,62,82].), reflecting the specific facets of their advanced chemical matter. More broadly, though, a general focus on ligand efficiency and maintenance of CNS drug-like physicochemical properties [83] appears to be important for achieving high brain BACE1 occupancy *in vivo* while minimizing off-target liabilities.

## 14.3 Conclusions

The emergence of BACE1 as a therapeutic target for the development of disease-modifying treatments for Alzheimer's disease has fueled intensive drug discovery efforts in which fragment-based approaches have played a key role. From several independent starting points identified through FBDD, a variety of design approaches provided a diverse set of lead structures that shared the nonplanar cyclic amidine minimal pharmacophore represented by general iminoheterocycle structure **1**. This novel chemotype for aspartyl protease inhibition provided direct, ligand-efficient access to both the contiguous S1–S3 subpockets and the non-prime side of BACE1. Even with key pharmacophore **1** as a foundation, there have been a number of innovative approaches to further optimization that have produced many structurally diverse inhibitors, as exemplified by compounds **12**, **20**, and **25**. It should be noted, however, that all of the disclosed compounds that have progressed into advanced preclinical testing or into clinical trials have remained relatively lean, occupying only two subsites of BACE1 with minimal rotatable bonds. Adherence to these design principles was critical for achieving high brain BACE1 occupancy *in vivo*, as evidenced by reduction of central Aβ levels in animal models. In this way, discovery of the iminoheterocycle class of BACE1 inhibitors transformed the field and, after a great deal of persistence and a high tolerance for setbacks, will now enable the most definitive test of the amyloid hypothesis to date in the clinic. The entry of multiple BACE1 inhibitors, all of which have likely arisen from or been influenced by fragment-based design, into clinical trials by several organizations is a testament to the progress of FBDD and its potential to impact the development of drugs for critical unmet medical needs.

## References

1 Hardy, J. and Selkoe, D.J. (2002) The amyloid hypothesis of Alzheimer's disease: progress and problems on the road to therapeutics. *Science*, **297**, 353–356.

2 Karran, E., Mercken, M., and De Strooper, B. (2011) The amyloid cascade hypothesis for Alzheimer's disease: an appraisal for the development of therapeutics. *Nature Reviews Drug Discovery*, **10**, 698–711.

3 Vassar, R., Kuhn, P.-H., Haass, C., Kennedy, M.E., Rajendran, L., Wong, P.C., and Lichtenthaler, S.F.J. (2014) Function, therapeutic potential and cell biology of BACE proteases: current status and future prospects. *Journal of Neurochemistry*, **130**, 4–28.

4 Lemere, C.A. (2013) Immunotherapy for Alzheimer's disease: hoops and hurdles. *Molecular Neurodegeneration*, **8**, 36–41.

5 Biogen Idec Presents Positive Interim Results from Phase 1B Study of Investigational Alzheimer's Disease Treatment Aducanumab (BIIB037) at 2015 AD/PD$^{TM}$ Conference. http://media.biogen.com/press-release/corporate/biogen-idec-presents-positive-interim-results-phase-1b-study-investigational (accessed 3 May 2015).

6 Yan, R. and Vassar, R. (2014) Targeting the β secretase BACE1 for Alzheimer's disease therapy. *Lancet Neurology*, **13**, 319–329.

7 Yuan, J., Venkatraman, S., Zheng, Y., McKeever, B.M., Lawrence, W., Dillard, L.W., and Singh, S.B. (2013) Structure-based design of β-site APP cleaving enzyme 1 (BACE1) inhibitors for the treatment of Alzheimer's disease. *Journal of Medicinal Chemistry*, **56**, 4156–4180.

8 Vassar, R., Bennett, B.D., Babu-Khan, S., Kahn, S., Mendiaz, E.A., Denis, P., Teplow, D.B., Ross, S., Amarante, P., Loeloff, R., Luo, Y., Fisher, S., Fuller, J., Edenson, S., Lile, J., Jarosinski, M.A., Biere, A.L., Curran, E., Burgess, T., Louis, J.-C., Collins, F., Treanor, J., Rogers, G., and Citron, M. (1999) β-secretase cleavage of Alzheimer's amyloid precursor protein by the transmembrane aspartic protease BACE. *Science*, **286**, 735–741.

9 Yan, R., Bienkowski, M.J., Shuck, M.E., Miao, H.Y., Tory, M.C., Pauley, A.M., Brashler, J.R., Stratman, N.C., Mathews, W.R., Buhl, A.E., Carter, D.B., Tomasselli, A.G., Parodi, L.A., Heinrikson, R.L., and Gurney, M.E. (1999) Membrane-anchored aspartyl protease with Alzheimer's disease β-secretase activity. *Nature*, **402**, 533–537.

10 Sinha, S., Anderson, J.P., Barbour, R., Basi, G.S., Caccavello, R., Davis, D., Doan, M., Dovey, H.F., Frigon, N., Hong, J., Jacobson-Croak, K., Jewett, N., Keim, P., Knops, J., Lieberburg, I., Power, M., Tan, H., Tatsuno, G., Tung, J., Schenk, D., Seubert, P., Suomensaari, S.M., Wang, S., Walker, D., Zhao, J., McConlogue, L., and John, V. (1999) Purification and cloning of amyloid precursor protein β-secretase from human brain. *Nature*, **402**, 537–540.

11 Hussain, I., Powell, D., Howlett, D.R., Tew, D.G., Meek, T.D., Chapman, C., Gloger, I.S., Murphy, K.E., Southan, C.D., Ryan, D.M., Smith, T.S., Simmons, D.L., Walsh, F.S., Dingwall, C., and Christie, C. (1999) Identification of a novel aspartic protease (Asp 2) as β-secretase. *Molecular and Cellular Neurosciences*, **14**, 419–427.

12 Lin, X., Koelsch, G., Wu, S., Downs, D., Dashti, A., and Tang, J. (2000) Human aspartic protease memapsin 2 cleaves the β-secretase site of β-amyloid precursor protein. *Proceedings of the National Academy of Sciences of the United States of America*, **97**, 1456–1460.

13 Roberds, S.L., Anderson, J., Basi, G., Bienkowski, M.J., Branstetter, D.G., Chen, K.S., Freedman, S.B., Frigon, N.L., Games, D., Hu, K., Johnson-Wood, K., Kappenman, K.E., Kawabe, T.T., Kola, I., Kuehn, R., Lee, M., Liu, W., Motter, R., Nichols, N.F., Power, M., Robertson, D.W., Schenk, D., Schoor, M., Shopp, G.M., Shuck, M.E., Sinha, S., Svensson, K.A., Tatsuno, G., Tintrup, H., Wijsman, J., Wright, S., and McConlogue, L. (2001) BACE knockout mice are healthy despite lacking the primary β-secretase activity in brain: implications for Alzheimer's disease therapeutics. *Human Molecular Genetics*, **10**, 1317–1324.

14 Mullan, M., Crawford, F., Axelman, K., Houlden, H., Lilius, L., Winblad, B., and Lannfelt, L. (1992) A pathogenic mutation for probable Alzheimer's disease in the APP gene at the N-terminus of β-amyloid. *Nature Genetics*, **1**, 345–347.

15 Haass, C., Lemere, C.A., Capell, A., Citron, M., Seubert, P., Schenk, D., Lannfelt, L., and Selkoe, D.J. (1995) The Swedish mutation causes early-onset Alzheimer's disease by β-secretase cleavage within the secretory pathway. *Nature Medicine*, **1**, 1291–1296.

16 Jonsson, T., Atwal, J.K., Steinberg, S., Snaedal, J., Jonsson, P.V., Bjornsson, S., Stefansson, H., Sulem, P., Gudbjartsson, D., Maloney, J., Hoyte, K., Gustafson, A., Liu, Y., Lu, Y., Bhangale, T., Graham, R.R., Huttenlocher, J., Bjornsdottir, G., Andreassen, O.A., Jonsson, E.G., Palotie, A., Behrens, T.W., Magnusson, O.T., Kong, A., Thorsteinsdottir, U., Watts, R.J., and Stefansson, K. (2012) A mutation in APP protects against Alzheimer's disease and age-related cognitive decline. *Nature*, **488**, 96–99.

17 Hong, L., Koelsch, G., Lin, X., Wu, S., Terzyan, S., Ghosh, A.K., Zhang, X.C., and Tang, J. (2000) Structure of the protease domain of memapsin 2 (β-secretase) complexed with inhibitor. *Science*, **290**, 150–153.

18 Hong, L., Turner, R.T. 3rd, Koelsch, G., Shin, D., Ghosh, A.K., and Tang, J. (2002) Crystal structure of memapsin 2 (β-secretase) in complex with an inhibitor OM00-3. *Biochemistry*, **41**, 10963–10967.

19 Tamamura, H., Kato, T., Otaka, A., and Fujii, N. (2003) Synthesis of potent β-secretase inhibitors containing a hydroxyethylamine dipeptide isostere and their structure–activity relationship studies. *Organic and Biomolecular Chemistry*, **1**, 2468–2473.

20 Stachel, S.J., Coburn, C.A., Steele, T.G., Jones, K.G., Loutzenhiser, E.F., Gregro, A.R., Rajapakse, H.A., Lai, M.T., Crouthamel, M.C., Xu, M., Tugusheva, K., Lineberger, J.E., Pietrak, B.L., Espeseth, A.S., Shi, X.P., Chen-Dodson, E., Holloway, M.K., Munshi, S., Simon, A.J., Kuo, L., and Vacca, J.P. (2004) Structure-based design of potent and selective cell-permeable inhibitors of human β-secretase (BACE-1). *Journal of Medicinal Chemistry*, **47**, 6447–6450.

21 Ghosh, A.K. and Osswald, H.L. (2014) BACE1 (β-secretase) inhibitors for the treatment of Alzheimer's disease. *Chemical Society Reviews*, **43**, 6765–6813.

22 Hopkins, A.L., Groom, C.R., and Alex, A. (2004) Ligand efficiency: a useful metric for lead selection. *Drug Discovery Today*, **9**, 430–431.

23 Sankaranarayanan, S., Holahan, M.A., Colussi, D., Crouthamel, M.C., Devanarayan, V., Ellis, J., Espeseth, A., Gates, A.T., Graham, S.L., Gregro, A.R., Hazuda, D., Hochman, J.H., Holloway, K., Jin, L., Kahana, J., Lai, M.T., Lineberger, J., McGaughey, G., Moore, K.P., Nantermet, P., Pietrak, B., Price, E.A., Rajapakse, H., Stauffer, S., Steinbeiser, M.A., Seabrook, G., Selnick, H.G., Shi, X.P., Stanton, M.G., Swestock, J., Tugusheva, K., Tyler, K.X., Vacca, J.P., Wong, J., Wu, G., Xu, M., Cook, J.J., and Simon, A.J. (2009) First demonstration of cerebrospinal fluid and plasma Aβ lowering with oral administration of a β-site amyloid precursor protein-cleaving enzyme 1 inhibitor in nonhuman primates. *Journal of Pharmacology and Experimental Therapeutics*, **328**, 131–140.

24 Lerchner, A., Machauer, R., Betschart, C., Veenstra, S., Rueeger, H., McCarthy, C., Tintelnot-Blomley, M., Jaton, A.-L., Rabe, S., Desrayaud, S., Enz, A., Staufenbiel, M., Paganetti, P., Rondeau, J.-M., and Neumann, U. (2010) Macrocyclic BACE-1 inhibitors acutely reduce Aβ in brain after po application. *Bioorganic and Medicinal Chemistry Letters*, **20**, 603–607.

25 Dineen, T.A., Weiss, M.M., Williamson, T., Acton, P., Babu-Khan, S., Bartberger, M.D., Brown, J., Chen, K., Cheng, Y., Citron, M., Croghan, M.D., Dunn, R.T., Esmay, J., Graceffa, R.F., Harried, S.S., Hickman, D., Hitchcock, S.A., Horne, D.B., Huang, H., Imbeah-Ampiah, R., Judd, T., Kaller, M.R., Kreiman, C.R., La, D.S., Li, V., Lopez, P., Louie, S., Monenschein, H., Nguyen, T.T., Pennington, L.D., Miguel, T.S., Sickmier, E.A., Vargas, H.M., Wahl, R.C., Wen, P.H., Whittington, D.A., Wood, S., Xue, Q., Yang, B.H., Patel, V.F., and Zhong, W. (2012) Design and synthesis of potent, orally efficacious hydroxyethylamine derived β-site amyloid precursor protein cleaving enzyme (BACE1) inhibitors. *Journal of Medicinal Chemistry*, **55**, 9025–9044.

26 Kaller, M.R., Halrried, S.S., Albrecht, B., Amarante, P., Babu-Khan, S., Bartberger, M.D., Brown, J., Brown, R., Chen, K., Cheng, Y., Citron, M., Croghan, M.D., Graceffa, R., Hickman, D., Judd, T.,

Kriemen, C., La, D., Li, V., Lopez, P., Luo, Y., Masse, C., Monenschein, H., Nguyen, T., Pennington, L.D., Miguel, T.S., Sickmier, E.A., Wahl, R.C., Weiss, M.M., Wen, P.H., Williamson, T., Wood, S., Xue, M., Yang, B., Zhang, J., Patel, V., Zhong, W., and Hitchcock, S. (2012) A potent and orally efficacious, hydroxyethylamine-based inhibitor of β-secretase. *ACS Medicinal Chemistry Letters*, **3**, 886–891.

27 Weiss, M.M., Williamson, T., Babu-Khan, S., Bartberger, M.D., Brown, J., Chen, K., Cheng, Y., Citron, M., Croghan, M.D., Dineen, T.A., Esmay, J., Graceffa, R.F., Harried, S.S., Hickman, D., Hitchcock, S.A., Horne, D.B., Huang, H., Imbeah-Ampiah, R., Judd, T., Kaller, M.R., Kreiman, C.R., La, D.S., Li, V., Lopez, P., Louie, S., Monenschein, H., Nguyen, T.T., Pennington, L.D., Rattan, C., San Miguel, T., Sickmier, E.A., Wahl, R.C., Wen, P.H., Wood, S., Xue, Q., Yang, B.H., Patel, V.F., and Zhong, W. (2012) Design and preparation of a potent series of hydroxyethylamine containing β-secretase inhibitors that demonstrate robust reduction of central β-amyloid. *Journal of Medicinal Chemistry*, **55**, 9009–9024.

28 Congreve, M., Aharony, D., Albert, J., Callaghan, O., Campbell, J., Carr, R.A.E., Chessari, G., Cowan, S., Edwards, P.D., Fredrickson, M., McMenamin, R., Murray, C.W., Patel, S., and Wallis, N. (2007) Application of fragment screening by X-ray crystallography to the discovery of aminopyridines as inhibitors of β-secretase. *Journal of Medicinal Chemistry*, **50**, 1124–1132.

29 Patel, S., Vuillard, L., Cleasby, A., Murray, C. W., and Yon, J. (2004) Apo and inhibitor complex structures of BACE (β-secretase). *Journal of Molecular Biology*, **343**, 407–416.

30 Stachel, S.J. (2009) Progress toward the development of a viable BACE-1 inhibitor. *Drug Development Research*, **70**, 101–110.

31 Probst, G. and Xu, Y.-Z. (2012) Small-molecule BACE1 inhibitors: a patent literature review (2006–2011). *Expert Opinion on Therapeutic Patents*, **22**, 511–540.

32 Baker, M. (2013) Fragment-based lead discovery grows up. *Nature Reviews Drug Discovery*, **12**, 5–7.

33 Forman, M., Kleijn, H.-J., Dockendorf, M., Palcza, J., Tseng, J., Canales, C., Egan, M., Kennedy, M., Laterza, O., Ma, L., Scott, J., Tanen, M., Apter, J., Backonja, M., Ereshefsky, L., Gevorkyan, H., Jhee, S., Rynders, R., Zari, A., Bryan, E., Wagner, J., Troyer, M., and Stone, J. (2013) The novel BACE inhibitor MK-8931 dramatically lowers CSF beta-amyloid in patients with mild-to-moderate Alzheimer's disease. *Alzheimer's and Dementia*, **9** (4), Supplement P139.

34 Swahn, B.-M., Kolmodin, K., Karlström, S., von Berg, S., Söderman, P., Holenz, J., Berg, S., Lindström, J., Sundström, M., Turek, D., Kihlström, J., Slivo, C., Andersson, L., Pyring, D., Rotticci, D., Öhberg, L., Kers, A., Bogar, K., von Kieseritzky, F., Bergh, M., Olsson, L.-L., Janson, J., Eketjäll, S., Georgievska, B., Jeppsson, F., and Fälting, J. (2012) Design and synthesis of β-secretase (BACE1) inhibitors with *in vivo* brain reduction of β-amyloid peptides. *Journal of Medicinal Chemistry*, **55**, 9346–9361.

35 Jeppsson, F., Eketjäll, S., Janson, J., Karlström, S., Gustavsson, S., Olsson, L.-L., Radesäter, A.-C., Ploeger, B., Cebers, G., Kolmodin, K., Swahn, B.-M., von Berg, S., Bueters, T., and Fälting, J. (2012) Discovery of AZD3839, a potent and selective BACE1 clinical candidate for the treatment of Alzheimer's disease. *Journal of Biological Chemistry*, **287**, 41245–41257.

36 Quartino, A., Huledal, G., Sparve, E., Lüttgen, M., Bueters, T., Karlsson, P., Olsson, T., Paraskos, J., Maltby, J., Claeson-Bohnstedt, K., Lee, C.-M., Alexander, R., Fälting, J., and Paulsson, B. (2014) Population pharmacokinetic and pharmacodynamic analysis of plasma Aβ40 and Aβ42 following single oral doses of the BACE1 inhibitor AZD3839 to healthy volunteers. *Clinical Pharmacology in Drug Development*, **3**, 396–405.

37 Gravenfors, Y., Viklund, J., Blid, J., Ginman, T., Karlstrom, S., Kihlstrom, J., Kolmodin, K., Lindstrom, J., von Berg, S., von Kierseritzky, F., Slivo, C., Swahn, B.-M., Olsson, L.-L., Johansson, P., Eketjäll, S., Fälting, J., Jeppsson, F., Strömberg, K., Janson, J., and Rahm, F.

(2012) New aminoimidazoles as β-secretase (BACE-1) inhibitors showing amyloid-β (Aβ) lowering in brain. *Journal of Medicinal Chemistry*, **55**, 9297–9311.

38 May, P.C., Dean, R.A., Lowe, S.L., Martenyi, F., Sheehan, S.M., Boggs, L.N., Monk, S.A., Mathes, B.M., Mergott, D.J., Watson, B.M., Stout, S.L., Timm, D.E., Smith LaBell, E., Gonzales, C.R., Nakano, M., Jhee, S.S., Yen, M., Ereshefsky, L., Lindstrom, T.D., Calligaro, D.O., Cocke, P.J., Hall, D.G., Friedrich, S., Citron, M., and Audia, J.E. (2011) Central reduction of amyloid-β in humans with an orally available, non-peptidic β-secretase inhibitor. *Journal of Neuroscience*, **31**, 16507–16516.

39 May, P.C., Willis, B.A., Lowe, S.L., Dean, R.A., Monk, S.A., Cocke, P.J., Audia, J.E., Boggs, L.N., Borders, A.R., Brier, R.A., Calligaro, D.O., Day, T.A., Ereshefsky, L., Erickso, J.A., Gevorkyan, H., Gonzales, C.R., James, D.E., Jhee, S.S., Komjathy, S.F., Li, L., Lindstrom, T.D., Mathe, B.M., Martenyi, F., Sheehan, S.M., Stout, S.L., Timm, D.E., Vaught, G.M., Watson, B.M., Winneroski, L.L., Yang, Z., and Mergott, D.J. (2015) The potent BACE1 inhibitor LY2886721 elicits robust central Aβ pharmacodynamic responses in mice, dogs, and humans. *Journal of Neuroscience*, **35**, 1199–1210.

40 Oehlrich, D., Prokopcova, H., and Gijsen, H.J.M. (2014) The evolution of amidine-based brain penetrant BACE1 inhibitors. *Bioorganic and Medicinal Chemistry Letters*, **24**, 2033–2045.

41 Menting, K.W. and Claassen, J.A.H.R. (2014) β-secretase inhibitor; a promising novel therapeutic drug in Alzheimer's disease. *Frontiers in Aging Neuroscience*, **6**, 165.

42 Murray, C.W., Callaghan, O., Chessari, G., Cleasby, A., Congreve, M., Frederickson, M., Hartshorn, M.J., McMenamin, R., Patel, S., and Wallis, N. (2007) Application of fragment screening by X-ray crystallography to β-secretase. *Journal of Medicinal Chemistry*, **50**, 1116–1123.

43 Geschwindner, S., Olsson, L.-L., Albert, J.S., Deinum, J., Philip, D., Edwards, P.D., de Beer, T., and Folmer, R.H.A. (2007) Discovery of a novel warhead against β-secretase through fragment-based lead generation. *Journal of Medicinal Chemistry*, **50**, 5903–5911.

44 Edwards, P.D., Albert, J.S., Sylvester, M., Aharony, D., Andisik, D., Callaghan, O., Campbell, J.B., Carr, R.A., Chessari, G., Congreve, M., Frederickson, M., Folmer, R.H.A., Geschwindner, S., Koether, G., Kolmodin, K., Krumrine, J., Mauger, R.C., Murray, C.W., Olsson, L.-L., Patel, S., Spear, N., and Tian, G. (2007) Application of fragment-based lead generation to the discovery of novel, cyclic amidine β-secretase inhibitors with nanomolar potency, cellular activity, and high ligand efficiency. *Journal of Medicinal Chemistry*, **50**, 5912–5925.

45 Kuglstatter, A., Stahl, M., Peters, J., Huber, W., Stihle, M., Schlatter, D., Benz, J., Ruf, A., Roth, D., Enderle, T., and Hennig, M. (2008) Tyramine fragment binding to BACE-1. *Bioorganic and Medicinal Chemistry Letters*, **18**, 1304–1307.

46 Yang, W., Fucini, R.V., Fahr, B.T., Randal, M., Lind, K.E., Lam, M.B., Lu, W., Lu, Y., Cary, D.R., Romanowski, M.J., Colussi, D., Pietrak, B., Allison, T.J., Munshi, S.K., Penny, D.M., Pham, P., Sun, J., Thomas, A.E., Wilkinson, J.M., Jacobs, J.W., McDowell, R.S., and Ballinger, M.D. (2009) Fragment-based discovery of nonpeptidic BACE-1 inhibitors using tethering. *Biochemistry*, **48**, 4488–4496.

47 Godemann, R., Madden, J., Joachim Krämer, J., Smith, M., Fritz, U., Hesterkamp, T., Barker, J., Höppner, S., Hallett, D., Cesura, A., Ebneth, A., and Kemp, J. (2009) Fragment-based discovery of BACE1 inhibitors using functional assays. *Journal of Medicinal Chemistry*, **48**, 10743–10751.

48 Madden, J., Dod, J.R., Godemann, R., Kraemer, J., Smith, M., Biniszkiewicz, M., Hallett, D.J., Barker, J., Dyekjaer, J.D., and Hesterkamp, T. (2010) Fragment-based discovery and optimization of BACE1 inhibitors. *Bioorganic and Medicinal Chemistry Letters*, **20**, 5329–5333.

49 Wang, Y.-S., Strickland, C., Voigt, J.H., Kennedy, M.E., Beyer, B.M., Senior, M.M., Smith, E.M., Nechuta, T.L., Madison, V.S., Czarniecki, M., McKittrick, B.A., Stamford, A.W., Parker, E.M., Hunter, J.C., Greenlee, W.J., and Wyss, D.F. (2010)

Application of fragment-based NMR screening, X-ray crystallography, structure-based design, and focused chemical library design to identify novel μM leads for the development of nM BACE-1 (β-site APP cleaving enzyme 1) inhibitors. *Journal of Medicinal Chemistry*, **53**, 942–950.

50 Zhu, Z., Sun, Z.-Y., Ye, Y., Voigt, J., Strickland, C., Smith, E.M., Cumming, J., Wang, L., Wong, J., Wang, Y.-S., Wyss, D.F., Chen, X., Kuvelkar, R., Kennedy, M.E., Favreau, L., Parker, E., McKittrick, B.A., Stamford, A., Czarniecki, M., Greenlee, W., and Hunter, J.C. (2010) Discovery of cyclic acylguanidines as highly potent and selective β-site amyloid cleaving enzyme (BACE) inhibitors. Part I. Inhibitor design and validation. *Journal of Medicinal Chemistry*, **53**, 951–965.

51 Cheng, Y., Judd, T.C., Bartberger, M.D., Brown, J., Chen, K., Fremeau, R.T., Jr., Hickman, D., Hitchcock, S.A., Jordan, B., Li, V., Lopez, P., Louie, S.W., Luo, Y., Michelsen, K., Nixey, T., Powers, T.S., Rattan, C., Sickmier, E.A., St. Jean, D.J. Jr., Wahl, R.C., Wen, P.H., and Wood, S. (2011) From fragment screening to *in vivo* efficacy: optimization of a series of 2-aminoquinolines as potent inhibitors of beta-site amyloid precursor protein cleaving enzyme 1 (BACE1). *Journal of Medicinal Chemistry*, **54**, 5836–5857.

52 Efremov, I.V., Vajdos, F.F., Borzilleri, K.A., Capetta, S., Chen, H., Dorff, P.H., Dutra, J.K., Goldstein, S.W., Mansour, M., McColl, A., Noell, S., Oborski, C.E., O'Connell, T.N., O'Sullivan, T.J., Pandit, J., Wang, H., Wei, B.Q., and Withka, J.M. (2012) Discovery and optimization of a novel spiropyrrolidine inhibitor of β-secretase (BACE1) through fragment-based drug design. *Journal of Medicinal Chemistry*, **55**, 9069–9088.

53 Gerritz, S.W., Zhai, W., Shi, S., Zhu, S., Toyn, J.H., Meredith, J.E., Jr., Iben, L.G., Burton, C.R., Albright, C.F., Good, A.C., Tebben, A.J., Muckelbauer, J.K., Camac, D.M., Metzler, W., Cook, L.S., Padmanabha, R., Lentz, K.A., Sofia, M.J., Poss, M.A., Macor, J.E., and Thompson, L.A., III (2012) Acyl guanidine inhibitors of β-secretase (BACE-1): optimization of a micromolar hit to a nanomolar lead via iterative solid- and solution-phase library synthesis. *Journal of Medicinal Chemistry*, **55**, 9208–9223.

54 Woltering, T.J., Wostl, W., Hilpert, H., Rogers-Evans, M., Pinard, E., Mayweg, A., Göbel, M., Banner, D.W., Benz, J., Travagli, M., Pollastrini, M., Marconi, G., Gabellieri, E., Guba, W., Mauser, H., Andreini, M., Jacobsen, H., Power, E., and Narquizian, R. (2013) BACE1 inhibitors: a head group scan on a series of amides. *Bioorganic and Medicinal Chemistry Letters*, **23**, 4239–4243.

55 Hilpert, H., Guba, W., Woltering, T.J., Wostl, W., Pinard, E., Mauser, H., Mayweg, A.V., Rogers-Evans, M., Humm, R., Krummenacher, D., Muser, T., Schnider, C., Jacobsen, H., Ozmen, L., Bergadano, A., Banner, D.W., Hochstrasser, R., Kuglstatter, A., David-Pierson, P., Fischer, H., Polara, A., and Narquizian, R. (2013) β-Secretase (BACE1) inhibitors with high *in vivo* efficacy suitable for clinical evaluation in Alzheimer's disease. *Journal of Medicinal Chemistry*, **56**, 3980–3995.

56 Baxter, E.W., Conway, K.A., Kennis, L., Bischoff, F., Mercken, M.H., De Winter, H.L., Reynolds, C.H., Tounge, B.A., Luo, C., Scott, M.K., Huang, Y., Braeken, M., Pieters, S.M.A., Berthelot, D.J.C., Masure, S., Bruinzeel, W.D., Jordan, A.D., Parker, M.H., Boyd, R.E., Qy, J., Alexander, R.S., Brenneman, D.E., and Reitz, A.B. (2007) 2-Amino-3,4-dihydroquinazolines as inhibitors of BACE-1 (β-site APP cleaving enzyme): use of structure based design to convert a micromolar hit into a nanomolar lead. *Journal of Medicinal Chemistry*, **50**, 4261–4264. PDB: 2Q15.

57 Stamford, A. and Strickland, C. (2013) Inhibitors of BACE for treating Alzheimer's disease: a fragment-based drug discovery story. *Current Opinion in Chemical Biology*, **17**, 320–328.

58 Cumming, J.N., Smith, E.M., Wang, L., Misiaszek, J., Durkin, J., Pan, J., Iserloh, U., Wu, Y., Zhu, Z., Strickland, C., Voigt, J., Chen, X., Kennedy, M.E., Kuvelkar, R., Hyde, L.A., Cox, K., Favreau, L., Czarniecki, M.F., Greenlee, W.J., McKittrick, B.A., Parker, E.M., and

Stamford, A.W. (2012) Structure based design of iminohydantoin BACE1 inhibitors: identification of an orally available, centrally active BACE1 inhibitor. *Bioorganic and Medicinal Chemistry Letters*, **22**, 2444–2449.

59 Zhu, Z., McKittrick, B., Sun, Z.-Y., Ye, Y.C., Voigt, J.H., Strickland, C., Smith, E.M., Stamford, A., Greenlee, W.J., Wu, Y., Iserloh, U., Mazzola, R., Caldwell, J., Cumming, J., Wang, L., Guo, T., Le, T.X.H., Saionz, K.W., Babu, S.D., and Hunter, R.C. (2005) Preparation of heterocyclic aspartyl protease inhibitors for treating various diseases. *PCT Int. Appl.* PATN WO 2005058311 A1.

60 Wyss, D.F., Wang, Y.-S., Eaton, H.L., Strickland, C., Voigt, J.H., Zhu, Z., and Stamford, A.W. (2012) Combining NMR and X-ray crystallography in fragment-based drug discovery: discovery of highly potent and selective BACE-1 inhibitors. *Topics in Current Chemistry*, **317**, 83–114.

61 Stamford, A.W., Scott, J.D., Li, S.W., Babu, S., Tadesse, D., Hunter, R., Wu, Y., Misiaszek, J., Cumming, J.N., Gilbert, E.J., Huang, C., McKittrick, B.A., Hong, L., Guo, T., Zhu, Z., Strickland, C., Orth, P., Voigt, J.H., Kennedy, M.E., Chen, X., Kuvelkar, R., Hodgson, R., Hyde, L.A., Cox, K., Favreau, L., Parker, E.M., and Greenlee, W.J. (2012) Discovery of an orally available, brain penetrant BACE1 inhibitor that affords robust CNS Aβ reduction. *ACS Medicinal Chemistry Letters*, **3**, 897–902.

62 Mandal, M., Zhu, Z., Cumming, J.N., Liu, X., Strickland, S., Mazzola, R.D., Caldwell, J.P., Leach, P., Grzelak, M., Hyde, L., Zhang, Q., Terracina, G., Zhang, L., Chen, X., Kuvelkar, R., Kennedy, M.E., Favreau, L., Cox, K., Orth, P., Buevich, A., Voigt, J., Wang, H., Kazakevich, I., McKittrick, B.A., Greenlee, W., Parker, E.M., and Stamford, A.W. (2012) Design and validation of bicyclic iminopyrimidinones as beta amyloid cleaving enzyme-1 (BACE1) inhibitors: conformational constraint to favor a bioactive conformation. *Journal of Medicinal Chemistry*, **55**, 9331–9345.

63 Nowak, P., Cole, D.C., Aulabaugh, A., Bard, J., Chopra, R., Cowling, R., Fan, K.Y., Hu, B., Jacobsen, S., Jani, M., Jin, G., Lo, M.-C., Malamas, M.S., Manas, E.S., Narasimhan, R., Reinhart, P., Robichaud, A.J., Stock, J.R., Subrath, J., Svenson, K., Turner, J., Wagner, E., Zhou, P., and Ellingboe, J.W. (2010) Discovery and initialoptimization of 5,5′-disubstituted aminohydantoins as potent β-secretase (BACE1) inhibitors. *Bioorganic and Medicinal Chemistry Letters*, **20**, 632–635.

64 Malamas, M.S., Robichaud, A., Erdei, J., Quagliato, D., Solvibile, W., Zhou, P., Morris, K., Turner, J., Wagner, E., Fan, K., Olland, A., Jacobsen, S., Reinhart, P., Riddell, D., and Menelas, P. (2010) Design and synthesis of aminohydantoins as potent and selective human b-secretase (BACE1) inhibitors with enhanced brain permeability. *Bioorganic and Medicinal Chemistry Letters*, **20**, 6597–6605.

65 Kobayashi, N., Ueda, K., Itoh, N., Suzuki, S., Sakaguchi, G., Kato, A., Yukimasa, A., Hori, A., Koriyama, Y., Haraguchi, H., Yasui, K., and Kanda, Y. (2007) Preparation of 2-aminodihydrothiazine derivatives as beta-secretase inhibitors. *PCT Int. Appl.* WO 2007049532 A1.

66 Suzuki, Y., Motoki, T., Kaneko, T., Takaishi, M., Ishida, T., Takeda, K., Kita, Y., Yamamoto, N., Khan, A., and Dimopoulos, P. (2009) Condensed aminodihydrothiazine derivative. *PCT Int. Appl.* WO 2009091016 A1.

67 Audia, J.E., Mergott, D.J., Sheehan, S.M., and Watson, B.M. (2009) Aminodihydrothiazine derivatives as bace inhibitors for the treatment of Alzheimer's disease. *PCT Int. Appl.* WO 2009134617 A1.

68 Swahn, B.-M., Holenz, J., Kihlstrom, J., Kolmodin, K., Lindstrom, J., Plobeck, N., Rottici, D., Sehgelmeble, F., Sundstrom, M., von Berg, S., Fälting, J., Georgievska, B., Gustavsson, S., Neelissen, J., Ek, M., Olsson, L.-L., and Berg, S. (2012) Aminoimidazoles as BACE-1 inhibitors: the challenge to achieve *in vivo* brain efficacy. *Bioorganic and Medicinal Chemistry Letters*, **22**, 1854–1859.

69 Cumming, J.N., Scott, J.D., Li, S.W., Cartwright, M., Chen, X., Cox, K., Forman, M., Gilbert, E.J., Hodgson, R., Hyde, L., Jin, Y., Kazakevich, I., Kuvelkar, R., Liang, X., Mei, H., Misiaszek, J., Orth, P., Stone, J.,

Strickland, C., Voigt, J.H., Wang, H., Werner, B., Wong, J., Parker, E.M., Greenlee, W.J., Kennedy, M.E., and Stamford, A.W. (2015) Discovery of MK-8931: a BACE inhibitor in phase 3 clinical development for Alzheimer's disease. *Abstract MEDI* 246 *at 249th ACS National Meeting, Denver, CO, March 22–26, 2015.*

70 Forman, M., Tseng, J., Palcza, J., Leempols, J., Ramael, S., Krishna, G., Ma, L., Wagner, J., and Troyer, M. (2013) The novel BACE inhibitor MK-8931 dramatically lowers CSF Aβ peptides in healthy subjects: results from a rising single dose study. *Neurology*, **78**. doi: 10.1212/WNL.78.1_MeetingAbstracts.PL02.004.

71 Merck Advances Development Program for Investigational Alzheimer's Disease Therapy, MK-8931. http://www.mercknewsroom.com/news-release/prescription-medicine-news/merck-advances-development-program-investigational-alzheimer (accessed 3 May 2015).

72 An Efficacy and Safety Trial of MK-8931 in Mild to Moderate Alzheimer's Disease (P07738) (EPOCH). http://www.clinicaltrials.gov/ct2/show/NCT01739348 (accessed 3 May 2015).

73 Efficacy and Safety Trial of MK-8931 in Participants With Prodromal Alzheimer's Disease (MK-8931-019) (APECS). http://www.clinicaltrials.gov/ct2/show/NCT01953601 (accessed 3 May 2015).

74 Malamas, M., Erdei, J., Gunawan, I., Barnes, K., Johnson, M., Hui, Y., Turner, J., Hu, Y., Wagner, E., Fan, K., Olland, A., Bard, J., and Robichaud, A.J. (2009) Aminoimidazoles as potent and selective human β-secretase (BACE1) inhibitors. *Journal of Medicinal Chemistry*, **52**, 6314–6323.

75 Assessment of Safety, Tolerability and Blood Concentrations of Single Doses of AZD3839 in Healthy Volunteers. http://www.clinicaltrials.gov/ct2/show/NCT01348737 (accessed 3 May 2015).

76 AZD3293 – Alzforum. http://www.alzforum.org/therapeutics/azd3293-0 (accessed 22 July 2015).

77 Lilly Voluntarily Terminates Phase II Study for LY2886721, a Beta Secretase Inhibitor, Being Investigated as a Treatment for Alzheimer's Disease. investo.lilly.com/releasedetail.cfm?ReleaseID=771353 (accessed 3 May 2015).

78 E2609 – Alzforum. http://www.alzforum.org/therapeutics/e2609 (accessed 22 July 2015).

79 JNJ-5486191 – Alzforum. http://www.alzforum.org/therapeutics/jnj-54861911 (accessed 22 July 2015).

80 BACE Inhibitor Trial Put on Hold. http://www.alzforum.org/news/research-news/bace-inhibitor-trial-put-hold (accessed 3 May 2015).

81 RG7129 – Alzforum. http://www.alzforum.org/therapeutics/rg7129 (accessed 22 July 2015).

82 Dineen, T.A., Chen, K., Cheng, A.C., Derakhchan, K., Epstein, O., Esmay, J., Hickman, D., Kreiman, C.E., Marx, I.E., Wahl, R.C., Wen, P.H., Weiss, M.M., Whittington, D.A., Wood, S., Fremeau J Jr., R.T., White, R.D., and Patel, V.F. (2014) Inhibitors of β-site amyloid precursor protein cleaving enzyme (BACE1): identification of (*S*)-7-(2-fluoropyridin-3-yl)-3-((3-methyloxetan-3-yl)ethynyl)-5′*H*-spiro[chromeno[2,3-b]pyridine-5,4′-oxazol]-2′-amine (AMG-8718). *Journal of Medicinal Chemistry*, **57**, 9811–9831.

83 Wager, T.T., Chandrasekaran, R.Y., Hou, X., Troutman, M.D., Verhoest, P.R., Villalobos, A., and Will, Y. (2010) Defining desirable central nervous system drug space through the alignment of molecular properties, *in vitro* ADME, and safety attributes. *ACS Chemical Neuroscience*, **1**, 420–434.

# 15
# Epigenetics and Fragment-Based Drug Discovery

*Aman Iqbal and Peter J. Brown*

## 15.1
## Introduction

The field of epigenetics studies the dynamic nature of our genome, which is constantly altered during our lifetime by post-translational modifications (PTMs) to histones and by switching key genes on and off. Histones are protein complexes that order and package DNA into nucleosomes, and numerous proteins constantly modify these histone proteins. These post-translational modifications have been shown to regulate various intracellular events including transcription and protein interaction networks key to human biology (Table 15.1). The field is experiencing an exponential growth in the knowledge of epigenetic proteins and signaling pathways and their involvement in life-threatening diseases including cancer [1–4].

Research efforts carried out by organizations such as the Structural Genomics Consortium (SGC) have generated potent, selective, and cell active chemical probes to validate potential therapeutic targets in oncology. Many of these drug targets have been shown to be mutated or overexpressed in cancers ranging from AML to multiple myeloma. This information, coupled with some preliminary cell-based and patient data, makes these targets of high clinical relevance. At present, there are only a handful of drugs in the clinic against these therapeutic targets. Development of chemical probes against specific epigenetic proteins will enable the understanding of their individual mechanisms in cells and disease. There is also an unmet need to design targeted therapies with known mechanisms of action against these specific epigenetic proteins. Some early research findings indicate that coadministration of epigenetic therapeutics along with existing cancer drugs may help in better treatment and delayed progression of cancer. Although a comprehensive review of the background of epigenetics is out of the scope of this chapter, we aim to cover the druggable aspect of epigenetics targets, with a focus on the role of fragment-based drug discovery (FBDD) in generating chemical probes and ligands of these targets. We have reviewed all published FBDD studies on potential epigenetic targets up to the end of 2014. This will give the reader an overview of the challenges in targeting these proteins via an FBDD approach.

*Fragment-based Drug Discovery: Lessons and Outlook*, First Edition. Edited by Daniel A. Erlanson and Wolfgang Jahnke.

**Table 15.1** Overview of epigenetic target families.

| | Writers | | Erasers | | Readers | | |
|---|---|---|---|---|---|---|---|
| **Family** | **Histone acetyltransferases** | **Histone methyltransferases** | **Histone deacetylases** | **Lysine demethylases** | **Bromodomains** | **Tudor, chromo, MBT, and PWWP** | **PHD domains** |
| Members | 18 | 60 | 17 | 25 | 61 | 95 | 104 |
| Activity | Add acetyl to lysines | Add methyl to lysines and arginines | Removal of acetyl from lysines | Removal of methyl from lysines | Binding to acetyllysines | Binding to methylated lysines and arginines | Binding to acetylated lysines, methylated lysines, and arginines |

## 15.2
## Epigenetic Families and Drug Targets

The epigenetic family of proteins is divided into (a) readers, proteins that bind to (acetyl or methyl) histone marks; (b) writers, proteins that carry out addition of post-translational modifications (PTMs) of histones; and (c) erasers, proteins that remove these PTMs from histones. The epigenetic proteins that include readers, writers, and erasers of epigenetic marks drive disease via two primary mechanisms. First, via aberrant activity due to mutation or altered expression of epigenetic proteins that leads to or even drives and maintains disease states. Second, the readers, writers, and erasers work in concert with many other cellular proteins, including environmentally responsive DNA binding transcription factors, and they can mediate altered gene expression patterns driven by upstream signals [1]. All of the individual readers, writers, and erasers are further divided into subprotein families based on the individual domains that carry out the histone binding and catalytic events.

There are small molecule ligands reported for each of the three major classes of acetyl and methyl readers, writers, and erasers. Over the past decade, the knowledge pertaining to the biochemistry, substrate selectivity, and three-dimensional structures of these classes of proteins has grown multifold revealing common structural and mechanistic features of their active sites. However, there is still a lot of progress to be made to discover selective and cell active small molecule ligands for a number of these epigenetic proteins in each family [5–8]. Chemical probe discovery efforts by the SGC have led to the development of small molecules that are selective for a particular epigenetic protein/domain and are cell active. These chemical probes are extremely important in studying the role of individual proteins in cells and enabling our understanding of the nascent cellular biology of each epigenetic family member. These molecules also act as useful tool compounds to aid the development of small molecule drugs.

Histone deacetylases (HDACs) are epigenetic proteins that catalyze the hydrolysis of *N*-acetyl lysine residues of histones, which package chromosomal DNA. HDACs have emerged as very important therapeutic targets in cancer and are the most widely studied epigenetic targets. There has been considerable effort to develop HDAC inhibitors (HDACi). In preclinical studies several classes of HDACi have been found to have potent anticancer activities, with remarkable tumor specificity [9].

After HDACs, bromodomains (BRDs) are the most well-studied family of epigenetic targets. BRDs belong to the reader class of epigenetic targets and are protein interaction modules that exclusively recognize acetylation motifs. There are 61 BRDs in the human genome, and there are crystal structures available for more than 25 of these. Cocrystal structures with peptides have demonstrated that acetyl-lysine (Kac) is recognized by a central deep hydrophobic cavity, where it is anchored by a hydrogen bond to an asparagine residue present in most BRDs. There is already a large number of small molecule modulators of BRD4 and other BET bromodomains in the literature [10].

Histone methyltransferases (HMTs) constitute another family of epigenetic proteins ("writers") which enable the transfer of a methyl group from the cofactor *S*-adenosyl methionine (SAM) to a lysine or arginine on histones [11]. More than 50 HMTs have been discovered in the past decade, and data suggests that some of these are attractive drug targets [12,13]. Although no marketed drugs exist to modulate HMT activity, several small molecule inhibitors are in development and in clinical trials against some of the most validated HMT targets such as EZH2, DOT1L, and PRMT5.

Histone demethylases (KDMs) are a family of epigenetic proteins ("erasers") that catalyze the removal of methyl groups from lysine on histones. There are 30 epigenetic KDMs subdivided into families. Some chemical probes of KDMs have also been developed to validate these enzymes as potential therapeutic targets [11,14]. There are a few small molecule inhibitors in development and in clinical trials against some of the KDMs such as LSD1.

The "royal family" consists of protein domains (PWWP, MBT, TUDOR, PHD, and CHROMO) that are readers of the histone code that contain aromatic cages capable of recognizing specific sequences and lysine/arginine methylation states on histone tails. These binding modules play a key role in epigenetic signaling and are part of a larger group of epigenetic targets that are becoming increasingly attractive for drug discovery [15–17]. As of today, there is only one chemical probe and two weak antagonists against this important epigenetic family [18–20]. A lot more work needs to be done to develop small molecule probes to better understand their role in epigenetic mechanisms.

## 15.3 Epigenetics Drug Discovery Approaches and Challenges

Some of the key challenges in designing and developing small molecule ligands of epigenetic targets are as follows:

a) **Target and isoform selectivity:** A large number of SET domain containing PMTs all bind to the cofactor SAM, which is analogous to kinase ATP binding. Hence, targeting the SAM binding site alone may not result in target- and isoform-selective inhibitors.
b) **Displacement of Cofactor:** A number of chemical probes against the SET domain-containing PMTs were observed to bind to the peptide binding site with a few of them extending into the cofactor (SAM) binding pocket which results in competitive binding to the cofactor binding site. SAM has been shown to bind with varying strength (0.5–200 μM) to these PMTs. Hence, there is a possibility that standard HTS screens might miss out on hits, which are not potent enough to displace SAM.
c) **Targeting the peptide binding site:** While both substrate and cofactor binding sites of the majority of the epigenetic enzyme targets are expected to be druggable, as described above, most of the epigenetic chemical

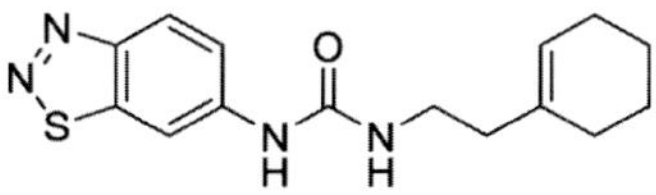

14u

**Figure 15.1** Chemical structure of 14u compound that binds to an allosteric pocket on PRMT3.

probes have been shown to bind in the lysine/substrate binding cleft [21]. However, since the lysine binding is shared by almost all epigenetic targets, it is possible that the HTS screens might not represent the best chemical diversity to identify novel chemical starting points in the substrate/peptide binding cleft. This is where fragment-based screening approaches can play an important role in identifying diverse chemical starting points binding to the substrate/peptide binding site, which can eventually be grown into leads traversing to the cofactor binding pocket.

d) **Identifying and targeting allosteric sites:** Mostly due to the efforts of the SGC, there is a large and ever-growing collection of high-resolution crystal structures of epigenetic proteins. However, this structural information needs to be exploited to discover ligands that bind to novel binding sites besides the cofactor and peptide binding sites, that is, allosteric sites. At present, there is only one allosteric inhibitor of an epigenetic target known [22]. Protein arginine methyltransferase 3 (PRMT3) catalyzes the dimethylation of GAR motifs of its main substrate: the 40 S ribosomal protein (rpS2) and also H4 (1–24) [23]. PRMT3 is known to be overexpressed in myocardial tissues from patients with atherosclerosis. Scientists at the SGC found a compound they called 14u (Figure 15.1) from diversity screening, which inhibited PRMT3 ($IC_{50}$ = 230 nM) in both a SAM and a peptide noncompetitive manner [22]. The cocrystal structure showed that the small molecule bound to a previously unknown allosteric pocket. As a result, the compound was selective for PRMT3 over 15 other PRMTs, PKMTs, and DNMT1. This study shows the potential of identifying and targeting allosteric sites on these epigenetic targets to discover and design novel selective inhibitors.

## 15.4 FBDD Case Studies

Although a majority of small molecule screening efforts against epigenetic targets still employ high-throughput screening (HTS) of proprietary libraries, there is emerging interest in the application of small molecule fragment-based screening (FBS) approaches to discover novel chemical starting points, allosteric binding sites, and expand chemical diversity. There are not many published examples of FBDD approaches in epigenetics. Below, we have reviewed all published case studies through the end of 2014 that employed FBDD to enable small molecule ligand design against epigenetic targets. As will be seen, the majority of FBDD programs were against "easy to target" bromodomains. There are FBDD projects running in pharmaceutical, biotechnology, and other private companies; the coming decade will be very exciting to see what role FBDD plays in developing small molecule therapeutics in the field of epigenetics.

### 15.4.1 BRD4 (Bromodomain)

BRD4 has been one of the most popular targets not only in epigenetics but also in drug discovery in general. There are innumerable publications, patents, and other research work around drug discovery of BRD4 antagonists in the public domain [24]. BRD4 has been characterized extensively as a key determinant in acute myeloid leukemia, multiple myeloma, Burkitt's lymphoma, NUT midline carcinoma, colon cancer, and inflammatory diseases [25,26].

Researchers at GSK were the first to report the use of FBDD to discover small molecule fragment hits against the BET bromodomain of BRD4 and to develop them into potent antagonists [27,28]. BET bromodomains bind to acetyllysine of the histone. Blocking the acetyl-lysine binding site of bromodomains prevents recognition of acetylated histone tails and alters the process of chromatin remodeling. Small molecules that prevent histone binding *in vitro* can produce profound *in vivo* biological responses. GSK reports preparing a focused fragment set with acetylysine (Kac) mimicking fragments resulting in 1376 fragments. Interestingly, the primary screening of the fragment set involved fluorescent anisotropy (FA), where the fragments were screened at 200 μM concentration each. The use of biochemical screening and high-concentration fragment screening as primary screening is common now. However, there are a number of limitations. First, the methodology works well only with proteins that have well-defined binding sites; secondly, the assay has to be robust such that it can identify weak (~>100 μM) fragment hits; and thirdly a lot of weakly binding fragments are missed, which could act as good starting points for lead discovery. The FA fragment screening resulted in 132 hits that showed >30% displacement in the assay. All 132 fragment hits were then screened by soaking in the apo-BRD4 crystals followed by X-ray diffraction. This resulted in 40 cocrystal structures. GSK identified three chemotypes that had not been crystallized in bromodomains before. The researchers then published a follow-up paper, which describes fragment-to-lead development of the phenylisoxazole sulphonamide series (Figure 15.2) [28].

Another paper describing fragment screening and lead development against BRD4 was reported by scientists at Constellation and Jubilant BioSys [29]. The researchers report fragment screening and identifying a micromolar amino isoxozle fragment hit for which they solved a cocrystal structure. GSK compared the binding mode of their fragment hit with those of JQ1 [30] and acetyl-lysine to show that the same hydrogen bonds are formed by all three compounds in the binding site, and developed the isoxazole fragment hit into an isoxazole azepine

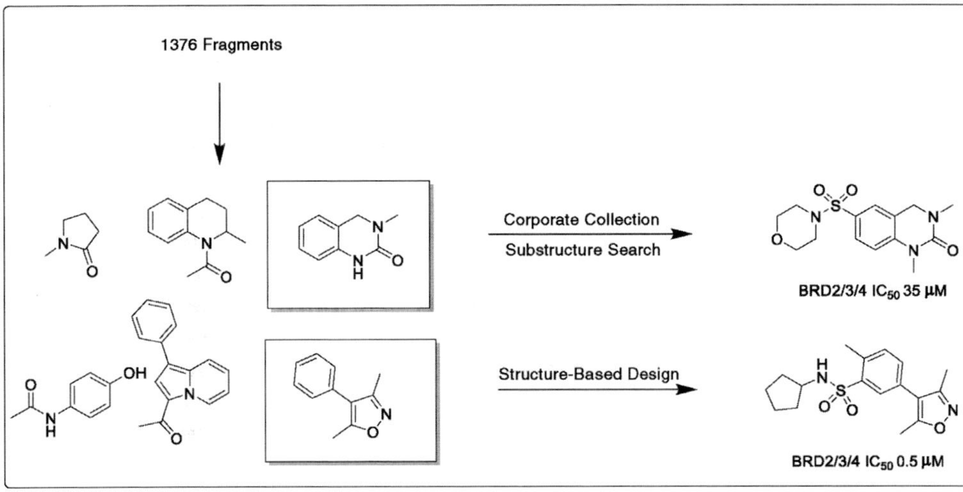


**Figure 15.2** Fragment-based screening and development of novel BRD4 inhibitors.

lead compound. When compared to the existing BRD4 inhibitors JQ1, I-BET151, and I-BET762 [30,31], their best lead compound was reported to have $IC_{50}$ of 26 nM (Figure 15.3). They also report dog and rat PK/PD experiments. Overall, this is nice work, which shows classical fragment-based lead development against a well-known target.

Another research publication from the University of Zurich discusses the use of *in silico* fragment screening against BRD4. Computational docking of fragments can be difficult, partly because the energetic differences between possible binding modes can be so small that it is impossible to select the best one. However, the researchers

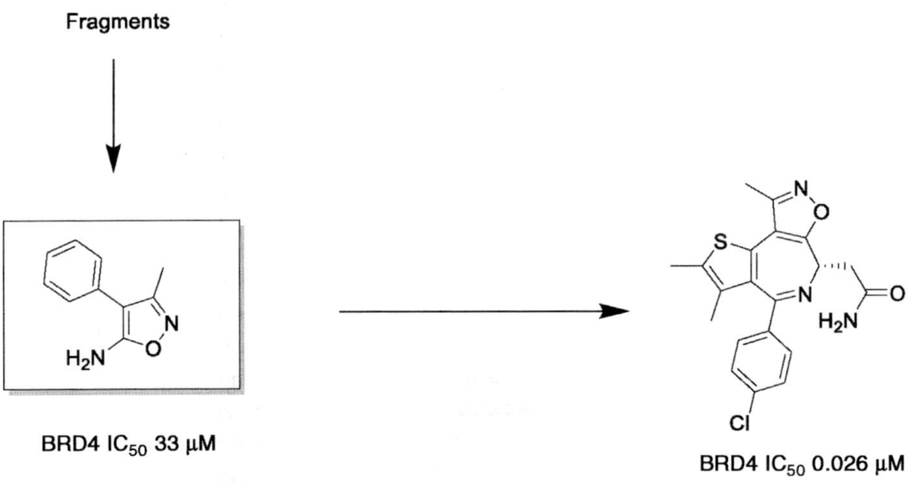


**Figure 15.3** Fragment to lead development of a novel isozazole azepine chemical series against BRD4.

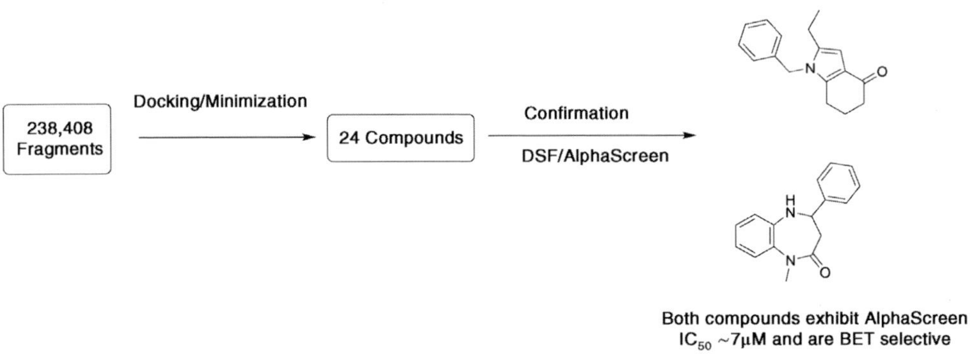


**Figure 15.4** Docking studies of fragment virtual library against BRD4.

describe the docking of 238 408 fragments [32], after a lot of initial filtering based on parameters such as molecular weight, rotatable bonds, and reactive features. Overall, they purchased 24 fragments after further CHARMM minimization and other filtering. Out of the 24 compounds that were tested in a BRD4 DSF assay, 4 fragments showed thermal shifts of more than 1 °C. These four fragments were further tested in the gold standard AlphaScreen assay resulting in the validation of two of the four fragment hits by obtaining $IC_{50}$s of 7 μM each (Figure 15.4).

Researchers at the Shanghai Institute of Material Medica describe the discovery of novel fragment hits against BRD4 and their conversion into small molecule leads with good metabolic stability and cell activity [33]. The researchers employed computational docking of fragments into the BRD4 acetyl-lysine binding pocket. They prepared an in-house fragment set of 487 fragments from the ZINC database of fragments and filtered them based on the traditional rule of three. Docking in BRD4 was carried out which resulted in the selection of top 41 fragment hits based on the mode of binding and chemical templates, followed by cocrystallization experiments of the 41 fragments with BRD4. Nine fragment hits were identified bound in the BRD4 acetyl-lysine binding pocket. Four out of the nine fragments bore different chemical moieties and yet bound to the same position as JQ-1. The researchers also identified an isoxazole-containing fragment hit as described above by the team at Constellation. The rest of the publication describes the conversion of the 2-thiazolidinone hit into potent lead molecules. The crystal structure of BRD4 with 2-thiazolidinone was a good starting point for medicinal chemistry. The 2-thiazolidinone fragment formed more hydrogen bonds in the binding pocket compared to the isoxazole fragment, in addition to the favorable position of the phenyl ring deep in the acetyl-lysine binding pocket of BRD4. Subsequent work involved using SAR studies in conjunction with some neat medicinal chemistry, which involved substituting the phenyl ring of the 2-thiazolidone fragment with functional groups that resulted in a steady increase in potency. Their best lead compounds had $IC_{50}$s of around 0.3 μM and showed exciting preliminary metabolic stability in liver microsomes and inhibition of HT-29 cells through a BRD4-dependent pathway.

Overall, this work demonstrates the power of structure-guided fragment-to-lead design. Assessment of *in vitro* DMPK profiles of early lead compounds (<1 μM) enabled the improvement physicochemical parameters during SAR development.

### 15.4.2
### EP300 (Bromodomain)

The CREB (cyclic-AMP response element binding protein) binding protein (CBP) and E1A binding protein (EP300) are pleiotropic lysine acetyltransferases that play key roles as transcriptional coactivators in human cells. It has emerged that antagonism of the CBP BRD, and therefore p53-mediated p21 activation, has potential clinical applications.

Researchers at SGC, published a paper describing fragment-based chemical probe design of novel ligands of EP300 BRD [34]. They report starting with a 3,5-dimethyl isoxazole molecule. Several reports describe 1,3-dimethylisoxazoles as potent ligands of the BET BRDs, including the benzimidazole compound A (Figure 15.5) [35]. Therefore, it was recognized that obtaining selectivity for CBP/p300 over the BET

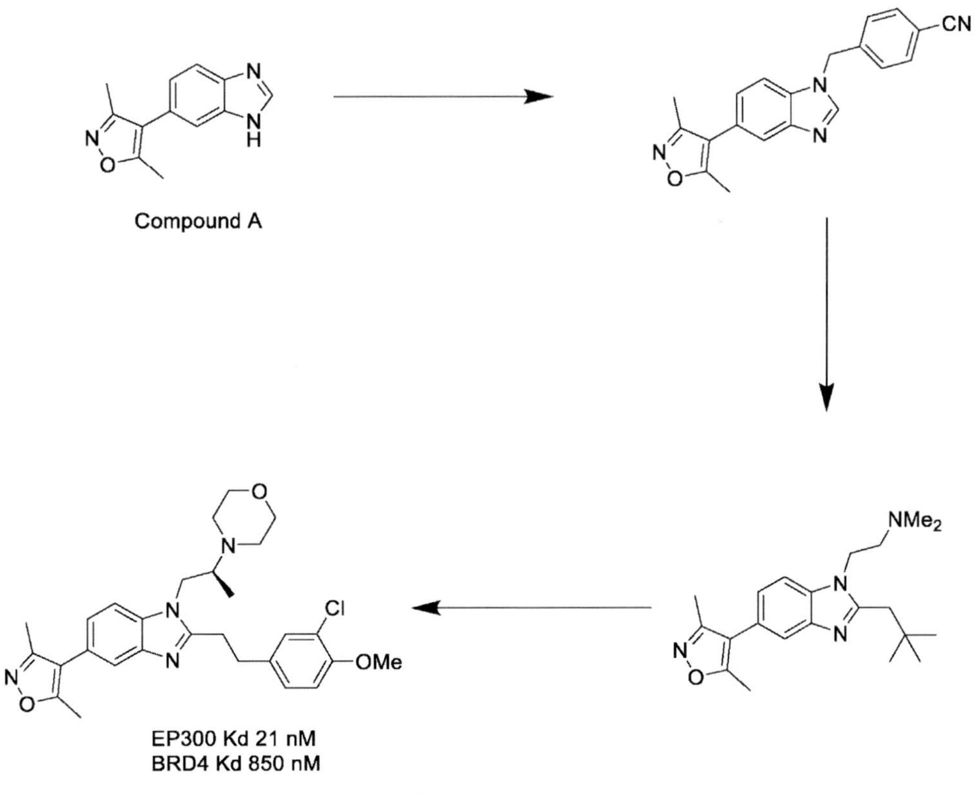


**Figure 15.5** Fragment-based small molecule lead development against the bromodomain of EP300.

BRDs could represent a substantial challenge. The fragment starting point seemed attractive to develop CBP BRD-selective inhibitors both because it had a low molecular weight (213 Da) and reasonable LipE (3.5) and ligand efficiency (0.45) for CBP and because it was found to have various points useful for the introduction of diversity to the core scaffold. The researchers then describe a structure-guided fragment-to-lead design, which involved significant medicinal chemistry efforts. All the compounds were screened by X-ray and Alphascreen against both EP300 and BRD4 simultaneously. This campaign resulted in a potent and selective small molecule antagonist of EP300 with a potency of 0.021 μM and a remarkable 40-fold selectivity over BRD4. In cells their lead compound inhibited CBP-mediated p53 activity in a luciferase-based reporter assay and showed low cytotoxicity. Overall, this story nicely illustrates the power of fragment-based screening and using structure-based design in guiding a target and domain-specific lead design campaign.

### 15.4.3 ATAD2 (Bromodomain)

Another successful fragment-to-lead design campaign against a bromodomain target is reported by Stephen Fesik and colleagues [36]. Overexpression of ATAD2, an epigenetic protein with bromodomain functionalities, has been linked to poor prognosis in prostate, lung, and triple-negative breast cancers, as well as in hepatocellular carcinoma [37]. Downregulation of ATAD2 via siRNA shows an increase in apoptotic activity, suggesting that a small-molecule inhibitor of ATAD2 could result in cancer cell death and tumor regression. Fesik and colleagues employed protein-observed (HSQC) NMR for primary fragment screening. Fesik and colleagues recognized ATAD2 as a challenging target, in terms of identifying hits via traditional high-throughput screening methods. They screened $^{15}$N-labeled ATAD2 against a fragment library of 13 800 fragments in pools of 12. Pool hits were then deconvoluted resulting in the identification of 65 fragments, corresponding to a hit rate of just 0.5%. The binding affinity of these fragments was assessed using the SOFAST-HMQC methodology [38] leading to the identification of 12 hits under 1 mM $K_d$. Most of the fragments bound in the affinity range of 300 μM–1 mM. The 65 hits represented 3 separate clusters from which four representative scaffolds were selected for crystallography to drive medicinal chemistry optimization. All fragments were reported to bind in the Kac binding pocket of ATAD2. The researchers comment that they offer these weakly binding fragments as potential starting points to develop them into novel chemical probes against ATAD2, in order to examine the biological impact of ATD2 bromodomain antagonism and to determine whether this is a viable strategy for treating human cancers.

### 15.4.4 BAZ2B (Bromodomain)

Another publication describes the use of fragment-based screening and lead development against a bromodomain target, BAZ2B, by researchers at SGC and

University of Cambridge [40]. BAZ2B is a non-BET bromodomain, which like other bromodomains on the phylogenetic tree has a shallow binding site. The biological role of BAZ2B is unclear; hence, a chemical probe against it would be useful.

A fragment library of 1300 fragments was screened against BAZ2B using Alphascreen as a primary screening assay, another example of a biochemical assay used for primary fragment screening. Fragment screening at 1 mM and subsequent $IC_{50}$ analysis resulted in the confirmation of 10 hits (0.8%). The $IC_{50}$s were in the range of 36–1000 μM. Nicely enough, all 10 fragments could be confirmed by orthogonal NMR-based techniques such as STD and water-LOGSY. Interestingly, when the same fragment library was screened against BRD2-BP1 and CREBBP, which are well-known druggable BRDs, the hit rate was much higher (1.8 and 6%, respectively). The authors solved cocrystal structures of BAZ2B with three of the fragments and describe fragment optimization for the most potent fragment hit tetrahydro-γ-carboline (THγC) (36 μM). The crystal structure showed that the fragment overlaid nicely on acetyl-lysine, which is the natural ligand. Upon successful optimization, a fragment was identified with an $IC_{50}$ of 8 μM, showing high selectivity over CREBBP and BRD2-BP1. This could act as an attractive starting point for antagonist design against BAZ2B bromodomain.

### 15.4.5 SIRT2 (Histone Deacetylase)

University of Minnesota researchers describe fragment-based lead development against SIRT2, a well-known HDAC. At the outset, the challenge was designing isoform-specific inhibitors with selectivity over SIRT1 and SIRT3. The researchers report preparation of focused fragment sets based on ligands known to bind to SIRTs including suramin and nicotinamide (NAD). The two focused fragment sets were screened using a biochemical assay against SIRTs 1–3 at 100 μM. This resulted in the identification of a number of fragment hits from the two sets. Biochemical screening is a popular primary tool for screening a large number of fragments quickly. However, the probability of missing out on weakly binding, yet useful, fragment hits is very high. Nevertheless, the researchers carried out medicinal chemistry in the absence of cocrystal structures to further develop these fragments into potent lead molecules. Docking of key fragments into the SIRT2 acetyl-lysine binding site enabled the development of two novel lead series against SIRT2, namely, naphthylamide sulfonic acids and naphthalene-benzamides. Biochemical screening enabled the combination of chemical features, leading to (5-benzamidonaphthalen-1/2-yloxy)-nicotinamide derivatives, resulting in compounds with superior anti-SIRT2 activity and excellent selectivity for SIRT2 over SIRT1 and SIRT3. Kinetic studies on their lead inhibitor also suggested that the compound is peptide competitive and noncompetitive with respect to the cofactor NAD.

Developing small molecule inhibitors that are selective toward an isoform of a target class is highly challenging. The authors nicely describe the role of fragment-based lead development to achieve selectivity over other isoforms. This is a common theme in epigenetics, where most of the targets in a family share very high similarity in their cofactor and peptide binding sites. Examples like those above show the potential of the application of FBDD approach in finding a solution to this problem.

### 15.4.6 Next-Generation Epigenetic Targets: The "Royal Family" and Histone Demethylases

So far, most of the efforts have been directed toward developing small molecule ligands of HDACs, HATs, and BRDs. This is primarily due to clear biological validation of these targets and their "easy-to-target" profiles. However, there is an unmet need to discover and develop small molecule probes against a number of epigenetics target classes such as KDMs, and methylysine/methylarginine readers. Reader proteins, which include TUDOR, MBT, Chromo, PWWP, and PHD (royal family) domain containing proteins, have been shown to be implicated in various cancers. However, as all these targets are protein–protein interaction complexes with histone proteins/peptides, they are deemed challenging for antagonist/probe design.

At SGC, there are ongoing efforts to develop chemical probes against the royal family using fragment-based drug discovery. Most of these royal family targets do not have reliable biochemical assays that can be used for HTS; hence, the application of FBDD is key in finding starting points for chemical probe design. At SGC, the most common primary screening techniques are X-ray crystallography and HSQC NMR followed by confirmation by ITC and DSF.

Researchers at the MRC laboratories in London and Cambridge report the discovery of the first small molecule binder to the PHD domain of the epigenetic protein Pygo [39]. The researchers identified Pygo-BcL9 as a promising target for inhibiting oncogenic beta-catenin by blocking protein–protein interaction interface and performed fragment screening using the rule of three-compliant Maybridge 1000 fragment set. Pools of five fragments were prepared and screened via HSQC NMR. A large number of pools showed chemical shifts, and deconvolution resulted in the validation of two out of seven pools. The two fragments identified were benzimidazole derivatives with almost identical structures. The cocrystal structures of these fragments with Pygo will be useful for future ligand design. These weakly binding small molecules were also able to displace the histone peptide (pentamer) from Pygo. This work shows that Pygo is druggable and future work can be carried out to develop small molecule ligands to study its role in cancer.

Chromodomains are epigenetic targets for which there are no antagonists reported in the literature. A recent publication describes a peptide-based approach, which resulted in the identification of a potent 200 nM peptidomimetic inhibitor of CBX7 [20]. Although the inhibitor still contains the trimethyl-lysine, this could act as a good template for future small molecule antagonist discovery against CBX7 using fragment-based approaches.

In another recent work, researchers at UNC North Carolina report the identification of the first small molecule antagonist against 53BP1, a TUDOR domain containing epigenetic target. 53BP1 plays a central role in DNA damage repair pathways, and is considered a therapeutic target of interest. Researchers carried out HTS and identified a fragment-like small molecule, UNC2170, which bound to 53BP1 with a $K_d$ of 22 μM and showed high selectivity over other TUDOR domain containing epigenetic targets. The authors report a cocrystal structure and future plans to further develop UNC2170 into a potent chemical probe against 53BP1 [41].

Histone demethylases (KDMs) as described above are extremely important epigenetic targets. There are very few small molecule inhibitors against KDM reported in literature. One of the most highly studied KDMs is LSD1. Classical HTS screening has not been successful in identifying novel hits against KDMs. Also, efforts to convert hits into potent selective leads have not been successful either. Fragment-based screening offers a great opportunity to identify novel small molecule fragment starting points that can be converted step by step into potent selective lead compounds.

## 15.5 Conclusions

There is rapid progress in the field of epigenetics to understand the role of various proteins in cellular mechanisms. Considerable efforts are ongoing to find small molecule ligands for many classes of epigenetic targets. Early research suggests that these targets may be tractable using conventional small molecule lead discovery approaches. However, early drug discovery endeavors in this area have been challenging, and much more work needs to be done before epigenetic systems can be more widely exploited for therapeutic uses. Fragment-based screening methodologies can have a significant impact on this area in the identification of novel chemical starting points. Small molecule fragments enable the exploration of much wider chemical space when compared to traditional HTS libraries. Moreover, the classical fragment-to-lead design approach has the potential of improving the potency and selectivity of these small molecule ligands. Targeting the cofactor binding sites of many of the epigenetic proteins is fraught with challenges of selectivity. The majority of epigenetic proteins bind to histone proteins/peptides hence making them challenging PPI targets. Finding distinct small-molecule replacements of acetyl-lysine and methyl-lysine/methylarginine moieties is difficult.

To summarize, the coming decade will see many exciting developments in the field of epigenetics biology. Target validation will enable the discovery of small molecule therapeutics to treat various diseases with high unmet medical need.

We strongly believe that fragment-based drug design will have an equally important impact in this field as it has in others.

## Abbreviations

| | |
|---|---|
| FBDD | fragment-based drug discovery |
| HMT | histone methyltransferase |
| PRMT/PMT | protein arginine methyltransferase |
| HDAC | histone deacetylase |
| BRD | bromodomain |
| SET | Su(var)3-9 and enhancer of zeste and trithorax |
| PWWP | Pro-Trp-Trp-Pro motif |
| PHD | plant homeodomain |
| MBT | malignant brain tumor domain |

## References


1 Arrowsmith, C.H., Bountra, C., Fish, P.V, Lee, K., and Schapira, M. (2012) Epigenetic protein families: a new frontier for drug discovery. *Nature Reviews Drug Discovery*, **11**, 384–400.

2 Helin, K. and Dhanak, D. (2013) Chromatin proteins and modifications as drug targets. *Nature*, **502**, 480–488.

3 Meaney, M.J. (2010) Epigenetics and the biological definition of Gene X environment interactions. *Child Development*, **81**, 41–79.

4 Spannhoff, A., Hauser, A.T., Heinke, R., Sippl, W., and Jung, M. (2009) The emerging therapeutic potential of histone methyltransferase and demethylase inhibitors. *ChemMedChem*, **4**, 1568–1582.

5 Bradner, J.E., West, N., Grachan, M.L., Greenberg, E.F., Haggarty, S.J., Warnow, T., and Mazitschek, R. (2010) Chemical phylogenetics of histone deacetylases. *Nature Chemical Biology*, **6**, 238–243.

6 Filippakopoulos, P., Qi, J., Picaud, S., Shen, Y., Smith, W.B., Fedorov, O., Morse, E.M., Keates, T., Hickman, T.T., Felletar, I., Philpott, M., Munro, S., McKeown, M.R., Wang, Y., Christie, A.L., West, N., Cameron, M.J., Schwartz, B., Heightman, T.D., La Thangue, N., French, C.A., Wiest, O., Kung, A.L., Knapp, S., and Bradner, J.E. (2010) Selective inhibition of BET bromodomains. *Nature*, **468**, 1067–1073.

7 Nicodeme, E., Jeffrey, K.L., Schaefer, U., Beinke, S., Dewell, S., Chung, C.W., Chandwani, R., Marazzi, I., Wilson, P., Coste, H., White, J., Kirilovsky, J., Rice, C.M., Lora, J.M., Prinjha, R.K., Lee, K., and Tarakhovsky, A. (2010) Suppression of inflammation by a synthetic histone mimic. *Nature*, **468**, 1119–1123.

8 Solomon, J.M., Pasupuleti, R., Xu, L., McDonagh, T., Curtis, R., DiStefano, P.S., and Huber, L.J. (2006) Inhibition of SIRT1 catalytic activity increases p53 acetylation but does not alter cell survival following DNA damage. *Molecular and Cellular Biology*, **26**, 28–38.

9 Bolden, J.E., Peart, M.J., and Johnstone, R.W. (2006) Anticancer activities of histone deacetylase inhibitors. *Nature Reviews Drug Discovery*, **5**, 769–784.

10 Muller, S., Filippakopoulos, P., and Knapp, S. (2011) Bromodomains as therapeutic targets. *Expert Reviews in Molecular Medicine*, **13**, e29.

11 Del Rizzo, P.A. and Trievel, R.C. (2014) Molecular basis for substrate recognition by lysine methyltransferases and demethylases. *Biochimica et Biophysica Acta*, **1839**, 1404–1415.

12 Schapira, M. and Ferreira de Freitas, R. (2014) Structural biology and chemistry of

protein arginine methyltransferases. *Medicinal Chemistry Communication*, **5**, 1779–1788.

13 Liu, Y., Liu, K., Qin, S., Xu, C., and Min, J. (2014) Epigenetic targets and drug discovery. Part 1: histone methylation. *Pharmacology and Therapeutics*, **143**, 275–294.

14 Thinnes, C.C., England, K.S., Kawamura, A., Chowdhury, R., Schofield, C.J., and Hopkinson, R.J. (2014) Targeting histone lysine demethylases: progress, challenges, and the future. *Biochimica et Biophysica Acta*, **1839**, 1416–1432.

15 Gayatri, S. and Bedford, M.T. (2014) Readers of histone methylarginine marks. *Biochimica et Biophysica Acta*, **1839**, 702–710.

16 Lu, R., Wang, G.G., and Tudor:, A. (2013) Versatile family of histone methylation "readers." *Trends in Biochemical Sciences*, **38**, 546–555.

17 Eissenberg, J.C. (2014) Structural biology of the chromodomain: form and function. *Gene*, **496**, 69–78.

18 James, L.I., Barsyte-Lovejoy, D., Zhong, N., Krichevsky, L., Korboukh, V.K., Herold, J.M., MacNevin, C.J., Norris, J.L., Sagum, C.A., Tempel, W., Marcon, E., Guo, H., Gao, C., Huang, X.-P., Duan, S., Emili, A., Greenblatt, J.F., Kireev, D.B., Jin, J., Janzen, W.P., Brown, P.J., Bedford, M.T., Arrowsmith, C.H., and Frye, S.V. (2013) Discovery of a chemical probe for the L3MBTL3 methyllysine reader domain. *Nature Chemical Biology*, **9**, 184–191.

19 Perfetti, M.T., Baughman, B.M., Dickson, B.M., Mu, Y., Cui, G., Mader, P., Dong, A., Norris, J.L., Rothbart, S.B., Strahl, B.D., Brown, P.J., Janzen, W.P., Arrowsmith, C.H., Mer, G., McBride, K.M., James, L.I., and Frye, S.V. (2015) Identification of a fragment-like small molecule ligand for the methyl-lysine binding protein, 53BP1. *ACS Chemical Biology*, **10** (4), 1072–1081.

20 Simhadri, C., Daze, K.D., Douglas, S.F., Quon, T.T.H., Dev, A., Gignac, M.C., Peng, F., Heller, M., Boulanger, M.J., Wulff, J.E., and Hof, F. (2014) Chromodomain antagonists that target the polycomb-group methyllysine reader protein chromobox homolog 7 (CBX7). *Journal of Medicinal Chemistry*, **57**, 2874–2883.

21 Vedadi, M., Barsyte-Lovejoy, D., Liu, F., Rival-Gervier, S., Allali-Hassani, A., Labrie, V., Wigle, T.J., Dimaggio, P.A., Wasney, G.A., Siarheyeva, A., Dong, A., Tempel, W., Wang, S.-C., Chen, X., Chau, I., Mangano, T.J., Huang, X.-P., Simpson, C.D., Pattenden, S.G., Norris, J.L., Kireev, D.B., Tripathy, A., Edwards, A., Roth, B.L., Janzen, W.P., Garcia, B.A., Petronis, A., Ellis, J., Brown, P.J., Frye, S.V, Arrowsmith, C.H., and Jin, J. (2011) A chemical probe selectively inhibits G9a and GLP methyltransferase activity in cells. *Nature Chemical Biology*, **7**, 566–574.

22 Liu, F., Li, F., Ma, A., Dobrovetsky, E., Dong, A., Gao, C., Korboukh, I., Liu, J., Smil, D., Brown, P.J., Frye, S.V, Arrowsmith, C.H., Schapira, M., Vedadi, M., and Jin, J. (2013) Exploiting an allosteric binding site of PRMT3 yields potent and selective inhibitors. *Journal of Medicinal Chemistry*, **56**, 2110–2124.

23 Guo, H., Wang, R., Zheng, W., Chen, Y., Blum, G., Deng, H., and Luo, M. (2014) Profiling substrates of protein arginine *N*-methyltransferase 3 with *S*-adenosyl-L-methionine analogues. *ACS Chemical Biology*, **9**, 476–484.

24 Garnier, J.-M., Sharp, P.P., and Burns, C.J. (2014) BET bromodomain inhibitors: a patent review. *Expert Opinion on Therapeutic Patents*, **24**, 185–199.

25 Zuber, J., Shi, J., Wang, E., Rappaport, A.R., Herrmann, H., Sison, E.A., Magoon, D., Qi, J., Blatt, K., Wunderlich, M., Taylor, M.J., Johns, C., Chicas, A., Mulloy, J.C., Kogan, S.C., Brown, P., Valent, P., Bradner, J.E., Lowe, S.W., and Vakoc, C.R. (2011) RNAi screen identifies Brd4 as a therapeutic target in acute myeloid leukaemia. *Nature*, **478**, 524–528.

26 Delmore, J.E., Issa, G.C., Lemieux, M.E., Rahl, P.B., Shi, J., Jacobs, H.M., Kastritis, E., Gilpatrick, T., Paranal, R.M., Qi, J., Chesi, M., Schinzel, A.C., McKeown, M.R., Heffernan, T.P., Vakoc, C.R., Bergsagel, P.L., Ghobrial, I.M., Richardson, P.G., Young, R.A., Hahn, W.C., Anderson, K.C., Kung, A.L., Bradner, J.E., and Mitsiades, C.S. (2011) BET bromodomain inhibition as a therapeutic strategy to target c-Myc. *Cell*, **146**, 904–917.

27 Chung, C.W., Dean, A.W., Woolven, J.M., and Bamborough, P. (20120) Fragment-based discovery of bromodomain inhibitors. Part 1. Inhibitor binding modes and implications for lead discovery. *Journal of Medicinal Chemistry*, **55**, 576–586.

28 Bamborough, P., Diallo, H., Goodacre, J.D., Gordon, L., Lewis, A., Seal, J.T., Wilson, D.M., Woodrow, M.D., and Chung, C.W. (2012) Fragment-based discovery of bromodomain inhibitors. Part 2. Optimization of phenylisoxazole sulfonamides. *Journal of Medicinal Chemistry*, **55**, 587–596.

29 Gehling, V.S., Hewitt, M.C., Vaswani, R.G., Leblanc, Y., Cote, A., Nasveschuk, C.G., Taylor, A.M., Harmange, J.C., Audia, J.E., Pardo, E., Joshi, S., Sandy, P., Mertz, J.A., Sims, R.J. 3rd, Bergeron, L., Bryant, B.M., Bellon, S., Poy, F., Jayaram, H., Sankaranarayanan, R., Yellapantula, S., Bangalore Srinivasamurthy, N., Birudukota, S., and Albrecht, B.K. (2013) Discovery, design, and optimization of isoxazole azepine BET inhibitors. *ACS Medicinal Chemistry Letters*, **4**, 835–840.

30 Matzuk, M.M., McKeown, M.R., Filippakopoulos, P., Li, Q., Ma, L., Agno, J.E., Lemieux, M.E., Picaud, S., Yu, R.N., Qi, J., Knapp, S., and Bradner, J.E. (2012) Small-molecule inhibition of BRDT for male contraception. *Cell*, **150**, 673–684.

31 Dawson, M.A., Prinjha, R.K., Dittmann, A., Giotopoulos, G., Bantscheff, M., Chan, W.-I., Robson, S.C., Chung, C., Hopf, C., Savitski, M.M., Huthmacher, C., Gudgin, E., Lugo, D., Beinke, S., Chapman, T.D., Roberts, E.J., Soden, P.E., Auger, K.R., Mirguet, O., Doehner, K., Delwel, R., Burnett, A.K., Jeffrey, P., Drewes, G., Lee, K., Huntly, B.J.P., and Kouzarides, T. (2011) Inhibition of BET recruitment to chromatin as an effective treatment for MLL-fusion leukaemia. *Nature*, **478**, 529–533.

32 Zhao, H., Gartenmann, L., Dong, J., Spiliotopoulos, D., and Caflisch, A. (2014) Discovery of BRD4 bromodomain inhibitors by fragment-based high-throughput docking. *Bioorganic and Medicinal Chemistry Letters*, **24**, 2493–2496.

33 Zhao, L., Cao, D., Chen, T., Wang, Y., Miao, Z., and Xu, Y. (2013) Fragment-based drug discovery of 2-thiazolidinones as inhibitors of the histone reader BRD4 bromodomain. *Journal of Medicinal Chemistry*, **56**, 3833–3851.

34 Hay, D.A., Fedorov, O., Martin, S., Singleton, D.C., Tallant, C., Wells, C., Picaud, S., Philpott, M., Monteiro, O.P., Rogers, C.M., Conway, S.J., Rooney, T.P., Tumber, A., Yapp, C., Filippakopoulos, P., Bunnage, M.E., Muller, S., Knapp, S., Schofield, C.J., and Brennan, P.E. (2011) Discovery and optimization of small-molecule ligands for the CBP/p300 bromodomains. *Journal of the American Chemical Society*, **136**, 9308–9319.

35 Hewings, D.S., Fedorov, O., Filippakopoulos, P., Martin, S., Picaud, S., Tumber, A., Wells, C., Olcina, M.M., Freeman, K., Gill, A., Ritchie, A.J., Sheppard, D.W., Russell, A.J., Hammond, E.M., Knapp, S., Brennan, P.E., and Conway, S.J. (2013) Optimization of 3,5-dimethylisoxazole derivatives as potent bromodomain ligands. *Journal of Medicinal Chemistry*, **56**, 3217–3227.

36 Harner, M.J., Chauder, B.A., Phan, J., and Fesik, S.W. (2014) Fragment-based screening of the bromodomain of ATAD2. *Journal of Medicinal Chemistry*, **57**, 9687–9692.

37 Filippakopoulos, P. and Knapp, S. (2014) Targeting bromodomains: epigenetic readers of lysine acetylation. *Nature Reviews Drug Discovery*, **13**, 337–356.

38 Schanda, P., Kupce, E., and Brutscher, B. (2005) SOFAST-HMQC experiments for recording two-dimensional heteronuclear correlation spectra of proteins within a few seconds. *Journal of Biomolecular NMR*, **33**, 199–211.

39 Miller, T.C., Rutherford, T.J., Birchall, K., Chugh, J., Fiedler, M., and Bienz, M. (2014) Competitive binding of a benzimidazole to the histone-binding pocket of the Pygo PHD finger. *ACS Chemical Biology*, **9**, 2864–2874.

40 Ferguson, F.M., Fedorov, O., Chaikuad, A., Philpott, M., Muniz, J.R., von Felletar, I., Delft, F., Heightman, T., Knapp, S., Abell, C., and Ciulli, A. (2013) Targeting low-druggability bromodomains: fragment based screening and inhibitor design against the BAZ2B bromodomain. *Journal of Medicinal Chemistry*, **56**, 10183–10187.

41 Perfetti, M.T., Baughman, B.M., Dickson, B.M., Mu, Y., Cui, G., Mader, P., Dong, A., Norris, J.L., Rothbart, S.B., Strahl, B.D., Brown, P.J., Janzen, W.P., Arrowsmith, C.H., Mer, G., McBride, K.M., James, L.I., and Frye, S.V. (2015) Identification of a fragment-like small molecule ligand for the methyl-lysine binding protein, 53BP1. *ACS Chemical Biology*, **10**(4), 1072–1081.

# 16
# Discovery of Inhibitors of Protein–Protein Interactions Using Fragment-Based Methods

*Feng Wang and Stephen W. Fesik*

## 16.1
## Introduction

There are approximately 20 687 protein-coding genes identified thus far from the whole human genome sequence [1] and an estimated 154 000–369 000 protein–protein interactions (PPIs) in humans [2]. Indeed, PPIs are intrinsic to almost every aspect of cellular function, including signal transduction, gene expression, cell metabolism, and cell fate determination [3,4]. Thus, as might be expected, many diseases are caused by the dysregulation of PPIs, and in principle, the proteins involved in PPIs could serve as promising drug targets. However, unlike enzymes, which have small molecules as the natural substrates and possess well-defined concave binding sites [5], PPIs are considered to be "undruggable" or at least very difficult to target with small molecules [6,7]. The challenges and problems associated with targeting PPIs include the following: (1) the interaction surfaces of PPIs are typically large (~1500–3000 $Å^2$) and generally flat without well-defined pockets [6,8], (2) PPIs do not have natural substrates that can be used as a starting point for inhibitor design [9], and (3) multiple noncontiguous interaction sites may be present [10].

Nevertheless, to expand the "druggable genome," much effort has been made toward discovering small molecule inhibitors of PPIs using a variety of different approaches. One of these approaches that has been used successfully to target PPIs is fragment-based methods in which low molecular weight compounds (fragments) are screened to identify hits that are subsequently optimized to generate high-affinity ligands (Figure 16.1) [11,12]. Compared to conventional approaches, fragment-based methods have several advantages. First, fragment-based screening can be used to explore chemical space more efficiently compared to conventional high-throughput screening (HTS), allowing a larger chemical space to be covered using a relatively small library. Although the fragment hits are generally weak in affinity, they typically have high ligand efficiencies (LE). Moreover, the smaller size of the fragments provides high-quality interactions for the PPI interface without the constraints from the rest of the molecule. Another advantage is that analogs of the fragment hits can be quickly obtained due to the

*Fragment-based Drug Discovery: Lessons and Outlook*, First Edition. Edited by Daniel A. Erlanson and Wolfgang Jahnke.

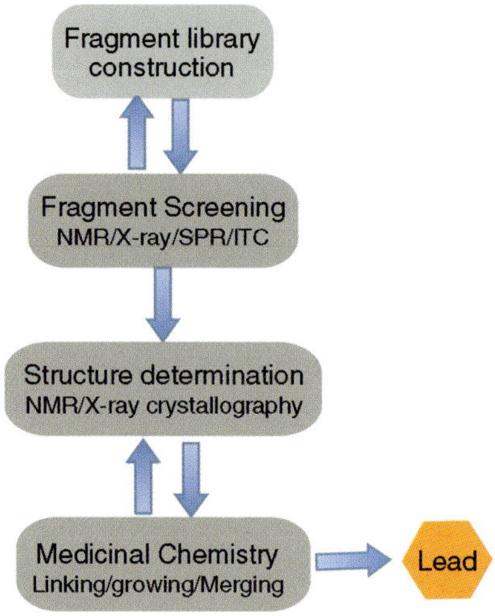


**Figure 16.1** Flow charts of fragment-based drug discovery process.

small size and low chemical complexity of the fragments, which has the advantage of making chemical synthesis of the analogs easier [13,14].

Fragment-based methods have been used in both academia and industry with a considerable amount of success [15]. Indeed, some of these compounds have progressed to clinical trials [16,17]. For example, Zelboraf (vemurafenib or PLX4032), a B-RAF inhibitor originating from a fragment-based screen, was approved by the FDA in 2011 for the treatment of late-stage melanoma [18]. Other drug candidates in clinic trials that were derived from fragment-based drug discovery (FBDD) include MK-8931 (Merck, BACE-1 inhibitor) [19], AT13387 (Astex, Hsp90 inhibitor) [20], and AT7519 (Astex, CDK2 inhibitor) [21]. For PPI inhibitors, ABT-737 and an orally bioavailable compound ABT-263 are potent small molecule inhibitors ($K_i \leq 1$ nM) of the antiapoptotic Bcl-2 family of proteins. They were discovered using fragment-based methods [22–24], and ABT-263 is currently being evaluated in clinical trials for the treatment of multiple cancers [25,26].

In this chapter, we describe the general strategies and practical considerations of FBDD, followed by some recent examples of how we are using these methods in our laboratory to target PPIs.

## 16.2 Fragment-Based Strategies for Targeting PPIs

### 16.2.1 Fragment Library Construction

The first step of FBDD is to obtain a fragment library that is suitable for conducting the screen. This library contains a special collection of small organic molecules that meet particular requirements. One popular selection criterion is the "rule of three" that corresponds to compounds with a molecular weight less than 300 Da, less than three H-bond donors and acceptors, less than three rotatable bonds, and a Clog*P* (the logarithm of partition coefficient between *n*-octanol and water) less than 3 [27]. Indeed, there are many commercially available fragment libraries that follow these rules [28]. In our own fragment library, we also used the rule of three but with slight variations. We used the following criteria: fragments must be commercially available, possess a molecular weight of 100–250 Da, contain up to four hydrogen bond donors, and have a Clog*P* of up to 3.5. To more specifically target PPIs, fragments with substructures that are known to bind frequently to proteins and are commonly found in known drugs were allowed to have looser cutoffs [12,29,30]. We did not include covalent modifiers, chelators, and those unwanted fragments that can nonspecifically bind to proteins, or cause protein aggregation or the denaturation of proteins. These fragments may contain functional groups such as Michael acceptors, hydrazines, hydroxyl amines, and free thiols [14]. In addition to removing compounds on the basis of a computational algorithm, our fragment library was further optimized by testing the compounds in three different screens, because not all bad compounds can be identified in advance. The "bad actors" that were identified in the screens were eliminated to further improve the library. Another important consideration is the size of the library. We prefer to use a relatively large library, and currently have a fragment library of approximately 15 000 compounds, which contains multiple chemically related compounds. Although the presence of several analogs seems redundant, the chance of identifying multichemotype hits is increased. Subtle modifications of the functional group may affect the low-affinity binders a lot, and make them under the detection limit. In addition, using a large library, we are able to obtain preliminary structure–activity relationship (SAR) from the initial screening that can provide direction for the follow-up analog screening [14].

### 16.2.2
### NMR-Based Fragment Screening Methods

Once the fragment library has been assembled, a suitable method needs to be used to conduct the screens. There are several screening methods that have been used to identify hits in a fragment screen. The low-affinity fragments generally require the use of sensitive biophysical assays for screening, although hits have also been identified using biochemical assays at high compound concentration [31,32]. The commonly used biophysical methods include NMR [33,34], X-ray crystallography [35], SPR [36], ITC [37], and LC–MS [38].

Of the NMR methods, the most robust approach is to monitor the chemical shift changes of isotopically labeled proteins upon the addition of the fragments. Using this approach, the ligand binding sites can be elucidated, and the initial structure–activity relationships can be obtained [33]. Thus, this method is not only able to identify hits but, if the assignments are known, can also be used to elucidate the small molecule binding mode to a certain extent. For smaller proteins (<30 kDa), $^{1}H/^{15}N$ heteronuclear single quantum correlation (HSQC) or SOFAST-HMQC can be used [39]. For larger proteins, TROSY experiments [40] or site selective labeling techniques [41] are available. These methods can potentially be effectively used for proteins with molecular weights as high as 100 kDa [42]. Protein detection is preferred, because it is more robust, has less false positives, and provides more binding information. However, it requires several hundred milligrams of isotopically labeled protein, and the size of the protein cannot be too large.

When protein detection methods are not feasible, detection of the ligands can be performed as an alternative. This approach consumes a much smaller amount of protein and does not have the size limitation. However, small molecule-based screening provides less structural information than protein-based screening, and it generates more false-positive and false-negative hits that need to be further validated. Currently, the commonly used ligand-based methods for screening are saturation transfer difference (STD) [43] and water-ligand observed via gradient spectroscopy (water-LOGSY) [44,45]. Both techniques are transfer-NOE-type experiments, capable of detecting fragments with affinities ranging from $10^{-8}$ to $10^{-3}$ M and actually work better for large proteins. Other methods, such as one-dimensional $^{1}H$ relaxation experiments [46], methods using a Carr–Purcell–Meiboom–Gill (CPMG) sequence [47], target-immobilized NMR screening [48], and Fluorine-19 relaxation [49] have also been employed.

### 16.2.3
### Structure Determination of Complexes

After completing the fragment-based screen, high-resolution 3D structures of protein–ligand complexes should be obtained to guide medicinal chemistry [50]. The availability of high-resolution structures of fragment hits bound to proteins dramatically increases the success of the hit-to-lead optimization in FBDD of PPIs. The complex structures can be determined by either X-ray crystallography or NMR spectroscopy.

X-ray crystallography is the preferred method to generate atomic resolution structures. If apo-protein crystals are amenable to high-concentration fragment soaking, the protein–fragment structures can be solved in a reproducible and rapid manner. It requires that the binding site of interest is unoccupied by the crystal contacts, and the crystals are tolerant to high concentrations of compounds and solvent. If crystals of apo-proteins cannot be obtained, protein engineering can be used to make proteins more rigid or cocrystallization conditions need to be found for each individual fragment [51–53]. The potency and solubility of the fragment hits are critical for success. To promote crystallization, the synthesis of more soluble and potent fragment analogs may be necessary.

Although labor-intensive and time-consuming, NMR can also be used for structure determination when X-ray structures cannot be obtained for small and mid-size proteins. Complete backbone and side-chain assignments and inter- and intra-molecular nuclear Overhauser effects (NOEs) are needed for a full NMR structure determination [54]. However, a full structure may not be required to obtain the structural information needed to guide hit-to-lead optimization. In some cases, a few intermolecular NOEs may be all that are required to generate a high-quality model of the protein/ligand complex that can be valuable for designing improved analogs [55].

### 16.2.4 Structure-Guided Hit-to-Lead Optimization

The structural information obtained from either NMR or crystallography gives essential insights into compound binding modes, and based on this structural information, a number of strategies can be used to generate more potent PPI inhibitors starting from relatively weak fragment hits. These include fragment linking, growing, and merging (Figure 16.2).

In our lab, fragment linking is the method of choice for converting fragment hits into high-affinity leads, when two fragments that bind to the protein in close proximity can be identified. In principle, the two fragments can be linked together to produce a linked molecule with a $K_d$ equal to the product of the $K_d$ values of the two fragments (Figure 16.2a) [56]. This concept was first demonstrated by linking together two fragments to produce a high-affinity ligand for FKBP. The method was dubbed "SAR by NMR" that stands for structure–activity relationship by nuclear magnetic resonance [33,57]. Other successful

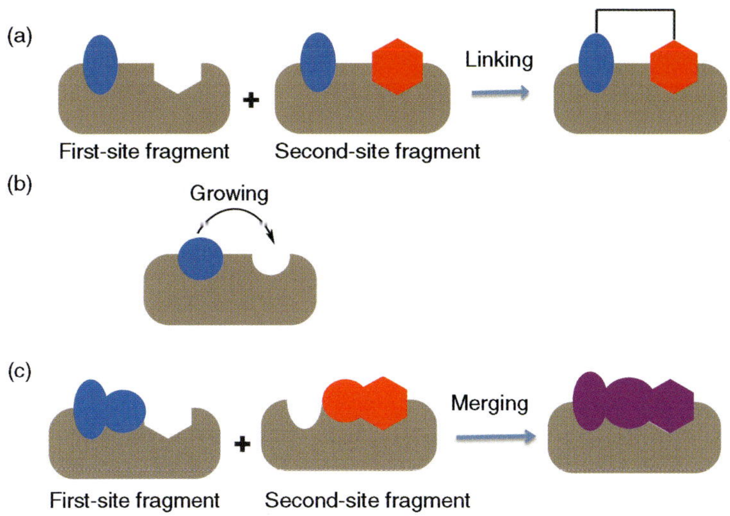


**Figure 16.2** Fragment-based hit-to-lead optimization strategies. (a) Fragment linking: two fragment hits are linked into a potent compound. (b) Fragment growing: a fragment hit is grown to the adjacent site. (c) Fragment merging: two compounds that share the same structural parts are combined into one molecule.

applications of fragment linking have been achieved on targets such as the matrix metalloproteinases (MMPs) [58,59], heat shock protein 90 (HSP90) [60], c-Src [61], thrombin [62], and Bcl-2 [63]. For the successful application of fragment linking, two fragment hits that bind close to each other must be identified and the structure of the ternary complex must be obtained. Thus, this approach is not always applicable. Although the first site hits are generally easy to obtain, the identification of second site hits can be a challenge. The second site screen should be conducted with the first site occupied, since the conformation of the second binding site can be altered in the presence of the first site ligand. The second site screen can be conducted using an HSQC experiment. Alternatively, the second site screen can be accomplished using a spin-labeled first site ligand [64] or by transferred NOE methods [65]. The paramagnetic relaxation enhancement (PRE) effect and interligand NOE (ILOE) can be detected when the first- and second-site ligands bind to the protein simultaneously [66].

Besides fragment linking, fragment growing has also been utilized for hit-to-lead optimization [14]. Fragment growing is often easier to implement than linking. The initial starting fragment hit is picked based on potency, LE, and synthetic tractability. Structural information on how the fragment binds to the protein can be used to guide the addition of functional groups to gain access to nearby unoccupied pockets to obtain additional interactions with the protein (Figure 16.2b). An example of the use of this strategy involves the optimization of an Hsp90 fragment hit, aminopyrimidine ($K_d = 20\,\mu M$) that was modified to produce a 60 nM inhibitor. This was accomplished using a fragment growing strategy that relied on high-throughput organic synthesis guided by X-ray structures of protein–ligand complexes [60].

Another approach that has been used to obtain more potent compounds is fragment merging in which the important parts of two fragments with overlapping binding sites are merged together into one compound (Figure 16.2c) [67]. More potent and more selective compounds can be synthesized using this approach by combining the key features for binding from multiple fragments. The application of fragment merging has been reported in the discovery of inhibitors of P450 [68], Renin [69], dengue protease [70], and Mcl-1 [71].

## 16.3
## Recent Examples from Our Laboratory

The focus of our lab is on cancer drug discovery using fragment-based methods and structure-based design. We are interested in attempting to target highly validated cancer targets that are known to be critical for the viability of many cancers. However, many of these proteins exert their function through protein–protein interactions and can be very difficult and challenging to target. Although risky, if successful, such compounds could have a dramatic impact for the treatment of cancer. In this section, we would like to introduce some of our recent work in which fragment-based methods were used in combination with other strategies to discover inhibitors of PPI targets. These targets include the N-terminal protein binding domain of the 70 kDa subunit of replication protein A (RPA), myeloid cell leukemia 1 (Mcl-1), and K-Ras.

## 16.3.1 Discovery of RPA Inhibitors

RPA is a heterotrimeric single-stranded DNA (ssDNA) binding protein composed of RPA70, RPA32, and RPA14 subunits. RPA is required for DNA replication, repair, and recombination as well as the checkpoint activation of the DNA damage response pathway [72,73]. The DNA damage response pathway effects are mediated through a common basic cleft in the N-terminal domain of RPA70 (RPA70 N) that interacts with many partner proteins, including p53, Rad9, and ATRIP [74–76]. Since cancer cells have higher levels of replicative stress than normal cells, inhibition of the DNA repair pathway could augment cancer cell killing in the presence of a DNA damaging agent. We sought to inhibit the DNA repair pathway by discovering a small molecule that selectively inhibits RPA70 N while leaving the other biological activities of RPA intact. To accomplish this, we conducted a fragment-based screen of RPA70 N using a fragment library consisting of ~15 000 compounds. In total, 149 hits were identified in the screen with $K_d$ values ranging from 600 μM to more than 2000 μM. The chemical shifts of particular residues (e.g., Ser55 and Thr60) that were located at opposite ends of the RPA70 N cleft were perturbed upon the addition of fragments. This suggested that there were two distinct fragment binding sites. The identification of compounds that bind to distinct nearby sites in the initial screen provided us with the opportunity to pursue a linking strategy to produce potent RPA70 N inhibitors [77]. To measure the binding affinity of the RPA70 N hits and their analogs, a high-throughput fluorescence polarization anisotropy (FPA) assay was developed and implemented in our lab that was much less time-consuming of an assay than using NMR methods [78]. In addition, in order to determine how the small molecules bind to RPA70 N, a mutant form of the protein (RPA70 N E7R) was used to alter key crystal lattice contacts, unmasking the ligand binding surface without affecting compound binding [79]. Using this mutant protein, crystal structures of RPA70 N when complexed to compounds in both sites were determined that were used to design linked compounds (Figure 16.3b) [77]. Based on the X-ray structures, interligand distances as low as 4.9 Å were observed between the closest heavy atoms. Based on this structural information, a two-atom ether-based linker was first used to connect two fragments, leading to a compound with an affinity of 26 μM (Figure 16.3a and c). Further modifications of this linked compound resulted in a 190 nM RPA70 N PPI inhibitor (Figure 16.3a and d).

The fragment-based hits and their analogs were also used to design potent stapled helix peptides [80,81]. Stapled helix peptides are locked into an active α-helical fold through site-specific all-hydrocarbon staples. This technology has

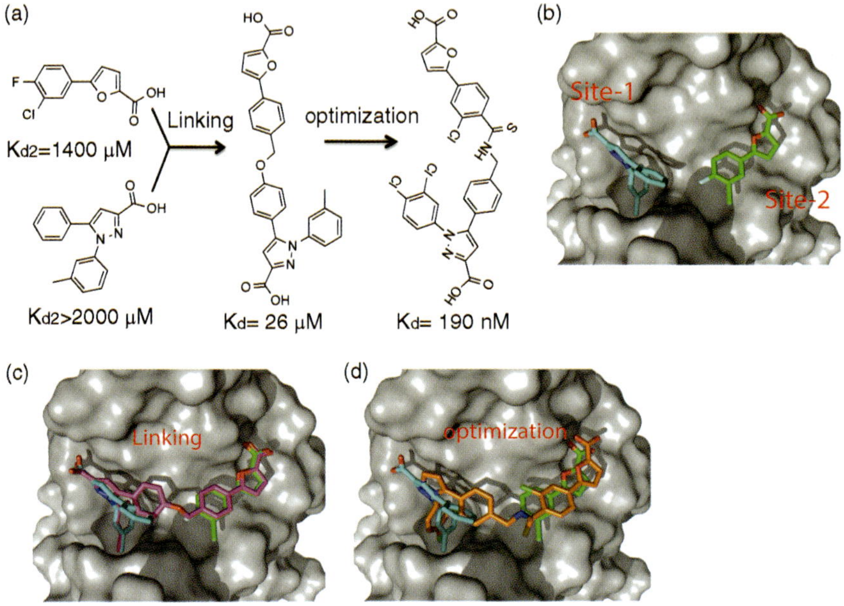


**Figure 16.3** Discovery of potent PPI inhibitors of RPA. (a) Overview of the linking and subsequent optimization strategies to design potent RPA70 N inhibitors. (b) X-ray structure (PDB: 4LUV) of RPA70 N in complex with two fragment ligands (site-1 cyan and site-2 green). (c) Overlaid structures of linked compound (PDB: 4LUZ, Purple) and the fragment hits. (d) Overlaid structures of optimized compound (PDB: 4LWC, orange) and the fragment hits.

also successfully been used to target PPIs [82]. Typically, stapled helix peptides are composed of the naturally occurring 20 amino acids, since the use of unnatural amino acids is very difficult and expensive. However, the SAR derived from fragment-based methods can facilitate the design of stapled helix peptides containing unnatural amino acids. Using the NMR-based fragment screen, the important binding "hot spots" on the PPI surface can be identified, which in the case of RPA70 N is a hydrophobic pocket near Ser55. During the lead optimization process, we found that a 3,4-dichlorosubstituted phenyl ring binds to this "hot spot" much better than the unsubstituted phenyl group. Based on these data, stapled helix peptides were designed that contained a 3,4-dichloro-substituted phenylalanine to replace a phenylalanine that binds to the same hydrophobic pocket near Ser55 as the small molecules. This modification resulted in a >100-fold improvement in affinity.

## 16.3.2
## Discovery of Potent Mcl-1 Inhibitors

Mcl-1 is a member of the Bcl-2 family of proteins that interacts with proapoptotic members of the same family and prevents cells from undergoing apoptosis [83,84].

The Mcl-1 protein is highly expressed in lung, breast, prostate, pancreatic, ovarian, and cervical cancers and leukemia [85,86], and when overexpressed, causes resistance to the Bcl-xL/Bcl-2 inhibitors ABT-737 and ABT-263 [87,88] as well as other anticancer therapies such as paclitaxel and vincristine [89]. Thus, Mcl-1 represents a promising target for a variety of cancers, and inhibition of Mcl-1 alone or in combination with other chemotherapies may serve as a potentially valuable anticancer therapy [90].

To discover a potent Mcl-1 inhibitor, we applied fragment-based methods and structure-based design [71,91]. From a screen of our fragment library against the recombinant, isotopically labeled Mcl-1, we identified 132 hits that were grouped into 11 distinct chemical classes. Among these hits, 93 hits inhibited Mcl-1 with $K_i < 500\,\mu M$, and more than 20 hits possessed a ligand efficiency greater than 0.25. Two classes of the hits exhibited different but overlapping chemical shift changes suggesting that they bind in a different manner. These include 5,6-ring-fused heterocyclic carboxylic acids (class I) and a group of hydrophobic aromatic compounds with a linker attached to a carboxylic acid (class II). We could not obtain X-ray structures of Mcl-1 complexed with these fragment ligands. However, we were able to obtain structural information on how they bind to Mcl-1 from NOE-derived distance restraints obtained from three-dimensional $^{13}C$-edited NOESY and $^{15}N$- and $^{13}C$-filter/edited NOESY spectra. These NMR-derived restraints were used to dock fragments from class I/II into a previously determined Mcl-1/Bim BH3 peptide complex structure (Figure 16.4). The hydrophobic groups of the two classes of fragments bind to distinct parts of a large pocket, and the carboxylic acids of the two classes of compounds were found to interact with Arg263 of Mcl-1. Based on the structural information, the two fragments were merged together to markedly improve binding affinities by more than 100-fold.

X-ray structures of the merged compounds were obtained that revealed that they maintained the favorable hydrophobic contacts of both fragments as well as the interaction between the common carboxylic acid and Arg263 of Mcl-1 (Figure 16.4). To further improve the binding affinity, the SAR of the merged compounds was explored using structure-based design. The most potent inhibitor that was reported from these studies had a $K_i$ value of 55 nM, and represents an excellent starting point for the design of improved analogs (Figure 16.4).

### 16.3.3
### Discovery of Small Molecules that Bind to K-Ras

The ras genes and their products are among the most well-studied cellular oncogenes [92]. K-Ras, a small GTPase, functions as a regulated switch, cycling between the GDP-bound (inactive) and GTP-bound (active) states [93–95]. The guanine nucleotide exchange factor (GEF) protein, son of sevenless (SOS), forms a complex with K-Ras and promotes the rate-limiting step in the activation of Ras by catalyzing the conversion of GDP to GTP [96,97]. The GTPase-activating proteins (GAPs) of K-Ras terminate signaling by inducing GTP hydrolysis [98].

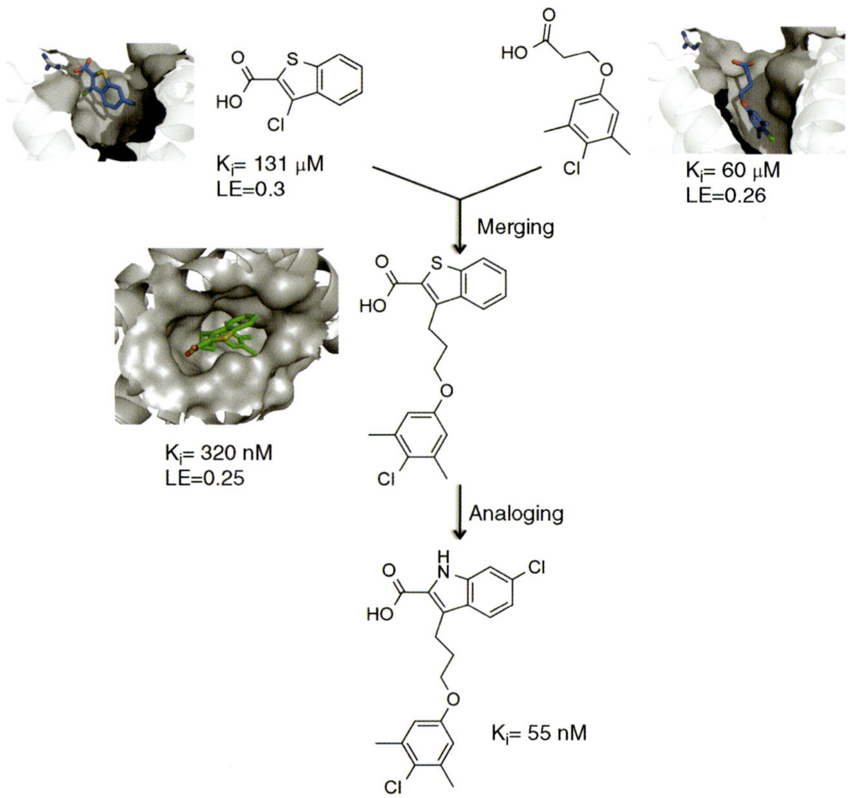


**Figure 16.4** Discovery of potent Mcl-1 inhibitors. Two compounds, benzothiophene (left) and tethered aromatic (right), were merged into a potent Mcl-1 inhibitor. Models of benzothiophene and tethered aromatic were calculated using NMR-derived distance restraints, and the X-ray structure of the merged compound complexed with Mcl-1 (PDB: 4HW3) is also shown. The merged compound was further modified to a more potent and selective Mcl-1 inhibitor.

Active K-Ras interacts with downstream effectors, including RAF, PI3 K, RIN, and PLC, and promotes cellular activities that represent the hallmarks of cancer [99,100]. K-Ras is mutated in ~30% of all cancers and is required for cancer cells to survive. K-Ras represents a highly validated target and has been the focus of many drug discovery efforts [101,102]. Most of this effort has focused on the discovery of inhibitors of Ras signaling either upstream or downstream of K-Ras [103], because the surface of K-Ras lacks well-defined binding pockets accessible for small molecules. Although very challenging, we attempted to target K-Ras directly. Using fragment-based methods, we discovered novel small molecules that bind directly to K-Ras between switch I and switch II. These molecules were shown to inhibit SOS-catalyzed K-Ras activation [104]. GDP-bound mutant K-Ras (G12D) was used in the initial NMR-based fragment screening, and about 140 hits out of 11 000 fragments were identified that can bind to

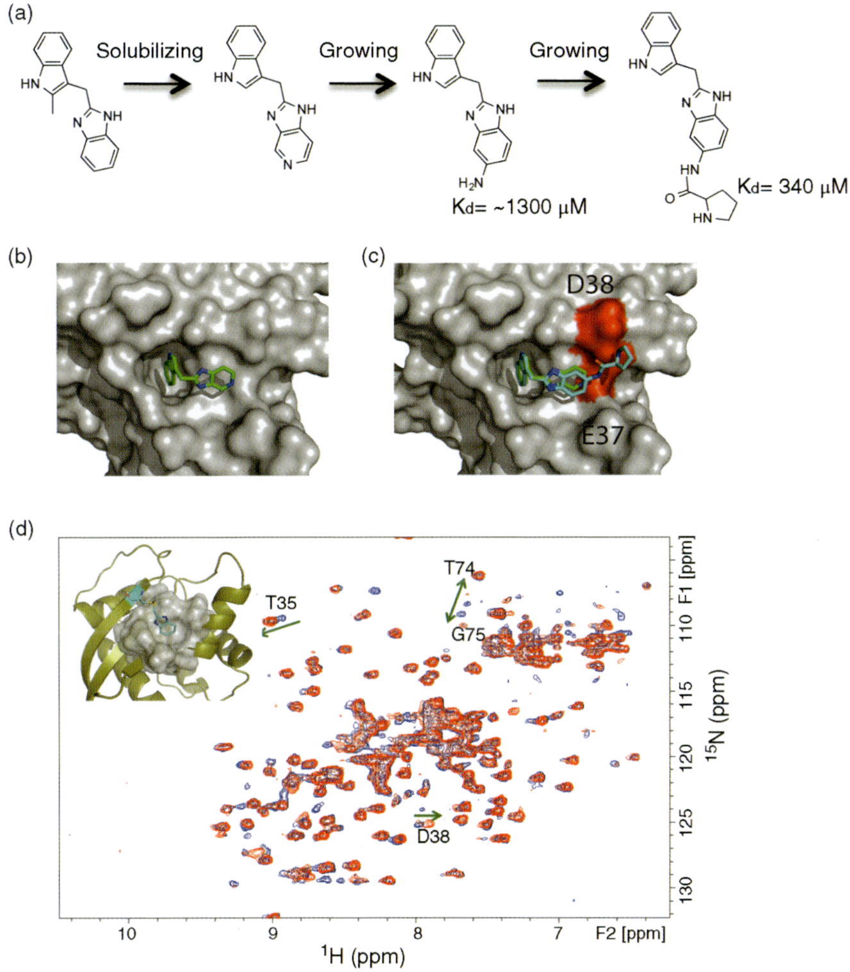


**Figure 16.5** Discovery of small molecules that bind to K-Ras. (a) Structures of K-Ras inhibitor compounds and the growing strategy are shown. (b) The compound was cocrystallized with GDP-K-Ras, and the structure (PDB: 4EPV) of binding cavity is shown. (c) Overlaid structures of initial (green) and grown (cyan) compounds. The negatively charged E37 and D38, which mediate the additional interaction, are labeled red. (d) Overlaid $^1H/^{15}N$ HMQC spectra of the uniformly $^{15}N$-labeled covalently modified K-Ras S39C mutant recoded in the absence (blue) and presence (red) of a second-site fragment hit. The structure of covalently modified K-Ras S39C mutant (PDB: 4PZZ) is also shown.

GDP-bound K-Ras (G12D) (hit rate = 1.3%). The initial fragment hits were not suitable for cocrystallization due to the low solubility, and more soluble analogs were generated for cocrystallization (Figure 16.5a). Cocrystal structures were obtained from wild-type and G12 V K-Ras and showed that the compounds bind between the α2 helix of switch II and the central β sheet (Figure 16.5b).

The binding affinities of these hits were weak with $K_d$ values in the millimolar range. However, a second binding cleft was generated upon the binding of the first-site ligands that contained two acidic residues (Glu37 and Asp38). To access this region of the protein and enhance the binding affinity, fragment growing was applied. The indole-benzimidazole fragment was functionalized with amino acids that contain positively charged amine groups. This resulted in compounds with a 10-fold increase in binding affinity (Figure 16.5a). The best compound showed $K_d$ value of 190 μM. The small molecules directly bind K-Ras at the Ras–SOS interface (Figure 16.5c). Besides our work, a group at Genentech independently discovered compounds that bound to the same site on K-Ras [105].

To obtain more potent inhibitors of K-Ras, we conducted a second-site screen against K-Ras to identify ligands that bind in close proximity to the first-site hits for the purpose of obtaining more potent inhibitors using linking or merging strategies [106]. However, our first-site ligands for K-Ras could not fully saturate the first-site pocket because of their weak binding and poor aqueous solubility. In order to saturate the first-site pocket, a panel of six cysteine mutants of K-Ras were generated and were used to covalently attach a first-site ligand. A small library of 32 thiol reactive first-site compounds was screened against the K-Ras mutants from which four mutant/compound combinations were identified. K-Ras S39C was selected in combination with a thiol reactive compound to perform the second-site fragment screen in which 20 hits were identified (Figure 16.5d). These hits are being used in structural studies to elucidate the binding modes for the possibility of linking to the first-site ligands.

## 16.4 Summary and Conclusions

Proteins that exert their activity by binding to other proteins represent very promising targets for the discovery of novel therapeutics for many diseases. In spite of the intrinsic difficulty of targeting PPIs, remarkable progress has been made in both academia and industry over the past few years [7]. One of the best methods for targeting PPIs is the fragment-based approach. In this chapter, we outlined some of the methods and strategies of FBDD. In our opinion, the quality of the fragment library is critical for FBDD and must be optimized and highly curated. In addition, we favor the use of NMR-based protein detection screening methods for identifying small molecules with weak affinity, mapping ligand binding sites, and obtaining preliminary SAR. Structural information on how the ligands bind to the protein obtained by X-ray crystallography or NMR spectroscopy is also critical. In some cases, it may be necessary to chemically modify the fragment hits to obtain ligands that are more suitable for conducting the structural studies. We illustrated the use of fragment linking, merging, and growing strategies in our efforts to target RPA, Mcl-1, and K-Ras, respectively. Moreover, we demonstrated fragment-based methods could be synergistically combined with other PPI targeting techniques to guide the design of stapled helix peptides containing unnatural amino acids.

For successful drug discovery, fragment-based methods can be a useful tool but requires the synergy of structural biology, biochemical assay development, bio-/chemo-informatics, and medicinal chemistry. Medicinal chemistry is very important even in the very early stage of fragment optimization. Medicinal chemistry modifications of low-affinity fragment hits can provide initial SAR and produce more soluble and potent ligands, facilitating the needed structural studies. For example, the synthesis of more soluble fragment analogs was crucial for the cocrystal structure determination of K-Ras. In addition, the SAR obtained from the fragments can also be easily applied on the linked or merged molecules that are much larger, as evidenced in the discovery of RPA70N inhibitors. Due to the successful applications of fragment-based methods in the discovery of PPI inhibitors, we foresee these methods are likely to be increasingly used in the future.

## Acknowledgments

We thank all the members in the Fesik lab for helpful discussions. This research was supported in part by these grants and awards to Stephen W. Fesik: the National Institutes of Health (NIH) Director's Pioneer Award (DP1OD006933/DP1CA174419), the NIH RO1 grant (RO1CA174887, RPA project), and Lustgarten Foundation Award (K-Ras project). This work was also supported by NIH ARRA stimulus grant to Lawrence J. Marnett (5RC2CA148375), an NCI SPORE grant in breast cancer to Carlos L. Arteaga (Grant P50CA098131, Mcl-1 project; Stephen W. Fesik as project leader), an NCI SPORE in GI Cancer to Robert J. Coffey (P50A095103-12, K-Ras project; Stephen W. Fesik as project leader), an NIH Fellowship Award to James D. Patrone (F32CA174315, Stephen W. Fesik as mentor), and a Deutscher Akademischer Austausch Dienst (DAAD) postdoctoral fellowship to Andreas O. Frank (Stephen W. Fesik as mentor). The use of the Advanced Photon Source, an Office of Science User Facility operated for the US Department of Energy (DOE) Office of Science by Argonne National Laboratory, was supported by the US DOE under Contract No. DE-AC02-06CH11357. The use of the LS-CAT Sector 21 was supported by the Michigan Economic Development Corporation and the Michigan Technology Tri-Corridor (Grant 085P1000817). The use of Vanderbilt NMR facility was supported in part by grants for NMR instrumentation from the US National Science Foundation (NSF) (0922862), NIH (S10 RR025677), and Vanderbilt University matching funds. A portion of the experiments described here used the Vanderbilt robotic crystallization facility, which was supported by NIH grant S10 RR026915.

## References


1 Consortium, E.P. (2012) An integrated encyclopedia of DNA elements in the human genome. *Nature*, **489**, 57–74.

2 Hart, G.T., Ramani, A.K., and Marcotte, E.M. (2006) How complete are current yeast and human protein-interaction networks? *Genome Biology*, **7**, 120.

3 Pawson, T. and Nash, P. (2000) Protein–protein interactions define specificity in signal transduction. *Genes and Development*, **14**, 1027–1047.

4 Lage, K. (2014) Protein–protein interactions and genetic diseases: the interactome. *Biochimica et Biophysica Acta*, **1842**, 1971–1980.

5 Bunnage, M.E. (2011) Getting pharmaceutical R&D back on target. *Nature Chemical Biology*, **7**, 335–339.

6 Hopkins, A.L. and Groom, C.R. (2002) The druggable genome. *Nature Reviews Drug Discovery*, **1**, 727–730.

7 Wells, J.A. and McClendon, C.L. (2007) Reaching for high-hanging fruit in drug discovery at protein–protein interfaces. *Nature*, **450**, 1001–1009.

8 Jones, S. and Thornton, J.M. (1996) Principles of protein–protein interactions. *Proceedings of the National Academy of Sciences of the United States of America*, **93**, 13–20.

9 Schramm, V.L. (1998) Enzymatic transition states and transition state analog design. *Annual Review of Biochemistry*, **67**, 693–720.

10 Wang, F., Marshall, C.B., and Ikura, M. (2013) Transcriptional/epigenetic regulator CBP/p300 in tumorigenesis: structural and functional versatility in target recognition. *Cellular and Molecular Life Sciences: CMLS*, **70**, 3989–4008.

11 Erlanson, D.A. (2012) Introduction to fragment-based drug discovery. *Topics in Current Chemistry*, **317**, 1–32.

12 Scott, D.E., Ehebauer, M.T., Pukala, T., Marsh, M., Blundell, T.L., Venkitaraman, A.R., Abell, C., and Hyvonen, M. (2013) Using a fragment-based approach to target protein–protein interactions. *ChemBioChem: A European Journal of Chemical Biology*, **14**, 332–342.

13 Winter, A., Higueruelo, A.P., Marsh, M., Sigurdardottir, A., Pitt, W.R., and Blundell, T.L. (2012) Biophysical and computational fragment-based approaches to targeting protein–protein interactions: applications in structure-guided drug discovery. *Quarterly Reviews of Biophysics*, **45**, 383–426.

14 Harner, M.J., Frank, A.O., and Fesik, S.W. (2013) Fragment-based drug discovery using NMR spectroscopy. *Journal of Biomolecular NMR*, **56**, 65–75.

15 Murray, C.W. and Rees, D.C. (2009) The rise of fragment-based drug discovery. *Nature Chemistry*, **1**, 187–192.

16 Valkov, E., Sharpe, T., Marsh, M., Greive, S., and Hyvonen, M. (2012) Targeting protein–protein interactions and fragment-based drug discovery. *Topics in Current Chemistry*, **317**, 145–179.

17 Turnbull, A.B., Boyd, S.M., and Walse, B. (2014) Fragment-based drug discovery and protein–protein interactions. *Research and Reports in Biochemistry*, **4**, 13–26.

18 Bollag, G., Tsai, J., Zhang, J., Zhang, C., Ibrahim, P., Nolop, K., and Hirth, P. (2012) Vemurafenib: the first drug approved for BRAF-mutant cancer. *Nature Reviews Drug Discovery*, **11**, 873–886.

19 Wang, Y.S., Strickland, C., Voigt, J.H., Kennedy, M.E., Beyer, B.M., Senior, M.M., Smith, E.M., Nechuta, T.L., Madison, V.S., Czarniecki, M., McKittrick, B.A., Stamford, A.W., Parker, E.M., Hunter, J.C., Greenlee, W.J., and Wyss, D.F. (2010) Application of fragment-based NMR screening, X-ray crystallography, structure-based design, and focused chemical library design to identify novel microM leads for the development of nM BACE-1 (beta-site APP cleaving enzyme 1) inhibitors. *Journal of Medicinal Chemistry*, **53**, 942–950.

20 Woodhead, A.J., Angove, H., Carr, M.G., Chessari, G., Congreve, M., Coyle, J.E., Cosme, J., Graham, B., Day, P.J., Downham, R., Fazal, L., Feltell, R., Figueroa, E., Frederickson, M., Lewis, J.,

McMenamin, R., Murray, C.W., O'Brien, M.A., Parra, L., Patel, S., Phillips, T., Rees, D.C., Rich, S., Smith, D.M., Trewartha, G., Vinkovic, M., Williams, B., and Woolford, A.J. (2010) Discovery of (2,4-dihydroxy-5-isopropylphenyl)-[5-(4-methylpiperazin-1-ylmethyl)-1,3-dihydrois oindol-2-yl]methanone (AT13387), a novel inhibitor of the molecular chaperone Hsp90 by fragment based drug design. *Journal of Medicinal Chemistry*, **53**, 5956–5969.

21 Wyatt, P.G., Woodhead, A.J., Berdini, V., Boulstridge, J.A., Carr, M.G., Cross, D.M., Davis, D.J., Devine, L.A., Early, T.R., Feltell, R.E., Lewis, E.J., McMenamin, R.L., Navarro, E.F., O'Brien, M.A., O'Reilly, M., Reule, M., Saxty, G., Seavers, L.C.A., Smith, D.M., Squires, M.S., Trewartha, G., Walker, M.T., and Woolford, A.J.A. (2008) Identification of *N*-(4-piperidinyl)-4-(2,6-dichlorobenzoylamino)-1*H*-pyrazole-3-carboxamide (AT7519), a novel cyclin dependent kinase inhibitor using fragment-based X-ray crystallography and structure based drug design. *Journal of Medicinal Chemistry*, **51**, 4986–4999.

22 Oltersdorf, T., Elmore, S.W., Shoemaker, A.R., Armstrong, R.C., Augeri, D.J., Belli, B.A., Bruncko, M., Deckwerth, T.L., Dinges, J., Hajduk, P.J., Joseph, M.K., Kitada, S., Korsmeyer, S.J., Kunzer, A.R., Letai, A., Li, C., Mitten, M.J., Nettesheim, D.G., Ng, S., Nimmer, P.M., O'Connor, J.M., Oleksijew, A., Petros, A.M., Reed, J.C., Shen, W., Tahir, S.K., Thompson, C.B., Tomaselli, K.J., Wang, B., Wendt, M.D., Zhang, H., Fesik, S.W., and Rosenberg, S.H. (2005) An inhibitor of Bcl-2 family proteins induces regression of solid tumours. *Nature*, **435**, 677–681.

23 Petros, A.M., Dinges, J., Augeri, D.J., Baumeister, S.A., Betebenner, D.A., Bures, M.G., Elmore, S.W., Hajduk, P.J., Joseph, M.K., Landis, S.K., Nettesheim, D.G., Rosenberg, S.H., Shen, W., Thomas, S., Wang, X., Zanze, I., Zhang, H., and Fesik, S.W. (2006) Discovery of a potent inhibitor of the antiapoptotic protein Bcl-xL from NMR and parallel synthesis. *Journal of Medicinal Chemistry*, **49**, 656–663.

24 Tse, C., Shoemaker, A.R., Adickes, J., Anderson, M.G., Chen, J., Jin, S., Johnson, E.F., Marsh, K.C., Mitten, M.J., Nimmer, P., Roberts, L., Tahir, S.K., Xiao, Y., Yang, X., Zhang, H., Fesik, S., Rosenberg, S.H., and Elmore, S.W. (2008) ABT-263: a potent and orally bioavailable Bcl-2 family inhibitor. *Cancer Research*, **68**, 3421–3428.

25 Rudin, C.M., Hann, C.L., Garon, E.B., de Oliveira, M.R., Bonomi, P.D., Camidge, D.R., Chu, Q., Giaccone, G., Khaira, D., Ramalingam, S.S., Ranson, M.R., Dive, C., McKeegan, E.M., Chyla, B.J., Dowell, B.L., Chakravartty, A., Nolan, C.E., Rudersdorf, N., Busman, T.A., Mabry, M.H., Krivoshik, A.P., Humerickhouse, R.A., Shapiro, G.I., and Gandhi, L. (2012) Phase II study of single-agent Navitoclax (ABT-263) and biomarker correlates in patients with relapsed small cell lung cancer. *Clinical Cancer Research*, **18**, 3163–3169.

26 Gandhi, L., Camidge, D.R., Ribeiro de Oliveira, M., Bonomi, P., Gandara, D., Khaira, D., Hann, C.L., McKeegan, E.M., Litvinovich, E., Hemken, P.M., Dive, C., Enschede, S.H., Nolan, C., Chiu, Y.L., Busman, T., Xiong, H., Krivoshik, A.P., Humerickhouse, R., Shapiro, G.I., and Rudin, C.M. (2011) Phase I study of Navitoclax (ABT-263), a novel Bcl-2 family inhibitor, in patients with small-cell lung cancer and other solid tumors. *Journal of Clinical Oncology: Official Journal of the American Society of Clinical Oncology*, **29**, 909–916.

27 Congreve, M., Carr, R., Murray, C., and Jhoti, H. (2003) A 'rule of three' for fragment-based lead discovery? *Drug Discovery Today*, **8**, 876–877.

28 Lepre, C.A. (2011) Practical aspects of NMR-based fragment screening. *Methods in Enzymology*, **493**, 219–239.

29 Hajduk, P.J., Bures, M., Praestgaard, J., and Fesik, S.W. (2000) Privileged molecules for protein binding identified from NMR-based screening. *Journal of Medicinal Chemistry*, **43**, 3443–3447.

30 Siegal, G., Ab, E., and Schultz, J. (2007) Integration of fragment screening and library design. *Drug Discovery Today*, **12**, 1032–1039.

31 Rishton, G.M. (2003) Nonleadlikeness and leadlikeness in biochemical screening. *Drug Discovery Today*, **8**, 86–96.

32 Erlanson, D.A., Braisted, A.C., Raphael, D.R., Randal, M., Stroud, R.M., Gordon, E.M., and Wells, J.A. (2000) Site-directed ligand discovery. *Proceedings of the National Academy of Sciences of the United States of America*, **97**, 9367–9372.

33 Shuker, S.B., Hajduk, P.J., Meadows, R.P., and Fesik, S.W. (1996) Discovering high-affinity ligands for proteins: SAR by NMR. *Science*, **274**, 1531–1534.

34 Hajduk, P.J., Meadows, R.P., and Fesik, S. W. (1999) NMR-based screening in drug discovery. *Quarterly Reviews of Biophysics*, **32**, 211–240.

35 Badger, J. (2012) Crystallographic fragment screening. *Methods in Molecular Biology*, **841**, 161–177.

36 Perspicace, S., Banner, D., Benz, J., Muller, F., Schlatter, D., and Huber, W. (2009) Fragment-based screening using surface plasmon resonance technology. *Journal of Biomolecular Screening*, **14**, 337–349.

37 Drinkwater, N., Vu, H., Lovell, K.M., Criscione, K.R., Collins, B.M., Prisinzano, T.E., Poulsen, S.A., McLeish, M.J., Grunewald, G.L., and Martin, J.L. (2010) Fragment-based screening by X-ray crystallography, MS and isothermal titration calorimetry to identify PNMT (phenylethanolamine *N*-methyltransferase) inhibitors. *The Biochemical Journal*, **431**, 51–61.

38 Duong-Thi, M.D., Bergstrom, M., Fex, T., Isaksson, R., and Ohlson, S. (2013) High-throughput fragment screening by affinity LC–MS. *Journal Of Biomolecular Screening*, **18**, 160–171.

39 Schanda, P., Kupce, E., and Brutscher, B. (2005) SOFAST-HMQC experiments for recording two-dimensional heteronuclear correlation spectra of proteins within a few seconds. *Journal of Biomolecular NMR*, **33**, 199–211.

40 Lescop, E., Kern, T., and Brutscher, B. (2010) Guidelines for the use of band-selective radiofrequency pulses in hetero-nuclear NMR: example of longitudinal-relaxation-enhanced BEST-type 1H-15N correlation experiments. *Journal of Magnetic Resonance*, **203**, 190–198.

41 Weigelt, J., van Dongen, M., Uppenberg, J., Schultz, J., and Wikstrom, M. (2002) Site-selective screening by NMR spectroscopy with labeled amino acid pairs. *Journal of the American Chemical Society*, **124**, 2446–2447.

42 Campos-Olivas, R. (2011) NMR screening and hit validation in fragment based drug discovery. *Current Topics in Medicinal Chemistry*, **11**, 43–67.

43 Mayer, M. and Meyer, B. (2001) Group epitope mapping by saturation transfer difference NMR to identify segments of a ligand in direct contact with a protein receptor. *Journal of the American Chemical Society*, **123**, 6108–6117.

44 Dalvit, C., Pevarello, P., Tato, M., Veronesi, M., Vulpetti, A., and Sundstrom, M. (2000) Identification of compounds with binding affinity to proteins via magnetization transfer from bulk water. *Journal of Biomolecular NMR*, **18**, 65–68.

45 Dalvit, C., Fogliatto, G., Stewart, A., Veronesi, M., and Stockman, B. (2001) WaterLOGSY as a method for primary NMR screening: practical aspects and range of applicability. *Journal of Biomolecular NMR*, **21**, 349–359.

46 Hajduk, P.J., Olejniczak, E.T., and Fesik, S.W. (1997) One-dimensional relaxation- and diffusion-edited NMR methods for screening compounds that bind to macromolecules. *Journal of the American Chemical Society*, **119**, 12257–12261.

47 Mashalidis, E.H., Sledz, P., Lang, S., and Abell, C. (2013) A three-stage biophysical screening cascade for fragment-based drug discovery. *Nature Protocols*, **8**, 2309–2324.

48 Vanwetswinkel, S., Heetebrij, R.J., van Duynhoven, J., Hollander, J.G., Filippov, D.V., Hajduk, P.J., and Siegal, G. (2005) TINS, target immobilized NMR screening: an efficient and sensitive method for ligand discovery. *Chemistry and Biology*, **12**, 207–216.

49 Dalvit, C., Fagerness, P.E., Hadden, D.T., Sarver, R.W., and Stockman, B.J. (2003) Fluorine-NMR experiments for high-throughput screening: theoretical aspects,

practical considerations, and range of applicability. *Journal of the American Chemical Society*, **125**, 7696–7703.

50 Murray, C.W. and Blundell, T.L. (2010) Structural biology in fragment-based drug design. *Current Opinion in Structural Biology*, **20**, 497–507.

51 Blundell, T.L., Jhoti, H., and Abell, C. (2002) High-throughput crystallography for lead discovery in drug design. *Nature Reviews Drug Discovery*, **1**, 45–54.

52 Hartshorn, M.J., Murray, C.W., Cleasby, A., Frederickson, M., Tickle, I.J., and Jhoti, H. (2005) Fragment-based lead discovery using X-ray crystallography. *Journal of Medicinal Chemistry*, **48**, 403–413.

53 Bottcher, J., Jestel, A., Kiefersauer, R., Krapp, S., Nagel, S., Steinbacher, S., and Steuber, H. (2011) Key factors for successful generation of protein-fragment structures requirement on protein, crystals, and technology. *Methods in Enzymology*, **493**, 61–89.

54 Wagner, G., Hyberts, S.G., and Havel, T.F. (1992) NMR structure determination in solution: a critique and comparison with X-ray crystallography. *Annual Review of Biophysics and Biomolecular Structure*, **21**, 167–198.

55 Meyer, B. and Peters, T. (2003) NMR spectroscopy techniques for screening and identifying ligand binding to protein receptors. *Angewandte Chemie*, **42**, 864–890.

56 Jencks, W.P. (1981) On the attribution and additivity of binding energies. *Proceedings of the National Academy of Sciences of the United States of America*, **78**, 4046–4050.

57 Hajduk, P.J., Meadows, R.P., and Fesik, S.W. (1997) Discovering high-affinity ligands for proteins. *Science*, **278**, 497–499.

58 Hajduk, P.J., Sheppard, G., Nettesheim, D.G., Olejniczak, E.T., Shuker, S.B., Meadows, R.P., Steinman, D.H., Carrera, G.M., Marcotte, P.A., Severin, J., Walter, K., Smith, H., Gubbins, E., Simmer, R., Holzman, T.F., Morgan, D.W., Davidsen, S.K., Summers, J.B., and Fesik, S.W. (1997) Discovery of potent nonpeptide inhibitors of stromelysin using SAR by NMR. *Journal of the American Chemical Society*, **119**, 5818–5827.

59 Wada, C.K. (2004) The evolution of the matrix metalloproteinase inhibitor drug discovery program at Abbott Laboratories. *Current Topics in Medicinal Chemistry*, **4**, 1255–1267.

60 Huth, J.R., Park, C., Petros, A.M., Kunzer, A.R., Wendt, M.D., Wang, X.L., Lynch, C.L., Mack, J.C., Swift, K.M., Judge, R.A., Chen, J., Richardson, P.L., Jin, S., Tahir, S.K., Matayoshi, E.D., Dorwin, S.A., Ladror, U.S., Severin, J.M., Walter, K.A., Bartley, D.M., Fesik, S.W., Elmore, S.W., and Hajduk, P.J. (2007) Discovery and design of novel HSP90 inhibitors using multiple fragment-based design strategies. *Chemical Biology and Drug Design*, **70**, 1–12.

61 Maly, D.J., Choong, I.C., and Ellman, J.A. (2000) Combinatorial target-guided ligand assembly: identification of potent subtype-selective c-Src inhibitors. *Proceedings of the National Academy of Sciences of the United States of America*, **97**, 2419–2424.

62 Howard, N., Abell, C., Blakemore, W., Chessari, G., Congreve, M., Howard, S., Jhoti, H., Murray, C.W., Seavers, L.C., and van Montfort, R.L. (2006) Application of fragment screening and fragment linking to the discovery of novel thrombin inhibitors. *Journal of Medicinal Chemistry*, **49**, 1346–1355.

63 Petros, A.M., Huth, J.R., Oost, T., Park, C.M., Ding, H., Wang, X.L., Zhang, H.C., Nimmer, P., Mendoza, R., Sun, C.H., Mack, J., Walter, K., Dorwin, S., Gramling, E., Ladror, U., Rosenberg, S.H., Elmore, S.W., Fesik, S.W., and Hajduk, P.J. (2010) Discovery of a potent and selective Bcl-2 inhibitor using SAR by NMR. *Bioorganic and Medicinal Chemistry Letters*, **20**, 6587–6591.

64 Jahnke, W. (2002) Spin labels as a tool to identify and characterize protein–ligand interactions by NMR spectroscopy. *ChemBioChem: A European Journal of Chemical Biology*, **3**, 167–173.

65 Li, D.W., DeRose, E.F., and London, R.E. (1999) The inter-ligand Overhauser effect: A powerful new NMR approach for mapping structural relationships of

macromolecular ligands. *Journal of Biomolecular NMR*, **15**, 71–76.

66 Jahnke, W., Florsheimer, A., Blommers, M.J., Paris, C.G., Heim, J., Nalin, C.M., and Perez, L.B. (2003) Second-site NMR screening and linker design. *Current Topics in Medicinal Chemistry*, **3**, 69–80.

67 Erlanson, D.A., McDowell, R.S., and O'Brien, T. (2004) Fragment-based drug discovery. *Journal of Medicinal Chemistry*, **47**, 3463–3482.

68 Hudson, S.A., McLean, K.J., Surade, S., Yang, Y.Q., Leys, D., Ciulli, A., Munro, A.W., and Abell, C. (2012) Application of fragment screening and merging to the discovery of inhibitors of the *Mycobacterium tuberculosis* cytochrome P450 CYP121. *Angewandte Chemie*, **51**, 9311–9316.

69 Ostermann, N., Ruedisser, S., Ehrhardt, C., Breitenstein, W., Marzinzik, A., Jacoby, E., Vangrevelinghe, E., Ottl, J., Klumpp, M., Hartwieg, J.C., Cumin, F., Hassiepen, U., Trappe, J., Sedrani, R., Geisse, S., Gerhartz, B., Richert, P., Francotte, E., Wagner, T., Kromer, M., Kosaka, T., Webb, R.L., Rigel, D.F., Maibaum, J., and Baeschlin, D.K. (2013) A novel class of oral direct renin inhibitors: highly potent 3,5-disubstituted piperidines bearing a tricyclic p3-p1 pharmacophore. *Journal of Medicinal Chemistry*, **56**, 2196–2206.

70 Behnam, M.A., Nitsche, C., Vechi, S.M., and Klein, C.D. (2014) C-terminal residue optimization and fragment merging: discovery of a potent peptide-hybrid inhibitor of dengue protease. *ACS Medicinal Chemistry Letters*, **5**, 1037–1042.

71 Friberg, A., Vigil, D., Zhao, B., Daniels, R. N., Burke, J.P., Garcia-Barrantes, P.M., Camper, D., Chauder, B.A., Lee, T., Olejniczak, E.T., and Fesik, S.W. (2013) Discovery of potent myeloid cell leukemia 1 (Mcl-1) inhibitors using fragment-based methods and structure-based design. *Journal of Medicinal Chemistry*, **56**, 15–30.

72 Wold, M.S. (1997) Replication protein A: a heterotrimeric, single-stranded DNA-binding protein required for eukaryotic DNA metabolism. *Annual Review of Biochemistry*, **66**, 61–92.

73 Oakley, G.G. and Patrick, S.M. (2010) Replication protein A: directing traffic at the intersection of replication and repair. *Frontiers in bioscience (Landmark edition)*, **15**, 883–900.

74 Xu, X., Vaithiyalingam, S., Glick, G.G., Mordes, D.A., Chazin, W.J., and Cortez, D. (2008) The basic cleft of RPA70N binds multiple checkpoint proteins, including RAD9, to regulate ATR signaling. *Molecular and Cellular Biology*, **28**, 7345–7353.

75 Bochkareva, E., Kaustov, L., Ayed, A., Yi, G.S., Lu, Y., Pineda-Lucena, A., Liao, J.C., Okorokov, A.L., Milner, J., Arrowsmith, C.H., and Bochkarev, A. (2005) Single-stranded DNA mimicry in the p53 transactivation domain interaction with replication protein A. *Proceedings of the National Academy of Sciences of the United States of America*, **102**, 15412–15417.

76 Ball, H.L., Myers, J.S., and Cortez, D. (2005) ATRIP binding to replication protein A-single-stranded DNA promotes ATR-ATRIP localization but is dispensable for Chk1 phosphorylation. *Molecular Biology of the Cell*, **16**, 2372–2381.

77 Frank, A.O., Feldkamp, M.D., Kennedy, J.P., Waterson, A.G., Pelz, N.F., Patrone, J.D., Vangamudi, B., Camper, D.V., Rossanese, O.W., Chazin, W.J., and Fesik, S.W. (2013) Discovery of a potent inhibitor of replication protein a protein–protein interactions using a fragment-linking approach. *Journal of Medicinal Chemistry*, **56**, 9242–9250.

78 Souza-Fagundes, E.M., Frank, A.O., Feldkamp, M.D., Dorset, D.C., Chazin, W.J., Rossanese, O.W., Olejniczak, E.T., and Fesik, S.W. (2012) A high-throughput fluorescence polarization anisotropy assay for the 70N domain of replication protein A. *Analytical Biochemistry*, **421**, 742–749.

79 Feldkamp, M.D., Frank, A.O., Kennedy, J.P., Patrone, J.D., Vangamudi, B., Waterson, A.G., Fesik, S.W., and Chazin, W.J. (2013) Surface reengineering of RPA70N enables cocrystallization with an

inhibitor of the replication protein A interaction motif of ATR interacting protein. *Biochemistry*, **52**, 6515–6524.

80 Walensky, L.D., Kung, A.L., Escher, I., Malia, T.J., Barbuto, S., Wright, R.D., Wagner, G., Verdine, G.L., and Korsmeyer, S.J. (2004) Activation of apoptosis *in vivo* by a hydrocarbon-stapled BH3 helix. *Science*, **305**, 1466–1470.

81 Frank, A.O., Vangamudi, B., Feldkamp, M.D., Souza-Fagundes, E.M., Luzwick, J.W., Cortez, D., Olejniczak, E.T., Waterson, A.G., Rossanese, O.W., Chazin, W.J., and Fesik, S.W. (2014) Discovery of a potent stapled helix peptide that binds to the 70N domain of replication protein A. *Journal of Medicinal Chemistry*, **57**, 2455–2461.

82 Walensky, L.D. and Bird, G.H. (2014) Hydrocarbon-stapled peptides: principles, practice, and progress. *Journal of Medicinal Chemistry*, **57**, 6275–6288.

83 Adams, J.M. and Cory, S. (1998) The Bcl-2 protein family: arbiters of cell survival. *Science*, **281**, 1322–1326.

84 Perciavalle, R.M. and Opferman, J.T. (2013) Delving deeper: MCL-1's contributions to normal and cancer biology. *Trends in Cell Biology*, **23**, 22–29.

85 Beroukhim, R., Mermel, C.H., Porter, D., Wei, G., Raychaudhuri, S., Donovan, J., Barretina, J., Boehm, J.S., Dobson, J., Urashima, M., Mc Henry, K.T., Pinchback, R.M., Ligon, A.H., Cho, Y.J., Haery, L., Greulich, H., Reich, M., Winckler, W., Lawrence, M.S., Weir, B.A., Tanaka, K.E., Chiang, D.Y., Bass, A.J., Loo, A., Hoffman, C., Prensner, J., Liefeld, T., Gao, Q., Yecies, D., Signoretti, S., Maher, E., Kaye, F.J., Sasaki, H., Tepper, J.E., Fletcher, J.A., Tabernero, J., Baselga, J., Tsao, M.S., Demichelis, F., Rubin, M.A., Janne, P.A., Daly, M.J., Nucera, C., Levine, R.L., Ebert, B.L., Gabriel, S., Rustgi, A.K., Antonescu, C.R., Ladanyi, M., Letai, A., Garraway, L.A., Loda, M., Beer, D.G., True, L.D., Okamoto, A., Pomeroy, S.L., Singer, S., Golub, T.R., Lander, E.S., Getz, G., Sellers, W.R., and Meyerson, M. (2010) The landscape of somatic copy-number alteration across human cancers. *Nature*, **463**, 899–905.

86 Andersen, M.H., Becker, J.C., and Straten, P.T. (2005) The antiapoptotic member of the Bcl-2 family Mcl-1 is a CTL target in cancer patients. *Leukemia*, **19**, 484–485.

87 Konopleva, M., Contractor, R., Tsao, T., Samudio, I., Ruvolo, P.P., Kitada, S., Deng, X., Zhai, D., Shi, Y.X., Sneed, T., Verhaegen, M., Soengas, M., Ruvolo, V.R., McQueen, T., Schober, W.D., Watt, J.C., Jiffar, T., Ling, X., Marini, F.C., Harris, D., Dietrich, M., Estrov, Z., McCubrey, J., May, W.S., Reed, J.C., and Andreeff, M. (2006) Mechanisms of apoptosis sensitivity and resistance to the BH3 mimetic ABT-737 in acute myeloid leukemia. *Cancer Cell*, **10**, 375–388.

88 Olberding, K.E., Wang, X., Zhu, Y., Pan, J., Rai, S.N., and Li, C. (2010) Actinomycin D synergistically enhances the efficacy of the BH3 mimetic ABT-737 by downregulating Mcl-1 expression. *Cancer Biology and Therapy*, **10**, 918–929.

89 Wertz, I.E., Kusam, S., Lam, C., Okamoto, T., Sandoval, W., Anderson, D.J., Helgason, E., Ernst, J.A., Eby, M., Liu, J.F., Belmont, L.D., Kaminker, J.S., O'Rourke, K.M., Pujara, K., Kohli, P.B., Johnson, A.R., Chiu, M.L., Lill, J.R., Jackson, P.K., Fairbrother, W.J., Seshagiri, S., Ludlam, M.J.C., Leong, K.G., Dueber, E.C., Maecker, H., Huang, D.C.S., and Dixit, V.M. (2011) Sensitivity to antitubulin chemotherapeutics is regulated by MCL1 and FBW7. *Nature*, **471**, 110–114.

90 Akgul, C. (2009) Mcl-1 is a potential therapeutic target in multiple types of cancer. *Cellular and Molecular Life Sciences*, **66**, 1326–1336.

91 Belmar, J. and Fesik, S.W. (2015) Small molecule Mcl-1 inhibitors for the treatment of cancer. *Pharmacology and Therapeutics*, **145**, 76–84.

92 Barbacid, M. (1987) Ras genes. *Annual Review of Biochemistry*, **56**, 779–827.

93 Katz, M.E. and McCormick, F. (1997) Signal transduction from multiple Ras effectors. *Current Opinion in Genetics and Development*, **7**, 75–79.

94 Reuther, G.W. and Der, C.J. (2000) The Ras branch of small GTPases: Ras family

members don't fall far from the tree. *Current Opinion in Cell Biology*, **12**, 157–165.

95 Takashima, A. and Faller, D.V. (2013) Targeting the RAS oncogene. *Expert Opinion on Therapeutic Targets*, **17**, 507–531.

96 Margarit, S.M., Sondermann, H., Hall, B.E., Nagar, B., Hoelz, A., Pirruccello, M., Bar-Sagi, D., and Kuriyan, J. (2003) Structural evidence for feedback activation by Ras. GTP of the Ras-specific nucleotide exchange factor SOS. *Cell*, **112**, 685–695.

97 Burns, M.C., Sun, Q., Daniels, R.N., Camper, D., Kennedy, J.P., Phan, J., Olejniczak, E.T., Lee, T., Waterson, A.G., Rossanese, O.W., and Fesik, S.W. (2014) Approach for targeting Ras with small molecules that activate SOS-mediated nucleotide exchange. *Proceedings of the National Academy of Sciences of the United States of America*, **111**, 3401–3406.

98 Bos, J.L., Rehmann, H., and Wittinghofer, A. (2007) GEFs and GAPs: critical elements in the control of small G proteins. *Cell*, **129**, 865–877.

99 Hanahan, D. and Weinberg, R.A. (2000) The hallmarks of cancer. *Cell*, **100**, 57–70.

100 Hanahan, D. and Weinberg, R.A. (2011) Hallmarks of cancer: the next generation. *Cell*, **144**, 646–674.

101 Podsypanina, K., Politi, K., Beverly, L.J., and Varmus, H.E. (2008) Oncogene cooperation in tumor maintenance and tumor recurrence in mouse mammary tumors induced by Myc and mutant Kras. *Proceedings of the National Academy of Sciences of the United States of America*, **105**, 5242–5247.

102 Baines, A.T., Xu, D.P., and Der, C.J. (2011) Inhibition of Ras for cancer treatment: the search continues. *Future Medicinal Chemistry*, **3**, 1787–1808.

103 Vasan, N., Boyer, J.L., and Herbst, R.S. (2014) A RAS renaissance: emerging targeted therapies for KRAS-mutated non-small cell lung cancer. *Clinical Cancer Research*, **20**, 3921–3930.

104 Sun, Q., Burke, J.P., Phan, J., Burns, M.C., Olejniczak, E.T., Waterson, A.G., Lee, T., Rossanese, O.W., and Fesik, S.W. (2012) Discovery of small molecules that bind to K-Ras and inhibit SOS-mediated activation. *Angewandte Chemie-International Edition in English*, **51**, 6140–6143.

105 Maurer, T., Garrenton, L.S., Oh, A., Pitts, K., Anderson, D.J., Skelton, N.J., Fauber, B.P., Pan, B., Malek, S., Stokoe, D., Ludlam, M.J., Bowman, K.K., Wu, J., Giannetti, A.M., Starovasnik, M.A., Mellman, I., Jackson, P.K., Rudolph, J., Wang, W., and Fang, G. (2012) Small-molecule ligands bind to a distinct pocket in Ras and inhibit SOS-mediated nucleotide exchange activity. *Proceedings of the National Academy of Sciences of the United States of America*, **109**, 5299–5304.

106 Sun, Q., Phan, J., Friberg, A.R., Camper, D.V., Olejniczak, E.T., and Fesik, S.W. (2014) A method for the second-site screening of K-Ras in the presence of a covalently attached first-site ligand. *Journal of Biomolecular NMR*, **60**, 11–14.

# 17
# Fragment-Based Discovery of Inhibitors of Lactate Dehydrogenase A

*Alexander L. Breeze, Richard A. Ward, and Jon Winter*

## 17.1
## Aerobic Glycolysis, Lactate Metabolism, and Cancer

It has been recognized since the 1950s that a fundamental shift in cellular metabolism toward highly elevated rates of glycolysis, even under aerobic conditions, is a particular hallmark of oncogenic transformation. The identification of this effect is attributed to Otto Warburg, who originally hypothesized that elevated aerobic glycolysis was the primary cause of cancer; however, the "Warburg effect" [1,2] is now understood in terms of an adaptive feature of oncogenic transformation rather than its origin. Cancer cells show rates of glycolysis as much as two orders of magnitude higher than those of normal, untransformed cells, but characteristically have impaired rates of mitochondrial oxidative phosphorylation via the tricarboxylic acid (TCA) or Krebs cycle. Although many advanced malignant tumors are poorly vascularized and exhibit correspondingly low oxygen tensions that are exacerbated by high cellular proliferation rates, this metabolic shift occurs in cancer cells even under conditions of relatively free oxygen availability, such as in peripheral or well-vascularized areas of the tumor, and indeed appears to be an early hallmark of malignant transformation [3–5]. The imbalance between accelerated glycolytic flux and impaired TCA cycle activity results in accumulation of pyruvate, which is then metabolized to lactate via the enzyme lactate dehydrogenase (LDH) [6,7].

The nature of the survival advantage conferred on cancer cells by metabolic transformation is still a matter of debate, but it seems likely that several factors conspire to favor the shift to aerobic glycolysis. First, low oxygen tensions in some tumor settings intrinsically favor metabolic adaptation toward lactate generation in order to maintain a positive energy balance. Second, the anabolic demands of fast-growing tumors necessitate plentiful availability of biosynthetic precursors to meet the demands of rapid cell proliferation and metabolism; these are effectively supplied via elevated flux through glycolysis, even in the presence of plentiful oxygen. Third, transformed cells often exhibit impaired mitochondrial function as a consequence of the selection pressure exerted against the role played by mitochondria in apoptosis. The conversion of excess pyruvate to

*Fragment-based Drug Discovery: Lessons and Outlook*, First Edition. Edited by Daniel A. Erlanson and Wolfgang Jahnke.

lactate by LDH (in particular, by the LDHA isoform; *vide infra*) has the additional advantage for the cancer cell of regenerating $NAD^+$ for availability upstream in glycolysis. The role of LDHA in the metabolic remodeling of cancer cells, and its strong link with overall prognosis, has prompted researchers to explore the potential for selective inhibition of LDHA activity to starve growing tumors of important sources of both energy and cellular building blocks. Much of the exciting recent progress made toward this goal has arisen from the application of fragment-based drug discovery (FBDD) approaches within a number of research groups in both academia and industry. In this chapter, we will describe in some detail the strategy adopted in AstraZeneca for our LDHA campaign, discuss some of the challenges we encountered, and explain the underlying rationale for our approach. We will also describe the trajectories charted by other groups to develop LDHA inhibitors using fragment-based methods, highlighting the differences in approach and interpretation since these are often the most illuminating learning points for future campaigns; we briefly touch on recently reported inhibitors derived from other strategies including high-throughput biochemical screening, and will reflect on the challenges and opportunities that lie ahead for LDHA inhibitors as they advance toward clinical testing.

## 17.2
## Lactate Dehydrogenase as a Cancer Target

L-lactate dehydrogenases are soluble, predominantly cytoplasmic NAD(H)-dependent oxidoreductases. They function as homo- or heterotetramers of catalytically independent subunits of the muscle (M) or heart (H) type, encoded by the *LDHA* or *LDHB* genes, respectively. Five distinct isoforms of LDH composed of different combinations of M and H subunits are distributed predominantly according to tissue type, with the 4M form (LDH-5, hereafter referred to as LDHA) being dominant in liver and striated muscle, while cardiac tissues are characterized by dominance of the 4H (LDH-1) isoform; alternative combinations give rise to the other isoforms that have specific distributions in normal tissues. The M and H subunits, while sharing around 75% sequence identity, preferentially catalyze either the pyruvate → lactate or lactate → pyruvate reactions, respectively (Figure 17.1a), in an ordered mechanism that involves initial binding of the cofactor, followed by that of the substrate, and dissociation of product and modified cofactor in reverse order [8].

Structurally, the human LDH monomer consists of 330 amino acid residues in a bilobal architecture of two domains, the larger of which adopts the canonical paired Rossman fold characteristic of NAD(H)-binding proteins [9]. Adjacent to the nicotinamide end of the NAD(H) pocket is the interface with the smaller "substrate binding" domain. A key feature of LDH that not only plays an important role both in the catalytic cycle and in substrate specificity but also has consequences for inhibitor design is the presence of an "active-site loop" (residues

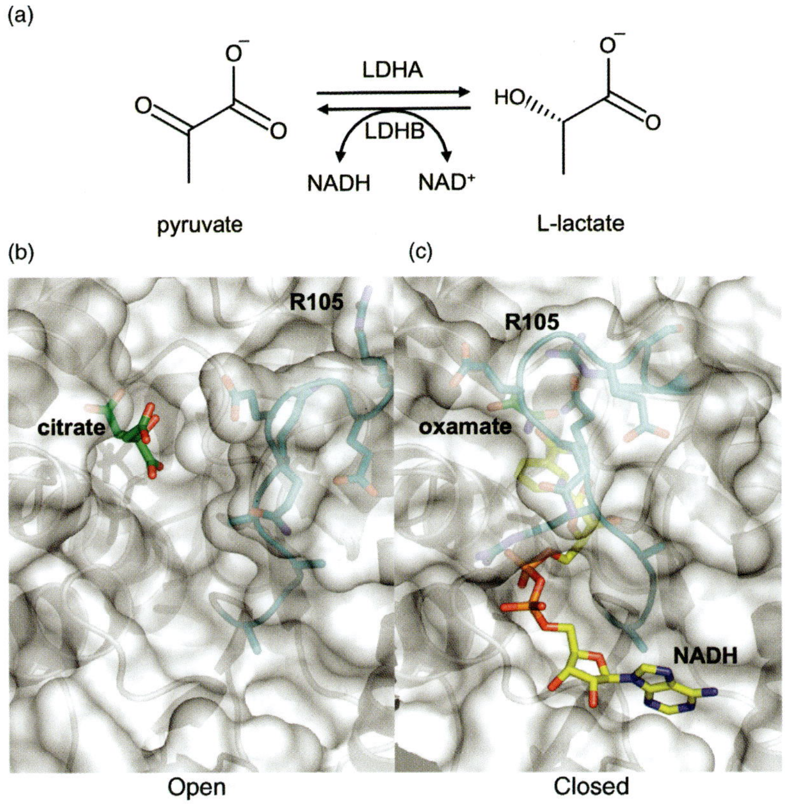


**Figure 17.1** (a) Interconversion of pyruvate and lactate catalyzed by lactate dehydrogenase isoforms. (b and c) X-ray crystal structures of human LDHA showing the active-site loop in "open" (b) and "closed" (c, from 1I10 [9]) conformations, highlighting the large excursion of the catalytic Arg105.

99–110) that alternately adopts "open" and "closed" conformations (Figure 17.1b and c). While the "open" conformation facilitates diffusion of the cofactor and substrate/product into and out of the active site, the "closed" conformation sequesters the loaded binding site from solvent water and furnishes an arginine side chain that is important for transition-state stabilization during the hydride transfer reaction.

The LDH-5 (LDHA) isoform that predominates in normal striated muscle and liver tissues displays the greatest catalytic efficiency in conversion of pyruvate to lactate [10]. It is also the isoform most strongly implicated in malignancy. LDHA gene expression is elevated in many cancers, and particularly so in invasive and hypoxic tumors [11], irrespective of the tissue of origin. Hypoxia-related transcription factors such as CREB and HIF-1, as well as signals from a number of growth factors that result in elevated levels of c-Myc, all function to promote LDHA gene expression via binding to response elements in the promoter region [12,13]. LDHA activity is subject to regulation via phosphorylation at a number of tyrosine residues including Tyr10, Tyr83, and Tyr238 that either enhance the catalytic activity of the enzyme or serve to target LDHA for translocation to the nucleus, where it may be implicated in transcription of a number of genes [14]. High LDHA expression levels are associated with poor prognosis, resistance to chemo- and radiotherapy, and an invasive phenotype [15–18]. Indeed, elevated LDHA activity and consequent export of lactate from cancer cells into the tumor environment may actively promote invasion and metastasis by triggering localized inflammatory responses [19] and through suppression of immune surveillance [20–24]. Consequently, LDHA has become a prognostic biomarker, with established value in the prediction of overall and progression-free survival across a number of malignancies [3].

Mounting evidence indicates that knockdown or silencing of *LDHA* inhibits cell growth and metastasis and can resensitize cancer cells to chemotherapeutic agents against which resistance had developed [25,26], a conclusion reinforced by recent inducible knockout studies in mice that have shown efficacy in non-small cell lung cancer models [27]. If the evidence that LDHA is a bona fide oncogene is increasingly compelling, what of the potential adverse consequences of its inhibition for normal physiology and the likelihood of unacceptable mechanism-based toxicity? A number of germ-line autosomal recessive mutations and deletions in the *LDHA* gene that reduce or abolish LDHA enzymatic activity are responsible for lactate dehydrogenase deficiency type A, also known as glycogen storage disease XI [28]. This condition, while linked to exertional myopathy, causes relatively few pathological effects under normal conditions. Taken together, these findings have served to highlight the attractiveness of LDHA inhibition as a potential therapeutic strategy to treat a wide range of cancers.

## 17.3
## "Ligandability" Characteristics of the Cofactor and Substrate Binding Sites in LDHA

The NADH cofactor occupies an extended cleft spanning the paired Rossman folds and abutting the substrate pocket that is furnished by the C-terminal substrate binding domain of the LDHA polypeptide chain. NADH adopts an extended conformation, spanning some 19 Å. The substrate (pyruvate for LDHA and lactate for LDHB) binds in a polar and charged region above the plane of the nicotinamide ring, poised for hydride transfer. The binding site is exceptionally diverse in its physicochemical character; both the adenine and nicotinamide groups occupy relatively lipophilic pockets, yet while the adenine pocket is rather large and solvent exposed, the nicotinamide pocket is more constrained and (along with the substrate pocket) is sequestered from solvent in the fully bound state by the enfolding of the active-site loop over the distal end of the cofactor groove. The ribose sugars of the adenosine and nicotinamide mononucleotide moieties occupy somewhat more polar environments, while the highly charged and polar central diphosphate is stabilized via salt-bridge interactions with an arginine donated by the active-site loop and by hydrogen bonds to backbone amides at the N-terminal end of the αB helix.

The potential mechanistic implications of exploiting particular binding sites or modes of inhibition should be considered when evaluating a putative drug target. In the case of enzyme targets, the question of whether binding modes that are competitive with substrate/product or cofactors are likely to be more or less attractive than uncompetitive, mixed, or noncompetitive mechanisms needs to be informed by an understanding of the relevant substrate, product, or cofactor concentrations and affinities under physiological conditions. Although reliable estimates of metabolite concentrations in relevant cellular compartments are not always easily obtained, these parameters clearly influence the level of occupancy of the target binding site, hence the difficulty of displacement with achievable *in situ* concentrations of the lead compound, and can inform the choice of format for assays used for hit-finding screens. NADH has a relatively high affinity for both LDHA and LDHB isoforms, with a $K_D$ ranging between 0.6 μM reported by Hewitt *et al.* [10] and 5 μM as determined in our laboratories by SPR (AstraZeneca, unpublished results); the oxidized ($NAD^+$) form of the cofactor binds significantly more weakly, however ($K_D \sim 200$ μM for human LDHA) [10]. Estimation of the free cytosolic concentrations of $NAD^+$ and NADH is challenging and the values much debated; although total cellular NAD/H concentration lies in the low millimolar range, the majority of this pool (and by far the majority of NADH) is believed to be protein bound [29]. The most credible estimates of combined free NAD/H concentration in the cytosol put the value somewhere around 100 μM, while the ratio of free $NAD^+$:NADH can be estimated from metabolite indicator measurements (e.g., via lactate:pyruvate ratios) to be >100 under normoxic conditions [30]. Taken together, these estimates suggest that both oxidized and reduced cofactor are probably present at free concentrations close to their respective $K_D$s, such that binding site occupancy under normal physiological conditions is likely to reside in the mid-range (between 10 and 90%). Conceptually, therefore, a competitive inhibition mode (with respect to NAD/H) should not present an insurmountable affinity barrier in the cellular context relative to $K_D$ values measured against cofactor-free enzyme *in vitro*. In light of this, we reasoned that it was legitimate to employ both cofactor-free LDHA for biophysical screening of compound libraries and activity-based hit-finding assays in which NADH is de facto present.

## 17.4 Previously Reported LDH Inhibitors

A number of inhibitors of LDH enzymes have been reported in the literature. They include substrate analogs, natural product compounds such as the polyphenol gossypol, and synthetic compounds derived from various virtual or experimental screens. None to date has exhibited either the potency or

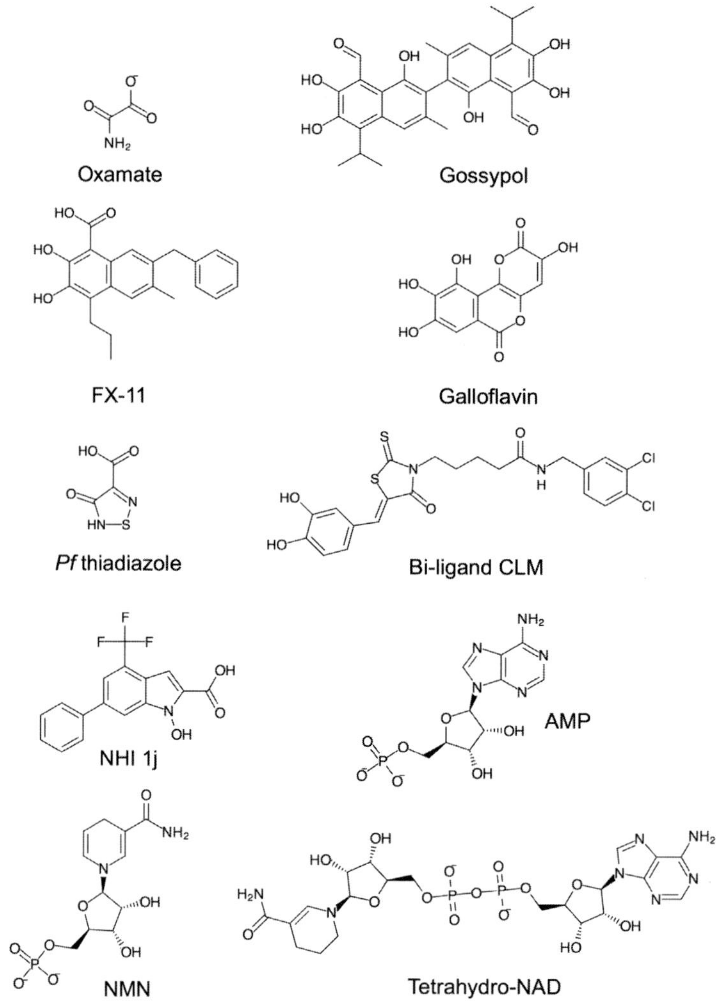


**Figure 17.2** Structures of published LDH inhibitors (refer to text for details and references).

physicochemical properties necessary for the ultimate clinical success as an LDH inhibitor. Structures of these compounds are illustrated in Figure 17.2.

Oxamic acid has long been known as a substrate analog inhibitor of LDH activity. It binds with a $K_D$ of ~20–30 μM to human LDHA [31] and is a competitive inhibitor of the enzyme with respect to substrate. Oxamate displays modest activity in cellular lactate production assays [32,33], likely limited by poor cell membrane permeability. Gossypol (AT-101) is a phenolic aldehyde compound isolated from seeds of the cotton plant and has been shown to inhibit a number of dehydrogenases by binding competitively with cofactor [34,35]; it is a non-competitive inhibitor of the spermatozoan lactate dehydrogenase isoform, LDH-X, with respect to lactate or pyruvate as substrates [35], but displays general cytotoxicity, limiting its potential as a therapeutic. The gossypol analog FX-11 was recently reported to inhibit progression of human lymphoma and pancreatic cancer xenografts in mice, via a mechanism that included suppression of cellular lactate production, increased oxygen consumption, production of reactive oxygen species (ROS), and triggering of apoptosis [36]. Galloflavin (GF), a gallic acid-derived compound, was disclosed as an LDHA inhibitor with a $K_i$ of around 5 μM against LDHA (and similar potency against LDHB) [37], and as an inhibitor of glycolysis and proliferation in breast cancer cell lines with a mean $IC_{50}$ of 125 μM [38]. GF was identified from a structure-based virtual screen of the NCI Open collection; interestingly, in a follow-up paper, the authors reported that GF inhibits the binding of single-stranded DNA (ssDNA) to LDHA, an interaction believed to be mutually exclusive with cofactor binding, suggesting that inhibition of the transcriptional activation role of LDHA could be an additional feature of the mode of action of NADH-competitive LDHA inhibitors [39].

Cameron *et al.* described a series of small, substrate-mimetic oxadiazole and thiadiazole acid inhibitors of *Plasmodium falciparum* LDH [40] that are selective over human LDHA and display a kinetic mechanism that is uncompetitive with respect to cofactor. Their potencies against *P. falciparum* LDH lie in the low-to-sub-micromolar range and they were reported to be active in mouse malarial infection models. A library of biligand compounds comprising a "common ligand mimic" (CLM) linked to a "specificity ligand" (SL) was elaborated by Sem *et al.* [41] based on mimicking the binding mode subclasses of the NAD/P/H cofactors of oxidoreductases; the authors reported 42 nM inhibition of LDH for one member of the library. The *N*-hydroxyindoles were disclosed by Granchi *et al.* as a novel series of LDHA inhibitors developed through structure-based design approaches [42]. The most potent of these compounds were shown to be competitive inhibitors of LDHA with respect to both NADH and pyruvate, achieving a $K_i$ of around 5 μM; *in silico* docking followed by molecular dynamics refinement suggested binding in the substrate pocket; however, our experimental evaluations using NMR and X-ray crystallography are more consistent with binding to the adenine pocket [43]. Cell-based studies of these compounds indicated inhibition of cell proliferation and suppression of lactate production across a range of cancer cell lines, albeit at high concentrations of inhibitor [42].

The constituent fragments of NADH, adenosine monophosphate (AMP) and nicotinamide mononucleotide (NMN), bind only weakly to LDHs and show commensurately little ability to inhibit enzymatic activity. On the other hand, an analog of NADH, tetrahydro-NAD, is a specific and reproducible inhibitor of LDH activity [44]. All three NADH-derived compounds proved very useful to us in establishing our fragment-based lead discovery platform.

None of these published inhibitors possess either physicochemical properties or *in vitro* cellular potencies typical of most licensed oral drug candidates. We profiled a number of compounds reported as LDHA inhibitors using our biophysical platform of 1D and 2D NMR, X-ray crystallography, and surface plasmon resonance (SPR) (see Section 17.5.3) prior to and during the course of our

LDHA lead generation program. Initially, the aim was to employ these as tool compounds with which to establish and validate our biophysical and activity-based assays, but as we became more confident in our assay platform we were, in turn, able to use our assays to evaluate and characterize reported inhibitors. While we could corroborate the binding of several of these compounds, there were also a number of reported LDHA inhibitors, including FX-11 [36] and the CLM-based biligand compounds [41] for which, in our hands, no specific binding or activity was detectable in any of our biophysical or activity-based assays, and/or we observed protein aggregation or superstoichiometric binding [43], suggestive of nonspecific compound aggregation as the likely origin of the LDHA inhibition observed in the original reports. Similar anomalous behavior for these same compounds was also noted by the Ariad group [45]. In retrospect, our observation of irreproducibility among reported inhibitors further underscores the challenging nature of LDHA as a drug target.

## 17.5
## Fragment-Based Approach to LDHA Inhibition at AstraZeneca

### 17.5.1
### High-Throughput Screening Against LDHA

In formulating a hit-finding strategy for LDHA, we initially embarked on a high-throughput screening (HTS)-based hit-finding approach, employing an LDHA enzymatic activity assay in the presence of 30 μM NADH with a resazurin/diaphorase readout couple to screen the full AstraZeneca compound deck at 10 μM concentration. In parallel with the development of our HTS assay, we established NMR-based methods for confirmation and characterization of the binding of compounds to LDHA, with the aim of using NMR for evaluation of compounds previously described in the literature, as well as hits from our HTS. Despite its large molecular mass (4 × ~35 kDa, that is, ca. 140 kDa), LDHA proved to yield $^1H-^{15}N$ TROSY-HSQC NMR spectra of relatively good quality when uniformly labeled with $^{15}N$ coupled with a high level of deuterium incorporation at nonlabile positions, and we were able to demonstrate the ability to monitor binding of both adenosine monophosphate (AMP) and nicotinamide mononucleotide (NMN) within the cofactor binding site. Furthermore, although we did not assign the backbone amide NMR resonances of LDHA, we could nevertheless identify subsets of amide peaks that underwent characteristic (albeit overlapping) patterns of chemical shift perturbation (CSPs) in response to the binding of known ligands to either the adenine or the nicotinamide subpockets of the cofactor binding site. This afforded us a means to provisionally classify arbitrary compounds according to binding site. We also established 1D "ligand-observed" $^1H$ NMR approaches using unlabeled LDHA, and again were able to establish proof of concept for the ability of this approach to detect compound binding. We employed NMR as part of a suite of structural and biophysical techniques to profile a number of previously described LDHA inhibitors from the literature, and were able to confirm specific binding for some but, notably, not all compounds tested ([43]; see above).

Despite identifying actives from our HTS screen, the hit rate was low (1% at a 30% inhibition cutoff, with 10 μM compound concentration) and we were unable to corroborate binding of any of the actives to human LDHA using 1D ligand-observed $^1$H NMR spectroscopy. We therefore concluded that the majority, if not all, the HTS actives were false positives, and subsequently established that silver contamination of many of the active compounds accounted for the inhibition of LDHA that we observed in the assay: LDHA is known to be highly sensitive to inhibition of activity by silver, via a mechanism presumed to involve covalent modification of the active site histidine [46], and compounds resynthesized via a silver-free route proved inactive when retested in our enzyme assay.

### 17.5.2
### Rationale and Strategy for Exploration of Fragment-Based Approaches

Having failed to identify attractive start points from our HTS, we reasoned that the chemical space encompassed by our corporate compound deck might be relatively sparse in areas suited to the nature of the binding pocket(s) available on the LDHA protein and that this might underlie the paucity of verifiable hit matter arising from the HTS. Computational studies have established that the number of potentially synthetically realizable compounds with "drug-like" characteristics increases approximately exponentially with the square of the number of heavy atoms [47]. Studies simulating the effect of molecular complexity on the likelihood of identifying a hit in a given binding site, based on the concept of matches or mismatches between molecular recognition features, have shown that reducing molecular "feature count" enhances the probability of binding (although the probability of detecting that binding is also a function of the assay sensitivity) [48]. Put another way, the molecular complexity of compounds in a screening library is negatively (and steeply) correlated with the efficiency with which chemical space can be sampled with a library comprising a given number of compounds. Hence, even typical large pharma compound decks exceeding millions of compounds of "drug-like" or "lead-like" [49] size (typically 350–450 Da) sample chemical space relatively sparsely, and may lack compounds capable of forming productive binding interactions with a target binding site at affinity levels detectable in a typical biochemical HTS assay. In the past decade and a half, FBDD (screening of much lower molecular mass, typically <200 Da, less complex molecules or "fragments," comprising minimally elaborated core molecular recognition elements) has become increasingly recognized as a means of improving the sampling of chemical space and enhancing lead generation success rates, even while screening numerically much smaller libraries (hundreds to low thousands of compounds) [50–54]. Having established NMR spectroscopy as a means to reliably detect, confirm, and characterize binding of compounds to LDHA, we initiated a screen of our generic fragment library using

NMR. At this point in our program, the primary objective was to establish the "ligandability" [55] of LDHA (in light of the failure of our HTS campaign) to facilitate portfolio decision making.

### 17.5.3
### Development of Our Biophysical and Structural Biology Platform

Our screening strategy combined 1D ligand-observed NMR (both waterLOGSY and $T_1\rho$ relaxation-filtered experiments) as the primary screening tool with follow-up confirmation by 2D $^1H-^{15}N$ TROSY-HSQC protein-observed NMR (employing uniformly $^{15}N$- and $^2H$-labeled LDHA). The former provided us with adequate throughput for screening our circa 1000-member diverse, generic fragment library, commensurate with modest protein consumption by means of "cocktailing" the compounds into mixtures of six. We chose to screen LDHA in a simple phosphate-based buffer system in the absence of bound NAD/H cofactor or substrate analogs. Our reasoning was that access to the full volume of the combined cofactor–substrate binding site would afford us the maximum opportunity to identify a diverse range of potential chemical start points that, with further elaboration, might provide the necessary affinity for a therapeutically effective inhibitor. In contrast to a biochemical activity-based assay, however, this approach is unlikely to find compounds that bind preferentially to the enzyme–cofactor or enzyme–cofactor–substrate complex; indeed, we employed competitive inhibition of the binding signal in the presence of cofactor analogs as a criterion for confirming specific binding of our fragment hits (*vide infra*). From our 1D NMR screen, we identified 44 hits (using the criterion of NADH competition as confirmation in this assay), representing a hit rate of 4.6%. Upon triage through 2D protein-observed NMR, we could quantify binding for only 13 of the original 44 fragments, as some of the 1D NMR hits gave CSPs too small to provide affinity estimates (implying very weak binding and/or limiting compound solubility). Thus, we obtained a confirmed hit rate of approximately 1.3%. Measurement of affinities from 2D TROSY-HSQC NMR titrations gave values between 300 and 4200 μM, with all but three hits having $K_D$ values >1000 μM. Figure 17.3 shows an example of the NMR binding data obtained for one of our hit fragments, **AZ9**. Ligand efficiencies (LE) [56], in this chapter given as $\Delta G$ (kcal mol$^{-1}$) per heavy atom, were also modest, averaging around 0.27, with a range between 0.23 and 0.32. According to the criteria we established in AstraZeneca through analysis of the hit-to-lead success rates of >35 fragment screening campaigns against targets for which we had also run HTS, we classified LDHA on this basis as having a "low" ligandability score [55]. Through analysis of CSP patterns in 2D TROSY-HSQC NMR experiments relative to those of AMP and NMN, we could also deduce that our hit fragments bound exclusively to the adenine subpocket. Therefore, despite the low confirmed hit rate, the nature of the hits that we obtained from NMR screening gave us enough encouragement that, contrary to what might have been inferred from the HTS outcome, LDHA was a potentially druggable, if challenging, target.

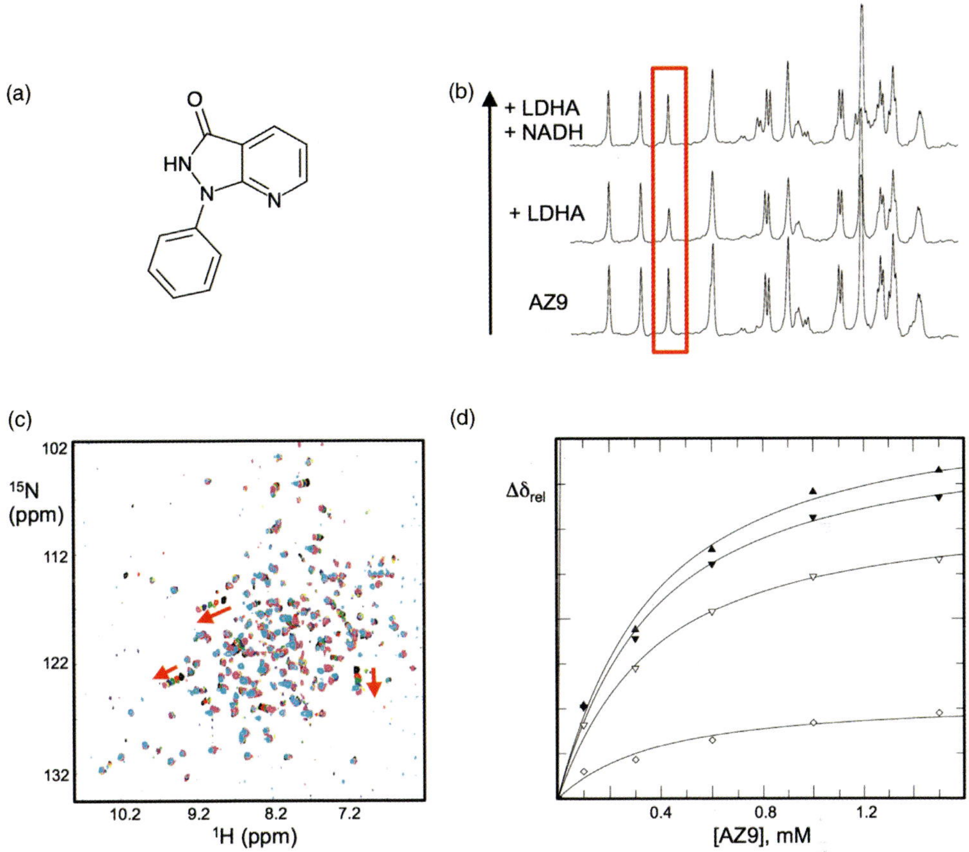


**Figure 17.3** (a) Structure of **AZ9**. (b) 1D $^1$H $T_1\rho$-filtered NMR binding data for fragment cocktail (6 compounds) including **AZ9**: bottom – 500 μM **AZ9**; middle – 500 μM **AZ9** + 15 μM human LDHA; and top – 500 μM **AZ9** + 15 μM human LDHA + 1 mM NADH. (c) $^1$H-$^{15}$N TROSY-HSQC spectra of human LDHA in the absence (black) and presence of 0.1 mM (red), 0.3 mM (green), 0.6 mM (blue), 1.0 mM (yellow), and 1.5 mM (magenta) **AZ9**, and 1.0 mM AMP (cyan). (d) Non-linear regression fits to a tight-binding model for measured CSPs of selected amide resonances from the NMR data shown in (c), yielding a calculated $K_D$ of 300 μM.

To provide an accurate means of $K_D$ determination, we developed an SPR direct binding assay [57,58] based on immobilization of recombinant LDHA to a sensor chip surface. Because of high nonspecific binding to the LDHA-bound chip surface, we found it necessary to employ a reference surface comprising acid-denatured LDHA as a control. Using our SPR assay, we were able to measure $K_D$ values for a number of our initial NMR fragment screen hits. While we found good agreement for some, there were also discrepancies; this is a commonly observed phenomenon in fragment-based lead discovery because of inherently different detection sensitivities of the techniques used to the weak binding phenomena being investigated, and the different susceptibilities to anomalous behavior that can lead to false positives or false negatives. It is our experience that fragment screens against targets of marginal ligandability, such as LDHA, tend to suffer higher rates of discrepancy between techniques than is the case for more highly ligandable targets; this is perhaps understandable given the lower affinities and ligand efficiencies of the fragment hits, and underscores the value of multiple orthogonal assays for fragment triage in providing the necessary degree of confidence to progress to the next step. In the case of LDHA, we found that a number of hits for which we had been able to measure $K_D$s by NMR (including the most ligand-efficient fragment by this measure) were too weak to be measurable in SPR. The approach that we adopted was thus to prioritize fragments that gave the most consistent behavior across the assays. Figure 17.4 shows a selection of our initial hits based on this prioritization mechanism. None of these initial fragments displayed any detectable activity in our

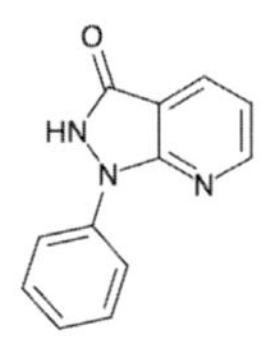


**AZ9**
$K_D$ (NMR) = 300 μM
LE (NMR) = 0.30
$K_D$ (SPR) = 1600 μM
LE (SPR) = 0.22

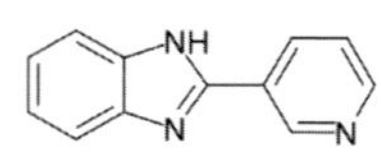


**AZ10**
$K_D$ (NMR) = 1400 μM
LE (NMR) = 0.26
$K_D$ (SPR) = weak
LE (SPR) = N/A

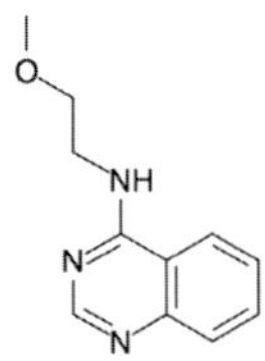


**AZ11**
$K_D$ (NMR) = 2400 μM
LE (NMR) = 0.23
$K_D$ (SPR) = 3000 μM
LE (SPR) = 0.23

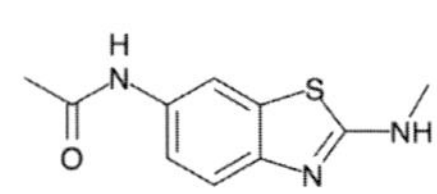


**AZ12**
$K_D$ (NMR) = 1000 μM
LE (NMR) = 0.27
$K_D$ (SPR) = 770 μM
LE (SPR) = 0.24

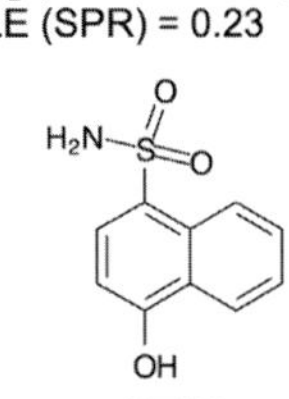


**AZ13**
$K_D$ (NMR) = 400 μM
LE (NMR) = 0.32
$K_D$ (SPR) = weak
LE (SPR) = N/A

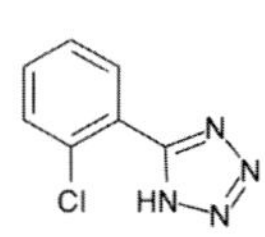


**AZ14**
$K_D$ (NMR) = 4200 μM
LE (NMR) = 0.27
$K_D$ (SPR) = weak
LE (SPR) = N/A

**Figure 17.4** Selection of prioritized hits from the initial AstraZeneca fragment screen.

enzymatic activity assay, as expected given the relatively high concentrations of NADH present.

In parallel with our establishment of NMR and SPR as robust biophysical assays for LDHA binding, we set about developing an X-ray crystallographic structure determination capability. Our experiences represent an object lesson in some of the challenges that can be encountered in developing a system that is suitable for driving structure-based elaboration of series derived from multiple fragment starting points, in that a crystal system that appears eminently well suited to providing structural data on one fragment series or binding subsite might be completely unsuited to a different series (or even a close analog), for sometimes nonobvious reasons. Our primary objective was to develop a soakable human LDHA-based crystal system, based on existing literature precedent (PDB code 1i10 [9]), but we were unable to adapt this precedent to the stringent demands imposed by the need to soak in fragments at high concentration: although we obtained apo crystals, they were fragile, diffracted to only moderate resolution in-house, and the catalytic loop was held in an open conformation by crystal contacts. Examination of LDHAs from other mammalian species revealed that the rat enzyme has a global sequence similarity to human LDHA of 94% (using the BestFit algorithm [59]) but is identical in the substrate and cofactor binding pockets; thanks to variation in surface residues that mediated crystal contacts, we were able to obtain robust, apo-form crystals of rat LDHA in a closed loop conformation into which we could readily soak compounds as large as NADH, and deemed this to be a suitable surrogate for driving structure-based optimization of hits. Using this crystal system, we established that all of the NMR hits for which we obtained structures bound to the adenine pocket of the cofactor binding site; this we had also corroborated based on their shift patterns in 2D TROSY-HSQC spectra. Figure 17.5 highlights key features of the interactions made by exemplar compounds **AZ12** and **AZ14** in this site. The

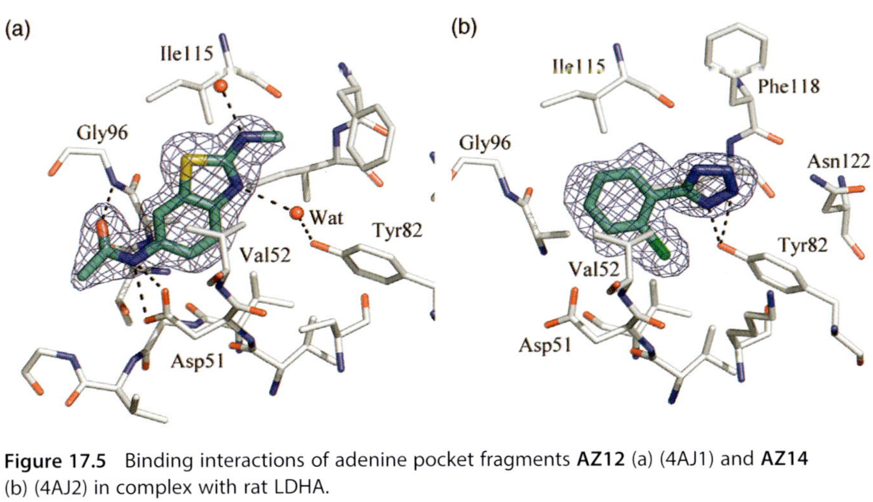


**Figure 17.5** Binding interactions of adenine pocket fragments **AZ12** (a) (4AJ1) and **AZ14** (b) (4AJ2) in complex with rat LDHA.

benzothiazole of **AZ12** binds in a predominantly lipophilic pocket, analogously to the adenine ring of the cofactor, and makes a water-mediated hydrogen bond to Tyr82 hydroxyl, while the amide group recapitulates hydrogen bonds made to the Asp51 side chain and to Gly96 by the ribose moiety of NADH. Fragment **AZ14** binds with its tetrazole ring in direct hydrogen-bond contact with Tyr82 and extends into a subpocket vacated by the movement of Phe118. Interestingly, we had anticipated that acidic fragments such as **AZ14** would bind in the substrate pocket, but we observed several such compounds to interact preferentially with the adenine site when soaked into our rat LDHA crystals; this might have reflected an innate bias toward this site in our crystal system through the occupation of the substrate pocket by malonate, which was present at high concentration (1.3–1.5 M) in the crystallization buffer.

Our development of a robust SPR-based assay for LDHA binding afforded us a higher throughput platform for fragment screening and characterization than was available from our NMR-based approach. This became particularly important in subsequent stages of the campaign, where we were able to exploit the relatively higher throughput and lower protein consumption of SPR, both to screen small libraries of analogs of our initial adenine pocket fragments and for screening of larger targeted fragment libraries in an attempt to identify binders in the substrate and nicotinamide subpockets. However, we continued to employ NMR as an orthogonal technique for confirmation and characterization of certain hits from our SPR screens.

### 17.5.4
### Elaboration of Adenine Pocket Fragments

With several structurally and biophysically validated adenine-site fragments in hand, our next step was to search for analogs that not only bound with higher affinity (and preferably superior LE) but also had bond vectors amenable to “growing” along the ribose channel and toward the diphosphate region of the cofactor binding site. From this point of view, fragments **AZ9**–**AZ12** appeared favorable, because their crystal structures all gave hints as to how they might be elaborated. **AZ14**, with its acidic tetrazole, was disfavored at this stage, as we anticipated needing an acid functionality for effective interaction with the substrate pocket and wanted to avoid an acid in the adenine pocket as well. To explore structure–activity relationships (SAR) around the initial hits, we performed a near-neighbor search of our corporate compound deck, mined commercial databases, and identified a number of analogs of compounds **AZ9**–**AZ12** that we tested for binding by NMR. Based on these results, we prioritized further analogs of compound **AZ12** that probed the effects of core substitution and of elaboration along the cofactor groove vector. In total, around 350 adenine site analogs were screened in our SPR assay, and of these we identified compound **AZ17** as the analog with the best overall affinity ($K_D = 40\,\mu M$ by SPR) and acceptable LE (0.29) (Figure 17.6). The close analog **AZ18** had slightly lower affinity and LE, but unlike **AZ17** it yielded unambiguous electron density when

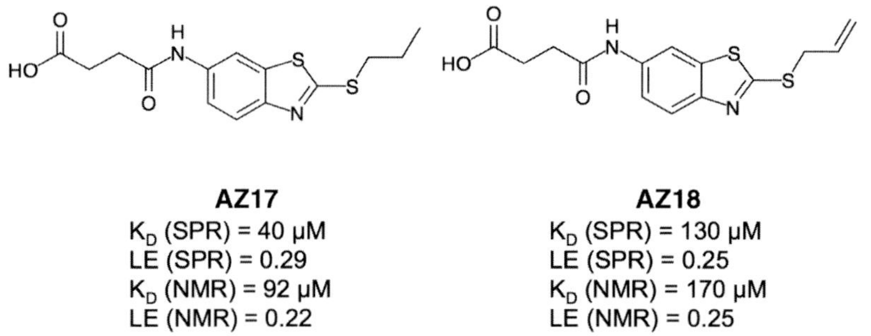


**Figure 17.6** Elaborated adenine pocket fragments **AZ17** and **AZ18**.

soaked into rat LDHA; this confirmed the conserved binding mode of **AZ12** and highlighted additional interactions with the phosphate binding groove that were picked up by the carboxylate interaction with the guanidinium of Arg98 and (via water) the backbone NH of Ala29. However, our best adenine pocket fragments had relatively poor lipophilic efficiency (LLE) [60], that of **AZ17** being 0.64 (calculated as $pK_D - cLogP$). Evidently, the rather lipophilic propyl thioether substituent on the benzothiazole provided a significant potency boost, but at the cost of poor solubility; still, we felt that there would be further scope for property modulation at a later point given the relatively solvent-exposed nature of the adenine subpocket. On the other hand, it was apparent by this stage of the campaign that, in order to achieve sufficient potency, our lead compounds would probably need to occupy the majority of the contiguous substrate–cofactor binding pocket and that a strategy based on linking our adenine pocket ligands to fragments binding at the distal end of the active site would be required.

### 17.5.5 Screening for Fragments Binding in the Substrate and Nicotinamide Pockets

Since we found very few fragments unequivocally binding in the nicotinamide or substrate subpockets in our initial generic fragment screen, we reasoned that a targeted fragment library built around mimicry of the nicotinamide and substrate would represent the most effective source of potential actives. We set about assembling a set derived from fragmentation of known NAD/H mimics [61–64] and combined this with substructural searches for carboxylic acids or amides coupled to five- or six-membered aromatic or aliphatic rings as putative substrate and nicotinamide mimics. In parallel, we added a series of fragments that contained a malonate substructure. NMR showed that malonate binds with a $K_D$ of 5 mM and a relatively high LE (0.45); as previously mentioned, our crystal system contained a bound malonate, and we were encouraged by the observation that (in contrast to oxamate) malonate's central methylene provides a potential vector from which to grow away from this pocket. Combined with supplementary sets derived from protein-based virtual screens

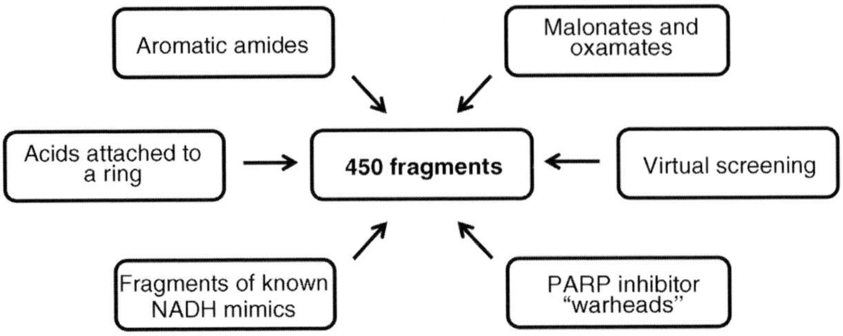


**Figure 17.7** AstraZeneca strategy for constructing fragment screening set to probe nicotinamide and substrate pockets.

against both apo and oxamate-bound LDHA structures, and after applying molecular property and visual filters in a pragmatic fashion, we obtained a screening set of 450 unique fragments with which to target the nicotinamide and substrate pockets. This strategy is summarized in Figure 17.7.

Before we were in a position to employ our rat LDHA crystal system to confirm binding mode for any hits we might obtain, we first had to confront the issue of the malonate occupying the site we were targeting. With a $K_D$ of 5 mM as measured by NMR, and a concentration of up to 1.5 M in the crystallization buffer, malonate in the substrate binding pocket would impose a greater than 300-fold $IC_{50}$ penalty (relative to $K_D$) on any fragment binding in the substrate pocket in competition with it – an impossible barrier to detection for what were expected in any case to be weakly binding compounds. In an effort to resolve this, we screened other precipitants and established that citrate could substitute for malonate while being capable of displacement by competing substrate pocket ligands.

Anticipating weaker interactions at this end of the target binding site, we elected to run our SPR screen at a relatively high fragment concentration of 2.5 mM and identified 40 initial hits. Despite a rather high nonspecific binding rate (as judged by the observation of binding to acid-denatured LDHA chip surfaces), and elimination of a significant number that did not provide reproducible $K_D$ data or that proved to bind in the adenine site, we were left with two confirmed malonate-derived hits (compounds **AZ20** and **AZ21**, Figure 17.8) for which we obtained X-ray crystal complex structures that unequivocally showed them to bridge both the substrate and the nicotinamide pockets. **AZ20** and **AZ21** exhibited good LE (0.33 and 0.27, respectively) and respectable affinities (214 and 280 μM), although that of **AZ20** was skewed by the lipophilic 4-bromo substituent, giving it a considerably worse LLE (2.1) than that of the 3,4-dimethoxy substituted **AZ21** (2.8). The carbon and oxygen linker atoms between the malonyl and phenyl portions of **AZ21** and **AZ20**, respectively, resulted in slight adjustments in binding mode, with the prochiral methylene of the malonyl group flipped between the two structures. The 4-methoxy group of **AZ21** appeared to provide a potentially convenient synthetic handle from which to project

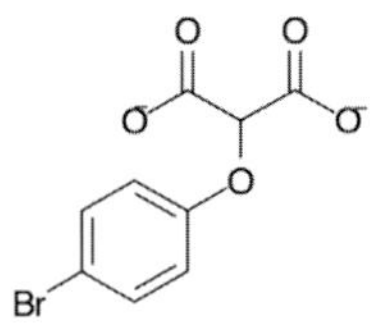

**AZ20**
$K_D$ (SPR) = 210 μM
LE (SPR) = 0.33
$K_D$ (NMR) = 470 μM
LE (NMR) = 0.30

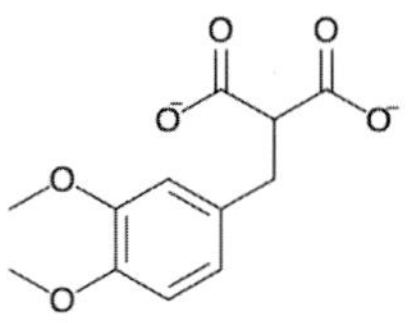

**AZ21**
$K_D$ (SPR) = 280 μM
LE (SPR) = 0.27
$K_D$ (NMR) = 1000 μM
LE (NMR) = 0.23

**Figure 17.8** Validated nicotinamide-substrate pocket fragment hits **AZ20** and **AZ21**.

additional functionality – or indeed, to link through to compounds binding in the adenosine-diphosphate pockets. Appreciating that the diacid was a liability for future achievement of cellular activity, we tested a few monoacid analogs of **AZ20**/**AZ21**, but these suffered precipitous losses in affinity to the extent that binding was barely detectable. Lacking significant numbers of additional malonate-containing analogs in our collection, we elected to turn our attention back to the adenine region, resolving to address the diacid question at a later stage with more elaborated and potent compounds.

### 17.5.6
### Reaching out Across the Void

With a distance of more than 7 Å separating **AZ12** and **AZ20** when overlayed in their respective LDHA crystal structures, it was felt that attempting to link via an extended methylene chain would be premature and unlikely to achieve much progress toward our goals. Therefore, we embarked on synthesizing a series of analog libraries based on the scaffolds of **AZ12, AZ17, AZ20**, and **AZ21** that aimed to extend these fragments across the intervening phosphate channel in order to reduce the gap to a more readily bridgeable distance. Because of the polar and solvent-exposed nature of the phosphate region, our chief aim was to extend our compounds while avoiding excessive loss of LE; we were not anticipating making substantial affinity gains at this stage. Encouragingly, among the compounds based on the **AZ12** scaffold tested in SPR that showed enhanced or preserved affinity (Figure 17.9), compound **AZ22** stood out as a potential breakthrough, in that its urea group not only makes productive interactions with the protein but also follows a trajectory that allows it to take a "shortcut" across the most polar part of the diphosphate region; instead, the urea carbonyl forms water-mediated interactions with this part of the binding groove, while its amide nitrogens form hydrogen bonds with Gly96. The affinity of **AZ22** was modest at 770 μM, but restoration of the *S*-propyl group (abandoned during the library synthesis because of poor compound solubility) boosted the $K_D$ value to an

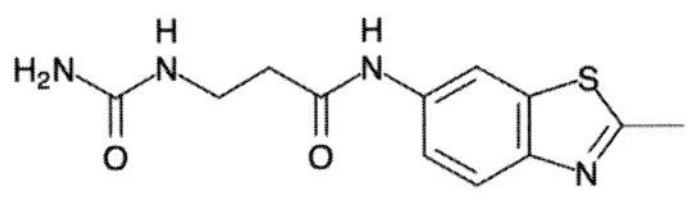


**AZ22**
$K_D$ (SPR) = 770 μM
LE (SPR) = 0.22
$K_D$ (NMR) = binds
LE (NMR) = N/A

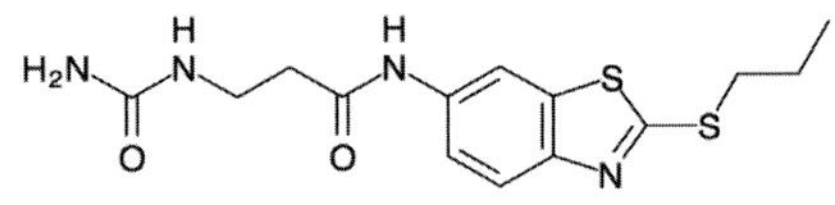


**AZ23**
$K_D$ (SPR) = 35 μM
LE (SPR) = 0.27
$K_D$ (NMR) = binds
LE (NMR) = N/A

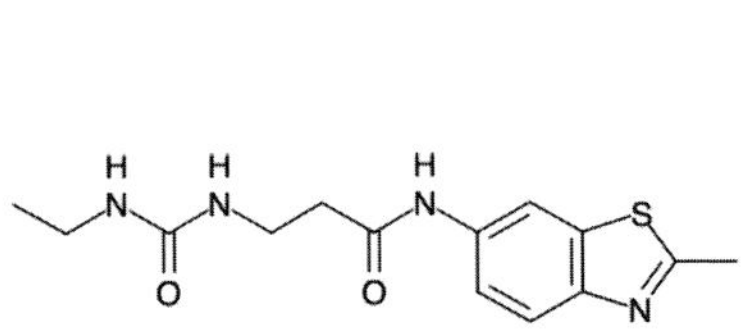


**AZ24**
$K_D$ (SPR) = 160 μM
LE (SPR) = 0.25
$K_D$ (NMR) = N/A
LE (NMR) = N/A

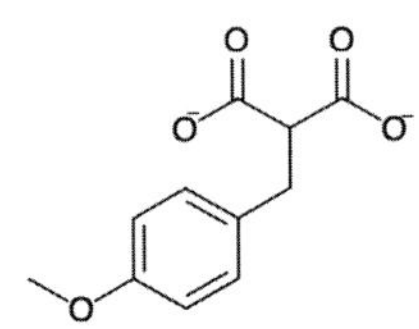


**AZ25**
$K_D$ (SPR) = 1100 μM
LE (SPR) = 0.25
$K_D$ (NMR) = 240 μM
LE (NMR) = 0.32

**Figure 17.9** Elaborated adenine pocket fragments **AZ22-AZ24**, together with prioritized nicotinamide-substrate pocket fragment **AZ25**.

encouraging 35 μM (compound **AZ23**), while **AZ24** demonstrated that impinging on the nicotinamide pocket from these adenine-site fragments via the *N*-ethyl substituent improved LE from 0.22 (compound **AZ12**) to 0.25.

The libraries based on the **AZ20** and **AZ21** scaffolds resulted in consistently superior affinities from carbon-linked malonate analogs, so these were prioritized. Compound **AZ25** (Figure 17.9) bound with a $K_D$ of 1100 μM by SPR and 240 μM by NMR (NMR LE = 0.32), and contained a single *p*-methoxy group that, by analogy with **AZ21**, we believed would form an excellent platform for linkage through to our adenine-site ureas. Superposition of the crystal structure of the complex of our urea-containing adenine-site fragment **AZ22** with that of the carbon-linked malonate **AZ21** (and subsequently a ternary complex generated by cosoaking, demonstrating the ability of both adenine and substrate pocket fragments to bind simultaneously) showed the distance between the terminal heavy atoms of the two ligands now to be just 3.1 Å, suggesting the possibility of bridging by as few as two intervening atoms.

### 17.5.7
### Fragment Linking and Optimization

Accordingly, we made our first linked compound **AZ26**, derived by merging the structures of **AZ24** and **AZ25**, based on computational modeling of the binding modes and comparative analysis of the probable linker geometry with the

Cambridge Structural Database (CSD). The synthetic routes involved in making linked compounds were complex and multistep with varying yields, involving formation of pendant amines on both partners, so we were keen to build confidence that linking was likely to be a productive strategy. Hence, we were very encouraged, on testing **AZ26** in our SPR assay, to find that the $K_D$ value had for the first time broken the μM barrier to achieve 130 nM, a gain of over 1000-fold on the starting fragments. Moreover, despite the well-known difficulties inherent in linking (e.g., through compromising the binding geometry of the starting fragments, conformational strain energy in the linker, or unfavorable binding site contacts in the linker region) [65–70], we had been able to maintain LE, at around 0.26. Even more gratifyingly, **AZ26** was the first compound (with the exception of tetrahydro-NAD) for which we could measure detectable activity in our enzyme assay ($IC_{50} = 4.2$ μM). Crystallography with rat LDHA confirmed that **AZ26** bound exactly as anticipated, with both adenine- and substrate-mimicking moieties closely recapitulating the binding modes of their progenitor fragments (Figure 17.10a). Again, we noted the "shortcut" taken by the urea across the diphosphate region compared to NADH, thus circumventing the need for a phosphate-mimic group (Figure 17.10b).

Our optimization of the linked inhibitors explored the effects of switching between central urea and central amide groups, as well as the effects of linker chain length and substitution of carbon for oxygen at the position equivalent to the

(a)

(b)

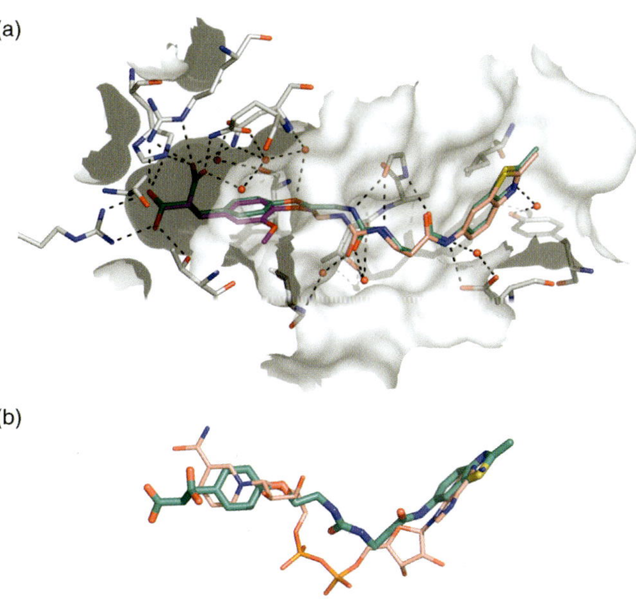

**Figure 17.10** (a) X-ray crystal structure showing binding mode of linked compound **AZ26** (cyan) in rat LDHA (4AL4), overlaid with parent fragments **AZ21** (purple) (4AJI) and **AZ24** (pink) (4AJL). (b) Bound conformation of **AZ26** (cyan) (4AL4) overlaid with that of NADH (1I10) [9] (pink).

4-methoxy of the substrate-nicotinamide fragments **AZ21** and **AZ25** and in the malonyl-phenyl linker. At this stage, we also revisited the SAR of the dicarboxylate in the substrate site, and the thiazole substituent in the adenine site. Broadly, the most potent combination was found to be a shortened (des-oxy) linker to the phenyl, an amide replacing the urea, a carbon-linked malonyl, and the *S*-propyl (rather than methyl) thiazole substituent. The most potent linked compound, incorporating all these features, was **AZ34**, with a $K_D$ (SPR) of 8 nM (LE = 0.29) and an $IC_{50}$ in our enzyme assay of 270 nM. **AZ34** was one of the three compounds that broke the μM barrier in the enzyme assay, and broadly there was a strong correlation for the linked compounds between $IC_{50}$ and $K_D$ measured by SPR ($R^2 = 0.86$), the ~30-fold offset being largely attributable to competition with NADH, present at 30 μM in the activity assay. The course we charted from initial fragment hits to our most potent lead compound is summarized in Figure 17.11. Corroborating our previous findings with the substrate-mimic fragments, the monocarboxylate analog of **AZ26** suffered around 25-fold loss in affinity ($K_D = 3.2$ μM).

Despite the good $K_D$ values we achieved with the best of our optimized linked compounds, we were not surprised that this level of binding affinity and enzyme inhibitory activity did not translate into detectable cellular activity, monitored via lactate levels in the culture medium of SKBR3 cells. Presuming the diacid to be the major liability in this regard, we explored the potential for methyl and ethyl mono- and diester derivatives to furnish the required cell membrane permeability while being hydrolysable to the active diacid by endogenous cytoplasmic esterases [71]. Monoesters were generally inactive, while dimethyl esters showed slightly superior cell potency to diethyl esters, with a relatively consistent drop-off between SPR affinity and cellular potency of around two to three orders of magnitude for those examples tested. Our most cell-active compound had a cell $IC_{50}$ of 4.2 μM, and is the dimethyl ester derivative of one of the highest affinity diacids, with methyl replacing the *S*-propyl of **AZ34** (SPR $K_D = 93$ nM and enzyme $IC_{50} = 0.5$ μM). This compound forms an attractive basis for further development of effective tool compounds with which to explore the effects of LDHA inhibition in different normal and transformed cell settings. Finally, initial results we obtained from affinity-based LDHB inhibition assays suggested a degree of selectivity for some of our linked diacid compounds. Further work is required to confirm and extend these findings in order to address the need for LDHA/LDHB selectivity in oncology-relevant model settings.

## 17.6 Fragment-Based LDHA Inhibitors from Other Groups

### 17.6.1 Nottingham

The first LDHA inhibitors reported to be inspired by a fragment-based lead generation philosophy were a series of bifunctional ligands elaborated by the group

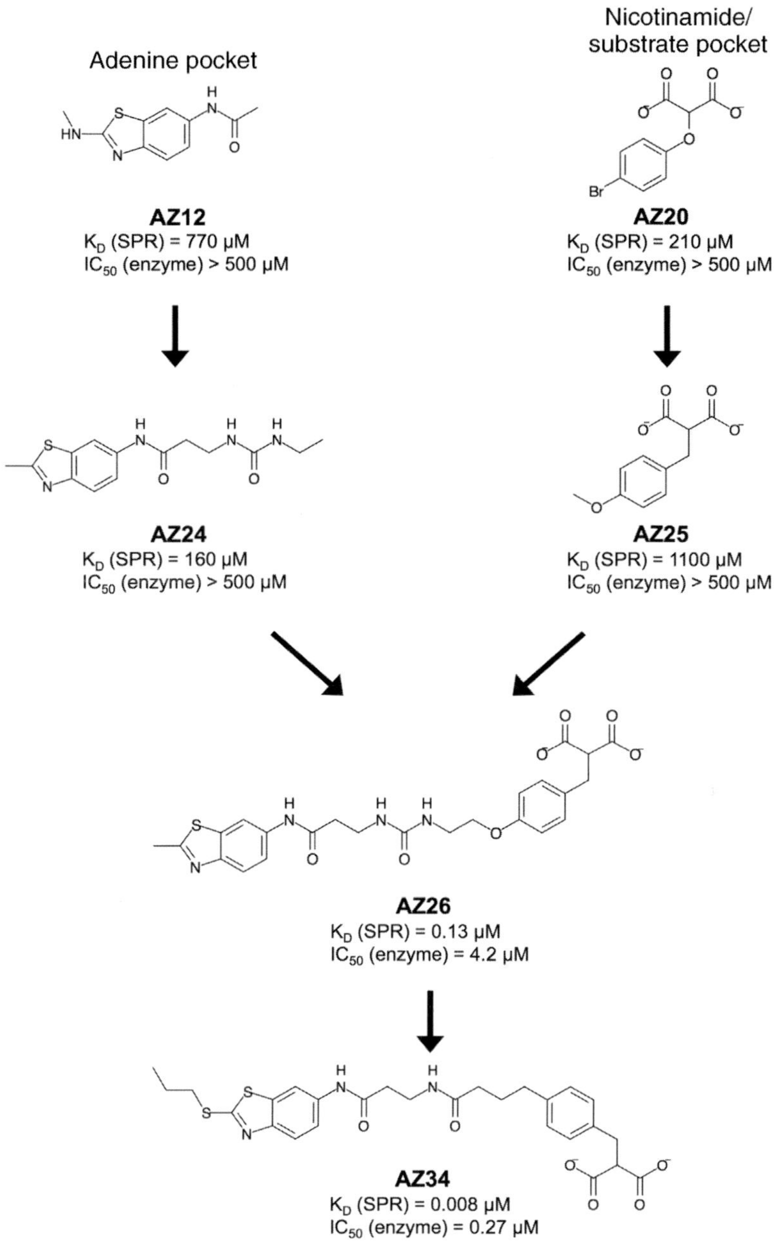


**Figure 17.11** Schematic illustration of evolution of potent AstraZeneca LDHA inhibitor **AZ34** from starting fragments **AZ12** and **AZ20**.

of Moses *et al.* at the University of Nottingham [72]. Based on analysis of the structure of human LDHA in complex with NADH and oxamate [9], this group reasoned that ligands spanning the entirety of the NADH and substrate sites could in principle be derived through linking compounds mimicking cofactor and substrate or product, respectively, and they identified copper-catalyzed "click" chemistry [73] as a means through which to achieve linkage of appropriate ligands. They initially identified the bis(indolyl)maleimides as candidate ligands for the adenine subpocket, based on precedent as adenosine mimics and inhibitors of sirtuins [74]; these compounds also presented opportunities for synthetic elaboration at a number of positions, including the incorporation of alkyne functionality to facilitate click linkage. Azido-derivatized *N*-alkyl oxamates were selected as synthetically exploitable candidates for the substrate binding pocket, on the basis that oxamate is a known inhibitor of LDHA. A hypothetical linked ligand, **N1** (Figure 17.12), was evaluated in modeling studies as being capable of forming the envisaged interactions, and on this basis a 31-member library of linked bifunctional ligands with varying chain lengths fused via a central 1,4-triazole was generated. The library members, along with their

**N1**

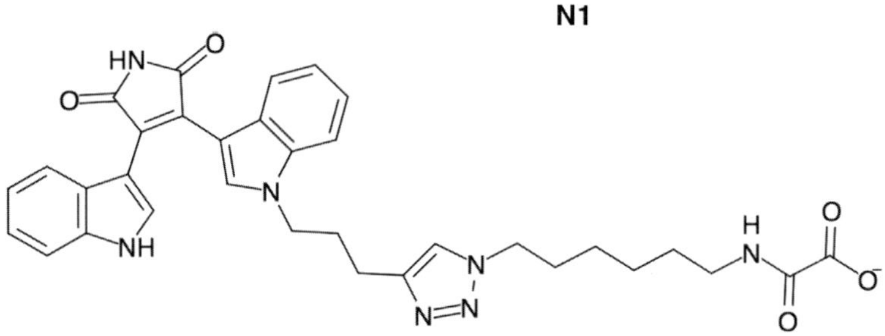

**N24**

$IC_{50}$ = 14.8 μM

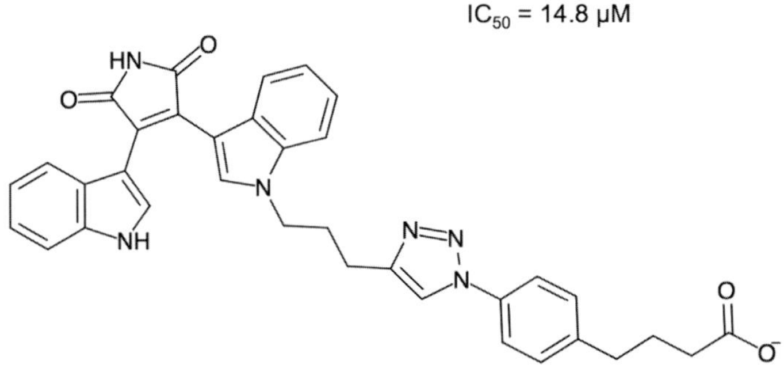

**Figure 17.12** Structures of Nottingham inhibitors: **N1** (hypothetical linked ligand) and the most potent (against human LDHA) library compound **N24**.

progenitor fragments, were tested for activity in an NADH depletion assay. Three compounds displayed inhibition equal to or greater than that of oxamate itself, with $IC_{50}$ values against human LDHA ranging from 15 μM (compound **N24**) to 174 μM (compared to 130 μM for oxamate) (Figure 17.12). In contrast, none of the unlinked oxamate-based fragments were active in the assay, while the bis(indolyl)maleimide fragments were insufficiently soluble in assay buffer. Docking studies suggested that the inhibitors bind as predicted from initial modeling and make favorable interactions with the cofactor and substrate binding sites. A potential challenge with this class of compounds is the presence of the electrophilic maleimide functionality. Encouragingly, however, none of the linked bifunctional ligands detectably inhibited LDHB, implying that selectivity between the two major isoforms of LDH is in principle possible, despite their high sequence similarity.

### 17.6.2
### Ariad

In a striking example of convergent evolution in drug discovery approaches to a challenging target, the group at Ariad adopted a very similar strategy in their fragment-based LDHA inhibitor campaign [45]. Notably, the Ariad team achieved similar levels of *in vitro* potency, yet with a structurally distinct chemical series (summarized in Figure 17.13). Interestingly, however, this group initially focused their attention on the substrate pocket, and this difference of initial emphasis shaped in an interesting and instructive way the divergent paths taken chemically by the Ariad and AstraZeneca groups. In this section, we will chiefly examine the differences between the two programs, since these provide the most enlightening glimpses into how professional drug discovery teams engaged in an FBDD campaign form hypotheses based on current understanding, acquire data whose information content is shaped by the experimental tools employed, and act on data as they emerge.

The Ariad team compiled a ~700-member diverse fragment library and screened them using saturation transfer difference (STD) NMR spectroscopy. Strikingly, of the 38 hits, 19 were carboxylic acids; these were in fact significantly enriched in the hit list compared to their representation in the library. This is an immediate difference from the AstraZeneca experience, where the majority of initial hits were neutral heterocycles. NADH competition in the STD NMR experiment was used to confirm specific binding (the group also ran a screen in the presence of NADH and obtained a very low hit rate, suggesting that the small substrate pocket did not readily accommodate the majority of the library compounds). For reasons similar to those of the AstraZeneca group, the team established a crystal system based on rabbit LDHA (93% sequence similarity to human LDHA) in complex with AMP; they then employed a back-soaking protocol to remove bound AMP from the crystals, leaving an empty cofactor and substrate site. With this crystal system, they were able to confirm that one of their initial carboxylic acid hits (**AR1**) bound to the substrate pocket via a salt

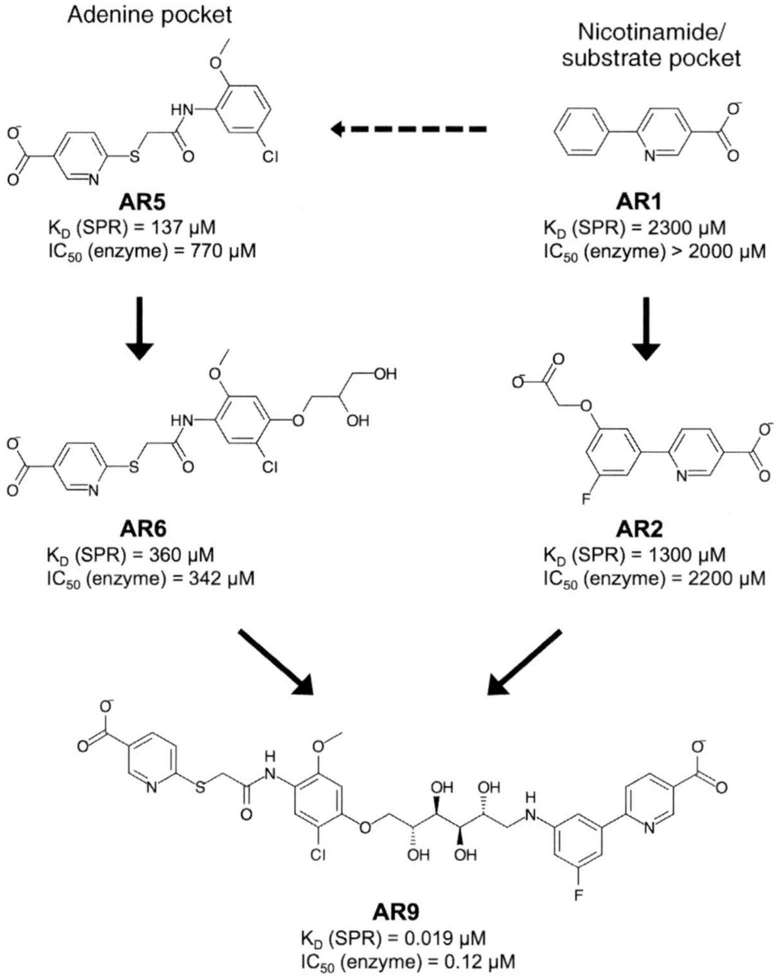


**Figure 17.13** Schematic illustration of evolution of potent Ariad LDHA inhibitor **AR9** from starting fragments **AR1** and **AR5**. The dashed arrow indicates the serendipitous derivation of adenine pocket fragments from substrate pocket-directed compound optimization.

bridge to Arg168. An LDHA SPR assay, established using carbonic anhydrase II as a reference chip surface, revealed that $K_D$ values for these initial hits ranged between 1 and 5 mM, with the majority, as expected, not displaying detectable inhibitory activity in the enzyme assay. Exploration of SAR around the initial phenyl pyridine carboxylic acid **AR1** ($K_D = 2.3$ mM; LE = 0.25) culminated in elaboration to compound **AR5**, which, with an SPR $K_D$ of 137 μM and an enzyme $IC_{50}$ of 770 μM, was the most potent compound thus far, with an LE of 0.23. Unexpectedly, however, on solving its structure in complex with rabbit LDHA, **AR5** was observed to bind in the adenine site. Unlike the AstraZeneca adenine-site fragments, **AR5** presented a carboxylic acid to the solvent-exposed side chain of Arg111, but analogously to AstraZeneca tetrazole **AZ4**, bound to the "Phe118-flipped" form of the site.

To explore the potential to grow from the adenine pocket start-point **AR5**, the Ariad team then appended groups to the 3-position of the phenyl ring. An oxy-propane 1,2-diol substituent at this position resulted in modest potency gain in the enzyme assay (fragment **AR6**; $IC_{50} = 342\,\mu M$), and crystallography showed that it mimicked NADH ribose interactions with the Asp51 side chain. Additional or alternative substitution patterns failed to achieve any potency gains and were often accompanied by a loss of LE. At this point, the team decided that significant potency and LE gains were only likely to be realized through linking the adenine- and substrate-site fragments. Cosoaking of fragments **AR2** and **AR6** revealed a distance of only 3.3 Å between terminal carbons of the fragments, providing confidence in the feasibility of a linking approach. The team scoped a range of seven- to nine-atom linkers including (as proof of concept) a simple ether link, which yielded a gain in enzyme potency to 59 μM and, for the first time, a submicromolar $K_D$ (850 nM; LE = 0.18) by SPR.

Optimization of the linker resulted in synthesis of compound **AR9**, the most potent in the series with an enzyme $IC_{50}$ of 120 nM, a $K_D$ of 19 nM, and LE = 0.21; this compound contains an eight-atom linker incorporating four hydroxyls, with an amine link to the fluorophenyl ring of the substrate-pocket fragment. The structure of the close analog **AR8**, with an oxygen replacing the amine, in complex with rabbit LDHA, is shown in Figure 17.14a overlaid on the parent fragments **AR2** and **AR6**, and shows that the polyhydroxylated linker makes a series of water-mediated hydrogen bonds to Gly28, Ala29, and Gly31, and that the linked compound preserves the binding mode of the constituent fragments. One additional water molecule is observed between the phenyl ring of the inhibitor and the backbone carbonyl atom of Thr247. These hydroxyl-mediated interactions were found to be very important in maintaining the high affinity of the optimized linked compounds.

Compound **AR9** was further modified in an attempt to enhance its ability to inhibit LDHA in a cellular context. As expected, the removal of either terminal carboxylate substantially decreased (by a factor of >20 at the "adenosine" end and >100 at the "substrate" end) *in vitro* inhibitory potency, but did result in measurable decreases in lactate levels in the medium of Burkitt's lymphoma cells. However, in contrast to the AstraZeneca compounds, conversion to the methyl esters at both carboxylates failed to elicit any cellular activity.

The Ariad and AstraZeneca compounds are fascinating examples of the highly divergent chemistries that can arise from two conceptually very similar lead generation approaches. That this is true even in the arena of fragment-based lead discovery – where it is sometimes posited that numerically small fragment libraries containing similar chemotypes will, ipso facto, give rise to rather indistinguishable lead molecules – is a testament to the astronomical number of degrees of freedom inherent in drug discovery as a preeminently human endeavor. The Nottingham compounds represent yet further degrees of

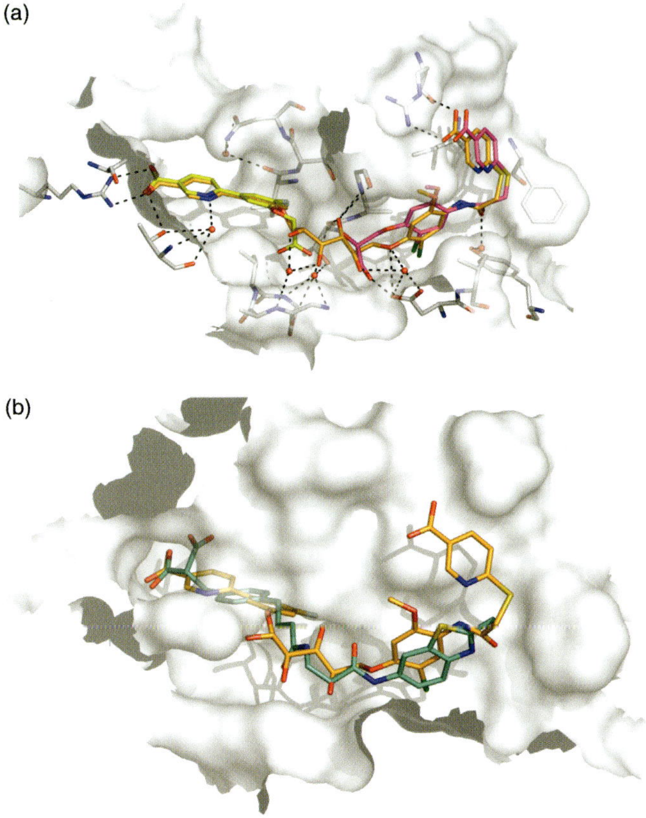

**Figure 17.14** (a) X-ray crystal structure showing binding mode and interactions of linked compound **AR8** (orange) (4I9 H) in rabbit LDHA, overlaid with parent fragments **AR2** (yellow) (4I9 N) and **AR6** (magenta) (4I9 N). (b) Overlaid binding mode of Ariad linked compound **AR8** (orange) (4I9 H) on AstraZeneca linked compound **AZ33** (cyan) bound to rat LDHA (gray surface) (4AJO).

divergence in the chemistries, albeit a result of distinct methodological and conceptual differences from those of the Ariad and AstraZeneca teams. Not only are the adenine and substrate pocket fragments of the three series chemically distinct but also the linker designs are highly divergent, with, for example, the Ariad compounds more closely following the path and interactions of the central diribose diphosphate moieties of the cofactor, while the AstraZeneca compounds take a "shortcut" across this region (Figure 17.14b). Retrospective analysis of the campaigns clearly reveals the decision points (and the data sets that informed those decisions) from which these divergent features emerged. One factor in common is the presence of two carboxylic acids, which in both cases is a clear liability for cellular activity, likely because of poor membrane permeability.

## 17.7
## Conclusions and Future Perspectives

The fragment campaigns we have discussed generated highly valuable understanding of the nature of the available small molecule binding sites in LDHA – understanding that in turn will undoubtedly facilitate the design of improved lead compounds with superior ligand efficiency, physical properties, and cell membrane permeability. Since the publication of fragment-based LDHA inhibitors from the Nottingham, AstraZeneca, and Ariad labs, two other groups (from Roche/Genentech and GlaxoSmithKline (GSK)) have reported discovery and characterization of new classes of LDHA inhibitors derived from high-throughput screening approaches [75–80] (summarized in Figure 17.15). Interestingly, in the case of the Roche/Genentech program, all the compounds (**RG26** [75], **RG35** [76], and **RG31** [78], from three distinct series) bind simultaneously with NADH but competitively with substrate; given the nature of the interactions, they most likely display uncompetitive inhibition kinetics with respect to cofactor, although data supporting this assumption (based on SPR measurements in the presence and absence of cofactor) are reported for only one of the series [77]. While the lipophilic acidic natures of the Roche/Genentech series do present some challenges in terms of physical properties, high *in vitro* enzyme potencies (down to 6 nM $IC_{50}$) and selectivity ranging up to 10-fold over LDHB were achieved with the most highly optimized compounds. High plasma protein binding is likely a factor in limiting the cell potency of these compounds, as modest effects on lactate production were seen only in the absence of added serum in cell assays [79].

Another series of compounds derived from HTS was reported in 2013 by Billiard *et al.* [80]. The GSK group identified quinoline 3-sulfonamides from screening of their corporate compound collection in high-throughput activity-based LDHA and LDHB assays followed by a thermal shift stabilization (differential scanning fluorimetry) assay [81] to confirm specific interaction with the enzyme. Following lead optimization, the most active examples of the series (e.g., GSK1, Figure 17.15) yielded enzyme $IC_{50}$ values in the single digit nM range and around 10–80 fold selectivity over LDHB, with an NADH-competitive mode of action. They also displayed potent inhibition of cellular lactate production in hepatocellular carcinoma cells although, as observed for other LDHA inhibitors, cellular $IC_{50}$s were significantly weaker than *in vitro* enzyme potencies, ranging between 400 and 600 nM. The GSK group investigated the origin of the drop-off and found high cytosolic LDHA concentrations exceeding 1 μM in many cancer cell lines, which would be expected to cause a shift in $IC_{50}$ for compounds with enzyme $K_i$ lower than the protein concentration; however, there was no correlation between cell potency and LDHA or LDHB expression levels. The encouraging levels of both potency and selectivity expressed by the quinoline 3-sulfonamides suggest a promising start point for further optimization, although suboptimal absorption, distribution, metabolism and excretion (ADME) properties and mitochondrial uncoupling effects at high doses render reported examples of the series unsuitable as *in vivo* probes.

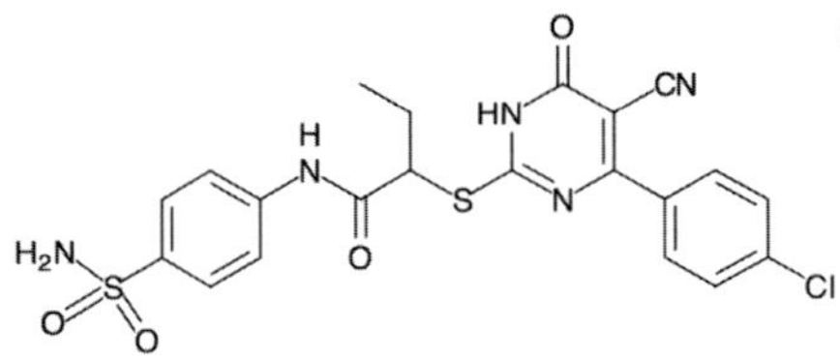

**RG26**
LDHA $IC_{50}$ = 0.65 μM
LDHB $IC_{50}$ = 2.4 μM

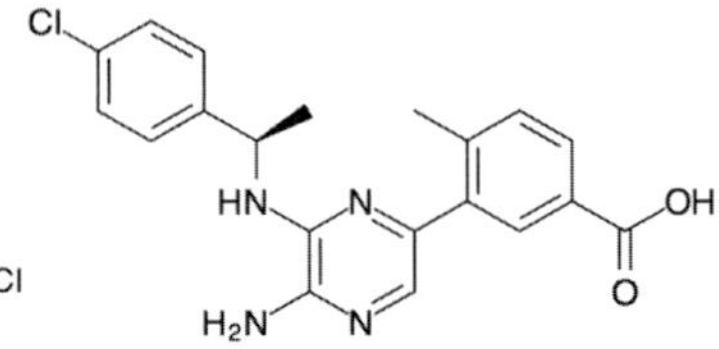

**RG35**
LDHA $IC_{50}$ = 0.5 μM
LDHB $IC_{50}$ = 2.0 μM

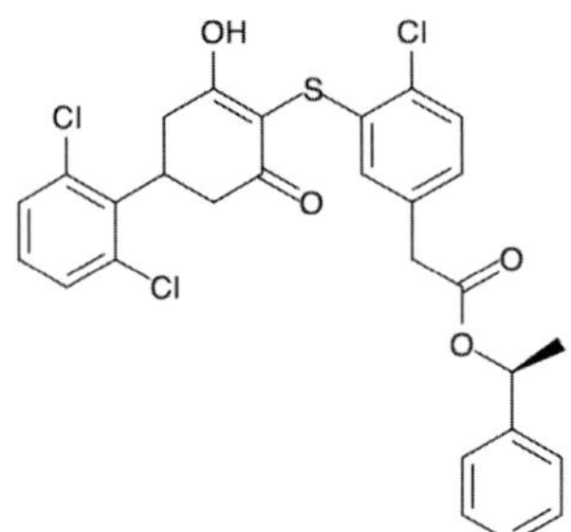

**RG31**
LDHA $IC_{50}$ = 0.006 μM
LDHB $IC_{50}$ ≈ 0.006 μM

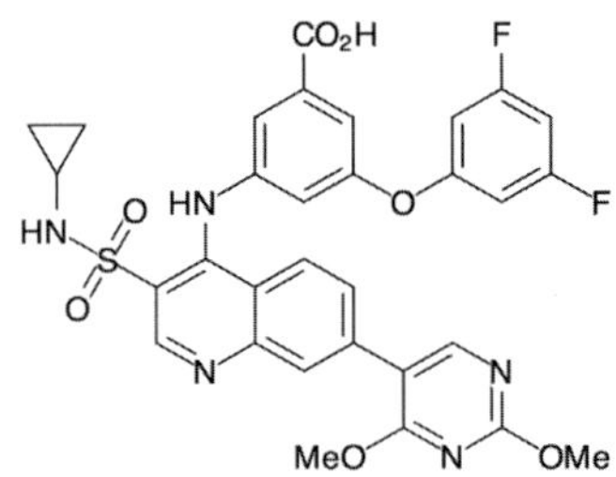

**GSK1**
LDHA $IC_{50}$ = 0.0026 μM
LDHB $IC_{50}$ = 0.043 μM

**Figure 17.15** Structures and potencies of Roche/Genentech inhibitors **RG26** [75], **RG35** [76], and **RG31** [78], and GlaxoSmithKline inhibitor **GSK1** [80].

LDHA remains a compelling cancer drug target because of its central role in metabolic remodeling of tumor cells, its strong association with malignant transformation, and the relatively modest pathology associated with congenital LDHA loss-of-function mutations. Whether or not high degrees of selectivity over LDHB will be required for clinical LDHA inhibitors remains to be established. At first sight, it might be assumed that LDHB inhibition should be avoided, on the basis of the predominance of the LDH-1 (4H) isoform in cardiac tissue and the physiological role of this isoform in lactate metabolism in cardiomyocytes. Despite the high sequence similarity between LDHA and LDHB, it does appear possible to obtain a measure of isoform selectivity. Another as yet unanswered question concerns the extent and duration of LDHA inhibition that will be required to achieve meaningful levels of *in vivo* efficacy. Development of sufficiently cell-active probe compounds with appropriate ADMET properties is thus a high priority. Achieving potent, cell-permeable inhibitors of LDHA has, however, proved highly challenging. Significant progress toward understanding how this goal might be achieved has arisen from the application of fragment-based biophysical screening strategies coupled with structure-based design. In addition to furnishing the first rigorously validated, nanomolar-potency ligands to be reported for LDHA, the fragment campaigns we have discussed generated highly valuable understanding of the molecular recognition landscapes of the cofactor and substrate sites in LDHA. The simultaneous emergence of new, rigorously profiled series from HTS [75–80] presents a real opportunity for oncology drug discovery: we now have within grasp the tantalizing possibility of hybridizing scaffolds from fragment-based and HTS-derived series to obtain a superior balance of potency and properties. The outlook for tackling this very challenging yet exceptionally high-value cancer target is as a result unquestionably brighter.

## References


1 Warburg, O. (1956) On the origin of cancer cells. *Science*, **123**, 309–314.

2 Kim, J. and Dang, C.V. (2006) Cancer's molecular sweet tooth and the Warburg effect. *Cancer Research*, **66**, 8927–8930.

3 Augoff, K., Hryniewicz-Jankowska, A., and Tabola, R. (2015) Lactate dehydrogenase 5: an old friend and a new hope in the war on cancer. *Cancer Letters*, **358**, 1–7.

4 Lewis, B., Morson, B.C., February, A.W., Jones, J.H., and Misiewicz, J.J. (1971) Abnormal lactic dehydrogenase isoenzyme patterns in ulcerative colitis with precancerous change. *Gut*, **12**, 16–19.

5 Yao, F., Zhao, T., Zhong, C., Zhu, J., and Zhao, H. (2013) LDHA is necessary for the tumorigenicity of esophageal squamous cell carcinoma. *Tumour Biology*, **34**, 25–31.

6 Everse, J. and Kaplan, N.O. (1973) Lactate dehydrogenases: structure and function. *Advances in Enzymology and Related Areas of Molecular Biology*, **37**, 61–133.

7 Goldman, R.D., Kaplan, N.O., and Hall, T. C. (1964) Lactic dehydrogenase in human neoplastic tissues. *Cancer Research*, **24**, 389–399.

8 Dunn, C.R., Wilks, H.M., Halsall, D.J., Atkinson, T., Clarke, A.R., Muirhead, H., and Holbrook, J.J. (1991) Design and synthesis of new enzymes based on the lactate dehydrogenase framework. *Philosophical Transactions of the Royal Society of London. Series B Biological Sciences*, **332**, 177–184.

9 Read, J.A., Winter, V.J., Eszes, C.M., Sessions, R.B., and Brady, R.L. (2001) Structural basis for altered activity of M- and H-isozyme forms of human lactate dehydrogenase. *Proteins: Structure, Function, and Genetics*, **43**, 175–185.

10 Hewitt, C.O., Eszes, C.M., Sessions, R.B., Moreton, K.M., Dafforn, T.R., Takei, J., Dempsey, C.E., Clarke, A.R., and Holbrook, J.J. (1999) A general method for relieving substrate inhibition in lactate dehydrogenases. *Protein Engineering*, **12**, 491–496.

11 Koukourakis, M.I., Giatromanolaki, A., Sivridis, E., Bougioukas, G., Didilis, V., Gatter, K.C., and Harris, A.L. (2003) Lactate dehydrogenase-5 (LDH-5) overexpression in non-small-cell lung cancer tissues is linked to tumour hypoxia, angiogenic factor production and poor prognosis. *British Journal of Cancer*, **89**, 877–885.

12 Kolev, Y., Uetake, H., Takagi, Y., and Sugihara, K. (2008) Lactate dehydrogenase-5 (LDH-5) expression in human gastric cancer: association with hypoxia-inducible factor (HIF-1alpha) pathway, angiogenic factors production and poor prognosis. *Annals of Surgical Oncology*, **15**, 2336–2344.

13 Shim, H., Dolde, C., Lewis, B.C., Wu, C.-S., Dang, G., Jungmann, R.A., Dalla-Favera, R., and Dang, C.V. (1997) c-Myc transactivation of LDH-A: implications for tumor metabolism and growth. *Proceedings of the National Academy of Sciences of the United States of America*, **94**, 6658–6663.

14 Cooper, J.A., Esch, F.S., Taylor, S.S., and Hunter, T. (1984) Phosphorylation sites in enolase and lactate dehydrogenase utilized by tyrosine protein kinases *in vivo* and *in vitro*. *Journal of Biological Chemistry*, **259**, 7835–7841.

15 Koukorakis, M.I., Pitiakoudis, M., Giatromanolaki, A., Tsarouha, A.,

Polychronidis, A., Sivridis, E., and Simopoulos, C. (2006) Oxygen and glucose consumption in gastrointestinal adenocarcinomas: correlation with markers of hypoxia, acidity and anaerobic glycolysis. *Cancer Science*, **97**, 1056–1060.

16 Zhuang, L., Scolyer, R.A., Murali, R., McCarthy, S.W., Zhang, X.D., Thompson, J.F., and Hersey, P. (2010) Lactate dehydrogenase 5 expression in melanoma increases with disease progression and is associated with expression of Bcl-XL and Mcl-1, but not Bcl-2 proteins. *Modern Pathology*, **23**, 45–53.

17 Koukourakis, M.I., Giatromanolaki, A., Winter, S., Leek, R., Sivridis, E., and Harris, A.L. (2009) Lactate dehydrogenase 5 expression in squamous cell head and neck cancer relates to prognosis following radical or postoperative radiotherapy. *Oncology*, **77**, 285–292.

18 Koukourakis, M.I., Giatromanolaki, A., Sivridis, E., Gatter, K.C., and Harris, A.L. (2006) Lactate dehydrogenase 5 expression in operable colorectal cancer: strong association with survival and activated vascular endothelial growth factor pathway: a report of the tumour angiogenesis research group. *Journal of Clinical Oncology*, **24**, 4301–4308.

19 Yabu, M., Shime, H., Hara, H., Saito, T., Matsumoto, M., Seya, T., Akazawa, T., and Inoue, N. (2011) IL-23-dependent and -independent enhancement pathways of IL-17A production by lactic acid. *International Immunology*, **23**, 29–41.

20 Shime, H., Yabu, M., Akazawa, T., Kodama, K., Matsumoto, M., Seya, T., and Inoue, N. (2008) Tumor-secreted lactic acid promotes IL-23/IL-17 proinflammatory pathway. *Journal of Immunology*, **180**, 7175–7183.

21 Dietl, K., Renner, K., Dettmer, K., Timischl, B., Eberhart, K., Dorn, C., Hellerbrand, C., Kastenberger, M., Kunz-Schughart, L.A., Oefner, P.J., Andreesen, R., Gottfried, E., and Kreutz, M.P. (2010) Lactic acid and acidification inhibit TNF secretion and glycolysis of human monocytes. *Journal of Immunology*, **184**, 1200–1209.

22 Fischer, K., Hoffmann, P., Voelkl, S., Meidenbauer, N., Ammer, J., Edinger, M., Gottfried, E., Schwarz, S., Rothe, G., Hoves, S., Renner, K., Timischl, B., Mackensen, A., Kunz-Schughart, L., Andreesen, R., Krause, S.W., and Kreutz, M. (2007) Inhibitory effect of tumor cell-derived lactic acid on human T cells. *Blood*, **109**, 3812–3819.

23 Gottfried, E., Kunz-Schughart, L.A., Ebner, S., Mueller-Klieser, W., Hoves, S., Andreesen, R., Mackensen, A., and Kreutz, M. (2006) Tumor-derived lactic acid modulates dendritic cell activation and antigen expression. *Blood*, **107**, 2013–2021.

24 Mendler, A.N., Hu, B., Prinz, P.U., Kreutz, M., Gottfried, E., and Noessner, E. (2012) Tumor lactic acidosis suppresses CTL function by inhibition of p38 and JNK/c-Jun activation. *International Journal of Cancer*, **131**, 633–640.

25 Wang, Z.Y., Loo, T.Y., Shen, J.G., Wang, N., Wang, D.M., Yang, D.P., Mo, S.L., Guan, X.Y., and Chen, J.P. (2012) LDH-A silencing suppresses breast cancer tumorigenicity through induction of oxidative stress mediated mitochondrial pathway apoptosis. *Breast Cancer Research and Treatment*, **131**, 791–800.

26 Sheng, S.L., Liu, J.J., Dai, Y.H., Sun, X.G., Xiong, X.P., and Huang, G. (2012) Knockdown of lactate dehydrogenase A suppresses tumor growth and metastasis of human hepatocellular carcinoma. *FEBS Journal*, **279**, 3898–3910.

27 Xie, H., Hanai, J.-I., Ren, J.-G., Kats, L., Burgess, K., Bhargava, P., Signoretti, S., Billiard, J., Duffy, K.J., Grant, A., Wang, X., Lorkiewicz, P.K., Schatzman, S., Bousamra, M., Lane, A.N., Higashi, R.M., Fan, T.W.M., Pandolfi, P.P., Sukhatme, V.P., and Seth, P. (2014) Targeting lactate dehydrogenase-A inhibits tumorigenesis and tumor progression in mouse models of lung cancer and impacts tumor-initiating cells. *Cell Metabolism*, **19**, 795–809.

28 Merkle, S., Favor, J., Graw, J., Hornhardt, S., and Pretsch, W. (1992) Hereditary lactate dehydrogenase A-subunit deficiency as cause of early postimplantation death of homozygotes in *Mus musculus*. *Genetics*, **131**, 413–421.

29 Lin, S.-J. and Guarente, L. (2003) Nicotinamide adenine dinucleotide, a metabolic regulator of transcription, longevity and disease. *Current Opinion in Cell Biology*, **15**, 241–246.

30 Houtkooper, R.H., Canto, C., Wanders, R. J., and Auwerx, J. (2010) The secret life of $NAD^+$: an old metabolite controlling new metabolic signaling pathways. *Endocrinology Reviews*, **31**, 194–223.

31 Eszes, C.M., Sessions, R.B., Clarke, A.R., Moreton, K.M., and Holbrook, J.J. (1996) Removal of substrate inhibition in a lactate dehydrogenase from human muscle by a single residue change. *FEBS Letters*, **399**, 193–197.

32 Zhai, X., Yang, Y., Wan, J., Zhu, R., and Wu, Y. (2013) Inhibition of LDH-A by oxamate induces G2/M arrest, apoptosis and increases radiosensitivity in nasopharyngeal carcinoma cells. *Oncology Reports*, **30**, 2983–2991.

33 Yang, Y., Su, D., Zhao, L., Zhang, D., Xu, J., Wan, J., Fan, S., and Chen, M. (2014) Different effects of LDH-A inhibition by oxamate in non-small cell lung cancer cells. *Oncotarget*, **5**, 11886–11896.

34 Vander Jagt, D.L., Deck, L.M., and Royer, R.E. (2000) Gossypol: prototype of inhibitors targeted to dinucleotide folds. *Current Medicinal Chemistry*, **7**, 479–498.

35 Olgiati, K.L. and Toscano, W.A. (1983) Kinetics of gossypol inhibition of bovine lactate dehydrogenase X. *Biochemical and Biophysical Research Communications*, **115**, 180–185.

36 Le, A., Cooper, C.R., Gouw, A.M., Dinavahi, R., Maitra, A., Deck, L.M., Royer, R.E., Vander Jagt, D.L., Semenza, G.L., and Dang, C.V. (2010) Inhibition of lactate dehydrogenase A induces oxidative stress and inhibits tumor progression. *Proceedings of the National Academy of Sciences of the United States of America*, **107**, 2037–2042.

37 Manerba, M., Vettraino, M., Fiume, L., Di Stefano, G., Sartini, A., Giacomini, E., Buonfiglio, R., Roberti, M., and Recanatini., M. (2012) Galloflavin (CAS 568-80-9): a novel inhibitor of lactate dehydrogenase. *ChemMedChem*, **7**, 311–317.

38 Farabegoli, F., Vettraino, M., Manerba, M., Fiume, L., Roberti, M., and Di Stefano, G. (2012) Galloflavin, a new lactate dehydrogenase inhibitor, induces the death of human breast cancer cells with different glycolytic attitude by affecting distinct signaling pathways. *European Journal of Pharmaceutical Sciences*, **47**, 729–738.

39 Fiume, L., Vettraino, M., Carnicelli, D., Arfilli, V., Di Stefano, G., and Brigotti, M. (2013) Galloflavin prevents the binding of lactate dehydrogenase A to single stranded DNA and inhibits RNA synthesis in cultured cells. *Biochemical and Biophysical Research Communications*, **430**, 466–469.

40 Cameron, A., Read, J., Tranter, R., Winter, V.J., Sessions, R.B., Brady, R.L., Vivas, L., Easton, A., Kendrick, H., Croft, S.L., Barros, D., Lavandera, J.L., Martin, J.J., Risco, F., Garcia-Ochoa, S., Gamo, F.J., Sanz, L., Leon, L., Ruiz, J.R., Gabarro, R., Mallo, A., and Gomez de las Heras, F. (2004) Identification and activity of a series of azole-based compounds with lactate dehydrogenase-directed anti-malarial activity. *Journal of Biological Chemistry*, **279**, 31429–31439.

41 Sem, D.S., Bertolaet, B., Baker, B., Chang, E., Costache, A.D., Coutts, S., Dong, Q., Hansen, M., Hong, V., Huang, X., Jack, R.M., Kho, R., Lang, H., Ma, C., Meininger, D., Pellecchia, M., Pierre, F., Villar, H., and Yu, L. (2004) Systems-based design of bi-ligand inhibitors of oxidoreductases: filling the chemical proteomic toolbox. *Chemistry and Biology*, **11**, 185–194.

42 Granchi, C., Roy, S., Giacomelli, C., Macchia, M., Tuccinardi, T., Martinelli, A., Lanza, M., Betti, L., Giannaccini, G., Lucacchini, A., Funel, N., Leon, L.G., Giovannetti, E., Peters, G.J., Palchaudhuri, R., Calvaresi, E.C., Hergenrother, P.J., and Minutolo, F. (2011) Discovery of *N*-hydroxyindole-based inhibitors of human lactate dehydrogenase isoform A (LDH-A) as starvation agents against cancer cells. *Journal of Medicinal Chemistry*, **54**, 1599–1612.

43 Ward, R.A., Brassington, C., Breeze, A.L., Caputo, A., Critchlow, S., Davies, G., Goodwin, L., Hassall, G., Greenwood, R.,

Holdgate, G.A., Mrosek, M., Norman, R.A., Pearson, S., Tart, J., Tucker, J.A., Vogtherr, M., Whittaker, D., Wingfield, J., Winter, J., and Hudson, K. (2012) Design and synthesis of novel lactate dehydrogenase A inhibitors by fragment-based lead generation. *Journal of Medicinal Chemistry*, **55**, 3285–3306.

44 Chapman, A.D.M., Cortes, A., Dafforn, T.R., Clarke, A.R., and Brady, R.L. (1999) Structural basis of substrate specificity in malate dehydrogenases: crystal structure of a ternary complex of porcine cytoplasmic malate dehydrogenase, α-ketomalonate and tetrahydroNAD. *Journal of Molecular Biology*, **285**, 703–712.

45 Kohlmann, A., Zech, S.G., Li, F., Zhou, T., Squillace, R.M., Commodore, L., Greenfield, M.T., Lu, X., Miller, D.P., Huang, W.S., Qi, J., Thomas, R.M., Wang, Y., Zhang, S., Dodd, R., Liu, S., Xu, R., Xu, Y., Miret, J.J., Rivera, V., Clackson, T., Shakespeare, W.C., Zhu, X., and Dalgarno, D.C. (2013) Fragment growing and linking lead to novel nanomolar lactate dehydrogenase inhibitors. *Journal of Medicinal Chemistry*, **56**, 1023–1040.

46 Javed, M.H. and Waqar, M.A. (1993) Effect of various metals and metabolites on the activity of lactate dehydrogenase. *Pakistan Journal of Pharmacology*, **10**, 7–14.

47 Fink, T. and Reymond, J.-L. (2007) Virtual exploration of the chemical universe up to 11 atoms of C, N, O, F: assembly of 26.4 million structures (110.9 million stereoisomers) and analysis for new ring systems, stereochemistry, physicochemical properties, compound classes, and drug discovery. *Journal of Chemical Information and Modeling*, **47**, 342–353.

48 Hann, M.M., Leach, A.R., and Harper, G. (2001) Molecular complexity and its impact on the probability of finding leads for drug discovery. *Journal of Chemical Information and Computer Science*, **41**, 856–864.

49 Oprea, T.I., Davis, A.M., Teague, S.J., and Leeson, P.D. (2001) Is there a difference between leads and drugs? A historical perspective. *Journal of Chemical Information and Computer Science*, **41**, 1308–1315.

50 Rees, D.C., Congreve, M., Murray, C.W., and Carr, R. (2004) Fragment-based lead discovery. *Nature Reviews Drug Discovery*, **3**, 660–672.

51 Erlanson, D.A., McDowell, R.S., and O'Brien, T. (2004) Fragment-based drug discovery. *Journal of Medicinal Chemistry*, **47**, 3463–3482.

52 Albert, J.S., Blomberg, N., Breeze, A.L., Brown, A.J.H., Burrows, J.N., Edwards, P. D., Folmer, R.H.A., Geschwindner, S., Griffen, E.J., Kenny, P.W., Nowak, T., Olsson, L., Sanganee, H., and Shapiro, A.B. (2007) An integrated approach to fragment-based lead generation: philosophy, strategy and case studies from AstraZeneca's drug discovery programmes. *Current Topics in Medicinal Chemistry*, **7**, 1600–1629.

53 Coyne, A.G., Scott, D.E., and Abell, C. (2010) Drugging challenging targets using fragment-based approaches. *Current Opinion in Chemical Biology*, **14**, 299–307.

54 Scott, D.E., Coyne, A.G., Hudson, S.A., and Abell, C. (2012) Fragment-based approaches in drug discovery and chemical biology. *Biochemistry*, **51**, 4990–5003.

55 Edfeldt, F.N.B., Folmer, R.H.A., and Breeze, A.L. (2011) Fragment screening to predict druggability (ligandability) and lead discovery success. *Drug Discovery Today*, **16**, 284–287.

56 Hopkins, A.L., Groom, C.R., and Alex, A. (2004) Ligand efficiency: a useful metric for lead selection. *Drug Discovery Today*, **9**, 430–431.

57 Bilitewski, U. (2006) Protein-sensing assay formats and devices. *Analytica Chimica Acta*, **568**, 232–247.

58 Cunningham, B.T. and Laing, L.G. (2008) Advantages and application of label-free detection assays in drug screening. *Expert Opinion on Drug Discovery*, **3**, 891–901.

59 Smith, T.F., Waterman, M.S., and Fitch, W.M. (1981) Comparative biosequence metrics. *Journal of Molecular Evolution*, **18**, 38–46.

60 Leeson, P.D. and Springthorpe, B. (2007) The influence of drug-like concepts on decision-making in medicinal chemistry.

*Nature Reviews Drug Discovery*, **6**, 881–890.

61 Karlberg, T., Hammarstrom, M., Schutz, P., Svensson, L., and Schuler, H. (2010) Crystal structure of the catalytic domain of human PARP2 in complex with PARP inhibitor ABT-888. *Biochemistry*, **49**, 1056–1058.

62 Woon, E.C.Y. and Threadgill, M.D. (2005) Poly(ADP-ribose)polymerase inhibition: where now? *Current Medicinal Chemistry*, **12**, 2373–2392.

63 Chen, L., Petrelli, R., Felczak, K., Gao, G., Bonnac, L., Yu, J.S., Bennett, E.M., and Pankiewicz, K.W. (2008) Nicotinamide adenine dinucleotide based therapeutics. *Current Medicinal Chemistry*, **15**, 650–670.

64 Burgos, E.S., Ho, M., Almo, S.C., and Schramm, V.L. (2009) A phosphoenzyme mimic, overlapping catalytic sites and reaction coordinate motion for human NAMPT. *Proceedings of the National Academy of Sciences of the United States of America*, **106**, 13748–13753.

65 Ichihara, O., Barker, J., Law, R.J., and Whittaker, M. (2011) Compound design by fragment-linking. *Molecular Informatics*, **30**, 298–306.

66 Barker, J.J., Barker, O., Courtney, S.M., Gardiner, M., Hesterkamp, T., Ichihara, O., Mather, O., Montalbetti, C.A.G.N., Muller, A., Varasi, M., Whittaker, M., and Yarnold, C.J. (2010) Discovery of a novel Hsp90 inhibitor by fragment linking. *ChemMedChem*, **5**, 1697–1700.

67 Hung, A.W., Silvestre, H.L., Wen, S., Ciulli, A., Blundell, T.L., and Abell, C. (2009) Application of fragment growing and fragment linking to the discovery of inhibitors of mycobacterium tuberculosis pantothenate synthetase. *Angewandte Chemie (International Edition in English)*, **48**, 8452–8456.

68 Nazare, M., Matter, H., Will, D.W., Wagner, M., Urmann, M., Czech, J., Schreuder, H., Bauer, A., Ritter, K., and Wehner, V. (2012) Fragment deconstruction of small, potent Factor Xa inhibitors: Exploring the superadditivity energetics of fragment linking in protein–ligand complexes. *Angew. Chem. Int. Ed.*, **51**, 905–911.

69 Fattori, D., Squarcia, A., and Bartoli, S. (2008) Fragment-based approach to drug lead discovery: overview and advances in various techniques. *Drugs R&D*, **9**, 217–227.

70 Crisman, T.J., Bender, A., Milik, M., Jenkins, J.L., Scheiber, J., Sukuru, S.C.K., Fejzo, J., Hommel, U., Davies, J.W., and Glick, M. (2008) "Virtual fragment linking": an approach to identify potent binders from low affinity fragment hits. *Journal of Medicinal Chemistry*, **51**, 2481–2491.

71 Beaumont, K., Webster, R., Gardner, I., and Dack, K. (2003) Design of ester prodrugs to enhance oral absorption of poorly permeable compounds: challenges to the discovery scientist. *Current Drug Metabolism*, **4**, 461–485.

72 Moorhouse, A.D., Spiteri, C., Sharm, P., Zloh, M., and Moses, J.E. (2011) Targeting glycolysis: a fragment based approach towards bifunctional inhibitors of hLDH-5. *Chemical Communications*, **47**, 230–232.

73 Kolb, H.C., Finn, M.G., and Sharpless, K.B. (2001) Click chemistry: diverse chemical function from a few good reactions. *Angewandte Chemie (International Edition in English)*, **40**, 2004–2021.

74 Trapp, J., Jochum, A., Meier, R., Saunders, L., Marshall, B., Kunick, C., Verdin, E., Goekjian, P., Sippl, W., and Jung, M. (2006) Adenosine mimetics as inhibitors of $NAD^+$-dependent histone deacetylases, from kinase to sirtuin inhibition. *Journal of Medicinal Chemistry*, **49**, 7307–7316.

75 Dragovich, P.S., Fauber, B.P., Corson, L.B., Ding, C.Z., Eigenbrot, C., Ge, H., Giannetti, A.M., Hunsaker, T., Labadie, S., Liu, Y., Malek, S., Pan, B., Peterson, D., Pitts, K., Purkey, H.E., Sideris, S., Ultsch, M., VanderPorten, E., Wei, B., Xu, Q., Yen, I., Yue, Q., Zhang, H., and Zhang, X. (2013) Identification of substituted 2-thio-6-*oxo*-1,6-dihydropyrimidines as inhibitors of human lactate dehydrogenase. *Bioorganic and Medicinal Chemistry Letters*, **23**, 3186–3194.

76 Fauber, B.P., Dragovich, P.S., Chen, J., Corson, L.B., Ding, C.Z., Eigenbrot, C., Giannetti, A.M., Hunsaker, T., Labadie, S., Liu, Y., Liu, Y., Malek, S., Peterson, D., Pitts, K., Sideris, S., Ultsch, M.,

VanderPorten, E., Wang, J., Wei, B., Yen, I., and Yue, Q. (2013) Identification of 2-amino-5-aryl-pyrazines as inhibitors of human lactate dehydrogenase. *Bioorganic and Medicinal Chemistry Letters*, **23**, 5533–5539.

77 Dragovich, P.S., Fauber, B.P., Boggs, J., Chen, J., Corson, L.B., Ding, C.Z., Eigenbrot, C., Ge, H., Giannetti, A.M., Hunsaker, T., Labadie, S., Li, C., Liu, Y., Liu, Y., Ma, S., Malek, S., Peterson, D., Pitts, K.E., Purkey, H.E., Robarge, K., Salphati, L., Sideris, S., Ultsch, M., VanderPorten, E., Wang, J., Wei, B., Xu, Q., Yen, I., Yue, Q., Zhang, H., Zhang, X., and Zhou, A. (2014) Identification of substituted 3-hydroxy-2-mercaptocyclohex-2-enones as potent inhibitors of human lactate dehydrogenase. *Bioorganic and Medicinal Chemistry Letters*, **24**, 3764–3771.

78 Labadie, S., Dragovich, P.S., Chen, J., Fauber, B.P., Boggs, J., Corson, L.B., Ding, C.Z., Eigenbrot, C., Ge, H., Ho, Q., Lai, K.W., Ma, S., Malek, S., Peterson, D., Purkey, H.E., Robarge, K., Salphati, L., Sideris, S., Ultsch, M., VanderPorten, E., Wei, B., Xu, Q., Yen, I., Yue, Q., Zhang, H., Zhang, X., and Zhou, A. (2015) Optimization of 5-(2,6-dichlorophenyl)-3-hydroxy-2-mercaptocyclohex-2-enones as potent inhibitors of human lactate dehydrogenase. *Bioorganic and Medicinal Chemistry Letters*, **25**, 75–82.

79 Fauber, B.P., Dragovich, P.S., Chen, J., Corson, L.B., Ding, C.Z., Eigenbrot, C., Labadie, S., Malek, S., Peterson, D., Purkey, H.E., Robarge, K., Sideris, S., Ultsch, M., Wei, B., Yen, I., Yue, Q., and Zhou, A. (2014) Identification of 3,6-disubstituted dihydropyrones as inhibitors of human lactate dehydrogenase. *Bioorganic and Medicinal Chemistry Letters*, **24**, 5683–5687.

80 Billiard, J., Dennison, J.B., Briand, J., Annan, R.S., Chai, D., Colón, M., Dodson, C.S., Gilbert, S.A., Greshock, J., Jing, J., Lu, H., McSurdy-Freed, J.E., Orband-Miller, L.A., Mills, G.B., Quinn, C.J., Schneck, J.L., Scott, G.F., Shaw, A.N., Waitt, G.M., Wooster, R.F., and Duffy, K.J. (2013) Quinoline 3-sulfonamides inhibit lactate dehydrogenase A and reverse aerobic glycolysis in cancer cells. *Cancer and Metabolism*, **1**, 19.

81 Pantoliano, M.W., Petrella, E.C., Kwasnoski, J.D., Lobanov, V.S., Myslik, J., Graf, E., Carver, T., Asel, E., Springer, B.A., Lane, P., and Salemme, F.R. (2001) High-density miniaturized thermal shift assays as a general strategy for drug discovery. *Journal of Biomolecular Screening*, **6**, 429–440.

# 18
# FBDD Applications to Kinase Drug Hunting

*Gordon Saxty*

## 18.1
## Introduction

The aim of this chapter is to describe what has happened in the kinase field of fragment-based drug discovery since the beginning of the decade.

The older kinase drugs from FBDD have been reviewed elsewhere and are highlighted in the conclusion [1–6]. We are delighted to see the launch of vemurafenib **(1)** (Zelboraf) – the first fragment-derived drug that received FDA approval for the treatment of late-stage melanoma in 2011. The 7-azaindole hinge-binding motif has recently seen as much interest as the quinazoline/cyanoquinolines experienced at the turn of the century, although at that time, publications related to this hinge binder were apparently less well known.

Interestingly, neither of the efficiency matrices, LE = 0.31, nor the LLE = 3.25 are outstanding for vemurafinib, but still the drug made it to the market (Figure 18.1). This was probably facilitated by Plexxikon's strategic decision to target the biologically relevant V600E mutation, while other companies had programs on wild-type B-RAF. In this chapter, we will discuss approaches and the results of recently published articles in the FBDD kinase literature. For the figures in this chapter, all ligands are drawn with a consistent view: as drawn in the binding mode to the kinase, the hinge is at the bottom, the gatekeeper is on

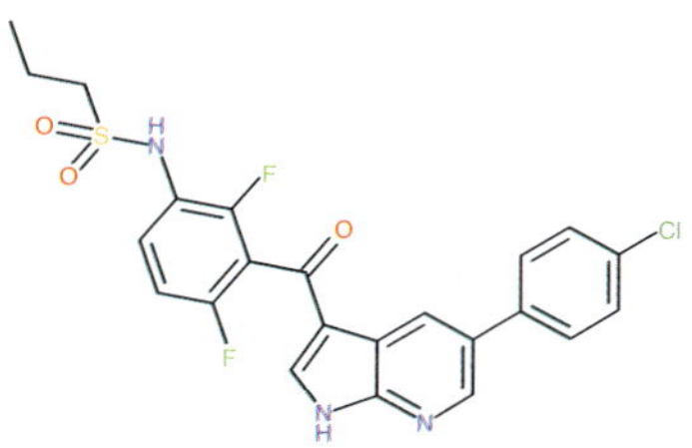


**(1)** BRAF V600E $IC_{50}$ = 0.031 μM
LE = 0.31 LLE 3.25

**Figure 18.1** Vemurafinib (1) the first, fragment derived, drug to be launched.

*Fragment-based Drug Discovery: Lessons and Outlook*, First Edition. Edited by Daniel A. Erlanson and Wolfgang Jahnke.

the left, and the "solvent exposed" area, if present, is toward the right-hand side of the scheme.

## 18.2 Virtual Screening and X-ray for PI3K

All isoforms of PI3K are targets, and each has known potential uses in different therapeutic areas. In a recent fragment screening campaign with PI3K p110β [7,8], the authors had a very clear understanding of the antitarget PI3K p110α because of the effects in insulin treatment. The project focussed on the p110β isoform subunit because it is associated with PTEN status and related to arterial thrombosis and oncogenesis.

The screening was carried out starting from a fragment collection, identified from the AstraZeneca compound bank. For this screening, 183 330 fragments were docked against a homology model of PI3K p110β (based on p110γ, pdb:2CHW). Selectivity of p110β over p110α at this stage was not envisaged or required. Poses were computationally analyzed. Fragments were ranked *in silico* to identify whether a minimum of 2 key hydrogen bonds were present in the active site of the docking. From 210 selected compounds that proceeded to biochemical assay, 18 hit fragments, in 5 chemotypes, were identified.

The authors observed an 8.6% hit rate from selected compounds screened in a bioassay ($IC_{50}$ range for p110β 3.6–81 μM; LE 0.35–0.53; LLE 3.7–7.0). The fragment selected for optimization was not the most potent from the screen.

In general, no selectivity of the fragment hits was observed versus the four isoforms: p110β, p110α, p110γ, and p110δ.

The key fragment hit **(2)**, published in the authors' second paper, also did not report those selectivity data (Scheme 18.1). We can only hypothesize that the compound was not completely soluble at the necessary concentrations. This second paper did report the isoform selectivity when a benzylamine **(3)** replaced the dimethylamine to give a desired sixfold selectivity for p110β over p110α. The fragment growth was based on a proposed docking pose of the morpholine, bound to the hinge region, and also of the pyrimidone oxygen to the protein. This growth strategy was expanded to a naphthalene **(4)** and gave good potency and efficiency.

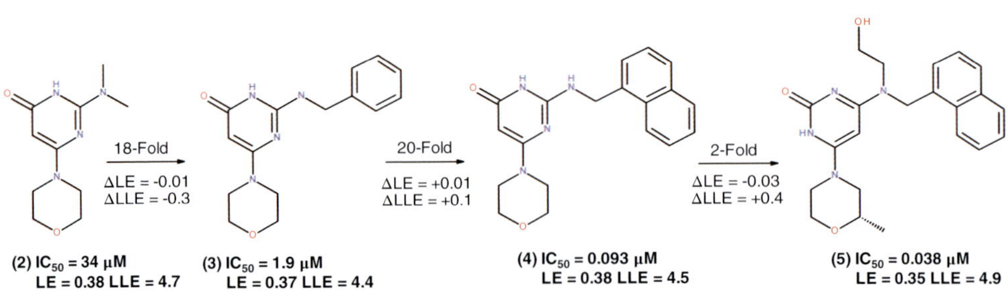


**Scheme 18.1** Fragment to lead for PI3K p110β.

In the next optimization step, the authors compromised potency for an improved measured log*D* by switching nitrogens around the central ring system. This aided solubility and metabolic stability. It also removed the propensity for reactive metabolites, which is obviously highly desirable.

In the second part of the paper, the authors describe the final optimization of this lead by the addition of a methyl to the morpholine and an ethanol motif onto an N–H. The final advanced molecule **(5)** is the first example of a potent p110β isoform inhibitor with 700-fold selectivity over p110α.

In conclusion, the authors reported what they set out to achieve: selectivity over the key antitarget, with the desired antithrombotic effect as demonstrated *in vivo* without the increased risk of bleeding and insulin resistance.

## 18.3
## High-Concentration Screening and X-ray for Rock1/2

The Lee Moffitt Cancer Centre used a combination of high-concentration screening (HCS) and structure-guided design to come up with moderately potent inhibitors of ROCK1/2 [9].

Because they did not have a fragment library, the authors had to synthesize a focused compound set before they progressed with limited resource.

In their library design, the authors used the known kinase binding motif – pyridine – and synthesized a library of seven compounds. They then synthesized an additional 10 compounds, using the known indazole kinase binding motif. A hit rate of 3/7 and 9/10, respectively, were observed where LE > 0.3 in ROCK1 and low μM potency were the criteria. The authors reported a 71% hit rate from HCS ($IC_{50}$ range for ROCK1 3.6–81 μM; LE 0.27–0.51).

All compounds synthesized at this stage were of a fragment size.

Because the initial fragment-to-lead efforts on pyridine were not fruitful, the authors used the known indazole template and improved potency. Then they applied the knowledge they had gained to produce compound (7), and confirmed the binding mode in ROCK1 (Scheme 18.2). This was elaborated into a

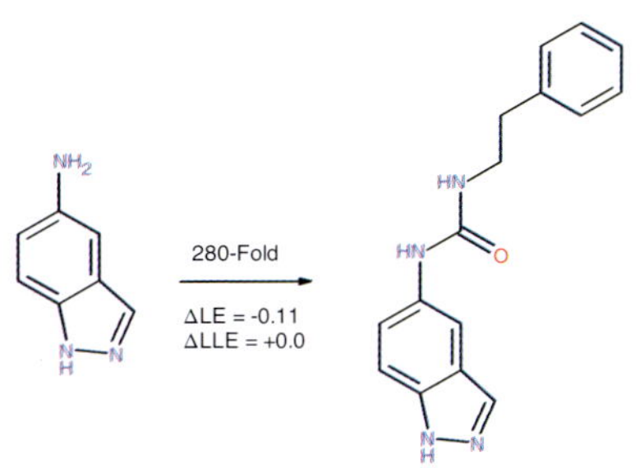


(6) $IC_{50}$ = 181 μM
LE = 0.51 LLE = 3.0

(7) $IC_{50}$ = 0.65 μM
LE = 0.40 LLE = 3.0

**Scheme 18.2** Fragment to lead for ROCK1.

1.69 μM ROCK1 and 0.10 μM ROCK2 inhibitor that worked in a relevant cell-based assay.

## 18.4
## Surface Plasmon Resonance for MAP4K4

The authors of this paper wanted to generate an inhibitor of MAP4K4 due to the cellular processes associated with neuronal degeneration. They set out to use the Genentech team's expertise in SPR technology [10]. They used a 2361 fragment collection (MW <350 Da) screening at a concentration of 100 μM in singletons and followed up with a 6-point $K_D$ determination to attain 225 confirmed hits [11]. The authors reported a 9.5% hit rate for SPR ($K_d$ range for MAP4K4 10–2000 μM; LE 0.30–0.67). The fragment reported was selected for elucidation based on LE and as a known drug scaffold for the kinase target class (substructure of the prodrug brivanib alaninate).

Unfortunately, this structurally enabled kinase did not afford a high-resolution protein X-ray crystal structure in complex with the fragment hits, although most other MAP4K4 protein–ligand structures were resolved to 2.2–2.6 Å. Therefore, the authors relied on a docking approach showing the nitrogen and two CH–O hydrogen bonds were presented to the hinge. At this point, there are two donor–acceptor–donor pharmacophores on the fragment. In addition, as the fragment is small enough, that is, with no blocking groups to clash with the gatekeeper, it also shows a symmetry element. There are at least four possible binding poses of the donor–acceptor–donor motif. It is possible that the mixture of binding mode may have caused the lack of clarity in the X-ray.

The authors proceeded with a strategy of using the docking mode and then growth. It would have been interesting if a more comprehensive fragment optimization strategy was reported at this point. For example, whether a "methyl scan" approach of several vectors would help prove or disprove their proposed docking mode.

In their paper, the authors discuss the addition of one heavy atom (methyl, compound **9**) and three heavy atoms (isopropyl, compound **10**) (Scheme 18.3). Both additions affected the LE and lost a minimal amount of potency. When there was an addition of six heavy atoms, potency was increased compared to the original fragment **8**. With the best 4-pyridyl group (compound **11**), there was a 10-fold increase of potency from **8** (Scheme 18.4). However, the LE was never achieved beyond the level found in symmetrical compound **8** from SPR and concluded the first fragment growth vector toward the DFG.

The second fragment growth of the pyrrolotriazine was toward the solvent exposed region, where it is known that groups give good potency increases. A good result at this stage was the addition of the six heavy-atom phenyl group.

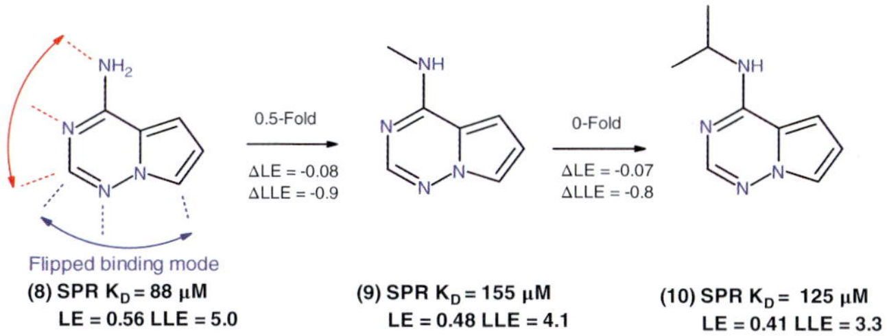


**Scheme 18.3** Initial fragment to lead for MAP4K4. Red and blue dotted lines represent potential binding modes to the hinge.

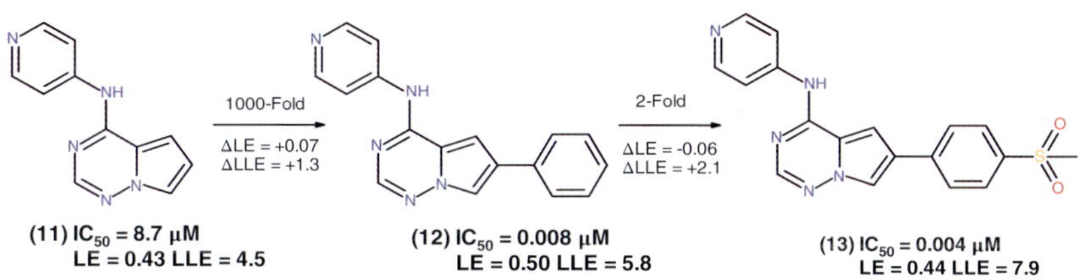


**Scheme 18.4** Fragment to lead for MAP4K4.

This addition gave a 1000-fold improvement in potency, as reported in a biochemical assay, and increased the LE and LLE (compound **12**).

The normal issues, plagued by the kinase small molecule field, of compound solubility and metabolic stability were highlighted. The authors immediately attempted to optimize these by lowering clog*P* by at least one unit. They chose three compounds for progression to PK. The results in PK showed that the "lipophilic kinase inhibitor" (log$D_{7.4}$ 3.5) with "solubilizing groups" added was poor in PK. However, a good oral bioavailability (72%F) was recorded with a nonsolubilizing sulfone group, compound **13**.

In conclusion, the authors reported the first fragment-to-lead campaign using SPR for MAP4K4 and achieved oral exposure in mice.

## 18.5 Weak Affinity Chromatography for GAK

This paper published by the SGC and the University of Linnaeus reports the relatively new technique of weak affinity chromatography (WAC) [12]. It is a technology validation paper, where the authors describe using WAC with cyclic G-associated kinase (GAK). This technology (also see Chapter 8) immobilized a protein of interest on to a capillary HPLC column. The immobilization occurred as a result of a reaction of the free protein surface lysine residues. These were immobilized onto aldehyde–silica residues using a reducing agent, sodium cyanoborohydride. A reference capillary HPLC column was also established, using ethanolamine to help mitigate any nonspecific stationary-phase binding.

The GAK protein was not structurally enabled at the time of publication [13]. Therefore, a homolog model was produced using five other proteins. From a library of 3200 fragments, the authors selected 170 compounds, based on an *in silico* Glide ranking in a virtual screen. It was important that the authors recognized that some fragments might be missed in this approach, because important water molecules would have been deleted from the protein active site before the virtual screening.

Then, the 170 compounds were screened in batch cocktails of 13–14 through both the GAK and the aminoethanol immobilized columns. Compounds were identified by a mass spectrum total ion count (MS TIC) with adenosine as a standard. The adenosine proved the activity of the GAK immobilized protein several weeks after immobilization. The retention time correlated with the $K_D$ value under standard conditions. More than 80% of the 170 fragments were detected by this methodology (i.e., quality control of the fragments occurred in the screen). From the retention time on the column, 78 fragments were estimated to have $K_D < 200\,\mu M$. The most potent fragment was retained for 117 min (GAK) versus 1.1 min (aminoethanol). This was confirmed by a thermal shift experiment.

The identified fragments were found to have a very good LE, up to 0.60, as measured by $K_D$. An additional advantage of this technique was the chiral resolution observed, where mixtures of enantiomers were screened.

The fragments obtained were biased to quercetin and adenosine. This is probably caused by the limitation and number of fragments chosen from the VS.

To conclude, the authors have shown the application of this technique to, at that time, a nonstructurally enabled kinase. They found several smaller fragments significantly more potent than adenosine. This is important work that addresses nonstructurally enabled targets in FBDD.

## 18.6 X-ray for CDK 4/6

In this paper, the authors set out to design CDK 4/6 inhibitors [14]. The biological rationale in oncology appears to be stronger for CDK4 inhibition. In addition, this paper shows that the CDK4 data appeared to be the priority enzyme to inhibit.

However, the protein–ligand structures of CDK4 were not reported in the paper, although the authors had published the structure 2 years earlier [15] and it is assumed that CDK6 was used as a surrogate.

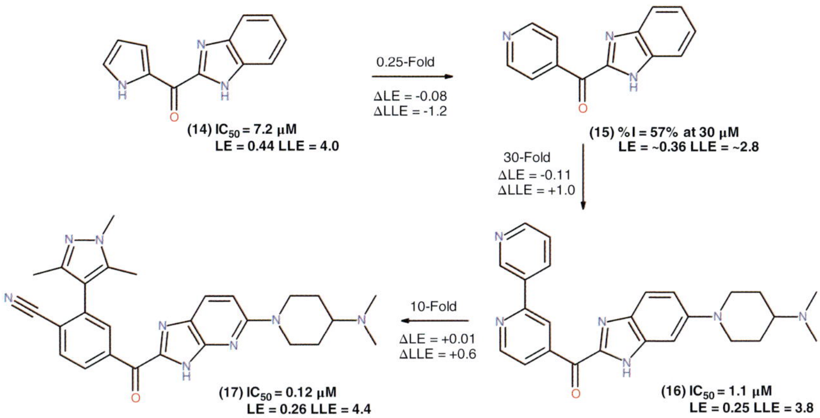


**Scheme 18.5** Fragment to lead for CDK6.

The authors make it clear, from the start of this project, that CDK1 or CDK2 inhibition was not acceptable. Their group already had a fragment-derived advanced CDK1/CDK2 phase II clinical asset [16].

The starting point in this project was a diarylketone that interacts with the hinge of CDK6. This is a particularly rare but not unknown chemotype to be found bound to the hinge, in the kinase literature. The statistics from the fragment screening campaign were not reported.

The fragment hit was optimized by changing the pyrrole to a pyridine where potency was affected, to the limit of solubility in the bioassay (compound **15**) (Scheme 18.5).

Compound **15** was then subjected to fragment growth in two directions, adding two ring systems to attain a 1.1 μM CDK6 inhibitor (compound **16**).

Compound **16** was designed to place a charged nitrogen motif in the protein. This would clash electrostatically with Lys89 in CDK2 (a Thr residue in CDK6) and begin to achieve selectivity against the antitarget.

Next, the authors fused an extra phenyl group on to the pyridine to obtain an isoquinoline compound that was 20-fold more active (not shown). From this point, optimization focused on keeping the molecular weight in a similar region to **16**.

The authors then used medicinal chemistry to optimize in three separate ways. The potent isoquinoline was changed to a multisubstituted pyrazole. A nitrogen atom replaced a carbon in the benzamidiazole. Finally, they modified the acceptor in the pyridine nitrogen to a nitrile (compound **17**) (Scheme 18.6).

This molecule was progressed through a biological cascade. The compound was 10-fold more potent for CDK4 than for CDK6. The compound showed a dose-dependent efficacy in a mantle cell lymphoma model that overexpressed CDK4's partner, cyclin D.

## 18.7
## High-Concentration Screening, Thermal Shift, and X-ray for CHK2

The van Montford group at the ICR start their paper describing this CHK2 fragment screen by stating the importance of orthogonal screening in FBDD. In this case study, they used a combination of an alpha screen and thermal shift techniques to reduce the false negative rate before X-ray crystallographic screening [17].

The authors then describe two series of hit expansion, using an in-house HTS library of 71 000 compounds. Their aim was to find structurally close analogs. Improvements in potency were found in the single-digit μM range.

In the first series, a quinoxaline amide **17** showed a small improvement in potency with the addition of two heavy atoms, compound **18**. There was also a similar improvement in a urea subseries (not shown).

In the second series **19**, the compound from the hit expansion phase improves potency, compound **20**. However, it is difficult to compare because the X-ray protein–ligand structures showed a shift in binding mode and an alternative hydrogen-bonding network to the hinge of the protein (Scheme 18.7).

To conclude, this screen shows a rich amount of X-ray data from several chemotypes, which would provide the discovery team with many options on growth and merging. The authors also comment that the two chemotypes described above are not reported in the CHK2 literature. The most potent fragment chemotype identified is significantly less well known (compound **21**) and, unless it

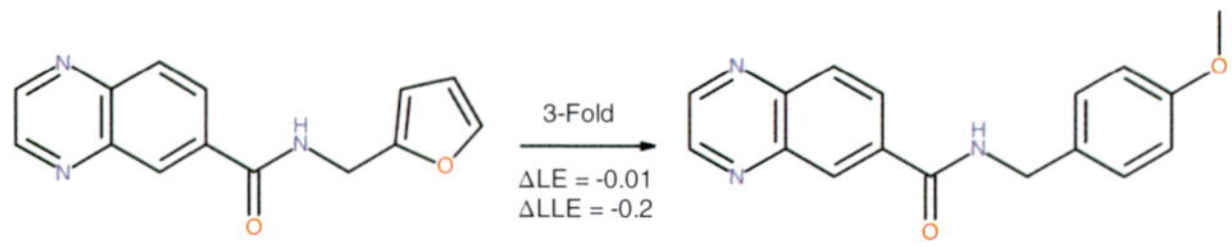


**(17)** $IC_{50}$ (ms) = 11.7 μM
LE = 0.36 LLE = 3.7

**(18)** $IC_{50}$ (ms) = 3.6 μM
LE = 0.35 LLE = 3.5

**Scheme 18.6** Fragment optimization for CHK2 using the reported mobile shift (ms) protocol.

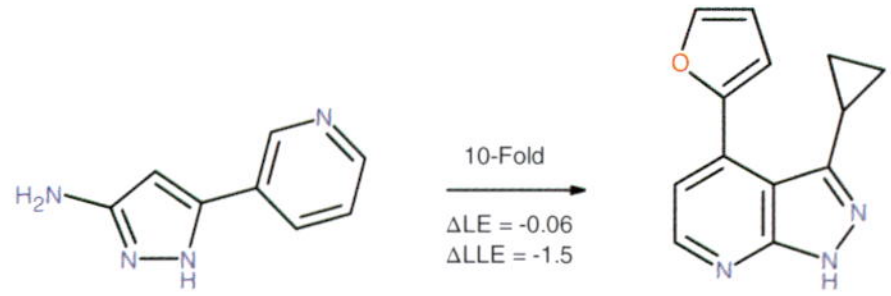


**(19)** $IC_{50}$ (ms) = 72 μM
LE = 0.48 LLE = 4.1

**(20)** $IC_{50}$ (ms) = 7.2 μM
LE = 0.42 LLE = 2.3

**Scheme 18.7** Fragment optimization for CHK2 using the mobile shift (ms) protocol.

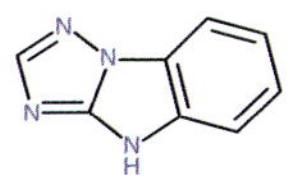

(21) $IC_{50}$(ms) = 2.7 μM
LE = 0.64 LLE = 4.4

**Scheme 18.8** Potent and LE fragment from screening against CHK2 using the mobile shift (ms) protocol.

had a known genotoxic flag, would be worthy of fragment-2-lead efforts given the LE and LLE (Scheme 18.8).

## 18.8 Virtual Screening and Computational Modeling for AMPK

The authors in this case study used computational models to aid *in silico* FBDD of AMPK inhibitors [18]. Their proprietary software enabled them to use grand canonical Monte Carlo (GCMC) fragment simulations of the 1785 small molecule fragments proposed by Moore [19].

The authors set up a docking model based on the protein AMPK (pdb code 2H6D) and used 22 residues of MARK1 (pdb 2HAK) to mimic the absent activation loop. Starting from nanomolar compound **22**, they describe the *in silico* driven FBDD replacement of three areas of **22**, followed by ADMET filtering of all the proposed molecules (Scheme 18.9).

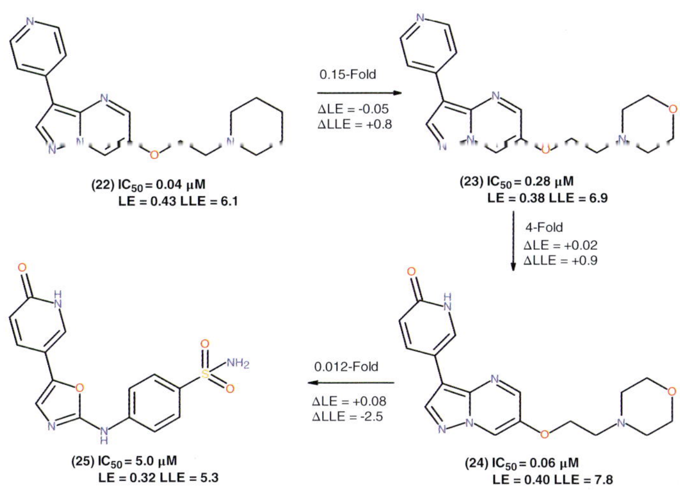


**Scheme 18.9** Computational fragment optimization for AMPK.

The first part of **22** to be replaced was the piperidine motif by a morpholine [20]. It did not appear to be fragment driven, because an inhibitory potency of sevenfold was lost. It was tried because the ADMET computational filter was expected to aid improved intestinal absorption.

Next, the structure–activity relationship of the compounds synthesized suggested that additional potency might be achievable in the replacement of the pyridyl motif. It pointed toward the aspartic acid residue of the DFG loop. The GCMC fragment simulations proposed several compounds, of which six were synthesized. The most encouraging *in silico* replacement of pyridine was a pyridone **24**. It recovered the LE and potency back to the original compound.

Finally, the last molecule designed and synthesized followed a dual GC–MC fragment simulations protocol. The initial replacement concentrated on the core hinge binder. Of the seven suggested cores, only three were synthesized because of the synthetic feasibility. This last GC–MC fragment simulation showed that phenylsulfonamide **25** would replace the ethyl piperidine moiety. Disappointingly, this computational approach afforded an 80-fold drop in potency from **24**.

The authors conclude that compound **24** was, in a relevant cell-based assay, more potent than compound **22**. However, each new compound reported from **22** to **24** showed a 1 log unit improvement in LLE, which is important.

## 18.9
## High-Concentration Screening, NMR, and X-ray FBDD for PDK1

The PDK1 story was one of the most exciting in the FBDD field. PDK1 was a target of great interest and seemed to be as popular in the field as HSP90 had been a few years earlier. Each major conference appeared to provide the next iteration of progress for the target. However, it is disappointing that we have not seen, to date, a fragment-derived clinical candidate from this target.

In this case study [21], GSK started with their 1065 fragments using an HCS. The authors describe 36 compounds that achieved the criteria of <400 μM and LE >0.40.

From these 36 compounds, 6 compounds were selected for progression to X-ray. Interestingly, only those that gave a positive result in NMR achieved a crystal structure.

The authors reported a 3.4% hit rate from HCS with the $IC_{50}$ range of concentrations 50–380 μM (LE 0.43–0.59). The authors chose fragments based on the criteria: LE >0.40, $IC_{50}$ <400 μM, whether or not they orthogonalized with STD NMR, and if the chemistry was thought to be tractable.

The approach of screening by catalog, referred to as "SAR by catalog," was achieved in this case by screening GSK's corporate library. Compounds were

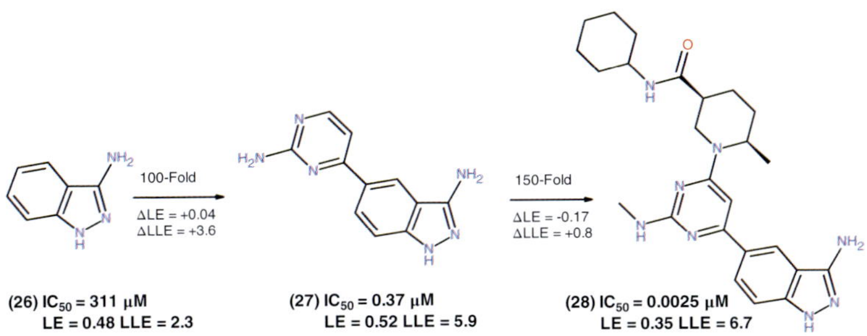


**Scheme 18.10** Fragment to lead for PDK1.

selected for progression if they were <300 Da and resulted in a good improvement in activity into the sub μM range.

Compound **27** was generally selective for a 226 Da fragment, but it had some contaminating activity versus AurA, CDK2, IKK1, JAK3, and KDR (Scheme 18.10).

It is arguable if this particular lead molecule **27** could have been identified by HTS of the GSK corporate collection as it was already present.

A following paper described further elaboration that afforded a very advanced compound **28** [22,23]. The conclusion of this paper was that **28** was only modestly efficacious, which required further elaboration. In summary, we observe an initial increase in LE and then a drop but the LLE has increased 4.4 log units, and finally 5 log units of potency was achieved during the optimization.

## 18.10 Tethering Mass Spectometry and X-ray for PDK1

This case study involves another example of FBDD against PDK1. It describes how the Erlanson group used the tethering-based mass spectrometry fragment-based approach [24].

The authors used the general and known kinase warhead molecule with the necessary "extender" attached. The electrophilic Michael acceptor reacted with the cysteine amino acid that the authors chose to mutate, within the protein active site.

The protein-extender complex was subjected to the disulfide fragment library. Several thousand compounds will afford a linked molecule. Only when good energy of binding is present will the mass spectrometry observe the linked library member and extender/protein. The hit compounds were confirmed by LC–MS determination. To determine and rank the hits, the small molecule inhibitors were synthesized. To achieve this, the covalent part was removed and disulfide replaced by carbons to give a stable small molecule inhibitor.

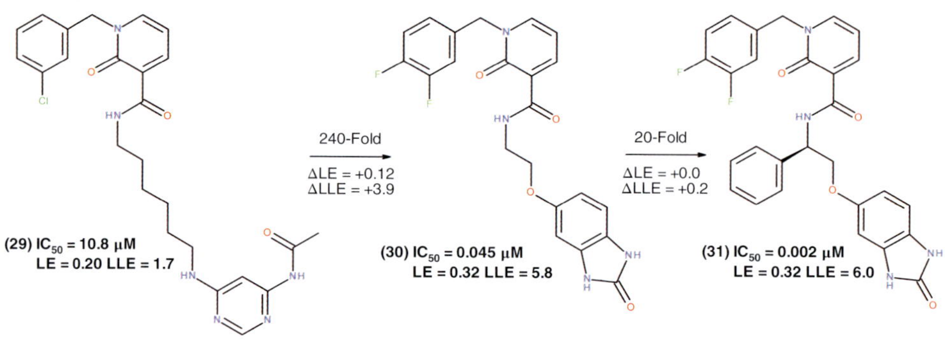


**Scheme 18.11** Fragment to lead for PDK1.

The hit compound **29** was 0.2 μM versus the unphosphorylated protein (Scheme 18.11). During this process, the original warhead and linker were optimized under standard medicinal chemistry conditions that mainly improved all interactions and locked down rotatable bonds.

Finally, the introduction of a chiral center afforded **31** that showed *in vivo* activity, when dosed by the intraperitoneal route. However, it also shows that the technique and technology can explore many pockets adjacent to a binding pocket. This example shows how large improvements in LE and LLE can be achieved.

## 18.11
## NMR and X-ray Case Study for Abl (Allosteric)

As the kinase field evolves, applications toward FBDD become more innovative and the work by Jahnke's group is a perfect example. The group goes into a high level of structural biology to produce modulators of ABL kinase. In this case study, the group adds significant value by describing allosteric inhibition and activation. This could potentially help with the generation of new classes of kinase drugs that can be tested in a clinical setting.

The clinicians need next-generation INDs and this makes sense as we see first-, second-, and now, for some oncology targets, even third-generation INDs are proceeding to clinical trials. What next after the *n*th clinical mutation occurs in the active site?

In the kinase field, the pharmaceutical or biotech company can either (1) design a new molecule to inhibit the relevant clinical mutation or (2) design a new molecule to be the best-in-class that inhibits the original kinase and all clinically relevant mutations, thereby hoping to supersede the current therapy.

In this case study [25], which was preceded by the known allosteric inhibitor GNF2, the authors apply FBDD to an allosteric pocket. They initiated the screening with the known ATP site inhibitor imatinib or STI571 that blocks binding in the site. A small library of 500 fragments was screened using NMR.

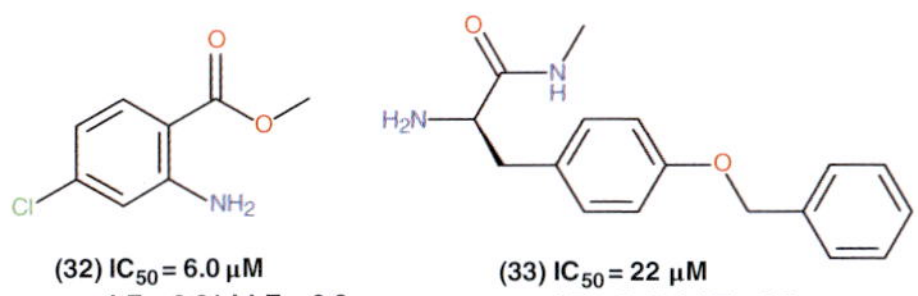


(32) $IC_{50} = 6.0\ \mu M$
LE = 0.61 LLE = 2.3

(33) $IC_{50} = 22\ \mu M$
LE = 0.31 LLE = 2.7

**Scheme 18.12** Fragment identification for ABL.

The screening identified several fragments but they disclosed only two compounds **32** and **33** (Scheme 18.12).

This case study is the example of FBDD delivering two new "novel" chemotypes that are significantly different from the known inhibitors or, in this case, modulators. The compounds have significantly potent $K_D$ values with good ligand efficiency. The authors also superimpose the natural ligand myristate to show that the ligands do not overlay and that a helix can "bend" toward the pocket.

The authors discuss in the paper the biological significance of this and they set up NMR protocols to observe the helix movement. They suggest that these fragments, depending on the orientation of the l-helix, can modulate antagonism or agonism.

Further medicinal chemistry evaluation of the fragments was not described in this paper, but it does show the power of FBDD screening, in this case by NMR. We must wait to see if activation of kinases will help design the kinase drugs of the future, perhaps in other therapeutic areas outside oncology.

## 18.12
## Review of Current Kinase IND's and Conclusions

In this last section, we would like to describe the current state of the kinase investigational new drugs (IND) in clinical trials.

In the early 1990s, everyone was highly skeptical that kinases, as a target class, could be inhibited effectively *in vivo*. There were two important perceived obstacles to a kinase drug: could they ever compete with ATP at 2 mM and could they gain enough selectivity against the ~600 closely related enzymes? However, over the past two decades, blockbuster breakthroughs have changed the pharmaceutical industry's opinion on the undrugable class to a drugable one.

The industry has produced many oncology drugs in the kinase area. When those patients on treatment relapse because of point mutations, the industry has another advantage: a clinically relevant target to inhibit. However, some hold the opinion that kinase drugs are "easy." It is clear that kinases, if structurally enabled, can easily find ligands to bind with, but making an IND is not easy. The kinase class of INDs suffered what was and still might be one of the worst failure rates in the clinic. That might also be associated with the link to the higher failure rate in oncology [26].

To review the success rates, we can describe the FBDD poster-child, vemurafenib. Plexxikon's HCS and X-ray follow up efforts have been reviewed on many occasions, as have all the other INDs listed in Table 18.1 [27]. Therefore, this

**Table 18.1** The list of compounds in clinical trials with physiochemical properties.

| | $IC_{50}$ | $pIC_{50}$ | ACDlog*P* | LE | LLE | Phase | Route |
|---|---|---|---|---|---|---|---|
| Vemurafenib[z] | 0.031 | 7.51 | 4.26 | 0.31 | 3.25 | Launched | Oral |
| 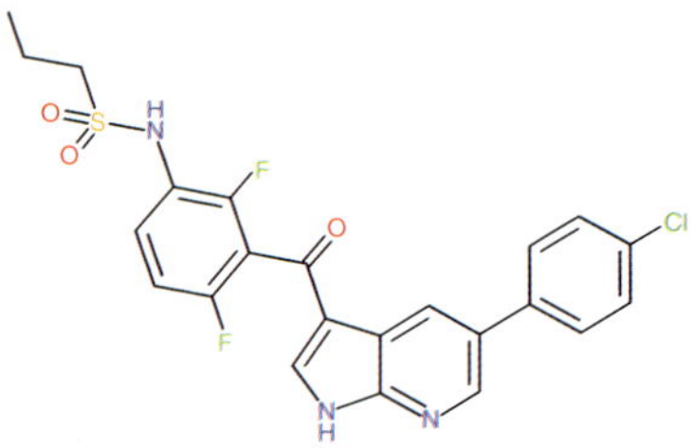 | | | | | | | |
| AT7519[x] | 0.047 | 7.33 | 0.95 | 0.4 | 6.38 | Phase II | IV |
| 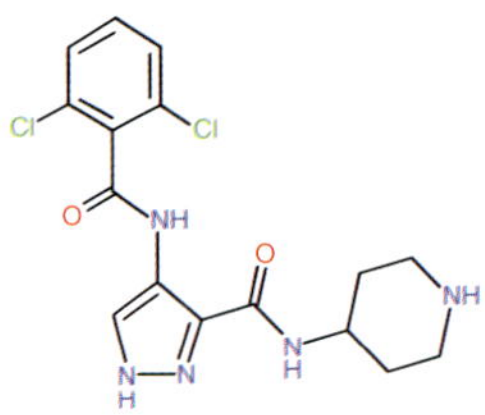 | | | | | | | |
| AT9283[x] | ~0.003 | 8.52 | 0.92 | 0.41 | 7.6 | Phase II | IV |
| 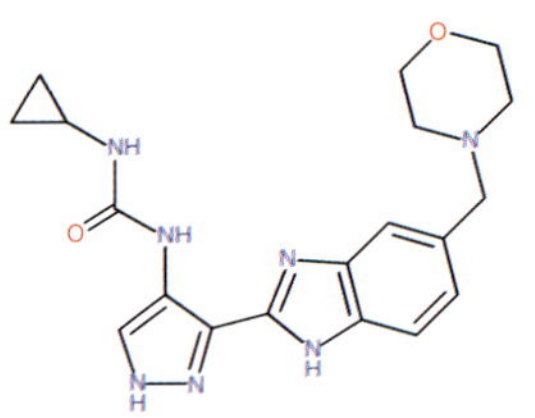 | | | | | | | |
| AZD5363[z] | 0.003 | 8.52 | 1.04 | 0.39 | 7.48 | Phase II | Oral |
| 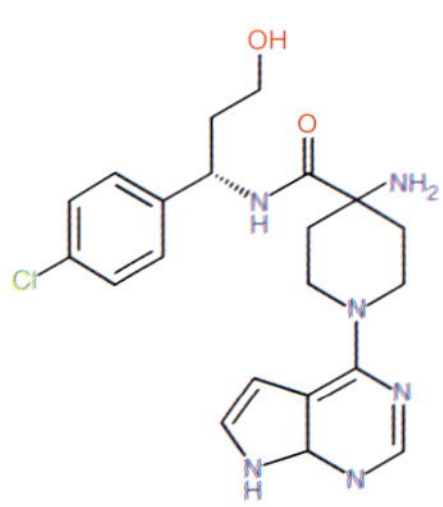 | | | | | | | |

**Table 18.1** (*Continued*)

| | $IC_{50}$ | $pIC_{50}$ | ACDlog*P* | LE | LLE | Phase | Route |
|---|---|---|---|---|---|---|---|
| Linifanib[y] 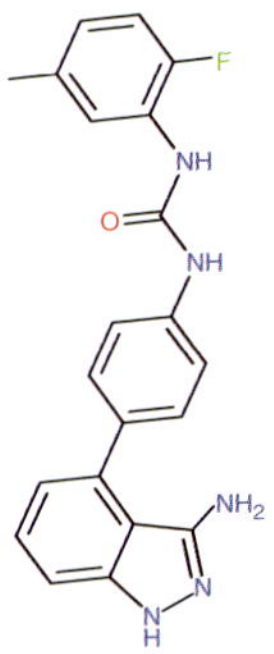 | 0.004 | 8.4 | 4.34 | 0.42 | 4.06 | Phase II | Oral |
| AT13 148[x] 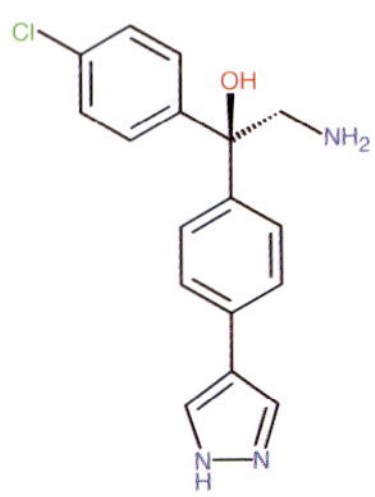 | 0.038 | 7.42 | 2.68 | 0.47 | 4.74 | Phase I | Oral |
| SNS-314[z] 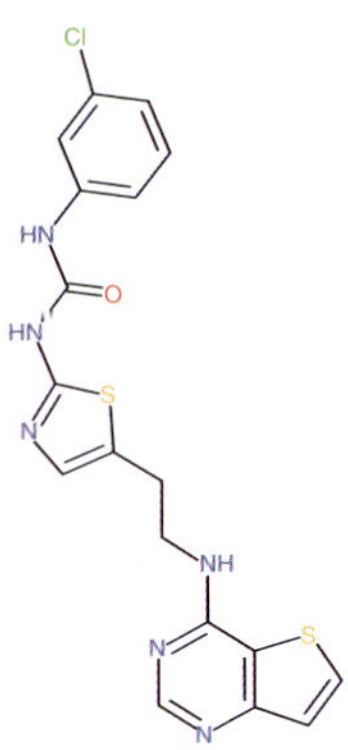 | 0.003 | 8.5 | 5.36 | 0.43 | 3.14 | Phase I | Oral |
| PLX3397 | Structure unknown. Status: phase II, oral. | | | | | | |
| JNJ-42756493 | Structure unknown. Status: phase I, oral. | | | | | | |
| SGX-393 | Status unknown (since partnered with Novartis) | | | | | | |

analysis will address the other issues we should consider within FBDD, when applied to kinases targets.

First, when concentrating on success, target consideration and patient selection is essential. Then, we must evaluate the novelty consideration versus the risk of proceeding with a new chemical chemotype that has never been used in a drug before. It is important to consider selectivity versus promiscuity and to decide whether we should rank LE/LLE or not and, given enough time and resources, perhaps have the time to fix LE/LLE. We should also decide whether we should have a low lipophilicity or pay less attention. Or alternatively, we could start at low lipophilicity and high ligand efficiency start points. Thus, lead optimization teams can then take advantage of the lower lipophilicity and molecular weight.

If we start this final review by considering novelty – the question is: *have we made "more novel" drugs?*

It is obvious that they are novel from a patent point of view. If we look at all the compounds in the table (Table 18.1), we can state that there are three main hinge-binding chemotypes (monocyclic pyrazole,[x] bicyclic indazole,[y] and the similar pyrrolopyrimidine/thienopyrimidine heterocycles[z]). These bind at the hinge region of the kinase. In clinical trials, we see only three: pyrazoles and the related fused bicycle indazole, plus a 7-azaindole class.

Note that scaffolding hopping by switching a carbon to a nitrogen or sulfur does not classify as something different in this analysis, if the binding to the hinge is the same.

We might view FBDD as delivering these three new chemotypes for the kinase field. We might decide that they are different, for example, from the very successful quinazoline class, where many HTS-derived kinase inhibitors came from, and were heavily patented (as are indazoles). If we analyze this view, we might argue that the HTS-derived quinazoline patents should significantly precede the FBDD-derived class. However, Chart 18.1 shows the annual distribution of all these hinge binders. They share approximately similar years of release. So, that dispels the myth, at this current point in time, that FBDD invented or popularized new chemotypes in the kinase field.

We hope that a full examination of the fragment chemical space, like the van Montfort work, will help popularize newer hinge binders in kinases.

Moving on to look at the potency of the specific molecules, we can see little discrimination and in part we do not have a large data set to prove or disprove these statements associated with Table 18.1. We can see that vemurafenib has one of the weaker potencies, although assay format and readouts are certainly different from the other inhibitors.

When we begin to consider clog*P*, we can see that the oral kinase inhibitors tend to be higher in lipophilicity with AZD5363 being the outlier. Vemurafenib has one of the highest calculated lipophilicity values. This brings us to the differing views about calculated properties. However, even the measured "real" properties in octanol and its relevance to drug hunting have been questioned [29,30]. Table 18.1 shows the most advanced oral inhibitor is one of the more lipophilic.

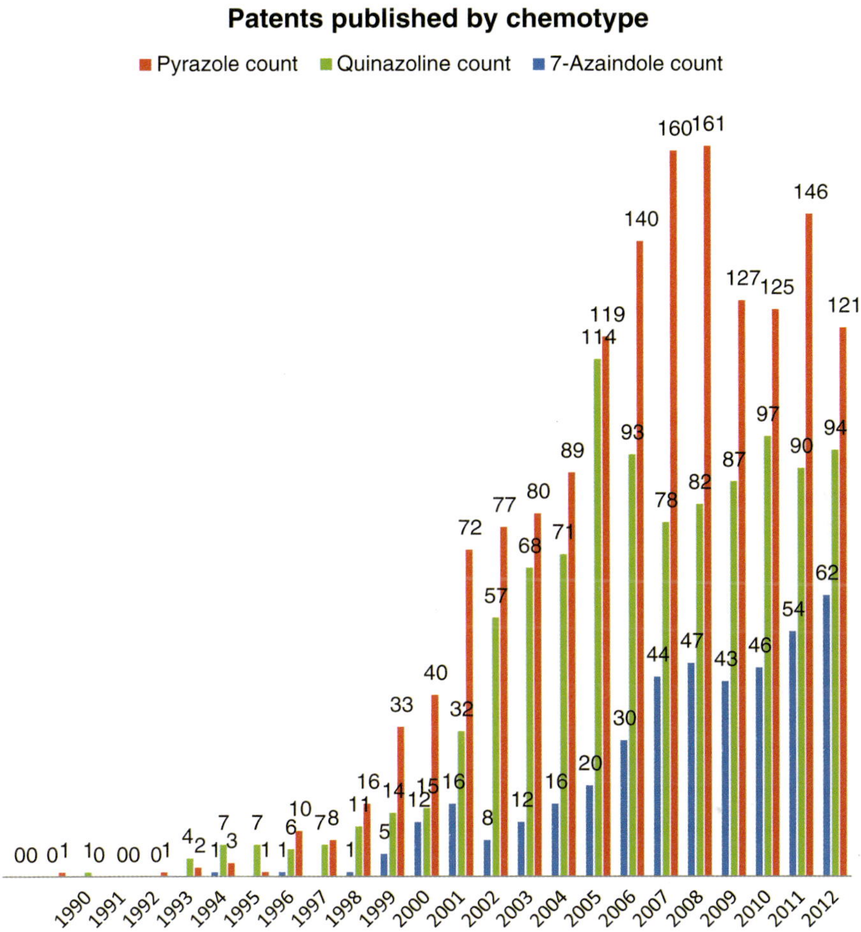


**Chart 18.1** Patents published for 3 kinase hinge binders [28].

Coupled with the weaker potency, it offers nearly the poorest LLE score. From a PFI point of view, it may not have been progressed [31].

The science community is full of very interesting views. On the one side, some say that poorer physiochemical properties will lead to higher failure in the clinic. This is undeniable. It is especially important if an organization is accounting for 20–30 clinical trials in the coming years and is aiming to save ~10% from the bottom line. If one takes the accountant paradigm shift and adds up all that loss in revenue, it looks like an awfully poor idea to set oneself up for failure. In contrast, on the other side, some say that if we all accept that there is a risk of higher failure, then outlier drugs just will not be invented. I am sure that in a few years' time we will be able to review the application of medicinal chemistry and if the trend will go bRO5 [32]. Only then can we decide which drugs should not have been advanced into clinical trials and which should be available to treat patients.

There is also a generalization that easier targets are now addressed leaving the undruggable targets that might require rule-breaker bRO5 small molecules to be invented.

We can illustrate this with the example of sorafenib (Nexavar). It was a true trailblazer, with all the *hearsay hideous LE and LLE properties*, yet it still proceeded to market launch in one of the shortest times for a kinase inhibitor (even with the delay and excellent work in pharmaceutical development) [33]. Our fragment community knew the drug was there for a significant time period, yet we did not come up with a fast-follower "nicely soluble" and "low log*P*" next-generation compound. Instead, the next-in-class IND, regorafenib (Stivarga) appeared, with projected sales in excess of $1 billion. It was as if sorafenib with an extra fluorine atom was treading in the similar *dreaded* physicochemical space. We can only ask why sorafenib was not superseded by a more drug-like compound.

So, where do we end up in the other paradigm? A biotech has one attempt for clinical trials. They might get a delay of 2 years for solubility and formulation, but they discharge the risk and manage to get human exposure, if they can afford to. With the typical status of biotech finances, the one or, if lucky, two chances in clinical trials will either be successful or inevitably cause *company attrition*, not to mention *drug attrition*.

We can conclude that the future for kinase drug discovery will still have an impact on standard ligands bound in the ATP site and all mutations that arise. It remains to be reported if they are bRO5.

What is interesting to report is the minimal real progress in the type II kinase area in this chapter, with the exception of the Erlanson's work. Especially, as some of the original kinase FBDD campaigns over 10 years ago found type II DFG-Out hits [34,35]. it will be satisfying to see more advances in type II inhibition from FBDD. Some of the most interesting work is done by Jahnke's group with newer mechanisms of kinase modulation. They may help the pharmaceutical industry to diversify away from oncology targets.

## References

1 Erlanson, D.A. (2009) *Kinase Inhibitor Drugs; Fragment-Based Drug Discovery of Kinase Inhibitors* (eds R. Li and J.A. Stafford), John Wiley & Sons, Inc., Hoboken.

2 Murray, C.W., Verdonk, M.L., and Rees, D.C. (2012) Experiences in fragment-based drug discovery. *Trends in Pharmacological Sciences*, **33**, 224–232.

3 Zartler, E.R. (2014) Fragonomics: the -omics with real impact. *ACS Medicinal Chemistry Letters*, **5**, 952–953.

4 Mazanetz, M., Law, R., and Whittaker, M. (2014) *Hit and Lead Identification from Fragments: De Novo Molecular Design* (ed. G. Schneider), Wiley-VCH Verlag GmbH, Weinheim.

5 Joseph-McCarthy, D., Campbell, A.J., Kern, G., and Moustakas, D. (2014) Fragment-based lead discovery and design. *Journal of Chemical Information and Modeling*, **54**, 693–704.

6 Toraskar, M.P., Singasane, N.S., Pichake, J.B., and Kadam, V. (2013) Fragment based drug discovery: a tool for drug

discovey. *International Journal of Drug Design and Discovery*, **4**, 1083–1092.

7 Giordanetto, F., Wållberg, A., Cassel, J., Ghosal, S., Kossenjans, M., Yuan, Z.-Q., Wang, X., and Liang, L. (2012) Discovery of 4-morpholino-pyrimidin-6-one and 4-morpholino-pyrimidin-2-one-containing phosphoinositide 3-kinase (PI3K) p110β isoform inhibitors through structure-based fragment optimisation. *Bioorganic and Medicinal Chemistry Letters*, **22**, 6665–6676.

8 Giordanetto, F., Kull, B., and Dellsén, A. (2011) Discovery of novel class 1 phosphatidylinositide 3-kinases (PI3K) fragment inhibitors through structure-based virtual screening. *Bioorganic and Medicinal Chemistry Letters*, **21**, 829–835.

9 Li, R., Martin, M.P., Liu, Y., Wang, B., Patel, R.A., Zhu, J.Y., Sun, N., Pireddu, R., Lawrence, N.J., Li, J., Haura, E.B., Sung, S.S., Guida, W.C., Schonbrunn, E., and Sebti, S.M. (2012) Fragment-based and structure-guided discovery and optimization of Rho kinase inhibitors. *Journal of Medicinal Chemistry*, **55**, 2474–2478.

10 Giannetti, A.M. (2011) *Experimental Design to Validated Hits: A Comprehensive Walk-through of Fragment Lead Identification Using Surface Plasmon Resonance; Methods in Enzymology: Fragment-Based Drug Design Tools, Practical Approaches, and Examples*, vol. **493** (ed. L.C. Kuo), Elsevier, San Diego, Burlington and London, pp. 169–218.

11 Wang, L., Stanley, M., Boggs, J.W., Crawford, T.D., Bravo, B.J., Giannetti, A.M., Harris, S.F., Magnuson, S.R., Nonomiya, J., Schmidt, S., Wu, P., Ye, W., Gould, S.E., Murray, L.J., Ndubaku, C.O., and Chen, H. (2014) Fragment-based identification and optimization of a class of potent pyrrolo[2,1-f][1,2,4]triazine MAP4K4 inhibitors. *Bioorganic and Medicinal Chemistry Letters*, **24**, 4546–4552.

12 Meiby, E., Knapp, S., Elkins, J.M., and Ohlson, S. (2012) Fragment screening of cyclin G-associated kinase by weak affinity chromatography. *Analytical and Bioanalytical Chemistry*, **404**, 2417–2425.

13 Chaikuad, A., Keates, T., Vincke, C., Kaufholz, M., Zenn, M., Zimmermann, B., Gutiérrez, C., Zhang, R.-G., Hatzos-Skintges, C., Joachimiak, A., Muyldermans, S., Herberg, F.W., Knapp, S., and Müller, S. (2014) Structure of cyclin G-associated kinase (GAK) trapped in different conformations using nanobodies. *Biochemical Journal*, **459**, 59.

14 Cho, Y.S., Angove, H., Brain, C., Hiu-Tung Chen, C., Cheng, H., Cheng, R., Chopra, R., Chung, K., Congreve, M., Dagostin, C., Davis, D.J., Feltell, R., Giraldes, J., Hiscock, S.D., Kim, S., Kovats, S., Lagu, B., Lewry, K., Loo, A., Lu, Y., Luzzio, M., Maniara, W., McMenamin, R., Mortenson, P.N., Benning, R., O'Reilly, M., Rees, D.C., Shen, J., Smith, T., Wang, Y., Williams, G., Woolford, A.J.-A., Wrona, W., Xu, M., Yang, F., and Howard, S. (2012) Fragment-based discovery of 7-azabenzimidazoles as potent, highly selective, and orally active CDK4/6 inhibitors. *ACS Medicinal Chemistry Letters*, **3**, 445–449.

15 Day, P.J., Cleasby, A., Tickle, I.J., O'Reilly, M., Coyla, J.E., Holding, F.P., McMenamin, R.L., Yon, J., Choprab, R., Lengauerc, C., and Jhoti, H. (2009) Crystal structure of human CDK4 in complex with a D-type cyclin. *PNAS*, **106**, 4166–4170.

16 Wyatt, P.G., Woodhead, A.J., Berdini, V., Boulstridge, J.A., Carr, M.G., Cross, D.M., Davis, D.J., Devine, L.A., Early, T.R., Feltell, R.E., Lewis, E.J., McMenamin, R.L., Navarro, E.F., O'Brien, M.A., O'Reilly, M., Reule, M., Saxty, G., Seavers, L.C.A., Smith, D.-M., Squires, M.S., Trewartha, G., Walker, M.T., and Woolford, A.J.-A. (2008) Identification of *N*-(4-piperidinyl)-4-(2,6-dichlorobenzoylamino)-1*H*-pyrazole-3-carboxamide (AT7519), a novel cyclin dependent kinase inhibitor using fragment-based X-ray crystallography and structure based drug design. *Journal of Medicinal Chemistry*, **51** 4986–4999.

17 Silva-Santisteban, M.C., Westwood, I.M., Boxall, K., and Brown, N. (2013) Fragment-based screening maps inhibitor

interactions in the ATP-binding site of checkpoint kinase 2. *PLoS ONE*, **8** (6), e65689.
18 Machrouhi, F., Ouhamou, N., Laderoute, K., Calaoagan, J., Bukhtiyarova, M., Ehrlich, P.J., and Klon, A.E. (2010) The rational design of a novel potent analogue of the 5′-AMP-activated protein kinase inhibitor compound C with improved selectivity and cellular activity. *Bioorganic and Medicinal Chemistry Letters*, **20**, 6394–6399.
19 Moore, W.R. Jr. (2005) Maximizing discovery efficiency with a computationally driven fragment approach. *Current Opinion in Drug Discovery and Development*, **8**, 355.
20 Viollet, B., Horman, S., Leclerc, J., Lantier, L., Foretz, M., Billaud, M., Giri, S., and Andreelli, F. (2010) AMPK inhibition in health and disease. *Critical Reviews in Biochemistry and Molecular Biology*, **45**, 276.
21 Medina, J.R., Blackledge, C.W., Heerding, D.A., Campobasso, N., Ward, P., Briand, J., Wright, L., and Axten, J.M. (2010) Aminoindazole PDK1 inhibitors: a case study in fragment-based drug discovery. *ACS Medicinal Chemistry Letters*, **1**, 439–442.
22 Medina, J.R., Becker, C.J., Blackledge, C. W., Duquenne, C., Feng, Y., Grant, S.W., Heerding, D., Li, W.H., Miller, W.H., Romeril, S.P., Scherzer, D., Shu, A., Bobko, M.A., Chadderton, A.R., Dumble, M., Gardiner, C.M., Gilbert, S., Liu, Q., Rabindran, S.K., Sudakin, V., Xiang, H., Brady, P.G., Campobasso, N., Ward, P., and Axten, J.M. (2011) *Journal of Medicinal Chemistry*, **54**, 1871–1895.
23 Medina, J.R. (2013) Selective 3-phosphoinositide-dependent kinase 1 (PDK1) inhibitors: dissecting the function and pharmacology of PDK1. *Journal of Medicinal Chemistry*, **56**, 2726–2737.
24 Erlanson, D.A., Arndt, J.W., Cancilla, M. T., Cao, K., Elling, R.A., English, N., Friedman, J., Hansen, S.K., Hession, C., Joseph, I., Kumaravel, G., Lee, W.C., Lind, K.E., McDowell, R.S., Miatkowski, K., Nguyen, C., Nguyen, T.B., Park, S., Pathan, N., Penny, D.M., Romanowski, M.J., Scott, D., Silvian, L., Simmons, R.L., Tangonan, B.T., Yang, W., and Sun, L. (2011) Discovery of a potent and highly selective PDK1 inhibitor via fragment-based drug discovery. *Bioorganic and Medicinal Chemistry Letters*, **21**, 3078–3083.
25 Jahnke, W., Grotzfeld, R.M., Pellé, X., Strauss, A., Fendrich, G., Cowan-Jacob, S.W., Cotesta, S., Fabbro, D., Furet, P., Mestan, J., and Marzinzik, A.L. (2010) Binding or bending: distinction of allosteric Abl kinase agonists from antagonists by an NMR-based conformational assay. *Journal of the American Chemical Society*, **132**, 7043–7048.
26 Kola, I. and Landis, J. (2004) Can the pharmaceutical industry reduce attrition rates? *Nature Reviews Drug Discovery*, **3**, 711–716.
27 Chessari, G. and Woodhead, A.J. (2009) From fragment to clinical candidate; a historical perspective. *Drug Discovery Today*, **14**, 668–675.
28 Analysis completed in SciFinder using kinase term and substructure of the hinge binder. Random analysis of hits confirmed the structures.
29 Kenny, P.W., Leitão, A., and Montanari, C. A. (2014) Ligand efficiency metrics considered harmful. *Journal of Computer-Aided Molecular Design*, **28**, 699–710.
30 Kenny, P.W., Montanari, C.A., and Prokopczyk, I.M. (2013) ClogPalk: a method for predicting alkane/water partition coefficient. *Journal of Computer-Aided Molecular Design*, **27**, 1–13.
31 Young, R.J., Green, D.V., Luscombe, C.N., and Hill, A.P. (2011) Getting physical in drug discovery II: the impact of chromatographic hydrophobicity measurements and aromaticity. *Drug Discovery Today*, **16**, 822–830.
32 Doak, B.C., Over, B., Giordanetto, F., and Kihlberg, J. (2014) Oral druggable space beyond the rule of 5: insights from drugs and clinical candidates. *Chemistry and Biology*, **21**, 1115–1142.
33 Analysis not shown.
34 Gill, A.L., Cleasby, A., and Jhoti, H. (2005) The discovery of novel protein kinase

inhibitors by using fragment-based high-throughput X-ray crystallography. *ChemBioChem*, **6**, 506–512.

35 Gill, A.L., Frederickson, M., Cleasby, A., Woodhead, S.J., Carr, M.G., Woodhead, A.J., Walker, M.T., Congreve, M.S., Devine, L.A., Tisi, D., O'Reilly, M., Seavers, L.C.A., Davis, D.J., Curry, J., Anthony, R., Padova, A., Murray, C.W., Carr, R.A.E., and Jhoti, H. (2005) Identification of novel p38α MAP kinase inhibitors using fragment-based lead generation. *Journal of Medicinal Chemistry*, **48**, 414–426.

# 19
# An Integrated Approach for Fragment-Based Lead Discovery: Virtual, NMR, and High-Throughput Screening Combined with Structure-Guided Design. Application to the Aspartyl Protease Renin.

*Simon Rüdisser, Eric Vangrevelinghe, and Jürgen Maibaum*

## 19.1
## Introduction

Fragment-based screening (FBS) has become an integral part of drug research [1] and several clinical-stage compounds originate from fragment-based lead discovery (FBLD) [2]. FBLD is conceptually different from traditional hit finding and hence the evolution of FBLD has required a change of mindset for both hit finding and hit evolution. It is now well recognized that fragments need to be approached differently compared to high-throughput screening (HTS) hits, not only experimentally but also in terms of chemical optimization strategies. FBLD is most successful when it is embedded in a broader array of additional discovery technologies strongly interconnected with virtual screening (VS), screening library design, HTS, biophysics, and structural biology [3]. As all of these technologies further evolve, FBLD constantly needs to be adapted to this changing landscape as seen over the last years. Fragment-based lead discovery applied to aspartyl proteases has been described for several clinically relevant targets. Renin as one example is described here; others are HIV-1 and BACE-1 (see also Chapter 14).

Biophysical methods for FBS usually have low-to-medium throughput [4], but with technologies using robust biochemical assays a significantly higher compound throughput is possible that allows screening of up to 10 000 fragments at concentrations as high as 1 mM [5]. Compared to the screening capacity of HTS, the number of screened fragments is still small, and compared to the vast chemical space of drug-like molecules, these numbers are negligible. However, a small collection of simple and small-size molecules is considered to cover more chemical space as a large number of more complex molecules with increased molecular weight, as was proposed based on combinatorial principles in the seminal paper by Hann *et al.* [6]. The chemical space covered by FBS is defined by the screening set that, even if well designed, will be limited in scope and diversity of chemical matter. In addition, to detect binding of weakly interacting fragments, often with dissociation constants ($K_D$) as high as 10 mM, a sensitive and robust assay is instrumental but not always available. It is therefore desirable to expand

*Fragment-based Drug Discovery: Lessons and Outlook*, First Edition. Edited by Daniel A. Erlanson and Wolfgang Jahnke.

the number of fragments screened by computational enrichment that is defined by *in-silico* screening of very large compound sets and in the second step by experimentally probing this selected smaller fragment collection. The principle of designing a focused fragment collection offers several advantages [7] and has been applied in our FBLD work on the aspartyl protease renin.

The discovery of inhibitors of BACE-1 as a potential new therapy for Alzheimer's disease, which is a particularly challenging target, has been seminal for how low-affinity fragments can evolve to drug-leads and eventually to drug candidates. In fact, most of the drug candidates for BACE-1 have been discovered by FBLD [2]. One of the most advanced inhibitors of BACE-1 is MK-8931, whose chemical structure is not disclosed, which is currently in clinical phase II/III trials. The starting point for the development of MK-8931 was a small fragment with an enzyme affinity of only 550 μM. Structural information was used to turn this fragment into a drug molecule that meets the stringent pharmacological criteria [8,9]. Eli Lilly identified isothiourea fragments that were developed into the candidate inhibitor LY-2811376 that has reached phase I (Figure 19.4). Because of off-target retinal pathology identified in preclinical toxicological studies, the drug was discontinued [10]. We refer to Chapter 14 in this volume for an in-depth discussion.

Human renin, an aspartyl protease, is an important drug target for the treatment of hypertension. Drug research efforts over more than 30 years culminated in the discovery of aliskiren (Tekturna®/Rasilez®; Figure 19.4), which is the only direct renin inhibitor currently on the market. More recently, we have applied

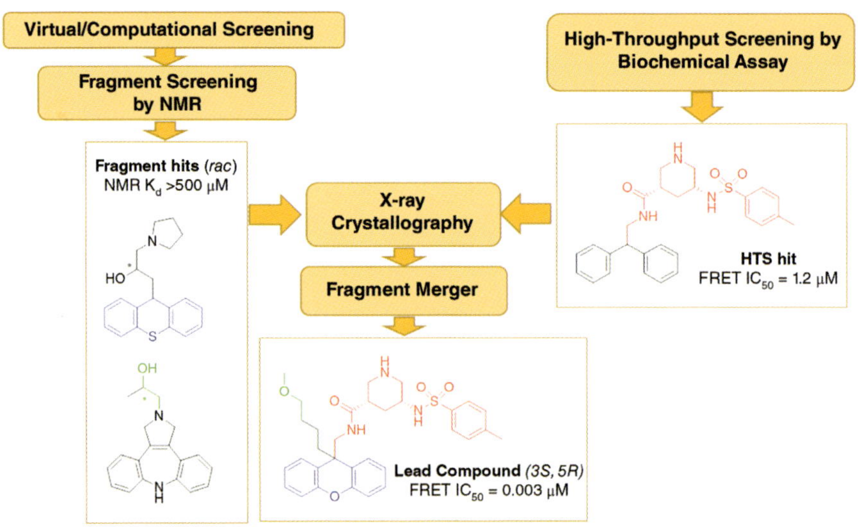


**Figure 19.1** An integrated, structure-based FBLD approach led to the discovery of a potent lead compound for renin. X-ray structures of hits from biophysical fragment screening and from a high-throughput biochemical assay were used in a structure guided effort to merge the individual chemical motifs.

FBLD in an integrated approach in order to discover inhibitors with improved pharmacological properties [11]. Our integrated efforts combine several elements that are described herein in detail. First, a renin-focused fragment library was specifically assembled by virtual screening in search for scaffolds interacting with the catalytic aspartates of the active site. Binding of the fragments was then probed by NMR using ligand observation techniques. Second, the Novartis compound archive was screened by high-throughput screening (HTS) using an enzymatic inhibition assay. Third, X-ray crystal structures of the enzyme–ligand complexes were generated and used to guide fragment hit-to-lead expansion. Crystallization trials were successful for two fragment hits and one HTS hit that allowed us to elaborate on their binding poses. Proximity of the two fragments to the catalytic dyad was not observed; instead, the hydrophobic $S_3$–$S_1$ patch and the $S_3^{sp}$ subpocket were occupied by the ligands. Merging of the fragments with the catalytic dyad-directed HTS hit resulted in a lead compound with good optimization opportunities (Figure 19.1). This combination of various screening technologies with X-ray crystallography allowed rapid progression of weakly binding fragments to a high-potency lead compound.

## 19.2
## Renin as a Drug Target

The renin–angiotensin–aldosterone system (RAAS) is a key regulatory pathway that controls blood pressure and the hemostasis of body fluid volume [12]. Renin enzymatic activity leads to the N-terminal cleavage of the alpha-globular protein angiotensinogen as its only known physiological substrate. This cleavage by renin generates the intermediate angiotensin I that is further processed by the Zn-metalloprotease angiotensin-converting enzyme (ACE) to form angiotensin II. This small peptide comprising an eight amino acid sequence is the principal mediator agent of the RAAS, which triggers multiple indirect and direct biological responses, such as constriction of blood vessels and retention of sodium and water in the kidney, through stimulation of the angiotensin receptor type-1 (AT-1). Pharmacological blockade of the RAAS by intervention at various levels has become a well-established therapeutic approach for the treatment of hypertension, a worldwide highly prevalent and important risk factor for myocardial infarction, stroke, heart, and renal failure, and other comorbidities [13]. Inhibitors of ACE and antagonists of the AT-1 receptor have emerged as conventional cornerstones of current antihypertensive practice. Renin inhibition at the first and rate-limiting step of the RAAS renders the entire pathway quiescent even under circumstances in which active plasma renin levels are significantly elevated as the result of compensatory negative feedback loop activation. Therefore, direct renin inhibition has been considered for a long time to be a highly attractive target in drug discovery, offering a potentially more advantageous and innovative option for the efficient management of hypertension and reduction of cardiovascular damage [14]. However, numerous early attempts to develop highly selective, orally bioavailable, small molecule renin inhibitors demonstrating *in-vivo* long-lasting efficacy met with little success [15]. Aliskiren (Tekturna®/Rasilez®; Figure 19.4), suitable for once-daily oral administration, was the first direct renin inhibitor to reach the market in 2007 and remains the only FDA approved drug of its class for the treatment of hypertension to date [16].

Aliskiren constitutes a fully nonpeptidic hydroxyethylene transition-state mimetic that tightly binds in a reversible, noncovalent manner to the enzyme cleft of human renin by spanning the $S_3$–$S_1$ 'superpocket' and the $S_{2'}$ recognition site, both flanking the catalytic dyad. This first-in-class drug has resulted from a unique structure-based topological design concept by exploiting the specific architecture of the contiguous hydrophobic $S_3$–$S_1$ 'superpocket' and its extension toward the nonsubstrate $S_3^{sp}$ cavity of human renin [15]. Retrospectively, the initial steps of this gradually evolving approach that eventually provided aliskiren may be considered as an early fragment growing strategy. By starting from a low-affinity ($IC_{50} = 30\,\mu M$) small ($S_1$–$S_{1'}$– $S_{2'}$)-spanning hydroxyethylene transition-state mimetic 'fragment', the hydrophobic $P_1$ position was further extended and grown into the large $S_3$–$S_1$ site considered as a hydrophobic hot spot for human renin and provided inhibitors with significantly improved binding affinities (Figure 19.2). Subsequent morphing of the $P_3$–$P_1$ pharmacophore and the discovery of the nonsubstrate $S_3^{sp}$ enzyme pocket by X-ray crystallography has then resulted in low nanomolar novel chemotype lead renin inhibitors. Continued efforts on the discovery of structurally distinct renin inhibitors with improved drug properties have been reported, and as a result several new development candidates have entered into early clinical trials more recently [19]. With the aim to identify a 'best-in-class' direct renin inhibitor, our research group at Novartis had initiated a broad multiple-approach screening program toward new drug-like lead chemotypes that included high-throughput screening (HTS) using an enzymatic and a competitive ligand binding assay format, *in-silico* three-dimensional pharmacophore searches [11,20], and combined NMR/X-ray fragment-based screen of a focused library as described herein.

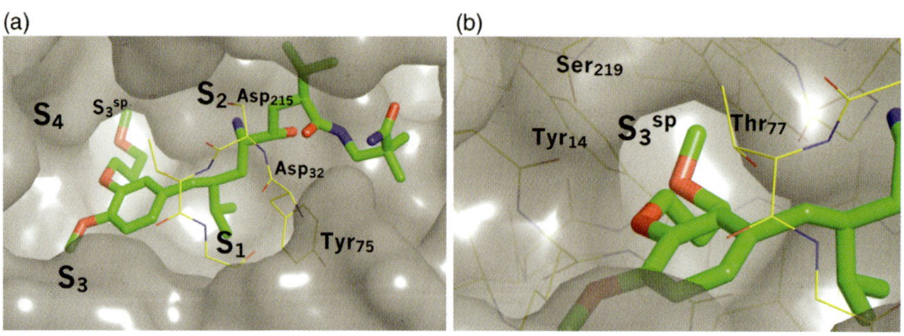


**Figure 19.2** (a and b) Aliskiren (Tekturna®/Rasilez®) is shown in complex with human renin to illustrate the topology of the binding pocket (pdb code 2V0Z.pdb) [17]. (b) Shows a close-up view of the $S_3$–$S_1$ hydrophobic patch focusing on the $S_3^{sp}$ subpocket. The figures of overlapping three-dimensional structures have been prepared using Pymol [18].

These integrated efforts in our hands turned out to be quite complementary leading to a diverse set of attractive starting leads from different sources well suited for further optimization.

## 19.3 The Catalytic Mechanism of Renin

The enzyme-catalyzed cleavage of the peptide substrate occurs through a general acid–general base mechanism in which one aspartyl carboxyl residue protonates the oxygen of the scissile bond, while the other aspartyl group deprotonates a conserved water molecule. For renin, the catalytic dyad residues are $Asp_{32}$ and $Asp_{215}$. The generated hydroxyl ion attacks the carbonyl carbon resulting in a gem-diol intermediate (Figure 19.3a). In the following step, protonation of the peptidic nitrogen by $Asp_{215}$ and deprotonation of the gem-diol by $Asp_{32}$ results in peptide bond cleavage [21,22]. A second conserved water molecule has a structural role due to its participation in a hydrogen-bond network that stabilizes the flap conformation during catalysis [23,24]. The structure and function of aspartyl proteases have been discussed in several review articles [23].

The protonation state of the catalytic dyad depends on the binding motifs of the inhibitor. Several computational methods were applied to identify the most probable dyad state. For instance, Rajamani and Reynolds used full linear-scaling quantum mechanical calculations to identify the preferred proton location for BACE-1. The calculations favor the monoprotonated state in the presence of

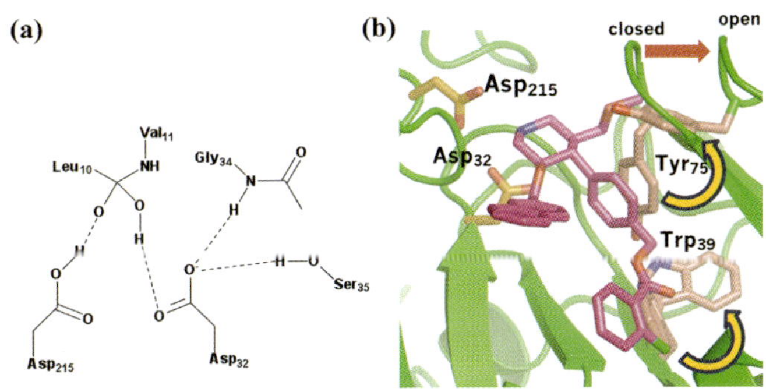


**Figure 19.3** (a) The proposed gem-diol transition state during proteolytic substrate cleavage. (b) The X-ray structure of renin in the closed flap conformation (2VOZ.pdb, aliskiren omitted) [35] is superposed with the open-flap conformation observed in complex with the piperidine derivative 6 (see Figure 19.6, $IC_{50} = 2$ nM, pdb code 1UHQ.pdb) [36,37]. The flap has moved by 4 Å ($Tyr_{75}C\alpha$, indicated by the red arrow) in order to accommodate the piperidine derivative, as the side chain of $Tyr_{75}$ (shown as sticks in salmon) swings outward to make room for the 4-phenyl-substituent. In the complex with the piperidine derivative, the cavity otherwise filled by the side chain of $Trp_{39}$ becomes occupied by the *o*-Cl-phenyl moiety of 6. The movements of the side chain of $Tyr_{75}$ and $Trp_{39}$ that are required for the accommodation of the piperidine derivative are indicated by orange arrows.

ligands and the diprotonated state in the absence of the ligand [25]. The influence of aspartate protonation on ligand binding and virtual ligand screening has been discussed recently [24]. It is now well recognized that the proper consideration of the protonation state is essential for hit finding [26].

The catalytic dyad of renin is monoprotonated in the unbound form and exists in two different states. In one state, the aspartates are coplanar with the proton centered in a symmetric conformation forming a low barrier hydrogen bond. In the other state, $Asp_{215}$ is protonated and this proton forms a hydrogen bond to the carbonyl oxygen of the scissile bond during catalysis [27]. The negative charge of the deprotonated $Asp_{32}$ is stabilized by a hydrogen bond to the NH of $Gly_{34}$ and the OH of $Ser_{35}$ [28]. Analysis of molecular dynamics simulation using explicit water indicates a noncoplanar orientation of the catalytic aspartates. Quantum mechanical calculations suggest furthermore that the noncoplanar orientation is energetically preferred before catalysis. The prevalence of coplanarity in X-ray structures was attributed to the low temperature, usually 100 K, at which diffraction data are collected [29].

The eukaryotic aspartyl proteases are comprised of two lobes, each lobe carrying one of the catalytic aspartates constituting the catalytic dyad. The central part of the substrate binding region is covered by a β-hairpin or flap region (Figure 19.3b). This flap domain can adopt multiple conformations and is usually in a closed position when an inhibitor is bound to the enzyme. The open flap conformation allows entrance of the substrate (or the inhibitor) and removal of the substrate products after proteolytic cleavage [23]. This receptor flexibility is one of the main complicating factors in VS and inhibitor design [30]. Detailed structural analysis and molecular dynamics (MD) simulations to generate representative structural ensembles are approaches to tackle this problem. This strategy has been tested on HIV-protease [31] and BACE-1 [24,32–34].

Beyond the catalytic water $W_1$, a structural water molecule $W_2$ is present that is involved in hydrogen bonds with the hydroxyl group of $Ser_{35}$ and $Tyr_{75}$. The indole NH of $Trp_{39}$ is in hydrogen bonding distance to $Tyr_{75}$, thereby further stabilizing the closed flap conformation. The series of hydrogen bonds $Asp_{32}$–$Ser_{35}$–$W_2$–$Tyr_{75}$–$Trp_{39}$ keeps the flap in close contact to the catalytic dyad. Explicit water molecular dynamics studies on renin indicate that the flap accommodates a closed, semiopen and open state in the unbound form. The semiopen conformation allows binding of the substrate and the catalytic water molecule [38].

## 19.4
## Virtual Screening

Virtual screening (VS) approaches have considerably advanced and are now an integral part of hit identification. In the early days of VS, compound screening by high-throughput assays was thought to become obsolete and to be entirely replaced by computational screening approaches. As of today, however, limitations in the predictive power of VS render the method complementary to experimental screening rather than a substitute. VS plays a role in hit finding by searching the vast chemical space computationally without the need to chemically prepare any molecules. As such, it is a relatively inexpensive way to enrich for active compounds or to act as a filter in order to reduce the number of compounds for screening in a biochemical activity or biophysical binding assay [39].

The first ligand-based VS methods applied two-dimensional (2D) or three-dimensional (3D) quantitative structure–activity relationship (QSAR) data, based on the hypothesis that there is a direct or strong relationship between the 2D or 3D structure of the ligand and its biological activity [40]. Another approach is the similarity method that is computationally inexpensive and retrieves compounds with chemical structures similar to known ligands. The characteristics of the molecules are encoded as fingerprints representing for instance the connectivity of atoms [41].

Structure-based VS uses the 3D structure of the target protein for enrichment of potentially active molecules. Coordinates generated by X-ray crystallography, NMR spectroscopy, or MD simulations can be used for this purpose. Ligands are docked to the target protein by optimizing intermolecular noncovalent interactions. This computational search results in a multitude of binding poses in different subpockets of the binding site. In order to identify the 'real' structure of the protein–ligand complex, either the binding free energy is calculated or a scoring function is applied for ranking [42]. It is very challenging to calculate the absolute binding free energy; however, this is often not required. Instead, a relative ranking of poses is possible by using simplified approximations to achieve the required computational efficiency. While currently available scoring functions are fast on the computational scale, they are not derived from well-defined physical models [43]. Even computationally expensive molecular mechanics with generalized Poisson–Boltzmann calculations fall short to discriminate inactives from active renin inhibitors [44]. To enhance the possibility of finding ligands and to alleviate the short-comings of the individual scoring functions, consensus scoring is often applied [45]. The process of ligand docking and scoring is further complicated by the fact that proteins may adopt a multitude of distinct conformations. Exploring backbone and side-chain flexibility is a challenge because such flexibility is difficult to treat computationally. It has been shown that incorporating protein flexibility in docking improves the predictive power of VS [46].

Both ligand- and protein structure-based VS methods can be applied either in a parallel or a sequential fashion. Parallel applications aim to combine the virtual hit lists in order to increase the probability of identifying a real binder. Sequential strategies are based on the following considerations: first, ligand-based VS, that is, 2D-, 3D-QSAR, or similarity methods, are computationally less demanding and can therefore serve as a preselection filter prior to target structure-based computations; second, in case certain 2D structural motifs are preferred to be present in the ligands of interest, undesired ligand structures can be removed from the library before conducting expensive structure-based computations [47].

VS methodologies to account for receptor plasticity have been implemented for BACE-1 [33,34]. In one study, five X-ray structures of BACE-1 in complex with distinct inhibitors were selected as the most relevant representatives for different flap conformational ensembles and a database of small molecules was docked using AutoDock4.2 (ensemble virtual screening – EVS). As an alternative approach, a grid of interaction energies was calculated and averaged for each grid point (virtual screening using energy weighted grids – EWVS) for the same five X-ray crystal complex structures [34]. Both methods were shown to have similar predictive power. However, EWVS is less demanding in terms of computational time rendering this method more suitable for screening of large molecular databases [34]. The previous example is a nice illustration of the importance to include dynamic conformational states derived from X-ray structures for VS. In the absence of relevant X-ray structures, these conformations need to be generated computationally. MD simulations may be used to compute conformations for docking. One example of docking to protein structures from MD simulations is the acylguanidine BACE-1 inhibitor WY-25105, resulting from FBS, which did not fit any known crystal structures at the time of its discovery. The complex structure that was obtained later revealed an open flap conformation previously not observed experimentally. MD simulations were used in a retrospective study to generate a BACE-1 conformation suitable for docking of WY-25105 [33].

Plasmepsins, which are related aspartyl proteases present in the food vacuole of the malaria parasite *Plasmodium falciparum*, are involved in the degradation of the host hemoglobin. It has been suggested that antimalarial drugs can be obtained by blocking the activity of the several plasmepsin isoforms. VS, including protein target flexibility, have been undertaken in search for plasmepsin inhibitors [48–51].

While several VS applications for renin have been published [44,52,53], we are not aware of VS studies for renin where protein flexibility has been considered explicitly. Selecting the relevant flap conformation is critical for structure generating programs, like GrowMol, which generate piperidine-based renin inhibitors if applied to the open flap conformation with the accompanied rotation of the $Trp_{39}$ and $Tyr_{75}$ side chain. These side chains movements open up new pockets that are then occupied by the nonpeptidic inhibitors. Inhibitors that bind by the stabilization of vacated side-chain pockets were termed group replacement-assisted binding peptidomimetics (GRAB peptidomimetics) [30]. Pharmacophore-based VS applications select ligands matching the pharmacophore model that is defined by the protein–ligand complex structure. Therefore, models based on the closed conformation will result in binders matching the corresponding binding geometry [54,55].

Virtual screening approaches have gained popularity in particular in academia with the aim to discover new chemotype hits for aspartyl proteases by screening publicly accessible compound collections, as documented by numerous reports describing the enrichment of active hits in biochemical assays by VS filters. However, many of these disclosures are not providing additional validation data generated either in orthogonal assays based on a different readout technology or by conducting biophysical experiments such as SPR or NMR, in order to demonstrate further evidence for true binding of identified hits to the target protein. In our experience, reliance on a single assay technology can often be misleading and we consider the comprehensive orthogonal validation of initial screening hits as true binders to be a prerequisite. Hence, the validity of conclusions drawn from docking and scoring performance alone in the absence orthogonal validation methods need to be considered with caution.

The concept of ligand efficiency (LE) has emerged recently and is particularly relevant in the context of FBLD. Although fragments are low affinity binders, their interactions with the target protein can nevertheless be highly efficient as only a few heavy atoms contribute strongly to the binding free energy of the fragment molecule. Therefore, the normalized binding free energy provides a meaningful metric to assess the quality of a given validated fragment hit. The thermodynamic definition of the binding free energy is $\Delta G = -RT\ln(K_D)$. In drug discovery projects, $IC_{50}$ values are typically obtained and the reasonable approximation $K_D \sim IC_{50}$ can be made. Dividing $\Delta G$ by the heavy atom count gives the LE that will be used here. Several other definitions of LE have been proposed in the literature by using different physicochemical properties as the normalizing factor. One example is the ligand lipophilicity index, which is an estimate of the specificity of the compound binding to the target relative to partitioning into 1-octanol [56,57].

## 19.5 Fragment-Based Lead Finding Applied to Renin and Other Aspartyl Proteases

Human, viral, and parasite aspartyl proteases have attracted much interest over the past decades as important drug discovery targets, and a number of low molecular weight aspartyl protease inhibitors are used clinically or are in late-stage development trials [58–61]. For example, several peptidomimetic inhibitors of the retroviral human immunodeficiency virus (HIV-1), which is the well-recognized causative agent for the acquired immunodeficiency syndrome (AIDS), are now available for drug treatment of patients. Initially, the target structure-based design of inhibitors has focused on the optimization of peptidomimetic scaffolds emulating the enzyme substrate binding conformation in the extended active site. Incorporation of secondary alcohols as stable transition-state motifs replacing the scissile peptide bond and mimicking an active site-bound substrate water molecule has been a very successful strategy to generate molecules with high binding affinities and selectivity for the specific target protease of interest.

Limited oral bioavailability, rapid *in-vivo* clearance and/or the low permeability for the blood–brain barrier (in the case of BACE-1) are often encountered with peptidomimetic inhibitors and have raised the interest in the identification of nonpeptidic inhibitor chemotypes with improved pharmacokinetic properties. Various approaches to screen large compound collections have been applied with the aim to identify new chemical matter as more promising starting points for further optimization. Conventional high-throughput biochemical assay screens and virtual screening have indeed resulted in the discovery of nonpeptide ligands of low molecular weight (<250–300 Da) for aspartyl proteases. The general concept of the fragment approach for lead discovery was introduced and

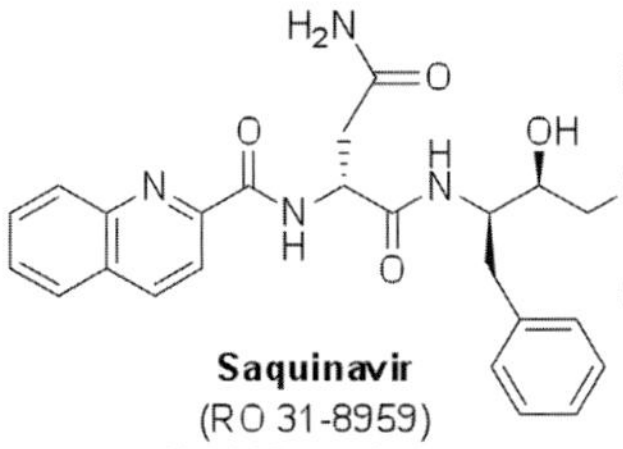

**Saquinavir**
(RO 31-8959)
HIV-1 $IC_{50}$ < 0.4 nM
$K_i$ < 0.12 nM
LE > 0.29

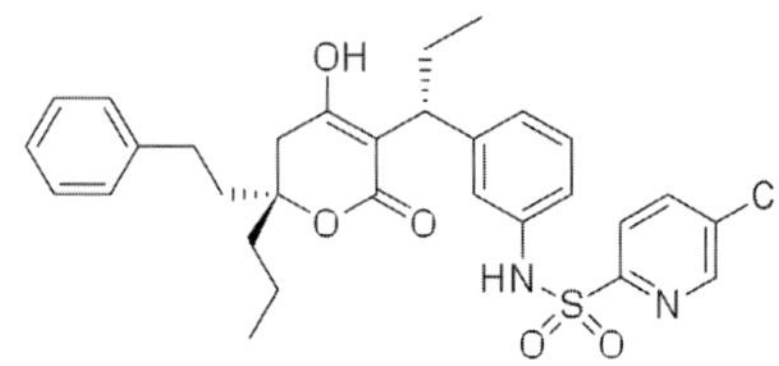

**Tipranavir**
(PNU-140690E)
HIV-1 $K_i$ = 8 pM
LE = 0.38

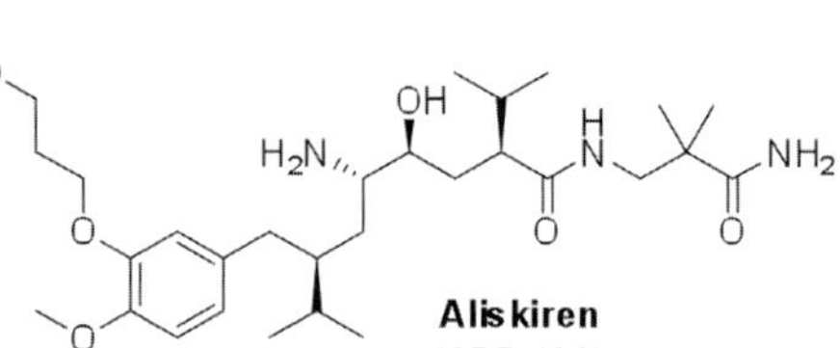

**Aliskiren**
(SPP-100)
hRenin $IC_{50}$ = 0.6 nM
LE = 0.32

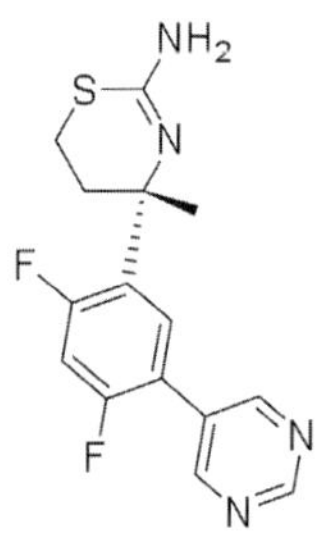

LY-2811376

BACE-1 $IC_{50}$ = 240 nM
Cell $EC_{50}$ = 300 nM
LE = 0.41

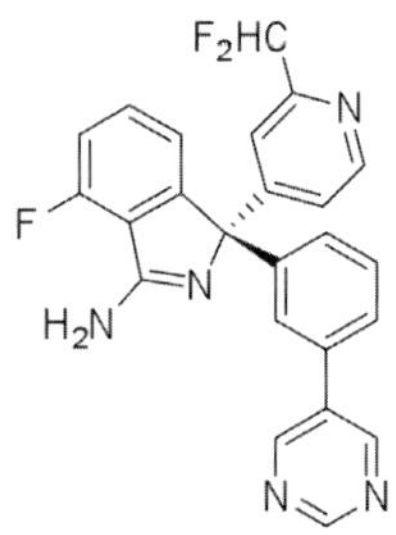

AZD-3839

BACE $IC_{50}$ = 35.1 nM
Cell $EC_{50}$ = 16.7 nM
LE = 0.32

**Figure 19.4** Aspartyl protease inhibitors in clinical use or in advanced stages of clinical trials. Saquinavir was one of the first marketed HIV-1 protease inhibitors and is based on a peptide analog transition-state mimetic concept [62]. Tipranavir represents a nonpeptidic HIV-1 protease inhibitor [63]. Aliskiren, a higly potent renin inhibitor lacking the extended peptide-like backbone of the first-generation renin inhibitors, has oral pharmacokinetic properties suitable for once daily dosing and is approved for the treatment of hypertension [35]. The BACE-1 inhibitor LY-2811376 resulted from a fragment screening initiative and reached clinical stage [10]. The aminoisoindole AZD-3839, developed by researchers at AstraZeneca, is another example of a BACE-1 inhibitor under clinical investigation [64].

implemented in an *in-silico* approach for ligand design in the computer program LUDI [65,66] and later in CombiDOCK [67]. Experimental screening of fragments was pioneered in the SHAPES [68] and needle [69] approaches and almost simultaneously the term ‘fragment screening’ became introduced [70].

As part of the more general conceptualization of fragment-based drug discovery (FBDD), both small generic and focused libraries of fragments were subsequently generated and screened against various aspartyl proteases. Interestingly, all the fragment or fragment-like hits reportedly discovered from both conventional and focused set screening campaigns were shown by X-ray crystallography to be positioned in close binding vicinity to the two catalytic aspartates at the center of the catalytic cleft. Furthermore, the structure of these small ligand molecules comprise either an OH-group-based recognition motif reminiscent of a classical hydroxyethylene dipeptide transition-state mimetic or more strikingly previously unprecedented basic amine structural motifs that can engage in hydrogen bonding and charge transfer interactions with the catalytic dyad.

At Dupont-Merck, a 3D *in-silico* search in the Cambridge Structural Database was conducted based on a three-point pharmacophore model derived from the crystal structure of a hydroxyethylene transition-state inhibitor bound to the HIV protease [71,72]. The poly-hydroxylated triphenyl analog **1** (Figure 19.5)

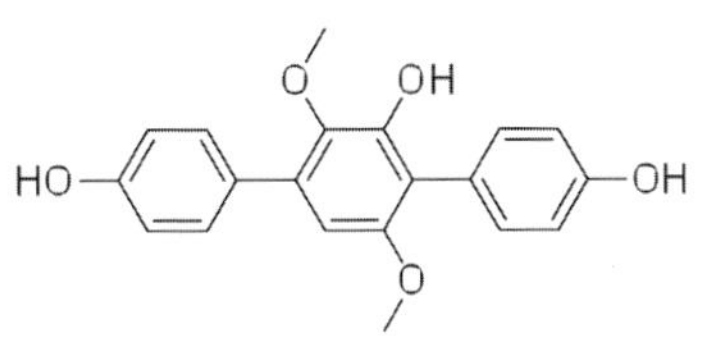

**1**
HIV-PR $K_i$ not reported
***Dupont-Merck 1994***

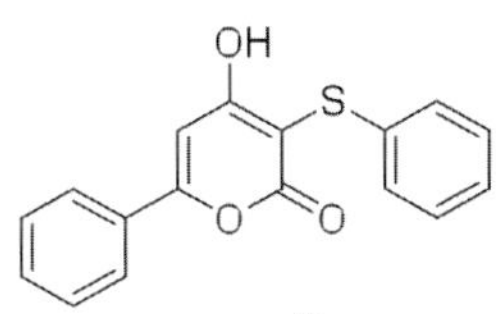

**2**
HIV-PR $K_i$ = 1.1 μM
HIV-PR $IC_{50}$ = 3 μM
LE = 0.36
***Parke-Davis, 1994***

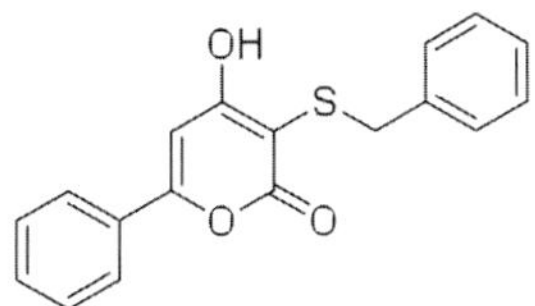

**3**
HIV-PR $K_i$ = 0.7 μM
HIV-PR $IC_{50}$ = 1.7 μM
LE = 0.36
X-ray validated

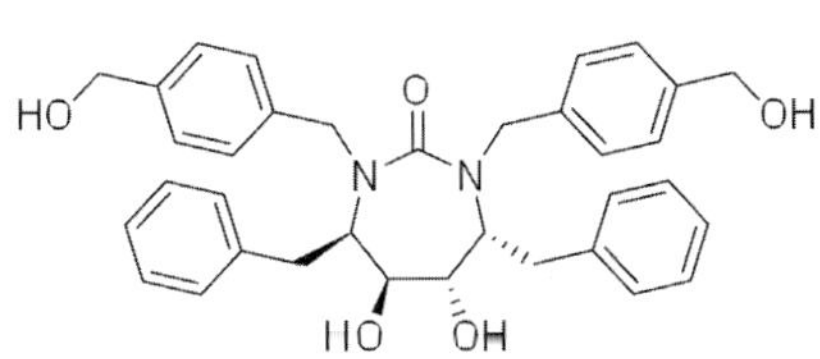

**DMP-323**
HIV-PR $K_i$ = 0.27 nM
LE = 0.32

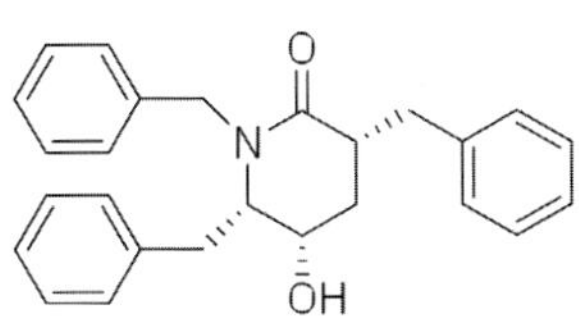

**Lucca, 1997**
HIV-PR $K_i$ = 9.4 μM

**Figure 19.5** Small molecule hits binding to the retroviral HIV-1 protease and the potent HIV-1 inhibitor DMP-323.

was considered by this research group as a 'hit' (no binding affinity data reported). In line with the defined basic spatial pharmacophore elements, molecular docking of **1** into the active site suggested both the hydroxy and the neighboring methoxy group at the center phenyl ring to be positioned in proximity to the catalytic aspartates. The two flanking phenolic residues were posed in the hydrophobic $S_1$ and $S_{1'}$ pockets, respectively, of the symmetric catalytic cleft. In addition, the second methoxy group at the opposite site of the central ring was predicted to displace the conserved structural water of the HIV protease flap region. This water molecule is unique to the retroviral protease and accepts two H-bonds from the backbone NH of an Ile, each from an identical subunit of the protease dimer ($Ile_{50}$ and $Ile_{50'}$). It also donates two H-bonds to the carbonyl oxygen of peptidomimetic HIV protease inhibitors, thereby keeping the flexible flap domain in a closed conformation. Hence, the incorporation of structural water mimetics into nonpeptide scaffolds was considered to provide selectivity for the viral versus human aspartyl protease. Based on this initial concept of key ligand–enzyme interactions and molecular modeling-guided structural morphing of the center scaffold, symmetric seven-membered ring cyclic ureas bearing a vicinal diol recognition motif evolved as a new chemotype of HIV protease inhibitors with improved antiviral potency. Several iterative molecular design and optimization studies ultimately resulted in highly potent and selective HIV protease inhibitors such as **DMP-323** (Figure 19.5).

High-volume biochemical screening of the compound collection at Parke-Davis identified the 4-hydroxylated pyran-2-one **2** (Figure 19.5) as a hit with micromolar inhibitory affinity to the HIV-1 protease [73]. Enzyme kinetic studies validated **2** as a competitive HIV protease active site binder. Confirmation of the binding pose for this chemotype in complex with the enzyme was accomplished subsequently by X-ray crystallography for the closely related analog **3** (Figure 19.5) bound to the active site: the enol hydroxyl group forms a H-bond network to the catalytic $Asp_{25}$ and $Asp_{125}$ as expected, while the critical structural water molecule is displaced by the lactone carbonyl and oxygen of the pyrone forming H-bonds to the two backbone NHs of the flap ($Ile_{50}$ and $Ile_{50'}$). Successful optimization of these early lead structures eventually culminated in the market introduction of tipranavir [Aptivus®, Boehringer-Ingelheim] in 2005 (Figure 19.4) [63].

A broad-scale screening program at Hoffmann-La Roche using the corporate compound library and a biochemical enzyme assay discovered racemic **4** to be a weak inhibitor of human renin, lacking inhibitory activity toward pepsin, cathepsin D, and HIV protease. Compound **4** represents a previously unknown chemotype for the inhibition of aspartyl proteases. The low resolution (ca. 3.5 Å) X-ray crystal structures of the active (*R,R*)-enantiomer ($IC_{50} = 25\,\mu M$) and its closely related analog **5** suggested that the piperidine nitrogen atom of these ligands is positioned near, and forming hydrogen bonds to, the two catalytic carboxylate residues of $Asp_{35}$ and $Asp_{215}$ [74]. Further interpretation of the electron density maps with respect to the orientation of the two piperidine substituents proved to be difficult and was based on the assumption that the enzyme active site

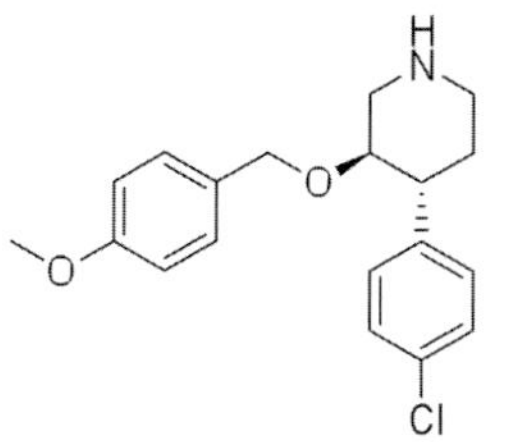

**4** *(rac)*
rh-Renin $IC_{50}$ = 50 µM
LE = 0.26
***Roche, 1999***

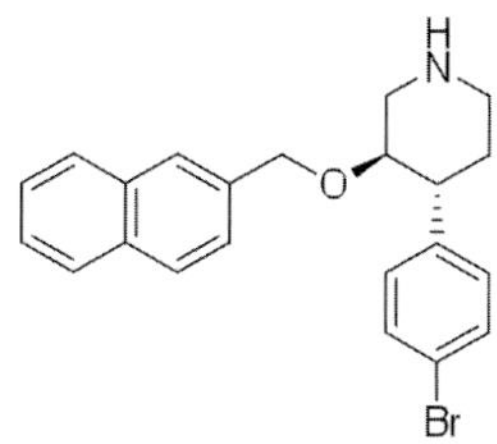

**5**
rh-Renin $IC_{50}$ = 5 µM
LE = 0.29
***X-ray***

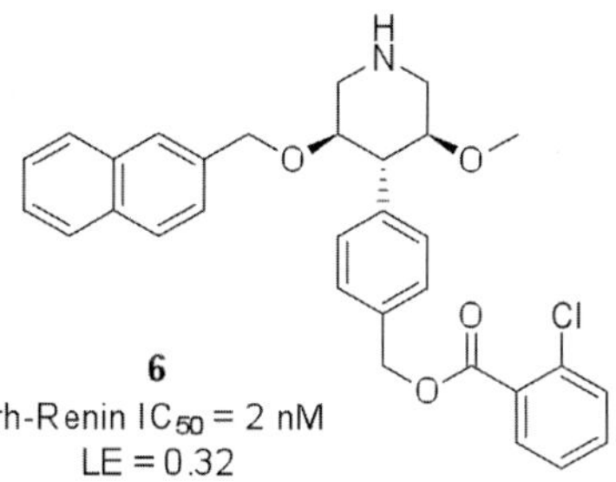

**6**
rh-Renin $IC_{50}$ = 2 nM
LE = 0.32
***X-ray***

**Figure 19.6** Fragment hits binding to recombinant human renin and inhibitor 6 binding to the open-flap pocket (see Figure 19.3).

conformation was unchanged compared to the renin conformation observed in complex with a potent peptidic transition-state mimetic inhibitor. The initial model of the crystal complex of **5** was later on revisited based on the X-ray crystal structure of a further optimized and more potent piperidine lead molecule (**6**, Figure 19.6) at higher resolution. Intriguingly, the extended lipophilic side chain of the *para*-phenyl residue of **6** occupied a newly opened binding pocket that is formed by the flexible flap domain of the protein adopting an open conformation. More specifically, the H-bond between the OH of $Tyr_{75}$ and the indole NH of the buried $Trp_{39}$ is disrupted as the consequence of major concerted side-chain rotations involving $Tyr_{75}$, $Leu_{73}$, and $Trp_{39}$ (see Figure 19.3). As a result, the *para*-phenyl ring of the ligand in the X-ray complex takes the position of the $Tyr_{75}$ side chain in the closed enzyme conformation, and the distal Cl-phenyl residue is accommodated by the 'deep flap pocket' generated by reorientation of the $Trp_{39}$ indole moiety. Retrospective analysis of the X-ray density of the low-affinity inhibitor **4** in complex with renin was compatible with a similar binding mode to a more open flap conformation of renin, though without the $Trp_{39}$ side chain flipping out of its position. In summary, the discovery of a weak-affinity fully nonpeptide small molecule hit has identified the basic piperidine scaffold as a viable transition-state surrogate that binds most likely in its protonated form to the aspartic acids of the catalytic dyad, and furthermore has

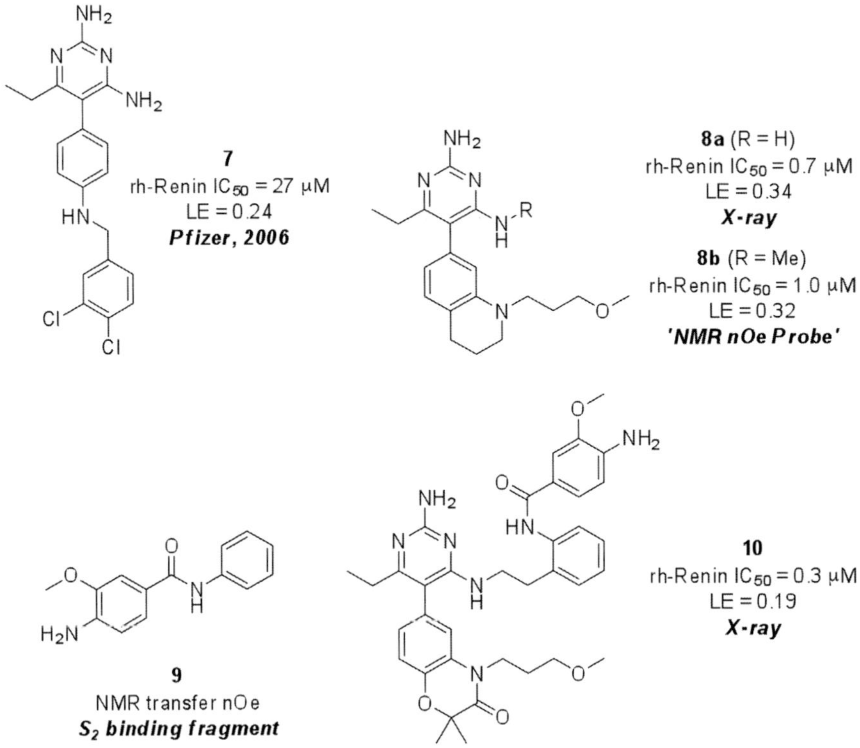


**Figure 19.7** Fragment-based second-site NMR screening approach applied to recombinant human renin.

uncovered unprecedented and 'ligandable' open flap conformational ensembles of human renin. These hallmark findings have provided new and highly attractive avenues for drug discovery efforts related to aspartyl proteases, and in particular for the design of new selective chemotype inhibitors of human renin with improved drug-like properties (*vide infra*) [37,75]. Similar design concepts based on mono- or bicyclic secondary amine transition-state surrogates in combination with structural elements binding to open flap conformations have been applied to the development of potent inhibitors of *Plasmodium* species as potential antimalarial agents [76].

The group at Pfizer has identified from an HTS the nonpeptidic and achiral 6-ethyl-2,4-diaminopyrimidine-based lead structure 7 (Figure 19.7) as a weakly active small molecule renin inhibitor [77]. The X-ray cocrystal structure of its 3,5-difluoro-substituted analog ($IC_{50} = 4\,\mu M$) demonstrated the ligand to bind to an active site conformation with a closed flap and moreover identified the planar 2,4-diaminopyrimidine head group as a new recognition element forming hydrogen bonds to the catalytic dyad. The small ethyl residue binds into the hydrophobic $S_1$ pocket and the extended side chain is accommodated by the $S_3$ and partially the $S_4$ pockets. Further SBDD efforts by removal of the distal benzylic portion and by incorporation of a rigid bicyclic $S_3$ template with the 3-methoxypropyl group binding into the nonsubstrate $S_3$ subpocket ($S_3^{sp}$) resulted in a significant potency improvement (compound **8a**, Figure 19.7). Targeting the large $S_2$ recognition site of renin favoring hydrophobic interactions was considered as one of the strategies to further improve the potency of this lead. A second-site NMR fragment screening approach based on saturation transfer nuclear Overhauser effect (nOe) was conducted to identify small molecules from a library of ~580 compounds (MW 100–150 Da) as potential $S_2$ binders. The *N*-methylated and $S_3$-modified analog **8b** was used as the coligand seeding probe in order to detect any interligand nOe signals at the putative vector position directing toward the $S_2$ site. The benzamide fragment **9** was detected as the only NMR hit exhibiting an interligand nOe with the probe **8b** and was also among the top ranking fragments from the same library that docked well in the $S_2$ pocket using GOLD [78]. Linking both the fragment **9** and the analog **8a** via a two-atom spacer provided compound **10** that was cocrystallized with rh-renin and demonstrated that the side-chain fragment indeed is positioned in the $S_2$ site. Further SAR optimization of the spacer and the $S_2$ binding motif improved the *in-vitro* potency to the low nanomolar $IC_{50}$ range [78,79].

Fragment-based lead generation was applied very early on in programs of several research groups to find small molecule drug-like inhibitors of β-secretase (BACE-1), which is considered as a key target for the treatment of Alzheimer's disease. A more detailed review is already provided in Chapter 14. Different approaches have been used with respect to the design and size of fragment libraries, applied screening technologies, and orthogonal validation of identified primary fragment hits. Several distinct chemical classes of novel active site binding fragments of BACE-1, mostly bearing cyclic or acyclic basic 'warheads' targeting the catalytic dyad, were identified as starting points for medicinal chemistry optimization. Noncanonical amine-containing BACE-1 fragments include, for example, aminopyridines [80], acylguanidines, (a) cyclic amidines [81], spiropyrrolidines, and structural variants of these chemotypes [82]. An overview of selected key representatives of fragment screen hits reported from different sources is illustrated in Figure 19.8.

At Astex, the platform of high-throughput X-ray crystallography for fragment-based screening referred to as the pyramid screening approach [83] has been successfully applied to the BACE-1 inhibitor program [80,84]. A small generic fragment library of 347 'drug fragments' was initially screened as mixtures of six, which identified the two low-affinity, high ligand efficacy hits **11**, **12** as well as the 2-aminopyridine analog **13** from a follow-up substructure search. A subsequent X-ray screen was conducted on a focused set of 60 fragments, which was generated by iterative virtual screening of a larger compound collection by using molecular properties filters and substructure filters for cyclic secondary amines recognized previously as transition-state surrogates binding to rh-renin. Final docking and ranking of the *in-silico* hits was conducted against different protein conformations of BACE-1. The structural features of the resulting fragment hit **14** (Figure 19.8) are reminiscent of classical aspartyl protease transition-state mimetics on the one hand, and in addition to the basic piperidine transition-

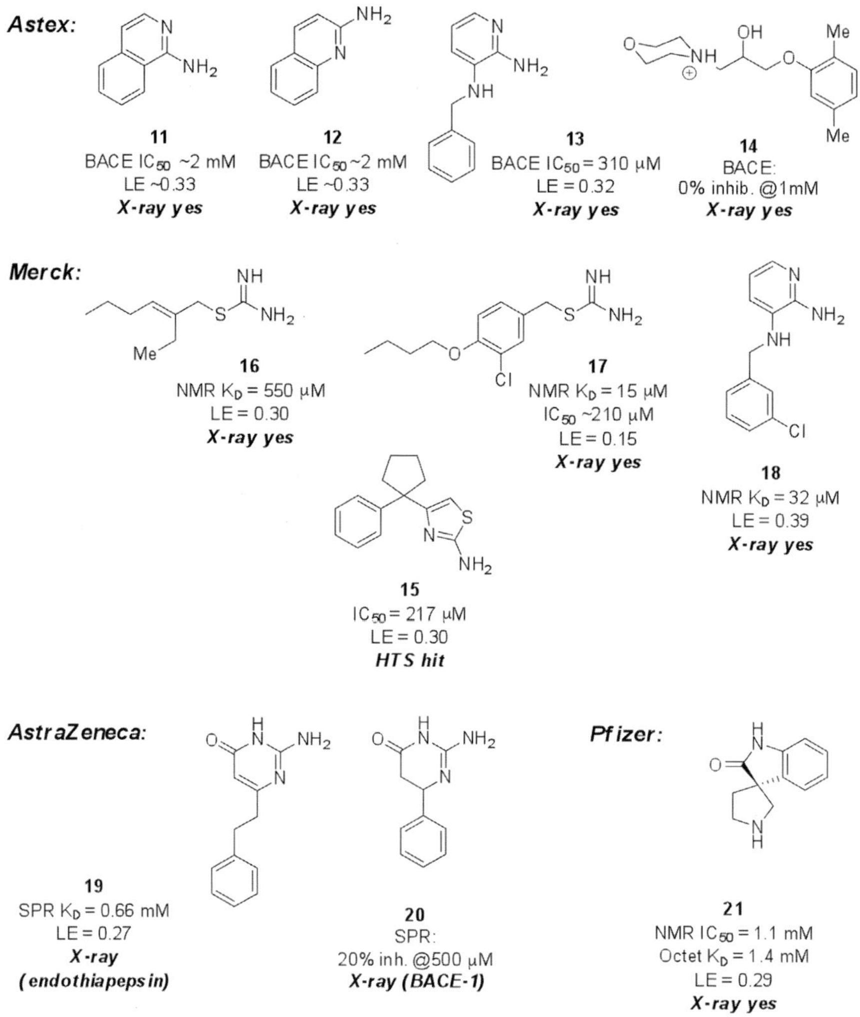


**Figure 19.8** Fragment-based screening hits for β-secretase (BACE-1).

state surrogate discovered for rh-renin. Indeed, the hydroxyl group of **14** was shown by X-ray (PDB 1FKN.pdb) to be positioned symmetrically between the two catalytic aspartates, while the morpholine nitrogen is in binding distance to the carbonyl of $Asp_{228}$. Inhibitory activities of the X-ray fragments were determined by a FRET-based enzymatic BACE-1 assay; however, further biophysical characterization of the binding kinetics and affinities of these fragments to the enzyme would be required, as was suggested by the encountered high false positive rate in the enzymatic assay at very high concentrations [80].

Interestingly, the fragment-like 2-amino-1,3-thiazole derivative **15** (Figure 19.8), which is structurally related to previously described $NH_2$-pyridine-type fragments, was identified as a weak inhibitor of BACE-1 from a HTS screening effort at Merck. A more than 200-fold improvement in inhibitory potency toward BACE-1 was accomplished by optimizing the residues at the phenyl ring and concomitant replacement of the cyclopentyl group with a 2,5-disubstituted benzyl moiety. Inhibitors bearing the weakly basic 2-amino-1,3-thiazole 'warhead' (calculated p$K_a$ ~5.5) were shown to be completely inactive in a biochemical assay when tested under higher pH conditions (pH 6.5 vs. 4.5) insufficient to protonate the heteroaromatic portion.

As part of a multiple hit-finding effort at Merck, the fully processed large-scale screening of a customized library of >10 000 fragments (either rule-of-three compliant or compatible with 'reduced complexity' rules) by $^{15}$N-HSQC NMR using a soluble version of isotope-labeled BACE-1 identified several distinct chemical classes of active site binders, as determined by NMR binding site mapping experiments [8]. The primary isothiourea fragment **16** triggered substructure searches that identified analog **17** with weak activity in an enzymatic assay but high ligand efficacy [57,85] based on NMR $K_D$ of 15 μM (Figure 19.8). Its X-ray co-crystal structure with BACE-1 revealed a H-bonding network between the weakly basic isothioura motif (p$K_a$ ~ 9) and the active site aspartates, while the hydrophobic substituted aryl portion occupied the contiguous $S_1$ and $S_3$ pockets. This work was extended to a 2D structure-based focused NMR fragment screening of the corporate library for chemically more stable heterocyclic bioisosteres including 2-aminopyridines (cf. fragment **18**, p$K_a$ ~ 7.2; Figure 19.8), 2-amino-imidazoles, and benzimidazoles. The planar nature of the 2-amino-pyridine core scaffold and its binding pose at the pivotal center of the catalytic dyad, however, was one limiting factor preventing the design of potent BACE-1 inhibitors with lead-like properties in this series. On the other hand, the aspartate-binding amidino motif of the NMR fragment hits **16**, **17** served as the conceptual basis for the design of related monocyclic iminopyrimidinone and iminohydantoin frameworks that upon extensive structural optimization ultimately delivered BACE-1 inhibitors with high potency, selectivity, and excellent pharmacokinetics properties resulting in CNS efficacy *in-vivo* [8,9]. At AstraZeneca, fragment screening by 1D and 2D NMR was performed on a medium-sized generic fragment library of ~2000 compounds at pH 4.5, accompanied by competition NMR experiments in the absence and presence of a known potent binder to BACE-1 in order to exclude false positives. Further validation of weak affinity hits was conduction by NMR water-LOGSY experiments and by SPR using the BIAcore technology. In lack of soakable BACE-1 crystal forms, the fungal aspartyl protease endothiapepsin was used as a surrogate enzyme (crystals obtained at pH 4.5, high protein homology with BACE-1 around the catalytic site) for elucidation of the binding characteristics of identified hits. A series of 6-substituted isocytosines such as **19** (Figure 19.8) were discovered as novel scaffolds binding to BACE-1 with very low affinity. In the X-ray complex with endothiapapsin, the exocyclic primary amine and one of the neighboring ring nitrogen atoms form a H-bonding network to the two catalytic aspartates. In addition, a weaker H-bond interaction was observed between the inhibitor oxygen and the backbone NH of $Asp_{81}$ of the flap domain. Further optimization in collaboration with Astex included structural modification of the original 'warheads' by merging with the dihydroisocytosine fragment **20** found by substructure search in the corporate compound collection afforded a BACE-1 inhibitor lead series with nanomolar potency [81,84].

Fragment screening by X-ray crystallography (soaking method) was performed by the Pfizer group using a proprietary Global Fragment Initiative (GFI) library comprising 340 compounds (tested as mixtures of 4). Compound **21** (Figure 19.8) characterized by a spiropyrrolidine core structure was identified as an unprecedented single hit. Several orthogonal methods were developed to correlate the structural information with binding affinity and functional inhibitory activity. Compound **21** was further investigated by biophysical methods and by an NMR-based functional assay and shown to be a reversible and competitive binder to BACE-1. Noteworthy, the X-ray crystal complex structure revealed an out-of-plane rotation of the carboxylate side chain of the catalytic $Asp_{32}$ by ~90° due to a steric clash with the pyrrolidine ring of the ligand. This significant conformational movement involving the catalytic dyad is very unusual to BACE-1 and in particular to other aspartyl protease structures, in which both aspartate residues are oriented in an almost coplanar position. Further ligand optimization efforts heavily involved crystallographic data and structure-based design, and resulted in a ~1000-fold potency improvement while maintaining high ligand efficacy [82].

Finally, an interesting enzyme active-site binding fragment discovery approach-based on tethering technology and screening of a library of >15 000 thiol-containing fragments in a panel of BACE-1 active site cysteine mutants under redox-controlled conditions has been reported [86]. Tethering of fragments was achieved by introducing cysteine mutations in the active site and screening a thiol reactive fragment library under controlled redox conditions. Fragments in the pool compete for disulfide formation and tethered fragments were detected by mass spectrometry. Those with inherent affinity were enriched and their X-ray structures were more easily obtained due to the covalent linkage. The structures of identified hits with weak binding affinities were clustered into a few chemotype scaffolds most of them comprising cyclic amines such as carboxamide-substituted piperidines and piperazines. The piperidine carboxamide core motif of both tethered and noncovalent lead ligands was shown by X-ray crystallography to bind to BACE-1 in an open flap conformation, displaying a conserved hydrogen-bonding pattern distinct from other BACE-1 inhibitors.

## 19.6 Renin Fragment Library Design

The strategy for designing a fragment library generally depends on several factors such as protein target knowledge as well as the overall hit finding and optimization approach [87–89]. In case the aim is to probe the binding pocket with unprecedented chemical matter, generic libraries are most suitable for screening.

Generic screening sets are assembled from small molecules by neglecting available specific knowledge about the target. These unbiased fragment collections have highly diversified molecular properties, such as the shape and number of hydrogen bond donor and/or acceptor groups. Other reasons for a generic compound selection are the flexibility of hot spots and shortcomings on the reliability of predicting the binding affinity. Focused screening sets, on the other hand, are designed by using information on both the ligand and the structure of the protein of interest, or alternatively of the target protein family. Screening of a focused fragment collection is appropriate when specific ligand chemotypes are preferred and when enrichment for binders is envisaged.

Screening a smaller number of compounds results in a decreased number of hits with a given hit rate unchanged. Low-to-medium throughput screening, which is typically applied to fragments, therefore, benefits from enriching for binders prior to experimental testing [88]. *In-silico* enrichment of fragments for the design of a focused screening library can be achieved in several ways. Ligand-based selection methods start from prior knowledge of chemical motifs known to interact preferentially with the target protein. These motifs are either based on known drugs from which key pharmacophore elements are extracted or on the mode of action of the enzyme. Enzyme mechanism-based principles include, for example, metal binding motifs for metalloproteases and electrophilic functional groups for cysteine and serine proteases. Such a target family approach increases the likelihood of detecting respective ligands; however, this will be at the cost of missing unexpected binding motifs [7].

Both structure-based and knowledge-based library design methods, the latter including privileged scaffolds, have been demonstrated to be successful [7,87]. We have used both approaches including molecular docking and scoring to identify fragments that preferentially bind to the active site of renin. Docking and scoring of fragments is regarded as particularly challenging. The reasoning is based on the observation that fragments often adopt multiple binding modes with similar binding free energies. Scoring functions still lack the required accuracy, and moreover the difference in the binding free energies for various poses is often within the error of the prediction method [90].

Fragment hits characterized as low molecular weight entities with relatively simple structures are often perceived to lack on-target selectivity. In particular, enzyme mechanism-based fragments bearing a reversible noncovalent recognition motif targeting the canonical catalytic dyad of an aspartyl protease are expected to bind in a conserved or likewise manner with relatively similar affinities to other members of this enzyme family. Very limited *in-vitro* selectivity data has been reported in the past for primary fragment hits and their close analogs binding to aspartyl proteases. Several reasons may account for this observation, for example, the lack of sensitive biochemical assays or biophysical methods to determine very weak binding affinities to off-target aspartyl proteases. On the other hand, small fragments interfering as transition-state surrogates with the catalytic mechanism constitute starting leads selective for one specific aspartyl protease. This could be due to differences in the pH optimum for enzyme activity that may impact the protonation state and potential conformational flexibility of the two aspartate carboxylic acid residues involved in catalyzing substrate cleavage, differences in the protein structural environment in the direct vicinity to the catalytic dyad, the presence of additional structural water molecules, and/or different spatial constraints at the center of the enzyme. All these factors have an impact on the binding poses of a fragment and thereby on the direction of suitable connectivity vectors for fragment growing and structural decoration toward the enzyme-specific recognition pockets. The dynamics of the protein enables a distinct set of conformations that provide additional binding sites for a given enzyme that are not present in other members of the target family. Fragments could bind to and stabilize a specific preexisting enzyme conformation that is sparsely populated and in equilibrium with other conformations in the ensemble and therefore generate a selectivity hot spot. Furthermore, we reasoned that improvement of the binding affinity for renin could be likely accomplished during fragment hit-to-lead exploration by targeting the renin-specific recognition sites in proximity to the catalytic dyad by appropriately decorating the fragment scaffold.

Our strategy focused on the discovery of fragment scaffolds interacting with the catalytic center of renin, comprising the two carboxylates of $Asp_{32}$ and $Asp_{215}$. Therefore, we restricted our selection to 'privileged' chemical motifs known to preferentially bind to the catalytic dyad. Hydroxyl compounds [91], statins [92], and phosphinates [93] are known active principals for aspartyl proteases. Furthermore, we included the piperidine motif and related cyclic amines in the renin fragment library (RFL) design since these constitute nontypical transition-state mimetics [37,75]. Additional nonpeptidic scaffolds included in the RFL are based on the pyrone structure first discovered by HTS for HIV protease inhibition at Parke-Davis [73]. Interestingly, the lactone carbonyl replaces a structural water molecule that is generally present in HIV protease inhibitor complexes [94]. We speculated that the pyrone and related motifs may bind in a similar mode to renin. At the time of this study, our laboratory was investigating the potential binding of related pyrone analogs to BACE-1. We had strong evidence that these compounds indeed do bind to the active site of BACE-1, suggesting that such ligands could also interact with renin.

For the selection of fragments potentially interacting with the catalytic site of renin, a diversity set of generic inhibitor substructures was generated, including 33 hydroxyl-based transition-state mimetics, 25 hydroxypyrone-related and 21 piperidine-related scaffolds (see Figure 19.9). These scaffolds were then used to conduct 79 substructures searches within our corporate archive. The results were merged and compounds having a molecular weight superior to 350 Da were discarded. A library of 4175 low molecular weight compounds was thus obtained.

To enrich this library with compounds having potentially a different mode of action, several similarity [95] searches were also conducted based on 247 nonpeptidic aspartyl protease inhibitors reported in the MDDR002.2 database [96] as reference. As for the mechanism-based selection, the same molecular weight cutoff was applied, and as a result a total of 375 additional compounds were identified and combined with the 4175 fragments from the substructure search.

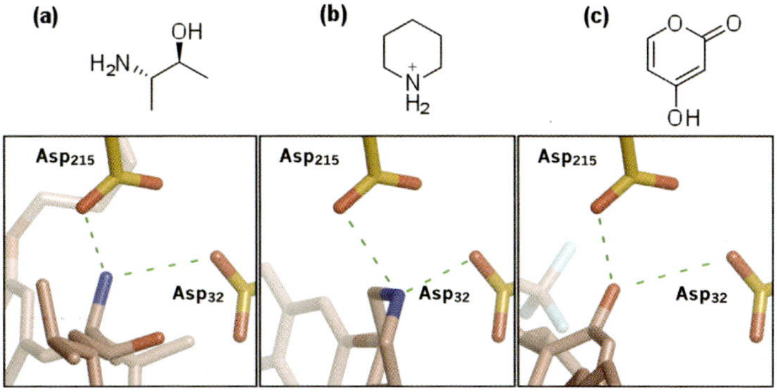


**Figure 19.9** Active principles for aspartyl protease inhibition ('priviledged motifs'): (a) transition-state mimetics, (b) piperidines, and (c) hydroxypyrones. The lower panels show the complex structures (ligands are shown as sticks in salmon color) of aliskiren (pdb code 2ZOV.pdb) [35], the piperidine derivate 6 (pdb code 1PR8.pdb, see Figure 19.6) [36,37], and tipranavir (pdb code 1D4Y.pdb) [63]. The catalytic aspartates are shown with the carbon atoms in yellow and the closest contacts with the inhibitor are indicated by dotted lines. The figures of three-dimensional structures have been prepared using Pymol [18].

3D coordinates of the selected compounds were generated using Corina (Molecular Networks Inc.). The coordinates of human renin in complex with a polyhydroxymonoamide inhibitor in the closed flap conformation of the enzyme (pdb code 1HRN.pdb; [97]) were used for docking after removing the inhibitor atoms. As this was a first pilot approach on renin, we did not consider additional human renin conformations representing various open flap binding sites initially. Also, similar open flap conformations described for human renin had not yet been reported for other aspartyl proteases such as BACE-1 and the plasmepsins at the time of this study. Choosing the appropriate protonation of the catalytic aspartates is crucial and we hypothesized that protonation of the inner $Asp_{32}$ oxygen is the most relevant state for docking of catalytic dyad-directed active principles. The hypothesis that protonation of the inner oxygen of Asp is preferred in complex with a transition-state mimetic inhibitor for BACE-1 was supported by molecular dynamics simulations [98]. Considering that the geometry of the catalytic machinery of BACE-1 and human renin is closely related, the same protonation state was assumed for human renin.

The library of 4550 compounds was screened virtually using GOLD 2.0 [99,100] without applying any constraints. Up to 20 poses were calculated by GOLD for each compound and the 15 000 top scored poses, according to the GOLD score, were selected for further analysis. A possible approach to improve scoring results is to use consensus scoring for postdocking processing. Combining multiple scoring functions for consensus scoring has been shown to improve the discrimination between active and inactive compounds [45]. Five different scoring functions available from the CScore module of Tripos [101] were thus applied. These are the D-Score (DOCK) [102], the PMF-Score [103], the G-Score (GOLD) [104], the ChemScore [105], and the F-Score (FlexX) [106]. The different scores were scaled and combined to generate a consensus score $Z$ for each pose that is defined as $Z = \sum z_i$ , where $z_i$ is the normal deviate score $z$ for a pose for the scoring function $i$. The sum is taken over the five scoring functions of Cscore and $z$ is defined as $z = (x - \overline{x})/\sigma$ where $x$ is the raw score, $\overline{x}$ is the average raw score of all the 15 000 poses, and σ is the standard deviation of the raw score for all the 15 000 poses. For compounds with several docking poses only the pose with the highest consensus score was considered, leading to a final set of 400 compounds (Figure 19.10). This

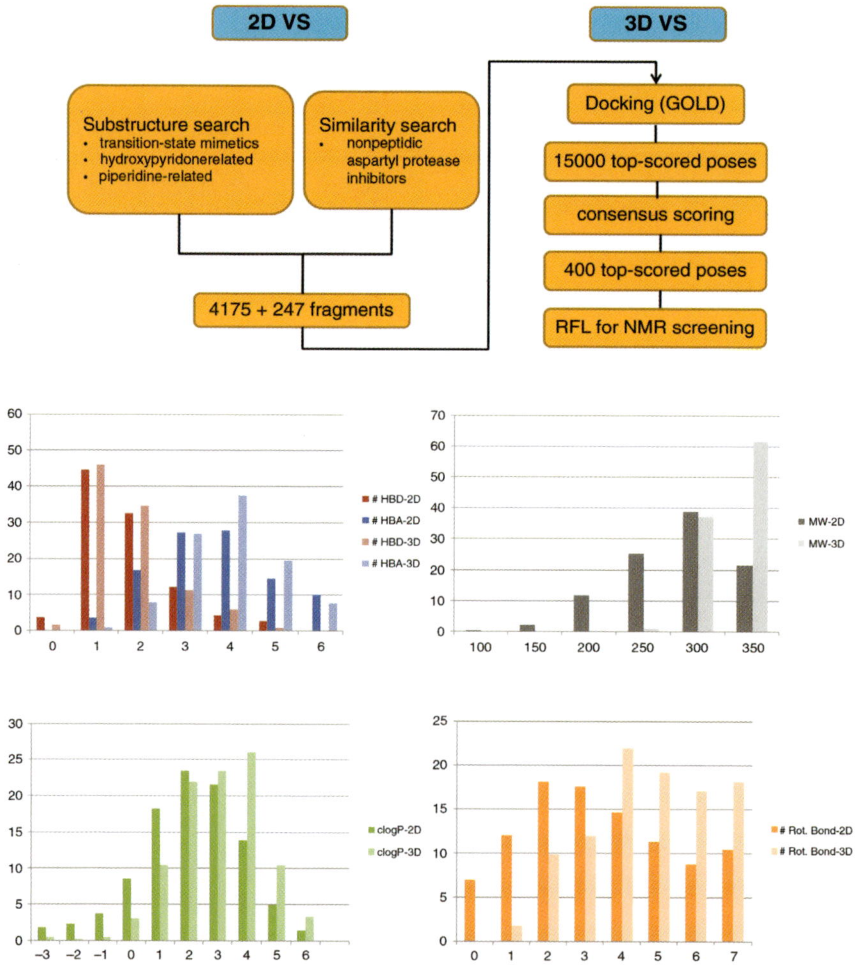


**Figure 19.10** The flowchart for the design of the RFL is shown in the top panel. The four panles below are the histograms displaying the distributions of the following molecular properties: number of hydrogen bond donors (HBDs), number of hydrogen bond acceptors (HBAs), molecular mass, clog*P*, and the number of rotatable bonds. The properties are displayed for the fragments after 2D VS and after 3D VS. Docking and consensus scoring enriches for fragments with higher molecular mass and clog*P* as indicated by the shift of the distribution toward higher values.

was considered to be the maximum size of an FBS library for NMR screening within a reasonable time frame. No additional filtering or prioritization was applied; hence the selection of 400 fragments was solely based on the consensus score.

Visual inspection of the docked structures showed that indeed 211 out of 400 potential aspartate binding motifs were actually in contact with the carboxylic acid side chains of the catalytic dyad. Most of the fragments were placed, at least in part, within the large hydrophobic $S_3$–$S_1$ pocket distant from the catalytic center. This indicated that van der Waals contacts are the main driving force for ligand binding and that the $S_3$–$S_1$ recognition site is a binding hot spot for hydrophobic interactions. Based on our experience, the GOLD-scoring function is well balanced between hydrophobic interactions and hydrogen bonds.

Several more general criteria are often applied for the selection of fragment screening collections. These include, for example, the molecular mass, number of nonhydrogen atoms, lipophilicity, and number of rotatable bonds, among others. The group at Astex has proposed the 'Rule-of-three', which is a set of physicochemical properties typically observed for fragments [107]. These are molecular weight $<300\,\mathrm{g\ mol^{-1}}$, number of hydrogen bond donors <3, number of hydrogen bond acceptors <3, and $\mathrm{clog}P < 3$. More practical aspects in library design apply as well, such as compound availability, purity, solubility, and stability. While the socalled 'Rule-of-three' was not applied in the design of this RFL, the profiles of the different properties of the RFL were very close to the current criteria applied to select lead-like structures that are found to correlate with good adsorption, distribution, metabolism, and excretion (ADME) properties [108].

Low-affinity fragments need to be tested at high concentrations that require high solubility of compounds in aqueous buffer solution. Fragment concentrations of 1 mM are frequently used for screening purposes. High compound concentrations can give rise to false positives in case the test probe is contaminated with reactive impurities. Some fragments react covalently with the protein without being immediately recognized as a potential reactive modifier from the chemical structure. These are referred to as pan assay interference compounds, or PAINS, and a list of moieties to watch out for has been published recently [109–111].

## 19.7 Fragment Screening by NMR T1ρ Ligand Observation

NMR spectroscopy has been the pioneering technology for FBS [112] and is still the method of choice in case the target protein can be readily produced in sufficient quantities [113]. The various NMR screening methods are categorized according to the observed molecular species, that is, small ligand molecule observation or target protein observation. In this study, small molecule observation has been used that is often applied for targets that are too large for protein observation NMR [113].

For the physical assembly of the RFL, the 400 top scoring compounds were selected as described in the previous section (Figure 19.10). The solubility, purity, and chemical structures of the fragments were checked by $^{1}$H NMR and 113 compounds had a solubility of at least 200 μM that is required for FBS. Furthermore, the $^{1}$H NMR spectra of the individual compounds were used to design mixtures of fragments (pools of 5–10 compounds) with minimal spectral overlap for ligand observation $^{1}$H NMR screening.

Samples with a 10 times molar excess of fragments over rh-renin were prepared and 1D $^{1}$H experiments with a relaxation filter delay were recorded. By this technique, changes in relaxation rates of small molecules that occur upon complex formation with the macromolecule are detected. Signals for the protein itself are not observed and therefore the method is not limited by the protein size [114]. The method is practical for rh-renin with a molecular mass of about 40 kDa that is too large for protein observation NMR in the absence of isotopic labeling.

A ligand-to-protein binding event is detected by an increase in the $^{1}$H relaxation rate of the small molecule. In practice, a series of $^{1}$H $^{1}$D spectra is recorded after applying spin-lock filters of increasing durations [115]. The intensity of the proton signals decrease with increasing spin-lock time and the decay rate can be fitted to the relaxation dispersion curve. The increase of relaxation rates upon interaction with the macromolecule is governed by the exchange rate between free and bound ligand and the modulation of the chemical environment of the corresponding proton. It provides, therefore, only a qualitative measure for the affinity and provides no indication on the binding site [116].

Figures (d) and (e) show the decrease in signal intensity of aromatic proton signals with increasing relaxation filter delays for **Hit 1** (at 7.44 ppm) and **Hit 2** (at 7.31 ppm), respectively. Data have been acquired in the absence of rh-renin (black circles), in the presence of rh-renin (red circles) and in the presence of rh-renin and aliskiren (blue circles). The solid curves are best fits to the data points using the relation $I(t) = \exp(-t/T1\rho)$. The relaxation dispersion indicates interaction of **Hit 1** and **Hit 2** with the enzyme. **Hit 2** showed complete recovery of the signal in the presence of aliskiren, that is, full competition. For **Hit 1**, the signal intensities were not fully recovered in the presence of aliskiren that indicates the presence of one or more binding sites for **Hit 1** outside of the aliskiren binding pocket.

We defined an initial hit in compound pools to show a decrease in signal intensity by at least 50% at a relaxation delay of 250 ms. In order to confirm specific binding, T1ρ relaxation data were acquired for single compounds and competition experiments with aliskiren, a tight-binding active site-directed renin inhibitor ($IC_{50} = 0.6$ nM) [35], were performed. Partial competition was considered to be indicative for the ligand interacting at the active site and at least partially overlapping with that of aliskiren. The competition experiments were recorded on samples containing single fragments with a 2.5-fold molar excess of aliskiren over rh-renin (Figure 19.11). As a result, out of the 113 fragments screened by applying the T1ρ relaxation filter, a total of 6 ligands qualified as NMR-confirmed hits (Figure 19.12). For two fragments, **Hit 2** and **Hit 3**, complete displacement in the presence of aliskiren was observed (Figure 19.11), and the other four hits showed partial displacement.

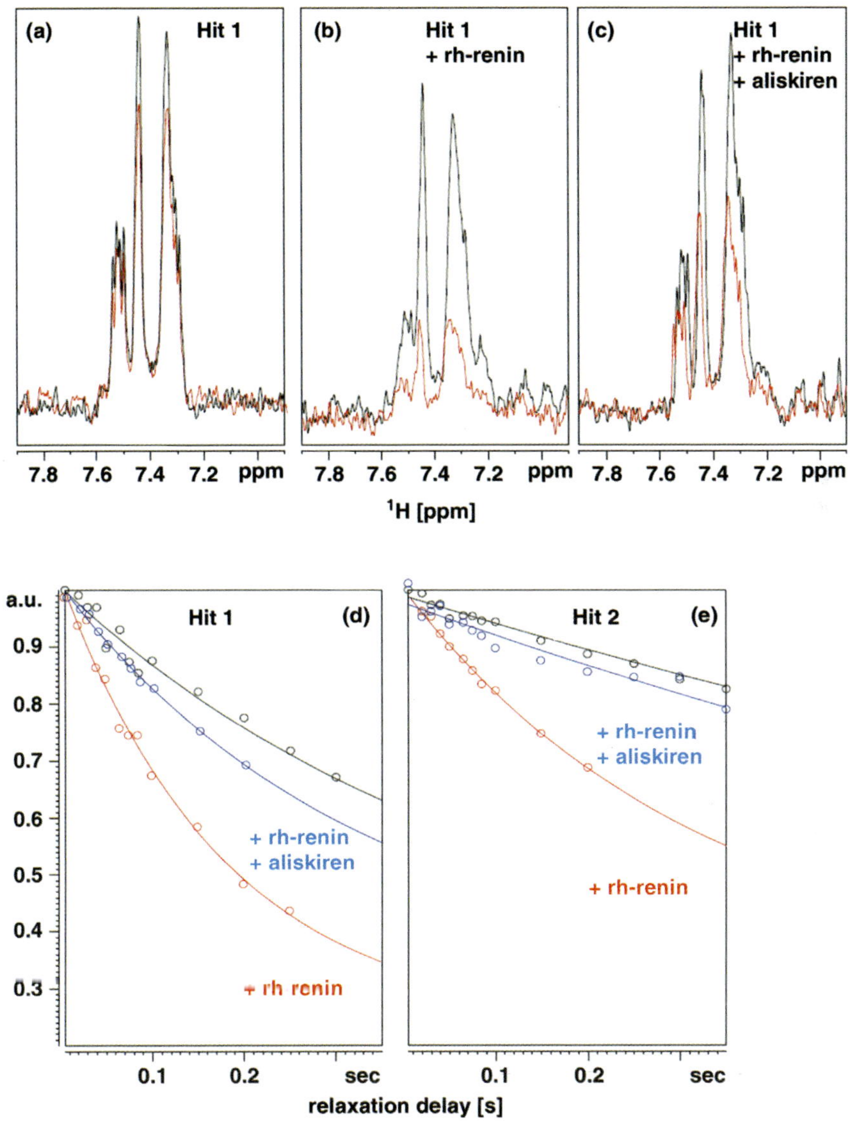


**Figure 19.11** $^{1}$H T1ρ spectra are shown for **Hit 1** at a relaxation delay of 5 ms (black curves) and 450 ms (red curves): (a) a moderate-intensity decrease at 450 ms relaxation delay versus 5 ms is observed for the sample without rh-renin; (b) strong signal intensity reduction in the presence of rh-renin indicates interaction of **Hit 1** with the macromolecule; and (c) partial recovery of the $^{1}$H signal upon addition of the competitior aliskiren were observed and indicated an overlap of the binding sites of both ligands. The datas have been acquired using the following concentrations: 100 μM **Hit 1**, 10 μM rh-renin and 25 μM aliskiren.

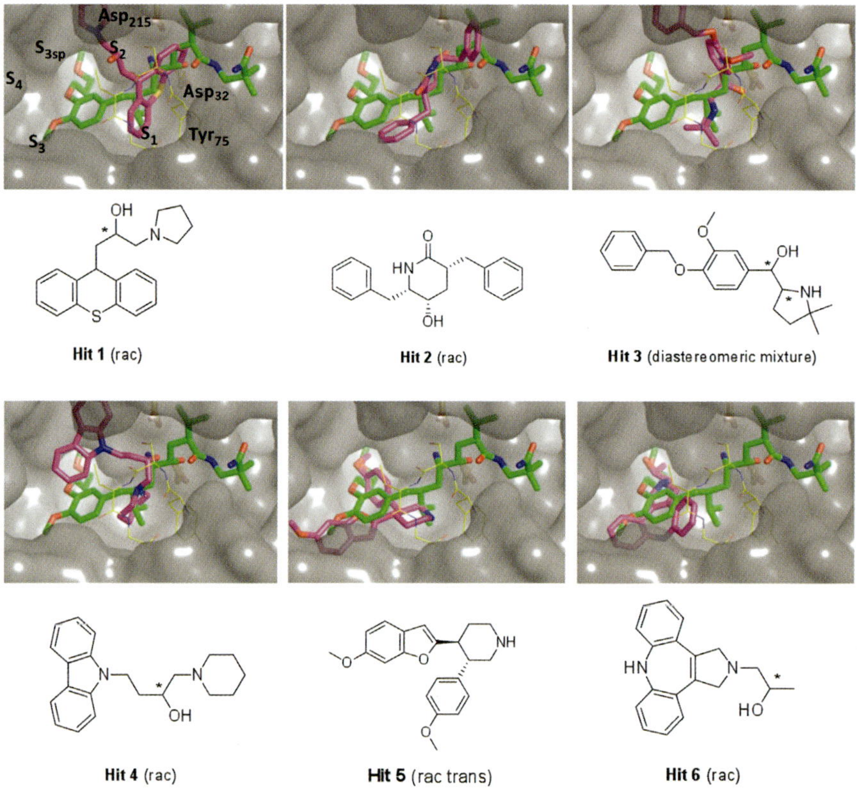


**Figure 19.12** T1ρ active hits identified from the NMR screen of the RFL and their predicted binding modes based on GOLD docking. The docking pose with the highest consensus score is shown for each ligand. The fragment structure models are superimposed with the transition-state mimetic inhibitor aliskiren (green color) bound to human renin (pdb code 2ZOV.pdb) [35]. The oxygen and nitrogen atoms of the ligands are colored in red and blue, respectively. Shown is the solvent-accessible surface of the active site of renin. Residues of the flap β-hairpin ($Tyr_{75}$, $Ser_{76}$, $Thr_{77}$, and $Gly_{78}$) in a closed conformation are depicted in yellow as thin lines and have been omitted from the protein surface calculation.

Two examples for which the consensus scoring preferred a direct binding of the 'priviledged' motifs to the catalytid dyad are **Hit 2** and **Hit 4**. The calculated structure of **Hit 2** showed the OH group symmetrically placed in the pivotal position between $Asp_{32}$ and $Asp_{215}$. This compound is reminiscent of a structurally closely related tribenzylated 4-OH-substituted δ-lactam ([117]; Figure 19.5) derivative that has been reported to be a micromolar inhibitor of the HIV protease. The OH group of **Hit 4** was docked equidistant between the outer oxygen of $Asp_{32}$ and $Asp_{215}$ in a similar position to the primary $NH_2$ group of the transition-state mimetic aliskiren, while the *N*-alkylated piperidine ring filled the hydrophobic $S_1$ pocket without any binding interactions to the aspartates (Figure 19.12). For the calculated structures of the other four confirmed NMR hits, the 'privileged' chemical motifs supposed to bind to the catalytic dyad did not show such an interaction as would have been expected. One example for non-catalytic dyad-directed binding is the docking pose of **Hit 5** (Figure 19.12) that showed the basic piperidine nitrogen to be about 5 Å distant from the center nitrogen of the piperidines such as **4–6** (Figure 19.6) binding to the open flap conformation of renin. In the closed flap conformation that we have used in our docking calculations, placement of the piperidine scaffold of **Hit 5** in a similar way as observed in the open flap conformation is sterically not possible since the C-4 benzofuran residue would clash with the side chain of $Tyr_{75}$. The NMR **Hits 1–Hit 6** have been tested in the biochemical inhibition assay but do not show inhibition with results in an $IC_{50} > 100\,\mu M$. Unfortunately, we were not successful in obtaining X-ray crystal structures of the NMR-confirmed **Hit 2**, **Hit 3**, **Hit 4**, and **Hit 5** in complex with rh-renin.

## 19.8 X-Ray Crystallography

The fragment optimization strategy greatly benefits from structural information of the protein–ligand complex. We, therefore, embarked on protein crystallography for the six NMR-validated hits in order to guide molecular design efforts. A robust crystallization system was established to obtain apo-protein crystals of recombinant glycosylated human renin suitable for subsequent soaking experiments. A screening protocol using the hanging drop vapor diffusion method was established to identify the best crystallization conditions in aqueous buffer systems. Crystals grew in the orthorhombic space group $P2_12_12_1$; this space group has not been reported previously for rh-renin, with sufficient diffraction quality. X-ray diffraction data were collected at the Swiss Light Source. Structures were solved by molecular replacement with the program MOLREP [118] using the coordinates of renin in complex with aliskiren (pdb code 2VOZ.pdb) [17] as a search model.

To prepare protein–ligand complexes, the crystals were incubated with 10–50 mM ligand in the presence of 10% DMSO. For all six validated NMR hits soaking experiments were conducted. Only for two co-crystals electron density in the active site could be observed. The densities were unambiguously assigned to the corresponding ligands and the X-ray structures for the two NMR fragment screening hits **Hit 1** and **Hit 6** (Figure 19.12) could be solved. An additional 20 NMR hits, that show ambiguous binding in the T1ρ experiments and/or no competition with aliskiren, were furthermore subjected to cocrystallization experiments that did not result in complex structures [11].

The cocrystal structures of **Hit 1** revealed an unexpected binding mode to the nonprime site of the enzyme. Both the 9*H*-thioxanthene moiety of **Hit 1** and the 5*H*-dibenzoazepine scaffold of **Hit 6** were found to be in close van der Waals contacts to the contiguous hydrophobic $S_3$–$S_1$ binding site completely filled by the ligand, and the flap β-hairpin (residues $Tyr_{75}$, $Ser_{76}$, $Thr_{77}$, and $Gly_{78}$) was in a closed conformation. Furthermore, both the hydroxylated side chain of **Hit 1** and

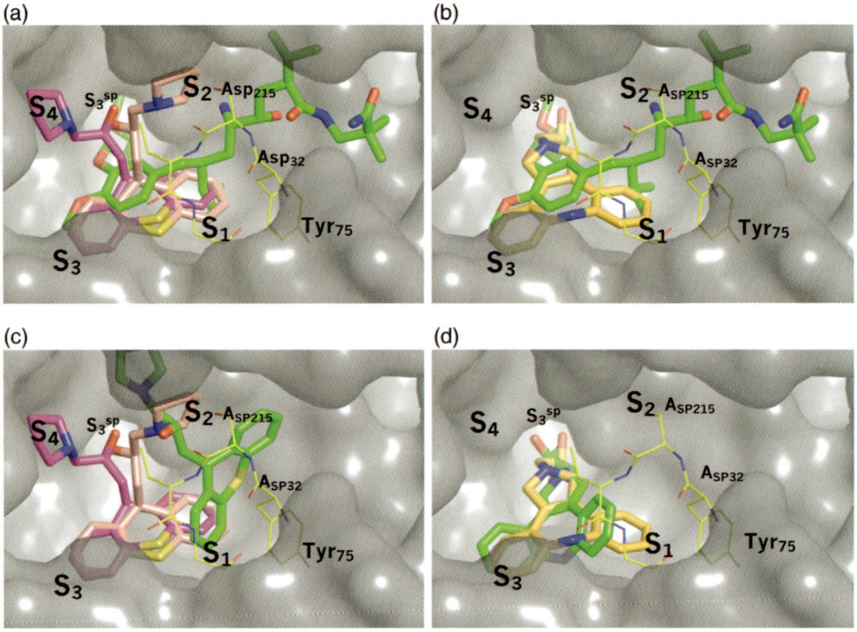


**Figure 19.13** X-ray cocrystal structures of (a) the (*S*)-configured **Hit 1** (shown as stick model in magenta color) superimposed with the (*R*)-configured **Hit 1** (salmon color) and (b) the 5*H*-dibenzoazepine **Hit 6** (orange color) in complex with renin. The ligand electron density of the crystal structure obtained with *rac*-**Hit 6** was only compatible with the (*R*)-configuration (as shown). The fragment structures are superimposed with the transition-state mimetic inhibitor aliskiren (green color) bound to renin (pdb code 2V0Z.pdb) [17]. Panels (c) and (d) show the overlay of the X-ray structures with the corresponding docking pose for **Hit 1** and **Hit 6**, respectively. The oxygen and nitrogen atoms of the ligands are colored in red and blue, respectively. Shown is the solvent-accessible surface of the active site of renin. Residues of the flap β-hairpin (residues $Tyr_{75}$, $Ser_{76}$, $Thr_{77}$, and $Gly_{78}$) in a closed conformation are depicted in yellow as thin lines and have been omitted from the protein surface calculation.

**Hit 6** and the pyrrolidine nitrogen atom of both enantiomers of **Hit 1** were in remote positions relative to the $Asp_{32}$ and $Asp_{215}$ carboxylates (Figure 19.13). This was surprising to us in view of the mechanism-biased fragment library design based on the query for privileged transition-state surrogate motifs for aspartic proteases. The X-ray data suggested that the binding of **Hit 1** and **Hit 6** is mainly driven by hydrophobic interactions to the hydrophobic $S_3$–$S_1$ cavity of renin.

While the predicted binding mode for **Hit 1** is different from the experimentally observed orientation (heavy atom RMSD = 5.9 Å for the corresponding enantiomer), the key binding features for **Hit 6** have been predicted with good accuracy (heavy atom RMSD = 1.6 Å). Remarkably, the consensus score docking pose generated with GOLD of **Hit 6** is very close to the experimentally observed binding pose, and even predicted the orientation of the OH-bearing side chain to the nonsubstrate binding pose. The 2-hydroxypropyl side chain of **Hit 6** extended into the nonsubstrate $S_3^{sp}$ cavity by extruding the ordered water of apo renin and forming a hydrogen bond to the carbonyl of $Thr_{216}$ at the bottom of the rigid channel. This was the first time that such a hydrogen bond involving $Thr_{216}$ had been observed. As a result, fragment screening by NMR followed by X-ray has resulted in the 'rediscovery' of two key design principles that have been critical for the conception of the topological transition-state mimetic inhibitor aliskiren [15,119], that is, targeting the $S_1$–$S_3$ hydrophobic hot spot and exploitation of the $S_3^{sp}$ binding site of human renin. In case of **Hit 1** bound to rh-renin, the very narrow and conformationally rigid nonsubstrate $S_3^{sp}$ channel [119] is not occupied by the sterically more demanding branched OH-bearing side chain. Instead, the terminal pyrrolidine residue of **Hit 1** in its (*S*)-configuration was bound to the $S_4$ pocket in one renin crystal monomer, while the pyrrolidine ring of the (*R*)-configured **Hit 1** in the second crystal monomer was pointing toward the $S_2$ pocket.

## 19.9 Renin Fragment Hit-to-Lead Evolution

With the three $S_3$–$S_1$ binding fragments **Hit 1** (*R*- and *S*-configured) and **Hit 6** (*R*-configured) and their cocrystal complex structures with rh-renin in hand, the next steps were to select a single best fragment with the most promising potential as starting lead, and to develop the chemistry strategy for structural modification by a fragment growing approach. In this context, several key aspects were taken into consideration, among others: the topology of the renin active site including different conformational ensembles relevant to drug design, the accessibility of specific recognition sites considered as 'binding hot spots', the presence of suitable ligand attachment points in a proper position and spatial orientation ('exit vector') for fragment growing toward targeted recognition sites, morphing of the fragment hit core prior to growing, and the complexity of chemical synthesis to allow for efficient SAR exploration. As an important notion, all previously known potent renin inhibitors were shown to tightly interact with the catalytic dyad of the enzyme by involving H-bonding networks or charge–charge interactions to the carboxylate side chains of $Asp_{32}$/$Asp_{215}$. We, therefore, sensed that expansion of the identified $S_3$–$S_1$ fragments toward the catalytic center would be very likely a prerequisite for improving the binding affinity by several orders of magnitude.

The terminal pyrrolidine ring of the *R*-configured **Hit 1** was found by X-ray to be positioned close to the entrance of the otherwise unoccupied $S_2$ site of rh-renin. However, growing into the $S_2$ pocket of rh-renin was not considered by us to be an attractive design concept for various reasons including the general topographic characteristics of this potentially more promiscuous binding site [119]. More specifically, it was argued that growing from an extended and flexible spacer into a rather large and hydrophobic area would afford a substantial increase in molecular size in order to gain significantly in potency, which in turn was projected to reduce substantially the ligand binding efficacy.

Both the large contiguous $S_3$–$S_1$ binding site, which is tightly filled by the tricyclic scaffolds of the fragment hits, and the narrow nonsubstrate $S_3^{sp}$ channel extending in a perpendicular orientation toward the center of the enzyme, have been recognized as key "ligand binding hot spots" of rh-human renin. Accordingly, merging the OH-containing $S_3^{sp}$ binding residue of **Hit 6** with the 9*H*-thioxanthene template of **Hit 1** was regarded as an attractive option to generate a structurally less complex fragment for further optimization. Initial computational modeling-assisted design work explored the feasibility of growing the identified fragments toward the catalytic center of rh-renin by superimposing their renin-bound poses to that of the potent inhibitor aliskiren (Figure 19.13). The C-2 position neighboring the basic tertiary amine of the annelated pyrroline ring of **Hit 6** appeared to be a suitable exit vector for tethering known recognition motifs that could potentially bind to the catalytic aspartate side chains. Several cyclic structural elements linked by a linear two-to-three atom spacer and bearing both a basic amine and a hydroxyl functional group in proper positions and spatial orientation relative to the catalytic dyad were proposed to closely overlap with the center 1,2-aminoalkohol backbone of aliskiren (C. Ehrhard, Novartis, unpublished data). Similarly, related aminoalcohol binding motifs tethered by a short spacer to the center C-6 position of **Hit 1** were investigated.

## 19.10
## Integration of Fragment Hits and HTS Hits

While our initial exploratory fragment-growing efforts were in progress, we identified a micromolar rh-renin inhibitor, compound **22** (Figure 19.14), from a high-throughput biochemical screen of the corporate compound collection using a miniaturized fluorescence polarization enzyme assay. Racemic **22** was the only validated hit that originated from a small split-and-mix combinatorial library of ~950 compounds, all bearing a 3,5-disubstituted center piperidine scaffold, which was specifically designed to potentially target aspartyl proteases [11]. The active (3*S*,5*R*)-enantiomer binds in an extended conformation by spanning the prime and nonprime sites of rh-renin in the closed flap conformation, as shown by X-ray crystallography (Figure 19.15). The basic 3,5-disubstituted center piperidine is in a pivotal position symmetrically between $Asp_{32}$ and $Asp_{215}$, thereby forming charge–charge binding interactions. The overlay of the rh-renin bound conformation of the HTS hit **22** with the fragment **Hit 1** was very instructive and helped to refine our early concept of linking a basic amine 'priviledged' transition-state surrogate to the $S_3$–$S_1$ scaffold (Figure 19.15). Particularly appealing was the almost ideal overlay of the diphenylmethane portion of **22** with the rigid tricyle of **Hit 1**, providing a strong rationale for merging the piperidine-carboxamide portion with the fragment scaffold. These findings prompted us to synthesize the racemic 9*H*-xanthene analog **23** that, however, was found to lack any inhibitory activity toward rh-renin in a FRET-based enzymatic assay at a 100 μM concentration. Despite its apparently very low intrinsic binding affinity, a crystal complex structure of **23**

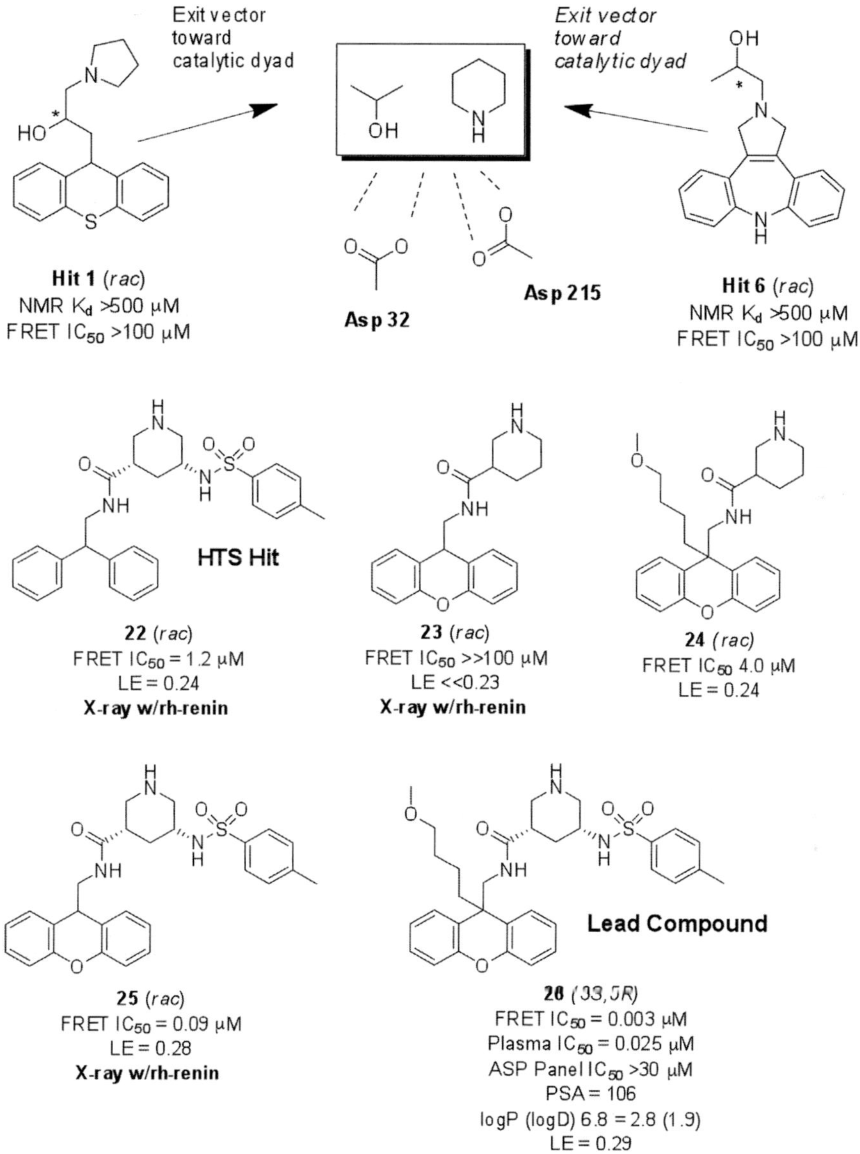


**Figure 19.14** Structures representing the key progression steps of a combined structure-based fragment growing and fragment–HTS hit merging approach, resulting in the novel potent and selective lead rh-renin inhibitor 26 (see text for details).

bound to the renin active site was obtained by soaking and resolved to 2.75 Å resolution (pdb code 4GJB.pdb). The experimental binding pose of the extended $S_3$–$S_1$ fragment closely overlapped with the corresponding portions of the HTS hit **22** and involved very similar specific binding interactions of the piperidine portion to

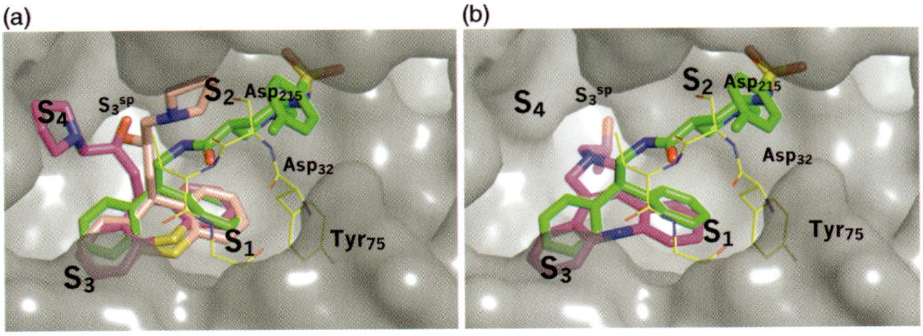


**Figure 19.15** Crystal structure of HTS hit **22** (pdb code 4GJA.pdb) [11] superimposed with (a) **Hit 1** and (b) the 5*H*-dibenzoazepine **Hit 6**. The (*S*)-configuration of **Hit 1** is shown as stick model (magenta color) together with the (*R*)-configuration (salmon color). The overlay provided a strong rationale for merging the piperidine-carboxamide with the tricycle of the fragment **Hit 1**. Merging of 22 with the 5*H*-dibenzoazepine of **Hit 6** appeared less appealing mainly for reasons of increased synthetic complexity.

the catalytic aspartates, as well as the carboxamide linker to the flap β-hairpin loop. Hence, the crystallographic data nicely validated at the structural level the principal design concept of growing the starting fragment out of the $S_3$–$S_1$ cavity toward the enzyme catalytic dyad. However, it was surprising to note that this did not result in a significant improvement in the inhibitory potency of the ligand.

Introduction of the methoxybutyl side chain at the tricycle C-6 position of **23** was predicted, based on superimposing the X-ray structures of **23** and aliskiren and modeling refinement, to properly fill the nonsubstrate $S_3^{sp}$ cavity and to potentially induce a hydrogen-bond interaction to $Tyr_{14}$ at the bottom of this binding site. Compound **24** was found to be a low micromolar inhibitor of rh-renin in the FRET assay and hence to be by far greater than 100-fold more potent compared to **23**. Ligand binding efficacy of the still weak inhibitor **24** (LE = 0.24) based on biochemical $IC_{50}$ data was significantly improved over that estimated for the starting fragment **Hit 1** and **Hit 6**. Overall, this demonstrated once again in an impressive manner that the deeply buried, rigid, and small $S_3^{sp}$ pocket of rh-renin constitutes a very important ligand binding site for the design of potent inhibitors [119].

Various strategies were then considered to further optimize the tethered fragments **23**, **24** for on-target potency by addressing either the closed or the open flap conformation of rh-renin [119]. For example, disubstituted piperidine analogs of **23** bearing mono- or bicyclic aryl residues at the C-4 position of the center transition-state surrogate scaffold were conceptualized to target potential interactions to the large and hydrophobic '$Tyr_{75}$–$Trp_{39}$ open-flap-pocket' discovered previously by researchers at Roche (GRAB peptidomimetics, *vide supra*). First prototype analogs bearing small nonoptimized C-4 aryl residues and lacking the $P_3^{sp}$ side chain were indeed found to have single-digit micromolar activity toward rh-renin, thus being more than two-orders of magnitude potent compared to **23** (Novartis, data unpublished). At this stage, however, the concern was raised that a significant enhancement of the predominantly hydrophobic van der Waals interactions to the deep-flap pocket would be required in order to gain substantially in potency, which in turn was expected to provide inhibitor molecules with potentially unfavorable physicochemical and pharmacokinetics properties, as was experienced previously with many GRAB peptidomimetic renin inhibitors [119].

Targeting the prime-site substrate recognition pockets for further optimization appeared to be more appealing [20]. This notion was in particular fueled by the X-ray structure of the rh-renin–**22** complex that revealed an unprecedented binding interaction of the arylsulfonamide portion to the $S_{1'}$ pocket as well as a previously unrecognized nonsubstrate binding site located between the proline-rich loop ($Met_{289}$–$Thr_{298}$) and the flap region ($Thr_{72}$–$Ser_{81}$) [20] (Figure 19.15). Introduction of the tolylsulfonamide motif to the piperidine C-3 position of **23** and **24**, and with the preferred (5*R*)-absolute configuration in place, resulted in a more than 1000-fold potency increase determined for inhibitors **25** and **26**, respectively, in the biochemical assay. The crystal structure of rh-renin bound inhibitor **26** confirmed the expected binding pose perfectly overlapping with the prime-site tolylsulfonamide portion of **26** and the tricyclic $P_3$–$P_1$ scaffold of fragment **Hit 1**. Moreover, the optimized flexible methoxybutyl side chain was shown to penetrate deeply into the $S_3^{sp}$ cavity in an extended conformation and to form a H bond to the $Tyr_{14}$ backbone nitrogen. Interestingly, the impact of introducing the optimal $P_3^{sp}$ methoxyalkyl chain appeared to be less pronounced for analogs bearing a high-affinity prime-site binding motif, as compound **26** was found to be only ~20-fold more potent compared to **25**. Intriguingly, the ligand binding efficacies (LE) calculated for analogs **24**, **25**, and **26** are in a very narrow range of 0.24–0.29, suggesting almost equal binding contributions of the flexible alkoxyalkyl $P_3^{sp}$ side chain as well as the prime-site sulfonamide moiety, when normalized by the number of heavy atoms added to the core scaffold.

Inhibitor **26** was broadly characterized for its *in-vitro* and *in-vivo* properties and served as a preclinical lead candidate. The compound showed moderate absolute oral bioavailability and significant blood-pressure lowering efficacy in a telemetered hypertensive double-transgenic rat model. Further extensive X-ray-guided SAR optimization efforts involved major structural modifications of the prime and non-prime site portions, addition of an OH group to the C-4 position at the center piperidine scaffold, and adjustment of the absolute stereochemistry depending on the nature of the prime-site residue. This work resulted ultimately in a novel candidate renin inhibitor with high potency, selectivity, and excellent pharmacokinetic properties to achieve sustained oral blood-pressure lowering efficacy *in-vivo* [120].

## 19.11
## Conclusions

Aspartyl proteases are important drug targets with ongoing drug discovery efforts in search for treatments for indications such as Alzheimer's disease. FBS and FBLD contribute to these efforts, ideally in an integrated approach combined with virtual, biophysical, biochemical screening as well as structure-guided drug design. While aspartyl proteases have proven to be challenging targets for docking due to their inherent conformational flexibility and due to the inaccuracy of scoring functions, virtual screening remains a valuable technique for hit discovery. Conformational flexibility of the target can be taken into account by including an ensemble of protein conformations in the 3D virtual screening calculations. Biochemical or biophysical FBS on aspartyl proteases, followed by FBLD, has resulted in drugs and drug candidates for HIV-1 protease, BACE-1, and renin. In all cases, protein–ligand complex structures have been critical both to elucidate the fragment binding modes and for guiding the design efforts.

In the work presented herein, we have used an integrated lead-finding approach based on the screening of a small and target-focused fragment library by ligand-observed NMR, determination of the binding pose of identified hits to rh-renin by X-ray crystallography, and subsequent hit-to-lead generation by structure-guided fragment merging with a novel prime-site chemotype renin inhibitor discovered by high-throughput screening. The fragment screening library was built by *in-silico* enrichment by applying in the first step a 2D ligand-based substructure and similarity filter based on a set of canonical, or 'privileged', recognition motifs for the catalytic dyad of aspartyl proteases, followed by docking and final compound selection by consensus scoring. The goal of this work was to uncover fragments that would interact with the catalytic dyad consisting of $Asp_{32}$ and $Asp_{215}$ at the center of the substrate binding cleft, which were suitable for growing into specific binding pockets of renin in order to gain inhibitory potency. A total of six structurally distinct fragments, indeed representing all known active site-recognition elements, were validated by NMR experiments to bind to the active site of rh-renin. For two of these hits, the X-ray complex structures were solved and show binding into the large $S_3$–$S_1$ pocket, a previously recognized 'hydrophobic hot-spot' for renin, but without direct interactions with the catalytic dyad. Most notably, the nonsubstrate binding channel $S_3^{sp}$, which has played a key role in the design of aliskiren, was 're-discovered' as an important target site for ligand binding by the experimental binding pose of one newly discovered fragment. The successful elucidation of the enzyme-bound ligand structures by X-ray is remarkable in view of their very low binding affinities. In summary, this work represents an impressive illustration of the power of biophysical screening approaches for active site mapping by low-affinity ligands of renin as a member of the aspartic protease family. Furthermore, the accomplishments made by this complementary and integrated approach within a relatively short time frame emphasize the benefit of exploiting the key strengths of individual lead-finding approaches in a combined effort.

## References

1 Murray, C.W., Verdonk, M.L., and Rees, D.C. (2012) Experiences in fragment-based drug discovery. *Trends in Pharmacological Sciences*, **33** (5), 224–232.

2 Baker, M. (2013) Fragment-based lead discovery grows up. *Nature Reviews Drug Discovery*, **12** (January), 5–7.

3 Kumar, A., Voet, A., and Zhang, K.Y.J. (2012) Fragment based drug design: from experimental to computational approaches. *Current Medicinal Chemistry*, **19**, 5128–5147.

4 Leonardo, H., Blundell, T.L., Abell, C., and Ciulli, A. (2013) Integrated biophysical approach to fragment screening and validation for fragment-based lead discovery. *Proceedings of the National Academy of Sciences of the United States of America*, **110** (32), 12984–12985.

5 Boettcher, A., Ruedisser, S., Erbel, P. *et al.* (2010) Fragment-based screening by biochemical assays: systematic feasibility studies with trypsin and MMP12. *Journal of Biomolecular Screening*, **15** (9), 1029–1041.

6 Hann, M.M., Leach, A.R., and Harper, G. (2001) Molecular complexity and its impact on the probability of finding leads for drug discovery. *Journal of Chemical Information and Computer Science*, **41** (3), 856–864.

7 Orry, A.J.W., Abagyan, R.A., and Cavasotto, C.N. (2006) Structure-based development of target-specific compound libraries. *Drug Discovery Today*, **11** (5), 261–266.

8 Wang, Y., Strickland, C., Voigt, J.H. *et al.* (2010) Application of fragment-based NMR screening, X-ray crystallography, structure-based design, and focused chemical library design to identify novel μ M leads for the development of nM BACE-1 (β-Site APP cleaving enzyme 1) inhibitors. *Journal of Medicinal Chemistry*, **53**, 942–950.

9 Zhu, Z., Sun, Z., Ye, Y. *et al.* (2010) Discovery of cyclic acylguanidines as highly potent and selective β -site amyloid cleaving enzyme (BACE) inhibitors: Part I – inhibitor design and validation. *Journal of Medicinal Chemistry*, **53**, 951–965.

10 Oehlrich, D., Prokopcova, H., and Gijsen, H.J.M. (2014) The evolution of amidine-based brain penetrant BACE1 inhibitors. *Bioorganic and Medicinal Chemistry Letters*, **24** (9), 2033–2045.

11 Ostermann, N., Ruedisser, S., Ehrhardt, C. *et al.* (2013) A novel class of oral direct renin inhibitors: highly potent 3,5-disubstituted piperidines bearing a tricyclic P3–P1 pharmacophore. *Journal of Medicinal Chemistry*, **11** (56), 2196–2206.

12 Zaman, M.A., Oparil, S., and Calhoun, D.A. (2002) Drugs targeting the renin–angiotensin–aldosterone system. *Nature Reviews Drug Discovery*, **1** (8), 621–636.

13 Ong, K.L., Cheung, B.M.Y., Man, Y.B. *et al.* (2007) Prevalence, awareness, treatment, and control of hypertension among United States adults 1999–2004. *Hypertension*, **49** (1), 69–75.

14 Bonanni, L. and Dalla Vestra, M. (2012) Oral renin inhibitors in clinical practice: a perspective review. *Therapeutic Advances in Chronic Disease*, **3** (4), 173–181.

15 Maibaum, J. and Feldman, D.L. (2009) Case history on Tekturnas (R)/Rasilez(R) (Aliskiren), a highly efficacious direct oral renin inhibitor as a new therapy for hypertension, in *Annual Reports in Medicinal Chemistry*, vol. 44, Elsevier, San Diego, CA 92101-4495, USA, pp. 105–127.

16 Jensen, C., Herold, P., and Brunner, H.R. (2008) Aliskiren: the first renin inhibitor for clinical treatment. *Nature Reviews Drug Discovery*, **7** (5), 399–410.

17 Rahuel, J., Rasetti, V., Maibaum, J. *et al.* (2000) Structure-based drug design: the discovery of novel nonpeptide orally active inhibitors of human renin. *Chemistry and Biology*, **7** (7), 493–504.

18 The PyMOL Molecular Graphics System, Version 1.5.0.5 Schrödinger, LLC.

19 Yokokawa, F. (2013) Recent progress on the discovery of non-peptidic direct renin inhibitors for the clinical management of hypertension. *Expert Opinion on Drug Discovery*, **8** (6), 673–690.

20 Lorthiois, E., Breitenstein, W., Cumin, F. *et al.* (2013) The discovery of novel potent *trans*-3,4-disubstituted pyrrolidine inhibitors of the human aspartic protease renin from *in silico* three-dimensional (3D) pharmacophore searches. *Journal of Medicinal Chemistry*, **56** (6), 2207–2217.

21 Blundell, T., Sibanda, B.L., and Pearl, L. (1983) Three-dimensional structure, specificity and catalytic mechanism of renin. *Nature*, **304**, 273–275.

22 Coates, L., Tuan, H., Tomanicek, S. *et al.* (2008) The catalytic mechanism of an aspartic proteinase explored with neutron and X-ray diffraction. *Journal of the American Chemical Society*, **130**, 7235–7237.

23 Andreeva, N.S. and Rumsh, L.E.V.D. (2001) Analysis of crystal structures of aspartic proteinases: on the role of amino acid residues adjacent to the catalytic site of pepsin-like enzymes. *Protein Science*, **10**, 2439–2450.

24 Gorfe, A.A. and Caflisch, A. (2005) Functional plasticity in the substrate binding site of β-secretase. *Structure*, **13** (10), 1487–1498.

25 Rajamani, R. and Reynolds, C.H. (2004) Modeling the protonation states of the catalytic aspartates in β-secretase. *Journal of Medicinal Chemistry*, **47**, 5159–5166.

26 Sussman, F., Villaverde, M.C., Domínguez, J.L., and Danielson, U.H. (2013) On the active site protonation state in aspartic proteases: implications for drug design. *Current Pharmaceutical Design*, **19**, 4257–4275.

27 Piana, S., Bucher, D., Carloni, P., and Rothlisberger, U. (2004) Reaction mechanism of HIV-1 protease by hybrid car-parrinello/classical MD simulations. *Journal of Physical Chemistry B*, **108** (30), 11139–11149.

28 Brás, N.F., Ramos, M.J., and Fernandes, P.A. (2012) The catalytic mechanism of mouse renin studied with QM/MM calculations. *Physical Chemistry Chemical Physics*, **14** (36), 12605–12613.

29 Friedman, R. and Caflisch, A. (2010) On the orientation of the catalytic dyad in aspartic proteases. *Proteins*, **78** (6), 1575–1582.

30 Rich, D.H., Bursavich, M.G., and Estiarte, M.A. (2002) Discovery of nonpeptide, peptidomimetic peptidase inhibitors that target alternate enzyme active site conformations. *Biopolymers*, **66** (2), 115–125.

31 Meagher, K.L. and Carlson, H.A. (2004) Incorporating protein flexibility in structure-based drug discovery: using HIV-1 protease as a test case. *Journal of the American Chemical Society*, **126** (41), 13276–13281.

32 Kacker, P., Masetti, M., Mangold, M. *et al.* (2012) Combining dyad protonation and active site plasticity in BACE-1 structure-based drug design. *Journal of Chemical Information and Modeling*, **52**, 1079–1085.

33 Ghemtio, L. and Muzet, N. (2013) Retrospective molecular docking study of WY-25105 ligand to β-secretase and bias of the three-dimensional structure flexibility. *Journal of Molecular Modeling*, **19** (8), 2971–2979.

34 Cosconati, S., Marinelli, L., Saverio, F. *et al.* (2012) Protein flexibility in virtual screening: the BACE-1 case study. *Journal of Chemical Information and Modeling*, **52**, 2697–2704.

35 Wood, J.M., Maibaum, J., Rahuel, J. *et al.* (2003) Structure-based design of aliskiren, a novel orally effective renin inhibitor. *Biochemical and Biophysical Research Communications*, **308** (4), 698–705.

36 The X-ray structure for compound 6 (1uhq.pdb) has been made publicly available by the Protein Data Bank during a restricted time period and later retracted for unknown reasons.

37 Oefner, C., Binggeli, A., Breu, V. *et al.* (1999) Renin inhibition by substituted piperidines: a novel paradigm for the inhibition of monomeric aspartic proteinases? *Chemistry and Biology*, **6** (3), 127–131.

38 Brás, N.F., Fernandez, P.A., and Ramos, M.J. (2014) Molecular dynamics studies on both bound and unbound renin protease. *Journal of Biomolecular Structure and Dynamics*, **32** (3), 351–363.

39 Tanrikulu, Y., Krüger, B., and Proschak, E. (2013) The holistic integration of virtual screening in drug discovery. *Drug Discovery Today*, **18** (7–8), 358–364.

40 Verma, J., Khedkar, V.M., and Coutinho, E.C. (2010) 3D-QSAR in drug design: a review. *Current Topics in Medicinal Chemistry*, **10** (1), 95–115.

41 Willett, P. (2006) Similarity-based virtual screening using 2D fingerprints. *Drug Discovery Today*, **11** (23–24), 1046–1053.

42 Lionta, E., Spyrou, G., Vassilatis, D.K., and Cournia, Z. (2014) Structure-based virtual screening for drug discovery:

principles, applications and recent advances. *Current Topics in Medicinal Chemistry*, **14**, 1923–1938.

43 Singh, N. and Warshel, A. (2010) Absolute binding free energy calculations: on the accuracy of computational scoring of protein–ligand interactions. *Proteins*, **78** (7), 1705–1723.

44 Subramanian, G. and Rao, S.N. (2013) Comprehending renin inhibitor's binding affinity using structure-based approaches. *Bioorganic and Medicinal Chemistry Letters*, **23** (24), 6667–6672.

45 Charifson, P.S., Corkery, J.J., Murcko, M. A., and Walters, W.P. (1999) Consensus scoring: a method for obtaining improved hit rates from docking databases of three-dimensional structures into proteins. *Journal of Medicinal Chemistry*, **42**, 5100–5109.

46 Fischer, M., Coleman, R.G., Fraser, J.S., and Shoichet, B.K. (2014) Incorporation of protein flexibility and conformational energy penalties in docking screens to improve ligand discovery. *Nature Chemistry*, **6** (7), 575–583.

47 Drwal, M.N. and Griffith, R. (2013) Combination of ligand- and structure-based methods in virtual screening. *Drug Discovery Today: Technologies*, **10** (3), e395–401.

48 Luksch, T., Chan, N.-S., Brass, S. *et al.* (2008) Computer-aided design and synthesis of nonpeptidic plasmepsin II and IV inhibitors. *ChemMedChem*, **3** (9), 1323–1336.

49 Luksch, T., Blum, A., Klee, N. *et al.* (2010) Pyrrolidine derivatives as plasmepsin inhibitors: binding mode analysis assisted by molecular dynamics simulations of a highly flexible protein. *ChemMedChem*, **5** (3), 443–454.

50 Degliesposti, G., Kasam, V., A., DaCosta. *et al.* (2009) Design and discovery of plasmepsin II inhibitors using an automated workflow on large-scale grids. *ChemMedChem*, **4** (7), 1164–1173.

51 Kasam, V., Zimmermann, M., Maa, A. *et al.* (2007) Design of new plasmepsin inhibitors: a virtual high throughput screening approach on the EGEE grid. *Journal of Chemical Information and Modeling*, **47** (5) 1818–1828.

52 Xiang, Y., Song, J., and Zhang, Z. (2014) Topomer CoMFA and virtual screening studies of azaindole class renin inhibitors. *Combinatorial Chemistry and High Throughput Screening*, **17** (5), 458–472.

53 Politi, A., Durdagi, S., Moutevelis-Minakakis, P. *et al.* (2009) Application of 3D QSAR CoMFA/CoMSIA and *in silico* docking studies on novel renin inhibitors against cardiovascular diseases. *European Journal of Medical Chemistry*, **44** (9), 3703–3711.

54 Moorthy, N.S.H.N., Brás, N.F., Ramos, M.J., and Fernandes, P.A. (2014) Binding mode prediction and identification of new lead compounds from natural products as renin and angiotensin converting enzyme inhibitors. *RSC Advances*, **4** (38), 19550.

55 Thangapandian, S., John, S., Sakkiah, S., and Lee, K.W. (2011) Potential virtual lead identification in the discovery of renin inhibitors: application of ligand and structure-based pharmacophore modeling approaches. *European Journal of Medical Chemistry*, **46** (6), 2469–2476.

56 Kuntz, I.D., Hen, K.C., Harp, K.A.S., and Ollman, P.A.K. (1999) The maximal affinity of ligands. *Proceedings of the National Academy of Sciences of the United States of America*, **96** (August), 9997–10002.

57 Hopkins, A.L., Keserü, G.M., Leeson, P.D. *et al.* (2014) The role of ligand efficiency metrics in drug discovery. *Nature Reviews Drug Discovery*, **13** (2), 105–121.

58 Klebe, G. (ed.) (2013) *Aspartic Protease Inhibitors. Drug Design*, Springer, Berlin, pp. 533–564.

59 Eder, J., Hommel, U., Cumin, F. *et al.* (2007) Aspartic proteases in drug discovery. *Current Pharmaceutical Design*, **13** (3), 271–285.

60 Lahiri, D.K., Maloney, B., Long, J.M., and Greig, N.H. (2014) Lessons from a BACE1 inhibitor trial: off-site but not off base. *Alzheimer's and Dementia*, **10**, 1–9.

61 Luo, X. and Yan, R. (2010) Inhibition of BACE1 for therapeutic use in Alzheimer's disease. *Journal of Clinical and Experimental Pathology*, **3** (6), 618–628.

62 Roberts, N.A., Martin, J.A., Kinchington, D. *et al.* (1990) Rational design of

peptide-based HIV proteinase inhibitors. *Science*, **248**, 358–361.
63 Thaisrivongs, S. and Strohbach, J.W. (1999) Structure-based discovery of tipranavir disodium (PNU-140690E): a potent, orally bioavailable, nonpeptidic HIV protease inhibitor. *Biopolymers*, **51** (1), 51–58.
64 Swahn, B., Kolmodin, K., Von Berg, S. *et al.* (2012) Design and synthesis of β-site amyloid precursor protein cleaving enzyme (BACE1) inhibitors with *in vivo* brain reduction of β-amyloid peptides. *Journal of Medicinal Chemistry*, **55**, 9346–9361.
65 Böhm, H.-J. (1992) The computer program LUDI: a new method for the *de novo* design of enzyme inhibitors. *Journal of Computer-Aided Molecular Design*, **6**, 61–78.
66 Böhm, H.-J. (1992) LUDI: rule-based automatic design of new substituents for enzyme inhibitor leads. *Journal of Computer-Aided Molecular Design*, **6** (6), 593–606.
67 Sun, Y., Ewing, T.J.A., Skillman, A.G., and Kuntz, I.D. (1998) CombiDOCK: structure-based combinatorial docking and library design. *Journal of Computer-Aided Molecular Design*, **12**, 597–604.
68 Fejzo, J., Lepre, C.A., Peng, J.W. *et al.* (1999) The SHAPES strategy: an NMR-based approach for lead generation in drug discovery. *Chemistry and Biology*, **6** (10), 755–769.
69 Boehm, H., Boehringer, M., Bur, D. *et al.* (2000) Novel inhibitors of DNA gyrase: 3D structure based biased needle screening, hit validation by biophysical methods, and 3D guided optimization. A promising alternative to random screening. *Journal of Medicinal Chemistry*, **43**, 2664–2674.
70 Hajduk, P.J., Gomtsyan, A., Didomenico, S. *et al.* (2000) Design of adenosine kinase inhibitors from the NMR-based screening of fragments. *Journal of Medicinal Chemistry*, **43**, 4781–4786.
71 Lam, P.Y.S., Jadhav, P.K., Eyermann, C.J. *et al.* (1994) Rational design of potent, bioavailable, nonpeptide cyclic ureas as HIV protease inhibitors. *Science*, **263** (18), 380–383.
72 Rayner, M.M., Cordova, B.C., Meade, R.P. *et al.* (1994) DMP 323, a nonpeptide cyclic urea inhibitor of human immunodeficiency virus (HIV) protease, specifically and persistently blocks intracellular processing of HIV gag polyprotein. *Antimicrobial Agents and Chemotherapy*, **38** (7), 1635–1640.
73 Prasad, J.V.N.V., Para, K.S., Lunney, E.A. *et al.* (1994) Novel series of achiral, low molecular weight, and potent HIV-1 protease inhibitors. *Journal of the American Chemical Society*, **116**, 6989–6990.
74 Vieira, E., Binggeli, A., Breu, V. *et al.* (1999) Substituted piperidines: highly potent renin inhibitors due to induced fit adaptation of the active site. *Bioorganic and Medicinal Chemistry Letters*, **9**, 1397–1402.
75 Güller, R., Binggeli, A., Breu, V. *et al.* (1999) Piperidine-renin inhibitors compounds with improved physicochemical properties. *Bioorganic and Medicinal Chemistry Letters*, **9**, 1403–1408.
76 Bursavich, M.G., West, C.W., Rich, D.H., and Highland, A.V. (2001) From peptides to non-peptide peptidomimetics: design and synthesis of new piperidine inhibitors of aspartic peptidases. *Organic Letters*, **3** (15), 2317–2320.
77 Olsworth, D.D., Beliotti, T., Bryant, J. *et al.* (2006) Discovery of 6-ethyl-2,4-diaminopyrimidine-based small molecule renin inhibitors. Abstr. Pap. 231st Natl. Meet. Am. Chem. Soc. Atlanta GA, March 26–30, 2006, Am. Chem. Soc. Washington DC, 2006, MEDI-87.
78 Bury, M., Cody, W.L., Emerson, D. *et al.* (2006) NMR auxillary binding screen for lead optimization: design of novel renin inhibitors that access both the S2 and S3-S3' pockets of rennin. Abstr. Pap. 231st Natl. Meet. Am. Chem. Soc. Atlanta GA, March 26–30, 2006, Am. Chem. Soc. Washington DC, 2006, MEDI-86.
79 Sarver, R.W., Peevers, J., Cody, W.L. *et al.* (2007) Binding thermodynamics of substituted diaminopyrimidine renin inhibitors. *Analytical Biochemistry*, **360** (1), 30–40.
80 Congreve, M., Aharony, D., Albert, J. *et al.* (2007) Application of fragment

screening by X-ray crystallography to the discovery of aminopyridines as inhibitors of β-secretase. *Journal of Medicinal Chemistry*, **50**, 1124–1132.

81 Geschwindner, S., Olsson, L., Albert, J.S. *et al.* (2007) Discovery of a novel warhead against β-secretase through fragment-based lead generation. *Journal of Medicinal Chemistry*, **50**, 5903–5911.

82 Efremov, I.V., Vajdos, F.F., Borzilleri, K.a. *et al.* (2012) Discovery and optimization of a novel spiropyrrolidine inhibitor of β-secretase (BACE1) through fragment-based drug design. *Journal of Medicinal Chemistry*, **55** (21), 9069–9088.

83 Davies, T.G. and Tickle, I.J. (2012) Fragment screening using X-ray crystallography. *Current Topics in Medicinal Chemistry*, **317**, 33–59.

84 Edwards, P.D., Albert, J.S., Sylvester, M. *et al.* (2007) Application of fragment-based lead generation to the discovery of novel, cyclic amidine β-secretase inhibitors with nanomolar potency, cellular activity, and high ligand efficiency. *Journal of Medicinal Chemistry*, **50**, 5912–5925.

85 Hopkins, A.L., Groom, C.R., and Alex, A. (2004) Ligand efficiency: a useful metric for lead selection. *Drug Discovery Today*, **9** (10), 430–431.

86 Yang, W., Fucini, R.V., Fahr, B.T. *et al.* (2009) Fragment-based discovery of nonpeptidic BACE-1 inhibitors using tethering. *Biochemistry*, **48** (21), 4488–4496.

87 Blomberg, N., Cosgrove, D.A., Kenny, P.W., and Kolmodin, K. (2009) Design of compound libraries for fragment screening. *Journal of Computer-Aided Molecular Design*, **23** (8), 513–525.

88 Barelier, S., Eidam, O., Fish, I. *et al.* (2014) Increasing chemical space coverage by combining empirical and computational fragment screens. *ACS Chemical Biology*, **9** (7), 1528–1535.

89 Schuffenhauer, A., Ruedisser, S., Marzinzik, A.L. *et al.* (2005) Library design for fragment based screening. *Current Topics in Medicinal Chemistry*, **5** (8), 751–762.

90 Chen, Y. and Shoichet, B.K. (2009) Molecular docking and ligand specificity in fragment-based inhibitor discovery. *Nature Chemical Biology*, **5** (5), 358–364.

91 Dann, J.G., Stammers, D.K., Harris, C.J. *et al.* (1986) Human renin: a new class of inhibitors. *Biochemical and Biophysical Research Communications*, **134** (1), 71–77.

92 Boger, J., Lohr, N.S., Ulm, E.H. *et al.* (1989) Novel renin inhibitors containing the amino acid statine. *Nature*, **303** (3), 81–84.

93 Allen, M.C., Fuhrer, W., Tuck, B. *et al.* (1989) Renin inhibitors. Synthesis of transition-state analogue inhibitors containing phosphorus acid derivatives at the scissile bond. *Journal of Medicinal Chemistry*, **32**, 1652–1661.

94 Erickson, J.W. (1993) Design and structure of symmetry-based inhibitors of HIV-1 protease. *Perspectives in Drug Discovery and Design*, **1** (1), 109–128.

95 Unity Fingerprints, UNITY, 4.2.1.

96 MDL Drug Data Report, MDL ISIS/HOST, 2002.2.

97 Tong, L., Pav, S., Lamarre, D. *et al.* (1995) High resolution crystal structures of recombinant human renin in complex with polyhydroxymonoamide inhibitors. *Journal of Molecular Biology*, **250** (2), 211–222.

98 Park, H., Lee, S., and Science, M. (2003) Determination of the active site protonation state of β-secretase from molecular dynamics simulation and docking experiment: implications for structure-based inhibitor design. *Journal of the American Chemical Society*, (4), 16416–16422.

99 Jones, G., Willett, P., and Glen, R. (1995) Molecular recognition of receptor sites using a genetic algorithm with a description of desolvation. *Journal of Molecular Biology*, **245** (1), 43–53.

100 Jones, G., Willett, P., and Glen, R. (1997) Development and validation of a genetic algorithm for flexible docking. *Journal of Molecular Biology*, **267**, 727–748.

101 "The SYBYL software.".

102 Ewing, T.J.A. and Kuntz, I.D. (1997) Critical evaluation of search algorithms for automated molecular docking and database screening. *Journal of Computational Chemistry*, **18** (9), 1175–1189.

103 Muegge, I. and Martin, Y.C. (1999) A general and fast scoring function for protein–ligand interactions: a simplified potential approach. *Journal of Medicinal Chemistry*, **42**, 791–804.

104 Jones, G., Willett, P., and Glen, R.C. (1995) A genetic algorithm for flexible molecular overlay and pharmacophore elucidation. *Journal of Computer-Aided Molecular Design*, **9** (6), 532–549.

105 Eldridge, M.D., Murray, C.W., Auton, T.R. *et al.* (1997) Empirical scoring functions. I. The development of a fast empirical scoring function to estimate the binding affinity of ligands in receptor complexes. *Journal of Computer-Aided Molecular Design*, **11**, 425–445.

106 Rarey, M., Kramer, B., Lengauer, T., and Klebe, G. (1996) A fast flexible docking method using an incremental construction algorithm. *Journal of Molecular Biology*, **261** (3), 470–489.

107 Congreve, M., Robin, C., Murray, C., and Harren, J. (2003) A "rule of three " for fragment-based lead. *Drug Discovery Today*, **8** (19), 876–877.

108 Verheij, H.J. (2006) Leadlikeness and structural diversity of synthetic screening libraries. *Molecular Diversity*, **10** (3), 377–388.

109 Baell, J.B. and Holloway, G.a. (2010) New substructure filters for removal of pan assay interference compounds (PAINS) from screening libraries and for their exclusion in bioassays. *Journal of Medicinal Chemistry*, **53** (7), 2719–2740.

110 Baell, J. and Walters, M.A. (2014) Chemical con artists foil drug discovery. *Nature*, **513**, 481–483.

111 Baell, J.B. (2010) Observations on screening-based research and some concerning trends in the literature. *Future Medicinal Chemistry*, **2** (10), 1529–1546.

112 Shuker, S.B., Hajduk, P.J., Meadows, R.P., and Fesik, S.W. (1996) Discovering high-affinity ligands for proteins: SAR by NMR. *Science*, **274** (5292), 1531–1534.

113 Harner, M.J., Frank, A.O., and Fesik, S.W. (2013) Fragment-based drug discovery using NMR spectroscopy. *Journal of Biomolecular NMR*, **56** (2), 65–75.

114 Hajduk, P.J., Olejniczak, E.T., Fesik, S.W. *et al.* (1997) One-dimensional relaxation- and diffusion-edited NMR methods for screening compounds that bind to macromolecules. *Journal of the American Chemical Society*, **7863** (9), 12257–12261.

115 Meiboom, S. and Gill, D. (1958) Modified spin-echo method for measuring nuclear relaxation times. *Review of Scientific Instruments*, **29** (8), 688.

116 Davis, D.G., Perlman, M.E., and London, R.E. (1994) Direct measurements of the dissociation-rate constant for inhibitor-enzyme complexes via the T1rho and T2 (CPMG) methods. *Journal of Magnetic Resonance Series B*, **104**, 266–275.

117 Lucca, G.V. (1997) Synthesis and evaluation of imidazolidinones as nonpeptidic HIV-protease inhibitors. *Bioorganic and Medicinal Chemistry Letters*, **7** (5), 495–500.

118 Vagin, A. and Teplyakov, A. (1997) MOLREP: an automated program for molecular replacement. *Journal of Applied Crystallography*, **30**, 1022–1025.

119 Webb, R.L., Schiering, N., Sedrani, R., and Maibaum, J. (2010) Direct renin inhibitors as a new therapy for hypertension. *Journal of Medicinal Chemistry*, **53** (21), 7490–7520.

120 Ehara, T., Irie, O., Kosaka, T. *et al.* (2014) Structure-based design of substituted piperidines as a new class of highly efficacious oral direct renin inhibitors. *Medicinal Chemistry Letters*, **5**, 787–792.

# Index



*Fragment-based Drug Discovery: Lessons and Outlook*, First Edition. Edited by Daniel A. Erlanson and Wolfgang Jahnke.

***q***



***r***